Introduction to

Finite and Spectral Element Methods Using MATLAB®

Second Edition

Introduction to
Finite and Spectral Element Methods Using MATLAB®

Second Edition

C. Pozrikidis

University of Massachusetts
Amherst, USA

CRC Press
Taylor & Francis Group
Boca Raton   London   New York

CRC Press is an imprint of the
Taylor & Francis Group, an **informa** business

A CHAPMAN & HALL BOOK

MATLAB® is a trademark of The MathWorks, Inc. and is used with permission. The MathWorks does not warrant the accuracy of the text or exercises in this book. This book's use or discussion of MATLAB® software or related products does not constitute endorsement or sponsorship by The MathWorks of a particular pedagogical approach or particular use of the MATLAB® software.

First published in paperback 2024

First published 2014
by CRC Press
2385 NW Executive Center Drive, Suite 320, Boca Raton FL 33431

and by CRC Press
4 Park Square, Milton Park, Abingdon, Oxon, OX14 4RN

CRC Press is an imprint of Taylor & Francis Group, LLC

© 2014, 2024 Taylor & Francis Group, LLC

Library of Congress Cataloging-in-Publication Data

Pozrikidis, C. (Constantine), 1958-
  Introduction to finite and spectral element methods using MATLAB / Constantine Pozrikidis. -- Second edition.
      pages cm
  Includes bibliographical references and index.
  ISBN 978-1-4822-0915-0 (hardcover : alk. paper)
  1. Finite element method--Data processing. 2. MATLAB. I. Title.

TA347.F5P7 2014
518'.25028553--dc23   2014007196

ISBN: 978-1-4822-0915-0 (hbk)
ISBN: 978-1-03-291750-4 (pbk)
ISBN: 978-0-429-16303-6 (ebk)

DOI: 10.1201/b17067

Visit the Taylor & Francis Web site at
http://www.taylorandfrancis.com

and the CRC Press Web site at
http://www.crcpress.com

Contents

Preface   xiii

FSELIB software library   xvii

Frequently asked questions   xxiii

1 The finite element method in one dimension   1
   1.1 Steady diffusion with linear elements   1
        1.1.1 Linear element interpolation   3
        1.1.2 Element grading   4
        1.1.3 Galerkin projection   7
        1.1.4 Formulation of a linear algebraic system   11
        1.1.5 Flux at the Dirichlet end   14
        1.1.6 Galerkin finite element equations via the Dirac delta function   16
        1.1.7 Relation to the finite difference method   20
   1.2 Finite element assembly   21
        1.2.1 Assembly of a linear system   25
        1.2.2 Thomas algorithm for a tridiagonal system   25
        1.2.3 Finite element code   27
        1.2.4 Convection (Robin or mixed) boundary condition   34
   1.3 Variational formulation and weighted residuals   35
        1.3.1 Homogeneous Dirichlet boundary conditions   36
        1.3.2 Inhomogeneous Dirichlet boundary conditions   39
        1.3.3 Dirichlet/Neumann boundary conditions   41
        1.3.4 Neumann/Dirichlet boundary conditions   43
   1.4 Helmholtz equation   45
   1.5 Steady diffusion with quadratic elements   51
        1.5.1 Element nodes and global nodes   51
        1.5.2 Galerkin finite element equations   53
        1.5.3 Thomas algorithm for pentadiagonal system   57
        1.5.4 Element matrices   58
        1.5.5 Finite element code   62
        1.5.6 Node condensation   67
        1.5.7 Arbitrary interior nodes   73
   1.6 Steady diffusion with quadratic modal expansions   78

2 Further applications in one dimension   **87**
  2.1 Unsteady diffusion 87
      2.1.1 Galerkin projection 88
      2.1.2 Integrating ODEs 90
      2.1.3 Forward Euler method 90
      2.1.4 Numerical stability 91
      2.1.5 Finite element code 96
      2.1.6 Crank–Nicolson integration 101
  2.2 Convection 105
      2.2.1 Linear elements 106
      2.2.2 Numerical dispersion due to spatial discretization 109
      2.2.3 Quadratic elements 110
      2.2.4 Integrating ODEs 110
      2.2.5 Nonlinear convection 113
  2.3 Convection–diffusion 114
      2.3.1 Steady linear convection–diffusion 114
      2.3.2 Nonlinear convection–diffusion 119
  2.4 Beam bending 120
      2.4.1 Euler–Bernoulli beam 120
  2.5 Finite element methods for beam bending 125
      2.5.1 Hermitian elements 126
      2.5.2 Galerkin projection 128
      2.5.3 Element stiffness and mass matrices 131
      2.5.4 One-element cantilever beam 132
      2.5.5 Cantilever beam with nodal loads 136
  2.6 Beam buckling 140
      2.6.1 Tip compression 142
      2.6.2 Buckling under a compressive tip force 143
      2.6.3 Buckling of a heavy vertical column 145

3 High-order and spectral elements in one dimension   **153**
  3.1 Element nodal sets 154
      3.1.1 Lagrange interpolation 155
      3.1.2 Evenly spaced nodes 156
      3.1.3 Element matrices 158
      3.1.4 C^0 continuity and shared element nodes 159
  3.2 Change of element nodal sets 161
  3.3 Spectral interpolation 165
      3.3.1 Lobatto nodal base 166
      3.3.2 Discretization code 174
      3.3.3 Legendre polynomials 174
      3.3.4 Chebyshev second-kind nodal base 178
  3.4 Lobatto interpolation and element matrices 180
      3.4.1 Lobatto mass matrix 181
      3.4.2 Lobatto integration quadrature 182

	3.4.3	Computation of the Lobatto mass matrix	183
	3.4.4	Computation of the Lobatto diffusion matrix	191
3.5	Spectral element code for steady diffusion	196
	3.5.1	Spectral accuracy	201
	3.5.2	Helmholtz equation	206
	3.5.3	Node condensation	206
3.6	Modal expansion	212
	3.6.1	Relation to the nodal expansion	213
	3.6.2	Implementation	215
3.7	Lobatto modal expansion	216
	3.7.1	Element diffusion matrix	216
	3.7.2	Element mass matrix	220
	3.7.3	Modal spectral element method	224
3.8	Arbitrary nodal sets	224
3.9	Unsteady diffusion	231
	3.9.1	Crank–Nicolson discretization	233
	3.9.2	Forward Euler discretization	240

4 The finite element method in two dimensions   **243**

4.1	Convection–diffusion in two dimensions	244
	4.1.1	Boundary conditions	246
	4.1.2	Galerkin projection	246
	4.1.3	Domain discretization and interpolation	248
	4.1.4	Galerkin finite element equations	250
	4.1.5	Implementation of the Dirichlet boundary condition	251
	4.1.6	Split nodes	253
	4.1.7	Variational formulation	253
4.2	Three-node triangles	257
	4.2.1	Element matrices	261
	4.2.2	Computation of the element diffusion matrix	263
	4.2.3	Computation of the element mass matrix	263
	4.2.4	Proof of the integration formula (4.2.36)	265
	4.2.5	Computation of the element advection matrix	267
4.3	Grid generation	270
	4.3.1	Successive subdivisions	270
	4.3.2	Delaunay triangulation	274
	4.3.3	Generalized connectivity matrices	279
	4.3.4	Element and node labeling schemes	282
4.4	Laplace's equation with the Dirichlet boundary condition	285
4.5	Eigenvalues of the Laplacian operator	293
4.6	Convection–diffusion with the Dirichlet boundary condition	297
4.7	Helmholtz's equation with the Neumann boundary condition	302
4.8	Laplace's equation with arbitrary boundary conditions	308
4.9	Surface elements	314
4.10	Bilinear quadrilateral elements	317

5 Quadratic and spectral elements in two dimensions   **321**

5.1 Six-node triangular elements 321
    5.1.1 Integral over a triangle 326
    5.1.2 Isoparametric interpolation and element matrices 326
    5.1.3 Element matrices and integration quadratures 328
    5.1.4 Elements with straight edges 335
5.2 Grid generation 340
    5.2.1 Circular disk 340
    5.2.2 Square 347
    5.2.3 L-shaped domain 347
    5.2.4 Square with a square or circular hole 347
    5.2.5 A rectangle with a circular hole 350
5.3 Laplace and Poisson equations 359
    5.3.1 Laplace equation 359
    5.3.2 Eigenvalues of the Laplacian operator 365
    5.3.3 Poisson equation 365
5.4 Convection–diffusion with the Dirichlet boundary condition 374
5.5 High-order triangle expansions 380
    5.5.1 Computation of the node interpolation functions 383
    5.5.2 The Lebesgue constant 387
    5.5.3 Node condensation 387
5.6 Appell polynomial base 387
    5.6.1 Incomplete biorthogonality 390
    5.6.2 Incomplete orthogonality 391
    5.6.3 Generalized Appell polynomials 391
5.7 Proriol polynomial base 392
    5.7.1 Orthogonality 394
    5.7.2 Orthogonal expansion 395
5.8 High-order node distributions 396
    5.8.1 Node distribution based on a one-dimensional master grid . . 396
    5.8.2 Uniform grid 398
    5.8.3 Lobatto grid on the triangle 402
    5.8.4 The Fekete set 405
    5.8.5 Further nodal distributions 406
5.9 Modal expansions in a triangle 407
    5.9.1 Implementation of the modal expansion 412
    5.9.2 Properties of the modal expansion 412
5.10 Surface elements 413
    5.10.1 Surface gradient 414
    5.10.2 Grid generation 415
5.11 High-order quadrilateral elements 416
    5.11.1 Eight-node serendipity elements 417
    5.11.2 12-node serendipity elements 419
    5.11.3 Grid nodes via tensor-product expansions 422
    5.11.4 Modal expansion 424

6 Applications in mechanics   **429**
  6.1 Elements of elasticity theory 429
      6.1.1 Deformation and constitutive equations 431
      6.1.2 Linear elasticity 432
  6.2 Plane stress and plane strain analysis 434
      6.2.1 Plane stress analysis 434
      6.2.2 Plane strain analysis 439
      6.2.3 Finite element formulation 440
  6.3 Finite element plane stress analysis 443
      6.3.1 Deformation due to an edge force 445
      6.3.2 Deformation due to a body force 460
  6.4 Plate bending 470
      6.4.1 Equilibrium equations 474
      6.4.2 Boundary conditions 475
      6.4.3 Constitutive and governing equations 477
      6.4.4 Circular plate 480
  6.5 Hermite triangles 482
  6.6 Morley's triangle 493
  6.7 Conforming triangles 496
      6.7.1 Six-node, 21-dof triangle 497
      6.7.2 The Hsieh–Clough–Tocher (HCT) element 498
  6.8 Finite element methods for plate bending 500
      6.8.1 Formulation as a biharmonic equation 506
      6.8.2 Formulation as a system of Poisson equations 518
  6.9 Buckling and wrinkling 523

7 Viscous flow   **541**
  7.1 Governing equations 541
  7.2 Finite element formulation 544
      7.2.1 Galerkin projections 544
      7.2.2 Discrete equations 546
  7.3 Stokes flow 547
      7.3.1 Governing equations 547
      7.3.2 Galerkin finite element equations 548
      7.3.3 Triangularization 552
  7.4 Stokes flow in a rectangular cavity 553
  7.5 Navier–Stokes flow 562
      7.5.1 Steady state 564
      7.5.2 Time integration 564
      7.5.3 Formulation based on the pressure Poisson equation 566

8 Finite and spectral element methods in three dimensions   **569**
  8.1 Convection–diffusion in three dimensions 569
      8.1.1 Boundary conditions 570
      8.1.2 Domain discretization 571

  8.1.3 Galerkin projection 571
  8.1.4 Galerkin finite element equations 572
  8.1.5 Element matrices 573
  8.1.6 Implementation of the Dirichlet boundary condition 574
 8.2 Tetrahedral elements 574
  8.2.1 Parametric representation 576
  8.2.2 Integral over the volume of a tetrahedron 578
  8.2.3 Element subdivision into eight tetrahedra 579
  8.2.4 Element subdivision into 12 tetrahedra 583
  8.2.5 Isoparametric interpolation 583
  8.2.6 Element diffusion matrix 583
  8.2.7 Element mass matrix 594
  8.2.8 Proof of the integration formula (8.2.36) 595
  8.2.9 Element advection matrix 597
 8.3 Domain discretization into four-node tetrahedra 598
  8.3.1 Delaunay tessellation 598
 8.4 Finite element codes with four-node tetrahedra 605
  8.4.1 Laplace's equation 605
  8.4.2 Eigenvalues of the Laplacian operator 612
 8.5 Orthogonal polynomials over a tetrahedron 615
  8.5.1 Karniadakis and Sherwin polynomials 616
  8.5.2 Orthogonal expansion 618
 8.6 High-order and spectral tetrahedral elements 620
  8.6.1 Uniform node distributions 621
  8.6.2 Arbitrary node distributions 625
  8.6.3 Spectral node distributions 625
  8.6.4 Gradient of the element node interpolation functions 628
  8.6.5 Numerical integration 628
 8.7 10-node quadratic tetrahedra 629
  8.7.1 Node interpolation functions 631
  8.7.2 Element diffusion and mass matrices 632
  8.7.3 Domain discretization 632
  8.7.4 Laplace's equation 651
  8.7.5 Eigenvalues of the Laplacian operator 651
 8.8 Modal expansions in a tetrahedron 661
 8.9 Hexahedral elements 664
  8.9.1 Parametric representation 665
  8.9.2 Integral over the volume of the hexahedron 666
  8.9.3 High-order and spectral hexahedral elements 667
  8.9.4 Modal expansion 668

Appendices   **673**

A Mathematical supplement   **675**
 A.1 Index notation 675

A.2 Kronecker's delta 675
A.3 Alternating tensor 676
A.4 Two- and three-dimensional vectors 676
A.5 Del or nabla operator 677
A.6 Gradient and divergence 677
A.7 Vector identities 678
A.8 Gauss divergence theorem 679
A.9 Gauss divergence theorem in the plane 680
A.10 Stokes's theorem 681

B Orthogonal polynomials **683**
B.1 Definitions and basic properties 683
    B.1.1 Orthogonality against lower-degree polynomials 684
    B.1.2 Roots of orthogonal polynomials 686
    B.1.3 Discrete orthogonality 686
    B.1.4 Gram polynomials 687
    B.1.5 Recursion relation 687
    B.1.6 Evaluation as the determinant of a tridiagonal matrix 689
    B.1.7 Clenshaw's algorithm 690
    B.1.8 Gram–Schmidt orthogonalization 690
    B.1.9 Orthonormal polynomials 692
    B.1.10 Christoffel–Darboux formula 692
B.2 Gaussian integration quadratures 694
    B.2.1 Evaluation of the integration weights 695
    B.2.2 Standard Gaussian quadratures 696
B.3 Lobatto integration quadrature 697
B.4 Chebyshev integration quadrature 699
B.5 Legendre polynomials 700
B.6 Lobatto polynomials 702
B.7 Chebyshev polynomials 703
B.8 Jacobi polynomials 706

C Linear solvers **709**
C.1 Gauss elimination 709
    C.1.1 Pivoting 711
    C.1.2 Implementation 712
    C.1.3 Symmetric matrices 715
    C.1.4 Computational cost 715
    C.1.5 Gauss elimination code 715
    C.1.6 Multiple right-hand sides 715
    C.1.7 Computation of the inverse 721
    C.1.8 Gauss–Jordan reduction 721
C.2 Iterative methods based on matrix splitting 722
    C.2.1 Jacobi's method 723
    C.2.2 Gauss–Seidel method 723

C.2.3   Successive over-relaxation (SOR) 724
C.2.4   Operator- and grid-based splitting 724
C.3   Iterative methods based on path search 725
C.3.1   Symmetric and positive-definite matrices 725
C.3.2   General methods 733
C.4   Finite element system solvers 734

D Function interpolation   **735**
D.1   The interpolating polynomial 736
D.1.1   Vandermonde matrix 736
D.1.2   Generalized Vandermonde matrix 738
D.1.3   Newton interpolation 739
D.2   Lagrange interpolation 740
D.2.1   Cauchy relations 740
D.2.2   Representation in terms of a generating polynomial 741
D.2.3   First derivative and the node differentiation matrix 742
D.2.4   Representation in terms of the Vandermonde matrix 745
D.2.5   Lagrange polynomials corresponding to polynomial roots . . 748
D.2.6   Lagrange polynomials for Hermite interpolation 749
D.3   Error in polynomial interpolation 751
D.3.1   Convergence and the Lebesgue constant 752
D.4   Chebyshev interpolation 755
D.5   Lobatto interpolation 756
D.6   Interpolation in two and higher dimensions 760

E Element grid generation   **763**

F Glossary   **765**

G MATLAB primer   **767**
G.1   Programming in MATLAB 768
G.1.1   Grammar and syntax 768
G.1.2   Precision 769
G.1.3   MATLAB commands 771
G.1.4   Elementary examples 773
G.2   MATLAB functions 777
G.3   Numerical methods 780
G.4   MATLAB graphics 781

References   **789**

Index   **795**

Preface

Five general classes of numerical methods are available for solving ordinary and partial differential equations encountered in the various branches of science and engineering, including finite difference, finite volume, finite element, boundary element, and spectral and pseudo-spectral methods. The relation and relative merits of these methods are discussed briefly in the *Frequently Asked Questions* section preceding Chapter 1.

An important advantage of the finite element method, and the main reason for its popularity among academics and industrial developers, is the ability to handle solution domains with arbitrary geometry. An attractive feature is the ability to generate solutions to problems governed by linear as well as nonlinear differential equations. Moreover, the finite element method enjoys a firm theoretical foundation that is mostly free of *ad hoc* schemes and heuristic numerical approximations, thereby inspiring confidence in the physical relevance of the solution.

A current search of books in print reveals over 200 items bearing in their title some variation of the general theme *finite element method*. Many of these books are devoted to special topics, such as heat transfer, computational fluid dynamics (CFD), structural mechanics, and stress analysis in elasticity. Other books are written from the point of view of the applied mathematician or numerical analyst with emphasis on convergence, error analysis, and numerical accuracy. Many excellent texts are suitable for second reading, while other texts do a superb job in explaining the fundamentals but fall short in describing the development and practical implementation of algorithms for non-elementary problems. In computational science and engineering, the devil is in the details.

The purpose of this text is to offer a venue for the rapid learning of the theoretical foundation and practical implementation of the finite element method and its companion spectral element method. The discussion has the form of a self-contained course that introduces the fundamentals on a need-to-know basis and emphasizes the development of algorithms and the computer implementation of essential procedures. The audience of interest includes students in science and engineering, practicing scientists and engineers, computational scientists, applied mathematicians, and scientific computing enthusiasts.

Consistent with the introductory nature of this text and its intended usage as a practical guide on finite and spectral element methods, error analysis and convergence studies are altogether ignored, and only the fundamental procedures and their implementation are discussed in sufficient detail. Specialized topics, such as Lagrangian formulations, free-boundary problems, infinite domains, and discontinuous Galerkin methods, are mentioned briefly in Appendix F entitled *Glossary*, which is meant to complement the subject index.

This text has been written to be used for self-study and is suitable as a textbook in a variety of courses in science and engineering. Since scientists and engineers of any discipline are familiar with the fundamental concepts of heat and mass transfer governed by the convection–diffusion equation, the main discourse relies on this paradigm. Once basic concepts have been explained and algorithms have been developed, problems in solid mechanics, fluid mechanics, and structural mechanics are discussed as implementations of the basic approach.

FSELIB

The importance of gaining simultaneous hands-on experience while learning the implementation of the finite element method, or any other numerical method, cannot be overstated. To achieve this goal, this book is accompanied by a library of user-defined MATLAB® functions and complete finite and spectral element codes encapsulated in the software library FSELIB.

Nearly all functions and complete codes of FSELIB are listed in the text. A few lookup tables, ancillary graphics functions, and slightly modified codes are listed in abbreviated form or have been omitted in the interest of space. The main codes of FSELIB are tabulated after the Contents.

The owner of this book can download and use the library freely subject to the conditions of the GNU public license from the Web site:

http://dehesa.freeshell.org/FSELIB

For instructional reasons and to reduce the overhead time necessary for learning how to run the codes, the library is almost completely free of `.dat` files containing data structures. All necessary parameters are defined in the main code and finite element grids are generated by automatic triangulation determined by the level of refinement for specific geometries. With this book as a user guide, the reader will be able to immediately run the codes as given on a modest computer and graphically display solutions to a variety of elementary and advanced problems.

MATLAB

The MATLAB computing environment was chosen primarily because of its ability to integrate numerical computation and computer graphics visualization, and also because of its popularity among students and professionals. For convenience, a brief MATLAB primer is included in Appendix G. Translation of a MATLAB code to another computer language is both straightforward and highly recommended. For clarity of exposition and to facilitate this translation, hidden operations embedded in the intrinsic MATLAB functions are deliberately avoided as much as possible in the FSELIB codes. Thus, two vectors are added explicitly, component by component, rather than implicitly by issuing symbolic vector addition. Similarly, the entries of a matrix are often initialized and manipulated in a double loop running over the indices, even though this makes for a longer code.

Second Edition

The Second Edition considerably extends the First Edition by including new topics and original material. Clarifications, explanations, detailed proofs, original derivations, new schematic illustrations and graphs have been inserted, and additional solved problems have been added in numerous places. The accommodate the new material, the text has been rearranged into two additional chapters with respect to the First Edition.

The discussion of computer programs in the Second Edition refers to the second release of the software library FSELIB hosting improved and new computer functions and complete finite and spectral element codes. Most important, complete finite element codes incorporating domain discretization modules in three dimensions are provided in the second release of FSELIB.

Further information, selected links to finite element recourses, and updates of the FSELIB library can be found at the book Web site:

http://dehesa.freeshell.org/FSEM2

Comments and corrections from readers are welcome and will be communicated to the audience with due credit at the book Web site.

C. Pozrikidis

FSELIB *software library*

The FSELIB software library accompanying this book contains scripts, miscellaneous user-defined functions, and complete finite and spectral element codes written in MATLAB, including modules for domain discretization, system assembly and solution, and graphics visualization. The codes are arranged in directories (folders), corresponding to the book chapters and appendices. The owner of this book can download the library freely from the Internet site:

<code>http://dehesa.freeshell.org/FSELIB</code>

Complete FSELIB finite and spectral element codes for problems in one, two, and three dimensions, corresponding Chapters 1–8 and appendices, are presented in the following tables. Each directory (folder) contains a *Readme* text file where the complete contents of the directory are listed.

Copyright © 2005 C. Pozrikidis

FSELIB, Finite and Spectral Element Library

The FSELIB software resides in the public domain and should be used strictly under the terms of the GNU General Public License, as stated at the GNU Internet page cited below.

This library is free software; you can redistribute it and/or modify it under the terms of the current GNU General Public License version 3, Gplv3, as published by the Free Software Foundation:

<code>http://www.gnu.org/copyleft/gpl.html</code>

This library is distributed in the hope that it will be useful, but *Without Any Warranty*; without even the implied warranty of *Merchantability of Fitness for a Particular Purpose*. See the GNU General Public License for more details.

Chapter 1: The finite element method in one dimension

Directory: 01_1D

Code	Problem	Element type
hlml	Helmholtz equation	linear
sdl	Steady diffusion	linear
sdl_robin	Steady diffusion	linear
sdq_beta	Steady diffusion	quadratic arbitrary interior node
sdq_cnd	Steady diffusion, condensed formulation	quadratic
sdq	Steady diffusion	quadratic mid-point interior node
sdq_modal	Steady diffusion	quadratic modal

Chapter 2: Further applications in one dimension

Directory: 02_1D

Code	Problem	Element type
beam	Cantilever beam bending	Hermitian
buckle_tree	Buckling of a heavy vertical column	Hermitian
udl	Unsteady diffusion	linear
scdl	Steady convection–diffusion	linear

Chapter 3: High-order and spectral elements in one dimension

Directory: *03_1D*

Code	Problem	Element type
hlms_lob	Helmholtz equation	spectral Lobatto
sds_any	Steady diffusion	arbitrary nodes
sds_lob_cnd	Steady diffusion, condensed system	spectral Lobatto
sds_lob	Steady diffusion	spectral Lobatto
sds_modal	Steady diffusion	modal expansion
uds_lob_cn	Unsteady diffusion, Crank–Nicolson method	spectral
uds_lob_fe	Unsteady diffusion, forward Euler method	spectral

Chapter 4: The finite element method in two dimensions

Directory: *04_2D*

Code	Problem	Element type
hlm3_n	Helmholtz's equation in a disk-like domain with the Neumann boundary condition	3-node triangle
lapl3_d	Laplace's equation in a disk-like domain with the Dirichlet boundary condition	3-node triangle
lapl3_dn	Laplace's equation in a disk-like domain with Dirichlet and Neumann boundary conditions	3-node triangle
lapl3_eig	Eigenvalues of the laplacian operator in a disk-like domain	3-node triangle
lapl3_dn_sqr	Laplace's equation in a square domain with Dirichlet and Neumann boundary conditions	3-node triangle
scd3_d	Steady convection–diffusion in a disk-like domain with the Neumann boundary condition	3-node triangle

Chapter 5: Quadratic and spectral elements in two dimensions

Directory: 05_2D

Code	Problem	Element type
lapl6_d_L	Laplace's equation in an L-shaped domain with the Dirichlet boundary condition	6-node triangle
lapl6_d	Laplace's equation in a disk-like domain with the Dirichlet boundary condition	6-node triangle
lapl6_d_rc	Laplace's equation in a rectangular domain with a circular hole with the Dirichlet boundary condition	6-node triangle
lapl6_d_sc	Laplace's equation in a square domain with a circular hole with the Dirichlet boundary condition	6-node triangle
lapl6_d_ss	Laplace's equation in a square domain with a square hole with the Dirichlet boundary condition	6-node triangle
lapl6_eig	Eigenvalues of the Laplacian operator	6-node triangle
scd6_d_rc	Steady convection–diffusion in a rectangular domain with a circular hole with the Neumann boundary condition	6-node triangle

Chapter 6: Applications in mechanics

Directory: 06_MECH

Code	Problem	Element type
bend_HCT	Bending of a clamped plate based on the biharmonic equation	HCT triangle
buckle_HCT	Buckling of a clamped plate based on the biharmonic equation	HCT triangle
psa6	Plane stress analysis in a rectangular domain possibly with a circular hole	6-node triangle
psaM	Plane stress analysis of a membrane patch under a uniform body force	6-node triangle

Chapter 7: Viscous flow

Directory: 07_FLUIDS

Code	Problem	Element type
cvt6	Stokes flow in a rectangular cavity	6-node triangle

Chapter 8: Finite and spectral element methods in three dimensions

Directory: *08_3D*

Code	Problem	Element type
laplt10_d	Laplace's equation	10-node tetrahedra
laplt10_eig	Eigenvalues of the Laplacian operator	10-node tetrahedra
laplt4_d	Laplace's equation	4-node tetrahedra
laplt4_eig	Eigenvalues of the Laplacian operator	4-node tetrahedra

Appendix C: Linear solvers

Directory: *AC_LIN*

Code	Problem
gel	Solution of an arbitrary linear system by Gauss elimination
cg	Solution of a symmetric system by the method of conjugate gradients

Frequently asked questions

The basic idea behind the finite element method and its relation to other methods are best explained by answering frequently asked questions.

- *What is the finite element method (FEM)?*

The finite element method is a numerical method for solving ordinary and partial differential equations encountered in the various branches of mathematical physics and engineering. Examples include Laplace's equation, Poisson's equation, Helmholtz's equation, the convection–diffusion equation, the equations of potential and viscous fluid flow, the equations of electrostatics and electromagnetics, and the equations of elastostatics and elastodynamics.

- *When was the finite element method conceived?*

The finite element method was developed in the mid-1950s for problems in stress analysis under the auspices of linear elasticity. In later years, the method was generalized and adapted to solve a broad range of differential equations with applications ranging from fluid mechanics, to structural dynamics, to electrostatics.

- *What is the Galerkin finite element method (GFEM)?*

The GFEM is a particular and most popular implementation of the FEM where algebraic equations are derived from the governing differential equations in a procedure called the Galerkin projection. The method is discussed extensively and exclusively in this book.

- *What are the advantages of the finite element method?*

The most significant practical advantage of the FEM is the ability to handle solution domains with arbitrary geometry. Another important advantage is that transforming the governing differential equations to a system of algebraic equations is performed in a way that is both theoretically sound and free of heuristic schemes and numerical approximations. Moreover, the finite element method is built on a rigorous theoretical foundation. Specifically, for a certain class of differential equations, it can be shown that the finite element method is equivalent to a properly posed functional minimization method.

- *What is the origin of the terminology "finite element?"*

In the finite element method, the solution domain is discretized into elementary units called finite elements. In the case of a two-dimensional domain, the finite elements can be triangular or quadrilateral elements. The discretization is typically unstructured, meaning that new elements can be added or removed without affecting an existing element structure and without requiring a global element or node relabeling.

• *Is there a restriction on the type of differential equation that the finite element method is not able to handle?*

In principle, the answer is negative. In practice, the finite element method is most suitable for diffusion-dominated problems and has been criticized for its weakness in handling convection–dominated problems occurring, for example, in high-speed flows. However, modifications of the basic procedure can be made to overcome this limitation and improve the performance of the method.

• *How does the FEM compare to the finite difference method (FDM)?*

In the finite difference method, a grid is introduced, the differential equation is applied at a grid node, and the derivatives are approximated with finite differences to obtain a system of algebraic equations. Because some grid nodes must be located at boundaries where boundary conditions are specified, the finite difference method is restricted to solution domains with simple geometry, or else requires the use of cumbersome boundary-fitted coordinates and artificial body forces for smearing out the boundary location.

• *How does the FEM compare to the finite volume method (FVM)?*

In the finite volume method, the solution domain is also discretized into elementary units called finite volumes. The differential equation is then integrated over the individual volumes and the divergence theorem is applied to derive equilibrium equations. In the numerical implementation, solution values are defined at the vertices, faces, or centers of the individual volumes, and undefined values are computed by neighbor averaging. Although the finite volume method is also able to handle domains with arbitrary geometrical complexity, the required *ad hoc* averaging is a serious drawback.

• *How does the FEM compare to the boundary element method (BEM)?*

The boundary element method (BEM) requires discretizing only the boundaries of a solution domain (e.g., Pozrikidis [48]). In contrast, the finite element method requires discretizing the whole of the solution domain, including the boundaries. In three dimensions, the BEM employs surface elements, whereas the FEM employs volume elements. From this reason, the boundary element method is superior to the finite element method. However, the BEM is primarily applicable to linear differential equations with constant coefficients. Despite claims to the contrary, its implementation for more general types of differential equations is both cumbersome and computationally demanding. The need to evaluate singular and hypersingular integrals is another weakness of the BEM.

• *How does the FEM compare to the spectral and pseudo-spectral methods?*

In one class of spectral and pseudo-spectral methods, the solution is expanded in a series of orthogonal basis functions, the expansion is substituted into the differential equation, and the coefficients of the expansion are computed by projection or collocation. These methods are suitable for solution domains with simple geometry.

• *What should one know before one is able to understand the theoretical foundation of the FEM?*

The basic concepts are discussed in a self-contained manner in this book. Prerequisites are college-level calculus, numerical methods, and a general familiarity with computer programming.

• *What should one know before one is able to write FEM code?*

Prerequisites are general-purpose numerical methods, including numerical linear algebra, function interpolation, and function integration. All necessary topics are discussed in this book and summaries are given in appendices. Familiarity with a computer programming language is another essential prerequisite.

• *What is the spectral element method?*

The spectral element method is an advanced implementation of the finite element method in which the solution over each element is expressed in terms of *a priori* unknown values at carefully selected spectral nodes. The advantage of the spectral element method is that stable solution algorithms and high accuracy can be achieved with a low number of elements under a broad range of conditions.

• *How can one keep up with developments in the finite and spectral element method?*

Several Web sites provide current information on various aspects of the finite and spectral element methods. Links are provided at the book's Web site given in the Preface.

The finite element method in one dimension

1

In the first chapter, we introduce the theoretical notions and practical concepts underlying the finite element method in one dimension by discussing the numerical solution of the steady diffusion equation in the presence of a distributed source. We begin by addressing the simplest implementation of the finite element method where the solution domain is divided into a number of intervals, called finite elements, and a function of interest is approximated with a linear function over each element. The union of the linear element functions provides us with a continuous, piecewise linear function representing a known or requisite field. Implementations for quadratic and high-order element functions arise as straightforward extensions of the basic formulation. Further applications of the finite element method in one dimension are discussed in Chapter 2.

In the most general formulation, the solution over each element is approximated with a polynomial of arbitrary degree defined by an appropriate set of interpolation nodes or expansion modes. Requiring a high interpolation accuracy subject to the available degrees of freedom leads us to the spectral element method discussed in Chapter 3.

1.1 Steady diffusion with linear elements

Consider heat conduction through a circular rod of length L at steady state, in the presence of a distributed source of heat due, for example, to a chemical reaction, as illustrated in Figure 1.1.1. To derive the governing equation, we place the x axis along the rod centerline and assume that the heat flux along the rod is given by Fick's law,

$$q(x) = -k \frac{\mathrm{d}f}{\mathrm{d}x}, \tag{1.1.1}$$

where k is the thermal conductivity of the rod material and $f(x)$ is the temperature distribution. A heat balance over an infinitesimal section of the rod, δx, requires that

$$q(x) - q(x + \delta x) + s(x)\, \delta x = 0, \tag{1.1.2}$$

1

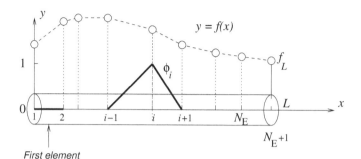

First element

FIGURE 1.1.1   Conduction through a rod extending from $x = 0$ to L at steady state.
The temperature distribution is computed by solving equation (1.1.4) subject to
the boundary conditions (1.1.5) and (1.1.6) using a finite element method with
linear interpolation functions. The function ϕ_i is the global interpolation function
of the ith node.

where $s(x)$ is the linear rate of heat production. If $s(x)$ is negative, heat escapes
from the rod across the cylindrical surface. Substituting Fick's law into equation
(1.1.2) and dividing by δx, we obtain

$$-k\,\frac{1}{\delta x}\,\Big[\,\Big(\frac{\mathrm{d}f}{\mathrm{d}x}\Big)_x - \Big(\frac{\mathrm{d}f}{\mathrm{d}x}\Big)_{x+\delta x}\,\Big] + s(x) = 0. \qquad (1.1.3)$$

Now taking the limit $\delta x \to 0$, we find that the temperature distribution satisfies
the steady-state heat conduction equation

$$k\,\frac{\mathrm{d}^2 f}{\mathrm{d}x^2} + s(x) = 0, \qquad (1.1.4)$$

which is a second-order ordinary differential equation (ODE).

Neumann and Dirichlet boundary conditions

In our discussion, we will assume that the boundary conditions specify: (a) the heat
flux at the left end of the rod located at $x = 0$,

$$q_0 \equiv -k\,\Big(\frac{\mathrm{d}f}{\mathrm{d}x}\Big)_{x=0}, \qquad (1.1.5)$$

where q_0 is a given constant, and (b) the temperature at the right end of the rod
located at $x = L$,

$$f(x = L) \equiv f_L, \qquad (1.1.6)$$

where f_L is another given constant. Equation (1.1.5) expresses a *Neumann* boundary condition, that is, a condition on the derivative of the unknown function,
whereas equation (1.1.6) expresses a *Dirichlet* boundary condition, that is, a condition on the value of the unknown function. Other possible boundary conditions
will be considered in later sections of this chapter.

1.1.1 Linear element interpolation

We begin implementing the finite element method by dividing the one-dimensional solution domain, $0 \leq x \leq L$, into N_E intervals, called finite elements, defined by element end-nodes located at positions x_i for $i = 1, \ldots, N_E + 1$, where $x_1 = 0$ and $x_{N_E+1} = L$, as shown in Figure 1.1.1. Next, we approximate the temperature distribution over the individual elements with linear functions whose union yields a continuous, piecewise linear function, drawn with the dashed polygonal line in Figure 1.1.1.

Global interpolation functions

To derive the mathematical representation of the polygonal approximation, we introduce the piecewise linear, tent-line, global interpolation functions, $\phi_i(x)$ for $i = 1, \ldots, N_E + 1$. The ith function is supported by the interval $[x_{i-1}, x_{i+1}]$, takes the value of unity at the ith node, drops linearly to zero at the adjacent nodes numbered $i - 1$ and $i + 1$, and remains zero outside the supporting interval $[x_{i-1}, x_{i+1}]$, as illustrated in Figure 1.1.1. By construction, the global interpolation functions satisfy the cardinal interpolation property

$$\phi_i(x_j) = \delta_{i,j}, \tag{1.1.7}$$

where $\delta_{i,j}$ is Kronecker's delta representing the identity matrix, that is,

$$\delta_{i,j} = \left\{ \begin{array}{ll} 1 & \text{if} \quad i = j, \\ 0 & \text{if} \quad i \neq j. \end{array} \right. \tag{1.1.8}$$

The piecewise linear approximation of the solution is represented by the global finite element expansion

$$f(x) = \sum_{j=1}^{N_E+1} f_j \, \phi_j(x), \tag{1.1.9}$$

where

$$f_j \equiv f(x = x_j) \tag{1.1.10}$$

is the *a priori* unknown value of the solution at the jth node. The Dirichlet boundary condition (1.1.6) at the right end requires that

$$f_{N_E+1} = f_L, \tag{1.1.11}$$

where f_L is a specified constant.

Weak formulation

The piecewise linear approximation expressed by (1.1.9) cannot possibly be made to satisfy the governing second-order differential equation. Taking the second derivative of the right-hand side of (1.1.9), we are faced with the zero function interrupted by a sequence of Dirac delta functions centered at the element nodes, as discussed in Section 1.1.6. However, while this is undoubtedly true, the difficulty will be overcome by seeking a solution of the *weak formulation* associated with the Galerkin projection of the differential equation, as discussed in Section 1.1.3. When this is done, the order of the differential equation is effectively lowered by one unit and the condition on the existence of the second derivative, which is implicit in the statement of the second-order differential equation, is replaced with a condition on the existence of the first derivative, which is satisfied by the piecewise linear approximation.

C^m *functions*

A continuous function with discontinuous first derivative at isolated points, such as the function described by the global finite element expansion (1.1.9), is called a C^0 function. A continuous function with continuous first derivative and discontinuous second derivative at isolated points is called a C^1 function. An infinitely differentiable function is called a C^∞ function. We will see that finite element solutions typically employ C^0 or C^1 functions, as the need arises, depending on the order of the differential equations.

1.1.2   Element grading

Two important features of the finite element method are: (*a*) ability to ensure adequate spatial resolution by varying the element size manually or adaptively, and (*b*) ease of accommodating irregular geometrical shapes. In the case of the one-dimensional problem presently considered, because the solution domain is a line, only the first feature is relevant.

Function elm_line1

FSELIB function *elm_line1*, listed in Table 1.1.1, discretizes an interval of the x axis into a graded mesh of N elements so that the element length increases or decreases from left to right in a geometrical fashion. Specifically, if Δx_1 is the size of the first element, and Δx_N is the size of the last element, then the ratio

$$\frac{\Delta x_N}{\Delta x_1} \equiv r \tag{1.1.12}$$

has a specified value, denoted as *ratio* in the code.

For the geometrical sequence, $\Delta x_2 = \alpha \, \Delta x_1$, where α is the element size expansion coefficient. By recursion,

$$\Delta x_k = \alpha^{k-1} \, \Delta x_1 \tag{1.1.13}$$

for $k = 1, \dots, N$. By definition,

$$r = \alpha^{N-1} \tag{1.1.14}$$

```
function xe = elm_line1 (x1,x2,n,ratio)

%===================================================
% Unsymmetrical discretization of a line segment
% into a graded mesh of n elements subtended
% between the left end-point x1
% and the right end-point x2
%
% The variable "ratio" is the ratio
% of the length of the last
% element to the length of the first element.
%
% xe: element end-nodes
%===================================================

%------------
% one element
%------------

if(n==1)
   xe(1) = x1; xe(2) = x2; return;
end

%--------------
% many elements
%--------------

if(ratio==1)
   alpha = 1.0;
   factor = 1.0/n;
else
   texp = 1/(n-1);
   alpha = ratio^texp;
   factor = (1.0-alpha)/(1.0-alpha^n);
end

deltax = (x2-x1) * factor;   % length of the first element

xe(1) = x1;                  % first point

for i=2:n+1
   xe(i)  = xe(i-1) + deltax;
   deltax = deltax*alpha;
end

%-----
% done
%-----

return;
```

TABLE 1.1.1   Function *elm_line1* performs the unsymmetrical discretization of a line segment into a graded mesh of elements.

(*a*)

(*b*)

(*c*)

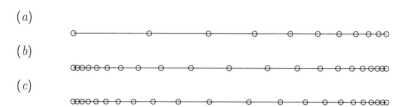

FIGURE 1.1.2   Graded element distributions generated by the FSELIB function (*a*)
elm_line1, (*b*) elm_line2, and (*c*) elm_line3.

and the total length of the interval is

$$L \equiv X_2 - X_1 = (1 + \alpha + \alpha^2 + \cdots + \alpha^{N-1}) \Delta x_1 = \frac{1 - \alpha^N}{1 - \alpha} \Delta x_1, \qquad (1.1.15)$$

where X_1 and X_2 are the left and right interval end-points. Rearranging (1.1.14),
we derive an expression for the expansion coefficient, α, in terms of the ratio, r,
and number of elements, N,

$$\alpha = r^{1/(N-1)}. \qquad (1.1.16)$$

Rearranging (1.1.15), we obtain an expression for the size of the first element,

$$\Delta x_1 = \frac{1 - \alpha}{1 - \alpha^N} L. \qquad (1.1.17)$$

Special provisions can be made to accommodate the special cases $N = 1$ or $r = 1$,
the latter corresponding to evenly spaced elements.

FSELIB script *elm_line1_dr*, listed in Table 1.1.2, drives the FSELIB function
elm_line1 to perform the element discretization and generate a graph of the element
end-points. An element distribution generated by this script is shown in Figure
1.1.2(*a*).

Functions elm_line2 and elm_line3

FSELIB functions *elm_line2* and *elm_line3*, listed in Tables 1.1.3 and 1.1.4, discretize an interval into a graded mesh of elements so that the element distribution
is symmetric with respect to the interval mid-point. Function *elm_line2* discretizes
the interval into an *even* number of elements, and function *elm_line3* discretizes the
interval into an *odd* number of elements. In both cases, the element length increases
or decreases in a geometrical fashion by a specified factor from the left end-point to
the interval mid-point (Problem 1.1.1). The ratio of the size of either one of the two
elements joining at the interval mid-point to the size of the first element is denoted
as *ratio* in function *elm_line2*. The ratio of the size of the element centered at the
interval mid-point to the size of the first element is denoted as *ratio* in the function
elm_line3.

```
%=======================================================
% Code elm_line1_dr
%
% Driver of the discretization function: elm_line1
%=======================================================

%------------
% input data
%------------

x1 = 0.0; x2 = 1.0;  %  end-points
n = 10;              %  number of elements
ratio = 0.1;         %  element geometrical ratio

%-----------
% discretize
%-----------

xe = elm_line1 (x1,x2,n,ratio);

%-----
% plot
%-----

ye = zeros(n+1,1);
plot (xe, ye,'-o');

%-----
% done
%-----
```

TABLE 1.1.2   Code *elm_line1_dr* is a driver for discretizing a line segment into a graded mesh of elements based on the FSELIB function *elm_line1*.

Typical element distributions generated by the corresponding FSELIB driver codes *elm_line2_dr* and *elm_line3_dr*, not listed in the text, are shown in Figure 1.1.2(*b, c*).

1.1.3   Galerkin projection

Following the general blueprint of the Galerkin finite element method (GFEM), we multiply both sides of the governing differential equation (1.1.4) by each one of the global interpolation functions ϕ_i, for $i = 1, \ldots, N_E$, and integrate the product over the solution domain to obtain the Galerkin projection equation

$$\int_0^L \phi_i(x) \left(k \frac{d^2 f}{dx^2} + s(x) \right) dx = 0. \tag{1.1.18}$$

Because the Dirichlet boundary condition (1.1.6) is imposed at the right end of the solution domain, the last global interpolation function, corresponding to the node

```
function xe = elm_line2 (x1,x2,n,ratio);

%========================================================
% Symmetric discretization of a line segment into a
% graded mesh of n elements subtended
% between the left end-point x1
% and the right end-point x2
%
% n is 1 or even: 2,4,6,...
%
% The variable "ratio" is the ratio of the size
% of the mid-elements to the size of the first element
%========================================================

%------------
% one element
%------------

if(n==1)
   xe(1) = x1;
   xe(2) = x2;
   return;
end

%-------------
% two elements
%-------------

if(n==2)
   xe(1) = x1;
   xe(2) = 0.5*(x1+x2);
   xe(3) = x2;
   return;
end

%--------------
% many elements
%--------------

xe = zeros(n+1,1);
nh = n/2;
xh = 0.5*(x1+x2);    % mid-point

if(ratio==1)
   alpha = 1.0; factor = 1.0/nh;
else
   texp = 1.0/(nh-1.0);
   alpha = ratio^texp;
   factor = (1.0-alpha)/(1.0-alpha^nh);
end

deltax = (xh-x1) * factor;   % length of first element

xe(1) = x1;                  % first point
```

TABLE 1.1.3   Function *elm_line2* (Continuing →)

```
%----------
% discretize
%----------

for i=2:nh+1
  xe(i)  = xe(i-1)+deltax;
  deltax = deltax*alpha;
end

deltax = deltax/alpha;

for i=nh+2:n+1;
  xe(i)  = xe(i-1)+deltax;
  deltax = deltax/alpha;
end

%-----
% done
%-----

return;
```

TABLE 1.1.3 ($\to$ Continued) Function *elm_line2* performs the symmetric discretization of a line segment into a graded mesh with an *even* number of elements.

labeled $i = N_E + 1$, has been excluded from the Galerkin projection. In Section
1.1.5, we will see that the projection of this function provides us with a method of
computing the unknown flux at the Dirichlet end.

In the next step, we use the rules of product differentiation to recast the integral
on the left-hand side of (1.1.18) into the form

$$\int_0^L \left(k \frac{d}{dx}\left(\phi_i \frac{df}{dx}\right) - k \frac{d\phi_i}{dx}\frac{df}{dx} + \phi_i\, s \right) dx = 0. \tag{1.1.19}$$

Applying the fundamental theorem of calculus to evaluate the integral of the first
term on the left-hand side, we obtain

$$-k\left(\phi_i \frac{df}{dx}\right)_{x=0} + k\left(\phi_i \frac{df}{dx}\right)_{x=L} - k\int_0^L \frac{d\phi_i}{dx}\frac{df}{dx}\, dx + \int_0^L \phi_i\, s\, dx = 0. \tag{1.1.20}$$

Finally, we use the cardinal interpolation property (1.1.7) to compute the end-
point values, $\phi_i(x = 0)$ and $\phi_i(x = L)$, and obtain

$$-\delta_{i,1}\, k \left(\frac{df}{dx}\right)_{x=0} + \delta_{i,N_E+1}\, k \left(\frac{df}{dx}\right)_{x=L} - k\int_0^L \frac{d\phi_i}{dx}\frac{df}{dx}\, dx + \int_0^L \phi_i\, s\, dx = 0,$$

$$\tag{1.1.21}$$

```
function xe = elm_line3 (x1,x2,n,ratio);

%==========================================================
% Symmetric discretization of a line segment into a
% graded mesh of n elements subtended
% between the left end-point x1
% and the right end-point x2
%
% n is odd: 1,3,5,...
%
% The variable "ratio" is the ratio of the size
% of the mid-element to the size of the first element
%==========================================================

%------------
% one element
%------------

if(n==1)
   xe(1) = x1;
   xe(2) = x2;
   return;
end

%--------------
% many elements
%--------------

if(ratio==1)
   alpha = 1.0;
   factor = 1.0/n;
else
   texp = 2.0/(n-1.0);
   alpha = ratio^texp;
   tmp1 = (n+1.0)/2.0;
   tmp2 = (n-1.0)/2.0;
   factor = (1.0-alpha)/(2.0-alpha^tmp1-alpha^tmp2);
end

deltax = (x2-x1) * factor;    % length of first element

%----------
% discretize
%----------

xe(1) = x1;    % first point

for i=2:(n+3)/2
   xe(i)  = xe(i-1)+deltax;
   deltax = deltax*alpha;
end

deltax = deltax/(alpha^2);
```

TABLE 1.1.4   Function *elm_line3* (Continuing →)

```
for i=(n+5)/2:n+1
  xe(i) = xe(i-1)+deltax;
  deltax = deltax/alpha;
end

%-----
% done
%-----

return;
```

TABLE 1.1.4 (→ Continued) Function *elm_line3* performs the symmetric discretization of a line segment into a graded mesh with an *odd* number of elements.

where $\delta_{i,j}$ is Kronecker's delta. Because we have specified that $i = 1, \ldots, N_E$, that is, we have excluded the last node, the second term on the left-hand side of equation (1.1.21) makes a zero contribution.

Implementing the boundary condition (1.1.5) in the first term on the left-hand side, we derive the Galerkin finite element equations

$$\delta_{i,1}\, q_0 - k \int_0^L \frac{\mathrm{d}\phi_i}{\mathrm{d}x} \frac{\mathrm{d}f}{\mathrm{d}x}\, \mathrm{d}x + \int_0^L \phi_i\, s\, \mathrm{d}x = 0, \tag{1.1.22}$$

which can be rearranged into the preferred form

$$\int_0^L \frac{\mathrm{d}\phi_i}{\mathrm{d}x} \frac{\mathrm{d}f}{\mathrm{d}x}\, \mathrm{d}x = \frac{q_0}{k}\, \delta_{i,1} + \frac{1}{k} \int_0^L \phi_i\, s\, \mathrm{d}x \tag{1.1.23}$$

for $i = 1, \ldots, N_E$.

1.1.4   Formulation of a linear algebraic system

In the next step, we substitute the finite element expansion (1.1.9) into the left-hand side of (1.1.23). For convenience, we assume that the source term is continuous and introduce the corresponding expansion

$$s(x) \simeq \sum_{j=1}^{N_E+1} s_j \phi_j(x), \tag{1.1.24}$$

where $s_j \equiv s(x = x_j)$ is the value of the source at the jth node. Rearranging the resulting equation, we derive a system of N_E linear equations for the N_E unknown values, f_j,

(*a*)

$$D_{ij} \equiv \int_0^L \frac{\mathrm{d}\phi_i}{\mathrm{d}x} \frac{\mathrm{d}\phi_j}{\mathrm{d}x}\,\mathrm{d}x = \begin{cases} \dfrac{1}{h_1} & \text{if } i = j = 1, \\[2mm] \dfrac{1}{h_{i-1}} + \dfrac{1}{h_i} & \text{if } i = j \neq 1 \text{ and } N_{\mathrm{E}} + 1, \\[2mm] \dfrac{1}{h_{N_{\mathrm{E}}}} & \text{if } i = j = N_{\mathrm{E}} + 1, \\[2mm] -\dfrac{1}{h_i} & \text{if } j = i + 1, \\[2mm] -\dfrac{1}{h_{i-1}} & \text{if } j = i - 1, \\[2mm] 0 & \text{otherwise.} \end{cases}$$

(*b*)

$$M_{ij} \equiv \int_0^L \phi_i\,\phi_j\,\mathrm{d}x = \begin{cases} \frac{1}{3} h_1 & \text{if } i = j = 1, \\[2mm] \frac{1}{3}\left(h_{i-1} + h_i\right) & \text{if } i = j \neq 1 \text{ and } N_{\mathrm{E}} + 1, \\[2mm] \frac{1}{3} h_{N_{\mathrm{E}}} & \text{if } i = j = N_{\mathrm{E}} + 1, \\[2mm] \frac{1}{6} h_i & \text{if } j = i + 1, \\[2mm] \frac{1}{6} h_{i-1} & \text{if } j = i - 1, \\[2mm] 0 & \text{otherwise.} \end{cases}$$

TABLE 1.1.5   Elements of the (*a*) global diffusion and (*b*) global mass matrix, where $h_i \equiv x_{i+1} - x_i$ is the ith element size.

$$\sum_{j=1}^{N_{\mathrm{E}}} \left(\int_0^L \frac{\mathrm{d}\phi_i}{\mathrm{d}x} \frac{\mathrm{d}\phi_j}{\mathrm{d}x}\,\mathrm{d}x \right) f_j = -\left(\int_0^L \frac{\mathrm{d}\phi_i}{\mathrm{d}x} \frac{\mathrm{d}\phi_{N+1}}{\mathrm{d}x}\,\mathrm{d}x \right) f_L + \frac{q_0}{k}\,\delta_{i,1}$$

$$+ \frac{1}{k} \sum_{j=1}^{N_{\mathrm{E}}+1} \left(\int_0^L \phi_i\,\phi_j\,\mathrm{d}x \right) s_j \qquad (1.1.25)$$

for $i = 1, \ldots, N_{\mathrm{E}}$. Because the Dirichlet boundary condition is specified at the right end, $x = L$, the terms multiplying the corresponding value, f_L, have been transferred to the right-hand side.

Detailed consideration of the global interpolation functions depicted in Figure 1.1.1, followed by analytical computation of the integrals on the left- and right-hand sides of (1.1.25), yields the integration formulas shown in Table 1.1.5. Substituting

$$\mathbf{D} \equiv \begin{bmatrix} \dfrac{1}{h_1} & -\dfrac{1}{h_1} & 0 & \cdots \\[2mm] -\dfrac{1}{h_1} & \dfrac{1}{h_1}+\dfrac{1}{h_2} & -\dfrac{1}{h_2} & \cdots \\[2mm] 0 & -\dfrac{1}{h_2} & \dfrac{1}{h_2}+\dfrac{1}{h_3} & \cdots \\[2mm] \vdots & \vdots & \vdots & \ddots \\[1mm] 0 & 0 & 0 & \cdots \\ 0 & 0 & 0 \\ 0 & 0 & 0 & \cdots & 0 \end{bmatrix} \longrightarrow$$

$$\begin{bmatrix} \cdots & 0 & & 0 & \\ \cdots & 0 & & 0 & \\ \cdots & 0 & & 0 & 0 \\ \ddots & \vdots & & \vdots & \vdots \\ \cdots & \dfrac{1}{h_{N_E-3}}+\dfrac{1}{h_{N_E-2}} & -\dfrac{1}{h_{N_E-2}} & & 0 \\[2mm] \cdots & -\dfrac{1}{h_{N_E-2}} & \dfrac{1}{h_{N_E-2}}+\dfrac{1}{h_{N_E-1}} & -\dfrac{1}{h_{N_E-1}} \\[2mm] \cdots & 0 & -\dfrac{1}{h_{N_E-1}} & \dfrac{1}{h_{N_E-1}}+\dfrac{1}{h_{N_E}} \end{bmatrix}$$

TABLE 1.1.6   Global square $N_E \times N_E$ diffusion matrix for linear elements, where $h_i \equiv x_{i+1} - x_i$ is the ith element size.

these values into (1.1.25), we derive a linear system,

$$\mathbf{D} \cdot \mathbf{f} = \mathbf{b}, \tag{1.1.26}$$

where $\mathbf{f}$ is the vector of the unknown temperatures at the nodes,

$$\mathbf{f} \equiv \begin{bmatrix} f_1, & f_2, & \cdots, & f_{N_E-1}, & f_{N_E} \end{bmatrix}^T, \tag{1.1.27}$$

and $\mathbf{D}$ is the $N_E \times N_E$ global *diffusion* or *conductivity* matrix, sometimes also called the *Laplacian* matrix, given in Table 1.1.6. In the remainder of this text, we refer to $\mathbf{D}$ as the global diffusion matrix.

Right-hand side

The right-hand side of the linear system is given by

$$\mathbf{b} \equiv \mathbf{c} + \frac{1}{k}\widetilde{\mathbf{M}} \cdot \mathbf{s}, \tag{1.1.28}$$

where $\widetilde{\mathbf{M}}$ is the $N_E \times (N_E + 1)$ *rectangular mass matrix* shown in Table 1.1.7(*a*), the nearly null N_E-dimensional vector $\mathbf{c}$ encapsulates the boundary conditions,

$$\mathbf{c} \equiv \begin{bmatrix} q_0/k, & 0, & \ldots, & 0, & f_L/h_{N_E} \end{bmatrix}^T, \tag{1.1.29}$$

and the $(N_E + 1)$-dimensional vector, $\mathbf{s}$, contains the nodal values of the source,

$$\mathbf{s} \equiv \begin{bmatrix} s_1, & s_2, & \ldots, & s_{N_E}, & s_{N_E+1} \end{bmatrix}^T. \tag{1.1.30}$$

Note that the global diffusion matrix has units of inverse length, whereas the global mass matrix has units of length. The N_E-dimensional vector arising from the matrix-vector product $\widetilde{\mathbf{M}} \cdot \mathbf{s}$ is shown in Table 1.1.7(*b*).

Tridiagonal algebraic system

Because the coefficient matrix of the linear system (1.1.26) is tridiagonal,[1] the solution can be found efficiently using the Thomas algorithm discussed in Section 1.2.2.

Neumann boundary condition

It is interesting to observe that the first equation of the linear system (1.1.26) takes the form

$$-\frac{f_2 - f_1}{h_1} = \frac{q_0}{k} + \frac{1}{6} h_1 (2 s_1 + s_2). \tag{1.1.31}$$

The fraction on the left-hand side expresses the forward difference approximation of the slope at the left end, $df/dx = -q_0/k$ at $x = 0$. The second term on the right-hand side contributes a correction based on the left end and adjacent nodal values of the source. This correction raises the accuracy of the method above the linear order associated with the two-point forward difference approximation.

1.1.5   Flux at the Dirichlet end

To compute the flux at the right end of the solution domain where the Dirichlet boundary condition (1.1.6) is imposed,

$$q_L \equiv k \left(\frac{df}{dx} \right)_{x=L}, \tag{1.1.32}$$

we apply the Galerkin projection expressed by equation (1.1.21) for $i = N_E + 1$, and obtain

$$q_L - k \int_0^L \frac{d\phi_{N_E+1}}{dx} \frac{df}{dx} \, dx + \int_0^L \phi_{N_E+1} \, s \, dx = 0. \tag{1.1.33}$$

---

[1]The word *tridiagonal* derives from the Greek words $\tau\rho\iota\alpha$, which means *three*, and $\delta\iota\alpha\gamma\omega\nu\iota\sigma\varsigma$, which means *diagonal*.

(a)

$$\widetilde{\mathbf{M}} \equiv \begin{bmatrix} \frac{1}{3}h_1 & \frac{1}{6}h_1 & 0 & 0 & \dots \\ \frac{1}{6}h_1 & \frac{1}{3}h_1 + \frac{1}{3}h_2 & \frac{1}{6}h_2 & 0 & \dots \\ 0 & \frac{1}{6}h_2 & \frac{1}{3}h_2 + \frac{1}{3}h_3 & 0 & \dots \\ \vdots & 0 & \dots & \dots & \dots \\ \vdots & \vdots & \vdots & \vdots & \vdots \\ 0 & \dots & \dots & \dots & \dots \end{bmatrix} \longrightarrow$$

$$\begin{bmatrix} 0 & 0 & 0 & 0 \\ 0 & \dots & 0 & \vdots \\ \dots & 0 & 0 & \vdots \\ \frac{1}{3}h_{N_\mathrm{E}-3} + \frac{1}{3}h_{N_\mathrm{E}-2} & \frac{1}{6}h_{N_\mathrm{E}-2} & 0 & \vdots \\ \frac{1}{6}h_{N_\mathrm{E}-2} & \frac{1}{3}h_{N_\mathrm{E}-2} + \frac{1}{3}h_{N_\mathrm{E}-1} & \frac{1}{6}h_{N_\mathrm{E}-1} & 0 \\ 0 & \frac{1}{6}h_{N_\mathrm{E}-1} & \frac{1}{3}h_{N_\mathrm{E}-1} + \frac{1}{3}h_{N_\mathrm{E}} & \frac{1}{6}h_{N_\mathrm{E}} \end{bmatrix}$$

(b)

$$\widetilde{\mathbf{M}} \cdot \mathbf{s} = \begin{bmatrix} \frac{1}{3}h_1 s_1 + \frac{1}{6}h_1 s_2 \\[2mm] \frac{1}{6}h_1 s_1 + \left(\frac{1}{3}h_1 + \frac{1}{3}h_2\right) s_2 + \frac{1}{6}h_2 s_3 \\[2mm] \vdots \\[2mm] \frac{1}{6}h_{N_\mathrm{E}-2} s_{N_\mathrm{E}-2} + \left(\frac{1}{3}h_{N_\mathrm{E}-2} + \frac{1}{3}h_{N_\mathrm{E}-1}\right) s_{N_\mathrm{E}-1} + \frac{1}{6}h_{N_\mathrm{E}-1} s_{N_\mathrm{E}} \\[2mm] \frac{1}{6}h_{N_\mathrm{E}-1} s_{N_\mathrm{E}-1} + \left(\frac{1}{3}h_{N_\mathrm{E}-1} + \frac{1}{3}h_{N_\mathrm{E}}\right) s_{N_\mathrm{E}} + \frac{1}{6}h_{N_\mathrm{E}} s_{N_\mathrm{E}+1} \end{bmatrix}$$

TABLE 1.1.7 (a) Global rectangular $N_\mathrm{E} \times (N_\mathrm{E} + 1)$ mass matrix and (b) its contribution to the right-hand side of the linear system for steady one-dimensional diffusion with linear elements.

Substituting expansions (1.1.9) and (1.1.24) and rearranging, we obtain

$$q_L = k \left(\int_{x_{N_E}}^{x_{N_E+1}} \frac{\mathrm{d}\phi_{N_E+1}}{\mathrm{d}x} \frac{\mathrm{d}\phi_{N_E}}{\mathrm{d}x} \, \mathrm{d}x \right) f_{N_E} + k \left(\int_{x_{N_E}}^{x_{N_E+1}} \frac{\mathrm{d}\phi_{N_E+1}}{\mathrm{d}x} \frac{\mathrm{d}\phi_{N_E+1}}{\mathrm{d}x} \, \mathrm{d}x \right) f_{N_E+1}$$

$$- \left(\int_{x_{N_E}}^{x_{N_E+1}} \phi_{N_E+1} \, \phi_{N_E} \, \mathrm{d}x \right) s_{N_E} - \left(\int_{x_{N_E}}^{x_{N_E+1}} \phi_{N_E+1} \, \phi_{N_E+1} \, \mathrm{d}x \right) s_{N_E+1}. \qquad (1.1.34)$$

Finally, we perform the integrations using the expressions given in Table 1.1.5 and find that

$$q_L = k \, \frac{f_{N_E+1} - f_{N_E}}{h_{N_E}} - \frac{1}{6} h_{N_E} \left(s_{N_E} + 2 \, s_{N_E+1} \right), \qquad (1.1.35)$$

which is the counterpart of the corresponding expression for the flux at the Neumann end displayed in (1.1.31). Formula (1.1.35) allows us to compute the flux at the Dirichlet end after the solution of the linear system (1.1.26) has been carried out.

Uniform source

It is instructive to consider the predictions of (1.1.35) for the special case of a uniform source function, $s = s_0$, where s_0 is a constant. For simplicity, we assume that the flux is zero at the Neumann end, $q_0 = 0$, and the boundary value at the Dirichlet end is $f_L = -s_0 L^2/(2k)$. The exact solution is readily found to be a quadratic function,

$$f(x) = -\frac{s_0}{2k} \, x^2. \qquad (1.1.36)$$

Denoting for convenience $h_{N_E} = h$, and substituting the expressions

$$f_{N_E+1} = f_L = -\frac{s_0}{2k} L^2, \qquad f_{N_E} = -\frac{s_0}{2k} (L - h)^2 \qquad (1.1.37)$$

into equation (1.1.35), we find that

$$q_L = -s_0 \, \frac{L^2 - (L - h)^2}{2h} - \frac{1}{2} s_0 \, h = -s_0 L, \qquad (1.1.38)$$

which is precisely the exact solution.

1.1.6 Galerkin finite element equations via the Dirac delta function

It is illuminating to rederive the Galerkin finite element equations compiled in the linear system (1.1.26) directly from the primary Galerkin projection (1.1.18). We begin by considering the graphs of the piecewise linear global finite element expansion and its piecewise constant first derivative, drawn with the broken and bold solid lines, respectively, in Figure 1.1.3. Both graphs have been extended so that $f(x)$ is a linear function with slope $\mathrm{d}f/\mathrm{d}x = -q_0/k$ beyond the left end of the solution domain, $x < 0$.

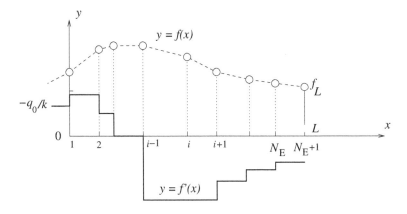

FIGURE 1.1.3 Graphs of the piecewise linear approximation (dashed line) and its first derivative (staircase bold line).

Dirac delta function

It is useful to introduce the Dirac delta function in one dimension, $\delta(x - x_0)$, physically representing a unit impulse applied at a point, x_0. The induced field is distinguished by the following properties:

1. $\delta(x - x_0)$ vanishes everywhere, except at the singular point $x = x_0$ where it becomes infinite.

2. The integral of $\delta(x - x_0)$ with respect to x over an interval $\mathcal{I}$ that contains the point x_0 is equal to unity,

$$\int_{\mathcal{I}} \delta(x - x_0) \, dx = 1. \tag{1.1.39}$$

   This property reveals that the delta function with argument of length, x, has units of inverse length.

3. The integral of the product of an arbitrary function $q(x)$ and the delta function over an interval $\mathcal{I}$ that contains the point x_0 is equal to the value of the function at the singular point,

$$\int_{\mathcal{I}} \delta(x - x_0) \, q(x) \, dx = q(x_0). \tag{1.1.40}$$

4. The integral of the product of an arbitrary function $q(x)$ and the delta function over an interval that does *not* contain the point x_0 is zero.

5. If a function $q(x)$ undergoes a discontinuity from a left value, q_-, to a right value, q_+, at a point, x_0, then the derivative at that point can be expressed

in terms of the delta function as

$$q'(x_0) = (q_+ - q_-)\,\delta(x - x_0) + \cdots,\qquad(1.1.41)$$

where the prime denotes the derivative and the dots denote non-singular terms.

6. In formal mathematics, $\delta(x - x_0)$ arises from the family of test functions

$$g(x - x_0) = \frac{1}{L}\left(\frac{\lambda}{\pi}\right)^{1/2}\exp\left[-\lambda\left(\frac{x - x_0}{L}\right)^2\right],\qquad(1.1.42)$$

in the limit as the dimensionless parameter λ tends to infinity, where L is an arbitrary length. Other test functions are available.

Graphs of the dimensionless function $G \equiv g\,L$ plotted against the dimensionless distance $X \equiv (x - x_0)/L$ are shown in Figure 1.1.4 for $\lambda = 1$ (dashed line), 2, 5, 10, 20, 50, 100, and 500 (dotted line). The maximum height of each graph is inversely proportional to its width, so that the area underneath each graph is equal to unity,

$$\int_{-\infty}^{\infty} g(x)\,\mathrm{d}x = \left(\frac{\lambda}{\pi L^2}\right)^{1/2}\int_{-\infty}^{\infty}\exp(-\lambda\frac{x^2}{L^2})\,\mathrm{d}x = 1.\qquad(1.1.43)$$

To prove this identity, we recall the definite integral associated with the error function,

$$\frac{2}{\sqrt{\pi}}\int_0^{\infty}\exp(-w^2)\,\mathrm{d}w = 1,\qquad(1.1.44)$$

and substitute $w = \lambda^{1/2}x/L$.

Galerkin projection

Using the fifth of the aforementioned properties of the Dirac delta function, we find that the second derivative of the global finite element expansion admits a generalized representation in terms of the delta function,

$$\frac{\mathrm{d}^2 f}{\mathrm{d}x^2} = \sum_{j=1}^{N_{\mathrm{E}}}(m_j - m_{j-1})\,\delta(x - x_j)\qquad(1.1.45)$$

for $0 \le x < L$, where

$$m_j = \frac{f_{j+1} - f_j}{h_j}\qquad(1.1.46)$$

is the slope of the approximate solution over the jth element, with the understanding that $m_0 = -q_0/k$, as required by the Neumann boundary condition at the left end. Now substituting (1.1.45) into the ith Galerkin projection (1.1.18), we obtain

$$\sum_{j=1}^{N_{\mathrm{E}}}(m_j - m_{j-1})\int_0^L \phi_i(x)\,\delta(x - x_j)\,\mathrm{d}x + \int_0^L \phi_i(x)\,s(x)\,\mathrm{d}x = 0.\qquad(1.1.47)$$

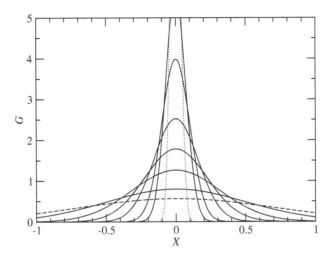

FIGURE 1.1.4   A family of dimensionless test functions $G \equiv gL$ described by (1.1.42), plotted against the dimensionless distance $X \equiv (x - x_0)/L$ for $\lambda = 1$ (dashed line), 2, 5, 10, 20, 50, 100, and 500 (dotted line). In the limit as the dimensionless parameter λ tends to infinity, we recover Dirac's delta function in one dimension.

Using the distinguishing properties of the Dirac delta function to evaluate the first integral, we obtain

$$\sum_{j=1}^{N_E} \phi_i(x_j)(m_j - m_{j-1}) + \int_0^L \phi_i(x)\, s(x)\, \mathrm{d}x = 0 \tag{1.1.48}$$

or

$$\sum_{j=1}^{N_E} \delta_{ij}(m_j - m_{j-1}) + \int_0^L \phi_i(x)\, s(x)\, \mathrm{d}x = 0, \tag{1.1.49}$$

yielding

$$m_i - m_{i-1} + \int_0^L \phi_i(x)\, s(x)\, \mathrm{d}x = 0. \tag{1.1.50}$$

Substituting the definition of the slope, m_i, from (1.1.46), we recover precisely the linear system (1.1.26).

We have demonstrated that integration by parts is not an essential step following the Galerkin projection, and should be regarded as a venue for bypassing the potentially cumbersome device of Dirac's delta function. It is sometimes erroneously stated that a critical step leading us from the strong to the weak formulation is integration by parts. In fact, the defining step is the Galerkin projection.

1.1.7 Relation to the finite difference method

When the elements are spaced evenly with uniform size h,

$$h_1 = h_2 = \cdots = h_{N_E} \equiv h, \tag{1.1.51}$$

the square global diffusion matrix shown in Table 1.1.6 takes the simpler form

$$\mathbf{D} = \frac{1}{h} \begin{bmatrix} 1 & -1 & 0 & \cdots & 0 & 0 & 0 \\ -1 & 2 & -1 & \cdots & 0 & 0 & 0 \\ 0 & -1 & 2 & \cdots & 0 & 0 & 0 \\ \vdots & \vdots & \vdots & \ddots & \vdots & \vdots & \vdots \\ 0 & 0 & 0 & \cdots & 2 & -1 & 0 \\ 0 & 0 & 0 & \cdots & -1 & 2 & -1 \\ 0 & 0 & 0 & \cdots & 0 & -1 & 2 \end{bmatrix}, \tag{1.1.52}$$

and the rectangular mass matrix shown in Table 1.1.7(a) takes the simpler form

$$\widetilde{\mathbf{M}} \equiv h \frac{1}{6} \left[\begin{array}{cccccccc|c} 2 & 1 & 0 & \cdots & 0 & 0 & 0 & 0 \\ 1 & 4 & 1 & \cdots & 0 & 0 & 0 & 0 \\ 0 & 1 & 4 & \cdots & 0 & 0 & 0 & 0 \\ \vdots & \vdots & \vdots & \ddots & \vdots & \vdots & \vdots & \vdots \\ 0 & 0 & 0 & \cdots & 4 & 1 & 0 & 0 \\ 0 & 0 & 0 & \cdots & 1 & 4 & 1 & 0 \\ 0 & 0 & 0 & \cdots & 0 & 1 & 4 & 1 \end{array} \right], \tag{1.1.53}$$

where the vertical line separates the last column. Apart from the first and last rows, the matrix on the right-hand side of (1.1.52) is recognized as the negative of the central differentiation matrix for the second derivative associated with the finite difference formula

$$f_i'' \simeq \frac{f_{i-1} - 2f_i + f_{i+1}}{h^2}. \tag{1.1.54}$$

The same behavior is encountered in the case of arbitrary element sizes with reference to the finite difference formula

$$f_i'' \simeq \frac{2}{h_i + h_{i-1}} \left(\frac{f(x_{i+1}) - f(x_i)}{h_i} - \frac{f(x_i) - f(x_{i-1})}{h_{i-1}} \right), \tag{1.1.55}$$

which is applicable for arbitrary node distributions (e.g., Pozrikidis [47]).

We conclude that the finite element method with uniform linear elements is equivalent to the finite difference method implemented with the second-order central difference approximation for the interior nodes. However, the rectangular mass matrix (1.1.53) enforces a neighbor averaging of the source term on either side of the ith node, thereby smearing the local contributions.

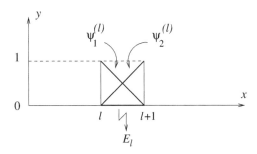

FIGURE 1.2.1 Illustration of the local linear interpolation functions $\psi_1^{(l)}$ and $\psi_2^{(l)}$ associated with the lth element. The function $\psi_1^{(l)}$ drops from 1 at the lth node to 0 at the $l + 1$ node. The function $\psi_2^{(l)}$ rises from 0 at the lth node to 1 at the $l + 1$ node.

PROBLEM

1.1.1 *Grid generation*

(a) Explain in narrative form the discretization algorithms implemented in the FSELIB function *elm_line2*.

(b) Repeat(a) for the function *elm_line3*.

(c) Execute codes *elm_line1_dr*, *elm_line2_dr*, and *elm_line3_dr* to generate and display element distributions of your choice.

1.2   Finite element assembly

To generate the coefficient matrix and right-hand side of the linear system (1.1.26), we used the *global* integration formulas shown in Table 1.1.5. In the practical implementation of the finite element method, the integrals arising from the Galerkin projection are assembled from corresponding integrals over the elements in a systematic way that facilitates the bookkeeping.

To demonstrate the process, we consider the Galerkin projection expressed by equation (1.1.23), introduce expansions (1.1.9) and (1.1.24), and replace the integrals over the solution domain $[0, L]$ with sums of integrals over the individual elements. The result is

$$k \sum_{j=1}^{N_E+1} \left(\sum_{l=1}^{N_E} \int_{E_l} \frac{\mathrm{d}\phi_i}{\mathrm{d}x} \frac{\mathrm{d}\phi_j}{\mathrm{d}x} \, \mathrm{d}x \right) f_j = q_0 \, \delta_{i,1} + \sum_{j=1}^{N_E+1} \left(\sum_{l=1}^{N_E} \int_{E_l} \phi_i \, \phi_j \, \mathrm{d}x \right) s_j \quad (1.2.1)$$

for $i = 1, \ldots, N_E$, where E_l stands for the lth element subtended between nodes x_l and x_{l+1}, as shown in Figure 1.2.1.

A key observation is that, because the support of the global interpolation functions is compact, that is, the global interpolation functions are non-zero only over two elements, the diffusion integral on the left-hand side of (1.2.1) can be recast into the form

$$\int_{E_l} \frac{\mathrm{d}\phi_i}{\mathrm{d}x} \frac{\mathrm{d}\phi_j}{\mathrm{d}x} \, \mathrm{d}x = \delta_{i,l} \int_{E_l} \frac{\mathrm{d}\phi_l}{\mathrm{d}x} \frac{\mathrm{d}\phi_j}{\mathrm{d}x} \, \mathrm{d}x + \delta_{i,l+1} \int_{E_l} \frac{\mathrm{d}\phi_{l+1}}{\mathrm{d}x} \frac{\mathrm{d}\phi_j}{\mathrm{d}x} \, \mathrm{d}x \qquad (1.2.2)$$

and then

$$\int_{E_l} \frac{\mathrm{d}\phi_i}{\mathrm{d}x} \frac{\mathrm{d}\phi_j}{\mathrm{d}x} \, \mathrm{d}x = \delta_{i,l} \left(\delta_{j,l} \int_{E_l} \frac{\mathrm{d}\phi_l}{\mathrm{d}x} \frac{\mathrm{d}\phi_l}{\mathrm{d}x} \, \mathrm{d}x + \delta_{j,l+1} \int_{E_l} \frac{\mathrm{d}\phi_l}{\mathrm{d}x} \frac{\mathrm{d}\phi_{l+1}}{\mathrm{d}x} \, \mathrm{d}x \right)$$

$$+ \delta_{i,l+1} \left(\delta_{j,l} \int_{E_l} \frac{\mathrm{d}\phi_{l+1}}{\mathrm{d}x} \frac{\mathrm{d}\phi_l}{\mathrm{d}x} \, \mathrm{d}x + \delta_{j,l+1} \int_{E_l} \frac{\mathrm{d}\phi_{l+1}}{\mathrm{d}x} \frac{\mathrm{d}\phi_{l+1}}{\mathrm{d}x} \, \mathrm{d}x \right), \qquad (1.2.3)$$

where $\delta_{p,q}$ is Kronecker's delta.

Element interpolation functions

In compact notation, we have

$$\int_{E_l} \frac{\mathrm{d}\phi_i}{\mathrm{d}x} \frac{\mathrm{d}\phi_j}{\mathrm{d}x} \, \mathrm{d}x = \delta_{il}\delta_{jl} \, A_{11}^{(l)} + \delta_{il}\delta_{j,l+1} \, A_{12}^{(l)}$$

$$+ \delta_{i,l+1}\delta_{jl} \, A_{21}^{(l)} + \delta_{i,l+1}\delta_{j,l+1} \, A_{22}^{(l)}, \qquad (1.2.4)$$

where $A_{ij}^{(l)}$ is the lth-element diffusion matrix with components

$$A_{11}^{(l)} \equiv \int_{E_l} \frac{\mathrm{d}\phi_l}{\mathrm{d}x} \frac{\mathrm{d}\phi_l}{\mathrm{d}x} \, \mathrm{d}x \quad = \int_{E_l} \frac{\mathrm{d}\psi_1^{(l)}}{\mathrm{d}x} \frac{\mathrm{d}\psi_1^{(l)}}{\mathrm{d}x} \, \mathrm{d}x,$$

$$A_{12}^{(l)} = A_{21}^{(l)} \equiv \int_{E_l} \frac{\mathrm{d}\phi_l}{\mathrm{d}x} \frac{\mathrm{d}\phi_{l+1}}{\mathrm{d}x} \, \mathrm{d}x \quad = \int_{E_l} \frac{\mathrm{d}\psi_1^{(l)}}{\mathrm{d}x} \frac{\mathrm{d}\psi_2^{(l)}}{\mathrm{d}x} \, \mathrm{d}x, \qquad (1.2.5)$$

$$A_{22}^{(l)} \equiv \int_{E_l} \frac{\mathrm{d}\phi_{l+1}}{\mathrm{d}x} \frac{\mathrm{d}\phi_{l+1}}{\mathrm{d}x} \, \mathrm{d}x \quad = \int_{E_l} \frac{\mathrm{d}\psi_2^{(l)}}{\mathrm{d}x} \frac{\mathrm{d}\psi_2^{(l)}}{\mathrm{d}x} \, \mathrm{d}x,$$

and $\psi_1^{(l)}$, $\psi_2^{(l)}$ are the first and second element-node interpolation functions shown in Figure 1.2.1.

Carrying out the integrations, we obtain

$$\mathbf{A}^{(l)} = \begin{bmatrix} A_{11}^{(l)} & A_{12}^{(l)} \\ A_{21}^{(l)} & A_{22}^{(l)} \end{bmatrix} = \frac{1}{h_l} \begin{bmatrix} 1 & -1 \\ -1 & 1 \end{bmatrix}, \qquad (1.2.6)$$

where $h_l \equiv x_{l+1} - x_l$ is the lth element size.

Working in a similar fashion with the integral on the right-hand side of (1.2.1) defining the mass matrix, we obtain

$$\int_{E_l} \phi_i\,\phi_j\,\mathrm{d}x = \delta_{il}\,\delta_{jl}\,B_{11}^{(l)} + \delta_{il}\,\delta_{j,l+1}\,B_{12}^{(l)}$$

$$+\delta_{i,l+1}\,\delta_{jl}\,B_{21}^{(l)} + \delta_{i,l+1}\,\delta_{j,l+1}\,B_{22}^{(l)}, \tag{1.2.7}$$

where $B_{ij}^{(l)}$ is the lth-element mass matrix with components

$$B_{11}^{(l)} \equiv \int_{E_l} \phi_l\,\phi_l\,\mathrm{d}x \quad = \int_{E_l} \psi_1^{(l)}\psi_1^{(l)}\mathrm{d}x,$$

$$B_{12}^{(l)} = B_{21}^{(l)} \equiv \int_{E_l} \phi_l\,\phi_{l+1}\,\mathrm{d}x \quad = \int_{E_l} \psi_1^{(l)}\psi_2^{(l)}\mathrm{d}x, \tag{1.2.8}$$

$$B_{22}^{(l)} \equiv \int_{E_l} \phi_{l+1}\,\phi_{l+1}\,\mathrm{d}x \quad = \int_{E_l} \psi_2^{(l)}\psi_2^{(l)}\mathrm{d}x.$$

Carrying out the integrations, we obtain

$$\mathbf{B}^{(l)} = \begin{bmatrix} B_{11}^{(l)} & B_{12}^{(l)} \\ B_{21}^{(l)} & B_{22}^{(l)} \end{bmatrix} = h_l\,\frac{1}{6}\begin{bmatrix} 2 & 1 \\ 1 & 2 \end{bmatrix}, \tag{1.2.9}$$

where h_l is the lth element size.

Substituting formulas (1.2.4) and (1.2.7) into equation (1.2.1), switching the summation order with respect to j and l, and using the distinguishing properties of Kronecker's delta to simplify the resulting expression, we find that

$$\sum_{l=1}^{N_E} \left[\delta_{il}\left(A_{11}^{(l)}\,f_l + A_{12}^{(l)}\,f_{l+1}\right) + \delta_{i,l+1}\left(A_{21}^{(l)}\,f_l + A_{22}^{(l)}\,f_{l+1}\right)\right] \tag{1.2.10}$$

$$= \frac{q_0}{k}\,\delta_{i,1} + \frac{1}{k}\sum_{l=1}^{N_E}\left[\delta_{il}\left(B_{11}^{(l)}\,s_l + B_{12}^{(l)}\,s_{l+1}\right) + \delta_{i,l+1}\left(B_{21}^{(l)}\,s_l + B_{22}^{(l)}\,s_{l+1}\right)\right].$$

The linear system that arises by applying the finite element Galerkin equation (1.2.10) for $i = 1,\ldots,N_E$ is identical to system (1.1.26). The only new feature is that the global diffusion matrix and the rectangular mass matrix are expressed in terms of the individual element matrices. Specifically, the global diffusion matrix and the rectangular mass matrix take the forms shown in Table 1.2.1. The vector $\mathbf{c}$ defined in (1.1.29) is given by

$$\mathbf{c} = \begin{bmatrix} q_0/k, & 0, & \ldots, & 0, & -A_{12}^{(N_E)}\,f_L \end{bmatrix}^T, \tag{1.2.11}$$

where $A_{12}^{(N_E)}$ is the off-diagonal component of the last element diffusion matrix.

(a)

$$\mathbf{D} = \begin{bmatrix}
A_{11}^{(1)} & A_{12}^{(1)} & 0 & 0 & \cdots \\
A_{21}^{(1)} & A_{22}^{(1)} + A_{11}^{(2)} & A_{12}^{(2)} & 0 & \cdots \\
0 & A_{21}^{(2)} & A_{22}^{(2)} + A_{11}^{(3)} & A_{12}^{(3)} & \cdots \\
\vdots & \vdots & \vdots & \ddots & \vdots \\
0 & 0 & 0 & \cdots & \cdots \\
0 & 0 & 0 & \cdots & \cdots \\
0 & 0 & 0 & \cdots & \cdots
\end{bmatrix}$$

$$\begin{bmatrix}
0 & 0 & 0 \\
0 & 0 & 0 \\
0 & 0 & 0 \\
\vdots & \vdots & \vdots \\
A_{22}^{(N_E-3)} + A_{11}^{(N_E-2)} & A_{12}^{(N_E-2)} & 0 \\
A_{21}^{(N_E-2)} & A_{22}^{(N_E-2)} + A_{11}^{(N_E-1)} & A_{12}^{(N_E-1)} \\
0 & A_{21}^{(N_E-1)} & A_{22}^{(N_E-1)} + A_{11}^{(N_E)}
\end{bmatrix}$$

(b)

$$\widetilde{\mathbf{M}} = \begin{bmatrix}
B_{11}^{(1)} & B_{12}^{(1)} & 0 & 0 & \cdots \\
B_{21}^{(1)} & B_{22}^{(1)} + B_{11}^{(2)} & B_{12}^{(2)} & 0 & \cdots \\
0 & B_{21}^{(2)} & B_{22}^{(2)} + B_{11}^{(3)} & B_{12}^{(3)} & \cdots \\
\vdots & \vdots & \vdots & \ddots & \vdots \\
0 & 0 & 0 & \cdots & \cdots \\
0 & 0 & 0 & \cdots & \cdots
\end{bmatrix}$$

$$\begin{bmatrix}
\cdots & 0 & 0 & 0 \\
\vdots & \vdots & \vdots & \vdots \\
\cdots & 0 & 0 & 0 \\
\cdots & B_{12}^{(N_E-2)} & 0 & 0 \\
\cdots & B_{22}^{(N_E-2)} + B_{11}^{(N_E-1)} & B_{12}^{(N_E-1)} & 0 \\
0 & B_{21}^{(N_E-1)} & B_{22}^{(N_E-1)} + B_{11}^{(N_E)} & B_{12}^{(N_E)}
\end{bmatrix}.$$

TABLE 1.2.1   (a) Square $N_E \times N_E$ global diffusion matrix, $\mathbf{D}$, and (b) rectangular $N_E \times (N_E+1)$ global mass matrix, $\widetilde{\mathbf{M}}$, in terms of corresponding element matrices.

Element assembly

Cursory inspection of (1.2.10) reveals an important property: the lth element contributes only to the Galerkin-projection equations for $i = l$ and $l+1$. Thus, the first element corresponding to $l = 1$ contributes to the first and second equations associated with the projections of ϕ_1 and ϕ_2, and the second element corresponding to $l = 2$ contributes to the second and third equations associated with the projections of ϕ_2 and ϕ_3. Similar contributions are made by subsequent elements. Accordingly, the linear system can be assembled by scanning the elements one by one, while making appropriate additive contributions to the coefficient matrices on the left- and right-hand sides.

The element diffusion matrix is singular

It is instructive to observe that the determinant of the element diffusion matrix shown in (1.2.6) is identically zero, which reveals that the element diffusion matrix is singular. The reason is that, in the one-element implementation with the Neumann boundary condition at both ends and in the absence of a distributed source, the solution can be found only up to an arbitrary constant if the boundary fluxes are equal in magnitude and opposite in sign, or does not exist otherwise. Physically, steady state can be established only when the inward flux balances the outward flux to prevent accumulation. The absence of a unique solution is conclusive evidence of a singular coefficient matrix.

1.2.1   Assembly of a linear system

To develop an algorithm for assembling the linear system (1.1.26), we map the *element nodes* to *unique global nodes* by introducing a *connectivity matrix*, $c_{l,j}$ for $l = 1, \ldots, N_E$ and $j = 1, 2$, defined as follows:

- $c_{l,1} = l$ is the global label of the first node of the lth element.

- $c_{l,2} = l + 1$ is the global label of the second node of the lth element.

In terms of the connectivity matrix, the global diffusion matrix, $\mathbf{D}$, and right-hand side, $\mathbf{b}$, of (1.1.26) can be assembled according to the algorithm shown in Table 1.2.2.

1.2.2   Thomas algorithm for a tridiagonal system

The tridiagonal nature of the global diffusion matrix, $\mathbf{D}$, allows us to solve the linear system arising from the finite element formulation efficiently using the legendary Thomas algorithm. To formalize the algorithm in general terms, we consider a linear system for an N-dimensional unknown vector, $\mathbf{x}$, with a given $N \times N$ coefficient matrix, $\mathbf{T}$,

$$\mathbf{T} \cdot \mathbf{x} = \mathbf{r}, \tag{1.2.12}$$

Do $l = 1, \ldots, N_E$ *Run over the elements to define the connectivity matrix*
   $c(l, 1) = l$
   $c(l, 2) = l + 1$
End Do

Do $i = 1, \ldots, N_E$ *Initialize to zero*
   $b_i = 0.0$
   Do $j = 1, \ldots, N_E$
      $D_{ij} = 0.0$
   End Do
End Do

$b_1 = q_0/k$ *Neumann boundary condition*

Do $l = 1, \ldots, N_E - 1$ *Run over the first $N_E - 1$ elements*
   Compute the element matrices $\mathbf{A}^{(l)}$ and $\mathbf{B}^{(l)}$ using (1.2.6) and (1.2.9)
   $i_1 = c(l, 1)$
   $i_2 = c(l, 2)$
   $D_{i_1,i_1} = D_{i_1,i_1} + A_{11}^{(l)}$
   $D_{i_1,i_2} = D_{i_1,i_2} + A_{12}^{(l)}$
   $D_{i_2,i_1} = D_{i_2,i_1} + A_{21}^{(l)}$
   $D_{i_2,i_2} = D_{i_2,i_2} + A_{22}^{(l)}$
   $b_{i_1} = b_{i_1} + (B_{11}^{(l)} \times s_{i_1} + B_{12}^{(l)} \times s_{i_2})/k$
   $b_{i_2} = b_{i_2} + (B_{21}^{(l)} \times s_{i_1} + B_{22}^{(l)} \times s_{i_2})/k$
End Do

Compute the last-element matrix entries, $A_{11}^{(N_E)}$, $A_{12}^{(N_E)}$, $B_{11}^{(N_E)}$, and $B_{12}^{(N_E)}$,
      using (1.2.6) and (1.2.9)

$D_{N_E,N_E} = D_{N_E,N_E} + A_{11}^{(N_E)}$ *Process the last element*

$b_{N_E} = b_{N_E} + (B_{11}^{(N_E)} \times s_{N_E} + B_{12}^{(N_E)} \times s_{N_E+1})/k - A_{12}^{(N_E)} \times f_L$

TABLE 1.2.2 Algorithm for assembling the finite element linear system for steady one-dimensional diffusion, $\mathbf{D} \cdot \mathbf{f} = \mathbf{b}$, in the case of linear elements.

where the vector $\mathbf{r}$ on the right-hand side is specified. The matrix $\mathbf{T}$ has the tridiagonal form

$$
\mathbf{T} = \begin{bmatrix}
a_1 & b_1 & 0 & \cdots & 0 & 0 & 0 \\
c_2 & a_2 & b_2 & \cdots & 0 & 0 & 0 \\
0 & c_3 & a_3 & \cdots & 0 & 0 & 0 \\
\vdots & \vdots & \vdots & \ddots & \vdots & \vdots & \vdots \\
0 & 0 & 0 & \cdots & a_{N-2} & b_{N-2} & 0 \\
0 & 0 & 0 & \cdots & c_{N-1} & a_{N-1} & b_{N-1} \\
0 & 0 & 0 & \cdots & 0 & c_N & a_N
\end{bmatrix}, \tag{1.2.13}
$$

where a_i, b_i, and c_i are specified matrix components. Thomas's algorithm proceeds in two stages. In the first stage, the tridiagonal system (1.2.12) is reduced to a bidiagonal system,

$$
\mathbf{B} \cdot \mathbf{x} = \mathbf{y}, \tag{1.2.14}
$$

involving a bidiagonal coefficient matrix,

$$
\mathbf{B} = \begin{bmatrix}
1 & d_1 & 0 & \cdots & 0 & 0 & 0 \\
0 & 1 & d_2 & \cdots & 0 & 0 & 0 \\
0 & 0 & 1 & \cdots & 0 & 0 & 0 \\
\vdots & \vdots & \vdots & \ddots & \vdots & \vdots & \vdots \\
0 & 0 & 0 & \cdots & 1 & d_{N-2} & 0 \\
0 & 0 & 0 & \cdots & 0 & 1 & d_{N-1} \\
0 & 0 & 0 & \cdots & 0 & 0 & 1
\end{bmatrix}. \tag{1.2.15}
$$

In the second stage, the bidiagonal system (1.2.14) is solved by backward substitution, which involves solving the last equation for the last unknown, x_N, and moving upward to compute the rest of the unknowns in a sequential fashion. The algorithm is shown in Table 1.2.3 and implemented in the FSELIB function *thomas*, listed in Table 1.2.4.

Thomas's algorithm is a special implementation of the inclusive method of Gauss elimination, which is applicable to general linear systems, as discussed in Section C.1, Appendix C. The sequence of computations is designed to bypass idle operations involving zeros.

1.2.3   Finite element code

The computational modules developed previously in this section can be combined into an integrated finite element code that performs three main tasks: (*a*) domain discretization, (*b*) assembly of a linear system, and (*c*) solution of a linear system.

To economize computer memory, we store the diagonal, super-diagonal, and sub-diagonal elements of the global diffusion matrix in three vectors,

$$
a_i^t \equiv D_{i,i}, \qquad b_i^t \equiv D_{i,i+1}, \qquad c_i^t \equiv D_{i-1,i}, \tag{1.2.16}
$$

Reduction to bidiagonal:

$$\begin{bmatrix} d_1 \\ y_1 \end{bmatrix} = \frac{1}{a_1} \begin{bmatrix} b_1 \\ r_1 \end{bmatrix}$$

Do $i = 1, \ldots, N-1$

$$\begin{bmatrix} d_{i+1} \\ y_{i+1} \end{bmatrix} = \frac{1}{a_{i+1} - c_{i+1}d_i} \begin{bmatrix} b_{i+1} \\ r_{i+1} - c_{i+1}\, y_i \end{bmatrix}$$

End Do

Backward substitution:

$x_N = y_N$

Do $i = N-1, \ldots, 1$

$\quad x_i = y_i - d_i\, x_{i+1}$

End Do

TABLE 1.2.3   Thomas algorithm for solving the tridiagonal system of N linear equations shown in (1.2.12).

corresponding to the tridiagonal form shown in (1.2.13), where the superscript t indicates *tridiagonal.* With this notation, the general assembly algorithm shown in Table 1.2.2 simplifies into the compact algorithm shown in Table 1.2.5.

The simplified algorithm is implemented in the FSELIB function *sdl_sys*, listed in Table 1.2.6.[2]

FSELIB code *sdl*, listed in Table 1.2.7, implements the complete finite element code in the following six stages:

- Data input.
- Finite element discretization.
- Specification of the source function.
- Assembly of a linear system.
- Solution of the linear system.
- Graphical display of the solution.

---

[2] *sdl* is the acronym for steady **d**iffusion with **l**inear elements.

```
function x = thomas (n,a,b,c,r)

%===================================================
% Thomas algorithm for solving a tridiagonal system
%
% n:      system size
% a,b,c: diagonal, superdiagonal,
%         and subdiagonal elements
% r:      right-hand side
%===================================================

%--------
% prepare
%-------

na = n-1;

%-----------------------------
% reduction to upper bidiagonal
%-----------------------------

d(1) = b(1)/a(1);

y(1) = r(1)/a(1);

for i=1:na
   i1 = i+1;
   den = a(i1)-c(i1)*d(i);
   d(i1) = b(i1)/den;
   y(i1) = (r(i1)-c(i1)*y(i))/den;
end

%------------------
% back substitution
%------------------

x(n) = y(n);

for i=na:-1:1
  x(i) = y(i)-d(i)*x(i+1);
end

%-----
% done
%-----

return;
```

TABLE 1.2.4 Function *thomas* solves a tridiagonal system of algebraic equations using the Thomas algorithm.

Do $i = 1, \ldots, N_E$
$\quad a_i^t = 0.0$ *Initialize the coefficients of the tridiagonal system*
$\quad b_i^t = 0.0$ *and right-hand side to zero*
$\quad c_i^t = 0.0$
$\quad b_i = 0.0$
End Do

$b_1 = q_0/k$ *Neumann boundary condition*

Do $l = 1, \ldots, N_E - 1$ *Run over the first N_E-1 elements*
$\quad$ Compute the element matrices $\mathbf{A}^{(l)}$ and $\mathbf{B}^{(l)}$ using (1.2.6) and (1.2.9)
$\quad a_l^t = a_l^t + A_{11}^{(l)}$
$\quad b_l^t = b_l^t + A_{12}^{(l)}$
$\quad c_{l+1}^t = c_{l+1}^t + A_{21}^{(l)}$
$\quad a_{l+1}^t = a_{l+1}^t + A_{22}^{(l)}$
$\quad b_{i_1} = b_{i_1} + (B_{11}^{(l)} \times s_l + B_{12}^{(l)} \times s_{l+1})/k$
$\quad b_{i_2} = b_{i_2} + (B_{21}^{(l)} \times s_l + B_{22}^{(l)} \times s_{l+1})/k$
End Do

Compute the last-element matrix components, $A_{11}^{(N_E)}$, $A_{12}^{(N_E)}$, $B_{11}^{(N_E)}$, and $B_{12}^{(N_E)}$,
$\quad$ using (1.2.6) and (1.2.9)

$a_{N_E}^t = a_{N_E}^t + A_{11}^{(N_E)}$ *Process the last element*
$b_{N_E} = b_{N_E} + (B_{11}^{(N_E)} \times s_{N_E} + B_{12}^{(N_E)} \times s_{N_E+1})/k - A_{12}^{(N_E)} \times f_L$

TABLE 1.2.5 Compact assembly of a tridiagonal linear algebraic system for linear elements, $\mathbf{D} \cdot \mathbf{f} = \mathbf{b}$, where a^t, b^t, and c^t are the components of the tridiagonal global diffusion matrix, $\mathbf{D}$.

```
function [at,bt,ct,b] = sdl_sys (ne,xe,q0,fL,k,s)

%=======================================================
% Compact assembly of the tridiagonal linear system
% for steady diffusion with linear elements (sdl)
%=======================================================

%--------------
% element size
%--------------

for l=1:ne
  h(l) = xe(l+1)-xe(l);
end

%---------------------------------
% initialize the tridiagonal matrix
%---------------------------------

at = zeros(ne,1);
bt = zeros(ne,1);
ct = zeros(ne,1);

%----------------------------
% initialize the right-hand side
%----------------------------

b = zeros(ne,1);

b(1) = q0/k;

%----------------------------------
% loop over the first ne-1 elements
%----------------------------------

for l=1:ne-1

  A11 = 1/h(l); A12 =-A11;
  A21 = A12;      A22 = A11;

  B11 = h(l)/3.0; B12 = 0.5*B11;
  B21 = B12;        B22 = B11;

  at(l) = at(l) + A11;
  bt(l) = bt(l) + A12;

  ct(l+1) = ct(l+1) + A21;
  at(l+1) = at(l+1) + A22;

  b(l)   = b(l)   + (B11*s(l) + B12*s(l+1))/k;
  b(l+1) = b(l+1) + (B21*s(l) + B22*s(l+1))/k;

end
```

TABLE 1.2.6   Function *sdl_sys* (Continuing →)

```
%----------------------------
% the last element is special
%----------------------------

A11 = 1/h(ne); A12 =-A11;

B11 = h(ne)/3.0; B12 = 0.5*B11;

at(ne) = at(ne) + A11;

b(ne) = b(ne) + (B11*s(ne) + B12*s(ne+1))/k - A12*fL;

%-----
% done
%-----

return;
```

TABLE 1.2.6 ($\rightarrow$ Continued) Function *sdl_sys* implements the compact assembly of a tridiagonal linear system for one-dimensional diffusion with linear elements.

Accuracy

The solution generated by the code for the Gaussian source function

$$s(x) = s_0 \exp(-5\,x^2/L^2)$$ (1.2.17)

in the interval $[0, L]$ is shown in Figure 1.2.2, where s_0 is a constant. Calculations for a specified value of the element stretch ratio $ratio = 5$ and number of elements,

$$N_E = 8, \quad 16, \quad 32, \quad 64, \quad 128, \quad 256,$$ (1.2.18)

yield the left-end value

$$f(0) = 1.9746, \quad 1.9662, \quad 1.9644, \quad 1.9640, \quad 1.9639, \quad 1.9639.$$ (1.2.19)

A similar set of calculations with uniformly spaced elements, $ratio = 1$, yield the left-end value

$$f(0) = 1.9510, \quad 1.9606, \quad 1.9631, \quad 1.9637, \quad 1.9638, \quad 1.9639.$$ (1.2.20)

These results clearly indicate that $f(0) = 1.9639$ is the exact solution, accurate to the fourth decimal place, achieved with 128 or 256 elements. In the case of evenly spaced elements, the numerical error, $e(0) \equiv f(0) - 1.9639$, is found to be

$$e(0) = -0.0129, \quad -0.0032, \quad -0.0008, \quad -0.0002, \quad -0.0001, \quad 0.0000,$$ (1.2.21)

to shown accuracy.

```
%================================
% Code sdl
%
% Steady one-dimensional diffusion
% with linear elements
%================================

%-----------
% input data
%-----------

L = 1.0;
k = 1.0; q0 =-1.0; fL = 0.0;
ne = 16; ratio = 5.0;

%----------------
% grid generation
%----------------

xe = elm_line1 (0,L,ne,ratio);

%------------------
% specify the source
%------------------

for i=1:ne+1
 s(i) = 10.0*exp(-5.0*xe(i)^2/L^2);
end

%-----------------
% compact assembly
%-----------------

[at,bt,ct,b] = sdl_sys (ne,xe,q0,fL,k,s);

%-------------
% linear solver
%-------------

f = thomas (ne,at,bt,ct,b);

f(ne+1) = fL;

%-----
% plot
%-----

plot(xe, f,'-ko');

%-----
% done
%-----
```

TABLE 1.2.7   Finite element code *sdl* for steady diffusion with linear elements.

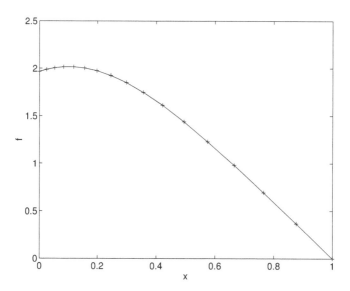

FIGURE 1.2.2   Finite element solution of the steady diffusion equation computed by
the FSELIB code *sdl* with 16 linear elements, $N_{\rm E} = 16$, and element stretch ratio
equal to 5. Other parameter values are $L = 1.0$, $k = 1.0$, $q_0 = -1.0$, $f(L) = 0.0$,
and $s_0 = 10.0$.

These numerical experiments suggest that when the number of elements is doubled, the numerical error decreases approximately by a factor of 4, which means
that the error is proportional to h^2, where h is a typical element size. We conclude
that the finite element method with linear elements for the steady diffusion problem with a smooth source term is second-order accurate in the element size. This
conclusion can be confirmed by performing a formal error analysis, keeping track of
the error introduced by the various numerical approximations (e.g., Karniadakis &
Sherwin [32]).

1.2.4   Convection (Robin or mixed) boundary condition

A convection boundary condition at the left end of the solution domain, also called
a *Robin* or *mixed* boundary condition, requires that the flux is related to the local
function value by a linear equation,

$$q_0 \equiv -k \left(\frac{\mathrm{d}f}{\mathrm{d}x} \right)_{x=0} = -h_T \left(f_0 - f_\infty \right), \qquad (1.2.22)$$

where h_T is a heat transfer coefficient and f_∞ is a specified left-end ambient temperature. The mixed condition is implemented in the finite element assembly algorithm shown in Table 1.2.5 by replacing the seventh line implementing the Neumann

boundary condition,

$$b_1 = q_0/k \tag{1.2.23}$$

with the following two lines:

$$b_1 = h_T f_\infty / k$$
$$a_1^t = h_T / k \tag{1.2.24}$$

The assembly algorithm is implemented in the FSELIB function *sdl_sys_robin*, not listed in the text. The finite element method itself is implemented in the FSELIB code *sdl_robin*, not listed in the text.

PROBLEMS

1.2.1 *Flux at the Dirichlet end*

Consider the implementation of the finite element method with linear elements, as discussed in the text. The flux at the Dirichlet end is defined as

$$q_L \equiv -k \left(\frac{\mathrm{d}f}{\mathrm{d}x} \right)_{x=L} . \tag{1.2.25}$$

Show that

$$q_L = -k \left(f_{N_E} A_{21}^{(N_E)} + f_{N_E+1} A_{22}^{(N_E)} \right) + s_{N_E} B_{21}^{(N_E)} + s_{N_E+1} B_{22}^{(N_E)} . \tag{1.2.26}$$

1.2.2 *Element diffusion and mass matrices*

Perform the integrations to derive the element diffusion and mass matrices shown in (1.2.6) and (1.2.9).

1.2.3 *Code sdl*

Execute the FSELIB code *sdl* with the same boundary conditions but a different source function of your choice and discuss the finite element solution.

1.2.4 *Convection (Robin or mixed) boundary condition*

Execute the FSELIB code *sdl_robin* and discuss the effect of the heat transfer coefficient, h_T.

1.3   Variational formulation and weighted residuals

For a certain class of differential equations, including the steady diffusion equation discussed in this chapter, the Galerkin projection method is equivalent to Ritz's implementation of Rayleigh's variational formulation discussed in this section. Moreover, the Galerkin finite element method itself can be regarded as a special implementation of the method of weighted residuals. These complementary viewpoints lend credence to the finite element method and offer an alternative venue for building the relevant mathematical framework.

Significance of the variational formulation

One advantage of the variational approach is that various physical effects can be incorporated readily in terms of energy functions that could be cumbersome to implement directly into the governing differential equations. In some disciplines of computational mechanics, the weak formulation based on the governing differential equations discussed in Section 1.1 is altogether abandoned and the finite element method is developed exclusively in the context of the variational formulation. In fact, some authors convey the misleading impression that the finite element method arises strictly in the context of the variational formulation. The truth is that the finite element method can be applied to a broader class of equations, even when a variational formulation is not available or apparent.

1.3.1 Homogeneous Dirichlet boundary conditions

To illustrate the connection with the variational formulation, we consider a linear differential operator, $\mathcal{L} < \omega >$, operating on the space of scalar functions, $\omega(x)$, defined in an interval, $[a, b]$, and satisfying the homogeneous Dirichlet boundary conditions

$$\omega(x = a) = 0, \qquad \omega(x = b) = 0. \tag{1.3.1}$$

Other boundary conditions at one or both ends will be considered later in this section.

Inner function production

To simplify the notation, we define the inner product of two functions, $\omega_1(x)$ and $\omega_2(x)$, as a curly bracket integral operation,

$$\{\omega_1, \omega_2\} \equiv \int_a^b \omega_1(x)\,\omega_2(x)\,\mathrm{d}x. \tag{1.3.2}$$

We further assume that the linear operator of interest is *self-adjoint*, which means that

$$\{\mathcal{L} < \omega_1 >,\, \omega_2\} = \{\mathcal{L} < \omega_2 >,\, \omega_1\}. \tag{1.3.3}$$

The counterparts of self-adjoint operators in matrix calculus are symmetric or Hermitian matrices.

Sturm–Liouville operator

An example of a self-adjoint operator is the Sturm–Liouville second-order operator

$$\mathcal{L} < \phi > = \frac{\mathrm{d}}{\mathrm{d}x}\left(p(x)\,\frac{\mathrm{d}\phi}{\mathrm{d}x}\right) - r(x)\,\phi(x), \tag{1.3.4}$$

where $p(x)$ and $r(x)$ are specified functions and $\phi(x)$ is an appropriate function. To demonstrate that this operator is self-adjoint, we must show that

$$\int_a^b \left[\frac{d}{dx}\left(p(x) \frac{d\omega_1}{dx} \right) - r(x)\,\omega_1 \right] \omega_2\,dx = \int_a^b \left[\frac{d}{dx}\left(p(x) \frac{d\omega_2}{dx} \right) - r(x)\,\omega_2 \right] \omega_1\,dx$$

$$(1.3.5)$$

for any two suitable functions, ω_1 and ω_2, that satisfy homogeneous boundary conditions, $\omega_1(a) = 0$, $\omega_1(b) = 0$, and $\omega_2(a) = 0$, $\omega_2(b) = 0$. The proof follows readily by integrating by parts the first term inside the integral on either side.

Rayleigh functional

Next, we introduce a source function, $s(x)$, and define the quadratic Rayleigh functional

$$\mathcal{F}<\omega(x)> \equiv \frac{1}{2}\left\{ \mathcal{L}<\omega(x)>, \omega(x) \right\} + \left\{ s(x), \omega(x) \right\},\qquad(1.3.6)$$

which maps a given function, $\omega(x)$, to a number, $\mathcal{F}<\omega(x)>$. We will demonstrate that the minimum or maximum value of the functional $\mathcal{F}<\omega(x)>$ is achieved for a function $f(x)$ that satisfies the linear differential equation

$$\mathcal{L}<f(x)> + s(x) = 0,\qquad(1.3.7)$$

subject to the boundary conditions $f(a) = 0$ and $f(b) = 0$.

Stated differently, minimizing or maximizing the functional $\mathcal{F}<\omega(x)>$ over a class of admissible functions that satisfy the boundary conditions $\omega(a) = 0$ and $\omega(b) = 0$ is equivalent to solving the differential equation (1.3.7) subject to the homogeneous Dirichlet boundary condition. In this sense, the differential equation (1.3.7) is the *Euler–Lagrange equation* of the functional (1.3.6).

To prove the equivalence, we perturb the solution, $f(x)$, to $f(x) + \epsilon\,v(x)$, where $v(x)$ is an appropriate disturbance function, also called a variation, required to satisfy the homogeneous Dirichlet boundary condition $v(a) = 0$ and $v(b) = 0$, and ϵ is a dimensionless number whose magnitude is much less than unity. Next, we consider the variation

$$\delta\mathcal{F} \equiv \mathcal{F}<f(x) + \epsilon\,v(x)> - \mathcal{F}<f(x)>,\qquad(1.3.8)$$

and use the definition of $\mathcal{F}$ to write

$$\mathcal{F}<f(x) + \epsilon\,v(x)> = \frac{1}{2}\left\{ \mathcal{L}<f(x)> + \epsilon\,\mathcal{L}<v(x)>, f(x) + \epsilon\,v(x) \right\}$$

$$+ \left\{ s(x), f(x) + \epsilon\,v(x) \right\}.\qquad(1.3.9)$$

Expanding the products inside the brackets, using the property

$$\left\{ \mathcal{L}<v(x)>, f(x) \right\} = \left\{ \mathcal{L}<f(x)>, v(x) \right\},\qquad(1.3.10)$$

and simplifying, we obtain

$$\delta \mathcal{F} = \epsilon \left\{ \mathcal{L} < f(x) > +s(x), \ v(x) \right\} + \frac{1}{2} \epsilon^2 \left\{ \mathcal{L} < v(x) >, v(x) \right\}, \qquad (1.3.11)$$

which shows that, if the first term on the right-hand side is zero for every x, then the functional is locally stationary and the minimum or maximum is achieved when $w(x) = f(x)$.

Ritz's method

In Ritz's method, the admissible functions $w(x)$ are expressed in a truncated series of a suitable set of basis functions, $\varphi_i(x)$, satisfying the homogeneous boundary conditions $\varphi_i(a) = 0$ and $\varphi_i(b) = 0$, as

$$w(x) = \sum_{i=1}^{N} c_i \, \varphi_i(x), \qquad (1.3.12)$$

where N is a specified truncation level. Substituting this expansion into the Rayleigh functional, we obtain

$$\mathcal{F} < w(x) > \equiv \mathcal{G}(c_1, \dots, c_N), \qquad (1.3.13)$$

where $\mathcal{G}$ is a quadratic function of the N coefficients, c_i. To compute these coefficients, we require the extremum condition

$$\frac{\partial \mathcal{G}}{\partial c_i} = 0 \qquad (1.3.14)$$

for $i = 1, \dots, N$, and solve the resulting system of linear algebraic equations to obtain the function $f(x)$.

Weighted residuals and weak formulation

Expression (1.3.11) shows that solving the differential equation (1.3.7) is equivalent to computing a function, $f(x)$, that satisfies the equation

$$\left\{ \mathcal{L} < f(x) > +s(x), \quad v(x) \right\} = 0 \qquad (1.3.15)$$

for any acceptable test function, $v(x)$, interpreted as a disturbance or variation. The left-hand side of (1.3.15) is a weighted residual of the differential equation. Replacing the differential equation with the integrated form (1.3.15) is the starting point in a class of approximate methods, known as methods of weighted residuals.

A solution procedure based on (1.3.15) is recognized as a *weak formulation*, whereas a solution procedure based on the primary differential equation is recognized as the *strong formulation*. The weak formulation carries the risk of introducing spurious components in the process of integration, playing the role of eigenfunctions.

In the Galerkin implementation of the method of weighted residuals, the solution is expanded as shown in (1.3.12), and the test functions, $v(x)$, are identified with each one of the basis functions, φ_i. When the operator, $\mathcal{L}$, satisfies the conditions stated earlier in this section, the Galerkin method becomes equivalent to Ritz's method based on the Rayleigh variational formulation. In the Galerkin finite element method, the basis functions, $\varphi_i(x)$, and the test functions, $v(x)$, are identified with the finite element global interpolation functions, ϕ_i, discussed in Section 1.1, as will be demonstrated later in this section by an example.

Rayleigh functional for one-dimensional diffusion

As an application, we note that the one-dimensional diffusion operator,

$$\mathcal{L} < \phi > = k \, \frac{\mathrm{d}^2 \phi}{\mathrm{d}x^2}, \qquad (1.3.16)$$

is a Sturm–Liouville operator with $p(x) = k$ and $r(x) = 0$. The variational formulation states that a solution of the differential equation

$$k \, \frac{\mathrm{d}^2 f}{\mathrm{d}x^2} + s(x) = 0, \qquad (1.3.17)$$

subject to homogeneous boundary conditions, $f(a) = 0$ and $f(b) = 0$, maximizes the quadratic functional

$$\mathcal{F} < \omega(x) > \equiv \frac{1}{2} k \int_a^b \frac{\mathrm{d}^2 \omega}{\mathrm{d}x^2} \, \omega(x) \, \mathrm{d}x + \int_a^b s(x) \, \omega(x) \, \mathrm{d}x \qquad (1.3.18)$$

over the set of all functions, $\omega(x)$, that satisfy the boundary conditions $\omega(a) = 0$ and $\omega(b) = 0$.

Integrating by parts the first term on the right-hand side of (1.3.18) and enforcing the homogeneous boundary conditions, we obtain the alternative expression

$$\mathcal{F} < \omega(x) > \equiv -\frac{1}{2} k \int_a^b \left(\frac{\mathrm{d}\omega}{\mathrm{d}x} \right)^2 \mathrm{d}x + \int_a^b s(x) \, \omega(x) \, \mathrm{d}x. \qquad (1.3.19)$$

Physically, the two integrals on the right-hand side express the thermal energy associated with the diffusive flux and the source. Maximizing this functional is equivalent to solving the steady diffusion equation (1.3.17) over the interval $[a, b]$, subject to the homogeneous Dirichlet conditions $f(a) = 0$ and $f(b) = 0$.

In the absence of a source term, $s(x) = 0$, the functional is negative and the maximum value of zero is clearly achieved for $\omega(x) = f(x) = 0$.

1.3.2   Inhomogeneous Dirichlet boundary conditions

Now we consider the steady diffusion equation (1.3.17) in an interval, $[a, b]$, subject to the inhomogeneous Dirichlet conditions

$$f(a) = f_a, \qquad f(b) = f_b. \qquad (1.3.20)$$

We will demonstrate that solving this equation is still equivalent to maximizing the functional given in (1.3.19) over all continuous functions, $w(x)$, that satisfy the specified Dirichlet boundary conditions, $w(a) = f_a$ and $w(b) = f_b$. Integrating by parts the first term on the right-hand side of (1.3.19), we obtain the functional

$$\mathcal{F} < w(x) >= \frac{1}{2} k \left[\int_a^b \frac{\mathrm{d}^2 w}{\mathrm{d}x^2} w \, \mathrm{d}x - \left(w \frac{\mathrm{d}w}{\mathrm{d}x} \right)_a^b \right] + \int_a^b s(x) w(x) \, \mathrm{d}x. \quad (1.3.21)$$

When homogeneous Dirichlet boundary conditions are prescribed, $w(a) = 0$ and $w(b) = 0$, the second term on the right-hand side does not appear.

To prove the equivalence to minimization, we perturb the solution of the forced diffusion equation from $f(x)$ to $f(x) + \epsilon\, v(x)$, and compute the difference in $\mathcal{F}$ based on (1.3.19), finding

$$\delta\mathcal{F} = -\epsilon k \int_a^b \frac{\mathrm{d}f}{\mathrm{d}x} \frac{\mathrm{d}v}{\mathrm{d}x} \, \mathrm{d}x + \epsilon \int_a^b s(x)\, v(x) \, \mathrm{d}x + O(\epsilon^2), \quad (1.3.22)$$

which can be restated as

$$\delta\mathcal{F} = \epsilon k \int_a^b \frac{\mathrm{d}^2 f}{\mathrm{d}x^2} v \, \mathrm{d}x - \epsilon k \left(v \frac{\mathrm{d}f}{\mathrm{d}x} \right)_a^b + \epsilon \int_a^b s(x)\, v(x) \, \mathrm{d}x + O(\epsilon^2). \quad (1.3.23)$$

Applying the boundary conditions $v(a) = 0$ and $v(b) = 0$, we obtain

$$\delta\mathcal{F} = \epsilon \left[\int_a^b \left(k \frac{\mathrm{d}^2 f}{\mathrm{d}x^2} + s(x) \right) v(x) \, \mathrm{d}x \right] + O(\epsilon^2). \quad (1.3.24)$$

If the function $f(x)$ satisfies (1.3.17) subject to the aforementioned boundary conditions, the first variation vanishes and the functional is stationary.

Alternatively, the variation in $\mathcal{F}$ can be computed based on (1.3.21), yielding

$$\delta\mathcal{F} = \epsilon \frac{1}{2} k \left[\int_a^b \left(\frac{\mathrm{d}^2 f}{\mathrm{d}x^2} v + \frac{\mathrm{d}^2 v}{\mathrm{d}x^2} f \right) \mathrm{d}x - \left(f \frac{\mathrm{d}v}{\mathrm{d}x} \right)_a^b \right] + \epsilon \int_a^b s(x)\, v(x) \, \mathrm{d}x + O(\epsilon^2). \quad (1.3.25)$$

Using Green's second identity in one dimension, we write

$$\int_a^b \frac{\mathrm{d}^2 v}{\mathrm{d}x^2} f \, \mathrm{d}x = \int_a^b \frac{\mathrm{d}^2 f}{\mathrm{d}x^2} v \, \mathrm{d}x + \left(f \frac{\mathrm{d}v}{\mathrm{d}x} - v \frac{\mathrm{d}f}{\mathrm{d}x} \right)_a^b, \quad (1.3.26)$$

and then apply the boundary conditions to recover precisely (1.3.24).

Rayleigh–Ritz method

In Ritz's method, the expansion (1.3.12) is endowed with a leading term, $\varphi_0(x)$, that satisfies the stipulated boundary conditions, $\varphi_0(a) = f_a$ and $\varphi_0(b) = f_b$, yielding

$$w(x) = \varphi_0(x) + \sum_{i=1}^N c_i\, \varphi_i(x). \quad (1.3.27)$$

The remaining basis functions, φ_i for $i = 1, \ldots, N$, satisfy homogeneous boundary conditions, $\varphi_i(a) = 0$ and $\varphi_i(b) = 0$. Substituting this expansion into the Rayleigh functional, we derive the quadratic algebraic form (1.3.13) and compute the coefficients, c_i, by requiring the extremum condition (1.3.14).

1.3.3 Dirichlet/Neumann boundary conditions

In the next case, we consider the steady diffusion equation (1.3.17), to be solved in an interval, $[a, b]$, subject to the Dirichlet boundary condition at the left end and the Neumann boundary condition at the right end,

$$f(a) = f_a, \qquad -k \left(\frac{\mathrm{d}f}{\mathrm{d}x} \right)_{x=b} = q_b, \qquad (1.3.28)$$

where f_a and q_b are prescribed values. We will demonstrate that computing the solution to this problem is equivalent to maximizing the functional

$$\mathcal{F} < w(x) > \equiv -\frac{1}{2} k \int_a^b \left(\frac{\mathrm{d}w}{\mathrm{d}x} \right)^2 \mathrm{d}x + \int_a^b s(x) \, w(x) \, \mathrm{d}x - w(x = b) \, q_b \qquad (1.3.29)$$

over all continuous functions, $w(x)$, that satisfy the left-end boundary condition alone, $w(a) = f_a$. The satisfaction of the Neumann boundary condition is enforced by the last term on the right-hand side of (1.3.29).

Integrating by parts the first term on the right-hand side of (1.3.29), we obtain

$$\mathcal{F} < w(x) > = \frac{1}{2} k \left[\int_a^b \frac{\mathrm{d}^2 w}{\mathrm{d}x^2} w \, \mathrm{d}x - \left(w \frac{\mathrm{d}w}{\mathrm{d}x} \right)_a^b \right] + \int_a^b s(x) \, w(x) \, \mathrm{d}x - w(x = b) \, q_b,$$
$$(1.3.30)$$

which can be restated as

$$\mathcal{F} < w(x) > = \frac{1}{2} k \int_a^b \frac{\mathrm{d}^2 w}{\mathrm{d}x^2} w \, \mathrm{d}x + \int_a^b s(x) \, w(x) \, \mathrm{d}x$$
$$+ \frac{1}{2} k f_a \left(\frac{\mathrm{d}w}{\mathrm{d}x} \right)_{x=a} - \frac{1}{2} w(x = b) \, q_b. \qquad (1.3.31)$$

When homogeneous boundary conditions are prescribed, $f_a = 0$ and $q_b = 0$, the last two terms on the right-hand side do not appear.

To prove the equivalence, we consider a perturbed solution, $f(x) + \epsilon \, v(x)$, where the disturbance function $v(x)$ satisfies the homogeneous left-end Dirichlet boundary condition, $v(a) = 0$, and compute the variation based on (1.3.29), finding

$$\delta \mathcal{F} = -\epsilon k \int_a^b \frac{\mathrm{d}f}{\mathrm{d}x} \frac{\mathrm{d}v}{\mathrm{d}x} \, \mathrm{d}x + \epsilon \int_a^b s(x) \, v(x) \, \mathrm{d}x - \epsilon \, v(x = b) \, q_b + O(\epsilon^2), \qquad (1.3.32)$$

which can be restated as

$$\delta \mathcal{F} = \epsilon \left[k \int_a^b \frac{\mathrm{d}^2 f}{\mathrm{d}x^2} v \, \mathrm{d}x - k \left(v \frac{\mathrm{d}f}{\mathrm{d}x} \right)_a^b + \int_a^b s(x) \, v(x) \, \mathrm{d}x - v(x = b) \, q_b \right] + O(\epsilon^2).$$

$$(1.3.33)$$

Enforcing the boundary conditions, we derive precisely (1.3.24). Thus, if the function $f(x)$ satisfies (1.3.17) subject to the aforementioned boundary conditions, the first variation vanishes and the functional is stationary.

Alternatively, the change in $\mathcal{F}$ can be computed based on (1.3.31), yielding

$$\delta \mathcal{F} = \epsilon \frac{1}{2} k \left[\int_a^b \left(\frac{\mathrm{d}^2 f}{\mathrm{d}x^2} v + \frac{\mathrm{d}^2 v}{\mathrm{d}x^2} f \right) \mathrm{d}x - \left(f \frac{\mathrm{d}v}{\mathrm{d}x} + v \frac{\mathrm{d}f}{\mathrm{d}x} \right)_a^b \right]$$

$$+ \epsilon \int_a^b s(x) \, v(x) \, \mathrm{d}x - \epsilon \, v(x = b) \, q_b + O(\epsilon^2). \qquad (1.3.34)$$

Using Green's second identity (1.3.26) to simplify the right-hand side, we derive (1.3.24).

Rayleigh–Ritz method

To implement Ritz's method, we introduce expansion (1.3.27), repeated here for convenience,

$$w(x) = \varphi_0(x) + \sum_{i=1}^{N} c_i \, \varphi_i(x), \qquad (1.3.35)$$

where the leading term, $\varphi_0(x)$, satisfies the specified Dirichlet condition at the left end, $\varphi_0(a) = f_a$. The rest of the basis functions, $\varphi_i(x)$, satisfy the homogeneous Dirichlet boundary condition, $\varphi_i(a) = 0$ for $i = 1, \ldots, N$. Substituting this expansion into the Rayleigh functional, we derive the algebraic form (1.3.13), and then require the extremum condition (1.3.14).

As an example, we consider the diffusion equation in the absence of source, $s = 0$. The Rayleigh functional takes the form

$$\mathcal{F} < w(x) > \equiv -\frac{1}{2} k \int_a^b \left(\frac{\mathrm{d}w}{\mathrm{d}x} \right)^2 \mathrm{d}x - w(x = b) \, q_b. \qquad (1.3.36)$$

Identifying $\varphi_0(x)$ with a constant function, $\varphi_0(x) = f_a$, truncating the expansion at the first term, $N = 1$, and choosing $\varphi_1(x) = x - a$, we obtain

$$w(x) = f_a + c_1 \, (x - a). \qquad (1.3.37)$$

A straightforward computation yields

$$\mathcal{F} < w(x) > \equiv \mathcal{G}(c_1) = -\frac{1}{2} k \, L \, c_1^2 - (f_a + c_1 L) \, q_b, \qquad (1.3.38)$$

where $L = b - a$ is the length of the solution domain. Requiring that

$$\frac{\partial \mathcal{G}}{\partial c_1} = -k\,L\,c_1 - L\,q_b = 0, \tag{1.3.39}$$

we obtain $c_1 = -q_b/k$, which can be substituted into (1.3.37) to produce the exact solution,

$$f(x) = f_a - \frac{q_b}{k}\,(x - a). \tag{1.3.40}$$

1.3.4 Neumann/Dirichlet boundary conditions

As a complement to the problem discussed in Section 1.3.3, we consider the steady diffusion equation (1.3.17), to be solved in an interval, $[a, b]$, subject to the Neumann boundary condition at the left end and the Dirichlet boundary condition at the right end,

$$-k\left(\frac{\mathrm{d}f}{\mathrm{d}x}\right)_{x=a} = q_a, \qquad f(b) = f_b, \tag{1.3.41}$$

where the values q_a and f_b are specified. Working as previously in this section, we find that computing the solution is equivalent to maximizing the functional

$$\mathcal{F} < w(x) > \equiv -\frac{1}{2}\,k \int_a^b \left(\frac{\mathrm{d}w}{\mathrm{d}x}\right)^2 \mathrm{d}x + \int_a^b s(x)\,w(x)\,\mathrm{d}x + w(x = a)\,q_a \tag{1.3.42}$$

over all continuous functions that satisfy the right-end boundary condition, $w(b) = f_b$. The satisfaction of the Neumann boundary condition at the left end is enforced by the last term on the right-hand side of (1.3.42).

To implement Ritz's method, we introduce expansion (1.3.35), where the leading term, $\varphi_0(x)$, satisfies the specified Dirichlet condition at the right end, $\varphi_0(b) = f_b$. The rest of the basis functions, $\varphi_i(x)$, satisfy the homogeneous Dirichlet boundary condition, $\varphi_i(b) = 0$ for $i = 1, \ldots, N$. Substituting this expansion into the Rayleigh functional, we derive the algebraic form (1.3.13), and then require the extremum condition (1.3.14).

To demonstrate the relation between the Rayleigh–Ritz and the Galerkin finite element method, we consider the steady diffusion problem discussed in Section 1.1, and set $a = 0$ and $b = L$. Next, we set the basis function $\varphi_0(x)$ proportional to the global interpolation function associated with the last node,

$$\varphi_0(x) = f_L\,\phi_{N_E+1}(x), \tag{1.3.43}$$

and identify the rest of the basis functions with the previous node interpolation functions,

$$\varphi_i(x) = \phi_i(x) \tag{1.3.44}$$

for $i = 1, \ldots, N_{\mathrm{E}}$, to obtain

$$\omega(x) = f_L \, \phi_{N_{\mathrm{E}}+1}(x) + \sum_{i=1}^{N_{\mathrm{E}}} \omega_i \, \phi_i(x) \tag{1.3.45}$$

and

$$\omega'(x) = f_L \, \phi'_{N_{\mathrm{E}}+1}(x) + \sum_{i=1}^{N_{\mathrm{E}}} \omega_i \, \phi'_i(x), \tag{1.3.46}$$

where a prime denotes a derivative with respect to x, and $\omega_i \equiv \omega(x_i)$. In addition, we express the source term in terms of the global interpolation functions as

$$s(x) = \sum_{i=1}^{N_{\mathrm{E}}+1} s_i \, \phi_i(x), \tag{1.3.47}$$

where $s_i \equiv s(x_i)$. Substituting these expansions into the Rayleigh functional (1.3.42), we obtain

$$\mathcal{F} < \omega(x) > \equiv \mathcal{G}(\omega_1, \omega_2, \ldots, \omega_{N_{\mathrm{E}}}) = -\frac{1}{2} k \sum_{i=1}^{N_{\mathrm{E}}+1} \left(\omega_i \sum_{j=1}^{N_{\mathrm{E}}+1} D_{ij} \, \omega_j \right)$$

$$+ \sum_{i=1}^{N_{\mathrm{E}}+1} \left(\omega_i \sum_{j=1}^{N_{\mathrm{E}}+1} M_{ij} \, s_j \right) + \omega_1 \, q_0, \tag{1.3.48}$$

where D_{ij} is the global diffusion matrix and M_{ij} is the global mass matrix defined in Table 1.1.5, and $\omega_{N_{\mathrm{E}}+1} = f_L$. In vector notation,

$$\mathcal{G}(\boldsymbol{\omega}) = -\frac{1}{2} k \, \boldsymbol{\omega} \cdot \mathbf{D} \cdot \boldsymbol{\omega} + \boldsymbol{\omega} \cdot \mathbf{M} \cdot \mathbf{s} + \omega_1 \, q_0. \tag{1.3.49}$$

To maximize the functional, we require that

$$\frac{\partial \mathcal{F}}{\partial \omega_i} = 0 \tag{1.3.50}$$

for $i = 1, \ldots, N_{\mathrm{E}}$, set $\omega_i = f_i$, and derive precisely the linear system (1.1.26).

PROBLEM

1.3.1 *Mixed boundary condition*

Develop the variational formulation of the steady diffusion equation subject to the Dirichlet boundary condition at the left end, $f(a) = f_a$, and the following mixed or Robin boundary condition at the right end,

$$q_b \equiv -k \left(\frac{\mathrm{d}f}{\mathrm{d}x} \right)_{x=b} = h_T \left(f_b - f_\infty \right), \tag{1.3.51}$$

where $f_b \equiv f(x = b)$, f_∞ is a specified ambient temperature, and h_T is a heat transfer coefficient.

1.4 Helmholtz equation

The Helmholtz equation in one dimension is a special case of the Poisson equation where the source is linear in the unknown function, $s = \alpha f$, yielding

$$\frac{\mathrm{d}^2 f}{\mathrm{d}x^2} + \alpha f = 0, \tag{1.4.1}$$

where α is a positive or negative constant with dimensions of inverse squared length. We will assume that the Neumann boundary condition (1.1.5) and the Dirichlet boundary condition (1.1.6) are specified at the ends of the solution domain, $[0, L]$, repeated below for convenience,

$$q_0 \equiv -k \left(\frac{\mathrm{d}f}{\mathrm{d}x}\right)_{x=0}, \qquad f(x = L) \equiv f_L, \tag{1.4.2}$$

where q_0 and f_L are two given constants.

Exact solution

When $\alpha > 0$, the exact solution is

$$f(x) = c_1 \sin(\sqrt{\alpha}x) + c_2 \cos(\sqrt{\alpha}x), \tag{1.4.3}$$

where the constants c_1 and c_2 are determined by the boundary conditions, yielding

$$c_1 = -\frac{q_0}{k\sqrt{a}}, \qquad c_2 = \frac{f_L - c_1 \sin(\sqrt{\alpha}L)}{\cos(\sqrt{\alpha}L)}. \tag{1.4.4}$$

When $\alpha < 0$, the exact solution is

$$f(x) = c_1 \exp(\sqrt{|\alpha|}\,x) + c_2 \exp(-\sqrt{|\alpha|}x), \tag{1.4.5}$$

where the constants c_1 and c_2 are determined by the boundary conditions, yielding

$$c_1 = \frac{f_L - \hat{q}_0 \exp(-\sqrt{|\alpha|}L)}{\exp(\sqrt{|\alpha|}L) + \exp(-\sqrt{|\alpha|}L)} \tag{1.4.6}$$

and

$$c_2 = \frac{f_L + \hat{q}_0 \exp(\sqrt{|\alpha|}L)}{\exp(\sqrt{|\alpha|}L) + \exp(-\sqrt{|\alpha|}L)}, \tag{1.4.7}$$

where

$$\hat{q}_0 \equiv \frac{q_0}{k\sqrt{|\alpha|}} \tag{1.4.8}$$

(see Problem 1.4.1). Approximations to these exact solutions can be obtained by the finite element method.

Finite element method

Following the discussion of the Galerkin finite element formulation with linear elements for the steady diffusion equation in Sections 1.1 and 1.2, we derive a tridiagonal system of linear equations for the nodal values of the solution,

$$(\mathbf{D} - \alpha \mathbf{M}) \cdot \mathbf{f} = \mathbf{b}, \tag{1.4.9}$$

where $\mathbf{f}$ is the vector of the unknown nodal temperatures,

$$\mathbf{f} \equiv \begin{bmatrix} f_1, & f_2, & \dots, & f_{N_E-1}, & f_{N_E} \end{bmatrix}^T, \tag{1.4.10}$$

$\mathbf{D}$ is the $N_E \times N_E$ global diffusion matrix shown in Table 1.1.6, and $\mathbf{M}$ is the $N_E \times N_E$ square global mass matrix shown in Table 1.4.1. Note that the square mass matrix $\mathbf{M}$ is identical to the rectangular mass matrix $\widetilde{\mathbf{M}}$ shown in Table 1.1.7(a), except that the last column has been discarded.

The right-hand side of the linear system is given by

$$\mathbf{b} = \begin{bmatrix} \dfrac{q_0}{k}, & 0, & \dots, & 0, & \left(\dfrac{1}{h_{N_E}} + \dfrac{1}{6} \alpha h_{N_E} \right) f_L \end{bmatrix}^T. \tag{1.4.11}$$

When $\alpha = 0$, the vector $\mathbf{b}$ reduces to the vector $\mathbf{c}$ shown in (1.1.29).

Finite element code

FSELIB function *hlml_sys*, listed in Table 1.4.2, assembles the linear system in the spirit of the function *sdl_sys* listed in Table 1.2.6.[3] The finite element method is implemented in the FSELIB code *hlml*, listed in Table 1.4.3. Results for a typical case are shown in Figure 1.4.1 where the nearly indiscernible dotted line represents the exact solution. The agreement between the numerical and the exact solution improves further as a higher number of elements is employed.

PROBLEMS

1.4.1 *Exact solution*

Derive the coefficients c_1 and c_2 in (1.4.6) and (1.4.7).

1.4.2 *Variational formulation*

Consider the one-dimensional Helmholtz equation (1.4.1) in an interval, $[a, b]$, subject to the Dirichlet boundary conditions $f(a) = f_a$ and $f(b) = f_b$. Show that solving this problem is equivalent to maximizing the Rayleigh functional

$$\mathcal{F} < \omega(x) >\equiv -\frac{1}{2} k \int_a^b \left(\frac{d\omega}{dx} \right)^2 dx + \frac{1}{2} \alpha \int_a^b \omega^2(x) \, dx, \tag{1.4.12}$$

---

[3] *hlml* is the acronym for **h**elmholtz equation with **l**inear e**l**ements.

(a)

$$\mathbf{M} = \begin{bmatrix} B_{11}^{(1)} & B_{12}^{(1)} & 0 & 0 & \cdots \\ B_{21}^{(1)} & B_{22}^{(1)} + B_{11}^{(2)} & B_{12}^{(2)} & 0 & \cdots \\ 0 & B_{21}^{(2)} & B_{22}^{(2)} + B_{11}^{(3)} & B_{12}^{(3)} & \cdots \\ \vdots & \vdots & \vdots & \ddots & \vdots \\ 0 & 0 & \cdots & \cdots & \cdots \\ 0 & 0 & \cdots & \cdots & \cdots \end{bmatrix} \longrightarrow$$

$$\begin{bmatrix} \cdots & 0 & 0 \\ \cdots & 0 & 0 \\ \cdots & 0 & 0 \\ \cdots & \cdots & \\ B_{21}^{(N_E-2)} & B_{22}^{(N_E-2)} + B_{11}^{(N_E-1)} & B_{12}^{(N_E-1)} \\ 0 & B_{21}^{(N_E-1)} & B_{22}^{(N_E-1)} + B_{11}^{(N_E)} \end{bmatrix}.$$

(b)

$$\mathbf{M} \equiv \begin{bmatrix} \frac{1}{3}h_1 & \frac{1}{6}h_1 & 0 & 0 & \cdots \\ \frac{1}{6}h_1 & \frac{1}{3}h_1 + \frac{1}{3}h_2 & \frac{1}{6}h_2 & 0 & \cdots \\ 0 & \frac{1}{6}h_2 & \frac{1}{3}h_2 + \frac{1}{3}h_3 & 0 & \cdots \\ \vdots & \vdots & \vdots & \ddots & \vdots \\ \vdots & \vdots & \cdots & \cdots & \cdots \\ 0 & \cdots & \cdots & \cdots & \cdots \end{bmatrix} \longrightarrow$$

$$\begin{bmatrix} 0 & 0 & 0 \\ 0 & \cdots & 0 \\ \cdots & 0 & 0 \\ \frac{1}{3}h_{N_E-3} + \frac{1}{3}h_{N_E-2} & \frac{1}{6}h_{N_E-2} & 0 \\ \frac{1}{6}h_{N_E-2} & \frac{1}{3}h_{N_E-2} + \frac{1}{3}h_{N_E-1} & \frac{1}{6}h_{N_E-1} \\ 0 & \frac{1}{6}h_{N_E-1} & \frac{1}{3}h_{N_E-1} + \frac{1}{3}h_{N_E} \end{bmatrix}$$

TABLE 1.4.1   (a) Global $N_E \times N_E$ mass matrix in terms of the element mass matrices. (b) Explicit expression in terms of the ith element size, $h_i \equiv x_{i+1} - x_i$.

```
function [at,bt,ct,b] = hlml_sys (ne,xe,q0,fL,k,alpha)

%=====================================================
% compact assembly of the tridiagonal linear system
% for the one-dimensional Helmholtz equation
%=====================================================

%-------------
% element size
%-------------

for l=1:ne
  h(l) = xe(l+1)-xe(l);
end

%---------------------------------
% initialize the tridiagonal matrix
%---------------------------------

at = zeros(ne,1);
bt = zeros(ne,1);
ct = zeros(ne,1);

%---------------------------------
% initialize the right-hand side
%---------------------------------

b = zeros(ne,1);
b(1) = q0/k;

%---------------------------------
% loop over the first ne-1 elements
%---------------------------------

for l=1:ne-1

  A11 = 1/h(l); A12 =-A11;
  A21 = A12;     A22 = A11;

  B11 = h(l)/3.0; B12 = 0.5*B11;
  B21 = B12;       B22 = B11;

  at(l) = at(l) + A11 - alpha * B11;
  bt(l) = bt(l) + A12 - alpha * B12;

  ct(l+1) = ct(l+1) + A21 - alpha * B21;
  at(l+1) = at(l+1) + A22 - alpha * B22;

end

%---------------------------
% the last element is special
%---------------------------
```

TABLE 1.4.2   Function *hlml_sys* (Continuing →)

```
A11 = 1.0/h(ne); A12 =-A11;
B11 = h(ne)/3.0; B12 = 0.5*B11;

at(ne) = at(ne) + A11 - alpha * B11;
b(ne) = b(ne) - (A12- alpha * B12)*fL;

%-----
% done
%-----

return;
```

TABLE 1.4.2   ($\rightarrow$ Continued) Function *hlml_sys* assembles a tridiagonal linear system for the one-dimensional Helmholtz equation with linear elements.

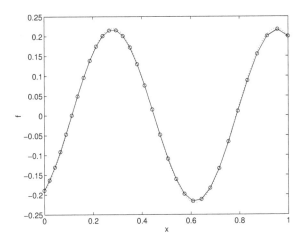

FIGURE 1.4.1   Finite element solution of the Helmholtz equation computed by the FSELIB code *hlml* with 16 linear elements, $N_E = 32$, element stretch ratio equal to 2, $L = 1.0$, $k = 1.0$, $q_0 = -1.0$, $f(L) = -0.2$, and $\alpha = 87.4$. The hardly discernible dotted line represents the exact solution.

over all continuous functions, $w(x)$, that satisfy the specified Dirichlet boundary conditions, $w(x = a) = f_a$ and $w(x = b) = f_b$.

1.4.3 *Finite element solution*

Plot the finite element solution of the Helmholtz equation for $N_E = 32$ elements, element stretch ratio equal to 2, $L = 1.0$, $k = 1.0$, $q_0 = 1.0$, $f(L) = 0.5$, and $\alpha = -54.2$. Confirm that the results are consistent with the exact solution.

```
%===============================================
% Code hlml
%
% code for the one-dimensional Helmholtz
% equation with linear elements
%
% f'' + alpha f = 0
%
% ne: number of elements
%===============================================

%-----------
% input data
%-----------

L = 1.0;  k = 1.0;
q0 = 1.0;  fL = 0.2;
alpha = 87.4;
ne = 32;  ratio = 2.0;

%----------------
% grid generation
%----------------

xe = elm_line1 (0,L,ne,ratio);

%-----------------
% compact assembly
%-----------------

[at,bt,ct,b] = hlml_sys (ne,xe,q0,fL,k,alpha);

%--------------
% linear solver
%--------------

f = thomas (ne,at,bt,ct,b);

f(ne+1) = fL;

%-----
% plot
%-----

plot(xe, f,'-ko');

%-----
% done
%-----
```

TABLE 1.4.3   Finite element code *hlml* for the one-dimensional Helmholtz equation
            with linear elements.

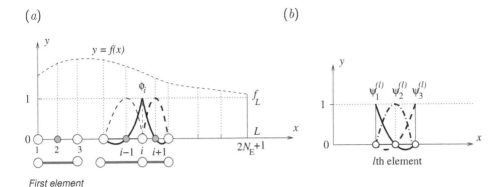

FIGURE 1.5.1 (*a*) Quadratic finite element interpolation in one dimension involving two shared element end-nodes and one interior node. The functions drawn with the heavy lines are the global interpolation functions. The solid line corresponds to the ith node, which is an element end-node, and the broken lines correspond to the $i \pm 1$ nodes, which are interior element nodes. (*b*) Graphs of three parabolic local interpolation functions over the lth element, $\psi_1^{(l)}$, $\psi_2^{(l)}$, and $\psi_3^{(l)}$.

1.5 Steady diffusion with quadratic elements

We can improve the accuracy of the finite element solution by using a higher number of elements, N_E, which is equivalent to reducing the element size, h. An alternative is to approximate the solution over the individual elements with quadratic functions, which amounts to increasing the polynomial order, p. The first method is classified as h-refinement, whereas the second method is classified as p-refinement, as discussed in Chapter 3. Combining the two methods yields the hp refinement.

1.5.1 Element nodes and global nodes

A quadratic polynomial (binomial) is defined by three coefficients and requires three element interpolation nodes. To ensure that the solution consisting of the union of the binomials is continuous over the entire solution domain, we use two *shared* element end-nodes, labeled as element nodes 1 and 3, and one *interior* element node, labeled as element node 2, as illustrated in Figure 1.5.1(*a*).

Adding to the $N_E + 1$ shared element end-nodes the N_E interior element nodes, we obtain

$$N_G = 2N_E + 1 \tag{1.5.1}$$

unique global nodes labeled 1 at the left end of the solution domain, located at $x = 0$, and N_G at the right end of the solution domain, located at $x = L$.

The global labels of the first end-node, interior node, and second end-node of the lth element are, respectively,

$$c_{l,1} = 2\,(l-1) + 1 = 2l - 1, \qquad c_{l,2} = 2\,(l-1) + 2 = 2l,$$
$$c_{l,3} = 2\,(l-1) + 3 = 2l + 1, \tag{1.5.2}$$

where $c_{l,j}$ is the connectivity matrix defined for $l = 1, \ldots, N_{\mathrm{E}}$ and $j = 1, 2, 3$.

Interpolation functions

Next, we introduce three parabolic interpolation functions for the lth-element, denoted by $\psi_1^{(l)}$, $\psi_2^{(l)}$, and $\psi_3^{(l)}$, as shown in Figure 1.5.1(b), defined as follows:

1. The first element function, $\psi_1^{(l)}$, takes the value of unity at the first element node and the value of zero at the second and third element nodes.

2. The second element function, $\psi_2^{(l)}$, takes the value of unity at the second element node and the value of zero at the first and third element nodes.

3. The third element function, $\psi_3^{(l)}$, takes the value of unity at the third element node, and the value of zero at the first and second element nodes.

Specific expressions for these functions will be derived later in this section for an arbitrary position of the interior node between the two end-nodes.

The union of (a) the end-node element interpolation functions corresponding to adjacent elements and (b) the interior node interpolation functions by themselves yields a set of *global* interpolation tent-like functions, $\phi_i(x)$. The global interpolation function of the ith global node, which happens to be an element end-node, is drawn with a heavy solid line in Figure 1.5.1(a). The global interpolation functions of the $i \pm 1$ global nodes, which happen to be interior element nodes, are drawn with heavy broken lines in Figure 1.5.1(a). Note that the global interpolation function associated with an end-node is supported by two elements, whereas that associated with an interior node is supported by one element alone.

Finite element expansion

The piecewise quadratic representation of the finite element solution is described by the expansion

$$f(x) = \sum_{j=1}^{N_{\mathrm{G}}} f_j\,\phi_j(x), \tag{1.5.3}$$

where $f_j \equiv f(x_j)$. The left-end Dirichlet boundary condition (1.1.6) requires that $f_{N_{\mathrm{G}}} = f_L$.

1.5.2 Galerkin finite element equations

To compute the $2N_{\mathrm{E}}$ unknown nodal values, f_j for $j = 1, \ldots, 2N_{\mathrm{E}}$, we work as in Section 1.1 for linear elements and derive the counterpart of the jth Galerkin projection equation (1.2.1),

$$k \sum_{j=1}^{N_{\mathrm{G}}} \left(\sum_{l=1}^{N_{\mathrm{E}}} \int_{E_l} \frac{\mathrm{d}\phi_i}{\mathrm{d}x} \frac{\mathrm{d}\phi_j}{\mathrm{d}x} \, \mathrm{d}x \right) f_j = q_0 \, \delta_{i,1} + \sum_{j=1}^{N_{\mathrm{G}}} \left(\sum_{l=1}^{N_{\mathrm{E}}} \int_{E_l} \phi_i \, \phi_j \, \mathrm{d}x \right) s_j \tag{1.5.4}$$

for $i = 1, \ldots, 2N_{\mathrm{E}}$, where E_l stands for the lth element. As in the case of linear elements, because the Dirichlet boundary condition is specified at the right end of the solution domain, the last node is excluded from the projection.

The counterpart of expression (1.2.4) for the diffusion matrix is

$$\int_{E_l} \frac{\mathrm{d}\phi_i}{\mathrm{d}x} \frac{\mathrm{d}\phi_j}{\mathrm{d}x} \, \mathrm{d}x = \delta_{i,c_{l,1}} \delta_{j,c_{l,1}} A_{11}^{(l)} + \delta_{i,c_{l,1}} \delta_{j,c_{l,2}} A_{12}^{(l)} + \delta_{i,c_{l,1}} \delta_{j,c_{l,3}} A_{13}^{(l)}$$
$$+ \delta_{i,c_{l,2}} \delta_{j,c_{l,1}} A_{21}^{(l)} + \delta_{i,c_{l,2}} \delta_{j,c_{l,2}} A_{22}^{(l)} + \delta_{i,c_{l,2}} \delta_{j,c_{l,3}} A_{23}^{(l)} \tag{1.5.5}$$
$$+ \delta_{i,c_{l,3}} \delta_{j,c_{l,1}} A_{31}^{(l)} + \delta_{i,c_{l,3}} \delta_{j,c_{l,2}} A_{32}^{(l)} + \delta_{i,c_{l,3}} \delta_{j,c_{l,3}} A_{33}^{(l)},$$

where $\mathbf{A}^{(l)}$ is the 3×3 element diffusion matrix whose components are given in Table 1.5.1(a). For the physical reasons discussed in Section 1.2, this matrix is singular irrespective of the location of the interior node.

The counterpart of expression (1.2.7) for the mass matrix is

$$\int_{E_l} \phi_i \, \phi_j \, \mathrm{d}x = \delta_{i,c_{l,1}} \delta_{j,c_{l,1}} B_{11}^{(l)} + \delta_{i,c_{l,1}} \delta_{j,c_{l,2}} B_{12}^{(l)} + \delta_{i,c_{l,1}} \delta_{j,c_{l,3}} B_{13}^{(l)}$$
$$+ \delta_{i,c_{l,2}} \delta_{j,c_{l,1}} B_{21}^{(l)} + \delta_{i,c_{l,2}} \delta_{j,c_{l,2}} B_{22}^{(l)} + \delta_{i,c_{l,2}} \delta_{j,c_{l,3}} B_{23}^{(l)} \tag{1.5.6}$$
$$+ \delta_{i,c_{l,3}} \delta_{j,c_{l,1}} B_{31}^{(l)} + \delta_{i,c_{l,3}} \delta_{j,c_{l,2}} B_{32}^{(l)} + \delta_{i,c_{l,3}} \delta_{j,c_{l,3}} B_{33}^{(l)},$$

where $\mathbf{B}^{(l)}$ is the 3×3 element mass matrix, given in Table 1.5.1(b).

The Galerkin finite element implementation results in a linear system,

$$\mathbf{D} \cdot \mathbf{f} = \mathbf{b}, \tag{1.5.7}$$

where

$$\mathbf{f} \equiv \left[\, f_1, \quad f_2, \quad \ldots, \quad f_{2N_{\mathrm{E}}-1}, \quad f_{2N_{\mathrm{E}}} \, \right]^T \tag{1.5.8}$$

is the $2N_{\mathrm{E}}$-dimensional solution vector encapsulating the unknown temperatures at all nodes, except at the last node where the Dirichlet boundary condition is specified.

(a)

$$A_{11}^{(l)} \equiv \int_{E_l} \frac{\mathrm{d}\phi_{c_{l,1}}}{\mathrm{d}x} \frac{\mathrm{d}\phi_{c_{l,1}}}{\mathrm{d}x} \, \mathrm{d}x = \int_{E_l} \frac{\mathrm{d}\psi_1^{(l)}}{\mathrm{d}x} \frac{\mathrm{d}\psi_1^{(l)}}{\mathrm{d}x} \, \mathrm{d}x$$

$$A_{12}^{(l)} = A_{21}^{(l)} \equiv \int_{E_l} \frac{\mathrm{d}\phi_{c_{l,1}}}{\mathrm{d}x} \frac{\mathrm{d}\phi_{c_{l,2}}}{\mathrm{d}x} \, \mathrm{d}x = \int_{E_l} \frac{\mathrm{d}\psi_1^{(l)}}{\mathrm{d}x} \frac{\mathrm{d}\psi_2^{(l)}}{\mathrm{d}x} \, \mathrm{d}x$$

$$A_{13}^{(l)} = A_{31}^{(l)} \equiv \int_{E_l} \frac{\mathrm{d}\phi_{c_{l,1}}}{\mathrm{d}x} \frac{\mathrm{d}\phi_{c_{l,3}}}{\mathrm{d}x} \, \mathrm{d}x = \int_{E_l} \frac{\mathrm{d}\psi_1^{(l)}}{\mathrm{d}x} \frac{\mathrm{d}\psi_3^{(l)}}{\mathrm{d}x} \, \mathrm{d}x$$

$$A_{22}^{(l)} \equiv \int_{E_l} \frac{\mathrm{d}\phi_{c_{l,2}}}{\mathrm{d}x} \frac{\mathrm{d}\phi_{c_{l,2}}}{\mathrm{d}x} \, \mathrm{d}x = \int_{E_l} \frac{\mathrm{d}\psi_2^{(l)}}{\mathrm{d}x} \frac{\mathrm{d}\psi_2^{(l)}}{\mathrm{d}x} \, \mathrm{d}x$$

$$A_{23}^{(l)} = A_{32}^{(l)} \equiv \int_{E_l} \frac{\mathrm{d}\phi_{c_{l,2}}}{\mathrm{d}x} \frac{\mathrm{d}\phi_{c_{l,3}}}{\mathrm{d}x} \, \mathrm{d}x = \int_{E_l} \frac{\mathrm{d}\psi_2^{(l)}}{\mathrm{d}x} \frac{\mathrm{d}\psi_3^{(l)}}{\mathrm{d}x} \, \mathrm{d}x$$

$$A_{33}^{(l)} \equiv \int_{E_l} \frac{\mathrm{d}\phi_{c_{l,3}}}{\mathrm{d}x} \frac{\mathrm{d}\phi_{c_{l,3}}}{\mathrm{d}x} \, \mathrm{d}x = \int_{E_l} \frac{\mathrm{d}\psi_3^{(l)}}{\mathrm{d}x} \frac{\mathrm{d}\psi_3^{(l)}}{\mathrm{d}x} \, \mathrm{d}x$$

(b)

$$B_{11}^{(l)} \equiv \int_{E_l} \phi_{c_{l,1}} \, \phi_{c_{l,1}} \, \mathrm{d}x = \int_{E_l} \psi_1^{(l)} \psi_1^{(l)} \, \mathrm{d}x$$

$$B_{12}^{(l)} = B_{21}^{(l)} \equiv \int_{E_l} \phi_{c_{l,1}} \, \phi_{c_{l,2}} \, \mathrm{d}x = \int_{E_l} \psi_1^{(l)} \psi_2^{(l)} \, \mathrm{d}x$$

$$B_{13}^{(l)} = B_{31}^{(l)} \equiv \int_{E_l} \phi_{c_{l,1}} \, \phi_{c_{l,3}} \, \mathrm{d}x = \int_{E_l} \psi_1^{(l)} \psi_3^{(l)} \, \mathrm{d}x$$

$$B_{22}^{(l)} \equiv \int_{E_l} \phi_{c_{l,2}} \, \phi_{c_{l,2}} \, \mathrm{d}x = \int_{E_l} \psi_2^{(l)} \psi_2^{(l)} \, \mathrm{d}x$$

$$B_{23}^{(l)} = B_{32}^{(l)} \equiv \int_{E_l} \phi_{c_{l,3}} \, \phi_{c_{l,2}} \, \mathrm{d}x = \int_{E_l} \psi_2^{(l)} \psi_3^{(l)} \, \mathrm{d}x$$

$$B_{33}^{(l)} \equiv \int_{E_l} \phi_{c_{l,3}} \, \phi_{c_{l,3}} \, \mathrm{d}x = \int_{E_l} \psi_3^{(l)} \psi_3^{(l)} \, \mathrm{d}x$$

TABLE 1.5.1   Components of the 3×3 (a) element diffusion matrix and (b) element mass matrix for quadratic elements.

The northwestern portion of the $2N_E \times 2N_E$ pentadiagonal[4] global diffusion matrix takes the form

$$
\mathbf{D} = \begin{bmatrix}
A_{11}^{(1)} & A_{12}^{(1)} & A_{13}^{(1)} & 0 & \cdots & \cdots & \cdots \\
A_{21}^{(1)} & A_{22}^{(1)} & A_{23}^{(1)} & 0 & 0 & \cdots & \cdots \\
A_{31}^{(1)} & A_{32}^{(1)} & A_{33}^{(1)} + A_{11}^{(2)} & A_{12}^{(2)} & A_{13}^{(2)} & 0 & \cdots \\
0 & 0 & A_{21}^{(2)} & A_{22}^{(2)} & A_{23}^{(2)} & 0 & \cdots \\
0 & 0 & A_{31}^{(2)} & A_{32}^{(2)} & A_{33}^{(2)} + A_{11}^{(3)} & A_{12}^{(3)} & \cdots \\
\vdots & \vdots & \vdots & \vdots & \vdots & \vdots & \vdots \\
0 & 0 & 0 & \cdots & \cdots & \cdots & \cdots \\
0 & 0 & 0 & \cdots & \cdots & \cdots & \cdots \\
0 & 0 & 0 & \cdots & \cdots & \cdots & \cdots
\end{bmatrix}, \qquad (1.5.9)
$$

which shows that $\mathbf{D}$ consists of partially overlapping 3×3 diagonal blocks.

The global diffusion matrix and right-hand side can be computed according to the algorithm given in Table 1.5.2, which is a generalization of the algorithm given in Table 1.1.1 for linear elements.

The right-hand side of the linear system (1.5.7) is given by

$$
\mathbf{b} \equiv \mathbf{c} + \frac{1}{k} \widetilde{\mathbf{M}} \cdot \mathbf{s}, \qquad (1.5.10)
$$

where the $2N_E$-dimensional vector

$$
\mathbf{c} \equiv \begin{bmatrix} q_0/k, & 0, & \cdots, & 0, & f_L/h_{N_E} \end{bmatrix} \qquad (1.5.11)
$$

encapsulates the boundary conditions, the $(N_E + 1)$-dimensional vector

$$
\mathbf{s} \equiv \begin{bmatrix} s_1, & s_2, & \cdots, & s_{2N_E}, & s_{2N_E+1} \end{bmatrix} \qquad (1.5.12)
$$

contains the nodal values of the source, and $\widetilde{\mathbf{M}}$ is a $2N_E \times (2N_E + 1)$ rectangular mass matrix,

$$
\widetilde{\mathbf{M}} = \begin{bmatrix}
B_{11}^{(1)} & B_{12}^{(1)} & B_{13}^{(1)} & 0 & \cdots & \cdots & \cdots \\
B_{21}^{(1)} & B_{22}^{(1)} & B_{23}^{(1)} & 0 & 0 & \cdots & \cdots \\
B_{31}^{(1)} & B_{32}^{(1)} & B_{33}^{(1)} + B_{11}^{(2)} & B_{12}^{(2)} & B_{13}^{(2)} & 0 & \cdots \\
0 & 0 & B_{21}^{(2)} & B_{22}^{(2)} & B_{23}^{(2)} & 0 & \cdots \\
0 & 0 & B_{31}^{(2)} & B_{32}^{(2)} & B_{33}^{(2)} + B_{11}^{(3)} & B_{12}^{(3)} & \cdots \\
\vdots & \vdots & \vdots & \vdots & \vdots & \vdots & \vdots \\
0 & 0 & 0 & \cdots & \cdots & \cdots & \cdots \\
0 & 0 & 0 & \cdots & \cdots & \cdots & \cdots
\end{bmatrix}. \qquad (1.5.13)
$$

---

[4]*Pentadiagonal* derives from the Greek words πεντε, which means *five*, and διαγωνιος, which means *diagonal*.

Do $l = 1, \ldots, N_E$ *Run over the elements to generate the connectivity matrix*
    $c(l, 1) = 2l - 1$
    $c(l, 2) = 2l$
    $c(l, 3) = 2l + 1$
End Do

Do $i = 1, \ldots, 2 N_E$ *Initialize the global diffusion matrix*
    $b_i = 0.0$ *and right-hand side to zero*
    Do $j = 1, \ldots, 2 N_E$
        $D_{ij} = 0.0$
    End Do
End Do

$b_1 = q_0/k$ *Neumann boundary condition*

Do $l = 1, \ldots, N_E$ *Run over all elements*

    Compute the element matrices $\mathbf{A}^{(l)}$ and $\mathbf{B}^{(l)}$ using the expressions given in Table 1.5.1

    $i_1 = c(l, 1), \qquad i_2 = c(l, 2), \qquad i_3 = c(l, 3)$

$$D_{i_1,i_1} = D_{i_1,i_1} + A_{11}^{(l)}, \qquad D_{i_1,i_2} = D_{i_1,i_2} + A_{12}^{(l)}, \qquad D_{i_1,i_3} = D_{i_1,i_3} + A_{13}^{(l)}$$
$$D_{i_2,i_1} = D_{i_2,i_1} + A_{21}^{(l)}, \qquad D_{i_2,i_2} = D_{i_2,i_2} + A_{22}^{(l)}, \qquad D_{i_2,i_3} = D_{i_2,i_3} + A_{23}^{(l)}$$
$$D_{i_3,i_1} = D_{i_3,i_1} + A_{31}^{(l)}, \qquad D_{i_3,i_2} = D_{i_3,i_2} + A_{32}^{(l)}, \qquad D_{i_3,i_3} = D_{i_3,i_3} + A_{33}^{(l)}$$

$$b_{i_1} = b_{i_1} + (B_{11}^{(l)} \times s_{i_1} + B_{12}^{(l)} \times s_{i_2} + B_{13}^{(l)} \times s_{i_3})/k$$
$$b_{i_2} = b_{i_2} + (B_{21}^{(l)} \times s_{i_1} + B_{22}^{(l)} \times s_{i_2} + B_{23}^{(l)} \times s_{i_3})/k$$
$$b_{i_3} = b_{i_3} + (B_{31}^{(l)} \times s_{i_1} + B_{32}^{(l)} \times s_{i_2} + B_{33}^{(l)} \times s_{i_3})/k$$

End Do

$b_{2N_E-1} = b_{2N_E-1} - A_{13}^{(N_E)} \times f_L$ *Process the last node*
$b_{2N_E} = b_{2N_E} - A_{23}^{(N_E)} \times f_L$

TABLE 1.5.2   Algorithm for the assembly of a linear system, $\mathbf{D} \cdot \mathbf{f} = \mathbf{b}$, for the quadratic elements shown in Figure 1.5.1.

Note that, given the element layout, the entries of the element diffusion and mass matrices and the finite element solution itself depend on the precise location of the element interior nodes.

Because the global diffusion matrix $\mathbf{D}$ shown in (1.5.9) is pentadiagonal, the solution of the linear system can be found efficiently using a modification of Thomas's algorithm for tridiagonal systems discussed in Section 1.2.2.

1.5.3   Thomas algorithm for pentadiagonal system

To formalize the Thomas algorithm for pentadiagonal systems in general terms, we consider an $N \times N$ linear system for an unknown vector, $\mathbf{x}$,

$$\mathbf{P} \cdot \mathbf{x} = \mathbf{r}, \tag{1.5.14}$$

subject to a specified vector, $\mathbf{r}$. The matrix $\mathbf{P}$ has the pentadiagonal form

$$\mathbf{P} = \begin{bmatrix} a_1 & b_1 & c_1 & 0 & 0 & \cdots & 0 & 0 & 0 \\ d_2 & a_2 & b_2 & c_2 & 0 & \cdots & 0 & 0 & 0 \\ e_3 & d_3 & a_3 & b_3 & c_3 & \cdots & 0 & 0 & 0 \\ \vdots & \vdots & \vdots & \vdots & \vdots & \vdots & \vdots & & \vdots \\ 0 & 0 & 0 & 0 & 0 & \cdots & d_{N-1} & a_{N-1} & b_{N-1} \\ 0 & 0 & 0 & 0 & 0 & \cdots & e_N & d_N & a_N \end{bmatrix}, \tag{1.5.15}$$

involving five vector sets, a_i, b_i, c_i, d_i, and e_i.

The modified Thomas algorithm proceeds in three stages. In the first stage, the pentadiagonal system (1.5.14) is reduced to a quadra-diagonal system,

$$\mathbf{Q} \cdot \mathbf{x} = \mathbf{r}', \tag{1.5.16}$$

involving the coefficient matrix

$$\mathbf{Q} = \begin{bmatrix} a_1' & b_1' & c_1' & 0 & 0 & \cdots & 0 & 0 & 0 \\ d_2' & a_2' & b_2' & c_2' & 0 & \cdots & 0 & 0 & 0 \\ 0 & 0 & d_3' & a_3' & b_3' & \cdots & 0 & 0 & 0 \\ \vdots & \vdots & \vdots & \vdots & \vdots & \vdots & \vdots & & \vdots \\ 0 & 0 & 0 & 0 & 0 & \cdots & d_{N-1}' & a_{N-1}' & b_{N-1}' \\ 0 & 0 & 0 & 0 & 0 & \cdots & 0 & d_N' & a_N' \end{bmatrix}. \tag{1.5.17}$$

In the second stage, the quadra-diagonal system (1.5.16) is reduced to the upper tridiagonal system

$$\mathbf{T} \cdot \mathbf{x} = \mathbf{r}'', \tag{1.5.18}$$

involving an upper tridiagonal coefficient matrix,

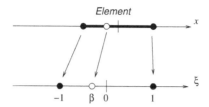

FIGURE 1.5.2   Mapping of an element from the x axis to the standard interval $[-1, 1]$
of the parametric ξ axis. The interior node is mapped to the point $\xi = \beta$. A
mid-point interior node is mapped to the origin of the ξ axis.

$$
\mathbf{T} =
\begin{bmatrix}
1 & b_1'' & c_1'' & 0 & 0 & \dots & 0 & 0 & 0 \\
0 & 1 & b_2'' & c_2'' & 0 & \dots & 0 & 0 & 0 \\
0 & 0 & 1 & b_3'' & c_3'' & \dots & 0 & 0 & 0 \\
\vdots & \vdots & \vdots & \vdots & \vdots & \vdots & \vdots & \vdots & \vdots \\
0 & 0 & 0 & 0 & 0 & \dots & 0 & 1 & b_{N-1}'' \\
0 & 0 & 0 & 0 & 0 & \dots & 0 & 0 & 1
\end{bmatrix}.
\tag{1.5.19}
$$

In the third stage, the upper tridiagonal system (1.5.18) is solved by backward
substitution, which involves solving the last equation for the last unknown, x_N, and
then moving upward to compute the rest of the unknowns in a sequential fashion.
The algorithm is outlined in Table 1.5.3 and implemented in the FSELIB function
penta, listed in Table 1.5.4.

1.5.4   *Element matrices*

To simplify the notation, we denote the lth-element nodes by

$$
x_{c_{l,1}} = x_1^{(l)}, \qquad x_{c_{l,2}} = x_2^{(l)}, \qquad x_{c_{l,3}} = x_3^{(l)}.
\tag{1.5.20}
$$

To compute the element diffusion and mass matrices, it is convenient to map each
element from the x axis to the standard interval $[-1, 1]$ of the parametric ξ axis
using the mapping function

$$
x = \frac{1}{2} \left(x_1^{(l)} + x_3^{(l)} \right) + \frac{1}{2} \left(x_3^{(l)} - x_1^{(l)} \right) \xi.
\tag{1.5.21}
$$

The first element node, $x_1^{(1)}$, is mapped to $\xi = -1$, the second element node, $x_2^{(l)}$, is
mapped to a point, $\xi = \beta$, and the third element node, $x_3^{(l)}$, is mapped to $\xi = 1$, as
shown in Figure 1.5.2, where β is a free parameter varying in the interval $(-1, 1)$.
Accordingly,

$$
dx = \frac{1}{2} h_l \, d\xi,
\tag{1.5.22}
$$

where $h_l = x_3^{(l)} - x_1^{(l)}$ is the element size.

Reduction to quadra-diagonal:

Do $i = 1, \ldots, N-2$
$\qquad c_i' = c_i$
End Do
$c_{N-1}' = 0.0, \qquad c_N' = 0.0$ $\qquad\qquad\qquad$ ← *Initialize for convenience*
$a_1' = a_1, \quad b_1' = b_1, \quad r_1' = r_1$
$d_2' = d_2, \quad a_2' = a_2, \quad b_2' = b_2, \quad r_2' = r_2$

Do $i = 2, 3, \ldots, N-1$

$$
\begin{bmatrix} a_{i+1}' \\ b_{i+1}' \\ d_{i+1}' \\ r_{i+1}' \end{bmatrix}
=
\begin{bmatrix} a_{i+1} \\ b_{i+1} \\ d_{i+1} \\ r_{i+1} \end{bmatrix}
- \frac{e_{i+1}}{d_i'}
\begin{bmatrix} b_i' \\ c_i' \\ a_i' \\ r_i' \end{bmatrix}
$$

End Do

Reduction to upper tridiagonal:

$$
\begin{bmatrix} b_1'' \\ c_1'' \\ r_1'' \end{bmatrix}
= \frac{1}{a_1'}
\begin{bmatrix} b_1'' \\ c_1'' \\ r_1'' \end{bmatrix}
$$

Do $i = 1, \ldots, N-1$

$$
\begin{bmatrix} b_{i+1}'' \\ c_{i+1}'' \\ r_{i+1}'' \end{bmatrix}
= \frac{1}{a_{i+1}' - d_{i+1}' b_i'}
\begin{bmatrix} b_{i+1} - d_{i+1}' c_i'' \\ c_{i+1}' \\ r_{i+1}' - d_{i+1}' r_i'' \end{bmatrix}
$$

End Do

Backward substitution:

$x_N = r_N''$
$x_{N-1} = r_{N-1}'' - b_{N-1}'' x_N$

Do $i = N-2, \ldots, 1$
$\qquad x_i = r_i'' - b_i'' x_{i+1} - c_i'' x_{i+2}$
End Do

TABLE 1.5.3   Modified Thomas algorithm for solving the pentadiagonal system of linear equations in (1.5.14).

```
function x = penta(n,a,b,c,d,e,r)

%==========================================
% Modified Thomas algorithm for solving
% a pentadiagonal system
%
% n:          system size
% a,b,c,d,e:  diagonal, super, super-super,
%             sub, and sub-sub elements
% r:          right-hand side
%==========================================

na = n-1;
nb = n-2;

%----------------------------
% reduction to quadra-diagonal
%----------------------------

for i=1:nb
  c1(i) = c(i);
end

c1(na) = 0.0; c1(n) = 0.0;

a1(1) = a(1); b1(1) = b(1); r1(1) = r(1);
d1(2) = d(2); a1(2) = a(2); b1(2) = b(2); r1(2) = r(2);

for i=2:na
  i1 = i+1;
  w = e(i1)/d1(i);
  a1(i1) = a(i1) - w*b1(i);
  b1(i1) = b(i1) - w*c1(i);
  d1(i1) = d(i1) - w*a1(i);
  r1(i1) = r(i1) - w*r1(i);
end

%----------------------------
% generate the system T x = r2
%----------------------------

b2(1) = b1(1)/a1(1);
c2(1) = c1(1)/a1(1);
r2(1) = r1(1)/a1(1);

for i=1:na
  i1 = i+1;
  den =  a1(i1)-d1(i1)*b2(i);
  b2(i1) = (b1(i1)-d1(i1)*c2(i))/den;
  c2(i1) =   c1(i1)/den;
  r2(i1) = (r1(i1)-d1(i1)*r2(i))/den;
end
```

TABLE 1.5.4 Function *penta* (Continuing →)

```
%------------------
% back substitution
%------------------

x(n) = r2(n);

x(n-1) = r2(n-1)-b2(n-1)*x(n);

for i=nb:-1:1    % step of -1
   x(i) = r2(i)-b2(i)*x(i+1)-c2(i)*x(i+2);
end

%-----
% done
%-----

return;
```

TABLE 1.5.4   ($\rightarrow$ Continued) FSELIB function *penta* for solving a pentadiagonal system of algebraic equations using the modified Thomas algorithm shown in Table 1.5.3.

Mid-point interior node

An interior node situated at the element mid-point, so that

$$x_2^{(l)} = \frac{1}{2}\left(x_1^{(l)} + x_3^{(l)}\right), \tag{1.5.23}$$

is mapped to the origin, $\beta = 0$. In Chapter 3, we will see that the midway position yields the highest possible interpolation accuracy for the available degrees of freedom. In terms of the canonical variable ξ, the element interpolation functions are found to be

$$\psi_1^{(l)}(\xi) = \frac{1}{2}\xi(\xi - 1), \qquad \psi_2^{(l)}(\xi) = 1 - \xi^2, \qquad \psi_3^{(l)}(\xi) = \frac{1}{2}\xi(\xi + 1), \tag{1.5.24}$$

where ξ is mapped to x by way of (1.5.21). Note that these interpolation functions are specific to the mid-point interior node, β. More general interpolation functions are given in Section 1.5.7.

Substituting the first interpolation function into the first expression shown in Table 1.5.1 and using (1.5.22), we obtain the first component of the element diffusion matrix,

$$A_{11}^{(l)} = \int_{E_l} \frac{d\psi_1^{(l)}}{dx}\frac{d\psi_1^{(l)}}{dx}\,dx = \frac{1}{2}h_l\int_{-1}^{1}\frac{2}{h_l}\frac{d\psi_1^{(l)}}{d\xi}\frac{2}{h_l}\frac{d\psi_1^{(l)}}{d\xi}\,d\xi. \tag{1.5.25}$$

Performing the integration, we obtain

$$A_{11}^{(l)} = \frac{1}{2\,h_l} \int_{-1}^{1} (2\,\xi - 1)^2 \, \mathrm{d}\xi = \frac{1}{h_l} \frac{7}{3}. \tag{1.5.26}$$

Working in a similar fashion with the rest of the elements, we derive the element diffusion matrix

$$\mathbf{A}^{(l)} = \frac{1}{3\,h_l} \begin{bmatrix} 7 & -8 & 1 \\ -8 & 16 & -8 \\ 1 & -8 & 7 \end{bmatrix}. \tag{1.5.27}$$

A straightforward computation shows that the determinant of this matrix is zero, which reveals that the matrix is singular (see also Problem 1.5.1(a)). The physical reason was discussed in Section 1.2.

The components of the element mass matrix are given by

$$B_{ij}^{(l)} = \int_{E_l} \psi_i^{(l)} \psi_j^{(l)} \, \mathrm{d}x = \frac{1}{2} h_l \int_{-1}^{1} \psi_i^{(l)} \psi_j^{(l)} \, \mathrm{d}\xi. \tag{1.5.28}$$

Performing the integrations, we obtain

$$\mathbf{B}^{(l)} = \frac{1}{30} h_l \begin{bmatrix} 4 & 2 & -1 \\ 2 & 16 & 2 \\ -1 & 2 & 4 \end{bmatrix}. \tag{1.5.29}$$

Note that the sum of all elements of this matrix is equal to the element size, h_l.

1.5.5   Finite element code

Now we may combine the elementary modules developed previously in this section into an integrated finite element code that performs three main functions: (a) domain discretization, (b) assembly of a linear system, and (c) solution of the linear system. To economize the computer memory, we store the non-zero diagonal components of the global diffusion matrix in five vectors,

$$e_i^p \equiv D_{i,i-2}, \quad d_i^p \equiv D_{i,i-1}, \quad a_i^p \equiv D_{i,i}, \quad b_i^p \equiv D_{i,i+1}, \quad c_i^p \equiv D_{i,i+2}, \tag{1.5.30}$$

corresponding to the pentadiagonal form shown in (1.5.15), where the superscript p denoted *pentadiagonal*.

FSELIB function *sdq_sys*, listed in Table 1.5.5, implements the assembly algorithm shown in Table 1.5.2 for mid-point interior nodes.[5] FSELIB code *sdq*, listed in Table 1.5.6, implements the complete finite element code in six familiar modules: data input; finite element discretization; specification of the source function; assembly of a linear system; solution of the linear system; plotting the solution.

---

[5] *sdq* is the acronym for steady **d**iffusion with **q**uadratic elements.

```
function [ap,bp,cp,dp,ep,b] = sdq_sys (ne,xe,q0,fL,k,s)

%========================================================
% Compact assembly of the pentadiagonal linear system
% for one-dimensional steady diffusion
% with quadratic elements (sdq)
%========================================================

%-------------
% element size
%-------------

for l=1:ne
  h(l) = xe(l+1)-xe(l);
end

%-----------------------------
% number of unique global nodes
%-----------------------------

ng = 2*ne+1;

%-----------------------------------
% initialize the pentadiagonal matrix
% and right-hand side
%-----------------------------------

ap = zeros(ng,1); bp = zeros(ng,1); cp = zeros(ng,1);
dp = zeros(ng,1); ep = zeros(ng,1);

b = zeros(ng,1); b(1) = q0/k;

%----------------------
% loop over all elements
%----------------------

for l=1:ne

  cf = 1.0/(3.0*h(l));

  A11 = 7.0*cf; A12 = -8.0*cf; A13 = cf;
  A21 = A12;    A22 = 16.0*cf; A23 = A12;
  A31 = A13;    A32 = A23;     A33 = A11;

  cf = h(l)/30.0;

  B11 = 4.0*cf; B12 =  2.0*cf; B13 = -1.0*cf;
  B21 = B12;    B22 = 16.0*cf; B23 = B12;
  B31 = B13;    B32 = B23;     B33 = B11;

  cl1 = 2*l-1; cl2 = 2*l; cl3 = 2*l+1;

  ap(cl1) = ap(cl1) + A11;
  bp(cl1) = bp(cl1) + A12;
  cp(cl1) = cp(cl1) + A13;
```

TABLE 1.5.5 Function *sdq_sys* (Continuing →)

```
      dp(cl2) = dp(cl2) + A21;
      ap(cl2) = ap(cl2) + A22;
      bp(cl2) = bp(cl2) + A23;

      ep(cl3) = ep(cl3) + A31;
      dp(cl3) = dp(cl3) + A32;
      ap(cl3) = ap(cl3) + A33;

      b(cl1) = b(cl1) + (B11*s(cl1) + B12*s(cl2) + B13*s(cl3))/k;
      b(cl2) = b(cl2) + (B21*s(cl1) + B22*s(cl2) + B23*s(cl3))/k;
      b(cl3) = b(cl3) + (B31*s(cl1) + B32*s(cl2) + B33*s(cl3))/k;

end

%------------------------------------
% implement the Dirichlet condition
%------------------------------------

b(ng-2) = b(ng-2) - cp(ng-2)*fL;
b(ng-1) = b(ng-1) - bp(ng-1)*fL;

%-----
% done
%-----

return;
```

TABLE 1.5.5 ($\rightarrow$ Continued) Function *sdq_sys* assembles of a pentadiagonal linear
   system for one-dimensional diffusion with quadratic elements and mid-point interior nodes.

Accuracy

A graph of the solution generated by the code with the Gaussian source function
defined in (1.2.17) is shown in Figure 1.5.3(a). Calculations with uniformly spaced
elements, $ratio=1$, and number of elements, $N_E = 4, 8, 16, 32$, and 64, corresponding
to system size $2N_E = 8, 16, 32, 64$, and 128, yield the left-end value

$$f(0) = 1.96339024, \quad 1.96383352, \quad 1.96385935, 1.96386094, \quad 1.963861036.$$

$$(1.5.31)$$

These results are more accurate than those listed in Section 1.2.3 for linear elements. A computation with a large number of elements reveals the left-end value,
1.9638 6104, to shown accuracy. The numerical error corresponding to the sequence
(1.5.31), defined as $\varepsilon(0) \equiv f(0) - 1.9638\ 6104$, is

$$\varepsilon(0) = -0.00047080, \quad -0.00002752, \quad -0.00000169, \quad -0.00000011, \quad -0.00000001.$$

$$(1.5.32)$$

```
%===================================
% Code sdq
%
% Steady one-dimensional diffusion
% with quadratic elements
%===================================

%-----------
% input data
%-----------

L = 1.0; k = 1.0; q0 =-1.0; fL = 0.0;

ne = 4; ratio = 2.0;

%----------------
% grid generation
%----------------

xe = elm_line1 (0,L,ne,ratio);

%-----------------------------
% number of unique global nodes
%-----------------------------

ng = 2*ne+1;

%--------------------------
% generate the global nodes
%--------------------------

Ic = 0;  % node counter

for i=1:ne
 Ic = Ic+1; xg(Ic) = xe(i);
 Ic = Ic+1; xg(Ic) = 0.5*(xe(i)+xe(i+1));
end

xg(ng) = xe(ne+1);

%-------------------
% specify the source
%-------------------

for i=1:ng
 s(i) = 10.0*exp(-5.0*xg(i)^2/L^2);
end

%-------------------------
% compact element assembly
%-------------------------

[ap,bp,cp,dp,ep,b] = sdq_sys(ne,xe,q0,fL,k,s);
```

TABLE 1.5.6   Finite element code *sdq* (Continuing →)

```
%--------------
% linear solver
%--------------

f = penta (ng-1,ap,bp,cp,dp,ep,b);

f(ng) = fL;

%-----
% plot
%-----

plot (xg, f,'-k+');

%-----
% done
%-----
```

TABLE 1.5.6   ($\rightarrow$ Continued) Finite element code *sdq* for steady one-dimensional diffusion with quadratic elements and mid-point interior nodes.

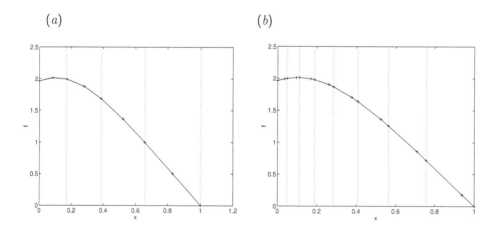

FIGURE 1.5.3   Numerical solution of the steady diffusion equation for $L = 1.0$, $k = 1.0$, $q_0 = -1.0$, and $f(L) = 0.0$. The elements are separated by the vertical dotted lines. (a) Solution with four quadratic elements, $N_E = 4$, element stretch ratio 2, and mid-point interior nodes ($\beta = 0$). The crosses represent the results of the FSELIB code *sdq*. (b) Solution with eight quadratic elements, $N_E = 8$, element stretch ratio 5, and interior nodes corresponding to $\beta = 0.5$. The crosses represent the results of the FSELIB code *sdq_beta*.

We observe that, as the number of elements is doubled, the numerical error is reduced approximately by a factor of 16, and this demonstrates that the error is proportional to h^4, where h is the element size. Thus, the finite element method with quadratic elements for the steady diffusion equation with a smooth source function, $s(x)$, is fourth-order accurate in the element size.

1.5.6 Node condensation

Inspection of the global diffusion matrix shown in (1.5.9) reveals that the unknowns corresponding to the interior element nodes can be eliminated in favor of the unknowns corresponding to the element end-nodes. Consequently, the size of the linear system can be reduced from $2N_E \times 2N_E$ to $N_E \times N_E$ following a process that is known as *node condensation*. Because the condensed system is tridiagonal, the solution can be found efficiently using the standard Thomas algorithm discussed in Section 1.2.2.

To implement node condensation, we perform the following computations:

- Multiply the second equation by the ratio $r_{12}^{(1)} \equiv A_{12}^{(1)}/A_{22}^{(1)}$ and subtract the outcome from the first equation.

- Multiply the second equation by the ratio $r_{32}^{(1)} \equiv A_{32}^{(1)}/A_{22}^{(1)}$ and subtract the outcome from the third equation.

- Repeat the calculation for all subsequent element blocks.

- Remove the Galerkin equations corresponding to the interior nodes.

In the end, we obtain an $N_E \times N_E$ condensed system,

$$\widehat{\mathbf{D}} \cdot \widehat{\mathbf{f}} = \widehat{\mathbf{b}}, \tag{1.5.33}$$

where

$$\widehat{\mathbf{D}} = \begin{bmatrix} A_{11}^{(1)} - r_{12}^{(1)} A_{21}^{(1)} & A_{13}^{(1)} - r_{12}^{(1)} A_{23}^{(1)} & & \cdots \\ A_{31}^{(1)} - r_{32}^{(1)} A_{21}^{(1)} & A_{33}^{(1)} - r_{32}^{(1)} A_{23}^{(1)} + A_{11}^{(2)} - r_{12}^{(2)} A_{21}^{(2)} & A_{13}^{(2)} - r_{12}^{(2)} A_{23}^{(2)} & \cdots \\ 0 & A_{31}^{(2)} - r_{32}^{(2)} A_{21}^{(2)} & \cdots & \cdots \\ \vdots & \vdots & \cdots & \cdots \\ 0 & 0 & \cdots & \cdots \end{bmatrix}$$

$$\tag{1.5.34}$$

is the condensed diffusion matrix and

$$\widehat{\mathbf{f}} \equiv \begin{bmatrix} f_1, & f_3, & \ldots, & f_{2N_E-3}, & f_{2N_E-1} \end{bmatrix}^T \tag{1.5.35}$$

is the vector of function values at the end-nodes. The first entry of the right-hand side of the condensed system is given by

$$\widehat{b}_1 = \frac{1}{k} \left[q_0 + \left(B_{11}^{(1)} - r_{12}^{(1)} B_{21}^{(1)} \right) s_1 + \left(B_{12}^{(1)} - r_{12}^{(1)} B_{22}^{(1)} \right) s_2 \right.$$
$$\left. + \left(B_{13}^{(1)} - r_{12}^{(1)} B_{23}^{(1)} \right) s_3 \right]. \qquad (1.5.36)$$

The second entry is given by

$$\widehat{b}_2 = \frac{1}{k} \left[q_0 \left(B_{31}^{(1)} - r_{32}^{(1)} B_{21}^{(1)} \right) s_1 + \left(B_{32}^{(1)} - r_{32}^{(1)} B_{22}^{(1)} \right) s_2 \right.$$
$$+ \left(B_{33}^{(1)} - r_{32}^{(1)} B_{23}^{(1)} + B_{11}^{(2)} - r_{12}^{(2)} B_{21}^{(2)} \right) s_3 \qquad (1.5.37)$$
$$\left. + \left(B_{12}^{(2)} - r_{12}^{(2)} B_{22}^{(2)} \right) s_4 + \left(B_{13}^{(2)} - r_{12}^{(2)} B_{23}^{(2)} \right) s_5 \right].$$

Subsequent entries have a similar structure. Once the solution of the condensed system for the element end-nodes has been found, the solution at the interior nodes can be recovered element by element by resorting to the removed equations.

The condensed global diffusion matrix shown in (1.5.34) consists of overlapping 2×2 condensed element matrices,

$$\widehat{\mathbf{A}}^{(l)} = \left[\begin{array}{c|c} A_{11}^{(l)} - r_{12}^{(l)} A_{21}^{(l)} & A_{13}^{(l)} - r_{12}^{(l)} A_{23}^{(l)} \\ \hline A_{31}^{(l)} - r_{32}^{(l)} A_{21}^{(l)} & A_{33}^{(l)} - r_{32}^{(l)} A_{23}^{(l)} \end{array} \right]. \qquad (1.5.38)$$

Accordingly, node condensation can be done on the element level, as implemented in the FSELIB function *sdq_cnd_sys*, listed in Table 1.5.7.[6]

FSELIB code *sdq_cnd*, listed in Table 1.5.8, implements the finite element method. The solution generated by this code with the frivolous source function defined in (1.2.17) is in perfect agreement with that shown in Figure 1.5.3(*a*).

Mid-point interior nodes

When the interior nodes is located at the element mid-points, we use (1.5.27) to find that

$$r_{12}^{(1)} = r_{32}^{(1)} = -\frac{1}{2} \qquad (1.5.39)$$

and thus

$$\widehat{\mathbf{A}}^{(l)} = \frac{1}{h_l} \left[\begin{array}{cc} 1 & -1 \\ -1 & 1 \end{array} \right], \qquad (1.5.40)$$

which is precisely the diffusion matrix with linear elements shown in (1.2.6). In fact, this expression holds true for any arbitrary position of the interior node.

---

[6] *sdq_cnd* is the acronym for **s**teady **d**iffusion with **q**uadratic elements using the **c**ondensed formulation.

```
function [at,bt,ct,b] = sdq_cnd_sys (ne,xe,q0,fL,k,s)

%===============================================================
% Compact assembly of a condensed tridiagonal linear system
% for one-dimensional steady diffusion
% with quadratic elements
%===============================================================

%-------------
% element size
%-------------

for l=1:ne
  h(l) = xe(l+1)-xe(l);
end

%--------------------------------
% initialize the tridiagonal matrix
%--------------------------------

at = zeros(ne+1,1); bt = zeros(ne+1,1); ct = zeros(ne+1,1);

%-------------------------------
% initialize the right-hand side
%-------------------------------

b = zeros(ne+1,1); b(1) = q0/k;

%----------------------
% loop over all elements
%----------------------

for l=1:ne

  cf = 1.0/(6.0*h(l));

  A11 = 14.0*cf; A12 =-16.0*cf; A13 = 2.0*cf;
  A21 = A12;     A22 = 32.0*cf; A23 = A12;
  A31 = A13;     A32 = A23;     A33 = A11;

  cf = h(l)/30.0;

  B11 = 4.0*cf; B12 =  2.0*cf;  B13 = -1.0*cf;
  B21 = B12;    B22 = 16.0*cf;  B23 = B12;
  B31 = B13;    B32 = B23;      B33 = B11;

  % condense the diffusion matrix:

  r12 = A12/A22;
  A11 = A11-r12*A21; A13= A13-r12*A23;

  r32=A32/A22;
  A31 = A31-r32*A21; A33= A33-r32*A23;

  at(l)   = at(l)   + A11; bt(l)   = bt(l)   + A13;
  ct(l+1) = ct(l+1) + A31; at(l+1) = at(l+1) + A33;
```

TABLE 1.5.7 Function *sdq_cnd_sys* (Continuing →)

```
% condense the right-hand side

cl1=2*l-1; cl2=2*l; cl3=2*l+1;

b(l)   = b(l)   +     (B11*s(cl1) + B12*s(cl2) + B13*s(cl3))/k;
b(l)   = b(l)   - r12*(B21*s(cl1) + B22*s(cl2) + B23*s(cl3))/k;
b(l+1) = b(l+1) +     (B31*s(cl1) + B32*s(cl2) + B33*s(cl3))/k;
b(l+1) = b(l+1) - r32*(B21*s(cl1) + B22*s(cl2) + B23*s(cl3))/k;

end

%---------------------------------
% implement the Dirichlet condition
%---------------------------------

b(ne) = b(ne)-A13*fL;

%-----
% done
%-----

return;
```

TABLE 1.5.7 ($\rightarrow$ Continued) Function *sdq_cnd_sys* implements the compact assembly of a condensed tridiagonal linear system for one-dimensional diffusion with quadratic elements and mid-point interior nodes.

Relation to the finite difference method

If the elements are spaced evenly, $h_1 = h_2 = \cdots = h_{N_E} = h$, and if the interior element nodes are located at the element mid-points, we may use the expressions in (1.5.29) to find that the right-hand side of the condensed linear system stated in (1.5.37) simplifies to

$$\widehat{b}_1 = \frac{q_0}{k} + \frac{h}{k}\frac{5}{30}\left(s_1 + 2s_2\right), \qquad \widehat{b}_2 = \frac{h}{k}\frac{1}{3}\left(s_2 + s_3 + s_4\right), \qquad \cdots \quad . \quad (1.5.41)$$

The ith equation of the condensed system takes the form

$$k\,\frac{f_{2i-3} - 2\,f_{2i-1} + f_{2i+1}}{h^2} + \frac{1}{3}\left(s_{2i-2} + s_{2i-1} + s_{2i}\right) = 0 \qquad (1.5.42)$$

for $i > 1$. For convenience, we set $j = 2i - 1$ and restate (1.5.42) as

$$k\,\frac{f_{j-2} - 2\,f_j + f_{j+2}}{h^2} + \frac{1}{3}\left(s_{j-1} + s_j + s_{j+1}\right) = 0, \qquad (1.5.43)$$

with the understanding that the index j is odd and greater than unity.

To demonstrate the connection with the finite difference method, we apply the governing equation (1.1.4) at nodes $j - 1$, j, and $j + 1$, and approximate the second

```
%===========================================
% Code sdq_cnd
%
% Code for steady one-dimensional diffusion
% with quadratic elements using the
% condensed formulation
%===========================================

%-----------
% input data
%-----------

L = 1.0; k = 1.0; q0 = -1.0; fL = 0.0;

ne = 4; ratio = 2.0;

%----------------
% grid generation
%----------------

xe = elm_line1 (0,L,ne,ratio);

%-----------------------------
% number of unique global nodes
%-----------------------------

ng = 2*ne+1;

%-------------------------
% generate the global nodes
%-------------------------

Ic = 0;   % counter

for i=1:ne
 Ic = Ic+1; xg(Ic) = xe(i);
 Ic = Ic+1; xg(Ic) = 0.5*(xe(i)+xe(i+1));
end

xg(ng) = xe(ne+1);

%-------------------
% specify the source
%-------------------

for i=1:ng
 s(i) = 10.0*exp(-5.0*xg(i)^2/L^2);
end

%-------------------------
% compact element assembly
%-------------------------

[at,bt,ct,b] = sdq_cnd_sys (ne,xe,q0,fL,k,s);

%-------------
% linear solver
%-------------
```

TABLE 1.5.8 Code *sdq_cnd* (Continuing →)

```
sol = thomas(ne,at,bt,ct,b);

%----------------------
% recover the solution
%----------------------

for i=1:ne
 f(2*i-1) = sol(i);
 f(2*i) = 0.0;      % ignore the mid-nodes
end

f(ng) = fL;

%-----
% plot
%-----

for i=1:ne+1
 yplot(i) = f(2*i-1);
end

plot(xe, yplot,'+');
hold on;
plot(xe, zeros(ne+1,1),'x');
```

TABLE 1.5.8 ($\rightarrow$ Continued) Finite element code *sdq_cnd* for steady one-dimensional diffusion with quadratic elements, using the condensed formulation.

derivative with finite differences involving five consecutive nodal values,

$$f_{j-2}, \qquad f_{j-1}, \qquad f_j, \qquad f_{j+1}, \qquad f_{j+2}. \tag{1.5.44}$$

Taking into consideration that these nodes are separated by the interval $\frac{1}{2}h$, we obtain the difference equations

$$\frac{k}{3h^2}\left(11\,f_{j-2} - 20\,f_{j-1} + 6\,f_j + 4\,f_{j+1} - f_{j+2}\right) + s_{j-1} = 0, \tag{1.5.45}$$

at the $j-1$ node,

$$\frac{k}{3h^2}\left(-f_{j-2} + 16\,f_{j-1} - 30\,f_j + 16\,f_{j+1} - f_{j+2}\right) + s_j = 0, \tag{1.5.46}$$

for the jth node, and

$$\frac{k}{3h^2}\left(-f_{j-2} + 4\,f_{j-1} + 6\,f_j - 20\,f_{j+1} + 11f_{j+2}\right) + s_{j+1} = 0 \tag{1.5.47}$$

for the $j+1$ node, The coefficients on the left-hand sides of equations (1.5.45) and (1.5.47) were generated using the FSELIB script *fdc*, not listed in the text, using the method of undetermined coefficients. The corresponding coefficients in the second equation are available in differentiation tables (e.g., Pozrikidis [47]). Adding (1.5.45)–(1.5.47) and simplifying, we recover precisely (1.5.43).

In summary, we have found that the finite element method with quadratic element functions and uniform elements is equivalent to the finite difference method implemented with the five-point difference approximation of the second derivative at the interior nodes.

1.5.7 Arbitrary interior nodes

When the interior node is mapped to the point $\xi = \beta$ in parameter space, the position of the second element node is

$$x_2(l) = \frac{1}{2}\left(x_1^{(l)} + x_3^{(l)}\right) + \frac{1}{2}\left(x_3^{(l)} - x_1^{(l)}\right)\beta, \qquad (1.5.48)$$

where $-1 < \beta < 1$ is a free parameter; the mid-point element position corresponds to $\beta = 0$. Conversely, equation (1.5.48) can be solved for β in terms of the three nodal positions.

Requiring the cardinal interpolation conditions, we find that the three quadratic element interpolation functions are given by

$$\psi_1^{(l)}(\xi) = \frac{1}{2(1+\beta)}\,(\xi - \beta)(\xi - 1), \qquad \psi_2^{(l)}(\xi) = \frac{1 - \xi^2}{1 - \beta^2},$$

$$\psi_3^{(l)}(\xi) = \frac{1}{2(1-\beta)}\,(\xi - \beta)(\xi + 1). \qquad (1.5.49)$$

When $\beta = 0$, we recover expressions (1.5.24) for the mid-point interior node.

Element matrices

The components of the element diffusion matrix are given in Table 1.5.9. When $\beta = 0$, we recover expressions (1.5.27) for the mid-point interior node. To circumvent lengthy algebra, the element mass matrix can be computed numerically using the four-point Lobatto integration quadrature, as discussed in Section 3.4.2. Alternatively, the element diffusion and mass matrices can be computed from those corresponding to the mid-point interior node using the transformation rules discussed in Section 3.2. Both methods are implemented in the function *sdq_beta_sys*, listed in Table 1.5.10.

Finite element code

FSELIB code *sdq_beta*, listed in Table 1.5.11, implements the complete finite element code. The numerical solution generated by the code with the frivolous source function defined in (1.2.17) is displayed with the crosses in Figure 1.5.3(*b*).

$$A_{11}^{(l)} = \frac{1}{3\,h_l}\,\frac{1}{(1+\beta)^2}\left(4 + 3\,(1+\beta)^2\right)$$

$$A_{12}^{(l)} = A_{21}^{(l)} = -\frac{1}{3\,h_l}\,\frac{8}{(1+\beta)(1-\beta^2)}$$

$$A_{13}^{(l)} = A_{31}^{(l)} = \frac{1}{3\,h_l}\,\frac{1}{(1-\beta^2)}\left(4 - 3\,(1-\beta^2)\right)$$

$$A_{22}^{(l)} = \frac{1}{3\,h_l}\,\frac{16}{(1-\beta^2)^2}$$

$$A_{23}^{(l)} = A_{32}^{(l)} = -\frac{1}{h_l}\,\frac{1}{3}\,\frac{8}{(1-\beta)(1-\beta^2)}$$

$$A_{33}^{(l)} = \frac{1}{h_l}\,\frac{1}{3}\,\frac{1}{(1-\beta)^2}\left(4 + 3\,(1-\beta)^2\right)$$

TABLE 1.5.9 Components of the element diffusion matrix for quadratic elements with arbitrary interior node determined by the parameter β.

PROBLEMS

1.5.1 *Evaluation of the diffusion and mass matrices*

(*a*) Derive the components of the element diffusion matrix shown in (1.5.27) and confirm that the matrix is singular. *Hint:* Observe that the sum of each row is zero.

(*b*) Derive the components of the element mass matrix shown in (1.5.29).

1.5.2 *Codes sdq and sdq_cnd*

(*a*) Execute the FSELIB code *sdq* with the same boundary conditions but a different source function of your choice and discuss the results.

(*b*) Repeat (*a*) for code *sdq_cnd* and confirm that the results remain unchanged.

1.5.3 *Arbitrary interior node*

(*a*) Verify that the element diffusion matrix is singular for any position of the interior node corresponding to an arbitrary value of the parameter β. *Hint:* Run script *edmq* of FSELIB.

(*b*) Execute the FSELIB code *sdq_beta* with parameters and conditions of your choice. Discuss the effect of the parameter β on the accuracy of the solution.

1.5.4 *Helmholtz equation*

Consider the Helmholtz equation in one dimension, equation (1.4.1) in a finite

```
function [ap,bp,cp,dp,ep,b] = sdq_beta_sys (ne,xe,beta,q0,fL,k,s)

%=======================================================
% Compact assembly of the pentadiagonal linear system
% for one-dimensional steady diffusion
% with quadratic elements (sdq)
% and arbitrary position of the interior node
% determined by the parameter "beta"
%=======================================================

%-------------
% element size
%-------------

for l=1:ne
  h(l) = xe(l+1)-xe(l);
end

%-----------------------------
% number of unique global nodes
%-----------------------------

ng = 2*ne+1;

%-----------------------------------
% initialize the pentadiagonal matrix
%-----------------------------------

ap = zeros(ng,1); bp = zeros(ng,1); cp = zeros(ng,1);
dp = zeros(ng,1); ep = zeros(ng,1);

%-----------------------------
% initialize the right-hand side
%-----------------------------

b = zeros(ng,1); b(1) = q0/k;

%----------------------
% loop over all elements
%----------------------

for l=1:ne

%----
% diffusion matrix analytically:
%----

  cf = 1.0/(3.0*h(l));

  A(1,1) =   1/(1+beta)^2 * (4+3*(1+beta)^2) * cf;
      A(1,2) = -8/((1+beta)*(1-beta^2)) * cf;
          A(1,3) =  1/(1-beta^2) * (4-3*(1-beta^2)) * cf;
```

TABLE 1.5.10 Function *sdq_beta_sys* (Continuing →)

```
  A(2,1) = A(1,2);
          A(2,2) = 16/(1-beta^2)^2 * cf;
          A(2,3) = -8/((1-beta)*(1-beta^2)) * cf;
  A(3,1) = A(1,3);
  A(3,2) = A(2,3);
  A(3,3) = 1/(1-beta)^2 * (4 + 3*(1-beta)^2) * cf;

%----
% alternative:
%
% diffusion matrix by the Vandermond matrix
% transformation
%----

% for beta=0:

  cf = 1.0/(3.0*h(1));
  A(1,1) = 7.0*cf; A(1,2) = -8.0*cf; A(1,3) = cf;
  A(2,1) = A(1,2);    A(2,2) = 16.0*cf; A(2,3) = A(1,2);
  A(3,1) = A(1,3);    A(3,2) = A(2,3);      A(3,3) = A(1,1);

%----
% vandermond matrix
%----

  vd(1,1) = 1; vd(1,2) = 0; vd(1,3) = 0;
  vd(2,1) = beta*(beta-1)/2; vd(2,2) = 1-beta^2;
  vd(2,3) = beta*(beta+1)/2;
  vd(3,1) = 0; vd(3,2)=0; vd(3,3)=1;

  invvd = inv(vd);

%----
% transform
%----

  A = invvd'*A*invvd;

%-------------------------------------------------
% mass matrix by the 4-point Lobatto quadrature:
%-------------------------------------------------

% Lobatto base points and weights

  xi(1) =-1.0; xi(2) =-1/sqrt(5); xi(3) = 1/sqrt(5); xi(4) = 1.0;
  w(1) = 1/6; w(2) = 5/6; w(3) = 5/6; w(4) = 1/6;

% initialize:

  for i=1:3
    for j=1:3
      B(i,j) = 0;
    end
  end
```

TABLE 1.5.10 Function *sdq_beta_sys* ($\rightarrow$ Continuing $\rightarrow$)

```
% Lobatto quadrature:

   for q=1:4

      psi(1) = (xi(q)-beta)*(xi(q)-1)/(2*(1+beta));
      psi(2) = (1-xi(q)^2)/(1-beta^2);
      psi(3) = (xi(q)-beta)*(xi(q)+1)/(2*(1-beta));

      B(1,1) = B(1,1) + psi(1)*psi(1)*w(q);
      B(1,2) = B(1,2) + psi(1)*psi(2)*w(q);
      B(1,3) = B(1,3) + psi(1)*psi(3)*w(q);
      B(2,2) = B(2,2) + psi(2)*psi(2)*w(q);
      B(2,3) = B(2,3) + psi(2)*psi(3)*w(q);
      B(3,3) = B(3,3) + psi(3)*psi(3)*w(q);

   end

   B(2,1) = B(1,2); B(3,1) = B(1,3); B(3,2) = B(2,3);

   B = h(l)/2.0*B;

%---------------
% mass matrix by the Vandermond matrix
% transformation
%---------------

% for beta=0:

   cf = h(l)/30.0;
   B(1,1) = 4.0*cf; B(1,2) =  2.0*cf; B(1,3) = -cf;
   B(2,1) = B(1,2); B(2,2) = 16.0*cf; B(2,3) = B(1,2);
   B(3,1) = B(1,3); B(3,2) = B(2,3);  B(3,3) = B(1,1);

%----
% transform
%----

   B = invvd'*B*invvd;

%---------
% assemble:
%---------

   cl1 = 2*l-1; cl2 = 2*l; cl3 = 2*l+1;

   ap(cl1) = ap(cl1) + A(1,1);
   bp(cl1) = bp(cl1) + A(1,2);
   cp(cl1) = cp(cl1) + A(1,3);

   dp(cl2) = dp(cl2) + A(2,1);
   ap(cl2) = ap(cl2) + A(2,2);
   bp(cl2) = bp(cl2) + A(2,3);

   ep(cl3) = ep(cl3) + A(3,1);
   dp(cl3) = dp(cl3) + A(3,2);
   ap(cl3) = ap(cl3) + A(3,3);
```

TABLE 1.5.10 Function *sdq_beta_sys* ($\rightarrow$ Continuing $\rightarrow$)

```
b(cl1) = b(cl1) + (B(1,1)*s(cl1) + B(1,2)*s(cl2) + B(1,3)*s(cl3))/k;
b(cl2) = b(cl2) + (B(2,1)*s(cl1) + B(2,2)*s(cl2) + B(2,3)*s(cl3))/k;
b(cl3) = b(cl3) + (B(3,1)*s(cl1) + B(3,2)*s(cl2) + B(3,3)*s(cl3))/k;

end

%-----------------------------------
% implement the Dirichlet condition
%-----------------------------------

b(ng-2) = b(ng-2) - cp(ng-2)*fL;
b(ng-1) = b(ng-1) - bp(ng-1)*fL;

%-----
% done
%-----

return;
```

TABLE 1.5.10 ($\rightarrow$ Continued) Function *sdq_beta_sys* assembles of a pentadiagonal
linear system for one-dimensional diffusion with quadratic elements and arbitrary
interior nodes.

interval, $[0, L]$, subject to the boundary conditions given by (1.4.2). Modify the code
sdq_beta into a code named *hlmq_beta* that solves the Helmholtz equation. Confirm
that the accuracy of the solution is insensitive to the position of the interior element
node.

1.6   Steady diffusion with quadratic modal expansions

The quadratic expansion of a function, $f(\xi)$, over the lth element takes the form

$$f(\xi) = f_1^{\mathrm{E}}\,\psi_1^{(l)}(\xi) + f_2^{\mathrm{E}}\,\psi_2^{(l)}(\xi) + f_3^{\mathrm{E}}\,\psi_3^{(l)}(\xi), \qquad (1.6.1)$$

where f_1^{E}, f_2^{E}, and f_3^{E} are element nodal values and ψ_i^{E} are the corresponding
quadratic element interpolation functions, as discussed in Section 1.5.

An alternative representation is provided by the quadratic *modal* expansion
expressed by

$$f(\xi) = f_1^{\mathrm{E}}\,\zeta_1^{(l)}(\xi) + c_2^{\mathrm{E}}\,\zeta_2^{(l)}(\xi) + f_3^{\mathrm{E}}\,\zeta_3^{(l)}(\xi), \qquad (1.6.2)$$

where

$$\zeta_1^{(l)}(\xi) = \frac{1}{2}\,(1 - \xi), \qquad \zeta_3^{(l)}(\xi) = \frac{1}{2}\,(1 + \xi) \qquad (1.6.3)$$

```
====================================
% Code sdq_beta
%
% steady one-dimensional diffusion
% with quadratic elements
% and arbitrary position of the
% interior node determined
% by the scalar parameter beta
%
% -1 < beta < 1
%
% beta=0 yields a mid-point node
%================================
%-----------
% input data
%-----------

L = 1.0; k = 1.0; q0 =-1.0; fL = 0.0;
ne = 8; ratio = 5.0;
beta = 0.5;

%----------------
% grid generation
%----------------

xe = elm_line1 (0,L,ne,ratio);

%-----------------------------
% number of unique global nodes
%-----------------------------

ng = 2*ne+1;

%-------------------------
% generate the global nodes
%-------------------------

Ic = 0;   % counter

for i=1:ne
 Ic = Ic+1;
 xg(Ic) = xe(i);
 Ic = Ic+1;
 xg(Ic) = 0.5*(xe(i)+xe(i+1)) + 0.5*(xe(i+1)-xe(i))*beta;
end

xg(ng) = xe(ne+1);

%-------------------
% specify the source
%-------------------

for i=1:ng
 s(i) = 10.0*exp(-5.0*xg(i)^2/L^2);
end
```

TABLE 1.5.11   Finite element code *sdq_beta* (Continuing →)

```
%------------------------
% compact element assembly
%------------------------

[ap,bp,cp,dp,ep,b] = sdq_beta_sys (ne,xe,beta,q0,fL,k,s);

%--------------
% linear solver
%--------------

f = penta (ng-1,ap,bp,cp,dp,ep,b);

f(ng) = fL;

%-----
% plot
%-----

plot(xg, f,'-k+');

for i=1:ne+1
 plot([xe(i),xe(i)], [0, 2.5],'k:');
end

%-----
% done
```

TABLE 1.5.11 ($\rightarrow$ Continued) Finite element code *sdq_beta* for steady one-dimensional diffusion with quadratic elements and arbitrary interior nodes.

are *linear* end-node interpolation functions independent of the location of the interior node, and c_2^{E} is a constant coefficient.

To satisfy the end-node interpolation conditions,

$$f(\xi = -1) = f_1^{\mathrm{E}}, \qquad f(\xi = 1) = f_3^{\mathrm{E}}, \qquad (1.6.4)$$

we require that the intermediate *quadratic* modal function, $\zeta_2^{(l)}(\xi)$, drops to zero at the end-nodes located at $\xi = \pm 1$, and set

$$\zeta_2^{(l)}(\xi) = 1 - \xi^2. \qquad (1.6.5)$$

The Galerkin finite element method of the quadratic modal expansion is formulated as discussed in Section 1.5 for the nodal expansion. The only difference is that the element-node interpolation functions, $\psi_i^{(l)}(\xi)$, are replaced by corresponding modal functions, $\zeta_i^{(l)}(\xi)$, and the associated interior nodal values, f_2^{E}, are replaced by the modal coefficients, c_2^{E}. Once f_1^{E}, c_2^{E}, and f_3^{E} are available, the solution can be reconstructed from the three-term modal expansion.

For example, the solution at the element mid-point, corresponding to $\xi = 0$, is recovered as

$$f_2^{\mathrm{E}} = f(\xi = 0) = \frac{1}{2} \left(f_1^{\mathrm{E}} + 2 c_2^{\mathrm{E}} + f_3^{\mathrm{E}} \right). \tag{1.6.6}$$

Galerkin implementation

In the Galerkin finite element implementation, the source term over the lth-element is approximated with a quadratic modal expansion,

$$s(\xi) = s_1^{\mathrm{E}} \, \psi_1^{(l)}(\xi) + d_2^{\mathrm{E}} \, \zeta_2^{(l)}(\xi) + s_3^{\mathrm{E}} \, \psi_3^{(l)}(\xi). \tag{1.6.7}$$

The coefficient d_2^{E} is evaluated by inverting the counterpart of (1.6.6) for the source,

$$d_2^{\mathrm{E}} = \frac{1}{2} \left(- s_1^{\mathrm{E}} + 2 \, s_2^{\mathrm{E}} - s_3^{\mathrm{E}} \right). \tag{1.6.8}$$

The global diffusion and mass matrices are assembled from corresponding element matrices defined with respect to the modal functions, $\zeta_i^{(l)}$.

Element matrices

The lth-element diffusion matrix is defined as

$$A_{ij}^{(l)} = \int_{E_l} \frac{\mathrm{d}\zeta_i^{(l)}}{\mathrm{d}x} \frac{\mathrm{d}\zeta_j^{(l)}}{\mathrm{d}x} \, \mathrm{d}x = \frac{1}{2} \, h_l \int_{-1}^{1} \frac{2}{h_l} \frac{\mathrm{d}\zeta_i^{(l)}}{\mathrm{d}\xi} \frac{2}{h_l} \frac{\mathrm{d}\zeta_j^{(l)}}{\mathrm{d}\xi} \, \mathrm{d}\xi, \tag{1.6.9}$$

yielding

$$A_{ij}^{(l)} = \frac{2}{h_l} \int_{-1}^{1} \frac{\mathrm{d}\zeta_i^{(l)}}{\mathrm{d}\xi} \frac{\mathrm{d}\zeta_j^{(l)}}{\mathrm{d}\xi} \, \mathrm{d}\xi. \tag{1.6.10}$$

Performing the integrations, we obtain

$$\mathcal{A}^{(l)} = \frac{1}{3 \, h_l} \begin{bmatrix} 3 & 0 & -3 \\ 0 & 16 & 0 \\ -3 & 0 & 3 \end{bmatrix}. \tag{1.6.11}$$

As in the case of the nodal expansion, this matrix is singular.

The element mass matrix is defined as

$$B_{ij}^{(l)} = \int_{E_l} \zeta_i^{(l)} \zeta_j^{(l)} \, \mathrm{d}x = \frac{1}{2} \, h_l \int_{-1}^{1} \zeta_i^{(l)} \zeta_j^{(l)} \, \mathrm{d}\xi. \tag{1.6.12}$$

Performing the integrations, we obtain

$$\mathcal{B}^{(l)} = h_l \, \frac{1}{30} \begin{bmatrix} 10 & 10 & 5 \\ 10 & 16 & 10 \\ 5 & 10 & 10 \end{bmatrix}. \tag{1.6.13}$$

Note that the modal element diffusion and mass matrices differ from the corresponding nodal matrices displayed in (1.5.27) and (1.5.29), with one exception. Since $\zeta_2^{(l)}(\xi) = \psi_2^{(l)}(\xi) = 1 - \xi^2$, the central values, $A_{22}^{(l)} = 16/(3\,h_l)$ and $B_{22}^{(l)} = 8h_l/15$, are consistent with those shown in (1.5.27) and (1.5.29).

Because of the sparseness of the element diffusion matrix, the global diffusion matrix takes the form

$$
\mathcal{D} =
\begin{bmatrix}
D_{11} & 0 & D_{13} & 0 & 0 & 0 & \cdots \\
0 & D_{22} & 0 & 0 & 0 & 0 & \cdots \\
D_{31} & 0 & D_{33} & 0 & D_{35} & 0 & \cdots \\
0 & 0 & 0 & D_{44} & 0 & 0 & \cdots \\
0 & 0 & D_{53} & 0 & D_{55} & 0 & \cdots \\
\vdots & \vdots & \vdots & \vdots & \vdots & \vdots & \ddots
\end{bmatrix}.
\tag{1.6.14}
$$

The superdiagonal and subdiagonal lines are zero, and every other element of the second superdiagonal and subdiagonal lines are also zero. Consequently, the odd-numbered unknowns representing the nodal values of the solution are decoupled from the even-number unknowns corresponding to the coefficients c_2^E of the modal functions $\zeta_2^{(l)}(\xi)$, and can be computed by solving a tridiagonal system of equations. The even-numbered unknowns can be recovered simply by dividing the corresponding entries on the right-hand side by the diagonal components. Thus, the modal expansion effectively implements node condensation.

Function *sdq_modal_sys*, listed in Table 1.6.1, generates a pentadiagonal system of equations for the vector of unknowns,

$$
\mathbf{f} \equiv \left[\, f_1^{(1)}, \quad c_2^{(1)}, \quad f_1^{(2)}, \quad c_2^{(2)}, \quad \ldots, \quad f_1^{(N_E)}, \quad c_2^{(N_E)} \,\right]^T.
\tag{1.6.15}
$$

Recall that, by C^0 continuity, $f_2^{(l)} = f_1^{(l+1)}$.

The finite element method is implemented in the FSELIB code *sdq_modal*, listed in Table 1.6.2. The results of the finite solution are identical to those obtained using the nodal expansion.

Modal compared to nodal expansion

The advantages of the modal expansion (1.6.2) will be discussed in Chapter 3 in the context of high-order and spectral element methods where an arbitrary number of interior modes, known as *bubble modes,* are added to the two linear end-node interpolation functions representing the so-called vertex modes displayed in (1.6.3). The bubble modes are multiplied by *a priori* unknown coefficients that are not necessarily associated with any interior nodes.

The main advantage of the modal approach is that, if the modes are chosen judiciously, the resulting global diffusion or mass matrices exhibit a desirable sparsity that promotes the well-conditioning and expedites the numerical solution of

```
function [ap,bp,cp,dp,ep,b] = sdq_modal_sys(ne,xe,q0,fL,k,s)

%=====================================================
% Compact assembly of a pentadiagonal linear system
% for one-dimensional steady diffusion
% with quadratic elements
% using the modal expansion
%=====================================================

%-------------
% element size
%-------------

for l=1:ne
   h(l) = xe(l+1)-xe(l);
end

%-----------------------
% number of global modes
%-----------------------

ng = 2*ne+1;

%-----------------------------------
% initialize the pentadiagonal matrix
%-----------------------------------

ap = zeros(ng,1); bp = zeros(ng,1); cp = zeros(ng,1);
dp = zeros(ng,1); ep = zeros(ng,1);

%-----------------------------
% initialize the right-hand side
%-----------------------------

b = zeros(ng,1);
b(1) = q0/k;

%----------------------
% loop over all elements
%----------------------

for l=1:ne

   cf = 1.0/h(l);

   A11 = cf;   A12 = 0;         A13 = -cf;
   A21 = A12;        A22 = 16.0*cf/3.0; A23 = 0;
   A31 = A13;        A32 = A23;         A33 = A11;

   cf = h(l)/6.0;

   B11 = 2.0*cf; B12 =   2.0*cf;  B13 = cf;
   B21 = B12;       B22 = 16.0/5.0*cf;  B23 = 2.0*cf;
   B31 = B13;       B32 = B23;          B33 = 2.0*cf;
```

TABLE 1.6.1   Function *sdq_modal_sys* (Continuing →)

```
   cl1 = 2*l-1; cl2 = 2*l; cl3 = 2*l+1;

   ap(cl1) = ap(cl1) + A11;
   bp(cl1) = bp(cl1) + A12;
   cp(cl1) = cp(cl1) + A13;

   dp(cl2) = dp(cl2) + A21;
   ap(cl2) = ap(cl2) + A22;
   bp(cl2) = bp(cl2) + A23;

   ep(cl3) = ep(cl3) + A31;
   dp(cl3) = dp(cl3) + A32;
   ap(cl3) = ap(cl3) + A33;

   b(cl1) = b(cl1) + (B11*s(cl1) + B12*s(cl2) + B13*s(cl3))/k;
   b(cl2) = b(cl2) + (B21*s(cl1) + B22*s(cl2) + B23*s(cl3))/k;
   b(cl3) = b(cl3) + (B31*s(cl1) + B32*s(cl2) + B33*s(cl3))/k;

end

%-----------------------------------
% implement the Dirichlet condition
%-----------------------------------

b(ng-2) = b(ng-2) - cp(ng-2)*fL;
b(ng-1) = b(ng-1) - bp(ng-1)*fL;

%-----
% done
%-----

return;
```

TABLE 1.6.1 (Continued →) Function *sdq_modal_sys* implements the compact assembly of a pentadiagonal linear system for one-dimensional diffusion with quadratic modal elements.

the final system of linear equations. High-order modal expansions are discussed in Sections 3.6 and 3.7.

PROBLEM

1.6.1 *Element diffusion and mass matrices*

(*a*) Derive the element diffusion matrix of the quadratic modal expansion shown in (1.6.11).

(*b*) Derive the element mass matrix of the quadratic modal expansion shown in (1.6.13).

```
%=================================
% Code sdq_modal
%
% steady one-dimensional diffusion
% with quadratic modal expansion
%=================================

%-----------
% input data
%-----------

L = 1.0;
k = 1.0; q0 =-1.0; fL = 0.0;

ne = 16; ratio = 5.0;

%----------------
% grid generation
%----------------

xe = elm_line1 (0,L,ne,ratio);

%-----------------------
% number of global modes
%-----------------------

ng = 2*ne+1;

%-------------------------
% generate the global nodes
%-------------------------

Ic = 0;   % counter

for i=1:ne
   Ic = Ic+1; xg(Ic) = xe(i);
   Ic = Ic+1; xg(Ic) = 0.5*(xe(i)+xe(i+1));
end

xg(ng) = xe(ne+1);

%-------------------
% specify the source
%-------------------

for i=1:ng
 s(i) = 10.0*exp(-5.0*xg(i)^2/L^2);
end

%-----------------------------------------------
% compute the modal coefficients of the source
% by collocation at the element mid-point
%-----------------------------------------------
```

TABLE 1.6.2   Code *sdq_modal* (Continuing →)

```
for i=1:ng
 smodal(i) = s(i);
end

for i=1:ne
 j=2*i;
 smodal(j) = (-s(j-1)+2.0*s(j)-s(j+1))/2.0;
end

%-------------------------------------------------------
% compact element assembly of the pentadiagonal system
%-------------------------------------------------------

[ap,bp,cp,dp,ep,rp] = sdq_modal_sys (ne,xe,q0,fL,k,smodal);

%------
% formulate and solve a tridiagonal system
%------

for i=1:ne
 j=2*i-1;
 at(i)=ap(j); bt(i)=cp(j); ct(i)=ep(j); rt(i)=rp(j);
end
sol = thomas (ne,at,bt,ct,rt);

for i=1:ne
 j = 2*i-1;
 f(j) = sol(i);
end

f(ng) = fL;

%------
% solve for the even-numbered unknowns
% to recover the mid-point values
%------

for i=1:ne
 j=2*i;
 f(j) = rp(j)/ap(j);  % modal
 f(j) = (f(j-1)+2.0*f(j)+f(j+1))/2.0; % nodal
end

%-----
% plot
%-----

plot(xg, f,'-k+');

%-----
% done
%-----
```

TABLE 1.6.2 (Continued →) Finite element code *sdq_modal* for steady diffusion with quadratic modal elements.

2

Further applications in one dimension

Building on our discussion of the finite element method for the one-dimensional steady diffusion equation in Chapter 1, we proceed to illustrate the implementation of the method to other ordinary and partial differential equations in one spatial dimension. We begin by discussing procedures for advancing in time the solution of the unsteady diffusion equation and the convection equation, and then study the solution of the convection–diffusion equation at steady state. In Sections 2.4–2.6, we consider applications in structural mechanics with reference to the bending and buckling of beams, governed by fourth-order differential equations. The union of the problems considered in this chapter provides us with an extended methodology that can be used with straightforward adaptations to solve general ordinary and partial differential equations.

2.1 Unsteady diffusion

Consider the unsteady version of the heat conduction problem in the presence of a distributed source discussed in Chapter 1. Performing a heat balance over an infinitesimal section of the rod illustrated in Figure 1.1.1, contained between two points located at x and $x + dx$, we find that the evolution of the temperature, $f(x,t)$, at a certain time, t, is governed by the unsteady heat conduction equation

$$\rho\, c_p\, \frac{\partial f}{\partial t} = k\, \frac{\partial^2 f}{\partial x^2} + s(x,t), \tag{2.1.1}$$

where ρ is the density of the rod material, c_p is the heat capacity under constant pressure, and k is the thermal conductivity. Dividing both sides of (2.1.1) by $\rho\, c_p$, we derive the alternative form

$$\frac{\partial f}{\partial t} = \kappa\, \frac{\partial^2 f}{\partial x^2} + \frac{s(x,t)}{\rho\, c_p}, \tag{2.1.2}$$

where

$$\kappa \equiv \frac{k}{\rho\, c_p} \tag{2.1.3}$$

is the *thermal diffusivity* with dimensions of length squared divided by time. Our goal is to compute a numerical solution of (2.1.2) subject to a given initial condition, $f(x, t = 0) = F(x)$, and two boundary conditions, one at each end.

2.1.1   Galerkin projection

To formulate the Galerkin finite element method, we follow the steps outlined in Chapter 1 for steady diffusion. The main difference is that the unknown nodal values, $f_i(t) \equiv f(x_i, t)$, are now time dependent. The ith Galerkin projection of (2.1.2) yields the counterpart of equation (1.1.20),

$$
\rho\, c_p \int_0^L \phi_i\, \frac{\partial f}{\partial t}\, \mathrm{d}x = -k\left(\phi_i\, \frac{\partial f}{\partial x}\right)_{x=0} + k\left(\phi_i\, \frac{\partial f}{\partial x}\right)_{x=L}
$$
$$
-k \int_0^L \frac{\partial \phi_i}{\partial x}\, \frac{\partial f}{\partial x}\, \mathrm{d}x + \int_0^L \phi_i\, s(x, t)\, \mathrm{d}x,
$$

(2.1.4)

where the index i runs over an appropriate number of nodes determined by the degree of the element interpolation functions and by the specified boundary conditions, Dirichlet versus Neumann, as discussed in Chapter 1.

Next, we substitute into (2.1.4) the finite element expansion of the numerical solution,

$$
f(x) = \sum_{j=1}^{N_G} f_j(t)\, \phi_j(x),
$$

(2.1.5)

and a corresponding expansion for the source term,

$$
s(x) = \sum_{j=1}^{N_G} s_j(t)\, \phi_j(x),
$$

(2.1.6)

where N_G is the number of unique global nodes. Rearranging and collecting all nodal projections, we derive a system of linear ordinary differential equations (ODEs),

$$
\mathbf{M} \cdot \frac{\mathrm{d}\mathbf{f}}{\mathrm{d}t} + \kappa\, \mathbf{D} \cdot \mathbf{f} = \kappa\, \mathbf{b},
$$

(2.1.7)

where $\mathbf{M}$ is a global square mass matrix, $\mathbf{D}$ is a global square diffusion matrix, and $\mathbf{b}$ is a properly constructed right-hand side.

Linear elements

In the case of the piecewise linear expansion (1.1.9), subject to the Neumann and Dirichlet boundary conditions (1.1.5) and (1.1.6), the system of ordinary differential equations (2.1.7) is the counterpart of the algebraic system (1.1.26), involving the square, tridiagonal, $N_E \times N_E$ global mass matrix $\mathbf{M}$ shown in Table 2.1.1.[1] The constant vector $\mathbf{b}$ on the right-hand side of (2.1.7) is defined in equations (1.1.28)–(1.1.30) and is given in Table 1.1.7(b). The square diffusion matrix, $\mathbf{D}$, is given in Table 1.1.6.

---

[1] The square mass matrix $\mathbf{M}$ arises by discarding the last column of the rectangular mass matrix $\widetilde{\mathbf{M}}$ shown in Table 1.1.7(a) and Table 1.2.1(b).

$$
\mathbf{M} \equiv
\begin{bmatrix}
\frac{1}{3} h_1 & \frac{1}{6} h_1 & 0 & 0 & \cdots \\
\frac{1}{6} h_1 & \frac{1}{3} h_1 + \frac{1}{3} h_2 & \frac{1}{6} h_2 & 0 & \cdots \\
0 & \frac{1}{6} h_2 & \frac{1}{3} h_2 + \frac{1}{3} h_3 & 0 & \cdots \\
\vdots & & 0 & \cdots & \cdots \quad \cdots \\
\vdots & & \vdots & & \vdots \quad \vdots \\
0 & \cdots & & \cdots & \cdots \quad \cdots
\end{bmatrix}
\longrightarrow
$$

$$
\begin{bmatrix}
0 & 0 & 0 \\
0 & \cdots & 0 \\
\cdots & 0 & 0 \\
\frac{1}{3} h_{N_{\mathrm{E}}-3} + \frac{1}{3} h_{N_{\mathrm{E}}-2} & \frac{1}{6} h_{N_{\mathrm{E}}-2} & 0 \\
\frac{1}{6} h_{N_{\mathrm{E}}-2} & \frac{1}{3} h_{N_{\mathrm{E}}-2} + \frac{1}{3} h_{N_{\mathrm{E}}-1} & \frac{1}{6} h_{N_{\mathrm{E}}-1} \\
0 & \frac{1}{6} h_{N_{\mathrm{E}}-1} & \frac{1}{3} h_{N_{\mathrm{E}}-1} + \frac{1}{3} h_{N_{\mathrm{E}}}
\end{bmatrix}
$$

TABLE 2.1.1   Global square $N_{\mathrm{E}} \times N_{\mathrm{E}}$ mass matrix for unsteady diffusion with linear elements.

Making substitutions, we find that the Galerkin finite element system (2.1.7) takes the form

$$
\frac{1}{3} h_1 \frac{\mathrm{d} f_1}{\mathrm{d} t} + \frac{1}{6} h_1 \frac{\mathrm{d} f_2}{\mathrm{d} t} + \kappa \sum_{j=1}^{N_{\mathrm{E}}} D_{1,j} f_j = \kappa b_1 \tag{2.1.8}
$$

and

$$
\frac{1}{6} h_{i-1} \frac{\mathrm{d} f_{i-1}}{\mathrm{d} t} + \frac{1}{3} (h_{i-1} + h_i) \frac{\mathrm{d} f_i}{\mathrm{d} t} + \frac{1}{6} h_i \frac{\mathrm{d} f_{i+1}}{\mathrm{d} t} + \kappa \sum_{j=1}^{N_{\mathrm{E}}} D_{ij} f_j = \kappa b_i \tag{2.1.9}
$$

for $i = 2, \ldots, N_{\mathrm{E}}$. Rearranging (2.1.9), we obtain

$$
\left(\frac{1}{3} \frac{h_{i-1}}{h_{i-1} + h_i} \right) \frac{\mathrm{d} f_{i-1}}{\mathrm{d} t} + \left(\frac{2}{3} \right) \frac{\mathrm{d} f_i}{\mathrm{d} t} + \left(\frac{1}{3} \frac{h_i}{h_{i-1} + h_i} \right) \frac{\mathrm{d} f_{i+1}}{\mathrm{d} t}
$$

$$
+ \kappa \frac{2}{h_{i-1} + h_i} \sum_{j=1}^{N_{\mathrm{E}}} D_{ij} f_j = \kappa \frac{2}{h_{i-1} + h_i} b_i \tag{2.1.10}
$$

for $i = 2, \ldots, N_{\mathrm{E}}$. The first three terms on the left-hand side of (2.1.10) contribute a weighted average of the time derivative at three neighboring nodes. The sum of the weights enclosed by the parentheses is equal to unity.

Uniform elements

When the element size is uniform, $h_1 = h_2 = \cdots = h_{N_E} \equiv h$, the Galerkin equations simplify to

$$\frac{1}{3}\frac{\mathrm{d}f_1}{\mathrm{d}t} + \frac{1}{6}\frac{\mathrm{d}f_2}{\mathrm{d}t} = \kappa\,\frac{f_2 - f_1}{h^2} + \frac{\kappa}{h}\,b_1 \tag{2.1.11}$$

and

$$\frac{1}{6}\frac{\mathrm{d}f_{i-1}}{\mathrm{d}t} + \frac{2}{3}\frac{\mathrm{d}f_i}{\mathrm{d}t} + \frac{1}{6}\frac{\mathrm{d}f_{i+1}}{\mathrm{d}t} = \kappa\,\frac{f_{i-1} - 2f_i + f_{i+1}}{h^2} + \frac{\kappa}{h}\,b_i \tag{2.1.12}$$

for $i = 2, \ldots, N_E$. The first term on the right-hand side of (2.1.12) is recognized as the central difference approximation of the second derivative, $\partial^2 f/\partial x^2$, evaluated at the ith node. The three terms on the left-hand side express a weighted average of the time derivative at the ith and adjacent nodes. A similar averaging of the source function is implicit in the constant term on the right-hand side, b_i.

2.1.2 Integrating ODEs

Equation (2.1.7) provides us with a coupled system of first-order, linear, ordinary differential equations (ODEs) in time, t, for the unknown node temperatures, $f_i(t)$. The coupling of nodal temperatures on the left-hand side, mediated through the global mass matrix, distinguishes the finite element method from the finite difference method. In practice, the system (2.1.7) is integrated in time using a standard numerical method for solving initial-value problems involving ordinary differential equations. Examples are the Euler method, the Crank–Nicolson method, and a Runge–Kutta method (e.g., Pozrikidis [47]).

2.1.3 Forward Euler method

In the simplest approach, the differential equation (2.1.7) is evaluated at a chosen time, t, and the time derivative on the left-hand side is approximated with a first-order forward finite difference, yielding

$$\mathbf{M} \cdot \frac{\mathbf{f}(t + \Delta t) - \mathbf{f}(t)}{\Delta t} + \kappa\,\mathbf{D} \cdot \mathbf{f}(t) = \kappa\,\mathbf{b}(t), \tag{2.1.13}$$

where Δt is a selected time step. The associated numerical error is on the order of Δt. Solving for the unknown vector $\mathbf{f}(t + \Delta t)$ on the left-hand side, we derive a tridiagonal system of equations

$$\mathbf{M} \cdot \mathbf{f}(t + \Delta t) = \mathbf{r}(t), \tag{2.1.14}$$

where

$$\mathbf{r}(t) \equiv \mathbf{M} \cdot \mathbf{f}(t) - \kappa\,\Delta t\,\big[\mathbf{D} \cdot \mathbf{f}(t) - \mathbf{b}(t)\big]. \tag{2.1.15}$$

The solution for $\mathbf{f}(t + \Delta t)$ can be found efficiently using the Thomas algorithm discussed in Section 1.2.2.

Mass lumping

Ideally, the global mass matrix $\mathbf{M}$ should be diagonal so that the linear system (2.1.14) can be solved by dividing each component of the right-hand side with the corresponding diagonal component of the coefficient matrix, yielding

$$f_i(t + \Delta t) = \frac{r_i}{M_{ii}}. \qquad (2.1.16)$$

In this ideal case, the forward Euler discretization provides us with an *explicit* numerical method.[2]

Later in this book, we will see that a diagonal mass matrix is particularly desirable in two and three dimensions where the global mass matrix is sparse but not tridiagonal or pentadiagonal, and the efficient Thomas algorithm may no longer be employed.

If we are not particularly interested in describing the transient evolution but only want to extract the solution established after a long period of time at steady state, we can simplify the algorithm by transferring the off-diagonal elements of the mass matrix $\mathbf{M}$ in each row to the corresponding diagonal element. In the finite element literature, this simplification is known as *mass lumping*. In Chapter 3, we will see that mass lumping also arises from the inexact integration of the element mass matrix implemented in a way that preserves the sum of the matrix elements in each row.

In the literature of finite element methods, the terminology *consistent formulation* implies the absence of mass lumping. Whether mass lumping is benign, tolerable, or detrimental must be assessed individually for each differential equation, numerical implementation, and physical application.

2.1.4 Numerical stability

In the case of linear elements with equal size, h, and mass lumping, the Galerkin projection (2.1.12) provides us with a coupled system of ordinary differential equations,

$$\frac{\mathrm{d} f_i}{\mathrm{d} t} = \kappa \frac{f_{i-1} - 2 f_i + f_{i+1}}{h^2} + \frac{\kappa}{h} b_i \qquad (2.1.17)$$

for $i = 2, \ldots, N_{\mathrm{E}}$. The fraction on the right-hand side is the central difference approximation of the second derivative at the ith node.

---

[2]The terminology *explicit* emphasizes that solving a system of linear algebraic equations is not required.

Applying the explicit Euler time discretization with time step Δt we obtain an algebraic equation,

$$\frac{f_i^{n+1} - f_i^n}{\Delta t} = \kappa \frac{f_{i-1}^n - 2f_i^n + f_{i+1}^n}{h^2} + \frac{\kappa}{h} b_i, \qquad (2.1.18)$$

where the superscript n denotes evaluation at a time level t_n, and the superscript $n + 1$ denotes evaluation at the next time level $t_{n+1} = t_n + \Delta t$. Rearranging, we derive an explicit formula,

$$f_i^{n+1} = \alpha\, f_{i-1}^n + (1 - 2\,\alpha)\, f_i^n + \alpha\, f_{i+1}^n + \frac{\kappa\,\Delta t}{h}\, b_i, \qquad (2.1.19)$$

where

$$\alpha \equiv \frac{\kappa\,\Delta t}{h^2} \qquad (2.1.20)$$

is the dimensionless diffusion number.

At steady state, $f_i^{n+1} = f_i^n \equiv \phi_i$ and the solution satisfies the difference equation

$$\phi_i = \alpha\, \phi_{i-1} + (1 - 2\,\alpha)\, \phi_i + \alpha\, \phi_{i+1} + \frac{\kappa\,\Delta t}{h}\, b_i. \qquad (2.1.21)$$

Simplifying, we derive a tridiagonal system of equations for the nodal values in the absence of the inconsequential time step, Δt. Subtracting (2.1.21) from (2.1.19), we find that the nodal deviation from the steady state, defined as

$$\varphi_i^n \equiv f_i^n - \phi_i, \qquad (2.1.22)$$

evolves according to the homogeneous difference equation

$$\varphi_i^{n+1} = \alpha\, \varphi_{i-1}^n + (1 - 2\,\alpha)\, \varphi_i^n + \alpha\, \varphi_{i+1}^n. \qquad (2.1.23)$$

In the case of linear elements with the Dirichlet boundary condition specified at *both ends*, equation (2.1.23) is applied for $i = 2, \ldots, N_{\mathrm{E}}$ to yield the vector equation

$$\boldsymbol{\varphi}^{(n+1)} = \mathbf{P} \cdot \boldsymbol{\varphi}^{(n)}, \qquad (2.1.24)$$

where

$$\boldsymbol{\varphi}^{(n+1)} \equiv \begin{bmatrix} \varphi_2^{n+1} \\ \varphi_3^{n+1} \\ \vdots \\ \varphi_{N-1}^{n+1} \\ \varphi_{N_{\mathrm{E}}}^{n+1} \end{bmatrix}, \qquad \boldsymbol{\varphi}^{(n)} \equiv \begin{bmatrix} \varphi_2^n \\ \varphi_3^n \\ \vdots \\ \varphi_{N_{\mathrm{E}}-1}^n \\ \varphi_{N_{\mathrm{E}}}^n \end{bmatrix} \qquad (2.1.25)$$

are solution vectors hosting $N = N_{\mathrm{E}} - 1$ nodes, and

$$
\mathbf{P} =
\begin{bmatrix}
1 - 2\,\alpha & \alpha & 0 & \cdots & 0 & 0 & 0 \\
\alpha & 1 - 2\,\alpha & \alpha & \cdots & 0 & 0 & 0 \\
0 & \alpha & 1 - 2\,\alpha & \cdots & 0 & 0 & 0 \\
\vdots & \vdots & \vdots & \ddots & \vdots & \vdots & \vdots \\
0 & 0 & 0 & \cdots & 1 - 2\,\alpha & \alpha & 0 \\
0 & 0 & 0 & \cdots & \alpha & 1 - 2\,\alpha & \alpha \\
0 & 0 & 0 & \cdots & 0 & \alpha & 1 - 2\,\alpha
\end{bmatrix}
\tag{2.1.26}
$$

is an $N \times N$ tridiagonal *projection matrix*. When a time step is made, the current deviation vector, $\varphi^{(n)}$, is multiplied by the projection matrix, $\mathbf{P}$, to yield an updated deviation vector, $\varphi^{(n+1)}$.

In the event that the time step Δt is constant, $t^n = (n - 1)\,\Delta t$, we obtain

$$
\varphi^{(n+1)} = \mathbf{P} \cdot \varphi^{(n)} = \mathbf{P} \cdot (\mathbf{P} \cdot \varphi^{(n-1)}) = \mathbf{P}^2 \cdot \varphi^{(n-1)}.
\tag{2.1.27}
$$

Repeating the process, we obtain

$$
\varphi^{(n+1)} = \mathbf{P}^{n+1} \cdot \varphi^{(0)}
\tag{2.1.28}
$$

for $n \geq 1$, where $\varphi^{(0)}$ corresponds to the initial state.

To determine the behavior of the solution vector, we introduce the eigenvectors of the projection matrix, $\mathbf{u}_m$, satisfying the equation

$$
\mathbf{P} \cdot \mathbf{u}_m = \lambda_m \,\mathbf{u}_m,
\tag{2.1.29}
$$

where λ_m are the corresponding eigenvalues. Next, we assume that $\mathbf{P}$ has N linearly independent eigenvectors and express the initial deviation from the steady state as a linear combination of these eigenvectors,

$$
\varphi^{(0)} = \sum_{m=1}^{N} c_m \,\mathbf{u}_m,
\tag{2.1.30}
$$

where c_m are appropriate coefficients. The first projection yields

$$
\mathbf{P} \cdot \varphi^{(0)} = \sum_{m=1}^{N} c_m \,\mathbf{P} \cdot \mathbf{u}_m = \sum_{m=1}^{N} c_m \lambda_m \,\mathbf{u}_m.
\tag{2.1.31}
$$

Repeating the projections, we obtain

$$
\varphi^{(n+1)} = \mathbf{P}^{(n+1)} \cdot \varphi^{(0)} = \sum_{m=1}^{N} c_m \lambda_m^{n+1} \,\mathbf{u}_m,
\tag{2.1.32}
$$

which shows that an arbitrary initial condition will grow in time if the magnitude of *at least one eigenvalue* is greater than unity. Since growth is physically unacceptable in a diffusion process, it must be attributed to numerical instability.

The formal criterion for numerical stability is that the spectral radius of the projection matrix is less than unity,

$$\max(|\lambda_m|) < 1. \tag{2.1.33}$$

By definition, the spectral radius of a matrix is the maximum of the magnitude of each eigenvalue.

The eigenvalues of the projection matrix (2.1.26) can be calculated exactly, and are given by

$$\lambda_m = 1 - 4\alpha \sin^2\left(\frac{m}{N+1}\frac{\pi}{2}\right) \tag{2.1.34}$$

for $m = 1, \ldots, N$ (e.g., Pozrikidis [47]). The eigenvector corresponding to the mth eigenvalue is

$$u_j^{(m)} = \sin\left(\frac{jm}{N+1}\pi\right) \tag{2.1.35}$$

for $m, j = 1, \ldots, N$ (Problem 2.1.1). Requiring that $|\lambda_m| < 1$ for any value of m or N, we derive the stability criterion

$$\alpha < \frac{1}{2}, \tag{2.1.36}$$

which can be restated as

$$\Delta t < \frac{h^2}{2\kappa}, \tag{2.1.37}$$

roughly requiring that the size of Δt is less than the size of h^2. Thus, if $h = 0.01$ in some units, Δt should be less than approximately 0.0001 in corresponding units. This stability constraint is extremely restrictive to be useful in practical applications.

Consistent formulation

The counterpart of the evolution equation (2.1.23) for the consistent formulation with elements of equal size is

$$\frac{1}{6}\varphi_{i-1}^{n+1} + \frac{2}{3}\varphi_i^{n+1} + \frac{1}{6}\varphi_{i+1}^{n+1} = \left(\frac{1}{6}+\alpha\right)\varphi_{i-1}^n + \left(\frac{2}{3}-2\alpha\right)\varphi_i^n + \left(\frac{1}{6}+\alpha\right)\varphi_{i+1}^n. \tag{2.1.38}$$

The counterpart of the vector equation (2.1.24) is

$$\mathbf{Q}\cdot\boldsymbol{\varphi}^{(n+1)} = (\mathbf{P}+\mathbf{Q}-\mathbf{I})\cdot\boldsymbol{\varphi}^{(n)}, \tag{2.1.39}$$

where $\mathbf{I}$ is the $N \times N$ identity matrix and

$$\mathbf{Q} = \frac{1}{6}\begin{bmatrix} 4 & 1 & 0 & \cdots & 0 & 0 & 0 \\ 1 & 4 & 1 & \cdots & 0 & 0 & 0 \\ 0 & 1 & 4 & \cdots & 0 & 0 & 0 \\ \vdots & \vdots & \vdots & \ddots & \vdots & \vdots & \vdots \\ 0 & 0 & 0 & \cdots & 4 & 1 & 0 \\ 0 & 0 & 0 & \cdots & 1 & 4 & 1 \\ 0 & 0 & 0 & \cdots & 0 & 1 & 4 \end{bmatrix} \tag{2.1.40}$$

is a tridiagonal matrix. Implementing mass lumping renders the matrix $\mathbf{Q}$ diagonal. Rearranging (2.1.39), we obtain

$$\varphi^{(n+1)} = \boldsymbol{\Pi} \cdot \varphi^{(n)}, \tag{2.1.41}$$

where

$$\boldsymbol{\Pi} \equiv \mathbf{Q}^{-1} \cdot (\mathbf{P} + \mathbf{Q} - \mathbf{I}) \tag{2.1.42}$$

is a new projection matrix.

The eigenvalues of the matrix $\mathbf{Q}$ can be calculated exactly, and are given by

$$\lambda_m^Q = 1 - \frac{2}{3}\sin^2\left(\frac{m}{N+1}\frac{\pi}{2}\right) \tag{2.1.43}$$

for $m = 1, \ldots, N$ (e.g., Pozrikidis [47]). Since the eigenvalues of the inverse of a matrix are equal to the inverses of the eigenvalues, the eigenvalues of the inverse matrix $\mathbf{Q}^{-1}$ are

$$\lambda_m^{Q^{-1}} = \frac{1}{1 - \frac{2}{3}\sin^2\left(\frac{m}{N+1}\frac{\pi}{2}\right)}. \tag{2.1.44}$$

The eigenvector corresponding to the mth eigenvalue is identical to the corresponding eigenvector of the matrix $\mathbf{P}$ defined in (2.1.35). Because the matrices $\mathbf{P}^{-1}$ and $\mathbf{P} + \mathbf{Q} - \mathbf{I}$ share eigenvectors, the eigenvalues of their product is equal to the product of their eigenvalues, and we may write

$$\lambda_m^{\Pi} = \lambda_m^{Q^{-1}}\left(\lambda_m^P + \lambda_m^Q - 1\right). \tag{2.1.45}$$

Substituting the eigenvalues, we find that

$$\lambda_m^{\Pi} = \frac{1 - (4\alpha + \frac{2}{3})\sin^2\left(\frac{m}{N+1}\frac{\pi}{2}\right)}{1 - \frac{2}{3}\sin^2\left(\frac{m}{N+1}\frac{\pi}{2}\right)} \tag{2.1.46}$$

for $m = 1, \ldots, N$.

Formula (2.1.46) for the eigenvalues can be verified for sample values of N and α by running the following MATLAB script entitled *proj_eig*:

```
N = 3;
al = 0.3;
P = [1-2*al al 0; al 1-2*al al; 0 al 1-2*al];
Q = [4 1 0; 1 4 1; 0 1 4]/6.0;
Pr = inv(Q)*(P+Q-eye(N));
eig(Pr)
```

The output can be confirmed to be consistent with the predictions of (2.1.46).

Now requiring that $|\lambda_i^{\mathbf{\Pi}}| < 1$, we derive the stability constraint

$$\alpha < \frac{1}{6}, \tag{2.1.47}$$

which can be restated as

$$\Delta t < \frac{h^2}{6\kappa}. \tag{2.1.48}$$

Comparing this expression with (2.1.37), we find that the consistent formulation imposes a somewhat more stringent constraint on the time step.

2.1.5   Finite element code

The forward Euler finite element method is implemented in the FSELIB code *udl*, listed in Table 2.1.2, according to the following steps:[3]

1. Data input and parameter definition.

2. Element node generation.

3. Definition of the source distribution along the rod.

4. Specification of the initial temperature distribution.

5. Compilation of the linear system (2.1.14).

6. Solution of the linear system using the Thomas algorithm.

7. Graphics display of the transient solution.

8. Return to Step 5 or termination of time stepping.

The linear system (2.1.14) is compiled using the FSELIB function *udl_sys*, listed in Table 2.1.3. The right-hand side, denoted as *lhs* in the code, is defined in (2.1.15). A graph of the transient solution is generated every *nplot* steps, where the parameter *nplot* is set in the input of the code. Once a plot is drawn, the counter *icount* is reset to zero.

Results in the interval $[0, 1]$ are shown in Figure 2.1.1 for ten evenly spaced elements with size $h = 0.1$, $\kappa = 1$, and two time steps $\Delta t = 0.0016$ or 0.0017,

---

[3] *udl* is the acronym for **u**nsteady **d**iffusion with **l**inear elements.

```
%=============================================
% Code   udl
%
% Finite element code for unsteady diffusion
% with linear elements using Euler's method
%=============================================

clear all
close all

%-----------
% input data
%-----------

L = 1.0;
k = 1.0; rho = 1.0; cp = 1.0;
q0 = -1.0; fL = 0.0;

Dt = 0.0017;

nsteps = 2000; nplot = 100;

ne = 10; ratio = 1.0;

%--------
% prepare
%--------

kappa = k/(rho*cp);

%----------------
% grid generation
%----------------

xe = elm_line1 (0,L,ne,ratio);

%---------------------------
% initial condition and source
%---------------------------

for i=1:ne+1
  f(1,i) = fL;
  s(i) = 10.0*exp(-5.0*xe(i)^2/L^2);
end

%-----
% plot
%-----

plot(xe, f,'x');
hold on;

icount = 0; % time step counter
```

TABLE 2.1.2   Code *udl* (Continuing →)

```
%--------------------
% begin time stepping
%--------------------

for irun=1:nsteps

%---------------------------------
% generate the tridiagonal system
%---------------------------------

[at,bt,ct,rhs] = udl_sys (ne,xe,q0,f,fL,k,kappa,s,Dt);

%--------------
% linear solver
%--------------

sol = thomas (ne,at,bt,ct,rhs);

f = [sol fL];      % include the value at the right end

%---------
% plotting
%---------

if(icount==nplot)
 plot(xe, f,'-o'); icount = 0;
end

icount = icount+1;

end

%--------------------
% end of time stepping
%--------------------

%-----
% done
%-----
```

TABLE 2.1.2 ($\to$ Continued) Finite element code *udl* for one-dimensional unsteady diffusion with linear elements using the forward Euler method.

corresponding to diffusion numbers $\alpha = 0.16$ and 0.17. A graph is generated every 100 time steps. The analysis in Section 2.1.4 suggested that the threshold for numerical stability is $\alpha = \frac{1}{6} \simeq 0.1667$. Consistent with the theoretical predictions, the first numerical solution for $\Delta t = 0.0016$ tends to the steady-steady solution smoothly, in the absence of visible oscillations. In spite of the smallness of the time step, the second numerical solution for $\Delta t = 0.0017$ develops an unphysical numerical oscillation dominated by a growing sawtooth wave. A catastrophic failure occurs at later times.

```
function [at,bt,ct,rhs] = udl_sys (ne,xe,q0,f,fL,k,kappa,s,Dt)

%=====================================================
% Compact assembly of the tridiagonal linear system
% for one-dimensional unsteady diffusion
% with linear elements (udl)
%=====================================================

%-------------
% element size
%-------------

for l=1:ne
  h(l) = xe(l+1)-xe(l);
end

%--------------------------------
% initialize the tridiagonal matrix
%--------------------------------

at = zeros(ne,1);
bt = zeros(ne,1);
ct = zeros(ne,1);

%--------------------------------
% initialize the right-hand side
%--------------------------------

rhs = zeros(ne,1); rhs(1) = Dt*q0/k;

%--------------------------------
% loop over the first ne-1 elements
%--------------------------------

for l=1:ne-1

  A11 = 1/h(l); A12 =-A11;        % diffusion matrix
  A21 = A12;       A22 = A11;
  B11 = h(l)/3.0; B12 = 0.5*B11; % mass matrix
  B21 = B12;       B22 = B11;
  at(l)    = at(l) + B11;            % tridiagonal matrix components
  bt(l)    = bt(l) + B12;
  ct(l+1) = ct(l+1) + B21;
  at(l+1) = at(l+1) + B22;
  rhs(l)    = rhs(l)    + B11*f(l) + B12*f(l+1);
  rhs(l+1) = rhs(l+1) + B21*f(l) + B22*f(l+1);
  rhs(l)    = rhs(l)    - Dt*kappa*(A11*f(l) + A12*f(l+1));
  rhs(l+1) = rhs(l+1) - Dt*kappa*(A21*f(l) + A22*f(l+1));
  rhs(l)    = rhs(l)    + Dt*kappa*(B11*s(l) + B12*s(l+1))/k;
  rhs(l+1) = rhs(l+1) + Dt*kappa*(B12*s(l) + B22*s(l+1))/k;

end
```

TABLE 2.1.3   Function *udl_sys* (Continuing →)

```
%---------------------------
% the last element is special
%---------------------------

A11 = 1.0/h(ne); A12 =-A11;
B11 = h(ne)/3.0; B12 = 0.5*B11;

at(ne) = at(ne) + B11;

rhs(ne) = rhs(ne) + (B11*f(ne) + B12*fL);
rhs(ne) = rhs(ne) - Dt*kappa*(A11*f(ne) + A12*fL);
rhs(ne) = rhs(ne) + Dt*kappa*(B11*s(ne) + B12*s(ne+1))/k;

%-----
% done
%-----

return;
```

TABLE 2.1.3   ($\rightarrow$ Continued) Function *udl_sys* assembles the linear system for one-dimensional unsteady diffusion with linear elements using the forward Euler method.

(a)   (b)

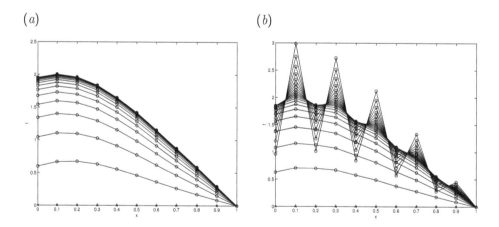

FIGURE 2.1.1   Solution of the unsteady diffusion equation generated by the finite element code *udl* for the source function shown in (1.2.17) with time step corresponding to diffusion number (a) $\alpha = 0.16$ and (b) 0.17. Theoretical analysis shows that the threshold for numerical stability is $\alpha = \frac{1}{6} = 0.1667$.

2.1.6 Crank–Nicolson integration

To circumvent the restriction on the time step, Δt, we may integrate the ordinary differential equations in time using an *implicit* numerical method.[4] In the Crank–Nicolson method, we select a time step, Δt, evaluate the differential equation (2.1.7) at a future time instant, $t + \frac{1}{2}\Delta t$, and approximate the time derivative on the left-hand side with a centered finite difference and the rest of the terms with averages to obtain

$$\mathbf{M} \cdot \frac{\mathbf{f}^{(n+1)} - \mathbf{f}^{(n)}}{\Delta t} + \frac{1}{2}\kappa\,\mathbf{D} \cdot \left(\mathbf{f}^{(n)} + \mathbf{f}^{(n+1)}\right) = \frac{1}{2}\kappa\left(\mathbf{b}^{(n)} + \mathbf{b}^{(n+1)}\right). \quad (2.1.49)$$

The associated error is on the order of Δt^2. Rearranging, we derive a linear algebraic system,

$$\mathbf{C} \cdot \mathbf{f}^{(n+1)} = \mathbf{r}. \quad (2.1.50)$$

The tridiagonal coefficient matrix on the left-hand side is given by

$$\mathbf{C} = \mathbf{B} + \frac{1}{2}\kappa\,\Delta t\,\mathbf{D}, \quad (2.1.51)$$

and the right-hand side is given by

$$\mathbf{r} = \left(\mathbf{M} - \frac{1}{2}\kappa\,\Delta t\,\mathbf{D}\right) \cdot \mathbf{f}^{(n)} + \frac{1}{2}\kappa\,\Delta t\left(\mathbf{b}^{(n)} + \mathbf{b}^{(n+1)}\right). \quad (2.1.52)$$

The numerical procedure involves solving the linear system (2.1.50) from a specified initial state for a sequence of steps with a constant or variable time step, Δt. The solution at every step can be found efficiently using the Thomas algorithm discussed in Section 1.2.2.

In the finite element implementation, the coefficient matrix, $\mathbf{C}$, and right-hand side, $\mathbf{b}$, are compiled by the FSELIB function *udl_sys_cn*, listed in Table 2.1.4.

A stability analysis, conducted as discussed in Section 2.1.4 for the explicit Euler method, shows that the spectral radius of the associated projection matrix is less than unity for any time step Δt, and the Crank–Nicolson method is unconditionally stable (e.g., Pozrikidis [47]).

PROBLEMS

2.1.1 *Eigenvalues and eigenvectors of the projection matrix*

Confirm the eigenvalues and eigenvectors of the matrix $\mathbf{P}$ given in (2.1.34) and (2.1.35).

---

[4]The terminology *implicit* emphasizes that solving a system of linear algebraic equations is required.

```
function [at,bt,ct,rhs] = udl_sys_cn (ne,xe,q0,f,fL,k,kappa,s,Dt)

%=====================================================
% Compact assembly of the tridiagonal linear system
% for one-dimensional unsteady diffusion
% with linear elements (udl) using the
% Crank-Nicolson method
%=====================================================

%-------------
% element size
%-------------

for l=1:ne
  h(l) = xe(l+1)-xe(l);
end

%---------------------------------
% initialize the tridiagonal matrix
%---------------------------------

at = zeros(ne,1);
bt = zeros(ne,1);
ct = zeros(ne,1);

%---------------------------------
% initialize the right-hand side
%---------------------------------

rhs = zeros(ne,1);
rhs(1) = Dt*q0/k;

%---------------------------------
% loop over the first ne-1 elements
%---------------------------------

for l=1:ne-1

  A11 = 1/h(l); A12 =-A11;        % diffusion matrix
  A21 = A12;    A22 = A11;

  B11 = h(l)/3.0; B12 = 0.5*B11;  % mass matrix
  B21 = B12;      B22 = B11;

  at(l)   = at(l)   + B11 + 0.5*Dt*kappa*A11;   % tridiagonal
  bt(l)   = bt(l)   + B12 + 0.5*Dt*kappa*A12;   % matrix components
  ct(l+1) = ct(l+1) + B21 + 0.5*Dt*kappa*A21;
  at(l+1) = at(l+1) + B22 + 0.5*Dt*kappa*A22;

  rhs(l)   = rhs(l)   + B11*f(l) + B12*f(l+1);
  rhs(l+1) = rhs(l+1) + B21*f(l) + B22*f(l+1);

  rhs(l)   = rhs(l)   - 0.5*Dt*kappa*(A11*f(l) + A12*f(l+1));
  rhs(l+1) = rhs(l+1) - 0.5*Dt*kappa*(A21*f(l) + A22*f(l+1));
```

TABLE 2.1.4   Function *udl_sys_cn* (Continuing →)

```
  rhs(1)   = rhs(1)   + Dt*kappa*(B11*s(1) + B12*s(1+1))/k;
  rhs(1+1) = rhs(1+1) + Dt*kappa*(B12*s(1) + B22*s(1+1))/k;

end

%------------------------
% the last element is special
%------------------------

A11 = 1.0/h(ne); A12 =-A11;
B11 = h(ne)/3.0; B12 = 0.5*B11;

at(ne) = at(ne) + B11 + 0.5*Dt*kappa*A11;

rhs(ne) = rhs(ne) + B11*f(ne);
rhs(ne) = rhs(ne) - 0.5*Dt*kappa*(A11*f(ne)+2.0*A12*fL);
rhs(ne) = rhs(ne) + Dt*kappa*(B11*s(ne)+B12*s(ne+1))/k;

%-----
% done
%-----

return;
```

TABLE 2.1.4   ($\rightarrow$ Continued) Function *udl_sys_cn* assembles the linear system for one-dimensional unsteady diffusion with linear elements using the Crank–Nicolson method.

2.1.2 *Code with mass lumping*

FSELIB function *udl_sys_lump*, not listed in the text, implements mass lumping. Run the code *udl* with the function *udl_sys_lump* and discuss the onset of numerical instability with reference to the theoretical predictions.

2.1.3 *Code for the Crank–Nicolson method*

Run the code *udl* with the Crank–Nicolson method and confirm that the algorithm is stable regardless of the size of the time step.

2.1.4 *Consistency analysis and hyperdiffusivity*

A consistency analysis is carried out working backwards from an algebraic difference equation to a modified differential equation (MDE). If the modified differential equation reduces to the governing partial differential equation as the time step and element size tend to zero independently, then the algebraic difference equation is consistent.

To carry out the consistency analysis of the difference equation (2.1.23), we express all discrete variables in terms of the value at the ith node at time level n.

Omitting the hats for simplicity, we obtain, for example,

$$\varphi_i^{n+1} = \varphi_i^n + \left(\frac{\partial \varphi}{\partial t}\right)_i^n \Delta t + \frac{1}{2}\left(\frac{\partial^2 \varphi}{\partial t^2}\right)_i^n \Delta t^2 + \cdots \tag{2.1.53}$$

and

$$\varphi_{i+1}^n = \varphi_i^n + \left(\frac{\partial \varphi}{\partial x}\right)_i^n \Delta x + \frac{1}{2}\left(\frac{\partial^2 \varphi}{\partial x^2}\right)_i^n \Delta x^2 + \cdots. \tag{2.1.54}$$

Substituting these expressions into equation (2.1.23) and simplifying, we obtain

$$\left(\frac{\partial \varphi}{\partial t}\right)_i^n \Delta t + \frac{1}{2}\left(\frac{\partial^2 \varphi}{\partial t^2}\right)_i^n \Delta t^2 + \cdots$$
$$= \alpha \left(\frac{\partial^2 \varphi}{\partial x^2}\right)_i^n \Delta x^2 + \frac{\alpha}{12}\left(\frac{\partial^4 \varphi}{\partial x^4}\right)_i^n \Delta x^4 + \cdots, \tag{2.1.55}$$

which can be rearranged into

$$\left(\frac{\partial \varphi}{\partial t}\right)_i^n = \kappa \left(\frac{\partial^2 \varphi}{\partial x^2}\right)_i^n - \frac{1}{2}\left(\frac{\partial^2 \varphi}{\partial t^2}\right)_i^n \Delta t + \frac{\kappa}{12}\Delta x^2 \left(\frac{\partial^4 \varphi}{\partial x^4}\right)_i^n + \cdots. \tag{2.1.56}$$

Because as $\Delta t \to 0$ and $\Delta x \to 0$ independently the modified differential equation (2.1.56) reduces to the unsteady diffusion equation, the numerical method is consistent.

The underlying continuous function φ satisfies the unsteady heat conduction equation in the absence of a source,

$$\frac{\partial \varphi}{\partial t} = \kappa \frac{\partial^2 \varphi}{\partial x^2}. \tag{2.1.57}$$

Differentiating with respect to time, we find that the solution also satisfies the high-order equation

$$\frac{\partial^2 \varphi}{\partial t^2} = \kappa^2 \frac{\partial^4 \varphi}{\partial x^4}. \tag{2.1.58}$$

Using this equation to eliminate the second derivative with respect to t on the right-hand side of (2.1.56), and rearranging, we obtain

$$\left(\frac{\partial \varphi}{\partial t}\right)_i^n = \kappa \left(\frac{\partial^2 \varphi}{\partial x^2}\right)_i^n - \kappa_4 \left(\frac{\partial^4 \varphi}{\partial x^4}\right)_i^n + \cdots, \tag{2.1.59}$$

where the coefficient

$$\kappa_4 \equiv \frac{1}{2}\kappa \Delta x^2 \left(\frac{1}{6} - \alpha\right) = \frac{1}{2}\kappa \left(\frac{1}{6}\Delta x^2 - \kappa \Delta t\right) \tag{2.1.60}$$

is called the *hyperdiffusivity*. Derive the hyperdiffusivity of the consistent formulation expressed by (2.1.38).

2.2 Convection

The unsteady diffusion equation discussed in Section 2.1 is complementary to the convection equation,

$$\frac{\partial f}{\partial t} + u \frac{\partial f}{\partial x} = 0, \tag{2.2.1}$$

where $f(x,t)$ is an unknown function of position, x, and time, t, and $u(x,f,t)$ is a prescribed convection velocity. The solution is to be found inside a specified interval, $a \leq x \leq b$, subject to a given initial condition, $f(x,t=0) = F(x)$, and *one* suitable boundary condition.

If the convection velocity, u, is independent of the convected field, f, that is $u(x,t)$ is a function of x and t explicitly but not implicitly through $f(x,t)$, the convection equation is quasi-linear. In that case, we say that the field $f(x,t)$ is *advected* with a specified position- and time-dependent advection velocity. In the special case where u is a constant, the convection equation is linear.

Finite element method

The implementation of the Galerkin finite element method follows the steps outlined in Section 2.1 for the unsteady diffusion equation. In the case of quasi-linear convection, the Galerkin projection takes the form

$$\int_a^b \phi_i \frac{\partial f}{\partial t} \, dx + \int_a^b u(x,t) \, \phi_i \frac{\partial f}{\partial x} \, dx = 0, \tag{2.2.2}$$

where the index i runs over an appropriate number of nodes. Substituting the finite element expansion of $f(x)$ involving N_G global interpolation functions,

$$f(x,t) = \sum_{j=1}^{N_G} f_j(t) \, \phi_j(x), \tag{2.2.3}$$

we obtain

$$\sum_{j=1}^{N_G} \left(\int_a^b \phi_i \, \phi_j \, dx \right) \frac{df_j}{dt} + \sum_{j=1}^{N_G} \left(\int_a^b u(x,t) \, \phi_i \frac{\partial \phi_j}{\partial x} \, dx \right) f_j = 0. \tag{2.2.4}$$

Rearranging, we obtain

$$\sum_{j=1}^{N_G} M_{ij} \frac{df_j}{dt} + \sum_{j=1}^{N_G} N_{ij} \, f_j = 0, \tag{2.2.5}$$

where M_{ij} is the global mass matrix and

$$N_{ij} \equiv \int_a^b u(x,t) \, \phi_i \frac{\partial \phi_j}{\partial x} \, dx \tag{2.2.6}$$

is the *global advection matrix*. Equation (2.2.5) provides us with a system of N_G linear ordinary differential equations,

$$\mathbf{M} \cdot \frac{d\mathbf{f}}{dt} + \mathbf{N} \cdot \mathbf{f} = \mathbf{0}, \qquad (2.2.7)$$

where the vector $\mathbf{f}$ encapsulates unknown nodal values.

To evaluate the global advection matrix, we express the convection velocity in the isoparametric form

$$u(x, t) = \sum_{k=1}^{N_G} u_k(t) \, \phi_k(x). \qquad (2.2.8)$$

Substituting this expansion into (2.2.6), we obtain

$$N_{ij} = \sum_{k=1}^{N_G} u_k \int_a^b \phi_k \, \phi_i \, \frac{\partial \phi_j}{\partial x} \, dx \equiv \sum_{k=1}^{N_G} u_k \, \Pi_{kij}, \qquad (2.2.9)$$

where

$$\Pi_{kij} \equiv \int_a^b \phi_k \, \phi_i \, \frac{\partial \phi_j}{\partial x} \, dx \qquad (2.2.10)$$

is a three-index global advection matrix. In terms of the matrix $\mathbf{\Pi}$, equation (2.2.7) takes the form

$$\mathbf{M} \cdot \frac{d\mathbf{f}}{dt} + \mathbf{u} \cdot \mathbf{\Pi} \cdot \mathbf{f} = \mathbf{0}. \qquad (2.2.11)$$

In practice, the matrix $\mathbf{\Pi}$ is assembled from the corresponding individual element advection matrices.

2.2.1   Linear elements

In the case of linear elements, the first derivative of the jth global interpolation function on the right-hand side of (2.2.6) is equal to $1/h_{j-1}$ over the $(j-1)$ element, $-1/h_j$ over the jth element, and zero over all other elements. Thus, the only non-zero components of the matrix N_{ij} are the tridiagonal elements $N_{i-1,j}$, $N_{i,j}$, and $N_{i+1,j}$. We find that

$$N_{i,i} = \frac{1}{h_{i-1}} \int_{x_{i-1}}^{x_i} (u_{i-1} \, \phi_{i-1} + u_i \, \phi_i) \, \phi_i \, dx - \frac{1}{h_i} \int_{x_i}^{x_{i+1}} (u_i \, \phi_i + u_{i+1} \, \phi_{i+1}) \, \phi_i \, dx,$$

$$N_{i-1,i} = -\frac{1}{h_{i-1}} \int_{x_{i-1}}^{x_i} (u_{i-1} \, \phi_{i-1} + u_i \, \phi_i) \, \phi_i \, dx, \qquad (2.2.12)$$

$$N_{i+1,i} = \frac{1}{h_i} \int_{x_i}^{x_{i+1}} (u_i \, \phi_i + u_{i+1} \, \phi_{i+1}) \, \phi_i \, dx.$$

The global advection matrix simplifies to

$$N_{ij} = (\delta_{i,j} - \delta_{i-1,j}) \frac{1}{h_{i-1}} \int_{x_{i-1}}^{x_i} (u_{i-1}\,\phi_{i-1} + u_i\,\phi_i)\,\phi_i\,dx$$

$$+(\delta_{i+1,j} - \delta_{i,j}) \frac{1}{h_i} \int_{x_i}^{x_{i+1}} (u_i\,\phi_i + u_{i+1}\,\phi_{i+1})\,\phi_i\,dx. \qquad (2.2.13)$$

Rearranging, we obtain

$$N_{ij} = u_{i-1} (\delta_{i,j} - \delta_{i-1,j}) \frac{1}{h_{i-1}} \int_{x_{i-1}}^{x_i} \phi_{i-1}\,\phi_i\,dx$$

$$+u_i \left[(\delta_{i,j} - \delta_{i-1,j}) \frac{1}{h_{i-1}} \int_{x_{i-1}}^{x_i} \phi_i^2\,dx + (\delta_{i+1,j} - \delta_{i,j}) \frac{1}{h_i} \int_{x_i}^{x_{i+1}} \phi_i^2\,dx \right]$$

$$+u_{i+1} (\delta_{i+1,j} - \delta_{i,j}) \frac{1}{h_i} \int_{x_i}^{x_{i+1}} \phi_{i+1}\,\phi_i\,dx, \qquad (2.2.14)$$

yielding

$$N_{ij} = u_{i-1} (\delta_{i,j} - \delta_{i-1,j}) \frac{1}{6} + u_i (\delta_{i+1,j} - \delta_{i-1,j}) \frac{1}{3} + u_{i+1} (\delta_{i+1,j} - \delta_{i,j}) \frac{1}{6}. \qquad (2.2.15)$$

The contribution of the advection term to the ith Galerkin projection is

$$\sum_{j=1}^{N_G} N_{ij} f_j = \frac{1}{6} \left[u_{i-1} (f_i - f_{i-1}) + 4\,u_i \frac{f_{i+1} - f_{i-1}}{2} + u_{i+1} (f_{i+1} - f_i) \right]. \qquad (2.2.16)$$

Note that the right-hand side involves a particular combination of backward, central, and forward differences.

Linear advection

If the convection velocity, u, is constant and equal to U, the global advection matrix takes the form

$$\mathbf{N} = U\widetilde{\mathbf{N}}, \qquad (2.2.17)$$

where

$$\widetilde{N}_{ij} \equiv \int_a^b \phi_i \frac{\partial \phi_j}{\partial x}\,dx \qquad (2.2.18)$$

is the dimensionless *damping matrix*.

In the case of linear elements presently considered, we find that

$$\widetilde{\mathbf{N}} = \frac{1}{2} \begin{bmatrix} -1 & 1 & 0 & 0 & \cdots & 0 & 0 & 0 \\ -1 & 0 & 1 & 0 & \cdots & 0 & 0 & 0 \\ 0 & -1 & 0 & 1 & \cdots & 0 & 0 & 0 \\ \vdots & \vdots & \vdots & \vdots & \vdots & \vdots & \vdots & \vdots \\ 0 & 0 & \cdots & \cdots & -1 & 0 & 1 & 0 \\ 0 & 0 & \cdots & \cdots & 0 & -1 & 0 & 1 \\ 0 & 0 & \cdots & \cdots & 0 & 0 & -1 & 1 \end{bmatrix}, \qquad (2.2.19)$$

independent of the element size. Apart from the first and last diagonal elements, the global advection matrix shown in (2.2.19) is skew-symmetric, that is, it is equal to the negative of its transpose. A perfectly skew-symmetric matrix has zero diagonal elements.

The global advection matrix shown in (2.2.19) consists of overlapping diagonal blocks of corresponding element matrices. The lth-element advection matrix is given by

$$\widetilde{\mathbf{C}}^{(l)} = \frac{1}{2} \begin{bmatrix} -1 & 1 \\ -1 & 1 \end{bmatrix}. \qquad (2.2.20)$$

Cancellation of the alternating 1 and -1 diagonal entries in the assembly process yields zeros along the diagonal, as shown on the right-hand side of (2.2.19).

Using the mass matrix displayed in Table 1.3.1(a), we find that the ith equation of the GFEM system (2.2.7) takes the form

$$\frac{h_{i-1}}{6} \frac{\mathrm{d} f_{i-1}}{\mathrm{d} t} + \frac{h_{i-1} + h_i}{3} \frac{\mathrm{d} f_i}{\mathrm{d} t} + \frac{h_i}{6} \frac{\mathrm{d} f_{i+1}}{\mathrm{d} t} + \frac{1}{2} U \left(f_{i+1} - f_{i-1} \right) = 0, \qquad (2.2.21)$$

which can be rearranged into

$$\left(\frac{1}{3} \frac{h_{i-1}}{h_{i-1} + h_i} \right) \frac{\mathrm{d} f_{i-1}}{\mathrm{d} t} + \left(\frac{2}{3} \right) \frac{\mathrm{d} f_i}{\mathrm{d} t} + \left(\frac{1}{3} \frac{h_i}{h_{i-1} + h_i} \right) \frac{\mathrm{d} f_{i+1}}{\mathrm{d} t} + U \frac{f_{i+1} - f_{i-1}}{h_{i-1} + h_i} = 0.$$
$$(2.2.22)$$

The first three terms on the left-hand side of (2.2.22) contribute a weighted average of the time derivative at three nodes. It is reassuring to observe that the sum of the weights enclosed by the parentheses is equal to unity, ensuring mass conservation. The last term on the left-hand side of (2.2.22) is the central difference approximation of the derivative $\partial f / \partial x$ at the ith node, computed over an interval of length $h_{i-1} + h_i$. Overall, (2.2.22) expresses a filtered or smeared finite difference approximation.

When the element size is uniform, $h_1 = h_2 = \cdots = h_{N_E} \equiv h$, we obtain the simplified form

$$\frac{1}{6} \frac{\mathrm{d} f_{i-1}}{\mathrm{d} t} + \frac{2}{3} \frac{\mathrm{d} f_i}{\mathrm{d} t} + \frac{1}{6} \frac{\mathrm{d} f_{i+1}}{\mathrm{d} t} + \frac{U}{2h} \left(f_{i+1} - f_{i-1} \right) = 0. \qquad (2.2.23)$$

When the convection velocity is not constant but varies with x, equation (2.2.21) can still be used with U on the left-hand side set equal to the nodal velocity, u_i, as an alternative to the primary form (2.2.16).

2.2.2 *Numerical dispersion due to spatial discretization*

Consider a solution in the form of a periodic wave with period λ, and assume that the solution of the discrete system (2.2.23) is given by the real or imaginary part of the function

$$f(x, t) = A \exp[-ik(x - ct)], \tag{2.2.24}$$

where i is the imaginary unit, $i^2 = -1$, $A = A_R + iA_I$ is the complex wave amplitude, the subscripts R and I denote the real and imaginary part, $k = 2\pi/\lambda$ is the wave number, and $c = c_R + ic_I$ is the complex phase velocity. The complex angular frequency of the oscillation is given by

$$\omega \equiv kc. \tag{2.2.25}$$

The exact solution is $c = U$ and $\omega = kU$.

Substituting (2.2.24) into (2.2.23), writing $x_i = (i - 1)h$, and simplifying, we find that

$$-ikc \left[\frac{1}{6} \exp(ikh) + \frac{2}{3} + \frac{1}{6} \exp(-ikh) \right] + \frac{U}{2h} \left[\exp(ikh) - \exp(-ikh) \right] = 0 \tag{2.2.26}$$

or

$$-ikc \left[\frac{2}{3} + \frac{1}{3} \cos(kh) \right] + i \frac{U}{h} \sin(kh) = 0. \tag{2.2.27}$$

Solving for the phase velocity, we obtain

$$\frac{c}{U} = \frac{\sin(kh)}{kh} \frac{3}{2 + \cos(kh)}. \tag{2.2.28}$$

As $kh \to 0$, corresponding to waves whose wavelength is long compared to the element size, the ratio c/U tends to unity.

For convenience, we assume that the length of the solution domain, $L = hN_{\mathrm{E}}$, is a multiple of the wavelength λ, that is,

$$L = n\lambda \tag{2.2.29}$$

for $n = 1, \ldots, N_{\mathrm{E}}$, where N_{E} is the number of available elements. The angular frequency, $\omega \equiv kc$, reduced by the time scale U/L, is given by

$$s \equiv \frac{\omega L}{U} = \frac{kcL}{U} = 2\pi n \frac{c}{U} = \frac{2\pi n}{kh} \sin(kh) \frac{3}{2 + \cos(kh)}. \tag{2.2.30}$$

Now, $kh = 2\pi h/\lambda$, and since $h = L/N_E$, $h/\lambda = n/N_E$ and

$$kh = \frac{n}{N_E} 2\pi \qquad (2.2.31)$$

for $n = 1, \ldots, N_E$. Consequently,

$$s \equiv \frac{\omega L}{U} = N_E \sin(kh) \frac{3}{2 + \cos(kh)}. \qquad (2.2.32)$$

When the ratio n/N_E is small, we may perform a Taylor series expansion to find that

$$s \equiv \frac{\omega L}{U} \simeq 2\pi n. \qquad (2.2.33)$$

In the lumped mass approximation, the fraction $3/[2 + \cos(kh)]$ is replaced by unity on the right-hand sides of (2.2.28) and (2.2.32) (Problem 2.2.2).

The exact solution of the linear convection equation states that the initial distribution travels with the specified velocity U, which means that $c = U$ and $s \equiv \omega L/U = 2\pi n$, consistent with the asymptotic result (2.2.33). Expression (2.2.32) reveals that the spatial discretization modifies the exact phase velocity by an amount that depends on the reduced wave number, kh, thereby introducing numerical dispersion.

The FSELIB script *dispersion*, listed in Table 2.2.1, generates a graph of $s/N_E \equiv \omega L/(U N_E)$ against the scaled wave number $n/N_E = kh/(2\pi)$ for the consistent and mass-lumped approximations. The results shown in Figure 2.2.1 for $N_E = 32$ suggest that the spatial discretization significantly affects the wave speed of moderately or poorly resolved waves, and thus introduces significant numerical dispersion.

2.2.3 Quadratic elements

The implementation of the Galerkin finite element method with quadratic elements follows the general guidelines described previously in this section for linear elements. In the case of quadratic elements with interior nodes situated at the element midpoints, the element advection matrix for constant advection velocity is given by

$$\widetilde{C}^{(l)} = \frac{1}{6} \begin{bmatrix} -3 & 4 & -1 \\ -4 & 0 & -4 \\ 1 & -4 & 3 \end{bmatrix}. \qquad (2.2.34)$$

The global advection matrix consists of overlapping diagonal blocks of the element advection matrix given in (2.2.34) (Problem 2.2.1).

2.2.4 Integrating ODEs

Equation (2.2.7) provides us with a coupled system of first-order linear ordinary differential equations for the unknown nodal values, $f_i(t)$. Implementing the simplest

```
ne = 32;
pi2 = 2*pi;

x     = zeros(ne,1);
s     = zeros(ne,1);
slump = zeros(ne,1);

for n=1:ne
    x(n) = n/ne;
    kh   = pi2*x(n);
    s(n) = sin(kh) * 3/(2+cos(kh));
    slump(n) = sin(kh);
end

figure(1);
plot(x, s,'o-');                           % consistent
plot(x, slump,'x-');                       % lumped
xe=[0 0.5]; ye=[0 0.5*pi2]; plot(xe,ye,'--'); % exact
axis([0 1 -2 2]);
xlabel('n/N_E'); ylabel('s/N_E');
```

TABLE 2.2.1   Script *dispersion* generates a graph of $s/N_E \equiv \omega L/(UN_E)$ against the scaled wave number $n/N_E = kh/(2\pi)$ for the consistent and mass-lumped approximations.

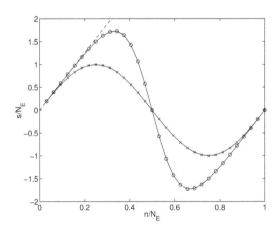

FIGURE 2.2.1   Dependence of the reduced angular frequency, $\omega L/(UN_E)$, on the scaled wave number, $kh/(2\pi)$, for $N_E = 32$ elements. The circles represent the predictions of (2.2.32), the $\times$ symbols correspond to the mass lumped approximation, and the dashed line represents the exact solution recovered for small values of kh.

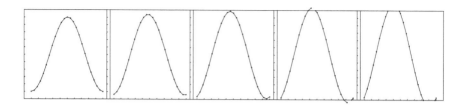

FIGURE 2.2.2   Evolution of a sinusoidal wave using the FTCS scheme with $N_{\mathrm{E}} = $ 24 elements and time step corresponding to $C = 0.05$. The depicted profiles correspond to scaled times $Ut/L = 0$, 1, 2, 3, and 3.

possible time discretization expressed by the forward Euler method, we write

$$\mathbf{M} \cdot \frac{\mathbf{f}(t + \Delta t) - \mathbf{f}(t)}{\Delta t} + \mathbf{N} \cdot \mathbf{f}(t) = \mathbf{0}, \tag{2.2.35}$$

where Δt is the time step. Rearranging, we derive a tridiagonal system of equations for the unknown vector $\mathbf{f}(t + \Delta t)$,

$$\mathbf{M} \cdot \mathbf{f}(t + \Delta t) = \left[\, \mathbf{M} - \Delta t\, \mathbf{N} \,\right] \cdot \mathbf{f}(t). \tag{2.2.36}$$

Implementing mass lumping to diagonalize the coefficient matrix on the left-hand side, we recover the explicit forward-time, centered-space (FTCS) finite difference equation. In the case of linear elements with uniform size h, the ith equation takes the form

$$f_i(t + \Delta t) = f_i(t) - C \left[\, f_{i+1}(t) - f_{i-1}(t) \,\right], \tag{2.2.37}$$

where

$$C \equiv \frac{U\Delta t}{h} \tag{2.2.38}$$

is the convection number.

Unfortunately, a stability analysis of the FTCS discretization reveals that the method is unstable for any value of C, that is, no matter how small the size of the time step (e.g., Pozrikidis [47]). A simulation of the evolution of a sinusoidal wave computed using the FTCS scheme with $N_{\mathrm{E}} = 24$ elements over a period L, corresponding to $n = 1$, and a time step corresponding to $C = 0.05$ is shown in Figure 2.2.2 at scaled times $Ut/L = 0$, 1, 2, 3, and 3. Two important features are evident: the unphysical growth of the wave amplitude, and a reduction in the phase speed with respect to U, consistent with our earlier discussion on numerical dispersion.

More generally, the non-diagonal structure of the advection matrix arising from the Galerkin finite element implementation causes serious or even incurable numerical instability in the case of explicit time integration, and significant numerical diffusivity in the case of implicit time integration. Moreover, the seemingly innocuous discretization (2.2.16) can be the source of a detrimental *aliasing instability* (e.g., Gresho & Sani [25], p. 59).

These difficulties underscore the sensitivity of the finite element method in problems dominated by convection. In practice, time integration is performed by a variety of explicit or semi-implicit numerical methods, such as the Runge–Kutta method (e.g., Gresho & Sani [25], Pozrikidis [47]).

2.2.5   Nonlinear convection

A prototypical nonlinear convection equation used to investigate the performance of numerical methods is the inviscid Burgers's equation

$$\frac{\partial f}{\partial t} + f \frac{\partial f}{\partial x} = 0 \tag{2.2.39}$$

(e.g., Pozrikidis [47]). Consider an initial condition where the function $f(x)$ is a sinusoidal wave. Because the convection velocity multiplying the spatial derivative, $\partial f/\partial x$, is equal to f, the larger the magnitude of the solution, the faster the magnitude of the local convection velocity. Accordingly, crests of a sinusoidal wave with zero mean travel to the right, troughs travel to the left, and the nodes remain stationary. The evolution leads to wave steepening followed by the eventual formation of a step discontinuity in the form of a shock wave.

Applying (2.2.11) with $\mathbf{f}$ in place of $\mathbf{u}$, we derive a nonlinear Galerkin finite element equation,

$$\mathbf{M} \cdot \frac{d\mathbf{f}}{dt} + \mathbf{f} \cdot \mathbf{\Pi} \cdot \mathbf{f} = \mathbf{0}. \tag{2.2.40}$$

The quadratic nonlinearity requires careful attention to prevent the onset of numerical instability and successfully obtain the structure of the solution at steady state.

PROBLEMS

2.2.1 *Quadratic elements*

Display the structure of the global advection matrix for linear convection with the quadratic elements discussed in Section 1.5.

2.2.2 *Dispersion with mass lumping*

Show that the second fraction on the right-hand sides of (2.2.28)–(2.2.32) are replaced by unity in the lumped mass approximation.

2.3 Convection–diffusion

The complementary cases of unsteady diffusion discussed in Section 2.1 and convection discussed in Section 2.2 can be synthesized into the unsteady convection–diffusion equation

$$\frac{\partial f}{\partial t} + u \frac{\partial f}{\partial x} = \kappa \frac{\partial^2 f}{\partial x^2} + \frac{s(x,t)}{\rho \, c_p}, \tag{2.3.1}$$

where $u(x, f, t)$ is a prescribed convection velocity and $\kappa = k/(\rho \, c_p)$ is the medium diffusivity. The solution of (2.3.1) follows a specified initial condition, $f(x, t = 0) = F(x)$, and satisfies a suitable number of boundary conditions.

2.3.1 *Steady linear convection–diffusion*

To illustrate the development of the Galerkin finite element method, we discuss the generalization of the steady diffusion problem introduced in Section 1.1, where the steady-state heat conduction equation (1.1.4) is replaced by the linear convection–diffusion equation

$$\rho \, c_p U \frac{\mathrm{d} f}{\mathrm{d} x} = k \frac{\mathrm{d}^2 f}{\mathrm{d} x^2} + s(x) \tag{2.3.2}$$

and U is a constant advection velocity. The importance of convection relative to diffusion is expressed by the dimensionless Péclet number,

$$\mathrm{Pe} \equiv \frac{\rho \, c_p U L}{k} = \frac{U L}{\kappa}, \tag{2.3.3}$$

where L is a characteristic length, presently identified with the length of the solution domain.

Applying the Galerkin finite element method with linear elements, we derive a $N_E \times N_E$ linear algebraic system similar to that shown in equation (1.1.26),

$$\mathbf{Q} \cdot \mathbf{f} = \mathbf{c} + \frac{1}{k} \widetilde{\mathbf{M}} \cdot \mathbf{s}, \tag{2.3.4}$$

where $\widetilde{\mathbf{M}}$ is the $N_E \times (N_E + 1)$ rectangular mass matrix shown in Table 1.1.7(a). When the elements are spaced evenly with uniform size h, the $N_E \times N_E$ coefficient matrix on the right-hand side of (2.3.4) is given in Table 2.3.1, where

$$\mathrm{Pe}_c \equiv \frac{\rho \, c_p U h}{k} = \frac{U h}{\kappa} \tag{2.3.5}$$

is the *cell Péclet number*. In the absence of convection, we set $\mathrm{Pe}_c = 0$ and recover the diffusion matrix shown in (1.1.52), $\mathbf{Q} = \mathbf{D}$. The N_E-dimensional vector $\mathbf{c}$ on the right-hand side of (2.3.4) is given by

$$\mathbf{c} \equiv \begin{bmatrix} q_0/k, & 0, & \cdots, & 0, & (1 + \tfrac{1}{2} \mathrm{Pe}_c) \frac{f_L}{h_{N_E}} \end{bmatrix}. \tag{2.3.6}$$

$$\mathbf{Q} = \frac{1}{h} \begin{bmatrix} 1 - \frac{1}{2}\mathrm{Pe}_c & -1 + \frac{1}{2}\mathrm{Pe}_c & 0 & 0 & \cdots \\ -1 - \frac{1}{2}\mathrm{Pe}_c & 2 & -1 + \frac{1}{2}\mathrm{Pe}_c & 0 & \cdots \\ 0 & -1 - \frac{1}{2}\mathrm{Pe}_c & 2 & -1 + \frac{1}{2}\mathrm{Pe}_c & 0 \\ \vdots & \vdots & \cdots & \cdots & \cdots \\ 0 & 0 & \cdots & \cdots & \cdots \\ 0 & 0 & \cdots & \cdots & \cdots \\ 0 & 0 & \cdots & \cdots & \cdots \end{bmatrix} \longrightarrow$$

$$\begin{bmatrix} \cdots & \cdots & 0 & 0 & 0 \\ \cdots & \cdots & 0 & 0 & 0 \\ \cdots & \cdots & 0 & 0 & 0 \\ \cdots & \cdots & \cdots & \vdots & \vdots \\ 0 & -1 - \frac{1}{2}\mathrm{Pe}_c & 2 & -1 + \frac{1}{2}\mathrm{Pe}_c & 0 \\ \cdots & 0 & -1 - \frac{1}{2}\mathrm{Pe}_c & 2 & -1 + \frac{1}{2}\mathrm{Pe}_c \\ \cdots & 0 & 0 & -1 - \frac{1}{2}\mathrm{Pe}_c & 2 + \frac{1}{2}\mathrm{Pe}_c \end{bmatrix},$$

TABLE 2.3.1   Coefficient matrix for linear convection–diffusion with evenly spaced elements.

As Pe_c increases, the coefficient matrix $\mathbf{Q}$ tends to obtain a skew-symmetric form,

$$\mathbf{Q} \rightarrow \frac{1}{\kappa} \mathbf{N}, \tag{2.3.7}$$

where $\mathbf{N}$ is the global advection matrix shown in (2.2.19). In this limit, the odd- and even-numbered unknowns are coupled only by the first equation represented by the first row on the right-hand side of the coefficient matrix $\mathbf{Q}$ shown in Table 2.3.1. From a practical standpoint, because the coefficient matrix is not diagonally dominant, numerical methods for solving the linear system either fail or become unreliable.

Finite element code

The assembly of the linear system (2.3.4) for linear elements is implemented in the FSELIB function *scdl_sys*, listed in Table 2.3.2. The function can accommodate the Neumann or the Robin boundary condition at the right end of the solution domain, as discussed in Section 1.3.2. FSELIB code *scdl*, listed in Table 2.3.3, implements the finite element code.

Results with the Robin boundary condition and parameter values $L = 1.0$, $k = 1.0$, $h_T = 1.0$, $f_\infty = 0.0$, $f_L = 0.0$, $\rho = 1.0$, $c_p = 1.0$, $N_E = 16$ elements of

```
function [at,bt,ct,b] ...
   ...
       = scdl_sys (ne,xe,q0,ht,finf,fL,k,U,rho,cp,s,leftbc)

%=========================================================
% Compact assembly of the tridiagonal linear system for
% one-dimensional steady convection-diffusion
% with linear elements (scdl)
%=========================================================

%-------------
% element size
%-------------

for l=1:ne
  h(l) = xe(l+1)-xe(l);
end

%-----------
% initialize
%-----------

at = zeros(ne,1);
bt = zeros(ne,1);
ct = zeros(ne,1);
b  = zeros(ne,1);

%--------------------
% boundary conditions
%--------------------

if(leftbc==1)         % Neumann
 b(1) = q0/k;
elseif(leftbc==2)     % Robin
 b(1) = ht*finf/k;
 at(1) = ht/k;
end

cf = 0.5*U*rho*cp/k;

C11 = -cf; C12 = cf; % advection matrix
C21 = -cf; C22 = cf;

%------------------------
% loop over ne-1 elements
%------------------------

for l=1:ne-1

  A11 = 1/h(l); A12=-A11; A22=A11;
  B11 = h(l)/3.0; B12=0.5*B11; B22=B11;

  at(l)   = at(l)   + A11 + C11;
  bt(l)   = bt(l)   + A12 + C12;
```

TABLE 2.3.2   Function *scdl_sys* (Continuing →)

```
  ct(l+1) = ct(l+1) + A12 + C21;
  at(l+1) = at(l+1) + A22 + C22;

  b(l)   = b(l)   + (B11*s(l) + B12*s(l+1))/k;
  b(l+1) = b(l+1) + (B12*s(l) + B22*s(l+1))/k;

end

%-------------
% last element
%-------------

A11 = 1/h(ne);   A12=-A11;
B11 = h(ne)/3.0; B12=0.5*B11;

at(ne) = at(ne) + A11+C11;
b(ne) = b(ne) + (B11*s(ne) + B12*s(ne+1))/k - (A12+C12)*fL;

%-----
% done
%-----

return;
```

TABLE 2.3.2 ($\rightarrow$ Continued) Function *scdl_sys* assembles a linear system for steady one-dimensional convection–diffusion with linear elements.

equal size, and $U = 0$ (highest curve), 20, 40, 60, 80, and 100 (lowest curve) are shown in Figure 2.3.1. As the convection velocity U becomes higher, and thus the Péclet number increases, the temperature distribution becomes increasingly steep and numerical oscillations appear. In practical applications, Pe can be on the order of 10^3 or even higher. This difficulty underlines the notion that the Galerkin finite element method performs best for conduction- or diffusion-dominated transport.

Petrov–Galerkin method

The poor performance of the Galerkin finite element method in problems with convection-dominated transport is remedied by the Petrov–Galerkin formulation where the differential equation to be solved is weighted by a linear combination of the global interpolation functions and their derivatives with position-dependent coefficients. (e.g., de Sampio [55], Yu & Heinrich [70]). This modification biases the difference equations resulting from the Petrov–Galerkin projection in a desired direction, thereby promoting the stability of the numerical method.

The Petrov–Galerkin method is the counterpart of the upwind differencing method in finite different implementations (e.g., Pozrikidis [47]).

```
%========================================================
% Code scdl
%
% Code for steady one-dimensional convection-diffusion
% with linear elements (scdl)
%========================================================

clear all
close all

%-----------
% input data
%-----------

L = 1.0; k = 1.0; q0 = -1.00; ht = 1.0; finf = 0.0; fL = 0.0;
rho = 1.0; cp = 1.0; U = 100.0; ne = 16; ratio = 1.0;

%----------------
% grid generation
%----------------

xe = elm_line1 (0,L,ne,ratio);

%-------------------
% specify the source
%-------------------

for i=1:ne+1
 s(i) = 10.0*exp(-5.0*xe(i)^2/L^2);
end

%-----------------
% element assembly
%-----------------

[at,bt,ct,b] = scdl_sys (ne,xe,q0,ht,finf,fL,k,U,rho,cp,s);

%--------------
% linear solver
%--------------

f = thomas (ne,at,bt,ct,b);

f(ne+1) = fL;

%-----
% plot
%-----

plot(xe, f,'-o'); xlabel('x'); ylabel('f');

%-----
% done
%-----
```

TABLE 2.3.3 Finite element code *scdl* for steady one-dimensional convection–
diffusion with linear elements.

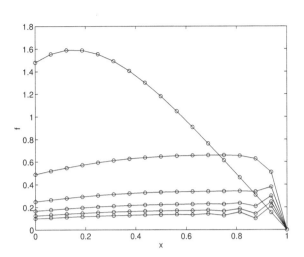

FIGURE 2.3.1   Finite element solution of the steady convection–diffusion equation
with linear elements computed with the FSELIB code *sdcl*, for $\mathrm{Pe} = 0$ (highest
curve), 20, 40, 60, 80, and 100. Artificial oscillations appear as the Péclet number
increases, yielding a convection-dominated transport.

2.3.2   Nonlinear convection–diffusion

A prototypical convection–diffusion equation incorporating the effect of convection
nonlinearity is the viscous Burgers's equation,

$$\frac{\partial f}{\partial t} + f \frac{\partial f}{\partial x} = \kappa \frac{\partial^2 f}{\partial x^2}, \qquad (2.3.8)$$

which is a generalization of the inviscid Burgers's equation shown in (2.2.39). Remarkably, when the solution domain extends over the entire x axis, that is, in the
absence of boundaries, the solution can be found exactly for any initial condition
by means of an integral transformation that reduces the viscous Burgers's equation
to the unsteady diffusion equation (e.g., Pozrikidis [47]).

The Galerkin finite element method provides us with a system of quadratic
ordinary differential equations (ODEs) for the time-dependent nodal solution vector,
$\mathbf{f}(t)$,

$$\mathbf{M} \cdot \frac{d\mathbf{f}}{dt} + \mathbf{f} \cdot \mathbf{\Pi} \cdot \mathbf{f} = -\mathbf{D} \cdot \mathbf{f} + \mathbf{b}, \qquad (2.3.9)$$

which is a generalization of (2.3.4) and (2.2.40), where $\mathbf{\Pi}$ is the three-index global
advection tensor defined in (2.2.10).

At steady state, we obtain a quadratic system of nonlinear equations,

$$\mathbf{f} \cdot \mathbf{\Pi} \cdot \mathbf{f} = -\mathbf{D} \cdot \mathbf{f} + \mathbf{b}, \tag{2.3.10}$$

which can be solved by iterative methods. In practice, the solution is found using Newton or modified Newton iterations, such as those implemented in Broyden's algorithms (e.g., Pozrikidis [47]).

PROBLEMS

2.3.1 *Convection dominated transport*

(a) Run the FSELIB code *scdl* with the Robin boundary condition at the left end of solution domain, parameter values $L = 1.0$, $k = 1.0$, $h_T = 1.0$, $f_\infty = 0.0$, $f_L = 0.0$, $\rho = 1.0$, $c_p = 1.0$, $U = 100$, and $N_E = 2, 4, 8, 16, 32$, and 64, evenly spaced elements. Discuss the behavior of the numerical solution.

(b) Repeat (a) with 16 elements and investigate the effect of element clustering near the right end of the computation domain where sharp gradients may arise.

2.3.2 *Neumann boundary condition*

Repeat Problem 2.3.1(a, b) with the Neumann boundary condition at the left end.

2.4   Beam bending

The finite element method finds important applications in the field of structural mechanics where it is used to describe the various modes of deformation and the dynamic response of beams, plates, and shells, subject to a permanent or transient load. The simplest member of a structure is a slender beam that bends under the influence of an imposed transverse load, while exhibiting insignificant twisting, elongation, and compression. Beams made of wood, steel, composite material, and concrete are familiar from everyday experience. In engineering design, the deformation and performance of beams are assessed by solving ordinary differential equations by numerical methods that are similar to those discussed previously in this chapter for the convection–diffusion equation. Because beam deflection is governed by a fourth-order differential equation, a new set of element interpolation functions must be introduced to ensure continuity of the derivatives of the finite element expansion.

2.4.1   Euler–Bernoulli beam

A straight beam with a uniform cross-section is called prismatic. For simplicity, we consider a prismatic beam whose cross-section is symmetric with respect to the xy plane, as shown in Figure 2.4.1(a). The shape of the beam before and after deformation is illustrated in Figure 2.4.1(b). The deflection is due to a distributed vertical load with force density, w, defined as the force divided by an infinitesimal

(a) (b)

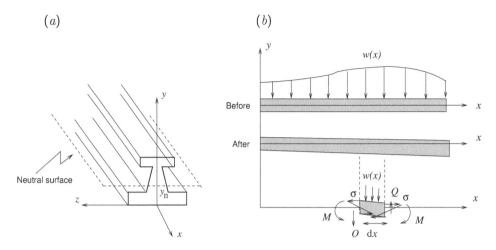

FIGURE 2.4.1 (*a*) Depiction of a prismatic beam showing the position of the neutral surface in the zx plane, located at $y = y_n$. (*b*) Illustration of a straight beam bending before and after deformation, showing the distribution of the axial stress over the cross-section, σ, and the transverse shear force, Q, for a given load, $w(x)$.

length along the x axis, reckoned to be positive when it is directed downward against the y axis.

The *axial stress* developing due to the deformation, σ, defined as the force divided by the cross-sectional area and the *transverse shear force*, Q, defined as the transverse shear stress integrated over the entire beam cross-section, are shown at the bottom of Figure 2.4.1(*b*).

To develop the equations governing the beam deflection, we perform a vertical force balance over an infinitesimal section of the beam with length dx, regarded as a free body. Under the sign conventions defined in Figure 2.4.1(*a*), we find that

$$Q(x) + w(x)\,\mathrm{d}x = Q(x + \mathrm{d}x). \tag{2.4.1}$$

Rearranging, we find that

$$w = \frac{\mathrm{d}Q}{\mathrm{d}x}. \tag{2.4.2}$$

The axial stress, σ, varies across the beam cross-section due to the stretching of axial fibers near the upper surface of the beam and the compression of axial fibers near the lower surface of the beam. The *neutral surface* of the beam is a material surface normal to the xy plane, located at $y = y_n$, that undergoes neither stretching nor compression, and is thus unstressed in the deformed configuration, $\sigma = 0$, as shown in Figure 2.4.1(*a*).

Variations of the axial stress over the beam cross-section give rise to a macroscopic bending moment about that z axis computed with respect to the neutral surface, given by

$$M = \iint \sigma \, \Delta y \, \mathrm{d}A, \tag{2.4.3}$$

where $\Delta y = y - y_N$ is the distance of a point from the neutral surface, and the integration is performed over the beam cross-section.

Performing a torque balance over an infinitesimal section of the beam with length $\mathrm{d}x$, we obtain

$$Q(x) \, \mathrm{d}x + M(x) = M(x + \mathrm{d}x). \tag{2.4.4}$$

Note that the neglected contribution from the vertical load, $w(x)$, is quadratic in $\mathrm{d}x$. Rearranging, we derive the differential relation

$$Q = \frac{\mathrm{d}M}{\mathrm{d}x}. \tag{2.4.5}$$

Substituting this expression into the force balance (2.4.2), we derive the governing equilibrium equation

$$\frac{\mathrm{d}^2 M}{\mathrm{d}x^2} = w(x). \tag{2.4.6}$$

Next, we relate the bending moment to the curvature of the neutral surface in the xy plane, denoted by κ, using the constitutive equation

$$M = EI\kappa, \tag{2.4.7}$$

where E is the modulus of elasticity of the beam material and I is the principal moment of inertia of the beam cross-section with respect to the z axis that is normal to the xy plane,

$$I \equiv \iint \Delta y^2 \, \mathrm{d}A. \tag{2.4.8}$$

To derive (2.4.7), we consider the simple configuration depicted in Figure 2.4.2 where a straight beam has been deformed into a circular arc with centerline radius, R. A fundamental assumption of the Euler–Bernoulli theory is that a material plane that is normal to the beam before deformation remains normal to the beam after deformation. The arc radius across the beam is $r = R + \Delta y$, and the axial stretch of fibers along the beam is $\epsilon = (R + \Delta y)/R$. Adopting a linear constitutive equation for the axial stress expressed by Hooke's law, we write

$$\sigma = E\left(\epsilon - 1\right) = E \frac{\Delta y}{R}. \tag{2.4.9}$$

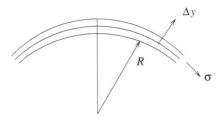

FIGURE 2.4.2 Exaggerated local view of a deformed beam used to derive a constitutive equation for the bending moment. Δy is the distance from the neutral surface.

Substituting this expression into (2.4.3), and setting $\kappa = 1/R$, we derive (2.4.7). Note that $\sigma = 0$ at the neutral surface, $\Delta y = 0$, as required.

For small vertical deflection, v, the curvature can be approximated with the negative of the second derivative,

$$\kappa \simeq -\frac{\mathrm{d}^2 v}{\mathrm{d}x^2}. \qquad (2.4.10)$$

For the bending configuration illustrated in Figure 2.4.2, the vertical deflection is negative, $v < 0$. Substituting (2.4.10) into the preceding expressions for the bending moment and transverse shear force, we obtain

$$M = -EI\frac{\mathrm{d}^2 v}{\mathrm{d}x^2}, \qquad Q = -\frac{\mathrm{d}}{\mathrm{d}x}\left(EI\frac{\mathrm{d}^2 v}{\mathrm{d}x^2}\right). \qquad (2.4.11)$$

Substituting further the expression for the bending moment into (2.4.6), we derive the governing differential equation

$$\frac{\mathrm{d}^2}{\mathrm{d}x^2}\left(EI\frac{\mathrm{d}^2 v}{\mathrm{d}x^2}\right) = -w(x). \qquad (2.4.12)$$

In the case of a homogeneous prismatic beam, the product EI is constant, independent of x. Rearranging, we obtain

$$\frac{\mathrm{d}^4 v}{\mathrm{d}x^4} = -\frac{w(x)}{EI}. \qquad (2.4.13)$$

To solve this fourth-order differential equation by the Galerkin finite element method, we must use at least a cubic polynomial expansion over the individual elements and require a C^1 solution. This means that the solution and its first derivative must be continuous across the element nodes. In contrast, in the case of heat conduction, governed by a second-order differential equation, we were able to obtain a solution using linear elements while requiring only C^0 continuity of the global expansion.

Boundary conditions

The beam bending equation must be solved subject to boundary conditions reflecting the physical circumstances of the problem under consideration. Since beam bending is governed by a fourth-order differential equation, we require two scalar boundary conditions at each end, adding up to a total of four boundary conditions.

When the end of a beam is *fixed*, *clamped*, or *built-in*, the deflection and its derivative are required to be zero,

$$v = 0, \qquad \frac{\partial v}{\partial x} = 0. \qquad\qquad (2.4.14)$$

An aircraft wing can be regarded as a beam that is built into the aircraft at one end.

When the end of a beam is *simply supported*, the deflection and the bending moment are required to be zero,

$$v = 0, \qquad M = 0. \qquad\qquad (2.4.15)$$

Physically, the end of a beam is simply supported when it rests on a ledge.

When the end of a beam is free, the transverse shear force and the bending moment are required to be zero,

$$Q = 0, \qquad M = 0. \qquad\qquad (2.4.16)$$

An aircraft wing can be regarded as a beam that is free at the far end.

A beam that is clamped at one end and is free at the other end is called a cantilever.

Variational formulation

In Section 1.3, we discussed the variational formulation of the steady diffusion equation in the presence of a distributed source. An analogous formulation is possible for the beam bending equation in the presence of vertical load. The variational formulation of the Euler–Bernoulli beam states that computing the solution of (2.4.12) over an interval, $[a, b]$, for a beam that is simply supported at both ends is equivalent to maximizing the functional

$$\mathcal{F} < w(x) > \equiv -\frac{1}{2} \int_a^b EI \left(\frac{d^2 w}{dx^2} \right)^2 dx + \int_a^b w(x) \, w(x) \, dx, \qquad (2.4.17)$$

over the set of all C^1 functions, $w(x)$, that satisfy the homogeneous boundary conditions

$$w(a) = 0, \qquad w''(a) = 0, \qquad w(b) = 0, \qquad w''(b) = 0, \qquad (2.4.18)$$

where a prime denotes a derivative with respect to x. The proof is carried out as discussed in Section 1.3 for the steady diffusion equation. Similar variational formulations are possible for other types of boundary conditions.

Timoshenko's theory

The Euler–Bernoulli theory is based on the fundamental assumption that a material plane that is normal to the beam before deformation remains normal to the beam after deformation. In Timoshenko's theory, this assumption is relaxed by introducing the displacement of point particles along the x and y axes, denoted as v_x and v_y, and stipulating that

$$v_x = u(x) - y \left(\frac{\partial v}{\partial x} - \gamma(x) \right), \qquad v_y = v(x) - y \left(\frac{\partial v}{\partial x} - \gamma(x) \right), \qquad (2.4.19)$$

where $u(x)$ and $v(x)$ are the x and y displacements of point particles in the neutral surface, and $\gamma(x)$ is the shear rotation (e.g., Popov [45]). The Euler–Bernoulli theory arises by setting $u(x) = 0$ and $\gamma(x) = 0$, computing the axial strain

$$\epsilon_{xx} \equiv \frac{\partial v_x}{\partial x} = -y \frac{\partial^2 v}{\partial x^2}, \qquad (2.4.20)$$

and then using Hooke's law to express the axial stress as

$$\sigma = E \, \epsilon_{xx} = -E \, y \frac{\partial^2 v}{\partial x^2}. \qquad (2.4.21)$$

Closure in Timoshenko's theory is achieved by relating $\gamma(x)$ to the transverse shear force, $Q(x)$, using an appropriate constitutive equation involving a shear modulus of elasticity, G, and an effective shear area, A_s, as $\gamma = Q/(GA_s)$. Finite element methods for the Timoshenko formulation are generalizations of those for the Euler–Bernoulli formulation discussed in this chapter.

PROBLEM

2.4.1 *Principal moment of inertia*

Present expressions for the principal moment of inertia of three beams with cross-sectional shapes of your choice.

2.5   Finite element methods for beam bending

To develop finite element methods for the beam bending problem formulated in Section 2.4, we divide the beam into N_E elements, where the lth element extends between the first element end-node, $x_1^{(l)}$, and the second element end-node, $x_2^{(l)}$. To ensure a C^1 finite element solution, we employ a cubic expansion over each element defined in terms of the element node vertical deflections and their derivatives expressing the slope,

$$v_1^{(l)}, \qquad v_1'^{(l)} \equiv \left(\frac{dv}{dx}\right)_1^{(l)}, \qquad v_2^{(l)}, \qquad v_2'^{(l)} \equiv \left(\frac{dv}{dx}\right)_2^{(l)}, \qquad (2.5.1)$$

where the subscript 1 or 2 denote the first or second element end-node. These values are shared by neighboring elements at corresponding positions.

2.5.1 Hermitian elements

It is convenient to introduce a dimensionless parameter, η, ranging in the interval $[0, 1]$, and express the x position over the lth element as

$$x = x_1^{(l)} + h_l\,\eta, \qquad (2.5.2)$$

where $h_l \equiv x_2^{(l)} - x_1^{(l)}$ is the element length. As η increases from 0 to 1, the field point x slides from $x_1^{(l)}$ to $x_2^{(l)}$. The cubic expansion over the lth element takes the form

$$v(\eta) = v_1^{(l)}\psi_1^{(l)}(\eta) + v_1'^{(l)}\psi_2^{(l)}(\eta) + v_2^{(l)}\psi_3^{(l)}(\eta) + v_2'^{(l)}\psi_4^{(l)}(\eta), \qquad (2.5.3)$$

where

$$\begin{array}{ll}
\psi_1^{(l)}(\eta) = 2\eta^3 - 3\eta^2 + 1, & \psi_2^{(l)}(\eta) = h_l\,\eta\,(\eta^2 - 2\eta + 1), \\[4pt]
\psi_3^{(l)}(\eta) = \eta^2\,(-2\eta + 3), & \psi_4^{(l)}(\eta) = h_l\,\eta^2\,(\eta - 1),
\end{array} \qquad (2.5.4)$$

are cubic *Hermitian* local interpolation functions, plotted in Figure 2.5.1. Note that the functions $\psi_1^{(l)}(\eta)$ and $\psi_3^{(l)}(\eta)$ are dimensionless, whereas the functions $\psi_2^{(l)}(\eta)$ and $\psi_4^{(l)}(\eta)$ have units of length.

Each of the functions $\psi_i^{(l)}$ satisfies a set of four boundary conditions. The first function satisfies the boundary conditions

$$\psi_1^{(l)}(\eta = 0) = 1, \qquad \left(\frac{d\psi_1^{(l)}}{d\eta}\right)_{\eta=0} = 0,$$

$$\psi_1^{(l)}(\eta = 1) = 0, \qquad \left(\frac{d\psi_1^{(l)}}{d\eta}\right)_{\eta=1} = 0. \qquad (2.5.5)$$

The second function satisfies the boundary conditions

$$\psi_2^{(l)}(\eta = 0) = 0, \qquad \left(\frac{d\psi_2^{(l)}}{d\eta}\right)_{\eta=0} = h_l,$$

$$\psi_2^{(l)}(\eta = 1) = 0, \qquad \left(\frac{d\psi_2^{(l)}}{d\eta}\right)_{\eta=1} = 0. \qquad (2.5.6)$$

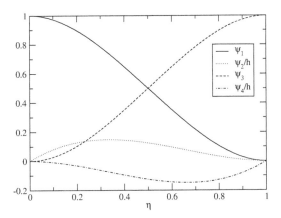

FIGURE 2.5.1   Graphs of the four Hermitian element interpolation functions over a cubic element of length h.

The third function satisfies the boundary conditions

$$\psi_3^{(l)}(\eta = 0) = 0, \qquad \left(\frac{\mathrm{d}\psi_3^{(l)}}{\mathrm{d}\eta}\right)_{\eta=0} = 0,$$

$$\psi_3^{(l)}(\eta = 1) = 1, \qquad \left(\frac{\mathrm{d}\psi_3^{(l)}}{\mathrm{d}\eta}\right)_{\eta=1} = 0. \tag{2.5.7}$$

The fourth function satisfies the boundary conditions

$$\psi_4^{(l)}(\eta = 0) = 0, \qquad \left(\frac{\mathrm{d}\psi_4^{(l)}}{\mathrm{d}\eta}\right)_{\eta=0} = 0,$$

$$\psi_4^{(l)}(\eta = 1) = 0, \qquad \left(\frac{\mathrm{d}\psi_4^{(l)}}{\mathrm{d}\eta}\right)_{\eta=1} = h_l. \tag{2.5.8}$$

An element supporting these interpolation functions is a cubic Hermitian element.

Evaluating (2.5.3) at $\eta = 0$ and 1, invoking the boundary conditions satisfied by the element interpolation functions, and writing

$$\frac{\mathrm{d}v}{\mathrm{d}x} = \frac{\mathrm{d}v}{\mathrm{d}\eta}\frac{\mathrm{d}\eta}{\mathrm{d}x} = \frac{\mathrm{d}v}{\mathrm{d}\eta}\frac{1}{h_l}, \tag{2.5.9}$$

we confirm the required properties

$$v(\eta = 0) = v_1^{(l)}, \qquad \left(\frac{\mathrm{d}v}{\mathrm{d}x}\right)_{\eta=0} = v_1'^{(l)},$$

$$v(\eta = 1) = v_2^{(l)}, \qquad \left(\frac{\mathrm{d}v}{\mathrm{d}x}\right)_{\eta=1} = v_2'^{(l)}, \tag{2.5.10}$$

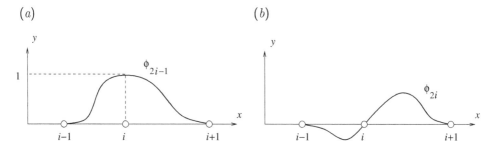

FIGURE 2.5.2   Illustration of the (a) deflection and (b) deflection-slope global interpolation functions associated with the ith global node.

where a prime denotes a derivative with respect to x.

Global interpolation functions

The union of the end-nodes of the N_E elements comprises a set of $N_G = N_E + 1$
unique global nodes. The ith global node hosts an odd-numbered deflection global
interpolation function, $\phi_{2i-1}(x)$, and an even-numbered deflection-slope global interpolation function, $\phi_{2i}(x)$, where $i = 1, \ldots, N_G$. The global interpolation functions consist of the union of corresponding element interpolation functions.

Since the ith cubic Hermitian element is subtended between the global nodes
numbered i and $i + 1$, the odd-numbered deflection global interpolation function,
$\phi_{2i-1}(x)$, is comprised of the union of $\psi_1^{(i)}(\eta)$ and $\psi_3^{(i-1)}(\eta)$, as illustrated in Figure
2.5.2(a). Similarly, the even-numbered deflection-slope global interpolation func-
tion, $\phi_{2i}(x)$, is comprised of the union of $\psi_2^{(i)}(\eta)$ and $\psi_4^{(i-1)}(\eta)$, as illustrated in
Figure 2.5.2(b).

In terms of the global interpolation functions, the finite element expansion over
the entire length of the beam is described by

$$v(x) = \sum_{j=1}^{N_G} v_j \, \phi_{2j-1}(x) + \sum_{j=1}^{N_G} v_j' \, \phi_{2j}(x), \qquad (2.5.11)$$

where v_j is the deflection and v_j' is the slope at the location of the jth global node.

2.5.2   Galerkin projection

To carry out the Galerkin projection, we multiply the governing equation (2.4.13)
by each one of the global interpolation functions and integrate the product over the
solution domain. For a beam extending from $x = 0$ to L, we write

$$\int_0^L \phi_i \, \frac{d^4 v}{dx^4} \, dx = -\frac{1}{EI} \int_0^L \phi_i \, w(x) \, dx. \qquad (2.5.12)$$

Integrating by parts on the left-hand side to lower the order of the fourth derivative, we find that

$$\int_0^L \phi_i \frac{d^4 v}{dx^4} dx = \int_0^L \phi_i \frac{d}{dx}\left(\frac{d^3 v}{dx^3}\right) dx = \left[\phi_i \frac{d^3 v}{dx^3}\right]_0^L - \int_0^L \frac{d\phi_i}{dx} \frac{d^3 v}{dx^3} dx, \quad (2.5.13)$$

where the square brackets signify the change of the enclosed expression between the upper and lower limits of integration.

Repeating the integration by parts for the last integral on the right-hand side, we obtain

$$\int_0^L \frac{d\phi_i}{dx} \frac{d^3 v}{dx^3} dx = \int_0^L \frac{d\phi_i}{dx} \frac{d}{dx}\left(\frac{d^2 v}{dx^2}\right) dx = \left[\frac{d\phi_i}{dx} \frac{d^2 v}{dx^2}\right]_0^L - \int_0^L \frac{d^2\phi_i}{dx^2}, \frac{d^2 v}{dx^2} dx.$$
$$(2.5.14)$$

Combining the preceding expressions and substituting the result into (2.5.12), we obtain

$$\int_0^L \frac{d^2\phi_i}{dx^2} \frac{d^2 v}{dx^2} dx + \left[\phi_i \frac{d^3 v}{dx^3} - \frac{d\phi_i}{dx} \frac{d^2 v}{dx^2}\right]_0^L = -\frac{1}{EI} \int_0^L \phi_i w(x) dx. \quad (2.5.15)$$

In terms of the transverse shear force, Q, and bending moment, M, given in (2.4.11), we obtain

$$EI \int_0^L \frac{d^2 v}{dx^2} \frac{d^2\phi_i}{dx^2} dx = \left[\phi_i Q - \frac{d\phi_i}{dx} M\right]_0^L - \int_0^L \phi_i w(x) dx \quad (2.5.16)$$

for $i = 1, \ldots, 2N_{\mathrm{G}}$.

Stiffness matrix

Substituting the finite element expansion (2.5.11) into (2.5.16) and compiling the resulting equations, we formulate a linear system of equations,

$$\mathbf{K} \cdot \mathbf{u} = \frac{1}{EI} \mathbf{b}, \quad (2.5.17)$$

where $\mathbf{K}$ is the global stiffness matrix with components

$$K_{ij} = \int_0^L \frac{d^2\phi_i}{dx^2} \frac{d^2\phi_j}{dx^2} dx \quad (2.5.18)$$

for $i, j = 1, \ldots, 2N_{\mathrm{G}}$, and

$$\mathbf{u} \equiv \begin{bmatrix} v_1, & v_1', & v_2, & v_2', & \ldots, & v_{N_{\mathrm{G}}}, & v_{N_{\mathrm{G}}}' \end{bmatrix}^T \quad (2.5.19)$$

is the vector of nodal deflections and slopes.

Right-hand side

The first two components of the vector $\mathbf{b}$ on the right-hand side of the linear system are given by

$$b_1 = \left[\phi_1 \, Q - \frac{\mathrm{d}\phi_1}{\mathrm{d}x} \, M \right]_0^L - \int_0^L \phi_1 \, w(x) \, \mathrm{d}x, \tag{2.5.20}$$

yielding

$$b_1 = -Q(0) - \int_0^L \phi_1 \, w(x) \, \mathrm{d}x, \tag{2.5.21}$$

and

$$b_2 = \left[\phi_2 \, Q - \frac{\mathrm{d}\phi_2}{\mathrm{d}x} \, M \right]_0^L - \int_0^L w(x) \, \phi_2 \, \mathrm{d}x, \tag{2.5.22}$$

yielding

$$b_2 = M(0) - \int_0^L \phi_2 \, w(x) \, \mathrm{d}x. \tag{2.5.23}$$

The intermediate components are given by

$$b_i = - \int_0^L \phi_i \, w(x) \, \mathrm{d}x \tag{2.5.24}$$

for $i = 3, \ldots, 2N_G - 2$. The penultimate component is given by

$$b_{2N_G - 1} = \left[\phi_{2N_G - 1} \, Q - \frac{\mathrm{d}\phi_{2N_G - 1}}{\mathrm{d}x} \, M \right]_0^L - \int_0^L \phi_{2N_G - 1} \, w(x) \, \mathrm{d}x, \tag{2.5.25}$$

yielding

$$b_{2N_G - 1} = Q(L) - \int_0^L \phi_{2N_G - 1} \, w(x) \, \mathrm{d}x. \tag{2.5.26}$$

The last component is given by

$$b_{2N_G} = \left[\phi_{2N_G} \, Q - \frac{\mathrm{d}\phi_{2N_G}}{\mathrm{d}x} \, M \right]_0^L - \int_0^L \phi_{2N_G} \, w(x) \, \mathrm{d}x, \tag{2.5.27}$$

yielding

$$b_{2N_G} = -M(L) - \int_0^L \phi_{2N_G} \, w(x) \, \mathrm{d}x. \tag{2.5.28}$$

Implementing the boundary conditions by specifying selected values of the generalized deflection vector $\mathbf{u}$, end-node transverse shear forces, $Q(0)$ and $Q(L)$, and bending moments, $M(0)$ and $M(L)$, we derive a final system of equations for the remaining unknowns.

If the load density function and its first derivative are continuous over the entire length of the beam, we may introduce the familiar Hermite expansion

$$w(x) = \sum_{j=1}^{N_{\mathrm{G}}} w_j \, \phi_{2j-1}(x) + \sum_{i=j}^{N_{\mathrm{G}}} w_j' \, \phi_{2j}(x), \qquad (2.5.29)$$

where w_j and w_j' are nodal values and slopes. Accordingly, we may write

$$\int_0^L \phi_i \, w(x) \, \mathrm{d}x = \sum_{j=1}^{N_{\mathrm{G}}} w_j \int_0^L \phi_i \, \phi_{2j-1} \, \mathrm{d}x + \sum_{i=j}^{N_{\mathrm{G}}} w_j' \int_0^L \phi_{2j} \, \phi_i \, \mathrm{d}x \qquad (2.5.30)$$

and identify the integrals on the right-hand side with the components of a generalized global mass matrix,

$$M_{ik} = \int_0^L \phi_i \, \phi_k \, \mathrm{d}x \qquad (2.5.31)$$

for $i, k = 1, \ldots, 2N_{\mathrm{G}}$.

2.5.3   Element stiffness and mass matrices

In practice, the global stiffness matrix, $\mathbf{K}$, is assembled from 4×4 element stiffness matrices with components

$$H_{ij}^{(l)} = \int_{E_l} \frac{\mathrm{d}^2 \psi_i^{(l)}}{\mathrm{d}x^2} \frac{\mathrm{d}^2 \psi_j^{(l)}}{\mathrm{d}x^2} \, \mathrm{d}x \qquad (2.5.32)$$

for $i, j = 1, 2, 3, 4$. Substituting the Hermitian element interpolation functions and carrying out the integrations, we obtain

$$\mathbf{H}^{(l)} = \frac{1}{h_l^3} \begin{bmatrix} 12 & 6\,h_l & -12 & 6\,h_l \\ 6\,h_l & 4\,h_l^2 & -6\,h_l & 2\,h_l^2 \\ -12 & -6\,h_l & 12 & -6\,h_l \\ 6\,h_l & 2\,h_l^2 & -6\,h_l & 4\,h_l^2 \end{bmatrix}. \qquad (2.5.33)$$

The assembled stiffness matrix, $\mathbf{K}$, takes the hexadiagonal[5] form shown in Table 2.5.1, with the understanding that the individual equations of the linear system are arranged in the order of the Galerkin projection with ϕ_1, ϕ_2, ϕ_3, ϕ_4, ..., $\phi_{2N_{\mathrm{G}}}$. The assembly is carried out according to the algorithm shown in Table 2.5.2, implemented in the FSELIB function *beam_sys*, listed in Table 2.5.3.

---

[5]The word *hexadiagonal* derives from the Greek words εξι, which means *six*, and διαγωνιος, which means *diagonal*.

$$\mathbf{K} = \begin{bmatrix}
H_{11}^{(1)} & H_{12}^{(1)} & H_{13}^{(l)} & H_{14}^{(1)} & 0 & 0 & \cdots \\
H_{21}^{(1)} & H_{22}^{(1)} & H_{23}^{(l)} & H_{24}^{(1)} & 0 & 0 & \cdots \\
H_{31}^{(1)} & H_{32}^{(1)} & H_{33}^{(1)}+H_{11}^{(2)} & H_{34}^{(1)}+H_{12}^{(2)} & H_{13}^{(2)} & H_{14}^{(2)} & \cdots \\
H_{41}^{(1)} & H_{42}^{(1)} & H_{43}^{(1)}+H_{21}^{(2)} & H_{44}^{(1)}+H_{22}^{(2)} & H_{23}^{(2)} & H_{24}^{(2)} & \cdots \\
0 & 0 & H_{31}^{(2)} & H_{32}^{(2)} & H_{33}^{(2)}+H_{11}^{(3)} & H_{34}^{(2)}+H_{12}^{(3)} & \cdots \\
0 & 0 & H_{41}^{(2)} & H_{42}^{(2)} & H_{43}^{(2)}+H_{21}^{(3)} & H_{44}^{(2)}+H_{22}^{(3)} & \cdots \\
\vdots & \vdots & \vdots & \vdots & \vdots & \vdots & \vdots \\
0 & 0 & 0 & 0 & 0 & 0 & \cdots \\
0 & 0 & 0 & 0 & 0 & 0 & \cdots \\
0 & 0 & 0 & 0 & 0 & 0 & \cdots
\end{bmatrix}$$

TABLE 2.5.1   Assembled global stiffness matrix for beam bending with Hermitian elements.

Similarly, the generalized global mass matrix can be assembled from 4×4 element mass matrices with components

$$B_{ij}^{(l)} = \int_{E_l} \psi_i^{(l)}\psi_j^{(l)}\,\mathrm{d}x \tag{2.5.34}$$

for $i,j = 1,2,3,4$. Substituting the Hermitian element interpolation functions and performing the integrations, we find that

$$\mathbf{B}^{(l)} = \frac{1}{420}\,h_l \begin{bmatrix}
156 & 22\,h_l & 54 & -13\,h_l \\
22\,h_l & 4\,h_l^2 & 13\,h_l & -3\,h_l^2 \\
54 & 13\,h_l & 156 & -22\,h_l \\
-13\,h_l & -3\,h_l^2 & -22\,h_l & 4\,h_l^2
\end{bmatrix}. \tag{2.5.35}$$

The assembled global mass matrix takes a hexadiagonal form similar to that shown in Table 2.5.1 for the stiffness matrix. The assembly can be carried by a straightforward modification of the algorithm shown in Table 2.5.2.

2.5.4   One-element cantilever beam

As an example, we consider the bending of a cantilever beam due to a vertical tip load of magnitude F, as shown in Figure 2.5.3. The beam load density can be described in terms of the one-dimensional Dirac delta function, $\delta(x - x_0)$, as

$$w(x) = F\,\delta(x - L). \tag{2.5.36}$$

Note that the one-dimensional delta function has units of inverse length. Accordingly, the load density has the expected units of force over length. Requiring the free-end boundary conditions

$$Q(L) = 0, \qquad\qquad M(L) = 0, \tag{2.5.37}$$

Specify the number of elements, N_{E}

$N_{\mathrm{G}} = N_{\mathrm{E}} + 1$ *Number of unique global nodes*

Do $i = 1, \ldots, 2N_{\mathrm{G}}$ *Initialize to zero*
   Do $j = 1, \ldots, 2N_{\mathrm{G}}$
      $K_{ij} = 0$
   End Do
End Do

Do $l = 1, \ldots, N_{\mathrm{E}}$ *Run over the elements*
   Compute the element stiffness matrix $\mathbf{H}^{(l)}$ using (2.5.33)
   $i_1 = 2l - 2$
   Do $i = 1, 2, 3, 4$
      Do $j = 1, 2, 3, 4$
         $K_{i_1+i,i_1+j} = K_{i_1+i,i_1+j} + H^{(l)}_{ij}$
      End Do
   End Do
End Do

TABLE 2.5.2   Algorithm for assembling the global stiffness matrix for beam bending with Hermitian elements.

FIGURE 2.5.3   Illustration of a beam subject to a tip load. The bold-faced numbers are the labels of the two global nodes. The one-element discretization provides us with the exact solution.

```
function [stiff] = beam_sys (ne,xe)

%==================================================
% Assembly of the stiffness matrix for beam bending
% with cubic Hermitian elements
%
% ne: number of elements,     xe: element end-points
% stiff: global stiffness matrix
%==================================================

ng = ne+1; % number of unique global nodes

%----------------------------------------
% element size and number of global nodes
%----------------------------------------

for l=1:ne
  h(l) = xe(l+1)-xe(l);
end

%-----------
% initialize
%-----------

stiff = zeros(2*ng,2*ng);

%-----------------------
% loop over the elements
%-----------------------

for l=1:ne
    esm(1,1) = 12.0; esm(1,2) = 6.0*h(l); esm(1,3) = -12.0;
    esm(1,4) = 6.0*h(l);
    esm(2,1) = esm(1,2); esm(2,2) = 4.0*h(l)^2; esm(2,3) = -6.0*h(l);
    esm(2,4) = 2.0*h(l)^2;
    esm(3,1) = esm(1,3); esm(3,2) = esm(2,3); esm(3,3) = 12.0;
    esm(3,4) = -6.0*h(l);
    esm(4,1) = esm(1,4); esm(4,2) = esm(2,4); esm(4,3) = esm(3,4);
    esm(4,4) = 4.0*h(l)^2;
    esm = esm/h(l)^3;   i1 = 2*(l-1);
    for i=1:4
      for j=1:4
        stiff(i1+i,i1+j) = stiff(i1+i,i1+j) + esm(i,j);
      end
    end
end

%-----
% done
%-----

return;
```

TABLE 2.5.3 Function *beam_sys* assembles the global stiffness matrix for the beam bending problem.

and using the distinctive properties of the delta function, we find that the right-hand side of the preliminary linear system simplifies to

$$b_1 = -Q(0), \qquad b_2 = M(0),$$

$$b_3 = -\int_0^L \phi_3\, w(x)\, \mathrm{d}x = -F\int_0^L \phi_3\, \delta(x - L)\, \mathrm{d}x = -F\,\phi_3(x = L) = -F,$$

$$b_4 = 0. \tag{2.5.38}$$

The Galerkin linear system takes the form

$$\frac{EI}{L^3}\begin{bmatrix} 12 & 6L & -12 & 6L \\ 6L & 4L^2 & -6L & 2L^2 \\ -12 & -6L & 12 & -6L \\ 6L & 2L^2 & -6L & 4L^2 \end{bmatrix}\cdot\begin{bmatrix} v_1 \\ v_1' \\ v_2 \\ v_2' \end{bmatrix} = \begin{bmatrix} -Q(0) \\ M(0) \\ -F \\ 0 \end{bmatrix}. \tag{2.5.39}$$

Specifying on physical grounds the built-in or clamped boundary conditions

$$v_1 = 0, \qquad\qquad v_1' = 0, \tag{2.5.40}$$

we obtain a system of four linear equations for four unknowns,

$$v_2, \qquad v_2', \qquad Q(0), \qquad M(0). \tag{2.5.41}$$

Because boundary conditions on the deflection and its slope are specified at the left end, only the projections with ϕ_3 and ϕ_4 expressed by the last two equations in (2.5.39) are necessary. In practice, it is expedient to sacrifice some computational cost and implement these boundary conditions after a complete preliminary system has been compiled. Replacing the redundant finite element equations with the specified boundary conditions reduces the preliminary system (2.5.39) to the final system

$$\frac{EI}{L^3}\begin{bmatrix} 1 & 0 & 0 & 0 \\ 0 & 1 & 0 & 0 \\ 0 & 0 & 12 & -6L \\ 0 & 0 & -6L & 4L^2 \end{bmatrix}\cdot\begin{bmatrix} v_1 \\ v_1' \\ v_2 \\ v_2' \end{bmatrix} = \begin{bmatrix} 0 \\ 0 \\ -F \\ 0 \end{bmatrix}, \tag{2.5.42}$$

which effectively encapsulates the 2×2 system

$$\frac{EI}{L^3}\begin{bmatrix} 12 & -6L \\ -6L & 4L^2 \end{bmatrix}\cdot\begin{bmatrix} v_2 \\ v_2' \end{bmatrix} = \begin{bmatrix} -F \\ 0 \end{bmatrix}. \tag{2.5.43}$$

The solution is

$$v_2 = -\frac{1}{3}\frac{FL^3}{EI}, \qquad v_2' = -\frac{1}{2}\frac{FL^3}{EI}, \tag{2.5.44}$$

which, in fact, is the exact solution of the problem under consideration.

The practice of implementing the boundary conditions after a complete preliminary finite element system has been compiled will be discussed further in this section and then again in Section 4.1.5 with reference to the finite element implementation in two dimensions.

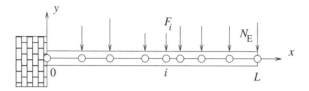

FIGURE 2.5.4   Finite element discretization of a beam subjected to nodal loads.

2.5.5   *Cantilever beam with nodal loads*

Consider a cantilever beam discretized into N_E elements, as illustrated in Figure 2.5.4, subject to concentrated nodal loads described by the load density function

$$w(x) = \sum_{i=1}^{N_G} F_i \, \delta(x - x_i), \qquad (2.5.45)$$

where $N_G = N_E + 1$ is the number of global nodes defining the element end-nodes, F_i is the ith nodal force, reckoned to be positive when directed downward, x_i is the position of the ith global node, and $\delta(x)$ is Dirac's one-dimensional delta function, having units of inverse length.

Requiring the free-end boundary conditions, $Q(L) = 0$ and $M(L) = 0$, and using the distinctive properties of the delta function, we find that the first component of the right-hand side of the preliminary linear system simplifies into

$$b_1 = -Q(0) - \int_0^L \phi_1 \, w(x) \, \mathrm{d}x = -Q(0) - F_1 \int_0^L \phi_1 \, \delta(x - x_1) \, \mathrm{d}x, \qquad (2.5.46)$$

yielding

$$b_1 = -Q(0) - F_1 \, \phi_1(x_1) = -Q(0) - F_1. \qquad (2.5.47)$$

The second component simplifies into

$$b_2 = M(0) - \int_0^L \phi_2 \, w(x) \, \mathrm{d}x = M(0). \qquad (2.5.48)$$

Subsequent components simplify into

$$
\begin{aligned}
b_3 &= -F_2, & b_4 &= 0, & \cdots \\
b_{2i-1} &= -F_i, & b_{2i} &= 0, & \cdots \\
b_{2N_G-1} &= -F_{N_G} & b_{2N_G} &= 0.
\end{aligned}
\qquad (2.5.49)
$$

FSELIB code *beam*, listed in Table 2.5.4, implements the finite element method. First, a preliminary linear system is assembled. The right-hand side is computed using the FSELIB function *beam_sys* listed in Table 2.5.3. Second, the left-end boundary conditions, $v_1 = 0$ and $v_1' = 0$, are implemented in the following three steps:

```
%=====================================
% Code beam
%
% Code for cantilever beam bending with
% Hermitian cubic elements, subject
% to nodal forcing
%=====================================

%-----------
% input data
%-----------

L = 1.0; E = 1.0; I = 1.0; ne = 16; ratio = 1.0;

%----------------
% grid generation
%----------------

xe = elm_line1 (0,L,ne,ratio);

ng = ne+1;   % number of unique global nodes

%----------------------
% specify the nodal load
%----------------------

for i=1:ng
 F(i) = 1.0/ne;
end

%-----------------------------
% compile the stiffness matrix
%-----------------------------

stiff = beam_sys (ne,xe); stiff = E*I*stiff;

%--------------------------
% preliminary right-hand side
%--------------------------

for i=1:ng
 b(2*i-1) = - F(i); b(2*i) = 0.0;
end

%-------------------------------------------
% implement the left-end boundary conditions
%-------------------------------------------

for i=1:2*ng
 stiff(1,i) = 0.0; stiff(2,i) = 0.0;
 stiff(i,1) = 0.0; stiff(i,2) = 0.0;
end

stiff(1,1) = 1.0; stiff(2,2) = 1.0;
```

TABLE 2.5.4   Code *beam* (Continuing $\rightarrow$)

```
b(1) = 0.0; b(2) = 0.0;

%--------------
% linear solver
%--------------

sol = b/stiff';

%-----------------------
% extract the deflections
%-----------------------

for i=1:ng
 v(i) = sol(2*i-1);
end

%----------------
% prepare to plot
%----------------

xlabel('x'); ylabel('y');
axis([0 1.1*L -0.2*L 0.10*L]);

%---------------
% plot the nodes
%---------------

plot (xe, zeros(ng),'-ko');

%----------------------
% plot the nodal forces
%----------------------

for i=1:ng
 plotx(1) = xe(i); ploty(1) = 0.0;
 plotx(2) = xe(i); ploty(2) = -F(i);
 plot(plotx, ploty,'k');
 plttx(1) = xe(i); pltty(1) = -F(i);
 plot(plttx, pltty,'kv');
end

%---------------------
% plot the deflections
%---------------------

plot(xe, v,'.-k');

%-----
% done
%-----
```

TABLE 2.5.4 ($\rightarrow$ Continued) Code *beam* solves the equations of beam bending subject to a discrete nodal load.

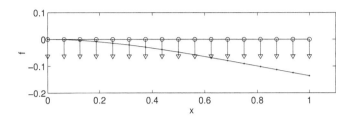

FIGURE 2.5.5   Beam deflection under a uniform transverse nodal load indicated by the vertical arrows computed by the FSELIB code *beam*.

1. All elements in the first and second columns and rows of the global stiffness matrix, $\mathbf{K}$, are set to zero.

2. The first two diagonal elements, K_{11} and K_{22}, are set to unity.

3. The first and second entries of the right-hand side, b_1 and b_2, are replaced by zeros.

In the case of one element, $N_E = 1$, the three stages amount to replacing the preliminary system (2.5.39) with the final system (2.5.42). This method of implementing the boundary conditions is desirable in that it preserves the symmetry of the stiffness matrix.

In the third stage, the linear system is solved using an internal function embedded in MATLAB, implemented in symbolic form by dividing a row vector by a matrix according to the statement:

$$\texttt{sol = b/stiff}'$$

which appears to defy the rules of matrix calculus. Because the solution, `sol`, is also a row vector, the transpose of the coefficient matrix, indicated by the prime, must be entered in the denominator according to MATLAB conventions, although in this case `stiff'` = `stiff` by the symmetry of the stiffness matrix. Methods for solving systems of linear equations in finite element applications are discussed in Appendix C.

A graph of the solution for uniform nodal loads and parameter values listed in the code is shown in Figure 2.5.5.

PROBLEMS

2.5.1 *Element stiffness matrix*

Confirm that the determinant of the element stiffness matrix shown in equation (2.5.33) is zero, indicating that the matrix is singular.

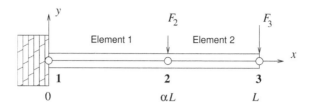

FIGURE 2.5.6   Two-element discretization of a beam subject to a nodal load. The
bold-faced numbers are the labels of the global nodes.

2.5.2 *Two-element cantilever beam*

Consider the two-element discretization of a cantilever beam subject to nodal load,
as shown in Figure 2.5.6. The length of the first element is αL, where α is a
dimensionless coefficient taking values in the interval $(0,1)$. Compile a 6×6 linear
system arising from the finite element formulation in terms of E, I, L, α, F_2, F_3,
$Q(0)$, and $M(0)$, before, and after the implementation of the boundary conditions
(2.5.40).

2.5.3 *Code beam*

Consider a cantilever beam where the load is zero at all nodes, except at the last
node located at the free end. Modify accordingly the FSELIB code *beam* and run
the modified code for several discretization levels. Compare and discuss the finite
element solutions.

2.6   Beam buckling

We continue the study of beams by considering deformation under the combined
action of a transverse load, as discussed in Sections 2.4 and 2.5, and a compressive
axial force, P, that is allowed to vary with axial position, as illustrated in Figure
2.6.1.

   Performing a force balance in the direction of the y axis, we recover the differential relation (2.4.2) for the transverse shear force. Performing an analogous torque
balance over the infinitesimal section with length dx, as shown in Figure 2.6.1, we
obtain

$$Q(x)\,dx + M(x) + P\,dv = M(x + dx), \qquad (2.6.1)$$

where v is the vertical deflection, Q is the transverse shear force, and M is the
bending moment. Rearranging, we obtain

$$Q = -P\frac{dv}{dx} + \frac{dM}{dx}. \qquad (2.6.2)$$

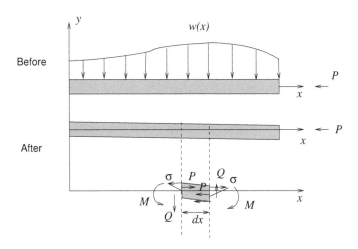

FIGURE 2.6.1   Illustration of combined beam bending and buckling under the influence of a transverse load and an axial compressive force, P, before and after
deformation. The distribution of the axial stress, σ, over a beam cross-section
and transverse shear force, Q, are shown at the bottom panel.

Substituting this expression into the force balance (2.4.2),

$$w = \frac{\mathrm{d}Q}{\mathrm{d}x},\tag{2.6.3}$$

we derive the equilibrium equation

$$\frac{\mathrm{d}^2 M}{\mathrm{d}x^2} - P\frac{\mathrm{d}v}{\mathrm{d}x}\left(P\frac{\mathrm{d}v}{\mathrm{d}x}\right) = w(x).\tag{2.6.4}$$

Finally, we express the bending moment in terms of the curvature by a linear
relation,

$$M = -EI\frac{\mathrm{d}^2 v}{\mathrm{d}x^2},\tag{2.6.5}$$

and derive the governing equation. In the case of a homogeneous prismatic beam
where the product EI is constant, we obtain a fourth-order equation,

$$EI\frac{\mathrm{d}^4 v}{\mathrm{d}x^4} + P\frac{\mathrm{d}v}{\mathrm{d}x}\left(P\frac{\mathrm{d}v}{\mathrm{d}x}\right) = -w(x).\tag{2.6.6}$$

When $P = 0$, we recover the governing equation (2.4.13) describing beam bending.

2.6.1 Tip compression

When the in-line load is due to tip compression, P is a constant, independent of x, and the beam deflection is governed by the linear equation

$$EI \frac{\mathrm{d}^4 v}{\mathrm{d}x^4} + P \frac{\mathrm{d}^2 v}{\mathrm{d}x^2} = -w(x). \tag{2.6.7}$$

The Galerkin finite element implementation provides us with the following generalization of (2.5.16),

$$EI \int_0^L \frac{\mathrm{d}^2 \phi_i}{\mathrm{d}x^2} \frac{\mathrm{d}^2 v}{\mathrm{d}x^2} \,\mathrm{d}x - P \int_0^L \frac{\mathrm{d}\phi_i}{\mathrm{d}x} \frac{\mathrm{d}v}{\mathrm{d}x} \,\mathrm{d}x$$

$$= \left[\phi_i \, Q - \frac{\mathrm{d}\phi_i}{\mathrm{d}x} \, M - \phi_i \frac{\mathrm{d}v}{\mathrm{d}x} \, P \right]_0^L - \int_0^L \phi_i \, w(x) \,\mathrm{d}x \tag{2.6.8}$$

for $i = 1, \ldots, 2N_{\mathrm{G}}$.

Substituting the finite element expansion (2.5.11) into (2.6.8) and compiling the resulting equations, we derive a linear system,

$$(EI \, \mathbf{K} - P \, \mathbf{D}) \cdot \mathbf{u} = \mathbf{b}, \tag{2.6.9}$$

where $\mathbf{K}$ is the global stiffness matrix defined in (2.5.18), $\mathbf{D}$ is the generalized global diffusion matrix, also called the global *geometrical stiffness matrix*, with components

$$D_{ij} = \int_0^L \frac{\mathrm{d}\phi_i}{\mathrm{d}x} \frac{\mathrm{d}\phi_j}{\mathrm{d}x} \,\mathrm{d}x \tag{2.6.10}$$

for $i, j = 1, \ldots, 2N_{\mathrm{G}}$, and the vector $\mathbf{u}$ is defined in (2.5.19).

The first two components of the vector $\mathbf{b}$ on the right-hand side of the linear system are given by

$$b_1 = \left[Q \, \phi_1 - M \frac{\mathrm{d}\phi_1}{\mathrm{d}x} - P \, \phi_1 \frac{\mathrm{d}v}{\mathrm{d}x} \right]_0^L - \int_0^L \phi_1 \, w(x) \,\mathrm{d}x, \tag{2.6.11}$$

yielding

$$b_1 = -Q(0) + P \, v_1' - \int_0^L \phi_1 \, w(x) \,\mathrm{d}x, \tag{2.6.12}$$

and

$$b_2 = \left[Q \, \phi_2 - M \frac{\mathrm{d}\phi_2}{\mathrm{d}x} - P \, \phi_2 \frac{\mathrm{d}v}{\mathrm{d}x} \right]_0^L - \int_0^L \phi_2 \, w(x) \,\mathrm{d}x, \tag{2.6.13}$$

yielding

$$b_2 = M(0) - \int_0^L \phi_2 \, w(x) \,\mathrm{d}x. \tag{2.6.14}$$

The intermediate components are given by

$$b_i = -\int_0^L \phi_i \, w(x) \, dx \tag{2.6.15}$$

for $i = 3, \ldots, 2N_G - 2$. The penultimate component is given by

$$b_{2N_G-1} = \left[\phi_{2N_G-1} \, Q - \frac{d\phi_{2N_G-1}}{dx} M - \phi_{2N_G-1} \frac{dv}{dx} P \right]_0^L - \int_0^L \phi_{2N_G-1} \, w(x) \, dx, \tag{2.6.16}$$

yielding

$$b_{2N_G-1} = Q(L) - P \, v'_{N_G} - \int_0^L \phi_{2N_G-1} \, w(x) \, dx. \tag{2.6.17}$$

The last component is given by

$$b_{2N_G} = \left[Q \, \phi_{2N_G} - \frac{d\phi_{2N_G}}{dx} M - \phi_{2N_G} \frac{dv}{dx} P \right]_0^L - \int_0^L \phi_{2N_G} \, w(x) \, dx, \tag{2.6.18}$$

yielding

$$b_{2N_G} = -M(L) - \int_0^L \phi_{2N_G} \, w(x) \, dx. \tag{2.6.19}$$

In practice, the global stiffness and geometrical stiffness matrices are assembled from corresponding 4×4 element matrices. The element stiffness matrix was given in (2.5.33). A detailed calculation shows that the lth-element geometrical stiffness matrix corresponding to $\mathbf{D}$ is given by

$$\mathbf{G}^{(l)} = \frac{1}{30 \, h_l} \begin{bmatrix} 36 & 3 \, h_l & -36 & 3 \, h_l \\ 3 \, h_l & 4 \, h_l^2 & -3 \, h_l & -h_l^2 \\ -36 & -3 \, h_l & 36 & -3 \, h_l \\ 3 \, h_l & -h_l^2 & -3 \, h_l & 4 \, h_l^2 \end{bmatrix}. \tag{2.6.20}$$

Implementing the boundary conditions by specifying selected values of the generalized deflection vector $\mathbf{u}$, end-node transverse shear forces $Q(0)$, $Q(L)$, and bending moments, $M(0)$, $M(L)$, provides us with the final system of equations for the remaining unknowns.

2.6.2   Buckling under a compressive tip force

In the absence of a transverse load, $w = 0$, the right-hand side of (2.6.6) is zero, yielding a linear homogeneous differential equation,

$$EI \frac{d^4 v}{dx^4} + P \frac{d^2 v}{dx^2} = 0. \tag{2.6.21}$$

A trivial solution is $v = 0$. Nontrivial solutions are possible for certain values of the compressive force, P, corresponding to different buckling modes.

To illustrate how these modes arise from the finite element formulation, we consider the one-element discretization providing us with four Galerkin equations encapsulated in the linear system

$$\mathbf{H} \cdot \begin{bmatrix} v_1 \\ v_1' \\ v_2 \\ v_2' \end{bmatrix} = \begin{bmatrix} -Q(0) + P\,v_1' \\ M(0) \\ Q(L) - P\,v_2' \\ -M(L) \end{bmatrix}, \qquad (2.6.22)$$

where

$$\mathbf{H} = \frac{EI}{L^3} \begin{bmatrix} 12 & 6L & -12 & 6L \\ 6L & 4L^2 & -6L & 2L^2 \\ -12 & -6L & 12 & -6L \\ 6L & 2L^2 & -6L & 4L^2 \end{bmatrix} - \frac{P}{30L} \begin{bmatrix} 36 & 3L & -36 & 3L \\ 3L & 4L^2 & -3L & -L^2 \\ -36 & -3L & 36 & -3L \\ 3L & -L^2 & -3L & 4L^2 \end{bmatrix}.$$

$$(2.6.23)$$

If the beam is simply supported at both ends, we require that

$$v_1 = 0, \qquad M(0) = 0, \qquad v_2 = 0, \qquad M(L) = 0 \qquad (2.6.24)$$

and obtain a system of linear equations for four unknowns,

$$v_1', \quad Q(0), \quad v_2', \quad Q(L). \qquad (2.6.25)$$

Physically, the beam is a toothpick pinched by two fingers at both ends. Implementing the boundary conditions (2.6.24), we find that the second and fourth equations in (2.6.22) provide us with a homogeneous system,

$$\begin{bmatrix} 4 & 2 \\ 2 & 4 \end{bmatrix} \cdot \begin{bmatrix} v_1' \\ v_2' \end{bmatrix} = \hat{P}\,\frac{1}{30} \begin{bmatrix} 4 & -1 \\ -1 & 4 \end{bmatrix} \cdot \begin{bmatrix} v_1' \\ v_2' \end{bmatrix}, \qquad (2.6.26)$$

where

$$\hat{P} \equiv \frac{PL^2}{EI} \qquad (2.6.27)$$

is the dimensionless compressive force.

Equation (2.6.26) has the standard form of a generalized eigenvalue problem expressed by the algebraic equation

$$\mathbf{A} \cdot \mathbf{x} = \lambda\,\mathbf{B} \cdot \mathbf{x}, \qquad (2.6.28)$$

where $\mathbf{A}$ and $\mathbf{B}$ are two matrices with the same dimensions, λ is an eigenvalue, and $\mathbf{x}$ is the corresponding eigenvector. A non-trivial solution of the 2×2 system

(2.6.26) exists only for two eigenvalues, $\hat{P}$. Once the eigenvalues have been found along with the corresponding eigensolutions, they can be substituted in the first and third equations in (2.6.22) to yield the transverse shear forces, $Q(0)$ and $Q(L)$.

The eigenvalues can be found expeditiously using the MATLAB function *eigen* in the following short session:

```
>> A = [4 2; 2 4];
>> B = [4 -1; -1 4]/30;
>> eig(A,B)

ans =
      12
      60
```

The exact solution predicts that the eigenvalues form an infinite sequence,

$$\hat{P}_n = n^2 \pi^2, \qquad (2.6.29)$$

where n is an integer. The numerical values of the one-element solution, 12 and 60, are fair approximations to the first two modes, $n = 1, 2$, corresponding to $\hat{P}_1 = 9.9$ and $\hat{P}_2 = 39.5$.

2.6.3 Buckling of a heavy vertical column

In the applications discussed in Sections 2.6.1 and 2.6.2, the compressive load is constant along the beam length. In applications encountered in practice, this is not always true. As an example, we consider the buckling of a heavy vertical beam before and after deformation, as illustrated in Figure 2.6.2(*a*). The axial compressive force is given by

$$P(x) = \rho g A (L - x), \qquad (2.6.30)$$

where A is the beam cross-sectional area, ρ is the beam density, and g is the acceleration of gravity. Repeating the earlier analysis, we find that the beam deflection is governed by the following modified version of the differential equation (2.6.21),

$$\frac{d^4 v}{dx^4} + \frac{1}{\ell^3} \frac{d}{dx} \left((L - x) \frac{dv}{dx} \right) = 0, \qquad (2.6.31)$$

where

$$\ell \equiv \left(\frac{EI}{\rho g A} \right)^{1/3} \qquad (2.6.32)$$

is a physical parameter with dimensions of length.

Integrating equation (2.6.31) once with respect to x, we derive a second-order equation for the linearized slope angle, $\theta \equiv dv/dx$,

$$\frac{d^2 \theta}{dx^2} + \frac{1}{\ell^3} (L - x) \theta = c, \qquad (2.6.33)$$

(a)   (b)

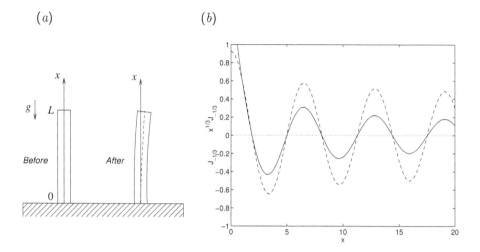

FIGURE 2.6.2   (*a*) Illustration of a heavy vertical beam buckling under the influence of its own weight. (*b*) Graphs of the Bessel function $J_{-1/3}(x)$ (solid line) and modulated Bessel function $x^{1/3}J_{-1/3}(x)$ (dashed line).

where c is a constant. Requiring the free-end condition $Q(L) = 0$, we find that $c = 0$ and derive the homogeneous equation

$$\frac{d^2\theta}{dx^2} = \frac{1}{\ell^3}\,(x - L)\,\theta. \tag{2.6.34}$$

In terms of the dimensionless spatial variable

$$\hat{x} \equiv \frac{L - x}{\ell}, \tag{2.6.35}$$

we obtain the Airy equation

$$\frac{d^2\theta}{d\hat{x}^2} + \hat{x}\,\theta = 0. \tag{2.6.36}$$

The general solution is

$$\theta(\xi) = \left(\frac{3}{2}\,\xi\right)^{1/3}\left(c_1\,J_{1/3}(\xi) + c_2\,J_{-1/3}(\xi)\right), \tag{2.6.37}$$

where c_1 and c_2 are two constants,

$$\xi \equiv \frac{2}{3}\,\hat{x}^{3/2}, \tag{2.6.38}$$

and $J_{\pm 1/3}$ are Bessel functions of fractional order (e.g., Abramowitz & Stegun [1], pp. 446–447). Graphs of the Bessel function $J_{-1/3}(x)$ and modulated Bessel function $x^{1/3}J_{-1/3}(x)$ are shown in Figure 2.6.2(*b*).

To evaluate the Bessel functions, $J_{\pm 1/3}(x)$, we may use the ascending series expansion

$$J_{\pm 1/3}(\xi) = (\frac{1}{2}\xi)^{\pm 1/3} \sum_{k=0}^{\infty} \frac{(-\frac{1}{4}\xi^2)^k}{k!\,\Gamma(\pm\frac{1}{3}+k+1)}, \tag{2.6.39}$$

where Γ is the Gamma function and an exclamation mark denotes the factorial (e.g., Abramowitz & Stegun [1], p. 360). Substituting this solution into (2.6.37), we obtain

$$\theta(\hat{x}) = c_1\,cA(\xi) + c_2\,\mathcal{B}(\xi), \tag{2.6.40}$$

where

$$A(x) = \frac{1}{\sqrt{3}}\,x\sum_{k=0}^{\infty} \frac{(-\frac{1}{9}x^9)^k}{k!\,\Gamma(\frac{1}{3}+k+1)} \tag{2.6.41}$$

and

$$\mathcal{B}(x) = \sqrt{3}\sum_{k=0}^{\infty} \frac{(-\frac{1}{9}x^9)^k}{k!\,\Gamma(-\frac{1}{3}+k+1)}. \tag{2.6.42}$$

It is significant to observe that $\mathcal{B}'(0) = 0$, whereas $A'(0) \neq 0$.

To evaluate the constants c_1 and c_2, we enforce the second free-end condition, $M(x = L) = 0$ or $\theta'(x = L) = 0$, and the fixed-end condition, $\theta(x = 0) = 0$. The former gives $c_1 = 0$ and the latter gives

$$J_{-1/3}(\frac{2}{3}\alpha^{3/2}) = 0, \tag{2.6.43}$$

where

$$\alpha \equiv \frac{L}{\ell} = L\left(\frac{\rho g A}{EI}\right)^{1/3} \tag{2.6.44}$$

is a dimensionless parameter. Thus, the eigenfunction is described by the dashed line in Figure 2.6.2(b). Buckling occurs when

$$\alpha^3 = L^3\frac{\rho g A}{EI} = \frac{9}{4}\varpi^2, \tag{2.6.45}$$

corresponding to the critical beam height

$$L_c = \left(\frac{9}{4}\varpi^2\frac{EI}{\rho g A}\right)^{1/3} \tag{2.6.46}$$

where ϖ is a zero of $J_{-1/3}$ (e.g., Cox & McCarthy [13]). The smallest zero is known to be $\varpi \simeq 1.86635$.

Finite element formulation

In the finite element formulation, solving the second-order differential equation (2.6.34) subject to the boundary conditions $\theta(0) = 0$ and $\theta'(L) = 0$ is reformulated as a generalized eigenvalue problem expressed by the linear system

$$\mathbf{D} \cdot \boldsymbol{\theta} = \frac{1}{\ell^3} \mathbf{E} \cdot \boldsymbol{\theta}. \tag{2.6.47}$$

The eigenvector $\boldsymbol{\theta}$ encapsulates the $N_{\mathrm{E}} - 1$ unknown nodal slopes,

$$\boldsymbol{\theta} = \begin{bmatrix} \theta_2, & \theta_3, & \dots, & \theta_{N_{\mathrm{E}}} \end{bmatrix}^T, \tag{2.6.48}$$

where N_{E} is the number of linear elements. The matrix $\mathbf{D}$ is the tridiagonal global diffusion matrix, and the matrix $\mathbf{E}$ is a modified tridiagonal mass matrix. The diagonal, superdiagonal, and subdiagonal entries of these matrices are generated by the FSELIB function *buckle_tree_sys*, listed in Table 2.6.1.

The solution of the generalized eigenvalue problem can be computed using the FSELIB function *buckle_tree* listed in Table 2.6.2. Calculations with 16 evenly spaced elements yield the small eigenvalue $\alpha^3 = 7.834$, which is in perfect agreement with the exact value, $9\varpi^2/4$, where ϖ is the aforementioned smallest zero of $\mathrm{J}_{-1/3}$.

PROBLEMS

2.6.1 *Element stiffness matrix*

Confirm that the determinant of the geometric element stiffness matrix displayed in (2.6.20) is zero, indicating that the matrix is singular.

2.6.2 *Buckling of a simply supported beam*

Derive analytical expressions for the eigenvalues and eigensolutions of (2.6.21) for a beam that is simply supported at both ends.

2.6.3 *Buckling of a beam with different types of support*

A beam that is simply supported at both ends is said to have a pin-pin support. Other types of support are the fixed-pin support, the fixed-fixed support, and the fixed-free support. Solving (2.6.21) by elementary analytical methods, we find that the eigenvalues are given by

$$\hat{P}_n = \left(\frac{n\pi}{\alpha} \right)^2, \tag{2.6.49}$$

where n is an integer and the coefficient α depends on the boundary conditions. For the pin-pin support, $\alpha = 1$, as discussed in the text.

(*a*) Show that, for a beam with a fixed-pin support, $\alpha = 0.7$. Derive and solve the one-element eigenvalue problem.

```
function [at,bt,ct,ar,br,cr] = buckle_tree_sys (ne,xe,L)

%=====================================================
% compact assembly of the tridiagonal linear system
% for the tree buckling equation
%
% at,bt,ct correspond to the diffusion matrix
% ar,br,cr correspond to the generalized mass matrix
%=====================================================

%--------------------------
% element size and mid-points
%--------------------------

for l=1:ne
  h(l) = xe(l+1)-xe(l);
  xm(l) = 0.5*(xe(l+1)+xe(l));
end

%----------------------------------
% initialize the tridiagonal matrices
%----------------------------------

at = zeros(ne,1);
bt = zeros(ne,1);
ct = zeros(ne,1);

ar = zeros(ne,1);
br = zeros(ne,1);
cr = zeros(ne,1);

%----------------------------
% the first element is special
% due to the Dirichlet condition
%----------------------------

A11 = 1.0/h(1); A12 =-A11;

B11 = h(1)/3.0; B12 = 0.5*B11;

at(1) = at(1) + A11;
xx = L-xm(1);
ar(1) = ar(1) + xx * B11;

%---------------------------------------
% loop over the subsequent ne-1 elements
%---------------------------------------

for l=2:ne

  A11 = 1/h(l); A12 =-A11;
  A21 = A12;    A22 = A11;
```

TABLE 2.6.1   Function *buckle_tree_sys* (Continuing →)

```
  at(1-1) = at(1-1) + A11;
  bt(1-1) = bt(1-1) + A12;
  ct(1) = ct(1) + A21;
  at(1) = at(1) + A22;

  B11 = h(1)/3.0;  B12 = 0.5*B11;
  B21 = B12;       B22 = B11;

  xx = L-xm(1);
  ar(1-1) = ar(1-1) + xx * B11;
  br(1-1) = br(1-1) + xx * B12;
  cr(1) = cr(1) + xx * B21;
  ar(1) = ar(1) + xx * B22;

end

%-----
% done
%-----

return;
```

TABLE 2.6.1 ($\rightarrow$ Continued) Function *buckle_tree_sys* generates two matrices pertinent to a generalized eigenvalue problem describing the buckling of a vertical prismatic beam under its own weight.

(*b*) Show that, for a beam with a fixed-fixed support, $\alpha = 0.5$. Derive and solve the one-element eigenvalue problem.

(*c*) Show that, for a beam with a fixed-free support, $\alpha = 2$. Derive and solve the one-element eigenvalue problem.

```
%==========================================
% Code  buckle_tree
%
% code for the column buckling equation
% with linear elements
%
% ne: number of elements
%==========================================

clear all
close all

%-----------
% input data
%-----------

L = 1.0;

ne = 16; ratio = 2.0;

%----------------
% grid generation
%----------------

xe = elm_line1 (0,L,ne,ratio);

%------------------
% compact assembly
%------------------

[at,bt,ct,ar,br,cr] = buckle_tree_sys (ne,xe,L);

A = zeros(ne,ne);
B = zeros(ne,ne);

for i=1:ne
 A(i,i) = at(i); B(i,i) = ar(i);
end

for i=1:ne-1
 A(i,i+1) = bt(i); A(i+1,i) = ct(i+1);
 B(i,i+1) = br(i); B(i+1,i) = cr(i+1);
end

eigenvalues = eig(A,B);

%-----
% done
%-----
```

TABLE 2.6.2   Code *buckle_tree* solves an eigenvalue problem describing the buckling of a vertical prismatic beam, regarded as a tree or telephone pole, deflecting under its own weight.

High-order and spectral elements in one dimension

3

In Chapters 1 and 2, we introduced the Galerkin finite element method in one spatial dimension with linear, quadratic, and cubic Hermitian elements. High-order finite element methods employ high-order polynomial expansions over the individual elements. The polynomial expansions themselves are defined by an appropriate number of element *interpolation* nodes that are generally distinct from the two shared *geometrical* element end-nodes.

Element and global nodes

In Chapter 1, we discussed quadratic elements defined by three interpolation nodes, including two shared geometrical end-nodes and one interior interpolation node, yielding a total of $N_G = 2N_E + 1$ global interpolation nodes, where N_E is the number of elements. Cubic elements are defined by four interpolation nodes, including two shared geometrical end-nodes and two interior interpolation nodes, yielding a total of $N_G = 3N_E + 1$ global interpolation nodes. More generally, pth-order elements are defined by $p + 1$ interpolation nodes, including two shared geometrical end-nodes and $p - 1$ interior interpolation nodes, yielding a total of $N_G = pN_E + 1$ global interpolation nodes.

h and p refinement

Improving the accuracy of the finite element solution by decreasing the element size, h, while holding the polynomial order fixed, is classified as an h-refinement. Typically, the error decreases like a power of h, where the exponent is determined by the polynomial order and smoothness of the solution. Improving the accuracy by raising the polynomial order, p, for a fixed element discretization, is classified as a p-refinement, typically associated with spectral convergence. The terminology *spectral* implies that the numerical error decreases faster than any power of $1/p$, where p is the order of the polynomial expansion. Combinations of the two refinement strategies yield the hp-refinement.

Galerkin projection

To implement the Galerkin finite element method for a high-order expansion, we integrate the product of the governing equation with each of the global interpolation functions associated with the global nodes, and thus derive a system of algebraic or differential equations. The system is subsequently modified by implementing the stipulated Neumann, Dirichlet, or other boundary conditions to obtain its final form, and then is solved by standard numerical methods.

Accuracy

In practice, we would like to use low-order expansions in regions where the solution is expected to vary smoothly, and high-order expansions in regions where the solution is expected to exhibit rapid spatial variations. Moreover, we would like to achieve the best possible accuracy for a given number of interpolation nodes. For linear and homogeneous differential equations, the node distribution is inconsequential to the numerical solution, and only determines the structure and standing of the global matrices quantified by the condition number. For inhomogeneous or nonlinear equations, the node distribution may play an important role in the accuracy and convergence of the solution.

Theoretical analysis of the interpolation error shows that, given the number of interpolation nodes to be distributed over an element, the highest interpolation accuracy is achieved when the interior nodes are distributed at positions corresponding to the zeros of certain families of orthogonal polynomials. When this is done, we obtain a spectral element expansion and associated spectral element method.

Summary

In this chapter, we discuss the theoretical foundation and practical implementation of high-order and spectral element methods in one spatial dimension. The theoretical development hinges on the notion of orthogonal polynomials, discussed in Appendix B, and on the theory of function interpolation, discussed in Appendix D. These two appendices should be reviewed, and preferably studied in detail, before continuing with this chapter. Readers who are not presently interested in the spectral element method can proceed without ramifications to the discussion on the finite element method in two dimensions, starting in Chapter 4. The spectral element method in two and three dimensions is discussed in Chapters 5 and 8.

3.1   Element nodal sets

Suppose that we have decided to approximate the solution over the lth element with a $p^{(l)}$-degree polynomial defined by $p^{(l)} + 1$ element interpolation nodes, including two geometrical end-nodes and $p^{(l)} - 1$ interpolation interior nodes. To simplify the notation, we denote $p^{(l)}$ by m, with the understanding that m may vary across the element assembly.

To facilitate the computations, it is helpful to map the lth element to the standard interval of the dimensionless ξ axis, $[-1, 1]$, as shown in Figure 3.1.1. The

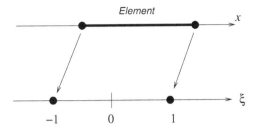

FIGURE 3.1.1   Mapping of an element from the x axis to the standard interval $[-1, 1]$
of the parametric ξ axis.

mapping is mediated by the function

$$x(\xi) = \frac{1}{2}\left(x_2^{(l)} + x_1^{(l)}\right) + \frac{1}{2}\left(x_2^{(l)} - x_1^{(l)}\right)\xi, \tag{3.1.1}$$

where $x_1^{(l)}$ is the first element end-node and $x_2^{(l)}$ is the second element end-node. As
ξ increases from -1 to 1, the field point x is shifted from the first element end-node
to the second element end-node.

The element nodes are deployed at positions ξ_i for $i = 1, \ldots, m + 1$ along the
ξ axis. The corresponding positions along the x axis arise from (3.1.1), and *vice
versa.*

3.1.1   Lagrange interpolation

The element interpolation functions can be conveniently identified with the mth-
degree Lagrange interpolating polynomials discussed in Section D.2, Appendix D,

$$\psi_i(\xi) = \frac{(\xi - \xi_1)\cdots(\xi - \xi_{i-1})(\xi - \xi_{i+1})\cdots(\xi - \xi_{m+1})}{(\xi_i - \xi_1)\cdots(\xi_i - \xi_{i-1})(\xi_i - \xi_{i+1})\cdots(\xi_i - \xi_{m+1})} \tag{3.1.2}$$

for $i - 1, \ldots, m + 1$. Cursory inspection reveals that these polynomials satisfy the
cardinal interpolation property

$$\psi_i(\xi_j) = \begin{cases} 1 & \text{if } i = j, \\ 0 & \text{if } i \neq j. \end{cases} \tag{3.1.3}$$

Stated differently, $\psi_i(\xi_j)$ is the identity matrix represented by Kronecker's delta,
$\psi_i(\xi_j) = \delta_{ij}$.

An equivalent representation is

$$\psi_i(\xi) = \frac{\Phi_{m+1}(\xi)}{(\xi - \xi_i)\,\Phi'_{m+1}(\xi_i)}, \tag{3.1.4}$$

where a prime denotes a derivative with respect to ξ and

$$\Phi_{m+1}(\xi) \equiv (\xi - \xi_1)(\xi - \xi_2) \cdots (\xi - \xi_m)(\xi - \xi_{m+1}) \tag{3.1.5}$$

is an $(m+1)$-degree Lagrange generating polynomial defined in terms of *all* element interpolation nodes, as discussed in Section D.2, Appendix D.

A function of interest, $f(\xi)$, defined over the lth element can be approximated with the mth-degree interpolating polynomial, $P_m(\xi)$, which can be constructed explicitly in terms of specified or *a priori* unknown nodal values, $f(\xi_i)$, and the Lagrange interpolating polynomials, as

$$f(\xi) \simeq P_m(\xi) = \sum_{i=1}^{m+1} f(\xi_i)\,\psi_i(\xi). \tag{3.1.6}$$

The aforementioned cardinal property of the Lagrange polynomials, $\psi_i(\xi_j) = \delta_{ij}$, ensures that

$$P_m(\xi_j) = \sum_{i=1}^{m+1} f(\xi_i)\,\psi_i(\xi_j) = \sum_{i=1}^{m+1} f(\xi_i)\,\delta_{ij} = f(\xi_j), \tag{3.1.7}$$

as required.

3.1.2 Evenly spaced nodes

Graphs of the node interpolation functions, ψ_i, in the canonical interval of the ξ axis, $[-1, 1]$, are shown in Figure 3.1.2 for sets of 5 ($m = 4$) and 12 ($m = 11$) evenly spaced nodes. These graphs were generated by the FSELIB function *lagrange_psi*, not listed in the text. As the number of interpolation nodes increases, oscillations arise near the ends of the interpolation domain as manifestations of the *Runge effect*. Unless a low-order polynomial approximation is employed, an even distribution of interpolation nodes is detrimental to the accuracy of the interpolation.

Runge function

To demonstrate explicitly the potential failure of the interpolation, we approximate the Runge function

$$f(\xi) = \frac{1}{1 + 25\,\xi^2} \tag{3.1.8}$$

in the interval $[-1, 1]$ with an mth-degree interpolating polynomial that passes through $m + 1$ data points, ξ_i for $i = 1, \ldots, m + 1$, where $\xi_1 = -1$ and $\xi_{m+1} = 1$. Graphs of several interpolating polynomials of increasing order m are shown in Figure 3.1.3 for evenly distributed interpolation nodes. The seemingly innocuous Runge function is plotted with the heavy solid line. As the number of data points, and thus the polynomial order, increases, the interpolation worsens near the two

(a)

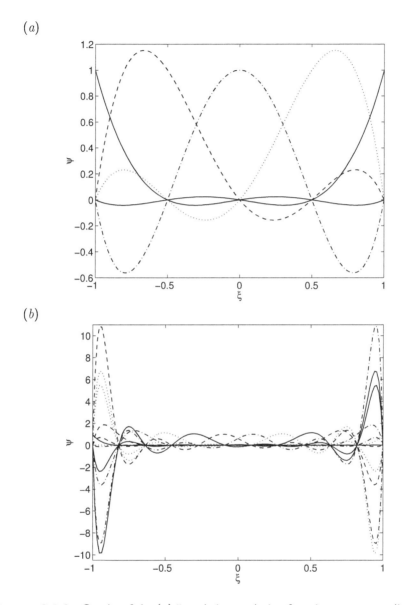

(b)

FIGURE 3.1.2   Graphs of the (a) 5-node interpolation functions corresponding to polynomial order $m = 4$, and (b) 12-node interpolation functions corresponding to
$m = 11$. In both cases, the nodes are spaced evenly in the interval between
$\xi_1 = -1$ and $\xi_{m+1} = 1$. As the size of the nodal set and thus the polynomial
order increases, Runge oscillations arise near the two ends of the interpolation
domain.

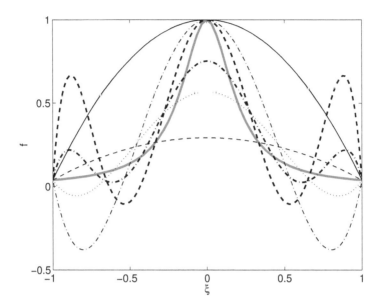

FIGURE 3.1.3   Lagrange interpolation of the Runge function, drawn with the heavy
    solid line. Interpolating polynomials are plotted for $m = 2$ (solid line), $m = 3$
    (dashed line), $m = 4$ (dash-dotted line), $m = 5$ (dotted line), $m = 6$ (heavy
    dashed line), and $m = 7$ (heavy dash-dotted line).

ends of the interpolation domain due to the Runge effect. Physically, because the
interpolated polynomial is reconstructed from evenly spaced data that pay equal
attention to the middle and to the two ends of the interpolation domain, insufficient
information is provided beyond the boundaries of the interpolation domain, $\xi < -1$
and $\xi > 1$, and this frustrates the interpolation.

In conclusion, as the polynomial order increases, a finite element solution based
on evenly spaced element interpolation nodes may become notably sensitive to the
numerical error or even fail. The potential failure can be circumvented by judiciously
deploying the interpolation nodes over each element to positions corresponding to
the zeros of orthogonal polynomials, as discussed in Section 3.2. For the moment, we
bypass the important issue of optimal node distribution and discuss the computation
of the element mass and diffusion matrices for a given element nodal set.

3.1.3   Element matrices

Expressing the physical variable x in terms of the canonical variable ξ using the
mapping function (3.1.1), and noting that

$$\mathrm{d}x = h_l \, \frac{1}{2} \, \mathrm{d}\xi, \tag{3.1.9}$$

where $h_l \equiv x_2^{(l)} - x_1^{(l)}$ is the element size, we find that the lth-element diffusion matrix is given by

$$A_{ij}^{(l)} \equiv \int_{X_1^{(l)}}^{X_2^{(l)}} \frac{\mathrm{d}\psi_i(x)}{\mathrm{d}x} \frac{\mathrm{d}\psi_j(x)}{\mathrm{d}x} \, \mathrm{d}x = \frac{2}{h_l} \int_{-1}^{1} \frac{\mathrm{d}\psi_i(\xi)}{\mathrm{d}\xi} \frac{\mathrm{d}\psi_j(\xi)}{\mathrm{d}\xi} \, \mathrm{d}\xi \qquad (3.1.10)$$

for $i, j = 1, \ldots, m+1$. For the physical reason discussed in Section 1.2, this matrix is singular irrespective of the number and location of the element interpolation nodes. Working in a similar fashion, we find that the element mass matrix is given by

$$B_{ij}^{(l)} \equiv \int_{X_1^{(l)}}^{X_2^{(l)}} \psi_i(x) \, \psi_j(x) \, \mathrm{d}x = \frac{1}{2} \, h_l \int_{-1}^{1} \psi_i(\xi) \, \psi_j(\xi) \, \mathrm{d}\xi \qquad (3.1.11)$$

for $i, j = 1, \ldots, m + 1$. The integrals with respect to ξ on the right-hand sides of (3.1.10) and (3.1.11) can be calculated by a combination of analytical and numerical methods, as will be discussed later in this chapter for specific node distributions.

Node differentiation matrix

Consider the computation of the derivatives of the element interpolation functions. Differentiating expression (3.1.4) using the rules of quotient differentiation, we find that

$$\frac{\mathrm{d}\psi_i(\xi)}{\mathrm{d}\xi} = \frac{\Phi'_{m+1}(\xi)(\xi - \xi_i) - \Phi_{m+1}(\xi)}{(\xi - \xi_i)^2 \, \Phi'_{m+1}(\xi_i)}. \qquad (3.1.12)$$

For future reference, we evaluate these derivatives at the interpolation nodes. Observing that $\Phi_{m+1}(\xi_j) = 0$, we derive the *node differentiation matrix*,

$$d_{ij} \equiv \left(\frac{\mathrm{d}\psi_i}{\mathrm{d}\xi} \right)_{\xi=\xi_j} = \begin{cases} \dfrac{\Phi'_{m+1}(\xi_j)}{(\xi_j - \xi_i) \, \Phi'(\xi_j)} & \text{if } \quad i \neq j, \\[4mm] \dfrac{\Phi''_{m+1}(\xi_i)}{2 \, \Phi'_{m+1}(\xi_i)} & \text{if } \quad i = j. \end{cases} \qquad (3.1.13)$$

The expression for $i = j$ arises by expanding $\Phi_{m+1}(\xi)$ and $\Phi'_{m+1}(\xi)$ in the numerator of (3.1.12) in Taylor series about the point ξ_i, and then taking the limit $\xi \, \rangle \, \xi_i$. An alternative is to use the l'Hôpital rule (Problem 3.1.1).

Further expressions for the derivatives of the element interpolation functions and node differentiation matrix are given in Section D.2.3, Appendix D.

3.1.4 C^0 continuity and shared element nodes

The implementation of the Galerkin finite element method for the steady heat conduction problem discussed in Chapter 1 culminates in a linear system, $\mathbf{D} \cdot \mathbf{f} = \mathbf{b}$, where $\mathbf{f}$ is the vector of unknown nodal temperatures, $\mathbf{D}$ is the global diffusion matrix, and $\mathbf{b}$ is a properly constructed right-hand side.

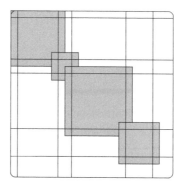

FIGURE 3.1.4   Structure of the global diffusion or mass matrix exhibiting overlapping
   diagonal blocks that host the element diffusion matrices.

The importance of employing a C^0 (continuous) global finite element expansion
implemented by shared element end-nodes is underscored by two key observations.
First, the Galerkin finite element equations were derived under the assumption of
continuous global interpolation functions. Second, if the elements did not share
end-nodes, the global diffusion matrix $\mathbf{D}$ would consist of non-overlapping diagonal
element blocks; consequently, the Galerkin equations corresponding to groups of
element nodes would be decoupled. Because the element diffusion matrix is singular,
the solution of the sub-blocks could be found only up to an arbitrary constant, and
the finite element formulation would provide us with an ill-posed problem.

To ensure continuity of the solution represented by the union of the element
expansions, we place the first interpolation node at the first geometrical element
end-node, and the last interpolation node at the second geometrical element end-
node,

$$\xi_1 = -1, \qquad \xi_{m+1} = 1, \qquad\qquad (3.1.14)$$

and require that the nodal values are shared by neighboring elements. The structure
of the global diffusion matrix, consisting of partially overlapping element diffusion
matrices, is illustrated in Figure 3.1.4. Note that overlap occurs only for one diagonal matrix element corresponding to the first or second element end-nodes.

The algorithm listed in Table 3.1.1 establishes a correspondence between element
and global nodes by means of a connectivity matrix, and counts the number of
unique global interpolation nodes, N_{G}.

PROBLEMS

3.1.1 *Node differentiation matrix*

Working as indicated in the text, prove the expression for $i = j$ shown in (3.1.13).

$Ic = 2$ *Initialize the global node counter*

Do $l = 1, \ldots, N_E$ *Run over the elements*
   $Ic = Ic - 1$ *Account for the common end-node*
   Do $i = 1, \ldots, N_p(l) + 1$ *Run over the element nodes*
      $c(l, i) = Ic$ *Evaluate the connectivity matrix*
      $Ic = Ic + 1$
   End Do
End Do

$N_G = Ic - 1$ *Total number of unique global nodes*

TABLE 3.1.1   Correspondence between element and global nodes, and counting of the total number of unique global nodes, N_G.

3.1.2 *Lebesgue constant for evenly spaced nodes*

The Lebesgue function is defined as

$$\mathcal{L}_m(\xi) \equiv \sum_{i=1}^{m+1} |\psi_i(\xi)|, \tag{3.1.15}$$

and the associated Lebesgue constant is defined as

$$\Lambda_m \equiv \max\big[\mathcal{L}(\xi)\big], \tag{3.1.16}$$

where $\max[\bullet]$ denotes the maximum value of the enclosed function over the interpolation interval $[-1, 1]$. Prepare a table of the Lebesgue constant against the polynomial order, m, for evenly spaced nodes and discuss its dependence on m.

3.2   Change of element nodal sets

Consider two different element nodal sets corresponding to the same polynomial order, m, denoted as sets I and II. Each set encapsulates $m + 1$ nodes, subject to the restriction that $\xi_1 = -1$ and $\xi_{m+1} = 1$.

The cardinal interpolation functions of the first set, $\psi_i^I(\xi)$, are related to the cardinal interpolation functions of the second set, $\psi_i^{II}(\xi)$, by the Lagrange interpolation formula

$$\psi_i^I(\xi) = \sum_{j=1}^{m+1} \psi_i^I(\xi_j^{II}) \, \psi_j^{II}(\xi) \tag{3.2.1}$$

for $i = 1, \ldots, m + 1$, without any approximation. Accordingly, the corresponding nodal interpolation function vectors are related by the linear transformation

$$\boldsymbol{\psi}^{\mathrm{I}}(\xi) = \mathbf{V}_{\mathrm{I,II}} \cdot \boldsymbol{\psi}^{\mathrm{II}}(\xi), \qquad (3.2.2)$$

where

$$\mathbf{V}_{\mathrm{I,II}} = \begin{bmatrix} \psi_1^{\mathrm{I}}(\xi_1^{\mathrm{II}}) & \psi_1^{\mathrm{I}}(\xi_2^{\mathrm{II}}) & \cdots & \psi_1^{\mathrm{I}}(\xi_m^{\mathrm{II}}) & \psi_1^{\mathrm{I}}(\xi_{m+1}^{\mathrm{II}}) \\ \psi_2^{\mathrm{I}}(\xi_1^{\mathrm{II}}) & \psi_2^{\mathrm{I}}(\xi_2^{\mathrm{II}}) & \cdots & \psi_2^{\mathrm{I}}(\xi_m^{\mathrm{II}}) & \psi_2^{\mathrm{I}}(\xi_{m+1}^{\mathrm{II}}) \\ \vdots & \vdots & \ddots & \vdots & \vdots \\ \psi_m^{\mathrm{I}}(\xi_1^{\mathrm{II}}) & \psi_m^{\mathrm{I}}(\xi_2^{\mathrm{II}}) & \cdots & \psi_m^{\mathrm{I}}(\xi_m^{\mathrm{II}}) & \psi_m^{\mathrm{I}}(\xi_{m+1}^{\mathrm{II}}) \\ \psi_{m+1}^{\mathrm{I}}(\xi_1^{\mathrm{II}}) & \psi_{m+1}^{\mathrm{I}}(\xi_2^{\mathrm{II}}) & \cdots & \psi_{m+1}^{\mathrm{I}}(\xi_m^{\mathrm{II}}) & \psi_{m+1}^{\mathrm{I}}(\xi_{m+1}^{\mathrm{II}}) \end{bmatrix} \qquad (3.2.3)$$

is a generalized Vandermonde matrix. Because the end-nodes are common in the two distributions, the first and last columns of $\mathbf{V}_{\mathrm{I,II}}$ are filled with zeros, except that the first and last diagonal elements that are equal to unity

$$\psi_i^{\mathrm{I}}(\xi_1^{\mathrm{II}}) = \delta_{i,1}, \qquad \psi_i^{\mathrm{I}}(\xi_{m+1}^{\mathrm{II}}) = \delta_{i,m+1}, \qquad (3.2.4)$$

yielding

$$\mathbf{V}_{\mathrm{I,II}} = \begin{bmatrix} 1 & \psi_1^{\mathrm{I}}(\xi_2^{\mathrm{II}}) & \cdots & \psi_1^{\mathrm{I}}(\xi_m^{\mathrm{II}}) & 0 \\ 0 & \psi_2^{\mathrm{I}}(\xi_2^{\mathrm{II}}) & \cdots & \psi_2^{\mathrm{I}}(\xi_m^{\mathrm{II}}) & 0 \\ \vdots & \vdots & \ddots & \vdots & \vdots \\ 0 & \psi_m^{\mathrm{I}}(\xi_2^{\mathrm{II}}) & \cdots & \psi_m^{\mathrm{I}}(\xi_m^{\mathrm{II}}) & 0 \\ 0 & \psi_{m+1}^{\mathrm{I}}(\xi_2^{\mathrm{II}}) & \cdots & \psi_{m+1}^{\mathrm{I}}(\xi_m^{\mathrm{II}}) & 1 \end{bmatrix}. \qquad (3.2.5)$$

As the first set tends to the second set, we obtain the identity matrix, $\mathbf{I}$.

Quadratic expansion

As an example, we identify the functions $\psi_i^{\mathrm{I}}(\xi)$ with the quadratic element interpolation functions corresponding to the mid-point interior node shown in (1.5.24), repeated below for convenience,

$$\psi_1^{\mathrm{I}}(\xi) = \frac{1}{2}\,\xi\,(\xi - 1), \qquad \psi_2^{\mathrm{I}}(\xi) = 1 - \xi^2, \qquad \psi_3^{\mathrm{I}}(\xi) = \frac{1}{2}\,\xi\,(\xi + 1), \qquad (3.2.6)$$

and the functions $\psi_i^{\mathrm{II}}(\xi)$ with the quadratic element interpolation functions corresponding to an arbitrary interior node located at $\xi_2^{\mathrm{II}} = \beta$; in both cases, $\xi_1 = -1$ and $\xi_3 = 1$. The associated 3×3 generalized Vandermonde matrix takes the form

$$\mathbf{V}_{\mathrm{I,II}} = \begin{bmatrix} 1 & \frac{1}{2}\,\beta\,(\beta - 1) & 0 \\ 0 & 1 - \beta^2 & 0 \\ 0 & \frac{1}{2}\,\beta\,(\beta + 1) & 1 \end{bmatrix}. \qquad (3.2.7)$$

The determinant of this matrix is equal to $1 - \beta^2$. When $\beta = 0$, we obtain the identity matrix, in agreement with physical intuition.

The second set of interpolation functions can be deduced from the first set by solving the linear system (3.2.2). The resulting expressions are identical to those displayed in (1.5.49), repeated below for convenience,

$$\psi_1^{\mathrm{II}}(\xi) = \frac{1}{2\,(1+\beta)}\,(\xi - \beta)\,(\xi - 1), \qquad \psi_2^{\mathrm{II}}(\xi) = \frac{1 - \xi^2}{1 - \beta^2},$$

$$\tag{3.2.8}$$

$$\psi_3^{\mathrm{II}}(\xi) = \frac{1}{2\,(1-\beta)}\,(\xi - \beta)\,(\xi + 1). \tag{3.2.9}$$

When $\beta = 0$, we recover the expressions in (3.2.6).

Converse transformation

Conversely, the second set of interpolation functions can be expressed in terms of the first set using the relations

$$\boldsymbol{\psi}^{\mathrm{II}}(\xi) = \mathbf{V}_{\mathrm{II,I}} \cdot \boldsymbol{\psi}^{\mathrm{I}}(\xi), \tag{3.2.10}$$

where

$$\mathbf{V}_{\mathrm{II,I}} = \begin{bmatrix} \psi_1^{\mathrm{II}}(\xi_1^{\mathrm{I}}) & \psi_1^{\mathrm{II}}(\xi_2^{\mathrm{I}}) & \cdots & \psi_1^{\mathrm{II}}(\xi_m^{\mathrm{I}}) & \psi_1^{\mathrm{II}}(\xi_{m+1}^{\mathrm{I}}) \\ \psi_2^{\mathrm{II}}(\xi_1^{\mathrm{I}}) & \psi_2^{\mathrm{II}}(\xi_2^{\mathrm{I}}) & \cdots & \psi_2^{\mathrm{II}}(\xi_m^{\mathrm{I}}) & \psi_2^{\mathrm{II}}(\xi_{m+1}^{\mathrm{I}}) \\ \vdots & \vdots & \ddots & \vdots & \vdots \\ \psi_m^{\mathrm{II}}(\xi_1^{\mathrm{I}}) & \psi_m^{\mathrm{II}}(\xi_2^{\mathrm{I}}) & \cdots & \psi_m^{\mathrm{II}}(\xi_m^{\mathrm{I}}) & \psi_m^{\mathrm{II}}(\xi_{m+1}^{\mathrm{I}}) \\ \psi_{m+1}^{\mathrm{II}}(\xi_1^{\mathrm{I}}) & \psi_{m+1}^{\mathrm{II}}(\xi_2^{\mathrm{I}}) & \cdots & \psi_{m+1}^{\mathrm{II}}(\xi_m^{\mathrm{I}}) & \psi_{m+1}^{\mathrm{II}}(\xi_{m+1}^{\mathrm{I}}) \end{bmatrix} \tag{3.2.11}$$

is another generalized Vandermonde matrix. For the reasons discussed earlier in this section,

$$\mathbf{V}_{\mathrm{II,I}} = \begin{bmatrix} 1 & \psi_1^{\mathrm{II}}(\xi_2^{\mathrm{I}}) & \cdots & \psi_1^{\mathrm{II}}(\xi_m^{\mathrm{I}}) & 0 \\ 0 & \psi_2^{\mathrm{II}}(\xi_2^{\mathrm{I}}) & \cdots & \psi_2^{\mathrm{II}}(\xi_m^{\mathrm{I}}) & 0 \\ \vdots & \vdots & \ddots & \vdots & \vdots \\ 0 & \psi_m^{\mathrm{II}}(\xi_2^{\mathrm{I}}) & \cdots & \psi_m^{\mathrm{II}}(\xi_m^{\mathrm{I}}) & 0 \\ 0 & \psi_{m+1}^{\mathrm{II}}(\xi_2^{\mathrm{I}}) & \cdots & \psi_{m+1}^{\mathrm{II}}(\xi_m^{\mathrm{I}}) & 1 \end{bmatrix}. \tag{3.2.12}$$

Comparing (3.2.2) with (3.2.10), we find that

$$\mathbf{V}_{\mathrm{II,I}} = \mathbf{V}_{\mathrm{I,II}}^{-1} \tag{3.2.13}$$

and *vice versa*.

Quadratic expansion

In the case of the quadratic elements discussed earlier in this section, $\xi_1^I = -1$, $\xi_2^I = 0$, and $\xi_3^I = 1$, yielding

$$
\mathbf{V}_{II,I} =
\begin{bmatrix}
1 & \dfrac{\beta}{2(\beta+1)} & 0 \\[2mm]
0 & \dfrac{1}{1-\beta^2} & 0 \\[2mm]
0 & \dfrac{\beta}{2(\beta-1)} & 1
\end{bmatrix},
\tag{3.2.14}
$$

which is the inverse of the matrix $\mathbf{V}_{I,II}$ shown in (3.2.7).

Relation between element matrices

Using the transformation (3.2.2), we find that the element diffusion and mass matrices of the first set are related to those of the second set by

$$
\mathbf{A}^I = \mathbf{V}_{I,II} \cdot \mathbf{A}^{II} \cdot \mathbf{V}_{I,II}^T, \qquad \mathbf{B}^I = \mathbf{V}_{I,II} \cdot \mathbf{B}^{II} \cdot \mathbf{V}_{I,II}^T
\tag{3.2.15}
$$

or

$$
\mathbf{A}^I = \mathbf{V}_{II,I}^{-1} \cdot \mathbf{A}^{II} \cdot \mathbf{V}_{II,I}^{-1^T}, \qquad \mathbf{B}^I = \mathbf{V}_{II,I}^{-1} \cdot \mathbf{B}^{II} \cdot \mathbf{V}_{II,I}^{-1^T},
\tag{3.2.16}
$$

where the superscript T denotes the matrix transpose.

Conversely, the element matrices of the second set are related to those of the first set by

$$
\mathbf{A}^{II} = \mathbf{V}_{II,I} \cdot \mathbf{A}^I \cdot \mathbf{V}_{II,I}^T, \qquad \mathbf{B}^{II} = \mathbf{V}_{II,I} \cdot \mathbf{B}^I \cdot \mathbf{V}_{II,I}^T
\tag{3.2.17}
$$

or

$$
\mathbf{A}^{II} = \mathbf{V}_{I,II}^{-1} \cdot \mathbf{A}^I \cdot \mathbf{V}_{I,II}^{-1^T}, \qquad \mathbf{B}^{II} = \mathbf{V}_{I,II}^{-1} \cdot \mathbf{B}^I \cdot \mathbf{V}_{I,II}^{-1^T}.
\tag{3.2.18}
$$

In theoretical analysis, these relations can be used to assess the sensitivity of the numerical solution to the choice of a nodal set.

Significance of the nodal set on the finite element solution

In the case of the Helmholtz equation, the finite element formulation with the second nodal set generates the element-level linear system

$$
(\mathbf{A}^{II} - \alpha\,\mathbf{B}^{II}) \cdot \mathbf{f}^{II} = \mathbf{0},
\tag{3.2.19}
$$

where α is a constant, with $\alpha = 0$ corresponding to Laplace's equation. Using (3.2.15), we derive the equivalent form

$$
(\mathbf{A}^I - \alpha\,\mathbf{B}^I) \cdot \mathbf{f}^I = \mathbf{0},
\tag{3.2.20}
$$

where

$$\mathbf{V}_{\mathrm{I,II}}^{T} \cdot \mathbf{f}^{\mathrm{I}} = \mathbf{f}^{\mathrm{II}}. \qquad (3.2.21)$$

Because this equation generates the II nodal values from the I nodal values by Lagrange interpolation, the choice of the nodal base is immaterial. We have reached the important conclusion that the finite element solution of the Helmholtz or Laplace equation is insensitive to the choice of nodal set for a fixed expansion order, m. However, the conditioning of the matrices involved in the final linear system do depend on the choice of nodal set.

In the case of Poisson's equation, the finite element formulation with the second nodal set culminates in a linear system,

$$\mathbf{A}^{\mathrm{II}} \cdot \mathbf{f}^{\mathrm{II}} = \mathbf{B}^{\mathrm{II}} \cdot \mathbf{s}^{\mathrm{II}}, \qquad (3.2.22)$$

where the vector $\mathbf{s}^{\mathrm{II}}$ contains the set II nodal values of the source. Using (3.2.15), we obtain

$$\mathbf{A}^{\mathrm{I}} \cdot \mathbf{f}^{\mathrm{I}} = \mathbf{B}^{\mathrm{I}} \cdot \mathbf{s}^{\mathrm{I}}, \qquad (3.2.23)$$

where

$$\mathbf{V}_{\mathrm{I,II}}^{T} \cdot \mathbf{f}^{\mathrm{I}} = \mathbf{f}^{\mathrm{II}}, \qquad \mathbf{V}_{\mathrm{I,II}}^{T} \cdot \mathbf{s}^{\mathrm{I}} = \mathbf{s}^{\mathrm{II}}. \qquad (3.2.24)$$

These equations provide us with the set II solution and source vectors from the corresponding set I vectors by Lagrange interpolation. Because the interpolated values of the source are generally different from the exact values, finite element solutions with different nodal sets will not necessarily be the same.

PROBLEM

3.2.1 *Change of nodal base*

Confirm that the matrix $\mathbf{V}_{\mathrm{II,I}}$ shown in (3.2.14) is the inverse of the matrix $\mathbf{V}_{\mathrm{I,II}}$ shown in (3.2.7).

3.3 Spectral interpolation

In the spectral element method, the element interpolation nodes are distributed at the zeros of an appropriate family of orthogonal polynomials over the canonical interval of the ξ axis, subject to the mandatory constraints that the first node is placed at $\xi_1 = -1$ and the last node is placed at $\xi_{m+1} = 1$, where m is the order of the polynomial expansion defined by $m + 1$ nodes. We recall that these constraints ensure the C^0 continuity of the finite element expansion.

3.3.1   Lobatto nodal base

The theory of polynomial interpolation, discussed in Appendix D, in conjunction with the theory of orthogonal polynomials, discussed in Appendix B, suggests that the intermediate $m-1$ nodes should be distributed at the zeros of the $(m-1)$-degree Lobatto polynomial, $\mathrm{Lo}_{m-1}(\xi)$, discussed in Section B.6, Appendix B. The first few Lobatto polynomials are listed in Table 3.3.1 along with a generating formula. Graphs of the first few Lobatto polynomials generated by the FSELIB function *lobatto_graphs* are shown in Figure 3.3.1.

By construction, any two Lobatto polynomials, Lo_i and Lo_j, satisfy the orthogonality property

$$\int_{-1}^{1} \mathrm{Lo}_i(\xi)\,\mathrm{Lo}_j(\xi)\,(1-\xi^2)\,\mathrm{d}\xi = \frac{2\,(i+1)(i+2)}{2i+3}\,\delta_{ij}, \qquad (3.3.1)$$

where δ_{ij} is Kronecker's delta representing the identity matrix. Further properties of the Lobatto polynomials are discussed in Section B.6, Appendix B.

Node interpolation functions

The main reason for adopting the zeros of the Lobatto polynomials is that the corresponding node interpolation functions are guaranteed to vary in the range $[-1,1]$, independent of the order of the polynomial approximation, m, that is,

$$|\psi_i(\xi)| \leq 1 \qquad (3.3.2)$$

for any i. The equality holds true only when $\xi = \xi_i$, as shown in Section D.5, Appendix D.

The quadratic element interpolation functions for polynomial order $m = 2$ are given by

$$\psi_1(\xi) = \frac{1}{2}\,\xi\,(\xi-1), \qquad \psi_2(\xi) = 1 - \xi^2, \qquad \psi_3(\xi) = \frac{1}{2}\,\xi\,(\xi+1). \qquad (3.3.3)$$

Graphs of the 4 cubic element interpolation functions corresponding to $m = 3$ and of the five quartic interpolation functions corresponding to $m = 4$, generated by the FSELIB function *lagrange_psi*, not listed in the text, are displayed in Figure 3.3.2. Note that each interpolation function achieves the maximum value of unity at the corresponding interpolation node, in agreement with inequality (3.3.2).

More generally, it can be shown that the $2m$-degree polynomial

$$G_{2m}(\xi) = \sum_{i=1}^{m+1} \psi_i^2(\xi) \qquad (3.3.4)$$

reaches the maximum value of unity only at the spectral nodes themselves, as discussed in Section D.5, Appendix D. The existence of this upper bound guarantees

$\text{Lo}_0(\xi) = 1$

$\text{Lo}_1(\xi) = 3\,\xi$

$\text{Lo}_2(\xi) = \frac{3}{2}\left(5\,\xi^2 - 1\right)$

$\text{Lo}_3(\xi) = \frac{5}{2}\left(7\,\xi^2 - 3\right)\xi$

$\text{Lo}_4(\xi) = \frac{15}{8}\left(21\,\xi^4 - 14\,\xi^2 + 1\right)$

$\text{Lo}_5(\xi) = \frac{1}{8}\left(693\,\xi^4 - 630\,\xi^2 + 105\right)\xi$

$\text{Lo}_6(\xi) = \frac{1}{16}\left(3003\,\xi^6 - 3465\,\xi^4 + 945\,\xi^2 - 35\right)$

$\cdots$

$$\text{Lo}_i(\xi) = \frac{1}{2^{i+1}(i+1)!}\frac{\mathrm{d}^{i+2}}{\mathrm{d}\xi^{i+2}}(\xi^2 - 1)^{i+1} = \frac{i\,(2(i+1))!}{2^{i+1}(i+1)!}\,\xi^i + \cdots$$

TABLE 3.3.1   The first few members of the triangular family of Lobatto polynomials defined in the interval $-1 \leq \xi \leq 1$. In the generating formula shown in the last entry, $i! = 1 \cdot 2 \cdot 3 \cdots i$, is the factorial. The Lobatto polynomials, Lo, are related to the Legendre polynomials, L, by $\text{Lo}_i(\xi) = L'_{i+1}(\xi)$, where a prime denotes a derivative.

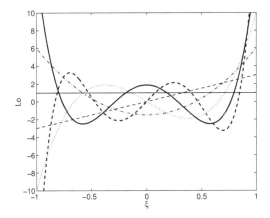

FIGURE 3.3.1   Graphs of the first six Lobatto polynomials, $\text{Lo}_m(\xi)$, in their native domain of definition, $[-1, 1]$ for $m = 0$ (solid line), $m = 1$ (dashed line), $m = 2$ (dash-dotted line), $m = 3$ (dotted line), $m = 4$ (heavy solid line), and $m = 5$ (heavy dashed line).

(a)   (b)

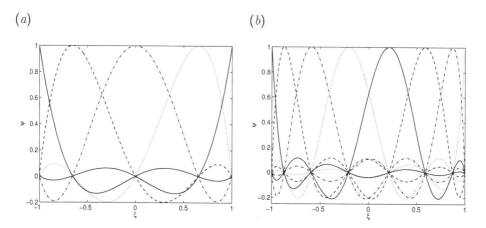

FIGURE 3.3.2   Graphs of the (a) 5-node interpolation functions corresponding to $m =$ 4, and (b) 8-node interpolation functions corresponding to $m = 7$ for the Lobatto nodal base.

that the magnitude of the global interpolation functions will not increase as the interpolation order increases, and thus guarantees rapid convergence. Because of this property, Runge oscillations are suppressed and the rate of convergence of the interpolation with respect to the interpolation order m is spectral, that is, it is faster than any power of $1/m$ for large m.

Lobatto polynomial zeros

Numerical values of the zeros of the first few Lobatto polynomials are shown in the second column of Table 3.3.2. The third column displays the corresponding integration weights used in the Lobatto integration quadrature discussed in Section 3.4.2 and in Section D.3, Appendix D. The FSELIB function *lobatto*, listed in Table 3.3.3, assigns numerical values to the roots of the Lobatto polynomials and evaluates the corresponding weights from a lookup table.

To compute the roots of $\mathrm{Lo}_4(\xi)$ using MATLAB, we may define the polynomial coefficient vector by issuing the statement:

```
>> c = [21, 0, -14, 0, 1]
```

where >> is the standard MATLAB prompt, and then invoke the MATLAB root finder function by typing:

```
>> roots(c)
```

MATLAB responds by printing the values:

	Zeros of Lo_i (Lobatto quadrature base points)	Integration weights
$i=1$	$t_1=0$	$\widehat{w}_1=\frac{4}{3}$
$i=2$	$t_1=-1/\sqrt{5}$ $t_2=-t_1$	$\widehat{w}_1=\frac{5}{6}$ $\widehat{w}_2=\widehat{w}_1$
$i=3$	$t_1=-\sqrt{3/7}$ $t_2=0.0$ $t_3=-t_1$	$\widehat{w}_1=49/60$ $\widehat{w}_2=32/45$ $\widehat{w}_3=\widehat{w}_1$
$i=4$	$t_1=-0.76505532392946$ $t_2=-0.28523151648064$ $t_3=-t_2$ $t_4=-t_1$	$\widehat{w}_1=0.37847495629785$ $\widehat{w}_2=0.55485837703549$ $\widehat{w}_3=\widehat{w}_2$ $\widehat{w}_4=\widehat{w}_1$
$i=5$	$t_1=-0.83022389627857$ $t_2=-0.46884879347071$ $t_3=0.0$ $t_4=-t_2$ $t_5=-t_1$	$\widehat{w}_1=0.27682604736157$ $\widehat{w}_2=0.43174538120986$ $\widehat{w}_3=0.48761904761905$ $\widehat{w}_4=\widehat{w}_2$ $\widehat{w}_5=\widehat{w}_1$
$i=6$	$t_1=-0.87174014850961$ $t_2=-0.59170018143314$ $t_3=-0.20929921790248$ $t_4=-t_3$ $t_5=-t_2$ $t_6=-t_1$	$\widehat{w}_1=0.21070422714350$ $\widehat{w}_2=0.34112269248350$ $\widehat{w}_3=0.41245879465870$ $\widehat{w}_4=\widehat{w}_3$ $\widehat{w}_5=\widehat{w}_2$ $\widehat{w}_6=\widehat{w}_1$

TABLE 3.3.2 Zeros of the first few Lobatto polynomials, Lo_i, and corresponding weights for the $(i+1)$-point Lobatto integration quadrature. The hat over the weights indicates that the node labels are shifted by one count when the Lobatto quadrature is applied.

```
function [Z, W] = lobatto(i)

%=================================================
% Zeros of the ith-degree Lobatto polynomial
% and corresponding weights for the Lobatto
% integration quadrature
%
% This table contains values for i = 1,2,..., 6
% The default value is i=6
%=================================================

%------
% check
%------

if(i>6)
   disp('-->'); disp(' lobatto: chosen Lobatto order');
   disp('   is not available; Will take i=6');
   i=6;
end

%-------
if(i==1)
%-------

Z(1) = 0.0; W(1) = 4.0/3.0;

%-----------
else if(i==2)
%-----------

Z(1) = -1.0D0/sqrt(5.0); Z(2) = -Z(1);
W(1) = 5.0/6.0; W(2) = W(1);

%------------
else if(i==3)
%------------

Z(1) = -sqrt(3.0/7.0); Z(2) = 0.0; Z(3) = -Z(1);
W(1) = 49.0/90.0; W(2) = 32.0/45.0; W(3) = W(1);

%------------
else if(i==4)
%------------

Z(1) = -0.76505532392946; Z(2) = -0.28523151648064;
Z(3) = -Z(2); Z(4) = -Z(1);
W(1) = 0.37847495629785; W(2) = 0.55485837703549;
W(3) = W(2); W(4) = W(1);

%------------
else if(i==5)
%------------
```

TABLE 3.3.3   Function *lobatto* (Continuing →)

```
Z(1) = -0.83022389627857; Z(2) = -0.46884879347071;
Z(3) = 0.0; Z(4) = -Z(2); Z(5) = -Z(1);
W(1) = 0.27682604736157; W(2) = 0.43174538120986;
W(3) = 0.48761904761905;
W(4) = W(2); W(5) = W(1);

%------------
else if(i==6)
%------------

Z(1) = -0.87174014850961; Z(2) = -0.59170018143314;
Z(3) = -0.20929921790248;
Z(4) = -Z(3); Z(5) = -Z(2); Z(6) = -Z(1);
W(1) = 0.21070422714350; W(2) = 0.34112269248350;
W(3) = 0.41245879465870;
W(4) = W(3); W(5) = W(2); W(6) = W(1);

%--
end
%--

%-----
% done
%-----

return;
```

TABLE 3.3.3 ($\rightarrow$ Continued) Function *lobatto* evaluates the zeros of the Lobatto polynomials and corresponding integration weights of the Lobatto integration quadrature.

```
                ans =
                    0.7651
                   -0.7651
                    0.2852
                   -0.2852
```

which are consistent with those listed in Table 3.3.2 for $i = 4$.

Lobatto interpolation nodes

With the Lobatto interpolation nodal base, the $m + 1$ element interpolation nodes are distributed at the scaled positions

$$\xi_1 = -1, \quad \xi_2 = t_1, \quad \xi_2 = t_2, \quad \ldots, \quad \xi_m = t_{m-1}, \quad \xi_{m+1} = 1, \quad (3.3.5)$$

where $t_1, t_2, \ldots, t_{m-1}$ are the zeros of the $(m - 1)$-degree Lobatto polynomial. The $m + 1$ interpolation nodes are thus the roots of the $(m+1)$-degree *completed Lobatto polynomial*, denoted by a hat and defined as

$$\widehat{\mathrm{Lo}}_{m+1}(t) \equiv (1 - t^2)\, \mathrm{Lo}_{m-1}(t). \tag{3.3.6}$$

The orthogonality property (3.3.1) ensures that

$$\int_{-1}^{1} \mathrm{Lo}_i(t)\, \widehat{\mathrm{Lo}}_j(t)\, \mathrm{d}t = \frac{2\,(i+1)(i+2)}{2i+3}\, \delta_{ij}, \tag{3.3.7}$$

where δ_{ij} is Kronecker's delta representing the identity matrix.

For example, having chosen $m = 1$, we place the interpolation nodes at

$$\xi_1 = -1, \qquad \xi_2 = 1 \tag{3.3.8}$$

and obtain the linear expansion discussed in Section 1.1.

Stepping up to $m = 2$, we note that $t_1 = 0$ is the single zero of $\mathrm{Lo}_1(\xi)$ and place the interpolation nodes at

$$\xi_1 = -1, \qquad \xi_2 = 0, \qquad \xi_3 = 1 \tag{3.3.9}$$

to obtain the quadratic expansion discussed in Section 1.3, where the interior element node lies at the element mid-point.

Stepping up once more to $m = 3$, we note that $t_1 = -1/\sqrt{5}$ and $t_2 = 1/\sqrt{5}$ are the two zeros of $\mathrm{Lo}_2(\xi)$ and place the interpolation nodes at the positions

$$\xi_1 = -1, \qquad \xi_2 = -\frac{1}{\sqrt{5}}, \qquad \xi_3 = \frac{1}{\sqrt{5}}, \qquad \xi_4 = 1, \tag{3.3.10}$$

to obtain a cubic expansion.

Runge function

To demonstrate the advantage of the Lobatto node distribution, we approximate the Runge function defined in (3.1.8) with an mth-degree interpolating polynomial that passes through $m + 1$ data with abscissas at the points ξ_i for $i = 1, \dots, m+1$, subject to the restriction that $\xi_1 = -1$ and $\xi_{m+1} = 1$. Graphs of the interpolating polynomials based on the Lobatto nodes generated by the FSELIB function *lagrange*, not listed in the text, are shown in Figure 3.3.3(*a*). As the number of interpolation base points and thus the polynomial order increases, the interpolating polynomial converges uniformly throughout the interpolation domain.

A further assessment of the interpolation accuracy is made in Figure 3.3.3(*b*) for the interpolated function

$$f(\xi) = \frac{1}{1 + 4\,\xi^2} \tag{3.3.11}$$

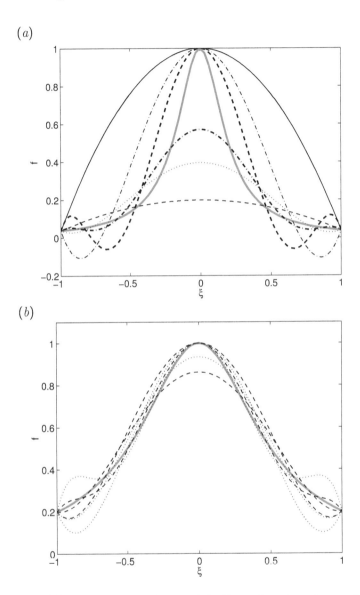

FIGURE 3.3.3   (*a*) Lagrange interpolation of the Runge function, drawn with the heavy
    solid line. Interpolating polynomials with Lobatto nodes are plotted for $m = 2$
    (solid line), $m = 3$ (dashed line), $m = 4$ (dash-dotted line), $m = 5$ (dotted line),
    $m = 6$ (heavy dashed line), and $m = 7$ (heavy dash-dotted line). (*b*) Comparison
    of the interpolating polynomials for the scaled Runge function $f(\xi) = 1/(1+4\xi^2)$
    with evenly spaced or Lobatto nodes. The dotted lines correspond to evenly spaced
    nodes and the dashed lines correspond to Lobatto interpolation nodes, in both
    cases for $m = 4, 5$, and 6.

FIGURE 3.3.4 Spectral node distribution on two elements generated by the
FSELIB code *discr_lob*. The circles mark the position of the shared element end-
nodes and the diamonds mark the position of the Lobatto interpolation interior
nodes.

in the interval $[-1, 1]$, which consists of a smaller portion of the Runge function
defined in (3.1.8). The graphs were generated by the FSELIB function *lagrange1*,
not listed in the text. In this case, the sixth-degree interpolation polynomial based
on the Lobatto nodes, drawn with a dashed line, faithfully approximates the interpolated function. In contrast, the interpolation polynomial based on evenly spaced
nodes, drawn with a dotted line, exhibits significant deviations at the ends of the
interpolation domain.

3.3.2 Discretization code

The FSELIB function *discr_lob*, listed in Table 3.3.4, defines the element end-points,
generates the Lobatto interpolation nodes, and returns the global nodes and connectivity matrix according to the algorithm shown in Table 3.1.1. The driver script
discr_lob_dr, listed in Table 3.3.5, calls *discr_lob* to display the generated node distribution. Results for a two-element arrangement, $N_E = 2$, and polynomial expansion
order $m = 6$ for the first element and $m = 4$ for the second element are shown in
Figure 3.3.4.

3.3.3 Legendre polynomials

The ith-degree Lobatto polynomial, Lo_i, derives from the $(i+1)$-degree Legendre
polynomial, $\mathrm{L}_{i+1}(\xi)$, as

$$\mathrm{Lo}_i(\xi) = \mathrm{L}'_{i+1}(\xi), \qquad (3.3.12)$$

where a prime denotes a derivative with respect to ξ, as discussed in Section B.6,
Appendix B. By construction, the ith-degree Legendre polynomial, $\mathrm{L}_i(\xi)$, satisfies
the second-order differential equation

$$\left[(\xi^2 - 1)\,\mathrm{L}'_i(\xi) \right]' = \left[(\xi^2 - 1)\,\mathrm{Lo}_{i-1}(\xi) \right]' = i\,(i+1)\,\mathrm{L}_i(\xi). \qquad (3.3.13)$$

An arbitrary pair of Legendre polynomials with degrees i and j satisfy the
orthogonality property

$$\int_{-1}^{1} \mathrm{L}_i(\xi)\,\mathrm{L}_j(\xi)\,\mathrm{d}\xi = \frac{2}{2i+1}\,\delta_{ij}, \qquad (3.3.14)$$

```
function [xe,xen,xien,xg,c,ng] = discr_lob (x1,x2,ne,ratio,np)

%========================================
% Discretize the interval (x1, x2) into "ne" elements
% and generate the np-order Lobatto element interpolation nodes
%
% xe:   element end-nodes
% xen:  element interpolation nodes
% xien: xi element interpolation nodes
% xg:   unique global nodes
% c:    connectivity matrix
% ng:   number of unique global nodes
%========================================

%-----------------------------
% define the element end-nodes
%-----------------------------

xe = elm_line1 (x1,x2,ne,ratio);

%----------------------------------------------
% define the element interpolation nodes (xen)
%----------------------------------------------

for l=1:ne

  m = np(l);

  xi(1) = -1.0;
  if(m>1)
   [tL, wL] = lobatto(m-1);
   for j=2:m
    xi(j) = tL(j-1);
   end
  end
  xi(m+1) = 1.0;

  mx = 0.5*(xe(l+1)+xe(l));
  dx = 0.5*(xe(l+1)-xe(l));

  for j=1:m+1
    xen(l,j) = mx + xi(j)*dx;
  end

  for j=1:m+1
    xien(l,j) = xi(j);
  end

end

%-------------------------------
% define the connectivity matrix
% and the global nodes
%-------------------------------
```

TABLE 3.3.4 Function *discr_lob* (Continuing →)

```
Ic = 2;      % global node counter

for l=1:ne
  Ic = Ic-1;
  for j=1:np(l)+1
    c(l,j) = Ic;
    xg(Ic) = xen(l,j);
    Ic = Ic+1;
  end
end

%----------------------
% total number of nodes
%----------------------

ng = Ic-1;

%-----
% done
%-----

return;
```

TABLE 3.3.4 ($\rightarrow$ Continued) Function *discr_lob* discretizes the solution domain into N_E elements with Lobatto element nodes. This function also counts the unique global nodes and generates the connectivity matrix.

where δ_{ij} is Kronecker's delta. The first 11 Legendre polynomials are shown in Table 3.3.6 along with a generating formula. Graphs of the Legendre polynomials generated using the FSELIB function *lagrange*, not listed in the text, are shown in Figure 3.3.5.

The Legendre polynomials can be computed using the recursion relation

$$L_{i+1}(\xi) = \frac{2i+1}{i+1}\,\xi\,L_i(\xi) - \frac{i}{i+1}\,L_{i-1}(\xi). \qquad (3.3.15)$$

The derivatives of the Legendre polynomials, and thus the Lobatto polynomials themselves, can be computed from the recursion relation

$$(1-\xi^2)\,L'_{i+1}(\xi) = (1-\xi^2)\,Lo_i(\xi) = (i+1)\left(L_i(\xi) - \xi\,L_{i+1}(\xi)\right). \qquad (3.3.16)$$

Replacing i with $i+1$ in (3.3.15) and rearranging, we find that

$$\xi\,L_{i+1}(\xi) = \frac{i+2}{2i+3}\,L_{i+2}(\xi) + \frac{i+1}{2i+3}\,L_i(\xi). \qquad (3.3.17)$$

Substituting this expression into the right-hand side of (3.3.16) and simplifying, we obtain an expression for the derivative,

$$(1-\xi^2)\,L'_{i+1}(\xi) = (1-\xi^2)\,Lo_i(\xi) = \frac{(i+1)(i+2)}{2i+3}\left(L_i(\xi) - L_{i+2}(\xi)\right). \qquad (3.3.18)$$

```
%======================================================================
% Code discr_lob_dr
%
% Driver script for discretizing the interval (0, L) into "ne"
% elements, and defining the Lobatto interpolation nodes
%======================================================================

%-----------
% input data
%-----------

L = 1.0;     % discretization length
ne = 2; ratio = 0.5;  % number of elements, stretch ratio
np(1) = 6; np(2) = 4;  % element polynomial orders

%-----------
% discretize
%-----------

[xe,xen,xien,xg,c,ng] = discr_lob (0,L,ne,ratio,np);

%-----
% plot
%-----

figure(1)
hold on;
yg = zeros(ng,1); plot (xg,yg,'-d');   % global nodes
ye = zeros(ne+1,1); plot (xe,ye,'o')   % end nodes
axis equal; axis ([ 0 L -0.10*L 0.10*L])

%-----
% done
%-----
```

TABLE 3.3.5   Driver script *discr_lob_dr* for discretizing the solution domain into N_E elements and defining the Lobatto interpolation nodes.

Finally, we use (3.3.13) and (3.3.18) and find that the completed Lobatto polynomials can be expressed in terms of the Legendre polynomials, L_i, as

$$\widehat{\text{Lo}}_{i+1}(\xi) \equiv (1 - \xi^2)\,\text{Lo}_{i-1}(\xi) = -i\,(i+1) \int_{-1}^{\xi} L_i(\xi')\,\mathrm{d}\xi'. \tag{3.3.19}$$

Performing the integration, we obtain

$$\widehat{\text{Lo}}_{i+1}(\xi) = \frac{i(i+1)}{2\,i+1}\left(L_{i-1}(\xi) - L_{i+1}(\xi) \right). \tag{3.3.20}$$

The relations presented in this section find applications in the computation of the element diffusion and mass matrices discussed in Section 3.4.

Legendre polynomials

$L_0(\xi) = 1$

$L_1(\xi) = \xi$

$L_2(\xi) = \frac{1}{2}(3\xi^2 - 1)$

$L_3(\xi) = \frac{1}{2}(5\xi^2 - 3)\xi$

$L_4(\xi) = \frac{1}{8}(35\xi^4 - 30\xi^2 + 3)$

$L_5(\xi) = \frac{1}{8}(63\xi^4 - 70\xi^2 + 15)\xi$

$L_6(\xi) = \frac{1}{16}(231\xi^6 - 315\xi^4 + 105\xi^2 - 5)$

$L_7(\xi) = \frac{1}{16}(429\xi^6 - 693\xi^4 + 315\xi^2 - 35)\xi$

$L_8(\xi) = \frac{1}{128}(6435\xi^8 - 12012\xi^6 + 6930\xi^4 - 1260\xi^2 + 35)$

$L_9(\xi) = \frac{1}{128}(12155\xi^8 - 25740\xi^6 + 18019\xi^4 - 4620\xi^2 + 315)\xi$

$L_{10}(\xi) = \frac{1}{256}(46189\xi^{10} - 109395\xi^8 + 90090\xi^6 - 30030\xi^4 + 3465\xi^2 - 63)$

$\cdots$

$L_i(\xi) = \frac{1}{2^i i!}\frac{d^i}{d\xi^i}(\xi^2 - 1)^i = \frac{(2i)!}{2^i i!}\xi^i + \cdots$

TABLE 3.3.6 The first 11 Legendre polynomials defined in the interval $-1 \leq \xi \leq 1$. In the generating formula shown in the last entry, $i! = 1 \cdot 2 \cdots i$, is the factorial. The Lobatto polynomials are the derivatives of the Legendre polynomials, $Lo_i(\xi) = L'_{i+1}(\xi)$, where a prime denotes a derivative.

3.3.4   Chebyshev second-kind nodal base

An alternative to the Lobatto node distribution is the Chebyshev second-kind distribution associated on the Chebyshev polynomials of the second kind, $\mathcal{T}_m(t)$, discussed in Appendix B. With this choice, the element interpolation nodes are distributed at the zeros of the $(m+1)$-degree *completed Chebyshev polynomial of the second kind*, denoted by a hat and defined as

$$\widehat{\mathcal{T}}_{m+1}(t) \equiv (1 - t^2)\,\mathcal{T}_{m-1}(t). \tag{3.3.21}$$

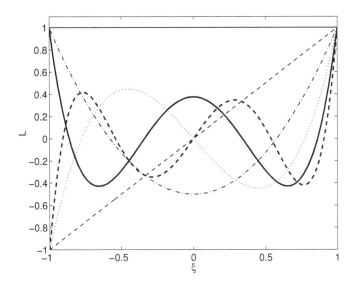

FIGURE 3.3.5 Graphs of the first six Legendre polynomials, $L_m(\xi)$, in their native domain of definition, $[-1, 1]$ for $m = 0$ (solid line), $m = 1$ (dashed line), $m = 2$ (dash-dotted line), $m = 3$ (dotted line), $m = 4$ (heavy solid line), and $m = 5$ (heavy dashed line). Note that the Legendre polynomials vary in the range $[-1, 1]$, while the Lobatto polynomials, plotted in Figure 3.3.1, vary over a wider range.

The nodal positions arise by projecting $m + 1$ evenly spaced nodes around the upper or lower half of the unit circle onto the horizontal ξ axis passing through the circle center, yielding

$$\xi_i = \cos\left(\frac{i-1}{m}\pi\right) \tag{3.3.22}$$

for $i = 1, \ldots, m + 1$.

Interpolating polynomials of several degrees for the narrow Runge function, $f(\xi) = 1/(1 + 4\xi^2)$, are compared in Figure 3.3.6 for the Lobatto and second-kind Chebyshev nodal sets. The quadratic polynomials for $m = 2$ are identical. Although higher-order polynomials differ, the differences are small. The second-kind Chebyshev nodal basis is favored in spectral methods for viscous flow (e.g., Peyret [44]).

PROBLEMS

3.3.1 *Polynomial root finder*

Use the MATLAB function *roots* to compute the zeros of the Lobatto polynomial $Lo_5(\xi)$. Confirm that the results with those listed in Table 3.3.2.

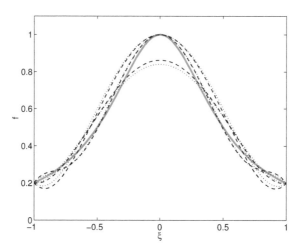

FIGURE 3.3.6 Interpolating polynomials for the scaled Runge function defined as
$f(\xi) = 1/(1 + 4\,\xi^2)$. The dotted lines are for Chebyshev interpolation nodes
and the dashed lines are for Lobatto interpolation nodes, in both cases for $m = 4$,
5, and 6.

3.3.2 *Lobatto interpolation base*

(a) Prepare graphs of the mth-degree interpolating polynomials for the function
$f(\xi) = 1/(1 + \xi^2)$ in the interval $[-1, 1]$, based on the Lobatto nodal base with
$m = 2, 3, 4$, and 5. Discuss the behavior of the interpolation error.

(b) Repeat part (a) for the second-kind Chebyshev nodal base.

3.4   Lobatto interpolation and element matrices

In the remainder of this chapter, we confine our attention to the Lobatto nodal set
introduced in Section 3.3.1. The Lagrange generating polynomial defined in (3.1.5)
takes the form

$$\Phi_{m+1}(\xi) = \frac{1}{c_{m-1}}\,(\xi^2 - 1)\,\mathrm{Lo}_{m-1}(\xi), \qquad (3.4.1)$$

where c_{m-1} is the coefficient of the highest power of the $(m-1)$-degree Lobatto
polynomial; for example, $c_3 = 3$.

Substituting (3.4.1) into the right-hand side of (3.1.4), we obtain the lth-element
interpolation functions

$$\psi_i(\xi) = \frac{1}{[(\xi^2 - 1)\,\mathrm{Lo}_{m-1}(\xi)]'_{\xi=\xi_i}}\,\frac{\xi^2 - 1}{\xi - \xi_i}\,\mathrm{Lo}_{m-1}(\xi), \qquad (3.4.2)$$

which are the Lagrange interpolating polynomials corresponding to the Lobatto nodes, complemented by the end-nodes, $\xi = \pm 1$. A more explicit expression arises by using property (3.3.13) to simplify the denominator of the first fraction on the right-hand side of (3.4.2), obtaining

$$\psi_i(\xi) = \frac{1}{m\,(m+1)\,L_m(\xi_i)} \frac{\xi^2 - 1}{\xi - \xi_i} \operatorname{Lo}_{m-1}(\xi). \tag{3.4.3}$$

The cardinal interpolation property $\psi_i(\xi_i) = 1$ requires that

$$\lim_{\xi \to \xi_i} \left[\frac{(\xi^2 - 1)\operatorname{Lo}_{m-1}(\xi)}{\xi - \xi_i} \right] = m\,(m+1)\,L_m(\xi_i). \tag{3.4.4}$$

As an example, we recall the mandatory choices $\xi_1 = -1$ and $\xi_{m+1} = 1$, and use (3.4.3) to compute the first and last interpolation functions

$$\psi_1(\xi) = \frac{1}{m\,(m+1)\,L_m(\xi = -1)}\,(\xi - 1)\operatorname{Lo}_{m-1}(\xi) \tag{3.4.5}$$

and

$$\psi_{m+1}(\xi) = \frac{1}{m\,(m+1)\,L_m(\xi = 1)}\,(\xi + 1)\operatorname{Lo}_{m-1}(\xi). \tag{3.4.6}$$

Choosing $m = 1$, and recalling that the first-degree Legendre polynomial is a linear function, $L_1(\xi) = \xi$, and the zeroth-degree Lobatto polynomial is equal to unity, $\operatorname{Lo}_0(\xi) = 1$, we obtain

$$\psi_1(\xi) = \frac{1}{2\,L_1(\xi = -1)}\,(\xi - 1)\operatorname{Lo}_0(\xi) = \frac{1}{2}\,(1 - \xi) \tag{3.4.7}$$

and

$$\psi_2(\xi) = \frac{1}{2\,\operatorname{Lo}_0(\xi = 1)}\,(1 + \xi)\operatorname{Lo}_0(\xi) = \frac{1}{2}\,(1 + \xi), \tag{3.4.8}$$

which are precisely the cardinal interpolation functions of the linear elements associated with the first and second element nodes.

3.4.1 Lobatto mass matrix

For convenience, we recast the expression for the element mass matrix given in (3.1.11) into the form

$$B_{ij}^{(l)} = \frac{1}{2}\,h_l\,\Theta_{ij} \tag{3.4.9}$$

for $i, j = 1, \ldots, m+1$, where

$$\Theta_{ij} \equiv \int_{-1}^{1} \psi_i(\xi)\,\psi_j(\xi)\,\mathrm{d}\xi \tag{3.4.10}$$

is a dimensionless mass matrix. Using expression (3.4.3) for the node interpolation functions, we find that

$$\Theta_{ij} \equiv \frac{1}{m^2 \, (m+1)^2 \, \mathrm{L}_m(\xi_j) \, \mathrm{L}_m(\xi_i)} \int_{-1}^{1} \mathcal{P}_{ij}(\xi) \, d\xi, \qquad (3.4.11)$$

where

$$\mathcal{P}_{ij}(\xi) \equiv \frac{(1-\xi)^2(1+\xi)^2}{(\xi - \xi_i)(\xi - \xi_j)} \, \mathrm{Lo}_{m-1}^2(\xi) = \frac{(\xi^2 - 1)^2}{(\xi - \xi_i)(\xi - \xi_j)} \, \mathrm{Lo}_{m-1}^2(\xi) \qquad (3.4.12)$$

is a $2m$-degree polynomial.

As ξ tends to -1, the numerator of the fraction in the last expression of (3.4.12) tends to zero quadratically and the fraction tends to zero, provided that $i \neq 1$ and $j \neq 1$. When $i = 1$ and $j = 1$, the fraction tends to the value of four. Similarly, as ξ tends to 1, the fraction tends to zero, provided that $i \neq m+1$ and $j \neq m+1$. When $i = m+1$ and $j = m+1$, the fraction tends to the value of four. Because of these properties, the matrices $\mathcal{P}_{ij}(\xi = \pm 1)$ are null, except for the first element, $\mathcal{P}_{1,1}(\xi = -1)$ and last element, $\mathcal{P}_{m+1,m+1}(\xi = 1)$. In compact notation,

$$
\begin{aligned}
\mathcal{P}_{ij}(\xi = -1) &= 4 \, \delta_{i1} \, \delta_{j1} \, \mathrm{Lo}_{m-1}^2(\xi = -1), \\
\mathcal{P}_{ij}(\xi = 1) &= 4 \, \delta_{i,m+1} \, \delta_{j,m+1} \, \mathrm{Lo}_{m-1}^2(\xi = 1),
\end{aligned}
\qquad (3.4.13)
$$

where δ_{ij} is Kronecker's delta.

3.4.2 Lobatto integration quadrature

In practice, the integral on the right-hand side of (3.4.11) is computed by numerical methods. The $(k+1)$-point Lobatto integration quadrature, discussed in Section B.3, Appendix B, provides us with the approximation

$$\int_{-1}^{1} f(\xi) \, d\xi \simeq \sum_{p=1}^{k+1} f(\xi = z_p) \, w_p, \qquad (3.4.14)$$

for a smooth function $f(\xi)$, where $k \geq 1$ is the selected quadrature order, z_p are quadrature base points, and w_p are the associated integration weights defined as follows:

- $z_1 = -1$ and $z_{k+1} = 1$.

- z_p for $p = 2, \ldots, k$ are the zeros of the $(k-1)$-degree Lobatto polynomial, Lo_{k-1}, denoted by t_i. Thus, $z_p = t_{p-1}$, which means that

$$z_2 = t_1, \qquad z_3 = t_2, \qquad \ldots, \qquad z_k = t_{k-1}. \qquad (3.4.15)$$

The integration weights are given by

$$w_1 = w_{k+1} = \frac{2}{k(k+1)}, \qquad w_p = \frac{2}{k(k+1)} \frac{1}{L_k^2(z_p)} \qquad (3.4.16)$$

for $p = 2, \ldots, k$, where L_k is a Legendre polynomial. Numerical values for the weights at the interior base points are given in the third column of Table 3.3.2, subject to the index shifting convention $w_p = \widehat{w}_{p-1}$ for $p = 2, \ldots, k$.

If the function $f(\xi)$ is a $(2k-1)$-degree polynomial, the numerical result obtained by applying the Lobatto quadrature is exact , as shown in Section B.3, Appendix B. Thus, if $k = 5$, the quadrature integrates exactly ninth-degree polynomials, even though it only employs six base points.

FSELIB function *int_lob*, listed in Table 3.4.1, implements the quadrature for an arbitrary integrand defined in the function. The integration base points and weights are imported from the FSELIB function *lobatto*, listed in Table 3.3.3. Executing the function *int_lob* with $k = 6$ to integrate the power function $f(\xi) = \xi^{10}$ returns the exact answer, in agreement with the theoretical predictions.

3.4.3 Computation of the Lobatto mass matrix

Applying the Lobatto quadrature to compute the integral on the right-hand side of (3.4.11), we obtain

$$\int_{-1}^{1} \mathcal{P}_{ij}(\xi) \, d\xi \simeq \sum_{p=1}^{k+1} \mathcal{P}_{ij}(\xi = z_p) \, w_p. \qquad (3.4.17)$$

Taking into consideration (3.4.13), we find that

$$\int_{-1}^{1} \mathcal{P}_{ij}(\xi) \, d\xi = \delta_{ij} \, \delta_{i1} \, 4 \, \mathrm{Lo}_{m-1}^2(\xi = -1) \, w_1 \qquad (3.4.18)$$

$$+ \sum_{p=2}^{k} \mathcal{P}_{ij}(\xi = z_p) \, w_p + \delta_{ij} \, \delta_{i(m+1)} \, , 4 \, \mathrm{Lo}_{m-1}^2(\xi = 1) \, w_{k+1}.$$

Since the integrand on the left-hand side is a $2m$-degree polynomial in ξ, the integration will be *exact* if $2m \leq 2k - 1$, which is true if

$$k > m, \qquad (3.4.19)$$

that is, if the order of the Lobatto quadrature is higher than the size of the nodal interpolation set by at least one unit.

FSELIB code *emm_lob*, listed in Table 3.4.2, computes the dimensionless mass matrix Θ introduced in (3.4.9) based on the Lobatto quadrature. Numerical values are presented in Table 3.4.3 for several polynomial orders, m. Note that the sum

```
function integral = int_lob(k)

%=======================================
% Numerical integration of a function
% in the interval [-1, 1]
% using the Lobatto quadrature
%=======================================

%--------------------------------------------------
% define the integration base points and weights
%--------------------------------------------------

z(1) = -1.0; z(k+1) = 1.0;

w(1) = 2/(k*(k+1)); w(k+1) = w(1);

if(k>1)

  [zL, wL] = lobatto(k-1);

  for i=2:k
   z(i) = zL(i-1);
   w(i) = wL(i-1);
   end

end

%-----------------------------
% generate the function values
%-----------------------------

for i=1:k+1
   xi = z(i);
   f(i) = xi^2;     % example, can also be generated by a function
end

%-----------------------
% perform the quadrature
%-----------------------

integral = 0.0;

for i=1:k+1
   integral = integral + f(i)*w(i);
end

%-----
% done
%-----

return
```

TABLE 3.4.1 Function *int_lob* implements the Lobatto integration quadrature expressed by the integration rule (3.4.14).

of all elements of Θ is equal to two for any value of m. The top and tail end of the FSELIB function *emm* that evaluates this matrix from a table is listed in Table 3.4.4. The complete function is included in FSELIB. This function will be used in Section 3.5 for building a spectral element code.

Karniadakis and Sherwin ([32], p. 44) present graphs of the condition number of the mass matrix for the Lobatto spectral nodes and for evenly spaced nodes. Their data confirm that the mass matrix for evenly spaced nodes is significantly worse conditioned for polynomial orders higher than five. A high condition number carries an increased risk of numerical instability and failure due to the growth of the computer round-off error.

Inexact integration and mass lumping

The inexact choice, $k = m$, has the practical advantage that $\mathcal{P}_{ij}(\xi = z_p) \neq 0$ only if $i = j = p$. The integration quadrature (3.4.18) simplifies to

$$\int_{-1}^{1} \mathcal{P}_{ij}(\xi)\,d\xi \simeq \delta_{ij}\left[\delta_{i1}\,4\,\mathrm{Lo}_{m-1}^2(\xi = -1)\,w_1\right.$$

$$\left. +\delta_{ip}\,\alpha_p\,w_p + \delta_{i(m+1)}\,4\,\mathrm{Lo}_{m-1}^2(\xi = 1)\,w_{m+1}\right] \tag{3.4.20}$$

for $p = 2, \ldots, m$. Using (3.4.4), we find that

$$\alpha_p \equiv \mathcal{P}_{ij}(\xi = z_p) = m^2(m+1)^2\,\mathrm{L}_m^2(z_p) \tag{3.4.21}$$

for $p = 2, \ldots, m$. We may now define

$$\alpha_1 \equiv 4\,\mathrm{Lo}_{m-1}^2(\xi = -1), \qquad \alpha_{m+1} \equiv 4\,\mathrm{Lo}_{m-1}^2(\xi = 1), \tag{3.4.22}$$

and recast (3.4.20) into the compact form

$$\int_{-1}^{1} \mathcal{P}_{ij}(\xi)\,d\xi \simeq \delta_{ij}\sum_{p=1}^{m+1}\delta_{ip}\,\alpha_p\,w_p. \tag{3.4.23}$$

We have found that inexact integration effectively diagonalizes the mass matrix by implementing mass lumping for any interpolation order, m.

Diagonalization of the mass matrix is especially desirable in time-dependent problems where the finite element method culminates in a system of ordinary differential equations for the nodal values, as will be discussed in Section 3.9. In these cases, forward time discretization leads to an explicit method that circumvents the need for solving a system of linear equations at each time step. However, because of numerical stability concerns, the explicit method is of interest primarily in convection-dominated transport.

Substituting (3.4.23) into (3.4.11), we find that the first diagonal entry of the inexact dimensionless mass matrix is given by

$$\Theta_{11}^{\mathrm{inexact}} = \frac{8\,\mathrm{Lo}_{m-1}^2(\xi = -1)}{m^3\,(m+1)^3\,\mathrm{L}_m^2(\xi = -1)}, \tag{3.4.24}$$

```
%=====================================
%  Code emm_lob
%
%  Evaluation of the dimensionless element mass matrix
%  Theta using the lobatto quadrature
%
%  m: order of polynomial expansion
%  k: order of Lobatto quadrature
%
%  setting m=k implements mass lumping
%=====================================

%---------------------------
% Lobatto interpolation nodes
%---------------------------

m = input('Enter the order of the polynomial expansion, m: ');

%---------------------------
% generate the spectral nodes
%---------------------------

xi(1) = -1.0;

if(m>1)
  [tL, wL] =  lobatto(m-1);
  for i=2:m
   xi(i) = tL(i-1);
  end
end

xi(m+1) = 1.0;

%-----------------------------------------
% generate the Lobatto integration points
% and corresponding weights
%-----------------------------------------

k = input('Enter the order of the Lobatto quadrature, k: ');

z(1) = -1.0;
w(1) = 2.0/(k*(k+1));

if(k>1)
  [zL, wL] = lobatto(k-1);
  for i=2:k
    z(i) = zL(i-1);    w(i) = wL(i-1);
  end
end

z(k+1) = 1.0;
w(k+1) = w(1);
```

TABLE 3.4.2 Code *emm_lob* (Continuing →)

```
%---------------------------
% compute the mass matrix
% by the Lobatto quadrature
%---------------------------

for ind=1:m+1    % loop over interpolation nodes, closes at echidna

   for jnd=1:m+1    % loop over interpolation nodes

   % Generate the integrand function values
   % consisting of the product of the interpolation functions
   % for nodes ind and jnd

      for j=1:k+1

        f(j) = 1.0;

        for i=1:m+1
         if(i ~= ind)
           f(j) = f(j)*(z(j)-xi(i))/(xi(ind)-xi(i));
         end
        end

        for i=1:m+1
         if(i ~= jnd)
           f(j) = f(j)*(z(j)-xi(i))/(xi(jnd)-xi(i));
         end
        end

      end

   % perform the (k+1)-point Lobatto quadrature
   % for the computed function values

      msmat = 0.0;

      for i=1:k+1
        msmat = msmat + f(i)*w(i);
      end

      mm(ind,jnd) = msmat;

   end

end    % echidna

%-----
% done
%-----
```

TABLE 3.4.2 ($\rightarrow$ Continued) Code *emm_lob* evaluates the dimensionless mass matrix Θ defined in (3.4.10).

Exact → Lumped

$m = 1$: $\dfrac{1}{3}\begin{bmatrix} 2 & 1 \\ 1 & 2 \end{bmatrix} \rightarrow \begin{bmatrix} 1 & 0 \\ 0 & 1 \end{bmatrix}$

$m = 2$: $\dfrac{1}{15}\begin{bmatrix} 4 & 2 & -1 \\ 2 & 16 & 2 \\ -1 & 2 & 4 \end{bmatrix} \rightarrow \dfrac{1}{3}\begin{bmatrix} 1 & 0 & 0 \\ 0 & 4 & 0 \\ 0 & 0 & 1 \end{bmatrix}$

$m = 3$:
$$\begin{bmatrix} 0.14285714 & 0.05323971 & -0.05323971 & 0.02380952 \\ 0.05323971 & 0.71428572 & 0.11904762 & -0.05323971 \\ -0.05323971 & 0.11904762 & 0.71428572 & 0.05323971 \\ 0.02380952 & -0.05323971 & 0.05323971 & 0.14285714 \end{bmatrix} \rightarrow$$

$$\rightarrow \begin{bmatrix} 0.16666667 & 0.0 & 0.0 & 0.0 \\ 0.0 & 0.83333333 & 0.0 & 0.0 \\ 0.0 & 0.0 & 0.83333333 & 0.0 \\ 0.0 & 0.0 & 0.0 & 0.16666667 \end{bmatrix}$$

$m = 4$:

$$\begin{bmatrix} 0.08888888 & 0.02592591 & -0.02962963 & 0.02592593 & -0.01111111 \\ 0.02592591 & 0.48395066 & 0.06913582 & -0.06049384 & 0.02592593 \\ -0.02962963 & 0.06913582 & 0.63209873 & 0.06913582 & -0.02962963 \\ 0.02592593 & -0.06049384 & 0.06913582 & 0.48395066 & 0.02592591 \\ -0.01111111 & 0.02592593 & -0.02962963 & 0.02592591 & 0.08888888 \end{bmatrix} \rightarrow$$

$$\rightarrow \begin{bmatrix} 0.1 & 0.0 & 0.0 & 0.0 & 0.0 \\ 0.0 & 0.54444444 & 0.0 & 0.0 & 0.0 \\ 0.0 & 0.0 & 0.71111111 & 0.0 & 0.0 \\ 0.0 & 0.0 & 0.0 & 0.54444444 & 0.0 \\ 0.0 & 0.0 & 0.0 & 0.0 & 0.1 \end{bmatrix}$$

TABLE 3.4.3 Tables of the dimensionless mass matrix Θ, defined in (3.4.11), for polynomial orders $m = 1$, 2, 3, and 4. The arrows point to the diagonalized mass-lumped approximation. Complete high-accuracy tables are included in the FSELIB function *emm_lob_tab*.

```
function elm_mm = emm_lob_tbl(m)

%=====================================================
% element mass matrix for an mth-order polynomial
% expansion, and m = 1-6
%=====================================================

%-----
% trap
%-----

if(m>5)
  disp('');
  disp(' elmm: Chosen number of points is not available');
  break;
end

%-------
if(m==1)
%-------

elm_mm = [2/3 1/3; 1/3 2/3];

%-----------
elseif(m==2)
%-----------

elm_mm = [
  0.266666666667   0.133333333333 -0.066666666667;
  0.133333333333   1.066666666667   0.133333333333;
 -0.066666666667   0.133333333333   0.266666666667];

%-----------
elseif(m==3)
%-----------

elm_mm =[
  0.14285714   0.05323971 -0.05323971   0.02380952;
  0.05323971   0.71428571   0.11904761 -0.05323971;
 -0.05323971   0.11904761   0.71428571   0.05323971;
  0.02380952 -0.05323971   0.05323971   0.14285714];

......

%--
end
%--

return;
```

TABLE 3.4.4 Function *emm_lob_tbl* contains tables of the dimensionless element mass
matrix Θ. The six dots near the end of the listing indicate unprinted code. Higher
accuracy data are implemented in the function included in FSELIB.

the last diagonal entry is given by

$$\Theta^{\text{inexact}}_{(m+1)(m+1)} = \frac{8\,\text{Lo}^2_{m-1}(\xi = 1)}{m^3\,(m+1)^3\,\text{L}^2_m(\xi = 1)}, \tag{3.4.25}$$

and the intermediate diagonal entries are given by

$$\Theta^{\text{inexact}}_{ii} = w_{i-1}$$

for $i = 2, \ldots, m$.

For example, when $m = 1$, corresponding to linear interpolation, we obtain the diagonal elements

$$\Theta^{\text{inexact}}_{11} = \frac{8\,\text{Lo}^2_0(\xi = -1)}{1 \cdot 8\,\text{L}^2_1(\xi = -1)} = \frac{8 \cdot (1)^2}{1 \cdot 8 \cdot (-1)^2} = 1 \tag{3.4.26}$$

and

$$\Theta^{\text{inexact}}_{22} = \frac{8\,\text{Lo}^2_0(\xi = 1)}{1 \cdot 8\,\text{L}^2_1(\xi = 1)} = \frac{8 \cdot (1)^2}{1 \cdot 8 \cdot (1)^2} = 1. \tag{3.4.27}$$

The inexact element mass matrix is thus given by

$$\Theta^{\text{inexact}} = \frac{1}{2}\,h_l \begin{bmatrix} 1 & 0 \\ 0 & 1 \end{bmatrix}. \tag{3.4.28}$$

Comparison with the exact matrix displayed in (1.2.6) confirms that inexact integration effectively lumps the off-diagonal elements of each row to the corresponding diagonal. Stated differently, the diagonal element of the inexact matrix is the sum of the elements in the corresponding row of the exact matrix.

In the case of quadratic interpolation, corresponding to $m = 2$, the diagonal elements are given by

$$\Theta^{\text{inexact}}_{11} = \frac{8\,\text{Lo}^2_1(\xi = -1)}{8 \cdot 27\,\text{L}^2_2(\xi = -1)} = \frac{8 \cdot (-3)^2}{8 \cdot 27 \cdot 1} = \frac{1}{3},$$

$$\Theta^{\text{inexact}}_{22} = w_1 = \frac{2}{2 \cdot 3}\,\frac{1}{\text{L}^2_2(\xi = 0)} = \frac{2}{2 \cdot 3}\,4 = \frac{4}{3}, \tag{3.4.29}$$

$$\Theta^{\text{inexact}}_{33} = \frac{8\,\text{Lo}_1(\xi = 1)^2}{8 \cdot 27\,\text{L}^2_2(\xi = 1)} = \frac{8 \cdot 3^2}{8 \cdot 27 \cdot 1} = \frac{1}{3}.$$

The inexact element mass matrix is thus given by

$$\Theta^{\text{inexact}} = \frac{1}{30}\,h_l \begin{bmatrix} 5 & 0 & 0 \\ 0 & 20 & 0 \\ 0 & 0 & 5 \end{bmatrix}. \tag{3.4.30}$$

Comparison with the exact result displayed in (1.5.29) confirms that inexact integration lumps the off-diagonal elements in each row to the diagonal entry. Numerical values of the lumped dimensionless mass matrix are shown in Table 3.4.3 for several polynomial degrees, m. FSELIB function *emm_lump*, not listed in the text, evaluates this matrix in the spirit of the function *emm*.

3.4.4   Computation of the Lobatto diffusion matrix

Next, we consider the computation of the element diffusion matrix given in (3.1.10). To isolate the effect of element size, we write

$$A_{ij}^{(l)} = \frac{2}{h_l}\,\Psi_{ij} \tag{3.4.31}$$

for $i, j = 1, \ldots, m + 1$, where Ψ is the dimensionless diffusion matrix,

$$\Psi_{ij} \equiv \int_{-1}^{1} \frac{d\psi_i}{d\xi}\frac{d\psi_j}{d\xi}\,d\xi. \tag{3.4.32}$$

Since the integrand is a $(2m - 2)$-degree polynomial in ξ, the integral can be computed exactly using the Lobatto integration quadrature (3.4.14) with $k = m$, yielding

$$\Psi_{ij} = \left(\frac{d\psi_i}{d\xi}\frac{d\psi_j}{d\xi}\right)w_1 + \sum_{p=2}^{m}\left[\left(\frac{d\psi_i}{d\xi}\frac{d\psi_j}{d\xi}\right)w_p\right]_{\xi=z_p} + \left(\frac{d\psi_i}{d\xi}\frac{d\psi_j}{d\xi}\right)_{\xi=1}w_{m+1}. \tag{3.4.33}$$

In terms of the differentiation matrix, d_{ij}, defined in (3.1.13), the integration quadrature reads

$$\Psi_{ij} = \sum_{p=1}^{m+1} d_{ip}\,d_{jp}\,w_p, \tag{3.4.34}$$

without any approximation.

Evaluation of the node differentiation matrix

To evaluate the differentiation matrix, d_{ip}, we resort to (3.1.13) and use property (3.3.13) to find that

$$d_{ip} = \frac{\mathrm{L}_m(\xi_p)}{\mathrm{L}_m(\xi_i)}\frac{1}{\xi_p - \xi_i} \tag{3.4.35}$$

for $i \neq p$, and

$$d_{ii} = \frac{\mathrm{L}'_m(\xi_i)}{2\,\mathrm{L}_m(\xi_i)} = \frac{\mathrm{Lo}_{m-1}(\xi_i)}{2\,\mathrm{L}_m(\xi_i)} \tag{3.4.36}$$

for $i = p$. Because ξ_i is a root of $\text{Lo}_{m-1}(\xi)$ for $i = 2, \ldots, m$,

$$d_{ii} = 0 \tag{3.4.37}$$

for $i = 2, \ldots, m$. The first and last diagonal values are found to be

$$
\begin{aligned}
d_{11} &= \frac{\text{Lo}_{m-1}(\xi_i = -1)}{2\,\text{L}_m(\xi_i = -1)} = -\frac{1}{4}\,m\,(m+1), \\
d_{m+1,m+1} &= \frac{\text{Lo}_{m-1}(\xi_i = 1)}{2\,\text{L}_m(\xi_i = 1)} = \frac{1}{4}\,m\,(m+1).
\end{aligned}
\tag{3.4.38}
$$

Expressions (3.4.35), (3.4.37), and (3.4.38) provide us with a complete definition of the node differentiation matrix, which can be used to evaluate the dimensionless diffusion matrix.

As an example, we consider the case $m = 1$, involving two interpolation nodes located at $\xi_1 = -1$ and $\xi_2 = 1$. Recalling that $\text{L}_1(\xi) = \xi$, we obtain the values

$$
\begin{aligned}
d_{1,1} &= -\frac{1}{2}, & d_{1,2} &= -\frac{1}{2}, \\
d_{2,1} &= \frac{1}{2}, & d_{2,2} &= \frac{1}{2},
\end{aligned}
\tag{3.4.39}
$$

which are consistent with the values obtained directly by differentiating (3.4.7) and (3.4.8) (Problem 3.4.4).

When $m = 2$, the interpolation nodes are located at $\xi_1 = -1$, $\xi_2 = 0$, and $\xi_3 = 1$. Recalling that $\text{L}_2(\xi) = \frac{1}{2}\,(3\,\xi^2 - 1)$, we obtain the values

$$
\begin{aligned}
d_{1,1} &= -\frac{3}{2}, & d_{1,2} &= -\frac{1}{2}, & d_{1,3} &= \frac{1}{2}, \\
d_{2,1} &= 2, & d_{2,2} &= 0, & d_{2,3} &= -2, \\
d_{3,1} &= -\frac{1}{2}, & d_{3,2} &= \frac{1}{2}, & d_{3,3} &= \frac{3}{2},
\end{aligned}
\tag{3.4.40}
$$

which are also consistent with the values obtained by directly differentiating (1.5.24) (Problem 3.4.4).

Alternatively, the node differentiation matrix can be evaluated using the general expressions (D.2.22) and (D.2.23) given in Appendix D for an arbitrary distribution of the element interpolation nodes,

$$d_{ij} = \frac{1}{\xi_j - \xi_i}\,\frac{(\xi_j - \xi_1)\cdots(\xi_j - \xi_{j-1})(\xi_j - \xi_{j+1})\cdots(\xi_j - \xi_{m+1})}{(\xi_i - \xi_1)\cdots(\xi_i - \xi_{i-1})(\xi_i - \xi_{i+1})\cdots(\xi_i - \xi_{m+1})} \tag{3.4.41}$$

for $i \neq j$, and

$$d_{ii} = \frac{1}{\xi_i - \xi_1} + \cdots + \frac{1}{\xi_i - \xi_{i-1}} + \frac{1}{\xi_i - \xi_{i+1}} + \cdots + \frac{1}{\xi_i - \xi_{m+1}} \tag{3.4.42}$$

for the diagonal components. We recall that all but the first and last diagonal components are precisely zero.

FSELIB code *edm_lob*, listed in Table 3.4.5, computes the dimensionless diffusion matrix $\mathbf{\Psi}$ based on the Lobatto quadrature shown in (3.4.34). The algorithm employs formulas (3.4.41) and (3.4.42) to evaluate the node differentiation matrix. Results are shown in Table 3.4.6.

The top and tail end of the FSELIB function *edm_lob_tbl* that evaluates this matrix from tabulated values is listed in Table 3.4.7. The complete function containing high-accuracy data is included in FSELIB. This function will be used in Section 3.5 for building a spectral element code.

PROBLEMS

3.4.1 *Quadratic element interpolation functions*

Apply formulas (3.4.3) for $m = 2$ to recover the quadratic element interpolation functions shown in (1.5.24).

3.4.2 *Lobatto quadrature*

Confirm by an example of your choice that the numerical result obtained by applying the $(k + 1)$-point Lobatto quadrature is exact if the integrated function, $f(\xi)$, is a $(2k - 1)$-degree polynomial.

3.4.3 *Evaluation of the element mass matrix*

Run the code *emm_lob* to evaluate the dimensionless mass matrix, $\mathbf{\Theta}$, for $m = 1, 2, 3$, and 4. Verify that the results are consistent with those shown in Table 3.4.3 for both the consistent and the mass-lumped formulation.

3.4.4 *Evaluation of the node differentiation matrix*

(a) Derive the components of the node differentiation matrix shown in (3.4.39) by directly differentiating the element interpolation functions.

(b) Repeat part (a) for (3.4.40).

3.4.5 *Evaluation of the element diffusion matrix*

(a) Run the code *edm_lob* to evaluate the dimensionless diffusion matrix, $\mathbf{\Psi}$, for $m = 1, 2, 3$, and 4. Verify that the results are consistent with those shown in Table 3.4.6.

(b) Repeat part (a), also computing the determinant and inverse of $\mathbf{\Psi}$ and discuss the results. The determinant of a matrix can be evaluated using the MATLAB function *det*, and the inverse of a matrix can be evaluated using the MATLAB function *inv*.

```
%================================================
%  Code edm_lob
%
%  Evaluation of the dimensionless
%  element diffusion matrix Psi
%  using the node differentiation matrix
%  and the Lobatto integration quadrature
%  The Lobatto base points and integration
%  weights are read from function "lobatto"
%  The node differentiation matrix is computed
%  using formulas (3.4.41) and (3.4.42)
%================================================

clear all

%-----------------------------
% Lobatto interpolation nodes
%-----------------------------

m = input('Enter the order of the polynomial expansion, m: ');

%------------------------------
% generate the Lobatto nodes and
% integration weights
%------------------------------

xi(1) = -1.0;
w(1)  = 2.0/(m*(m+1));

if(m>1)

  [tL, wL] = lobatto(m-1);

  for i=2:m
   xi(i) = tL(i-1);
   w(i)  = wL(i-1);
  end

end

xi(m+1) = 1.0;
w(m+1)  = w(1);

%-----------------------------------------
% compute the node differentiation matrix
% (ndm)
%-----------------------------------------

for i=1:m+1
  for j=1:m+1
```

TABLE 3.4.5   Code *edm_lob* (Continuing →)

```
      %---
      if(i ~= j)
      %---
         ndm(i,j) = 1.0/(xi(j)-xi(i));

         for l=1:m+1
            if(l ~= j) ndm(i,j) = ndm(i,j)*(xi(j)-xi(l)); end
            if(l ~= i) ndm(i,j) = ndm(i,j)/(xi(i)-xi(l)); end
         end

      %---
      else
      %---

         ndm(i,i) = 0.0;

         for l=1:m+1
            if(i ~= l) ndm(i,i) = ndm(i,i)+1.0/(xi(i)-xi(l)); end
         end

      %---
      end
      %---
   end
end

%----------------------------
% compute the diffusion matrix
% by the Lobatto quadrature
%----------------------------

for i=1:m+1     % loop over interpolation nodes

   for j=1:m+1          % loop over interpolation nodes

      dm(i,j) = 0.0;

      for p=1:m+1
         dm(i,j) = dm(i,j) + ndm(i,p)*ndm(j,p)*w(p);
      end

   end

end

%-----
% done
%-----
```

TABLE 3.4.5   ($\rightarrow$ Continued) Code *edm_lob* evaluates the dimensionless diffusion matrix Ψ, defined in (3.4.31), using the Lobatto integration quadrature.

$m = 1:$ $\frac{1}{2} \begin{bmatrix} 1 & -1 \\ -1 & 1 \end{bmatrix}$

$m = 2:$

$$\frac{1}{12} \begin{bmatrix} 14 & -16 & 2 \\ -16 & 32 & -16 \\ 2 & -16 & 14 \end{bmatrix}$$

$m = 3:$

$$\begin{bmatrix} 2.1666666 & -2.4392091 & 0.3558758 & -0.0833333 \\ -2.4392091 & 4.1666666 & -2.0833333 & 0.3558758 \\ 0.3558758 & -2.0833333 & 4.1666666 & -2.4392091 \\ -0.0833333 & 0.3558758 & -2.4392091 & 2.1666666 \end{bmatrix}$$

$m = 4:$

$$\begin{bmatrix} 3.500000 & -3.912885 & 0.533333 & -0.170448 & 0.050000 \\ -3.912885 & 6.351852 & -2.903703 & 0.635185 & -0.170448 \\ 0.533333 & -2.903703 & 4.740740 & -2.903703 & 0.533333 \\ -0.170448 & 0.635185 & -2.903703 & 6.351852 & -3.912885 \\ 0.050000 & -0.170448 & 0.533333 & -3.912885 & 3.500000 \end{bmatrix}$$

TABLE 3.4.6   Tables of the dimensionless diffusion matrix $\boldsymbol{\Psi}$, defined in (3.4.31), for polynomial orders $m = 1, 2, 3,$ and 4. Complete high-accuracy tables are included in the FSELIB function *edm*.

3.5 Spectral element code for steady diffusion

The implementation of the spectral element method for the steady diffusion problem discussed in Chapter 1 culminates in a linear system of algebraic equations,

$$\mathbf{D} \cdot \mathbf{f} = \mathbf{b}, \tag{3.5.1}$$

where the unknown vector, $\mathbf{f}$, contains the values of the solution at all nodes, expect at the very last node where the Dirichlet boundary condition is specified,

$$\mathbf{f} \equiv \begin{bmatrix} f_1, & f_2, & \cdots, & f_{N_G-2}, & f_{N_G-1} \end{bmatrix}^T, \tag{3.5.2}$$

and N_G is the number of global interpolation nodes. The $(N_G - 1) \times (N_G - 1)$ global diffusion matrix, $\mathbf{D}$, consists of partially overlapping element diffusion matrices, as illustrated in Figure 3.1.4.

```
function elm_dm = edm_lob_tbl(m)

%==========================================================
% element diffusion matrix for an mth order polynomial
% expansion, and m = 1-6
%==========================================================

%-----
% trap
%-----

if(m>5)
  disp('');
  disp(' edm: Chosen number of points is not available');
  break;
end

%-------
if(m==1)
%-------

elm_dm = [0.5 -0.5; -0.5 0.5];

%-----------
elseif(m==2)
%-----------

elm_dm = [
 1.16666667 -1.33333333  0.16666667;
-1.33333333  2.66666667 -1.33333333;
 0.16666667 -1.33333333  1.16666667];

%-----------
elseif(m==3)
%-----------

elm_dm =[
 2.1666666 -2.4392091  0.3558758 -0.0833333;
-2.4392091  4.1666666 -2.0833333  0.3558758;
 0.3558758 -2.0833333  4.1666666 -2.4392091;
-0.0833333  0.3558758 -2.4392091  2.1666666];

......

%--
end
%--

return;
```

Table 3.4.7   FSELIB function *edm_lob_tbl* contains tables of the dimensionless element
diffusion matrix Ψ. The six dots near the end of the listing denote unprinted
code. Higher accuracy data is implemented in the function included in FSELIB.

Assembly

The assembly algorithm shown in Table 3.5.1 is a generalization of that for linear elements, shown in Table 1.2.2, and quadratic elements, shown in Table 1.5.2. Linear and quadratic expansions arise, respectively, when $N_p(l) = 1$ or 2 for all elements, $l = 1, \ldots, N_E$, where $N_p(l)$ is the lth element polynomial order.

FSELIB function *sds_lob_sys*, listed in Table 3.5.2, implements the assembly algorithm shown in Table 3.5.1.[1] For programming simplicity, the function generates a full $N_G \times N_G$ global diffusion matrix, denoted as *gdm* in the code, and associated right-hand side, denoted as *b* in the code. The size of the final system to be solved is $(N_G - 1) \times (N_G - 1)$, corresponding to all but the last node where the Dirichlet boundary condition is specified.

Spectral element code

Code *sds_lob*, listed in Table 3.5.3, implements the spectral element code in the following five modules:

1. Data input and parameter definition.

2. Element and interpolation node definition.

3. System assembly.

4. Linear solver module.

5. Graphics module.

For simplicity, the input data are hard-coded in the main program. In a professional code, the data are read from an input file or entered through a graphical user interface (GUI). Function *discr_lob*, listed in Table 3.3.4, is employed to define the element interpolation nodes. The linear system is assembled using the function *sds_lob_sys*, listed in Table 3.5.2. The graphics module involves a standard xy plot.

The linear solver embedded in MATLAB is used to solve the compiled linear system. A MATLAB function is used to remove the extraneous last column and row of the global diffusion matrix as well as the last entry of the right-hand side, so as to reduce the dimension of the linear system to the proper size, $(N_G - 1) \times (N_G - 1)$. The solution of the linear system is found in symbolic form by dividing a vector by a matrix. Methods for solving systems of linear equations in finite element applications are discussed in Appendix C.

The solution generated by the code for the Gaussian source function displayed in (1.2.17) is displayed in Figure 3.5.1. Three elements, $N_E = 3$, with polynomial expansion orders $N_p(1) = 2$ for the first element, $N_p(2) = 2$ for the second element, and $N_p(3) = 4$ for the third element are used in this calculation. The crosses denote the computed nodal values at the interpolation nodes. The dotted vertical lines and circles on the x axis mark the location of the element end-nodes.

---

[1] *sds* is the acronym for steady **d**iffusion with **s**pectral elements.

Do $i = 1, \ldots, N_G - 1$ *Initialize to zero*
    $b_i = 0$
    Do $j = 1, \ldots, N_G - 1$
       $D_{ij} = 0$
    End Do
End Do

$b_1 = q_0/k$ *Neumann boundary condition at the left end*

Do $l = 1, \ldots, N_E$ *Run over the elements*

    Compute the element matrices $\mathbf{A}^{(l)}$ and $\mathbf{B}^{(l)}$

    Do $i = 1, \ldots, N_p(l) + 1$ *Run over the spectral element nodes*
       $i_1 = c(l, i)$
       Do $j = 1, \ldots, N_p(l) + 1$ *Run over the spectral element nodes*
          $i_2 = c(l, j)$
          $D_{i_1,i_2} = D_{i_1,i_2} + A_{ij}^{(l)}$
          $b_{i_1} = b_{i_1} + (B_{ij}^{(l)} \times s_{i_2})/k$
       End Do
    End Do

End Do

$m = N_p(N_E)$ *Process the last node*

Do $i = 1, \ldots, m$ *Run over the last element nodes*
    $i_1 = c(N_e, i)$
    $b_{i_1} = b_{i_1} - B_{i,m+1}^{(N_E)} \times f_L$
End Do

TABLE 3.5.1   Algorithm for assembling a $(N_G - 1) \times (N_G - 1)$ linear system, $\mathbf{D} \cdot \mathbf{f} = \mathbf{b}$, for spectral element expansion. $N_p(l)$ is a specified polynomial order over the lth element.

```
function [gdm,b] = sds_lob_sys (ne,xe,np,ng,c,q0,fL,k,s)

%===========================================================
% Assembly of the linear system for one-dimensional
% steady diffusion with spectral elements (sds)
%
% gdm: global diffusion matrix
% b:   right-hand side
% c:   connectivity matrix
%===========================================================

%-------------
% element size
%-------------

for l=1:ne
  h(l) = xe(l+1)-xe(l);
end

%-----------
% initialize
%-----------

gdm = zeros(ng,ng);
b = zeros(1,ng);

b(1) = q0/k;

%------------------------------------------------
% loop over the elements to compose gdm and b
%------------------------------------------------

for l=1:ne

  m = np(l);

  elm_mm = 0.5*h(l)*emm_lob_tbl(m);     % element mass matrix

  elm_dm = 2.0*edm_lob_tbl(m)/h(l);     % element diffusion matrix

  for i=1:m+1

    i1 = c(l,i);

    for j=1:m+1
      j1 = c(l,j);
      gdm(i1,j1) = gdm(i1,j1) + elm_dm(i,j);
      b(i1) = b(i1) + elm_mm(i,j)*s(j1);
    end

  end

end
```

TABLE 3.5.2   Function *sds_lob_sys* (Continuing →)

```
%-----------------------------------
% implement the Dirichlet condition
% at the last node
%-----------------------------------

m = np(ne);

for i=1:m
  i1 = c(ne,i);
  b(i1) = b(i1) - elm_dm(i,m+1)*fL
end

%-----
% done
%-----

return;
```

TABLE 3.5.2 ($\rightarrow$ Continued) Function *sds_sys* assembles a linear system for one-dimensional diffusion with spectral elements.

To assess whether implementing the high-order method is worthy of the additional analytical and programming effort, we compare the spectral element solution with the solution of the entry-level finite element method that employs linear elements, as discussed in Section 1.1. Properly scaled values of the solution at the left end of the solution domain, $f(x = 0)$, computed with (a) the consistent spectral element method, (b) the mass-lumped spectral element method, and (c) the linear element method, which is nearly identical to the central-differencing finite difference method, are shown in Table 3.5.4. In all cases, the elements are evenly spaced.

The results shown on the left part of Table 3.5.4 reveal that the effect of mass lumping is insignificant for all but the crudest discretization. Comparison of the numerical solutions for a fixed number of global nodes, N_G, clearly demonstrates the superiority of the spectral element method over the linear element method. For example, the spectral element solution with 16 global nodes, $N_G = 16$, is accurate to the fifth significant figure, whereas the finite element solution with 128 global nodes, $N_G = 128$, is accurate only to the fourth significant figure.

3.5.1   *Spectral accuracy*

The performance of the spectral element method can be quantified further with reference to the steady diffusion equation in the interval $[0, L]$, with source term

$$s(x) = -s_0 \, \exp(x/L), \tag{3.5.3}$$

```
%================================================
% Code sds_lob
%
% Spectral element code for steady diffusion
%================================================

%-----------
% input data
%-----------

L = 1.0; k = 1.0; q0 =-1.0; fL = 0.0;

ne = 3; ratio = 5.0;                % number of elements, stretch ratio
np(1) = 2; np(2) = 2; np(3) = 4;   % element polynomial order

%------------------------
% element node generation
%------------------------

[xe,xen,xien,xg,c,ng] = discr_lob (0,L,ne,ratio,np);

%-------------------
% specify the source
%-------------------

for i=1:ng
 s(i) = 10.0*exp(-5.0*xg(i)^2/L^2);
end

%-----------------
% element assembly
%-----------------

[gdm,b] = sds_lob_sys (ne,xe,np,ng,c,q0,fL,k,s);

%--------------
% linear solver
%--------------

gdm(:,ng) = [];   % remove the last (ng) column
gdm(ng,:) = [];   % remove the last (ng) row
b(:,ng) = [];     % remove the last (ng) element
f = b/gdm';       % solve the linear system
f = [f fL];       % include the value at the right end

%-----
% plot
%-----

plot(xg, f,'x-'); hold on;
ye= zeros(ne+1,1); plot(xe,ye,'o');
xlabel('x'); ylabel('f');
```

TABLE 3.5.3   Spectral element code *sds* for steady one-dimensional diffusion.

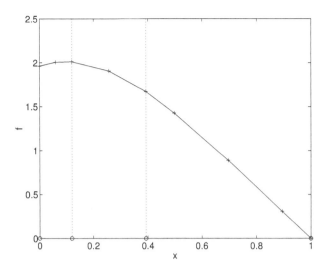

FIGURE 3.5.1 Solution generated by the FSELIB spectral element code *sds_lob* for steady diffusion in the presence of the source displayed in (1.2.17). In this computation, three elements are used, $N_E = 3$, with polynomial expansions orders $N_p(1) = 2$, $N_p(2) = 2$, and $N_p(3) = 4$. The crosses denote the computed values at the interpolation nodes. The dotted vertical lines and circles on the x axis mark the location of the element end-nodes.

| Spectral | | | consistent | lumped | Linear uniform | |
N_G	N_E	N_p	$f(0)$	$f(0)$	$N_E = N_G$	$f(0)$
3	2	1	1.8024	2.2163	2	1.8024
5	2	2	1.9512	1.9512	4	1.9140
7	2	3	1.9043	←	6	1.9412
9	2	4	1.9638	←	8	1.9510
11	2	5	1.9639	←	10	1.9556
16	3	5	1.9639	←	15	1.9602
					32	1.9632
					64	1.9637
					128	1.9638

TABLE 3.5.4 Accuracy and convergence of the spectral element method compared to the finite element method with linear, evenly spaced elements.

where s_0 is a constant. The boundary values,

$$q_0 = -L\,s_0, \qquad f_L = f_0\,e, \tag{3.5.4}$$

are designed so that the exact solution is the exponential function

$$f(x) = f_0\,\exp(x/L), \tag{3.5.5}$$

where $f_0 = s_0 L^2/k$.

The relative error in the left-end value, defined as

$$E \equiv |f(0)/f_0 - 1.0|, \tag{3.5.6}$$

is plotted in Figure 3.5.2 against the available degrees of freedom, N, on a log-log scale. In the case of the finite element method, N is the number of evenly spaced linear elements, N_{E}. In the case of the spectral element method, N is the expansion order, m, for the one-element solution, $N_{\mathrm{E}} = 1$.

As N increases, the slope of the dashed line corresponding to the linear elements tends to the value of -2, indicating that the numerical error behaves like $1/N^2$. As the polynomial degree increases, the slope of the dotted line tracing the results of the spectral element code continues to decrease, suggesting that the error decreases

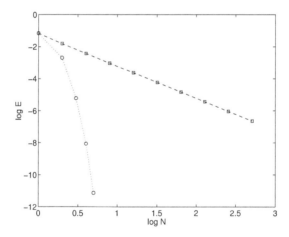

FIGURE 3.5.2   Error in the left-end value for a model problem, plotted against the available degrees of freedom, N, for linear elements (dashed line), and Lobatto nodes (dotted line). The slope of the dashed line corresponding to the linear elements tends to the value of -2, suggesting that the numerical error behaves like $1/N^2$. The slope of the dotted line tracing the results of the spectral code continues to decrease, indicating that the error decreases at a rate that is faster than any power of $1/N$.

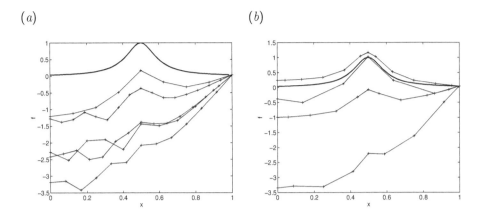

FIGURE 3.5.3   Numerical solutions of the steady diffusion equation with two elements of equal size, $N_E = 2$, and an increasing number of (a) evenly spaced or (b) spectral nodes. The exact solution is the Runge function represented by the smooth line in both graphs. The circles, squares, diamonds, triangles, and crosses in (a) correspond to polynomial orders $m = 4, 5, 6, 7$, and 8. The numerical solutions in (b) correspond to polynomial orders $m = 3, 4, 5$, and 6.

at a rate that is faster than any power of $1/N$, and may thus be classified as spectral. This behavior is typical of a p-type convergence.

Runge function

In a more stringent test, we consider the steady diffusion equation in the interval $[0, L]$ in the presence of the source function

$$s(x) = 200\, s_0\, \frac{1 - 75\, \hat{x}^2}{(1 + 25\, \hat{x}^2)^3}, \tag{3.5.7}$$

where s_0 is a constant and $\hat{x} = 2\,(x/L) - 1$ is a dimensionless parameter varying in the interval $[-1, 1]$. The boundary conditions, $q_0 = -100\, L\, s_0/26^2$ and $f_L = \gamma/26$, are designed so that the exact solution is the Runge function,

$$f(x) = \frac{\gamma}{1 + 25\, \hat{x}^2}, \tag{3.5.8}$$

where $\gamma = s_0 L^2/k$. Numerical solutions with two elements of equal size, $N_E = 2$, and an increasing number of (a) evenly spaced or (b) spectral nodes are shown in Figure 3.5.3. The exact solution is represented by the smooth line in both graphs.

The numerical solutions with evenly spaced nodes suffer from oscillations, exhibiting an apparent lack of convergence as the polynomial order increases from 4 (circles) to 8 (crosses). The numerical solutions with spectral nodes behave much better, exhibiting a uniform improvement as the polynomial order is raised from 3 (circles) to 6 (triangles).

3.5.2 Helmholtz equation

As a further application of the spectral element method, we consider the solution of the one-dimensional Helmholtz equation,

$$\frac{d^2 f}{dx^2} + \left(\frac{\pi}{L}\right)^2 f = 0, \qquad (3.5.9)$$

in the interval $[0, L]$, subject to the boundary conditions

$$(df/dx)_{x=0} = f_0', \qquad f(x = L) = f_L, \qquad (3.5.10)$$

where f_0' and f_L are two given constants. The exact solution for $f_L = 0$ is a sinusoidal function,

$$f = \gamma \sin(\pi x/L), \qquad (3.5.11)$$

where $\gamma = f_0' L/\pi$. The solution is found using the FSELIB code *hlm_lob*, not listed in the text. The pertinent linear system is assembled by the FSELIB function *hlm_lob_sys*, not listed in the text.

A graph of the error in the left-end value, $E \equiv |f(0)/\gamma|$, plotted against the available degrees of freedom, N, is shown in Figure 3.5.4 on a log-log scale. In the case of the finite element method, N is the number of global nodes with evenly spaced quadratic elements, $N_G = 2N_E + 1$. In the case of the spectral element method, N is the number of global nodes for the two-element solution, $N_E = 2$. As N increases, the slope of the solid line corresponding to the quadratic elements tends to the value of -4, indicating that the numerical error behaves like $1/N^4$. In this limit, the slope of the dotted line tracing the results of the spectral element code continues to decrease, revealing a spectral convergence.

It should be emphasized that, for the reasons discussed in Section 3.2, the results shown in Figure 3.5.4 are insensitive to the precise location of the interior nodes in the quadratic expansion as well as in the high-order element expansion. Thus, using evenly spaced element nodes would yield exactly the same answer.

3.5.3 Node condensation

In Section 1.5.6, we demonstrated that the unknowns corresponding to the interior nodes of quadratic elements can be eliminated in favor of those corresponding to the element end-nodes by the process of *node condensation*. As a result, the size of the final linear system is reduced nearly by a factor of 2. Because the condensed system is tridiagonal, the solution can be found efficiently using the standard Thomas algorithm discussed in Section 1.2.2.

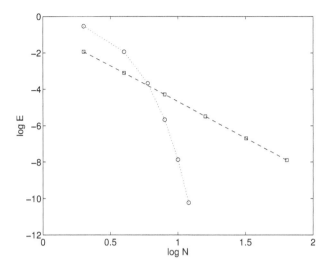

FIGURE 3.5.4   Error in the left-end value in the solution of the Helmholtz equation,
plotted against the available degrees of freedom, N, for quadratic elements (solid
line) and Lobatto nodes (circles). The slope of the solid line corresponding to
the quadratic elements is nearly equal to -4, suggesting that the numerical error
behaves like $1/N^4$. The slope of the dotted line representing the results of the
spectral element code continues to decrease, indicating that the error decreases
at a rate that is faster than any power of $1/N$.

Node condensation in the spectral element method can be implemented at the
element level by a slight modification of the method of Gauss–Jordan reduction used
for solving systems of linear equations, as discussed in Section C.1.8, Appendix C.
Consider the contribution of the lth element to the Galerkin equations expressed
by the constituent linear system

$$\sum_{j=1}^{m+1} A_{ij}^{(l)} f_j^{(l)} = b_i^{(l)} \tag{3.5.12}$$

for $i = 1, \ldots, m + 1$, where m is the order of the element expansion, $f_j^{(l)}$ are the
unknown nodal values of the solution, and $A_{ij}^{(l)}$ is the element diffusion matrix
represented by the shaded blocks in Figure 3.1.4. The right-hand side of (3.5.12) is
given by

$$b_i^{(l)} \equiv \frac{1}{k} \sum_{j=1}^{m+1} B_{ij}^{(l)} s_j^{(l)}, \tag{3.5.13}$$

where $s_j^{(l)}$ are the known nodal values of the source. To implement node condensation, we transform the dense system (3.5.12) into a bordered diagonal system,

$$\sum_{j=1}^{m+1} \mathcal{C}_{ij}^{(l)} f_j^{(l)} = \beta_i^{(l)}, \tag{3.5.14}$$

where the new coefficient matrix takes the form

$$\mathcal{C}^{(l)} = \begin{bmatrix} \mathcal{C}_{11}^{(l)} & 0 & 0 & \cdots & 0 & 0 & \mathcal{C}_{1,m+1}^{(l)} \\ \mathcal{C}_{21}^{(l)} & \mathcal{C}_{22}^{(l)} & 0 & \cdots & 0 & 0 & \mathcal{C}_{2,m+1}^{(l)} \\ \mathcal{C}_{31}^{(l)} & 0 & \mathcal{C}_{33}^{(l)} & \cdots & 0 & 0 & \mathcal{C}_{3,m+1}^{(l)} \\ \vdots & \vdots & \vdots & \ddots & \vdots & \vdots & \vdots \\ \mathcal{C}_{m-1,1}^{(l)} & 0 & 0 & \cdots & \mathcal{C}_{m-1,m-1}^{(l)} & 0 & \mathcal{C}_{m-1,m+1}^{(l)} \\ \mathcal{C}_{m,1}^{(l)} & 0 & 0 & \cdots & 0 & \mathcal{C}_{m,m}^{(l)} & \mathcal{C}_{m,m+1}^{(l)} \\ \mathcal{C}_{m+1,1}^{(l)} & 0 & 0 & \cdots & 0 & 0 & \mathcal{C}_{m+1,m+1}^{(l)} \end{bmatrix} . \tag{3.5.15}$$

Note that all elements, except for the elements along the diagonal and down the first and last columns, are zero. The redeeming feature of this reduction is that, because the interior nodes are decoupled from the end-nodes, they can be removed from the Galerkin system, yielding a condensed system where the end-nodes only appear.

To reduce the element equations into the form shown in (3.5.14), we multiply the second equation by the ratio $r_{12} \equiv A_{12}^{(l)}/A_{22}^{(l)}$, and subtract the outcome from the first equation to eliminate $f_2^{(l)}$. Repeating this calculation, we eliminate $f_2^{(l)}$ from the third, fourth, fifth, and all subsequent equations. In the next cycle, we eliminate $f_3^{(l)}$ from the first, second, fourth, fifth, and all subsequent equations. The process continues until we have eliminated $f_m^{(l)}$ from the first, second, third, all the way up to the $m-1$ equation, and then from the $m+1$ equation.

The condensed global diffusion matrix consists of overlapping 2×2 condensed element matrices,

$$\hat{\mathbf{A}}^{(l)} = \begin{bmatrix} \mathcal{C}_{11}^{(l)} & \mathcal{C}_{1,m+1}^{(l)} \\ \mathcal{C}_{m+1,1}^{(l)} & \mathcal{C}_{m+1,m+1}^{(l)} \end{bmatrix} = \frac{1}{h_l} \begin{bmatrix} 1 & -1 \\ -1 & 1 \end{bmatrix}, \tag{3.5.16}$$

which is precisely the diffusion matrix with linear elements shown in (1.2.6).

FSELIB function *sds_lob_sys_cnd*, listed in Table 3.5.5, implements an algorithm for assembling a condensed $N_E \times N_E$ system of equations. Code *sds_lob_cnd*, listed in Table 3.5.6, implements the spectral element code. For simplicity, the condensed system is solved using the MATLAB linear solver instead of Thomas's algorithm. The results are identical to those obtained with the non-condensed code *sds_lob*.

```
function [gdm,b] = sds_lob_sys_cnd (ne,xe,np,c,q0,fL,k,s)

%===============================================================
% Assembly of the condensed linear system for one-dimensional
% steady diffusion with spectral elements (sds)
%
% gdm: global diffusion matrix
% b:   right-hand side
% c:   connectivity matrix
%===============================================================

%-------------
% element size
%-------------

for l=1:ne
  h(l) = xe(l+1)-xe(l);
end

%-----------
% initialize
%-----------

gdm = zeros(ne+1,ne+1);
b = zeros(1,ne+1);

b(1) = q0/k;

%----------------------
% loop over the elements
%----------------------

for l=1:ne      % closes at echidna

   m = np(l);   % element polynomial expansion order

   elm_dm = 2.0*edm_lob_tbl(m)/h(l);          % element diffusion matrix

   elm_mm = 0.5*h(l)*emm_lob_tbl(m);          % element mass matrix

  % Compute the element-equations right-hand side

   for i=1:m+1
     belm(i) = 0.0;
     for j=1:m+1
      j1 = c(l,j)
      belm(i) = belm(i) + elm_mm(i,j)*s(j1);
     end
     belm(i) = belm(i)/k;
   end

   % Condense the element diffusion matrix
   % and right-hand side
```

TABLE 3.5.5 Function *sds_lob_sys_cnd* (Continuing $\longrightarrow$)

```
%-----------
   if(m > 1)        % Gauss-Jordan reduction; closes at squirrel
%-----------

   for i=2:m        % loop over interior nodes
      for j=1:m+1   % loop over all element nodes
%------
      if(i ~= j)    % skip the self node

         rji = elm_dm(j,i)/elm_dm(i,i);
         for p=1:m+1
          elm_dm(j,p) = elm_dm(j,p) - rji*elm_dm(i,p);
         end
         belm(j) = belm(j) - rji*belm(i);

      end
%------
   end
   end

%-----------
   end     % squirrel
%-----------

   % Assign element equations to global equations
   % The first and last element equations are assigned
   % to the l and l+1 global equations of the condensed system

   gdm(l,   l)   = gdm(l,l)       + elm_dm(1,   1);
   gdm(l,   l+1) = gdm(l,   l+1)  + elm_dm(1,   m+1);
   gdm(l+1,l)    = gdm(l+1,l)     + elm_dm(m+1,1);
   gdm(l+1,l+1)  = gdm(l+1,  l+1) + elm_dm(m+1,m+1);

   b(l)   = b(l)   + belm(1);
   b(l+1) = b(l+1) + belm(m+1);

end     % echidna

%-----------
% last node
%-----------

 b(ne) = b(ne) - elm_dm(1,m+1)*fL

%-----
% done
%-----

return;
```

TABLE 3.5.5 ($\to$ Continued) Function *sdsc_lob_sys_cnd* assembles a condensed linear system for one-dimensional diffusion with spectral elements.

```
%============================================================
% Code sds_lob_cnd
%
% Steady one-dimensional diffusion with spectral elements
% using the condensed formulation where only the end-nodes
% appear in the final system of equations
%============================================================

%-----------
% input data
%-----------

L = 1.0; k = 1.0; q0 = -100/26^2; fL = 1/26.0;

ne = 3; ratio = 1.0;                % number of elements, stretch ratio

np(1) = 3; np(2) = 3; np(3) = 2;   % element expansion order

%-----------------------
% element node generation
%-----------------------

[xe,xen,xien,xg,c,ng] = discr_lob(0,L,ne,ratio,np);

%------------------
% specify the source
%------------------

for i=1:ng
   xhat = 2*xg(i)/L-1;
   s(i) = 200*(1-75*xhat^2)/(1+25*xhat^2)^3 ;
end

%----------------
% element assembly
%----------------

[gdm,b] = sds_lob_sys_cnd (ne,xe,np,c,q0,fL,k,s);

%--------------
% linear solver
%--------------

gdm(:,ne+1) = [];    % remove the last (ne+1) column
gdm(ne+1,:) = [];    % remove the last (ne+1) row
b(:,ne+1) = [];      % remove the last (ne+1) node
f = b/gdm';          % solve the linear system
f = [f fL];          % add the value at the right end

%-----
% plot
%-----
```

TABLE 3.5.6 Code *sdsc_lob_cnd* (Continuing $\rightarrow$)

```
plot(xe, f,'x-'); hold on;

ye= zeros(ne+1,1); plot(xe,ye,'o');

xlabel('x'); ylabel('f');

%-----
% done
%-----
```

TABLE 3.5.6 ($\rightarrow$ Continued) Code *sdsc_lob_cnd* for steady one-dimensional diffusion with the condensed formulation.

PROBLEM

3.5.1 *Codes sds_)lob and sds_lob_cnd*

(*a*) Execute the FSELIB code *sds_lob* with the same boundary conditions but a different source function, using a number of elements and element nodes of your choice. Discuss the results of your computations.

(*b*) Repeat part (*a*) for the condensed code *sds_lob_cnd*. Verify that the results are identical to those generated by code *sd_lob*.

(*c*) Repeat part (*a*) and prepare the counterpart of Table 3.5.4. Discuss the accuracy of the spectral element solution with reference to the finite element solution with linear elements.

3.6   Modal expansion

The mth-order polynomial expansion over the lth element can be expressed in terms of a new set of $m + 1$ basis functions, $\zeta_i(\xi)$, with associated expansion coefficients, c_i, in the modal form

$$f(\xi) = c_1^{(l)} \, \zeta_1(\xi) + \sum_{i=2}^{m} c_i^{(l)} \, \zeta_i(\xi) + c_{m+1}^{(l)} \, \zeta_{m+1}(\xi), \qquad (3.6.1)$$

where

$$\zeta_1(\xi) = \frac{1}{2}(1 - \xi), \qquad \zeta_{m+1}(\xi) = \frac{1}{2}(1 + \xi) \qquad (3.6.2)$$

are the first- and last-node *linear* interpolation functions. The interior modes, also called bubble modes, $\zeta_i(\xi)$, are at most mth-degree polynomials anchored to zero at the end-nodes,

$$\zeta_i(\xi = \pm 1) = 0 \qquad (3.6.3)$$

for $i = 2, \ldots, m$. Evaluating (3.6.1) at $\xi = \pm 1$ and using (3.6.3), we find that

$$c_1^{(l)} = f_1^{(l)}, \qquad c_{m+1}^{(l)} = f_{m+1}^{(l)}. \tag{3.6.4}$$

To ensure C^0 continuity of the solution, we require that the end-node values $f_1^{(l)}$ and $f_{m+1}^{(l)}$ are shared by neighboring elements at the corresponding positions.

To guarantee the satisfaction of the bubble-mode conditions (3.6.3), we set

$$\zeta_i(\xi) = (1 - \xi^2) \, Q_{i-2}(\xi) \tag{3.6.5}$$

for $i = 2, \ldots, m$. A *hierarchical* modal expansion arises by specifying that the functions $Q_{i-2}(\xi)$ are $(i-2)$-degree polynomials.

For example, when $m = 2$, corresponding to the quadratic approximation, we obtain a single interior node,

$$\zeta_2(\xi) = (1 - \xi^2) \, Q_0(\xi). \tag{3.6.6}$$

Setting $Q_0(\xi) = c$ and choosing for convenience the value of the constant $c = 1$, we recover the quadratic modal expansion shown in (1.6.3).

3.6.1 Relation to the nodal expansion

The modal functions can be expressed in terms of a set of $m + 1$ nodal interpolation functions constructed with reference to an arbitrary set of element nodes, ξ_j, subject to the restrictions $\xi_1 = -1$ and $\xi_{m+1} = 1$. An example is the set of the Lobatto nodal interpolation functions displayed in (3.4.3).

Regarding the nodal interpolation functions as Lagrange polynomials, as discussed in Appendix D, we write

$$\zeta_i(\xi) = \sum_{j=1}^{m+1} \zeta_i(\xi_j) \, \psi_j(\xi). \tag{3.6.7}$$

For example, when $i = 1$, we find that

$$\zeta_1(\xi) = \frac{1}{2} \sum_{j=1}^{m+1} (1 - \xi_j) \, \psi_j(\xi). \tag{3.6.8}$$

When $i = m + 1$, we find that

$$\zeta_{m+1}(\xi) = \frac{1}{2} \sum_{j=1}^{m+1} (1 + \xi_j) \, \psi_j(\xi). \tag{3.6.9}$$

The modal function vector may then be related to the nodal function vector by the linear transformation

$$\boldsymbol{\zeta} = \mathbf{V} \cdot \boldsymbol{\psi}, \tag{3.6.10}$$

where

$$\mathbf{V} = \begin{bmatrix} 1 & \zeta_1(\xi_2) & \cdots & \zeta_1(\xi_m) & 0 \\ 0 & \zeta_2(\xi_2) & \cdots & \zeta_2(\xi_m) & 0 \\ \vdots & \vdots & \ddots & \vdots & \vdots \\ 0 & \zeta_m(\xi_2) & \cdots & \zeta_m(\xi_m) & 0 \\ 0 & \zeta_{m+1}(\xi_2) & \cdots & \zeta_{m+1}(\xi_m) & 1 \end{bmatrix} \tag{3.6.11}$$

is a generalized Vandermonde matrix. Note that the first and last columns of $\mathbf{V}$ are filled with zeros, with the exception of the first and last diagonal elements that are equal to unity. For $m = 1$, corresponding to the linear expansion, we find that $\mathbf{V}$ is the 2×2 identity matrix. For $m = 2$ with $\xi_1 = -1$, $\xi_2 = 0$, and $\xi_3 = 1$, and $Q_0 = 1$, we find that

$$\mathbf{V} = \begin{bmatrix} 1 & \frac{1}{2} & 0 \\ 0 & 1 & 0 \\ 0 & \frac{1}{2} & 1 \end{bmatrix}. \tag{3.6.12}$$

The element diffusion matrix associated with the modal expansion is defined as

$$\mathcal{A}_{ij}^{(l)} = \frac{2}{h_l} \int_{-1}^{1} \frac{d\zeta_i}{d\xi} \frac{d\zeta_j}{d\xi} \, d\xi. \tag{3.6.13}$$

Using (3.6.10), we obtain

$$\mathcal{A}^{(l)} = \mathbf{V} \cdot \mathbf{A}^{(l)} \cdot \mathbf{V}^T, \tag{3.6.14}$$

where $\mathbf{A}^{(l)}$ is the element diffusion matrix corresponding to a nodal expansion and the superscript T indicates the matrix transpose. Conversely,

$$\mathbf{A}^{(l)} = \mathbf{V}^{-1} \cdot \mathcal{A}^{(l)} \cdot \mathbf{V}^{-1^T}. \tag{3.6.15}$$

The element mass matrix associated with the modal expansion is defined as

$$\mathcal{B}_{ij}^{(l)} = \frac{1}{2} h_l \int_{-1}^{1} \zeta_i \zeta_j \, d\xi. \tag{3.6.16}$$

Using (3.6.10), we obtain

$$\mathcal{B}^{(l)} = \mathbf{V} \cdot \mathbf{B}^{(l)} \cdot \mathbf{V}^T, \tag{3.6.17}$$

where $\mathbf{B}^{(l)}$ is the element mass matrix of an arbitrary nodal expansion. Conversely,

$$\mathbf{B}^{(l)} = \mathbf{V}^{-1} \cdot \mathcal{B}^{(l)} \cdot \mathbf{V}^{-1^T}. \tag{3.6.18}$$

These relations allow us to compute the element nodal diffusion and matrices from the corresponding modal matrices, and *vice versa*, in terms of the Vandermonde matrix based on the modal functions.

3.6.2   Implementation

The Galerkin finite element method for the modal expansion is implemented as discussed previously in this chapter for the nodal expansion. The only difference is that the element-node interpolation functions, $\psi_i(\xi)$, are replaced by the non-cardinal modal functions, $\zeta_i(\xi)$, and the interior nodal values, $f_i^{(l)}$, are replaced by the expansion coefficients, $c_i^{(l)}$.

Expansion of the source

To implement the method for the steady diffusion equation discussed in Chapter 1, we approximate the source term over the lth element with an mth-degree polynomial, $\mathcal{S}_m$, expressed in the familiar modal form

$$\mathcal{S}_m(\xi) = \sum_{i=1}^{m+1} d_i^{(l)}\, \zeta_i(\xi), \tag{3.6.19}$$

and compute the $m + 1$ modal coefficients, $d_i^{(l)}$ for $i = 1, \ldots, m + 1$, by requiring the interpolation conditions $\mathcal{S}(\xi_j) = s(\xi_j)$ for $i = 1, \ldots, m + 1$, where ξ_j is a set of element nodes. Enforcing these conditions provides us with a system of linear equations,

$$\mathbf{V}^T \cdot \mathbf{d}^{(l)} = \mathbf{s}^{(l)}, \tag{3.6.20}$$

where $\mathbf{V}^T$ is the transpose of the generalized Vandermonde matrix defined in (3.6.11), given by

$$\mathbf{V}^T = \begin{bmatrix} 1 & 0 & \cdots & 0 & 0 \\ \zeta_1(\xi_2) & \zeta_2(\xi_2) & \cdots & \zeta_m(\xi_2) & \zeta_{m+1}(\xi_2) \\ \vdots & \vdots & \ddots & \vdots & \vdots \\ \zeta_1(\xi_m) & \zeta_2(\xi_m) & \cdots & \zeta_m(\xi_m) & \zeta_{m+1}(\xi_m) \\ 0 & 0 & \cdots & 0 & 1 \end{bmatrix}. \tag{3.6.21}$$

It is evident from the structure of this matrix that $d_1^{(l)} = s_1^{(l)}$ and $d_{m+1}^{(l)} = s_{m+1}^{(l)}$.

Nodal reconstruction

Once the finite element solution has been completed, the unknown function can be evaluated at the nodes using the counterpart of (3.6.20), yielding

$$\mathbf{f}^{(l)} = \mathbf{V}^T \cdot \mathbf{c}^{(l)}. \tag{3.6.22}$$

The lth-element equation block of the Galerkin finite element system with the modal expansion takes the form

$$\mathcal{A}^{(l)} \cdot \mathbf{c}^{(l)} = \mathcal{B}^{(l)} \cdot \mathbf{d}^{(l)}. \tag{3.6.23}$$

Restating this equation in terms of the element matrices for the nodal expansion, we obtain

$$\mathbf{V} \cdot \mathbf{A}^{(l)} \cdot \mathbf{V}^T \cdot \mathbf{c}^{(l)} = \mathbf{V} \cdot \mathbf{B}^{(l)} \cdot \mathbf{V}^T \cdot \mathbf{d}^{(l)} = \mathbf{V} \cdot \mathbf{B}^{(l)} \cdot \mathbf{s}^{(l)}. \tag{3.6.24}$$

Combining this equation with (3.6.22), we obtain

$$\mathbf{A}^{(l)} \cdot \mathbf{f}^{(l)} = \mathbf{B}^{(l)} \cdot \mathbf{s}^{(l)}, \tag{3.6.25}$$

which shows that the modal and nodal expansions are equivalent.

PROBLEM

3.6.1 *Lobatto modal expansion*

Display the structure of the Vandermonde matrix for evenly spaced nodes and $Q_p(\xi) = \xi^p$.

3.7   Lobatto modal expansion

For reasons that will become evident in hindsight, now we identify the polynomial $Q_{i-2}(\xi)$ introduced in (3.6.5) with the $(i-2)$-degree Lobatto polynomial $\mathrm{Lo}_{i-2}(\xi)$, yielding the interior modes

$$\zeta_i(\xi) = (1 - \xi^2) \, \mathrm{Lo}_{i-2}(\xi) \equiv \widehat{\mathrm{Lo}}_i(\xi) \tag{3.7.1}$$

for $i = 2, \ldots, m$, where $\widehat{\mathrm{Lo}}_i$ is a completed Lobatto polynomial (Patera [42]). Using the relations in (3.3.20), we find the alternative representations

$$\zeta_i(\xi) = -i \, (i-1) \int_{-1}^{\xi} \mathrm{L}_{i-1}(\xi') \, \mathrm{d}\xi' = -i \, \frac{i-1}{2i-1} \left(\mathrm{Lo}_i(\xi) - \mathrm{Lo}_{i-2}(\xi) \right) \tag{3.7.2}$$

(Szabó & Babuška [65], Karniadakis & Sherwin [32]). Explicitly, the first few modes are given by

$$\zeta_2(\xi) = 1 - \xi^2, \qquad \zeta_3(\xi) = 3 \, (1 - \xi^2) \, \xi,$$

$$\zeta_4(\xi) = \frac{3}{2} \, (1 - \xi^2) \, (5\,\xi^2 - 1), \qquad \ldots \tag{3.7.3}$$

for $m \geq 2$.

FSELIB function *vdm_modal_lob*, listed in Table 3.7.1 evaluates the generalized Vandermonde matrix defined in (3.6.11) up to a certain size.

3.7.1   *Element diffusion matrix*

The four corner entries of the element diffusion matrix defined in (3.6.14) can be evaluated readily using (1.2.6),

$$A_{11}^{(l)} = A_{m+1,m+1}^{(l)} = \frac{1}{h_l}, \qquad A_{1,m+1}^{(l)} = A_{m+1,1}^{(l)} = -\frac{1}{h_l}. \tag{3.7.4}$$

```
function vdm = vdm_modal_lob(m,xi)

%-----------------------------
% compute the vandermonde matrix
% for a given nodal set xi
% for the Lobatto modal expansion
%-----------------------------

  vdm = zeros(m+1:m+1); vdm(1,1)=1; vdm(m+1,m+1)=1;

  if(m==1) return; end

%---
  for j=2:m
%---

  x = xi(j);

  vdm(1,j)   = (1-x)/2;
  vdm(2,j)   =   1-x^2;
  vdm(m+1,j) = (1+x)/2;
  if(m>2)
    vdm(3,j) = 3*(1-x^2)*x;
  end
  if(m>3)
    vdm(4,j)=3*(1-x^2)*(5*x^2-1)/2;
  end
  if(m>4)
    vdm(5,j)=5*(1-x^2)*(7*x^2-3)*x/2;
  end
  if(m>5)
    vdm(6,j)=15*(1-x^2)*(21*x^4-14*x^2+1)/8;
  end
  if(m>6)
    vdm(7,j)=(1-x^2)*(693*x^4-630*x^2+105)*x/8;
  end
  if(m>7)
    vdm(8,j)=(1-x^2)*(3003*x^6-3465*x^4+945*x^2-35)/16;
  if(m>8)
    disp(' -->');
    error(' vdm_modal_lob: Sorry this high order not yet implemented');
  end
  end

%---
  end % over j
%---

%---
% done
%---

return
```

TABLE 3.7.1   Function *vdm_modal_lob* evaluates the Vandermonde matrix for the Lobatto modal expansion.

According to property (3.3.13),

$$\frac{\mathrm{d}\zeta_i}{\mathrm{d}\xi} = -(i-1)\,i\,\mathrm{L}_{i-1}(\xi) \tag{3.7.5}$$

for $i = 2, \ldots, m$, where L_{i-1} is a Legendre polynomial. Using this relation, we find that

$$A_{ij}^{(l)} = \frac{2}{h_l}\,(i-1)(j-1)\,i\,j\,\int_{-1}^{1} \mathrm{L}_{i-1}(\xi)\,\mathrm{L}_{j-1}(\xi)\,\mathrm{d}\xi \tag{3.7.6}$$

for $i, j = 2, \ldots, m$. Invoking the orthogonality condition (3.3.14), we obtain the final expression

$$A_{ij}^{(l)} = \frac{1}{h_l}\,\frac{4\,(i-1)^2\,i^2}{2i-1}\,\delta_{ij}, \tag{3.7.7}$$

for $i, j = 2, \ldots, m$. Moreover, we find that

$$A_{1j}^{(l)} = A_{j1}^{(l)} = \frac{2}{h_l}\int_{-1}^{1} \frac{\mathrm{d}\zeta_1}{\mathrm{d}\xi}\frac{\mathrm{d}\zeta_j}{\mathrm{d}\xi}\,\mathrm{d}\xi = \frac{j\,(j-1)}{h_l}\int_{-1}^{1} \mathrm{L}_{j-1}(\xi)\,\mathrm{d}\xi = 0 \tag{3.7.8}$$

for $j = 2, \ldots, m$, thanks to the orthogonality of the Legendre polynomials. Working in a similar fashion, we find that $A_{m+1,j}^{(l)} = A_{j,m+1}^{(l)} = 0$, for $j = 2, \ldots, m$.

Combining these results, we find that the $(m+1) \times (m+1)$ element diffusion matrix has a desirable nearly diagonal form,

$$\mathcal{A}^{(l)} = \frac{1}{h_l}
\begin{bmatrix}
1 & 0 & 0 & 0 & \cdots & 0 & 0 & -1 \\
0 & \frac{16}{3} & 0 & 0 & \cdots & 0 & 0 & 0 \\
0 & 0 & \frac{144}{5} & 0 & \cdots & 0 & 0 & 0 \\
\vdots & \vdots & \vdots & \ddots & \vdots & \vdots & \vdots & \vdots \\
0 & 0 & 0 & \cdots & \frac{4\,(i-1)^2\,i^2}{2i-1} & \cdots & 0 & 0 \\
\vdots & \vdots & \vdots & \vdots & \vdots & \ddots & \vdots & \vdots \\
0 & 0 & 0 & \cdots & 0 & \frac{4\,(m-2)^2\,(m-1)^2}{2m-3} & 0 & 0 \\
0 & 0 & 0 & \cdots & 0 & 0 & \frac{4\,(m-1)^2\,m^2}{2m-1} & 0 \\
-1 & 0 & 0 & \cdots & 0 & 0 & 0 & 1
\end{bmatrix}.
\tag{3.7.9}$$

For example, when $m = 3$, we obtain the 4×4 matrix

$$\mathcal{A}^{(l)} = \frac{1}{h_l}
\begin{bmatrix}
1 & 0 & 0 & -1 \\
0 & \frac{16}{3} & 0 & 0 \\
0 & 0 & \frac{144}{5} & 0 \\
-1 & 0 & 0 & 1
\end{bmatrix}. \tag{3.7.10}$$

```
function elm_dm =  edm_modal_lob(m,h)

%-------------------------------
% element diffusion matrix for the
% Lobatto modal expansion
%-------------------------------

    for i=1:m+1
      for j=1:m+1
        elm_dm(i,j)= 0;
      end
    end

    for i=2:m
      elm_dm(i,i) = 4*(i-1)^2 * i^2/((2*i-1)*h);
    end

    elm_dm(1,1)   = 1/h; elm_dm(1,m+1)   =-1/h;
    elm_dm(m+1,1)=-1/h; elm_dm(m+1,m+1)= 1/h;

%---
% done
%---

return
```

TABLE 3.7.2   Function *edm_modal_lob* evaluates the element diffusion matrix for the Lobatto modal expansion.

Applying the Laplace expansion of the determinant with respect to the first or last column or row, we find that the element diffusion matrix is singular.

In summary, the choice of Lobatto polynomials has the significant advantage of diagonalizing the portion of the element diffusion matrix corresponding to the interior modes.

The FSELIB function *edm_modal_lob*, listed in Table 3.7.2, evaluates the element diffusion matrix based on the formulas derived in this section. The input includes the expansion order, m, and the element size, h.

It should be noted that if instead we had used the alternate scaled modes

$$\widehat{\zeta}_i = -\left(\frac{2}{2i-1}\right)^{1/2} i\,(i-1)\,\zeta_i \tag{3.7.11}$$

for $i = 2,\ldots,m$, we would have found that the interior diagonal elements of the matrix on the right-hand side of (3.7.9) are all equal to 2 (Szabó & Babuška [65]).

3.7.2 *Element mass matrix*

Substituting into the element mass matrix given in (3.6.16) the modal functions for the interior modes, we obtain

$$B_{ij}^{(l)} = h_l \frac{1}{2} \int_{-1}^{1} (1 - \xi^2)^2 \, \mathrm{Lo}_{i-2}(\xi) \, \mathrm{Lo}_{j-2}(\xi) \, \mathrm{d}\xi \qquad (3.7.12)$$

for $i, j = 2, \ldots, m$. Using property (3.3.18), we obtain

$$B_{ij}^{(l)} = h_l \frac{1}{2} \frac{(i-1)i(j-1)j}{(2i-1)(2j-1)} \int_{-1}^{1} \left(L_{i-2} - L_i \right) \left(L_{j-2} - L_j \right) \mathrm{d}\xi. \qquad (3.7.13)$$

Because of the orthogonality condition (3.3.14), $B_{ij}^{(l)}$ is non-zero only when $j = i$ or $i \pm 2$.

The interior diagonal components are given by

$$B_{ii}^{(l)} = h_l \frac{1}{2} \frac{(i-1)^2 i^2}{(2i-1)^2} \int_{-1}^{1} \left(L_{i-2}^2 + L_i^2 \right) \mathrm{d}\xi, \qquad (3.7.14)$$

yielding

$$B_{ii}^{(l)} = h_l \frac{2(i-1)^2 i^2}{(2i-3)(2i-1)(2i+1)} \qquad (3.7.15)$$

for $i = 2, \ldots, m$.

The corresponding off-diagonal components are given by

$$B_{i(i+2)}^{(l)} = -h_l \frac{1}{2} \frac{(i-1)(i+1)i(i+2)}{(2i-1)(2i+3)} \int_{-1}^{1} L_i^2 \, \mathrm{d}\xi, \qquad (3.7.16)$$

yielding

$$B_{i(i+2)}^{(l)} = -h_l \frac{(i-1)i(i+1)(i+2)}{(2i-1)(2i+1)(2i+3)} \qquad (3.7.17)$$

for $i = 2, \ldots, m - 1$, and

$$B_{i(i-2)}^{(l)} = -h_l \frac{1}{2} \frac{(i-1)(i-3)i(i-2)}{(2i-1)(2i-5)} \int_{-1}^{1} L_{i-2}^2 \, \mathrm{d}\xi, \qquad (3.7.18)$$

yielding

$$B_{i(i-2)}^{(l)} = -h_l \frac{(i-3)(i-2)(i-1)i}{(2i-5)(2i-3)(2i-1)} \qquad (3.7.19)$$

for $i = 3, \ldots, m$.

Invoking the orthogonality of the quadratic and higher-degree Lobatto polynomials against the linear polynomials associated with the first- and last-node linear interpolation functions, we find that the $(m+1) \times (m+1)$ element mass matrix takes a nearly tridiagonal banded form, as shown in Table 3.7.3(a). The interior elements are computed using expressions (3.7.15)–(3.7.19), and the bordering elements are given in Table 3.7.3(b).

When $m = 2$, corresponding to a quadratic element expansion, we obtain the dense 3×3 element mass matrix

$$\mathcal{B}^{(l)} = \frac{1}{30} h_l \begin{bmatrix} 10 & 10 & 5 \\ 10 & 16 & 10 \\ 5 & 10 & 10 \end{bmatrix}, \tag{3.7.20}$$

also displayed in (1.6.13) with reference to the nodal expansion.

When $m = 3$, corresponding to a cubic expansion, we obtain

$$\mathcal{B}^{(l)} = \begin{bmatrix} \mathcal{B}_{11}^{(l)} & \mathcal{B}_{12}^{(l)} & \mathcal{B}_{13}^{(l)} & \mathcal{B}_{14}^{(l)} \\ \mathcal{B}_{21}^{(l)} & \mathcal{B}_{22}^{(l)} & 0 & \mathcal{B}_{24}^{(l)} \\ \mathcal{B}_{31}^{(l)} & 0 & \mathcal{B}_{33}^{(l)} & \mathcal{B}_{34}^{(l)} \\ \mathcal{B}_{41}^{(l)} & \mathcal{B}_{42}^{(l)} & \mathcal{B}_{43}^{(l)} & \mathcal{B}_{44}^{(l)} \end{bmatrix}, \tag{3.7.21}$$

where the filled elements are non-zero (Problem 3.7.2).

When $m = 4$, corresponding to the quartic expansion, we obtain

$$\mathcal{B}^{(l)} = \begin{bmatrix} \mathcal{B}_{11}^{(l)} & \mathcal{B}_{12}^{(l)} & \mathcal{B}_{13}^{(l)} & 0 & \mathcal{B}_{15}^{(l)} \\ \mathcal{B}_{21}^{(l)} & \mathcal{B}_{22}^{(l)} & 0 & \mathcal{B}_{24}^{(l)} & \mathcal{B}_{25}^{(l)} \\ \mathcal{B}_{31}^{(l)} & 0 & \mathcal{B}_{33}^{(l)} & 0 & \mathcal{B}_{35}^{(l)} \\ 0 & \mathcal{B}_{42}^{(l)} & 0 & \mathcal{B}_{44}^{(l)} & 0 \\ \mathcal{B}_{51}^{(l)} & \mathcal{B}_{52}^{(l)} & \mathcal{B}_{53}^{(l)} & 0 & \mathcal{B}_{55}^{(l)} \end{bmatrix}, \tag{3.7.22}$$

where the filled elements are non-zero. The matrices (3.7.21) and (3.7.22) differ from those of the corresponding nodal expansions.

FSELIB function *emm_modal_lob*, listed in Table 3.7.4, evaluates the element mass matrix based on the formulas derived in this section. The input includes the expansion order, m, and the element size, h.

As an aside, we note that to evaluate the integral

$$\int_{-1}^{1} \frac{1-\xi}{2} \left(1 - \xi^2\right) \mathrm{d}\xi \tag{3.7.23}$$

using MATLAB, we may issue the statements:

(a)

$$
\mathcal{B}^{(l)} =
\begin{bmatrix}
\mathcal{B}^{(l)}_{11} & \mathcal{B}^{(l)}_{12} & \mathcal{B}^{(l)}_{13} & 0 & 0 & 0 & 0 & 0 & \cdots \\
\mathcal{B}^{(l)}_{21} & \mathcal{B}^{(l)}_{22} & 0 & \mathcal{B}^{(l)}_{24} & 0 & 0 & 0 & 0 & \cdots \\
\mathcal{B}^{(l)}_{31} & 0 & \mathcal{B}^{(l)}_{33} & 0 & \mathcal{B}^{(l)}_{35} & 0 & 0 & 0 & \cdots \\
0 & \mathcal{B}^{(l)}_{42} & 0 & \mathcal{B}^{(l)}_{44} & 0 & \mathcal{B}^{(l)}_{46} & 0 & 0 & \cdots \\
\cdots & \cdots & \cdots & \cdots & \cdots & \cdots & \cdots & \cdots & \cdots \\
0 & 0 & 0 & 0 & 0 & 0 & 0 & 0 & \cdots \\
\mathcal{B}^{(l)}_{(m+1)1} & \mathcal{B}^{(l)}_{(m+1)2} & \mathcal{B}^{(l)}_{(m+1)3} & 0 & 0 & 0 & 0 & 0 & \cdots
\end{bmatrix}
$$

$$
\begin{bmatrix}
0 & 0 & \cdots & 0 & 0 & 0 & \mathcal{B}^{(l)}_{1(m+1)} \\
0 & 0 & \cdots & 0 & 0 & 0 & \mathcal{B}^{(l)}_{2(m+1)} \\
0 & 0 & \cdots & 0 & 0 & 0 & \mathcal{B}^{(l)}_{3(m+1)} \\
0 & 0 & \cdots & 0 & 0 & 0 & 0 \\
\cdots & \cdots & \cdots & \cdots & \cdots & \cdots & \cdots \\
0 & 0 & \cdots & \mathcal{B}^{(l)}_{m(m-2)} & 0 & \mathcal{B}^{(l)}_{mm} & 0 \\
0 & 0 & 0 & \cdots & 0 & 0 & \mathcal{B}^{(l)}_{(m+1)(m+1)}
\end{bmatrix}
$$

(b)

$$
\mathcal{B}^{(l)}_{11} = \mathcal{B}^{(l)}_{(m+1)(m+1)} = \frac{1}{3}\, h_l
$$

$$
\mathcal{B}^{(l)}_{1(m+1)} = \mathcal{B}^{(l)}_{(m+1)1} = \frac{1}{6}\, h_l
$$

$$
\mathcal{B}^{(l)}_{12} = \mathcal{B}^{(l)}_{21} = h_l \,\frac{1}{2} \int_{-1}^{1} \frac{1-\xi}{2}\,(1-\xi^2)\,\mathrm{d}\xi = \frac{1}{3}\, h_l
$$

$$
\mathcal{B}^{(l)}_{13} = \mathcal{B}^{(l)}_{31} = h_l \,\frac{1}{2} \int_{-1}^{1} \frac{1-\xi}{2}\,(1-\xi^2)\,3\,\xi\,\mathrm{d}\xi = -\frac{1}{5}\, h_l
$$

$$
\mathcal{B}^{(l)}_{2(m+1)} = \mathcal{B}^{(l)}_{(m+1)2} = h_l \,\frac{1}{2} \int_{-1}^{1} \frac{1+\xi}{2}\,(1-\xi^2)\,\mathrm{d}\xi = \frac{1}{3}\, h_l
$$

$$
\mathcal{B}^{(l)}_{3(m+1)} = \mathcal{B}^{(l)}_{(m+1)3} = h_l \,\frac{1}{2} \int_{-1}^{1} \frac{1+\xi}{2}\,(1-\xi^2)\,3\,\xi\,\mathrm{d}\xi = \frac{1}{5}\, h_l
$$

TABLE 3.7.3   (a) Structure of the element mass matrix for the Lobatto modal expansion and (b) evaluation of the non-zero elements in the first and last columns and rows.

```
function elm_mm = emm_modal_lob(m,h)

%----------------------------
% element mass matrix for the
% Lobatto modal expansion
%----------------------------

   for i=1:m+1
     for j=1:m+1
        elm_mm(i,j)= 0;
     end
   end

   for i=2:m
    elm_mm(i,i) = h*2*(i-1)^2*i^2/((2*i-3)*(2*i-1)*(2*i+1));
   end

   for i=2:m-1
    elm_mm(i,i+2) = - h*(i-1)*i*(i+1)*(i+2)/((2*i-1)*(2*i+1)*(2*i+3));
   end

   for i=3:m
    elm_mm(i,i-2) = - h*(i-3)*(i-2)*(i-1)*i/((2*i-5)*(2*i-3)*(2*i-1));
   end

   elm_mm(1,1)   =  h/3; elm_mm(m+1,m+1) =  h/3;
   elm_mm(1,2)   =  h/3; elm_mm(2,1)     =  h/3;
   elm_mm(1,3)   = -h/5; elm_mm(3,1)     = -h/5;

   elm_mm(1,m+1) = h/6; elm_mm(m+1,1) = h/6;
   elm_mm(2,m+1) = h/3; elm_mm(m+1,2) = h/3;
   elm_mm(3,m+1) = h/5; elm_mm(m+1,3) = h/5;

%---
% done
%---

return
```

TABLE 3.7.4 Function *emm_modal_lob* evaluates the element mass matrix of the Lobatto modal expansion.

```
>> syms x
>> int(0.5*(1-x)*(1-x^2),x,-1,1)
```

where the double angle bracket (>>) is the MATLAB command line prompt. The first statement declares x as a symbolic object. The second statement computes the indefinite integral with respect to x, and then evaluates the indefinite integral between the specified integration limits, -1 and 1.

3.7.3 Modal spectral element method

The modal spectral element method arises by identifying the element interpolation nodes for the source involved in (3.6.20) with the Lobatto spectral nodes. The algorithm is implemented in the FSELIB code *sds_modal*, listed in Table 3.7.5, for the steady diffusion problem discussed in Chapter 1.

FSELIB unction *discr_lob*, listed in Table 3.3.1, is called to generate the element spectral nodes. Function *vdm_modal_lob*, listed in Table 3.7.1, is called to generate the Vandermonde matrix on each element for the purpose of evaluating the expansion coefficients, $\mathbf{d}^{(l)}$. FSELIB function *sds_modal_sys*, listed in Table 3.7.6, is called to compile the global diffusion matrix and right-hand side of a linear system for the steady diffusion problem discussed in Chapter 1. The element diffusion and mass matrices are computed by the functions *edm_modal_lob* and *emm_modal_lob*, listed in Tables 3.7.3 and 3.7.4. The results are identical to those obtained for the Lobatto spectral nodal expansion.

PROBLEMS

3.7.1 *Expansion coefficients*

Evaluate the exact expansion coefficients, c_i, for the following functions defined in the interval $-1 \leq \xi \leq 1$, and expansion orders, m: (a) $f(\xi) = 1 + 4\xi$, $m = 1$, (b) $f(\xi) = 1 + 4\xi + \xi^2$, $m = 2$, and (c) $f(\xi) = 1 + 4\xi + \xi^2 + \xi^3$, $m = 3$.

3.7.2 *Element mass matrix*

Evaluate the 4×4 element mass matrix (3.7.21) corresponding to $m = 3$.

3.7.3 *Element diffusion matrix*

Evaluate the 5×5 element diffusion matrix for $m = 4$ and use the MATLAB function *det* to compute its determinant.

3.7.4 *Legendre modal expansion*

Discuss the structure of the element mass matrix for the Legendre modal expansion,

$$\zeta_i(\xi) = (1 - \xi^2) \, \mathrm{L}_{i-2}(\xi) \tag{3.7.24}$$

for $i = 2, \ldots, m$, where $\mathrm{L}_{i-2}(\xi)$ are the Legendre polynomials discussed in Section B.5, Appendix B.

3.8 Arbitrary nodal sets

The methods and procedures developed earlier in this chapter allow us to compute the element diffusion and mass matrices for *arbitrary nodal sets* over the elements based on the conversion formulas (3.6.15) and (3.6.18). This is done by combining the following three functions:

```
%======================================
% Code sds_modal
%
% modal expansion with spectral elements
%======================================

%-----------
% input data
%-----------

L = 1.0; k = 1.0; fL = 0; q0 =-1.0;

ne = 2; ratio = 1.0;      % number of elements, stretch ratio
np(1) = 3;                % element order expansions
np(2) = 3;

%------------------------
% element node generation
%------------------------

[xe,xen,xien,xg,c,ng] = discr_lob (0,L,ne,ratio,np);

%--------------------
% source coefficients
%--------------------

for l=1:ne

  m = np(l);

  for j=1:m+1
   s(j) = 10.0*exp(-5.0*xen(l,j)^2/L^2);
  end

  vdm = vdm_modal_lob(m,xien(l,:));

  d = s/vdm;

  for j=1:m+1
    dg(c(l,j)) = d(j);
  end

  clear s vdm d

end

%-----------------
% element assembly
%-----------------

[gdm,b] = sds_modal_sys (ne,xe,np,ng,c,q0,fL,k,dg);
```

TABLE 3.7.5   Code *sds_modal* (Continuing →)

```
%---------------
% linear solver
%---------------

gdm(:,ng) = [];   % remove the last (ng) column
gdm(ng,:) = [];   % remove the last (ng) row
b(:,ng) = [];     % remove the last (ng) element

cg = b/gdm';      % solve the linear system
cg = [cg fL];     % add the value at the right end

%---
% extract the element modal coefficients
%---

shift = 0;
for l=1:ne
 for j=1:np(l)+1
  c(l,j) = cg(shift+j);
 end
 shift = shift+np(l);
end

%---
% reconstruct the nodal values
%---

Ic = 0;

for l=1:ne

  vdm = vdm_modal_lob(m,xien(l,:));
  felm = vdm'*c(l,:)';

  for j=1:np(l)
   Ic = Ic+1;
   f(Ic) = felm(j);
  end

end

f = [f fL];       % add the value at the right end

%-----
% plot
%-----

plot(xg, f,'k-+')

%-----
% done
%-----
```

TABLE 3.7.5   ($\rightarrow$ Continued) Code *sds_modal* implements the modal spectral element method for steady diffusion.

```
function [gdm,b] = sds_modal_sys (ne,xe,np,ng,c,q0,fL,k,dg)

%---------------------------------------
% Assembly of a linear system
% for one-dimensional steady diffusion
% with the Lobatto modal expansion
%---------------------------------------

%-------------
% element size
%-------------

for l=1:ne
  h(l) = xe(l+1)-xe(l);
end

%----------
% initialize
%----------

gdm = zeros(ng,ng); % global diffusion matrix
b = zeros(1,ng);    % RHS

b(1) = q0/k;    % Neumann condition on the left

%----------------------
% loop over the elements
%----------------------

for l=1:ne

   m = np(l);

% element diffusion and mass matrix

   clear elm_dm elm_mm
   elm_dm =  edm_modal_lob(m,h(l));
   elm_mm =  edm_modal_lob(m,h(l));

    for i=1:m+1
       i1 = c(l,i);
       for j=1:m+1
       j1 = c(l,j);
       gdm(i1,j1) = gdm(i1,j1) + elm_dm(i,j);
       b(i1)      = b(i1)      + elm_mm(i,j)*dg(j1)/k;
       end
    end

%--
end   % end of loop over elements
%--

%----------
% last node
%----------

m = np(ne);
```

TABLE 3.7.6   Function *sds_modal_sys* (Continuing →)

```
for i=1:m
  i1 = c(ne,i);
  b(i1) = b(i1) - elm_dm(i,m+1)*fL;
end

%-----
% done
%-----

return;
```

TABLE 3.7.6 ($\rightarrow$ Continued) Function *sds_modal_sys* compiles a linear system for the modal spectral element method.

1. The Vandermonde matrix function *vdm_modal_lob*, listed in Table 3.7.1.

2. Function *edm_modal_lob*, listed in Table 3.7.2.

3. Function *emm_modal_lob*, listed in Table 3.7.4.

Assume that m is a specified element expansion order, h is the element size, and ξ_i for $i = 1, \ldots, m+1$ is a nodal set, where $\xi_1 = -1$ and $\xi_{m+1} = 1$. To compute the element diffusion and mass matrices, denoted as *elm_dm* and *elm_mm*, we insert the following lines in a MATLAB code:

```
vdm = vdm_modal_lob(m,xi);                  % vandermond matrix
elm_dm_modal = edm_modal_lob(m,h);          % modal diffusion matrix
elm_mm_modal = emm_modal_lob(m,h);          % modal mass matrix
elm_dm = inv(vdm)*elm_dm_modal*inv(vdm');   % nodal diffusion matrix
elm_mm = inv(vdm)*elm_mm_modal*inv(vdm');   % nodal mass matrix
```

A finite element code may then be built for arbitrary element distributions and corresponding nodal sets.

Arbitrary nodal set discretization

Function *discr_any*, listed in Table 3.8.1, defines arbitrary element node positions and global nodes. The input includes the x positions of the two ends of the solution domain, the number of elements, and the degree of the interpolating polynomial over each element. Uniform and Lobatto nodal distributions over each element are implemented in the code. A selection is made according to the choice index *idistr*. An error message is issued if a distribution is not available. Other distributions can be incorporated by straightforward modifications (Problem 3.8.1).

```
function [xe,xen,xien,xg,c,ng] = discr_any (x1,x2,ne,ratio,np,idistr)

%=========================================
% Discretize the interval (x1, x2) into "ne" elements
% and generate the np-order element interpolation nodes
%
% xe:    element end-nodes
% xen:   element x interpolation nodes
% xien:  element xi interpolation nodes
% xg:    unique global nodes
% c:     connectivity matrix
% ng:    number of unique global nodes
%=========================================

%----------------------------
% define the element end-nodes
%----------------------------

xe = elm_line1 (x1,x2,ne,ratio);

%--------------------------------------------------
% define the element interpolation nodes (xen)
%--------------------------------------------------

for l=1:ne

  m = np(l);

%---
  if(idistr==1)   % uniform
%---

    for j=1:m+1
       xi(j) = -1.0+2.0*(j-1)/m;
    end

%---
  elseif(idistr==2)   % Lobatto
%---

    xi(1) = -1.0;

    if(m>1)
     [tL, wL] = lobatto(m-1);
     for j=2:m
      xi(j) = tL(j-1);
     end
    end

    xi(m+1) = 1.0;
```

TABLE 3.8.1   Function *discr_any* (Continuing →)

```
%---
  else
%---

    error('discr_any: unknown option for idistr')

%---
  end
%---

  for j=1:m+1
    xix(l,j) = 0.5*(xe(l+1)+xe(l)) + 0.5*(xe(l+1)-xe(l))*xi(j);
  end

  for j=1:m+1
    xien(l,j) =  xi(j);
     xen(l,j) = xix(j);
  end

end

%-------------------------------
% define the connectivity matrix
% and the global nodes
%-------------------------------

Ic = 2;      % global node counter

for l=1:ne    % run over elements

  Ic = Ic-1;  % account for shared nodes

  for je=1:np(l)+1
    c(l,je) = Ic;
    xg(Ic) = xen(l,je);
    Ic = Ic+1;
  end

end

%----------------------
% total number of nodes
%----------------------

ng = Ic-1;

%-----
% done
%-----

return;
```

TABLE 3.8.1 ($\rightarrow$ Continued) Function *discr_any* defines the element node positions and compiles the global nodes.

Assembly of a linear system

The linear system originating from the finite element method is assembled by the FSELIB function *sds_any_sys*, listed in Table 3.8.2. The five function calls mentioned earlier in this section are included verbatim in the code.

Finite element code

Code *sds_any*, listed in Table 3.8.3, solves the steady diffusion equation. The source function and boundary conditions are designed so that the exact solution is the Runge function, given in (3.5.8). The code employs the function *discr_any* to define element and global nodes, and the function *sds_any_sys* to compile a linear system.

PROBLEMS

3.8.1 *Geometrically arranged nodes*

Modify the function FSELIB function *discr_any* so that the node distribution over each element is generated by the FSELIB function *elm_line2* or *elm_line3* discussed in Section 1.1, Chapter 1.

3.8.2 *Plotting*

Write a function that plots the entire finite element solution consisting of the union of the interpolating polynomials over each element.

3.9   Unsteady diffusion

To illustrate the implementation of the spectral element method in problems involving time dependence, we consider unsteady heat conduction in a rod, as discussed in Section 2.1. The distribution of the temperature along the rod, $f(x,t)$, is governed by equation (2.1.2), repeated below for convenience,

$$\frac{\partial f}{\partial t} = \kappa \frac{\partial^2 f}{\partial x^2} + \frac{s(x,t)}{\rho\, c_p}. \tag{3.9.1}$$

The solution is to be found subject to the Neumann boundary condition (1.1.5) and the Dirichlet boundary condition (1.1.6), also repeated below for convenience,

$$q_0 \equiv -k \left(\frac{\partial f}{\partial x}\right)_{x=0}, \qquad f(x=L) \equiv f_L. \tag{3.9.2}$$

For simplicity, we assume that the left-end flux, q_0, the right-end temperature, f_L, and the source term, $s(x)$, remain constant in time.

Since the temperature is prescribed at the right end of the rod, the Galerkin projection provides us with a system of $N_G - 1$ ordinary differential equations (ODEs) for the temperatures at all but the last node,

```
function [gdm,b] = sds_any_sys (ne,xe,np,ng,c,xien,q0,fL,k,s)

%-------------------------------------------------
% Assemble a linear system
% for one-dimensional steady diffusion
% for any arbitrary nodal distribution
% converted into the evenly spaced nodal base
%
% gdm: global diffusion matrix
% b:   right-hand side
% c:   connectivity matrix
%-------------------------------------------------

%-------------
% element size
%-------------

for l=1:ne
  h(l) = xe(l+1)-xe(l);
end

%----------
% initialize
%----------

gdm = zeros(ng,ng);

b   = zeros(1,ng);     % right-hand side

b(1) = q0/k;     % Neumann condition on the left

%----------------------
% loop over the elements
%----------------------

for l=1:ne

  m = np(l);

  clear xi

  for j=1:m+1
    xi(j) = xien(l,j);
  end

  vdm = vdm_modal_lob(m,xi);

  elm_dm_modal = edm_modal_lob(m,h(l));
  elm_mm_modal = emm_modal_lob(m,h(l));

  elm_dm = inv(vdm)*elm_dm_modal*inv(vdm');
  elm_mm = inv(vdm)*elm_mm_modal*inv(vdm');
```

TABLE 3.8.2 Function *sds_any_sys* (Continuing →)

```
% assemble:

   for i=1:m+1
      i1 = c(1,i);
      for j=1:m+1
      j1 = c(1,j);
      gdm(i1,j1) = gdm(i1,j1) + elm_dm(i,j);
      b(i1)      =  b(i1)     + elm_mm(i,j)*s(j1)/k;
      end
   end

%--
end    % end of loop over elements
%--

%----------
% last node
%----------

m = np(ne);

for i=1:m
   i1 = c(ne,i);
   b(i1) = b(i1) - elm_dm(i,m+1)*fL;
end

%-----
% done
%-----

return;
```

TABLE 3.8.2 ($\rightarrow$ Continued) Function *sds_any_sys* assembles a linear system for steady diffusion with an arbitrary nodal set.

$$\mathbf{M} \cdot \frac{d\mathbf{f}}{dt} + \kappa \mathbf{D} \cdot \mathbf{f} = \kappa \mathbf{b}, \tag{3.9.3}$$

where N_G is the number of global nodes.

FSELIB function *gdmm_lob*, listed in Table 3.9.1, generates the $N_G \times N_G$ global diffusion matrix, $\mathbf{D}$, and associated global mass matrix, $\mathbf{M}$, for the Lobatto spectral element discretization.

3.9.1   Crank–Nicolson discretization

The Crank–Nicolson discretization of the differential equation (3.9.3) provides us with a linear system of equations for a vector, $\mathbf{f}^{n+1}$, containing the unknown nodal

```
%==================================================
% Code sds_any
%
% Steady 1D diffusion for arbitrary element
% node distribution
%==================================================

clear all
close all

%-----------
% input data
%-----------

idistr = 1; % uniform
idistr = 2; % Lobatto

%-----------
% input data
%-----------

L = 1.0; k = 1.0;
q0 = -100/26^2;
fL = 1/26.0;

ne = 3; ratio = 5.0;              % number of elements, stretch ratio
np(1) = 2; np(2) = 2; np(3) = 4;  % element interpolation order

%-----
% trap
%-----

for i=1:ne

   if(np(i)>8)
   disp(' -->');
   disp(' max polynomial order is 8');
   error(' sds_any: Sorry this high order not yet implemented');
   end

end

%-----------
% element node generation
%-----------

[xe,xen,xien,xg,c,ng] = discr_any (0,L,ne,ratio,np,idistr);

%-------------------
% specify the source
%-------------------

for i=1:ng
  xhat = 2*xg(i)-1;
  s(i) = 200*(1-75*xhat^2)/(1+25*xhat^2)^3 ;
end
```

TABLE 3.8.3   Code *sds_any* (Continuing →)

```
%------------------
% element assembly
%------------------

[gdm,b] = sds_any_sys (ne,xe,np,ng,c,xien,q0,fL,k,s);

%--------------
% linear solver
%--------------

gdm(:,ng) = [];    % remove the last (ng) column
gdm(ng,:) = [];    % remove the last (ng) row
b(:,ng) = [];      % remove the last (ng) element

f = b/gdm';        % solve the linear system
f = [f fL];        % add the value at the right end

%-----
% plot
%-----

plot(xg, f,'k-+')

%-----
% done
%-----

return
```

TABLE 3.8.3 ($\rightarrow$ Continued) Code *sds_any* solves the steady diffusion equation with arbitrary element node distributions.

values at the $n + 1$ time level,

$$\mathbf{C} \cdot \mathbf{f}^{n+1} = \mathbf{r}, \tag{3.9.4}$$

as discussed in Section 2.1.6. The coefficient matrix on the left-hand side is given by

$$\mathbf{C} = \mathbf{M} + \frac{1}{2} \kappa \, \Delta t \, \mathbf{D} \tag{3.9.5}$$

and the right-hand side is given by

$$\mathbf{r} = \left(\mathbf{M} - \frac{1}{2} \kappa \, \Delta t \, \mathbf{D} \right) \cdot \mathbf{f}^n + \kappa \, \Delta t \, \mathbf{b}, \tag{3.9.6}$$

where Δt is the time step and the vector $\mathbf{f}^n$ contains the known nodal values at the nth time level.

FSELIB code *uds_lob_cn*, listed in Table 3.9.2, implements the spectral element method in the following steps:

```
function [gdm,gmm] = gdmm_lob (ne,xe,np,ng,c)

%================================================
% Assembly of the global diffusion matrix (gdm)
% and global mass matrix (gmm)
%
% ne:      number of elements
% np(l):   polynomial order over the lth element
% ng:      number of distinct global nodes
% c:       connectivity matrix
%================================================

%--------------
% element sizes
%--------------

for l=1:ne
  h(l) = xe(l+1)-xe(l);
end

%----------
% initialize
%----------

gdm = zeros(ng,ng);
gmm = zeros(ng,ng);

%----------------------
% loop over the elements
%----------------------

for l=1:ne

    m = np(l);
    elm_dm = 2.0*edm_lob_tbl(m)/h(l);        % element diffusion matrix
    elm_mm = 0.5*h(l)*emm_lob_tbl(m);        % element mass matrix

    for ip=1:m+1
       i1 = c(l,ip);
       for jp=1:m+1
         i2 = c(l,jp);
         gdm(i1,i2) = gdm(i1,i2) + elm_dm(ip,jp);
         gmm(i1,i2) = gmm(i1,i2) + elm_mm(ip,jp);
       end
    end

end

return;
```

TABLE 3.9.1   Function *gdmm_lob* assembles the global diffusion and mass matrices
        for one-dimensional unsteady diffusion with the Lobatto spectral element method.

```
%==================================================
% Code uds_lob_cn
%
% Spectral-element code for unsteady diffusion
% with the Crank-Nicolson method
%
% SYMBOLS:
%
% ne:  number of elements
% np(l):  polynomial order over the lth element
% ng:  number of distinct global nodes
% c:   connectivity matrix
%==================================================

%----------
% input data
%----------

L = 1.0;                         % length of the rod
rho = 1.0; cp = 1.0; k = 1.0;    % physical properties
q0 = -1.0; fL = 0.0;             % boundary conditions
ne = 3; ratio = 5.0;             % number of elements, stretch ratio
np(1) = 4; np(2) = 2; np(3) = 4; % polynomial element order
dt = 0.01; nsteps = 100;         % time stepping
nplot=10;                        % will display after nplot steps

%--------
% prepare
%--------

kappa = k/(rho*cp);        % thermal diffusivity

%------------------
% initial condition
%------------------

for l=1:ng
  f(1,l) = fL;
end

%-----------------------
% element node generation
%-----------------------

[xe,xen,xg,c,ng] = discr_lob (0,L,ne,ratio,np);

%-------------------------
% source (time independent)
%-------------------------

for l=1:ng
  s(i) = 10.0*exp(-5.0*xg(i)^2/L^2);
end
```

TABLE 3.9.2 Code *uds_lob_cn* (Continuing →)

```
%-----
% plot
%-----

plot(xg, f,'-x'); hold on;        %  spectral nodes
xlabel('x'); ylabel('f');
ye = zeros(ne+1,1); plot(xe,ye,'o');   % element end-nodes

%-----------------------------------
% global diffusion and mass matrices
%-----------------------------------

[gdm,gmm] = gdmm_lob (ne,xe,np,ng,c);

%-------------------------------
% generate the vector b on the rhs
%-------------------------------

b = zeros(1,ng-1);

b(1) = q0/k;

for ip=1:ng-1
  for jp=1:ng
    b(ip) = b(ip) + gmm(ip,jp)*s(jp);
  end
end

m = np(ne);

for i=1:m
  i1 = c(ne,i);
  b(i1) = b(i1) - gdm(i1,ng)*fL;
end

%-----------------------------------------
% generate the matrix on the left-hand side
%-----------------------------------------

lhs = zeros(ng-1,ng-1);

for ip=1:ng-1
  for jp=1:ng-1
    lhs(ip,jp) = gmm(ip,jp) + 0.5*dt*kappa*gdm(ip,jp);
  end
end

%--------------------
% begin time stepping
%--------------------

icount = 0;    % step counter
```

TABLE 3.9.2 Code *uds_lob_cn* ($\rightarrow$ Continuing $\rightarrow$)

```
for irun=1:nsteps

%----------------------------
% generate the right-hand side
%----------------------------

rhs = zeros(1,ng-1);

for ip=1:ng-1

  rhs(ip) = dt*kappa*b(ip);

  for jp=1:ng
    rhs(ip) = rhs(ip) + (gmm(ip,jp) ...
      ...
        - 0.5*dt*kappa*gdm(ip,jp))*f(jp);
  end

end

%--------------
% linear solver
%--------------

sol = rhs/lhs';     % solve the (ng-1)x(ng-1) linear system

f = [sol fL];       % add the value at the right end

%---------
% plotting
%---------

if(icount==nplot)
 plot(xg, f,'-o'); icount = 0;
end

icount = icount+1;

end

%---------------------
% end of time stepping
%

%-----
% done
%-----
```

TABLE 3.9.2 ($\rightarrow$ Continued) Spectral-element code *uds_lob_cn* for one-dimensional unsteady diffusion with the Crank–Nicolson method.

1. Data input and parameter definition.

2. Element and node definition by the function *discr_lob*, listed in Table 3.3.4.

3. Specification of the initial temperature distribution and definition of the source distribution along the rod.

4. Computation of the global diffusion and mass matrices using the function *gdmm_lob*, listed in Table 3.9.1.

5. Compilation of a linear algebraic system.

6. Initiation of time stepping.

7. Computation of the right-hand side of the linear system (3.9.4).

8. Solution of the linear system using the linear solver embedded in MATLAB.

9. Visualization module designed so that graph of the transient solution is generated every *Nplot* steps, where *Nplot* is a preset parameter.

10. Return to Step 7 or termination of time stepping.

The compilation of the linear system in Step 5 involves the computation of the coefficient matrix $\mathbf{C}$ defined in (3.9.5), denoted as *lhs* in the code, and the computation of the vector $\mathbf{b}$ on the right-hand side of (3.9.6). Because the vector $\mathbf{b}$ does not change in time, it is evaluated only once, outside the time-stepping loop. More generally, this vector will be updated inside the time-stepping loop.

The solution generated by the code is displayed in Figure 3.9.1(*a*). The right-end Dirichlet condition specifies that $f_L = 0$, the initial condition prescribes that $f(x, t = 0) = 0$, and other parameter values are defined in the code. The element end-nodes are marked by dotted vertical lines and the interpolation nodes are marked with circles. As time progresses, the transient solution smoothly tends to the steady-state solution, in agreement with physical intuition.

3.9.2   *Forward Euler discretization*

To demonstrate the importance of numerical stability, we consider time discretization according to the forward Euler method. Working in the familiar way, we derive the linear system

$$\mathbf{M} \cdot \mathbf{f}^{n+1} = \mathbf{r}, \tag{3.9.7}$$

where the vector on the right-hand side is given by

$$\mathbf{r} = \left(\mathbf{M} - \kappa \, \Delta t \, \mathbf{D} \right) \cdot \mathbf{f}^n + \kappa \, \Delta t \, \mathbf{b}. \tag{3.9.8}$$

The spectral element method is implemented in the FSELIB code *uds_fe*, not listed in the text. The numerical solution generated by the code for conditions identical to those listed in the text for the Crank–Nicolson code *uds__lob_cn*, except that the time step is $\Delta t = 0.00010785$, is shown in Figure 3.9.1(*b*). The developing numerical

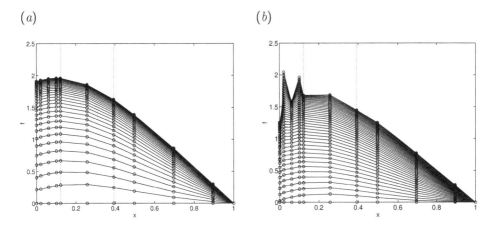

FIGURE 3.9.1 Solution of the unsteady diffusion equation generated by the FSELIB spectral element code (*a*) *uds_lob_cn* implementing the Crank–Nicolson method and (*b*) *uds_lob_fe* implementing the forward Euler method, for the source function given in (1.2.17).

instability in spite of the smallness of the time step underscores the significance of the implicit time discretization underlying the Crank–Nicolson method.

In the calculation shown in Figure 3.9.1(*b*), the solution domain is divided into ten intervals separated by the spectral nodes, and the size of the smallest interval is $h \simeq 0.02$. The stability criterion (2.1.48) imposes the approximate restriction

$$\Delta t < \frac{h^2}{6\kappa} \simeq 0.000066, \tag{3.9.9}$$

which is consistent with the results of the numerical simulation.

PROBLEMS

3.9.1 *Significance of mass lumping*

Modify the FSELIB code *uds_cn* to implement mass lumping. Discuss the significance of mass lumping by comparison with the consistent formulation.

3.9.2 *Forward Euler method*

Run the FSELIB code *uds_fe* for several sizes of the time step, Δt. Discuss the stability of the solution with reference to the theoretical criterion shown in (2.1.48).

3.9.3 *Node condensation for unsteady diffusion*

Discuss whether it is possible to implement node condensation in the case of unsteady diffusion.

The finite element method in two dimensions

4

Having discussed the fundamental concepts underlying the formulation and implementation of the Galerkin finite element method in one spatial dimension, we proceed to extend the methodology to two dimensions. The generalized framework developed in this chapter will allow us to build algorithms for solving the Laplace equation, the Poisson equation, the unsteady heat conduction equation, the convection equation, the convection–diffusion equation, the equations of linear elasticity in solid mechanics, and the equations of viscous flow in hydrodynamics. In the discourse, we will demonstrate one of the most powerful features of the finite element method, which is the ability to accommodate solution domains with arbitrary and even complex geometry.

The implementation of the Galerkin finite element method in two dimensions follows the general blueprint of that in one dimension discussed in Chapters 1 and 2, involving the following basic modules:

1. *Formulation of the Galerkin projection using global interpolation functions, and application of Gauss's divergence theorem to reduce the order of the highest-order derivative.*

2. *Domain discretization into finite elements defined by geometrical nodes, assignment of element interpolation nodes, and computation of node and element connectivity matrices, as required.*

   To ensure C^0 continuity of the finite element expansion, adjacent elements must share some interpolation nodes. Unless a physical discontinuity or a singularity is anticipated, the solution must be common at the shared element nodes.

3. *Derivation of a system of algebraic or differential equations by substituting the finite element expansion into the Galerkin projection.*

4. *Computation of the element diffusion, mass, and advection matrices, as required.*

   These computations are done mostly numerically with the aid of integration quadratures pertinent to the selected element shape.

5. *Assembly of the global linear system in terms of the element matrices.*

   The algorithm relies on a connectivity matrix that associates element inter-polation nodes to unique global nodes.

6. *Implementation of the boundary conditions.*

   This is done most conveniently by modifying the coefficient matrix and right-hand side of the linear system arising from the projection of the global interpo-lation functions corresponding to *all* global nodes, possibly with the addition of contour integrals implementing the Neumann boundary condition, as re-quired.

7. *Solution of the final algebraic system.*

   For simplicity, we will use the linear solver embedded in MATLAB. In prac-tice, because the dimension of the linear system can be large, custom-made algorithms are employed, as discussed in Appendix C.

8. *Time stepping.*

   In problems with time dependence, the algorithm is applied at a sequence of time steps to yield the evolution of the solution from a specified initial state.

In this chapter, the formulation and implementation of these basic procedures will be demonstrated by discussing selected modular cases of the unsteady convection–diffusion equation. Once the general methodology has been established, high-order and spectral element methods will arise as natural extensions, as discussed in Chapter 4. Applications in solid and structural mechanics are discussed in Chapter 5, and applications in fluid mechanics are discussed in Chapter 6.

4.1   Convection–diffusion in two dimensions

Consider unsteady heat transport in the xy plane in the presence of a distributed source due, for example, to a chemical reaction, as illustrated in Figure 4.1.1(a). The evolution of the temperature field, $f(x, y, t)$, is governed by the convection–diffusion equation

$$\rho\, c_p \left(\frac{\partial f}{\partial t} + u_x \frac{\partial f}{\partial x} + u_y \frac{\partial f}{\partial y} \right) = k \left(\frac{\partial^2 f}{\partial x^2} + \frac{\partial^2 f}{\partial y^2} \right) + s, \qquad (4.1.1)$$

where ρ, c_p, and k are the medium density, heat capacity, and thermal conductiv-ity, $u_x(x, y, f, t)$ and $u_y(x, y, f, t)$ are the x and y components of the velocity, and $s(x, y, f, t)$ is a distributed source. Note that equation (4.1.1) is the two-dimensional counterpart of the one-dimensional rod equation (2.3.1). For simplicity, the con-vection velocity and source are assumed to depend on x, y, and t explicitly, but not implicitly by way of $f(x, y, t)$. To signify this explicit dependence, we write $u_x(x, y, t)$, $u_y(x, y, t)$ and $s(x, y, t)$.

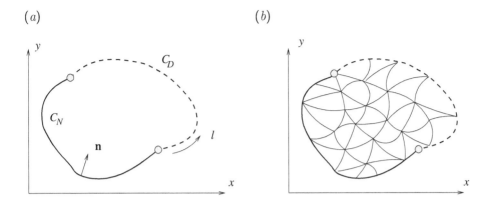

FIGURE 4.1.1   (*a*) Illustration of heat conduction in a plate with arbitrary geometry in the xy plane, showing the Dirichlet contour, C_D, and the Neumann contour, C_N. (*b*) Finite element discretization of the solution domain into triangular elements with curved sides.

In vector notation, the transport equation (4.1.1) takes the compact form

$$\frac{\partial f}{\partial t} + \mathbf{u} \cdot \nabla f = \kappa \nabla^2 f + \frac{s}{\rho \, c_p}, \tag{4.1.2}$$

which can be rewritten as

$$\frac{\partial f}{\partial t} + \mathbf{u} \cdot \nabla f = \kappa \left(\nabla^2 f + \frac{s}{k} \right), \tag{4.1.3}$$

where

$$\kappa \equiv \frac{k}{\rho \, c_p} \tag{4.1.4}$$

is the thermal diffusivity with dimensions of length squared over time,

$$\nabla f = \left(\frac{\partial f}{\partial x}, \ \frac{\partial f}{\partial y} \right) \tag{4.1.5}$$

is the two-dimensional gradient, and

$$\nabla^2 f \equiv \nabla \cdot \nabla f = \frac{\partial^2 f}{\partial x^2} + \frac{\partial^2 f}{\partial y^2} \tag{4.1.6}$$

is the two-dimensional Laplacian operator, equal to the divergence of the gradient. We may write

$$\mathbf{u} \cdot \nabla f = u_x \frac{\partial f}{\partial x} + u_y \frac{\partial f}{\partial y} = |\mathbf{u}| \frac{\partial f}{\partial l_u}, \tag{4.1.7}$$

where l_u is the arc length measured in the direction of $\mathbf{u}$.

4.1.1 Boundary conditions

Equation (4.1.2) is to be solved in a domain, D, that is enclosed by a contour, C, subject to two complementary boundary conditions: (*a*) the *Neumann* boundary condition specifying the inward or outward normal derivative of the unknown function, or (*b*) the *Dirichlet* boundary condition specifying the boundary distribution of the unknown function. Other boundary conditions, including the mixed boundary condition, also called a convection or Robin boundary condition, can be handled by similar methods.

Neumann boundary condition

In the present problem of heat transport, the Neumann boundary condition prescribes the heat flux along the Neumann portion of C, denoted as C_N, drawn with the solid line in Figure 4.1.1(*a*),

$$-k\,\mathbf{n}\cdot\nabla f \equiv -k\frac{\partial f}{\partial l_n} = q(l), \qquad (4.1.8)$$

where $\mathbf{n} = (n_x, n_y)$ is the unit vector normal to C pointing *into* the solution domain, $q(l)$ is a given function of arc length, l, along C, as shown in Figure 4.1.1(*a*), and

$$\mathbf{n}\cdot\nabla f = n_x\frac{\partial f}{\partial x} + n_y\frac{\partial f}{\partial y}. \qquad (4.1.9)$$

The notation $\partial/\partial l_n$ denotes the derivative with respect to the inward length normal to the boundary. If $q(l) > 0$, in which case $\mathbf{n}\cdot\nabla f < 0$, heat enters the solution domain; if $q(l) < 0$, in which case $\mathbf{n}\cdot\nabla f > 0$, heat escapes from the solution domain across C_N.

Dirichlet boundary condition

The Dirichlet boundary condition prescribes the temperature distribution along the Dirichlet portion of the boundary, denoted as C_D, drawn with the broken line in Figure 4.1.1(*a*),

$$f = g(l), \qquad (4.1.10)$$

where $C_D = C - C_N$ is the complement of C_N and $g(l)$ is a specified function.

4.1.2 Galerkin projection

To develop the Galerkin finite element method, we follow the general steps outlined in the introduction of this chapter. In the first step, we carry out the Galerkin projection of the governing differential equation (4.1.1) using as weighting functions the global interpolation functions, $\phi_i(x, y)$. The interpolation functions are defined according to the selected element types and location of the associated element interpolation nodes, as will be discussed in this section in great detail. For the moment, we assume that these functions are available.

Multiplying the right-hand side of (4.1.1) by $\phi_i(x, y)$, integrating the product over the solution domain, D, and manipulating the integrand to form a divergence, we obtain

$$\iint_D \phi_i \left(k \nabla^2 f + s \right) dx \, dy = \iint_D \left(k \phi_i \left(\nabla \cdot \nabla f \right) + \phi_i s \right) dx \, dy$$

$$= \iint_D \left(k \left(\nabla \cdot (\phi_i \nabla f) - \nabla \phi_i \cdot \nabla f \right) + \phi_i s \right) dx \, dy \qquad (4.1.11)$$

$$= k \iint_D \nabla \cdot (\phi_i \nabla f) \, dx \, dy + \iint_D \left(-k \nabla \phi_i \cdot \nabla f + \phi_i s \right) dx \, dy.$$

Next, we apply the Gauss divergence theorem under the assumption that the function ϕ_i is continuous throughout D, and obtain the equation

$$\iint_D \phi_i \left(k \nabla^2 f + s \right) dx \, dy \qquad (4.1.12)$$

$$= -k \oint_C \phi_i \, \mathbf{n} \cdot \nabla f \, dl + \iint_D \left(-k \nabla \phi_i \cdot \nabla f + \phi_i s \right) dx \, dy.$$

Substituting the definition of the boundary flux, $q \equiv -k \, \mathbf{n} \cdot \nabla f$, into the first integral on the right-hand side and rearranging, we obtain

$$\iint_D \phi_i \left(k \nabla^2 f + s \right) dx \, dy = -k \iint_D \nabla \phi_i \cdot \nabla f \, dx \, dy + Q_i + S_i, \qquad (4.1.13)$$

where

$$Q_i \equiv \oint_C \phi_i \, q \, dl, \qquad S_i \equiv \iint_D \phi_i \, s \, dx \, dy \qquad (4.1.14)$$

are boundary and domain integrals involving the boundary flux and distributed source, weighted by the ith global interpolation function. If the ith node associated with ψ_i is an interior node, ϕ_i is zero along the entire contour C, and $Q_i = 0$.

Next, we multiply the left-hand side of (4.1.1) by ϕ_i, integrate the product over the solution domain, D, set the resulting expression equal to the right-hand side of (4.1.13), and divide by $\rho \, c_p$ to obtain the Galerkin equation

$$\iint_D \phi_i \frac{\partial f}{\partial t} \, dx \, dy + \iint_D \phi_i \, \mathbf{u} \cdot \nabla f \, dx \, dy$$

$$= \frac{1}{\kappa} \left(-\iint_D \nabla \phi_i \cdot \nabla f \, dx \, dy + \frac{1}{k} (Q_i + S_i) \right), \qquad (4.1.15)$$

which provides us with a foundation for the Galerkin finite element method (GFEM).

4.1.3 Domain discretization and interpolation

To implement the finite element method, we discretize the solution domain, D, into a set of N_E elements, as illustrated in Figure 4.1.1(b). In practice, the elements have triangular or rectangular shapes defined by a small group of geometrical element nodes.

The collection of all geometrical element nodes comprises a set of N_{GN} *unique geometrical global nodes.* If an element node coincides with a neighboring element node, the two nodes are mapped to the same global node through a connectivity matrix, as will be discussed later in this section. Accordingly, the element nodes are assigned *global labels* ranging from 1 to N_{GN}.

If all elements are defined by the same number of nodes, N_{EN}, because of node sharing, the number of unique global nodes is less than $N_{EN}N_E$, where N_E is the number of elements. For example, if all elements are triangles with straight edges defined by the three vertices, $N_{EN} = 3$, then the number of unique global nodes is less than $3\,N_E$.

Interpolation nodes

To describe the *a priori* unknown solution in numerical form, we introduce interpolation element nodes, not all of which necessarily coincide with the geometrical element nodes. The collection of all interpolation element nodes comprises a set of N_G *unique global interpolation nodes.*

In the *isoparametric interpolation*, the set of interpolation element nodes coincides with the set of geometrical element nodes and the number of global interpolation nodes is equal to the number of global geometrical nodes, $N_G = N_{GN}$. In this chapter we discuss the isoparametric interpolation and in Chapter 5 we discuss super-parametric interpolation in the context of the spectral element method where $N_G > N_{GN}$.

Connectivity matrix and element flags

In the second stage of the implementation, boundary flags and a connectivity matrix are introduced with two main goals:

- *Map the local element nodes to the unique global nodes.*

- *Designate whether a certain element or global node is a boundary node and store the boundary conditions.*

A connectivity matrix is introduced, $c(i, j)$ for $i = 1, \ldots, N_E$, such that $c(i, j)$ is the global label of the jth node of the ith element. Thus, $c(i, j)$ takes values in the range $1, \ldots, N_G$, where N_G is the number of unique global nodes.

Assume that the ith element hosts m interpolation nodes. To specify the boundary conditions, an element-node flag is introduced, $efl(i, j)$ for $i = 1, \ldots, N_E$ and

$j = 1, \ldots, m$, such that

$$efl(i,j) = \begin{cases} 0 & \text{if the } j\text{th node is not a boundary node,} \\ \neq 0 & \text{if the } j\text{th node is a boundary node.} \end{cases} \qquad (4.1.16)$$

If the solution domain is enclosed by a single contiguous contour, we may set $efl(i,j) = 1$ for those nodes located at that contour. If the solution domain is enclosed by two distinct contours, we may set $efl(i,j) = 1$ for those nodes located at the first contour, and $efl(i,j) = 2$ for the nodes located at the second contour. Multiply connected domains enclosed by a higher number of distinct boundary contours are handled in a similar fashion.

In practice, the connectivity matrix and the boundary flags are generated in the process of triangulation, as discussed in Section 4.3.

Global node flags

Correspondingly, a global node flag is introduced, $gfl(i)$ for $i = 1, \ldots, N_G$, such that

$$gfl(i) = \begin{cases} 0 & \text{if the } i\text{th global node is not a boundary node,} \\ \neq 0 & \text{if the } i\text{th global node is a boundary node.} \end{cases} \qquad (4.1.17)$$

The type of boundary condition specified at the global nodes can be flagged by additional arrays. Companion data arrays can be introduced to hold Dirichlet, Neumann, or other boundary data, as necessary.

Isoparametric interpolation

In the third stage of the implementation, the requisite solution is expressed in terms of *a priori* unknown values at the global interpolation nodes, f_j, and associated *global cardinal interpolation functions*, ϕ_j, in the isoparametric form

$$f(x,y,t) = \sum_{j=1}^{N_G} f_j(t)\,\phi_j(x,y). \qquad (4.1.18)$$

By definition, the global interpolation function $\phi_j(x,y)$ takes the value of unity at the jth global interpolation node and the value of zero at all other global interpolation nodes, as illustrated in Figure 4.1.2. Specific expressions will be given in this chapter for triangular elements defined by three vertex nodes and in Chapter 5 for triangular elements defined by six nodes, including three vertex nodes and three edge nodes.

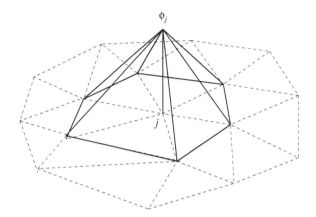

FIGURE 4.1.2 Illustration of a tent-like global interpolation function associated with the jth global node. By definition, $\phi_j(x,y)$ takes the value of unity at the jth global node and the value of zero at all other global nodes.

4.1.4 Galerkin finite element equations

Inserting expansion (4.1.18) and a similar expansion for the source term, s, into (4.1.15), we derive the Galerkin finite element equations

$$
\sum_{i=j}^{N_G} M_{ij}\,\frac{\mathrm{d}f_j}{\mathrm{d}t} + \sum_{i=j}^{N_G} N_{ij}\,f_j = \frac{1}{\kappa}\left(-\sum_{i=j}^{N_G} D_{ij}\,f_j + \frac{1}{k}\oint_C \phi_i\,q\,\mathrm{d}l + \frac{1}{k}\sum_{i=j}^{N_G} M_{ij}\,s_j \right),
$$

$$(4.1.19)$$

where

$$
D_{ij} \equiv \iint_D \nabla\phi_i \cdot \nabla\phi_j \,\mathrm{d}x\,\mathrm{d}y
\tag{4.1.20}
$$

is the global diffusion matrix,

$$
M_{ij} \equiv \iint_D \phi_i\,\phi_j \,\mathrm{d}x\,\mathrm{d}y
\tag{4.1.21}
$$

is the global mass matrix, and

$$
N_{ij} \equiv \iint_D \phi_i\,\mathbf{u} \cdot \nabla\phi_j \,\mathrm{d}x\,\mathrm{d}y
\tag{4.1.22}
$$

is the global advection matrix.

Applying (4.1.19) for the global interpolation functions associated with the interpolation nodes where the Dirichlet boundary condition is *not* prescribed, we obtain a system of ordinary differential equations for the unknown nodal values encapsulated in a vector, $\mathbf{f}$,

$$\mathbf{M} \cdot \frac{d\mathbf{f}}{dt} + \mathbf{N} \cdot \mathbf{f} = \kappa \left(-\mathbf{D} \cdot \mathbf{f} + \mathbf{b} \right), \tag{4.1.23}$$

where the vector $\mathbf{p}$ on the right-hand side incorporates the given source term and the boundary conditions,

$$b_i \equiv \frac{1}{k} \oint_C \phi_i\, q\, dl + \frac{1}{k} \sum_{j=1}^{N_G} M_{ij}\, s_j. \tag{4.1.24}$$

The first integral on the right-hand side of (4.1.24) is non-zero only if the ith node is a boundary node where the Neumann boundary condition is specified.

Element matrices

The domain integrals in (4.1.20), (4.1.21), and (4.1.22) are assembled in terms of corresponding element integrals, as discussed in Section 1.2. The lth-element diffusion matrix is given by

$$A_{ij}^{(l)} \equiv \iint_{E_l} \nabla \psi_i^{(l)} \cdot \nabla \psi_j^{(l)} \, dx\, dy, \tag{4.1.25}$$

the corresponding element mass matrix is given by

$$B_{ij}^{(l)} \equiv \iint_{E_l} \psi_i^{(l)} \, \psi_j^{(l)} \, dx\, dy, \tag{4.1.26}$$

and the corresponding element advection matrix is given by

$$C_{ij}^{(l)} \equiv \iint_{E_l} \psi_i^{(l)} \, \mathbf{u} \cdot \nabla \psi_j^{(l)} \, dx\, dy, \tag{4.1.27}$$

where the integration is performed over the element area, E_l, and the indices i and j run over the element nodes. The algorithm shown in Table 4.1.1 assembles the $N_G \times N_G$ global diffusion matrix assisted by a connectivity matrix, $c(i, j)$, that maps element nodes to global nodes. The mass and advection matrices are assembled in a similar fashion.

4.1.5   Implementation of the Dirichlet boundary condition

It is not necessary to carry out the Galerkin projection at a node where the Dirichlet boundary condition is imposed, that is, at a node that lies at a Dirichlet contour, C_D, where the unknown function is specified. In practice, it is expedient to make this exception after an extended preliminary $N_G \times N_G$ system has been compiled, at a minimal computational cost.

Do $i = 1, \ldots, N_G$ *Initialize to zero*
    Do $j = 1, \ldots, N_G$
        $D_{ij} = 0.0$
    End Do
End Do

Do $l = 1, \ldots, N_E$ *Run over the elements*
    Compute the element diffusion matrix $\mathbf{A}^{(l)}$
    Do $i = 1, \ldots, N(l)$ *Run over the element interpolation nodes*
        $i_1 = c(l, i)$
        Do $j = 1, \ldots, N(l)$ *Run over the element interpolation nodes*
           $i_2 = c(l, j)$
           $D_{i_1, i_2} = D_{i_1, i_2} + A_{ij}^{(l)}$
        End Do
    End Do

End Do

TABLE 4.1.1   Assembly of the $N_G \times N_G$ global diffusion matrix, $\mathbf{D}$, in terms of element diffusion matrices.

In the case of steady heat conduction, the system of ordinary differential equations (4.1.23) reduces into a linear algebraic system,

$$\mathbf{D} \cdot \mathbf{f} = \mathbf{b}. \tag{4.1.28}$$

Assume that the boundary conditions specify that $f = f_m$ at the mth global node, where f_m is a given value. In the finite element implementation, after a preliminary Galerkin system has been compiled for *all* nodes, the mth constituent equation is discarded and replaced with the Dirichlet boundary condition, working in four stages:

1. All entries on the right-hand side, b_i, are replaced with $b_i - D_{im} f_m$ for $i = 1, \ldots, N_G$.

2. All elements in the mth column and mth row of $\mathbf{D}$ are set to zero.

3. The diagonal element, D_{mm}, is set to unity.

4. The mth entry of the right-hand side, b_m, is replaced by f_m.

Although this procedure introduces some redundancy, the preserved symmetry of the coefficient matrix after implementing the boundary conditions is a significant advantage.

Do $m = 1, \ldots, N_{\mathrm{G}}$ *Run over all global nodes*

    If $(gfl(m){=}1)$ then   *Dirichlet boundary node*

        Do $i = 1, \ldots, N_{\mathrm{G}}$ *Run over all global nodes*

          $b_i = b_i - D_{im} * bcd(m)$

          $D_{im} = 0$

          $D_{mi} = 0$

        End Do

        $D_{mm} = 1.0$

        $b_m = bcd(m)$

    End If

End Do

TABLE 4.1.2   Implementation of the Dirichlet boundary condition for steady two-dimensional conduction. The vector component $bcd(m)$ contains the prescribed boundary value of the solution at the mth global node that is a boundary node.

The method is implemented in the algorithm shown in Table 4.1.2, subject to the convention that the vector component $bcd(m)$ contains the prescribed boundary value of the solution specified by the Dirichlet boundary condition at the position of the mth global node, which is a boundary node.

4.1.6   Split nodes

When a boundary node lies at a corner or junction where different conditions are applied on either side, or when a discontinuity is expected across a shared element node, the node develops a split personality. One way to handle this dichotomy is to map the dual node to two ghost global nodes, one corresponding to the left and the second to the right side of the boundary. By doing so, we effectively introduce an artificial crack extending up to the nearest interior node, as illustrated in Figure 4.1.3. Because of the implicit presence of the crack, the ghost nodes support two distinct global interpolation functions. Appropriate boundary conditions may then be enforced at the two ghost nodes, as discussed in Section 4.8.

4.1.7   Variational formulation

In Section 1.2, we demonstrated an intimate relationship between Galerkin's method and Ritz's implementation of Rayleigh's variational formulation for one-dimensional diffusion at steady state. Moreover, we showed that the finite element method is a

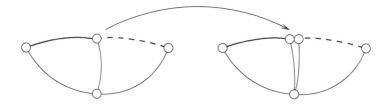

FIGURE 4.1.3   Illustration of node splitting to accommodate a discontinuity or account
for a corner.

specific implementation of the method of weighted residuals. In the case of steady
two-dimensional diffusion governed by a Poisson equation,

$$k \nabla^2 f + s(x,y) = 0, \tag{4.1.29}$$

subject to the boundary conditions (4.1.8) and (4.1.10), the variational formulation
states that computing the solution, f, is equivalent to maximizing the functional

$$\mathcal{F} < \omega(x,y) > \equiv -\frac{1}{2} k \iint_D |\nabla \omega|^2 \, dx \, dy + \iint_D s \omega \, dx \, dy + \int_{C_N} q(l) \, \omega \, dl \tag{4.1.30}$$

over all functions, $\omega(x,y)$, that satisfy the Dirichlet boundary condition (4.1.10),
where C_N in the last integral of (4.1.30) is the Neumann contour. The satisfaction
of the Neumann boundary condition is enforced by the last term on the right-hand
side of (4.1.30). Note that

$$|\nabla \omega|^2 \equiv \left(\frac{\partial \omega}{\partial x} \right)^2 + \left(\frac{\partial \omega}{\partial y} \right)^2, \tag{4.1.31}$$

which differs from the Laplacian operator. If the Dirichlet boundary condition is
specified around the entire boundary, C_N is null and the last term on the right-hand
side of (4.1.30) does not appear.

Alternative functional

Manipulating the first integral on the right-hand side of (4.1.30), we obtain

$$\iint_D |\nabla \omega|^2 \, dx \, dy = \iint_D \left[\frac{\partial}{\partial x}\left(\omega \frac{\partial \omega}{\partial x} \right) - \omega \frac{\partial^2 \omega}{\partial x^2} + \frac{\partial}{\partial y}\left(\omega \frac{\partial \omega}{\partial y} \right) - \omega \frac{\partial^2 \omega}{\partial y^2} \right] dx \, dy$$

$$= \iint_D \left[\nabla \cdot (\omega \, \nabla \omega) - \omega \, \nabla^2 \omega \right] dx \, dy. \tag{4.1.32}$$

Using the Gauss divergence theorem, we obtain

$$\iint_D |\nabla \omega|^2 \, dx \, dy = - \oint_C \omega \, \mathbf{n} \cdot \nabla \omega \, dl - \iint_D \omega \, \nabla^2 \omega \, dx \, dy, \tag{4.1.33}$$

where $\mathbf{n}$ is the unit vector normal to C, pointing *into* the solution domain, D. Substituting this expression into (4.1.30), we obtain

$$\mathcal{F} < \omega(x, y) >= \frac{1}{2} k \left(\iint_D \omega \nabla^2 \omega \, dx \, dy + \oint_C \omega \mathbf{n} \cdot \nabla \omega \, dl \right)$$
$$+ \iint_D s\omega \, dx \, dy + \int_{C_N} q(l) \, \omega \, dl, \tag{4.1.34}$$

which can be rearranged into

$$\mathcal{F} < \omega(x, y) >= \frac{1}{2} k \iint_D \omega \nabla^2 \omega \, dx \, dy + \iint_D s\omega \, dx \, dy$$
$$+ \frac{1}{2} k \int_{C_D} g(l) \, \mathbf{n} \cdot \nabla \omega \, dl + \frac{1}{2} \int_{C_N} q(l) \, \omega \, dl. \tag{4.1.35}$$

If the Dirichlet boundary condition is specified around the entire boundary, C_N is null and the last term on the right-hand side of (4.1.34) does not appear.

Proof of equivalence

To prove equivalence to minimization, we perturb the solution from $f(x, y)$ to $f(x, y) + \epsilon\, v(x, y)$, where the disturbance function, $v(x, y)$, also called a *variation*, is required to satisfy the homogeneous Dirichlet boundary condition $v = 0$ over the Dirichlet contour C_D, and ϵ is a dimensionless number whose magnitude is much less than unity. Next, we consider the difference

$$\delta \mathcal{F} \equiv \mathcal{F} < f(x, y) + \epsilon\, v(x, y) > - \mathcal{F} < f(x, y) >, \tag{4.1.36}$$

and use expression (4.1.30) to find that

$$\delta \mathcal{F} = \epsilon \left(-k \iint_D \nabla f \cdot \nabla v \, dx \, dy + \iint_D s(x, y) \, v \, dx \, dy + \int_{C_N} q(l) \, v \, dl \right) + O(\epsilon^2). \tag{4.1.37}$$

Concentrating on the first integral on the right-hand side, we write

$$\iint_D \nabla f \cdot \nabla v \, dx \, dy = \iint_D \left[\nabla \cdot (v \nabla f) - v \nabla^2 f \right] dx \, dy \tag{4.1.38}$$

and apply the divergence theorem to obtain

$$\iint_D \nabla f \cdot \nabla v \, dx \, dy = - \oint_C v \mathbf{n} \cdot \nabla f \, dl - \iint_D v \nabla^2 f \, dx \, dy. \tag{4.1.39}$$

Consequently,

$$\delta \mathcal{F} = \epsilon \iint_D \left(k \nabla^2 f + s \right) v \, dx \, dy + \epsilon k \oint_C v \mathbf{n} \cdot \nabla f \, dl + \epsilon \int_{C_N} q(l) \, v \, dl + O(\epsilon^2). \tag{4.1.40}$$

Finally, we recall that $v = 0$ on C_D and invoke the definition of the boundary flux, $q \equiv -k\,\mathbf{n}\cdot\nabla f$, to obtain

$$\delta\mathcal{F} = \epsilon\iint_D \left(k\,\nabla^2 f + s\right) v\,\mathrm{d}x\,\mathrm{d}y + O(\epsilon^2). \tag{4.1.41}$$

If a function, $f(x, y)$, satisfies (4.1.29) subject to the aforementioned boundary conditions, the first variation vanishes and the functional is stationary.

Alternatively, we may use (4.1.34) to compute the variation

$$\delta\mathcal{F} = \epsilon\frac{1}{2}k\left[\iint_D (v\,\nabla^2 f + f\,\nabla^2 v)\,\mathrm{d}x\,\mathrm{d}y + \oint_C (v\,\mathbf{n}\cdot\nabla f + f\,\mathbf{n}\cdot\nabla v)\,\mathrm{d}l\right]$$
$$+ \epsilon\left[\iint_D s(x, y)\,v\,\mathrm{d}x\,\mathrm{d}y + \int_{C_N} q(l)\,v\,\mathrm{d}l\right] + O(\epsilon^2). \tag{4.1.42}$$

Now using Green's second identity,

$$\iint_D f\,\nabla^2 v\,\mathrm{d}x\,\mathrm{d}y = \iint_D f\,\nabla^2 v\,\mathrm{d}x\,\mathrm{d}y - \oint_C (f\,\mathbf{n}\cdot\nabla v - v\,\mathbf{n}\cdot\nabla f)\,\mathrm{d}l, \tag{4.1.43}$$

we simplify the first integral on the right-hand side to obtain

$$\delta\mathcal{F} = \epsilon k\left(\iint_D v\,\nabla^2 f\,\mathrm{d}x\,\mathrm{d}y + \oint_C v\,\mathbf{n}\cdot\nabla f\,\mathrm{d}l\right)$$
$$+ \epsilon\iint_D s(x, y)\,v\,\mathrm{d}x\,\mathrm{d}y + \epsilon\int_{C_N} q(l)\,v\,\mathrm{d}l + O(\epsilon^2), \tag{4.1.44}$$

which reduces to (4.1.41).

Ritz and finite element methods

Ritz's implementation of the variational formulation and the Galerkin implementation of the method of weighted residuals are similar to those discussed in Section 1.3 in one dimension with a combination of the Dirichlet and Neumann boundary condition.

PROBLEM

4.1.1 *Variational formulation*

(a) Show that (4.1.30) is the two-dimensional counterpart of the one-dimensional form (1.3.29).

(b) Consider the linear equation

$$k\,\nabla^2 f + g(x, y)\,f + s(x, y) = 0, \tag{4.1.45}$$

where $g(x, y)$ is a specified function, subject to the boundary conditions (4.1.8) and (4.1.10). Show that the pertinent Rayleigh functional is given by (4.1.30), provided

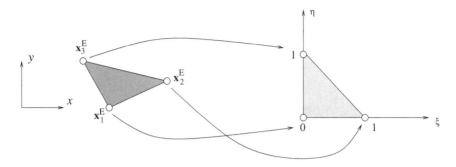

FIGURE 4.2.1   A three-node triangle with straight edges in the xy plane is mapped to a right isosceles triangle in the parametric $\xi\eta$ plane.

that the term

$$\frac{1}{2} \iint_D g(x, y)\, \omega^2 \, \mathrm{d}x\, \mathrm{d}y \qquad (4.1.46)$$

is added to the right-hand side.

4.2   Three-node triangles

Having outlined the basic principles of the Galerkin finite element method in two dimensions, we proceed to discuss the particulars of the implementation and build a complete finite element code.

In the simplest implementation, the solution domain in the xy plane is discretized into triangular elements with straight edges defined by 3 geometrical *vertex nodes*,

$$\mathbf{x}_i^{\mathrm{E}} = (\, x_i^{\mathrm{E}},\, y_i^{\mathrm{E}} \,) \qquad (4.2.1)$$

for $i = 1, 2, 3$, as illustrated on the left of Figure 4.2.1. Note that the vertices are numbered in the counterclockwise direction around the element contour. The superscript E emphasizes that these are local or *element* nodes, which can be mapped to unique *global* nodes in terms of a connectivity matrix.

Element node interpolation functions

To describe an element in the physical xy plane, we map it to a standard right isosceles triangle in the $\xi\eta$ parametric plane, as shown in Figure 4.2.1. The first element node is mapped to the origin, $\xi = 0, \eta = 0$, the second to the point $\xi = 1, \eta = 0$ on the ξ axis, and the third to the point $\xi = 0, \eta = 1$ on the η axis. The mapping from the physical to the parametric plane is mediated by the linear expansion

$$\mathbf{x} = \mathbf{x}_1^{\mathrm{E}}\, \psi_1(\xi, \eta) + \mathbf{x}_2^{\mathrm{E}}\, \psi_2(\xi, \eta) + \mathbf{x}_3^{\mathrm{E}}\, \psi_3(\xi, \eta), \qquad (4.2.2)$$

where $\psi_i(\xi, \eta)$ are element node interpolation functions satisfying a familiar cardinal property: $\psi_i = 1$ at the ith element node and $\psi_i = 0$ at the other two element nodes, so that

$$\psi_i(\xi_j, \eta_j) = \delta_{i,j} \tag{4.2.3}$$

for $i, j = 1, 2, 3$, where $\delta_{i,j}$ is Kronecker's delta, and

$$(\xi_1, \eta_1) = (0, 0), \qquad (\xi_2, \eta_2) = (1, 0), \qquad (\xi_3, \eta_3) = (0, 1) \tag{4.2.4}$$

are the coordinates of the vertices in the $\xi\eta$ plane.

To derive the node interpolation functions, we write

$$\psi_i(\xi, \eta) = a_i + b_i\,\xi + c_i\,\eta, \tag{4.2.5}$$

and compute the coefficients, a_i, b_i, and c_i, to satisfy the aforementioned cardinal interpolation conditions. For example, for node labeled 1, we require that

$$\begin{aligned}
\psi_1(\xi_1, \eta_1) &= a_1 + b_1 \times 0.0 + c_1 \times 0.0 = a_1 = 1, \\
\psi_1(\xi_2, \eta_2) &= a_1 + b_1 \times 1.0 + c_1 \times 0.0 = a_1 + b_1 = 0, \\
\psi_1(\xi_3, \eta_3) &= a_1 + b_1 \times 0.0 + c_1 \times 1.0 = a_1 + c_1 = 0,
\end{aligned} \tag{4.2.6}$$

which yields $a_1 = 1$, $b_1 = -1$, and $c_1 = -1$. Substituting these values into (4.2.5) and repeating the calculation for the other two nodes, we obtain the expressions

$$\psi_1(\xi, \eta) = \zeta, \qquad \psi_2(\xi, \eta) = \xi, \qquad \psi_3(\xi, \eta) = \eta, \tag{4.2.7}$$

where

$$\zeta = 1 - \xi - \eta. \tag{4.2.8}$$

The trio of variables (ξ, η, ζ) comprise triangle *barycentric coordinates*. Physically,

$$\zeta = A_1/A, \qquad \xi = A_2/A, \qquad \eta = A_3/A, \tag{4.2.9}$$

where A_1, A_2, and A_3 are the areas of the sub-triangles defined by the field point, $\mathbf{x}$, as shown in Figure 4.2.2.

A graph of the first interpolation function, ψ_1, over the area of the triangle in the $\xi\eta$ plane, generated by the FSELIB function *psi_3*, not listed in the text, is shown in Figure 4.2.3. The other two interpolation functions corresponding to the second or third node have similar shapes.

Substituting the interpolation functions given in (4.2.7) into (4.2.2), we obtain a mapping function that is a *complete linear function* in ξ and η, consisting of a constant term, a term that is linear in ξ, and a term that is linear in η,

$$\mathbf{x} = \mathbf{x}_1^{\mathrm{E}} + (\mathbf{x}_2^{\mathrm{E}} - \mathbf{x}_1^{\mathrm{E}})\,\xi + (\mathbf{x}_3^{\mathrm{E}} - \mathbf{x}_1^{\mathrm{E}})\,\eta. \tag{4.2.10}$$

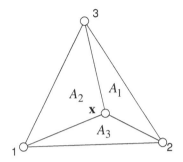

FIGURE 4.2.2 The barycentric coordinates are defined with respect to the areas of 3 triangles defined by an arbitrary field point, **x**, and two vertices.

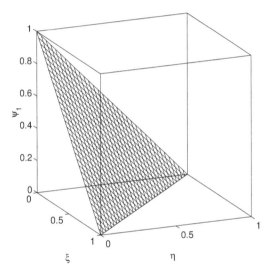

FIGURE 4.2.3 Graph of the element interpolation function ψ_1 associated with the first vertex node of a three-node triangle.

Explicitly, the x and y components of the mapping are given by

$$\begin{aligned}
x &= x_1^{\mathrm{E}} + (x_2^{\mathrm{E}} - x_1^{\mathrm{E}})\,\xi + (x_3^{\mathrm{E}} - x_1^{\mathrm{E}})\,\eta,\\
y &= y_1^{\mathrm{E}} + (y_2^{\mathrm{E}} - y_1^{\mathrm{E}})\,\xi + (y_3^{\mathrm{E}} - y_1^{\mathrm{E}})\,\eta.
\end{aligned} \tag{4.2.11}$$

For a specified location, (x, y), the corresponding coordinates (ξ, η) over the parametric triangle can be found by solving a system of two linear equations expressed by (4.2.11).

Integral of a function over a triangle

The Jacobian matrix of the mapping from the xy to the $\xi\eta$ plane is defined as

$$\mathbf{J} \equiv \begin{bmatrix} \dfrac{\partial x}{\partial \xi} & \dfrac{\partial x}{\partial \eta} \\[2mm] \dfrac{\partial y}{\partial \xi} & \dfrac{\partial y}{\partial \eta} \end{bmatrix}. \tag{4.2.12}$$

The determinant of the Jacobian matrix is the *surface metric coefficient,*

$$h_S \equiv \det(\mathbf{J}) = \frac{\partial x}{\partial \xi}\frac{\partial y}{\partial \eta} - \frac{\partial x}{\partial \eta}\frac{\partial y}{\partial \xi} = \left| \frac{\partial \mathbf{x}}{\partial \xi} \times \frac{\partial \mathbf{x}}{\partial \eta} \right|. \tag{4.2.13}$$

The last expression in (4.2.13) is the magnitude of the outer product of two vectors, $\partial\mathbf{x}/\partial\xi$ and $\partial\mathbf{x}/\partial\eta$ (Appendix A). Since these vectors lie in the xy plane, their outer product is a vector pointing along or against the z axis that is normal to the xy plane.

Using the linear mapping functions shown in (4.2.11), we find that

$$\frac{\partial \mathbf{x}}{\partial \xi} = \mathbf{x}_2^{\mathrm{E}} - \mathbf{x}_1^{\mathrm{E}}, \qquad \frac{\partial \mathbf{x}}{\partial \eta} = \mathbf{x}_3^{\mathrm{E}} - \mathbf{x}_1^{\mathrm{E}}, \tag{4.2.14}$$

calculate

$$\mathbf{J} = \begin{bmatrix} x_2^{\mathrm{E}} - x_1^{\mathrm{E}} & x_3^{\mathrm{E}} - x_1^{\mathrm{E}} \\ y_2^{\mathrm{E}} - y_1^{\mathrm{E}} & y_3^{\mathrm{E}} - y_1^{\mathrm{E}} \end{bmatrix}, \tag{4.2.15}$$

and find that

$$h_S = \left| (\mathbf{x}_2^{\mathrm{E}} - \mathbf{x}_1^{\mathrm{E}}) \times (\mathbf{x}_3^{\mathrm{E}} - \mathbf{x}_1^{\mathrm{E}}) \right|, \tag{4.2.16}$$

yielding

$$h_S = (x_2^{\mathrm{E}} - x_1^{\mathrm{E}})(y_3^{\mathrm{E}} - y_1^{\mathrm{E}}) - (x_3^{\mathrm{E}} - x_1^{\mathrm{E}})(y_2^{\mathrm{E}} - y_1^{\mathrm{E}}). \tag{4.2.17}$$

Invoking the geometrical interpretation of the outer vector product, we find that h_S is equal to the area of a rectangle with two sides coinciding with the vectors $\mathbf{x}_2^{\mathrm{E}} - \mathbf{x}_1^{\mathrm{E}}$ and $\mathbf{x}_3^{\mathrm{E}} - \mathbf{x}_1^{\mathrm{E}}$, which is equal to twice the area of the triangle in the physical xy plane. In conclusion, the surface metric coefficient of a three-node triangle is then

$$h_S = 2A, \tag{4.2.18}$$

and A is the area of the triangle in the physical xy plane.

Using elementary calculus, we find that the integral of a function $f(x,y)$ over the physical triangle in the xy plane can be expressed as an integral over the parametric triangle in the $\xi\eta$ plane as

$$\iint f(x,y)\, \mathrm{d}x\, \mathrm{d}y = \iint f(\xi,\eta)\, h_S\, \mathrm{d}\xi\, \mathrm{d}\eta = 2A \iint f(\xi,\eta)\, \mathrm{d}\xi\, \mathrm{d}\eta. \tag{4.2.19}$$

The integrals with respect to ξ and η can be computed analytically in simple cases or numerically in more involved cases, as will be discussed later in this section.

4.2.1   Element matrices

In the *isoparametric interpolation*, a function of interest defined over an element is expressed in a form that is analogous to that shown in (4.2.2),

$$f(x, y, t) = \sum_{i=1}^{3} f_i^{\mathrm{E}}(t)\, \psi_i(\xi, \eta), \tag{4.2.20}$$

with the understanding that the point $\mathbf{x} = (x, y)$ is mapped to (ξ, η), and *vice versa*, through (4.2.2).

Using the preceding integration formulas, we find that the *l*th element diffusion matrix is given by

$$A_{ij}^{(l)} \equiv \iint_{E_l} \nabla\psi_i \cdot \nabla\psi_j \, \mathrm{d}x\, \mathrm{d}y = 2A \iint \nabla\psi_i \cdot \nabla\psi_j \, \mathrm{d}\xi\, \mathrm{d}\eta \tag{4.2.21}$$

for $i, j = 1, 2, 3$, where

$$\nabla\psi_i = \left(\frac{\partial\psi_i}{\partial x},\ \frac{\partial\psi_i}{\partial y} \right). \tag{4.2.22}$$

The element mass matrix is given by

$$B_{ij}^{(l)} = \iint_{E_l} \psi_i\, \psi_j \, \mathrm{d}x\, \mathrm{d}y = 2A \iint \psi_i\, \psi_j \, \mathrm{d}\xi\, \mathrm{d}\eta, \tag{4.2.23}$$

and the element advection matrix is given by

$$C_{ij}^{(l)} = \iint_{E_l} \psi_i\, \mathbf{u} \cdot \nabla\psi_j \, \mathrm{d}x\, \mathrm{d}y = 2A \iint \psi_i\, \mathbf{u} \cdot \nabla\psi_j \, \mathrm{d}\xi\, \mathrm{d}\eta, \tag{4.2.24}$$

where $\mathbf{u} = (u_x, u_y)$ is the advection velocity.

Computation of the gradient

To compute the element diffusion and advection matrices, we require the gradients of the element interpolation functions, $\nabla\psi_i$ for $i = 1, 2, 3$. These can be found readily using the relations

$$\frac{\partial\mathbf{x}}{\partial\xi} \cdot \nabla\psi_i = \frac{\partial\psi_i}{\partial\xi}, \qquad \frac{\partial\mathbf{x}}{\partial\eta} \cdot \nabla\psi_i = \frac{\partial\psi_i}{\partial\eta}, \tag{4.2.25}$$

which state that the projection of the gradient vector in the direction of the ξ-line vector, $\partial\mathbf{x}/\partial\xi$, is the partial derivative of $\mathbf{x}$ with respect to ξ, and the projection of the gradient vector in the direction of the η-line vector, $\partial \mathbf{x}/\partial \eta$, is the partial derivative of $\mathbf{x}$ with respect to η.

Explicitly, the equations in (4.2.25) read

$$\frac{\partial x}{\partial \xi}\frac{\partial \psi_i}{\partial x} + \frac{\partial y}{\partial \xi}\frac{\partial \psi_i}{\partial y} = \frac{\partial \psi_i}{\partial \xi}, \qquad \frac{\partial x}{\partial \eta}\frac{\partial \psi_i}{\partial x} + \frac{\partial y}{\partial \eta}\frac{\partial \psi_i}{\partial y} = \frac{\partial \psi_i}{\partial \eta}, \tag{4.2.26}$$

which can be compiled into a linear system,

$$\mathbf{J}^T \cdot \nabla \psi_i = \begin{bmatrix} \dfrac{\partial \psi_i}{\partial \xi} \\ \dfrac{\partial \psi_i}{\partial \eta} \end{bmatrix}, \tag{4.2.27}$$

where $\mathbf{J}^T$ is the transpose of the Jacobian matrix defined in (4.2.12), given by

$$\mathbf{J}^T \equiv \begin{bmatrix} \dfrac{\partial x}{\partial \xi} & \dfrac{\partial y}{\partial \xi} \\ \dfrac{\partial x}{\partial \eta} & \dfrac{\partial y}{\partial \eta} \end{bmatrix}. \tag{4.2.28}$$

The determinant of $\mathbf{J}^T$ is equal to the determinant of $\mathbf{J}$, which is equal to the surface metric coefficient, $h_S = 2A$.

Using (4.2.14), we find that the equations in (4.2.25) take the specific form

$$(\mathbf{x}_2^E - \mathbf{x}_1^E) \cdot \nabla \psi_i = \frac{\partial \psi_i}{\partial \xi}, \qquad (\mathbf{x}_3^E - \mathbf{x}_1^E) \cdot \nabla \psi_i = \frac{\partial \psi_i}{\partial \eta}. \tag{4.2.29}$$

Accordingly,

$$\mathbf{J}^T = \begin{bmatrix} x_2^E - x_1^E & y_2^E - y_1^E \\ x_3^E - x_1^E & y_3^E - y_1^E \end{bmatrix}, \tag{4.2.30}$$

which is consistent with (4.2.15).

Substituting the specific expressions for the cardinal interpolation functions given in (4.2.7), we derive three linear systems,

$$\mathbf{J}^T \cdot \nabla \psi_1 = \begin{bmatrix} -1 \\ -1 \end{bmatrix}, \qquad \mathbf{J}^T \cdot \nabla \psi_2 = \begin{bmatrix} 1 \\ 0 \end{bmatrix}, \qquad \mathbf{J}^T \cdot \nabla \psi_3 = \begin{bmatrix} 0 \\ 1 \end{bmatrix}. \tag{4.2.31}$$

The solutions are

$$\nabla \psi_1 = \frac{1}{2A} \begin{bmatrix} -(y_3^E - y_2^E) \\ x_3^E - x_2^E \end{bmatrix}, \qquad \nabla \psi_2 = \frac{1}{2A} \begin{bmatrix} -(y_1^E - y_3^E) \\ x_1^E - x_3^E \end{bmatrix},$$

$$\nabla \psi_3 = \frac{1}{2A} \begin{bmatrix} -(y_2^E - y_1^E) \\ x_2^E - x_1^E \end{bmatrix}. \tag{4.2.32}$$

Note that the gradient $\nabla\psi_1$ is perpendicular to the side 23, the gradient $\nabla\psi_2$ is perpendicular to the side 31, and the gradient $\nabla\psi_3$ is perpendicular to the side 12. The expressions in (4.2.32) are used to evaluate the element diffusion and advection matrices.

4.2.2   Computation of the element diffusion matrix

Substituting (4.2.32) into (4.2.21), we find that the element diffusion matrix is given by

$$\mathbf{A}^{(l)} = \frac{1}{4A} \begin{bmatrix} |\mathbf{d}_{32}|^2 & \mathbf{d}_{32}\cdot\mathbf{d}_{13} & \mathbf{d}_{32}\cdot\mathbf{d}_{21} \\ \mathbf{d}_{32}\cdot\mathbf{d}_{13} & |\mathbf{d}_{13}|^2 & \mathbf{d}_{13}\cdot\mathbf{d}_{21} \\ \mathbf{d}_{32}\cdot\mathbf{d}_{21} & \mathbf{d}_{13}\cdot\mathbf{d}_{21} & |\mathbf{d}_{21}|^2 \end{bmatrix}, \tag{4.2.33}$$

where

$$\mathbf{d}_{32} = \mathbf{x}_3^E - \mathbf{x}_2^E, \qquad \mathbf{d}_{13} = \mathbf{x}_1^E - \mathbf{x}_3^E, \qquad \mathbf{d}_{21} = \mathbf{x}_2^E - \mathbf{x}_1^E. \tag{4.2.34}$$

The sum of the elements in any row or column is zero.

If the triangle in the xy plane is orthogonal with the right angle occurring at the first element node labeled 1, then $\mathbf{d}_{13}\cdot\mathbf{d}_{21} = 0$ and two off-diagonal elements of $\mathbf{A}^{(l)}$ are zero. Similar simplifications occur when the right angle occurs at one of the other two nodes. If all angles of the triangles are acute, all off-diagonal elements of the element diffusion matrix are negative. If one angle is obtuse, two corresponding elements of the element diffusion matrix are positive.

FSELIB function *edm3*, listed in Table 4.2.1, evaluates the element diffusion matrix from the coordinates of the three nodes based on the preceding expressions. It can be confirmed by direct evaluation that the determinant of this matrix is zero, which means that the matrix is singular (Problem 4.2.2). The physical reason was discussed in Section 1.2.

4.2.3   Computation of the element mass matrix

Substituting (4.2.7) into (4.2.23), we derive an expression for the element mass matrix,

$$\mathbf{B}^{(l)} = 2A \iint \begin{bmatrix} \zeta^2 & \zeta\xi & \zeta\eta \\ \xi\zeta & \xi^2 & \xi\eta \\ \eta\zeta & \eta\xi & \eta^2 \end{bmatrix} \mathrm{d}\xi\,\mathrm{d}\eta. \tag{4.2.35}$$

To compute the double integral, we use the integration formula

$$\iint \zeta^p\,\xi^q\,\eta^r\,\mathrm{d}\xi\,\mathrm{d}\eta = \frac{p!\,q!\,r!}{(p+q+r+2)!}, \tag{4.2.36}$$

```
function edm = edm3 (x1,y1,x2,y2,x3,y3)

%=============================================
% Evaluation of the element diffusion matrix
% for a three-node triangle
%=============================================

  d32x = x3-x2; d32y = y3-y2;
  d13x = x1-x3; d13y = y1-y3;
  d21x = x2-x1; d21y = y2-y1;

  A = 0.5*(d13x*d21y - d31y*d21x);    % element area

  A4 = 4.0*A;

  edm(1,1) = (d32x*d32x + d32y*d32y)/A4;
  edm(1,2) = (d32x*d13x + d32y*d13y)/A4;
  edm(1,3) = (d32x*d21x + d32y*d21y)/A4;

  edm(2,1) = (d13x*d32x + d13y*d32y)/A4;
  edm(2,2) = (d13x*d13x + d13y*d13y)/A4;
  edm(2,3) = (d13x*d21x + d13y*d21y)/A4;

  edm(3,1) = (d21x*d32x + d21y*d32y)/A4;
  edm(3,2) = (d21x*d13x + d21y*d13y)/A4;
  edm(3,3) = (d21x*d21x + d21y*d21y)/A4;

%-----
% done
%-----

return;
```

TABLE 4.2.1 Function *edm3* evaluates the element diffusion matrix from the coordinates of the vertices of a three-node triangle.

to be proven in Section 4.2.4, where p, q, and r are non-negative integers, and an exclamation mark denotes the factorial, $m! = 1 \cdot 2 \cdots m$ for any integer, m. A straightforward calculation yields

$$\mathbf{B}^{(l)} = A \frac{1}{12} \begin{bmatrix} 2 & 1 & 1 \\ 1 & 2 & 1 \\ 1 & 1 & 2 \end{bmatrix}. \qquad (4.2.37)$$

Note that the element mass matrix depends only on the element area, A, and is independent of the element shape. The sum of all entries of the element mass matrix is equal to the area of the triangle, A.

The mass-lumped diagonal element mass matrix, designated by a hat, is given by

$$\widehat{\mathbf{B}}^{(l)} = A \frac{1}{3} \begin{bmatrix} 1 & 0 & 0 \\ 0 & 1 & 0 \\ 0 & 0 & 1 \end{bmatrix}. \tag{4.2.38}$$

The trace of the lumped mass matrix is also equal to the area of the triangle, A.

Inexact integration and lumped mass matrix

It is instructive to note that the lumped form arises from the inexact integration of the right-hand side of (4.2.35) using the trapezoidal rule. In this approximation, the entries of the matrix inside the integral are assigned constant values that are equal to the arithmetic mean of the three values at the three vertices. This results in the value of $1/3$ for the diagonal elements and the value of zero for the off-diagonal elements, as shown in (4.2.38).

4.2.4   Proof of the integration formula (4.2.36)

To prove the integration formula (4.2.36), we map the standard triangle in the $\xi\eta$ plane to the standard square in the $\xi'\eta'$ parametric plane using Duffy's transformation, as shown in Figure 4.2.4. The mapping is mediated by the functions

$$\xi = \frac{1+\xi'}{2}\frac{1-\eta'}{2}, \qquad \eta = \frac{1+\eta'}{2}, \tag{4.2.39}$$

yielding

$$\zeta \equiv 1 - \xi - \eta = \frac{1-\xi'}{2}\frac{1-\eta'}{2}, \tag{4.2.40}$$

where $-1 < \xi' < 1$ and $-1 < \eta' < 1$. For future reference, we note that

$$1 - \eta = 1 - \frac{1+\eta'}{2} = \frac{1-\eta'}{2}. \tag{4.2.41}$$

The inverse transformation is given by

$$\xi' = \frac{2\xi}{1-\eta} - 1, \qquad \eta' = 2\eta - 1. \tag{4.2.42}$$

The integral of a function, $f(\xi, \eta)$, over the area of the triangle in the $\xi\eta$ plane can be expressed as an integral over the standard square as

$$\iint f(\xi, \eta)\, \mathrm{d}\xi\, \mathrm{d}\eta = \int_{-1}^{1}\int_{-1}^{1} f(\xi, \eta)\, h'_s\, \mathrm{d}\xi'\, \mathrm{d}\eta'. \tag{4.2.43}$$

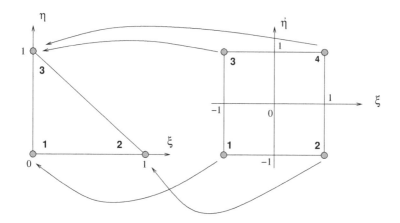

FIGURE 4.2.4   Mapping of the standard triangle to the standard square using Duffy's transformation. The bold-faced numbers are the element node labels.

The metric coefficient of the transformation is equal to the determinant of the Jacobian matrix of the transformation, given by

$$
h'_S = \det\!\left(\begin{bmatrix} \dfrac{\partial \xi}{\partial \xi'} & \dfrac{\partial \eta}{\partial \xi'} \\[2mm] \dfrac{\partial \xi}{\partial \eta'} & \dfrac{\partial \eta}{\partial \eta'} \end{bmatrix}\right) = \det\!\left(\begin{bmatrix} \tfrac{1}{4}(1-\eta') & 0 \\[2mm] -\tfrac{1}{4}(1+\xi') & \tfrac{1}{2} \end{bmatrix}\right) = \frac{1}{8}\,(1-\eta'). \quad (4.2.44)
$$

Thus,

$$
\iint f(\xi,\eta)\,\mathrm{d}\xi\,\mathrm{d}\eta = \frac{1}{8}\int_{-1}^{1}\int_{-1}^{1} f(\xi,\eta)\,(1-\eta')\,\mathrm{d}\xi'\,\mathrm{d}\eta'. \quad (4.2.45)
$$

The double integral over the standard square can be computed accurately by applying a Gaussian quadrature twice, once with respect to ξ' and the second time with respect to η'.

Now applying the integration formula (4.2.45) for the function $f(\xi,\eta) = \zeta^p\,\xi^q\,\eta^r$, we find that

$$
\iint \zeta^p\,\xi^q\,\eta^r\,\mathrm{d}\xi\,\mathrm{d}\eta = \frac{1}{8}\int_{-1}^{1}\int_{-1}^{1}\left(\frac{1-\xi'}{2}\right)^p\left(\frac{1-\eta'}{2}\right)^p\left(\frac{1+\xi'}{2}\right)^q
$$
$$
\times\left(\frac{1-\eta'}{2}\right)^q\left(\frac{1+\eta'}{2}\right)^r(1-\eta')\,\mathrm{d}\xi'\,\mathrm{d}\eta' \quad (4.2.46)
$$

or

$$
\iint \zeta^p\,\xi^q\,\eta^r\,\mathrm{d}\xi\,\mathrm{d}\eta = \frac{1}{4}\left[\int_{-1}^{1}\left(\frac{1-\xi'}{2}\right)^p\left(\frac{1+\xi'}{2}\right)^q\mathrm{d}\xi'\right]
$$
$$
\times\left[\int_{-1}^{1}\left(\frac{1-\eta'}{2}\right)^{p+q+1}\left(\frac{1+\eta'}{2}\right)^r\mathrm{d}\eta'\right]. \quad (4.2.47)
$$

The integral with respect to ξ', enclosed by the first square brackets on the right-hand side, can be evaluated by introducing a new variable, $\omega \equiv \frac{1}{2}(1 - \xi')$, writing

$$\mathcal{J} \equiv \int_{-1}^{1} \left(\frac{1-\xi'}{2}\right)^p \left(\frac{1+\xi'}{2}\right)^q d\xi' = 2 \int_{0}^{1} \omega^p (1 - \omega)^q \, d\omega, \qquad (4.2.48)$$

and then

$$\mathcal{J} \equiv 2\,\mathrm{B}(p+1, q+1) = 2\,\frac{p!\,q!}{(p+q+1)!}, \qquad (4.2.49)$$

where $\mathrm{B}(k, l)$ is the beta function (e.g., Abramowitz & Stegun [1]). The integral with respect to η' can be evaluated in a similar fashion, yielding

$$\int_{-1}^{1} \left(\frac{1-\eta'}{2}\right)^{p+q+1} \left(\frac{1+\eta'}{2}\right)^r d\eta' = 2\,\frac{(p+q+1)!\,r!}{(p+q+r+1)!}. \qquad (4.2.50)$$

Putting these results together, we derive the integration formula (4.2.36).

4.2.5   Computation of the element advection matrix

When the advection velocity, $\mathbf{u}$, is constant or can be approximated with a constant, $\mathbf{U} = (U_x, U_y)$, the element advection matrix simplifies to

$$C_{ij}^{(l)} = 2A\,\mathbf{U} \cdot \iint \psi_i \, \nabla \psi_j \, d\xi \, d\eta \equiv \alpha\,\beta_j, \qquad (4.2.51)$$

where α is a numerical coefficient given by

$$\alpha = \iint \psi_1 \, d\xi \, d\eta = \iint \psi_2 \, d\xi \, d\eta = \iint \psi_3 \, d\xi \, d\eta = \frac{1}{6}, \qquad (4.2.52)$$

β_j are dimensional coefficients given by

$$\beta_1 = \mathbf{V} \cdot \mathbf{d}_{32}, \qquad \beta_2 = \mathbf{V} \cdot \mathbf{d}_{13}, \qquad \beta_3 = \mathbf{V} \cdot \mathbf{d}_{21}, \qquad (4.2.53)$$

the vector $\mathbf{V}$ is defined as

$$\mathbf{V} = (U_y, -U_x), \qquad (4.2.54)$$

and the node distances $\mathbf{d}_{ij}$ are defined in (4.2.34). Note that, because the rows of the element advection matrix are identical, the determinant is zero and the matrix is singular with zero rank, that is, it has three null eigenvalues.

   FSELIB function *eam3*, listed in Table 4.2.2, computes the element advection matrix based on the preceding expressions. The input includes the velocity components, u_x and u_y, assumed to be uniform over the triangle, and the coordinates of the three element nodes.

```
function [eam] = eam3 (ux,uy,x1,y1,x2,y2,x3,y3)

%===========================================
% Evaluation of the element advection matrix
% for a three-node triangle
%===========================================

 d32x = x3-x2; d32y = y3-y2;
 d13x = x1-x3; d13y = y1-y3;
 d21x = x2-x1; d21y = y2-y1;

 eam(1,1) = (-ux*d32y + uy*d32x)/6.0;
 eam(1,2) = (-ux*d13y + uy*d13x)/6.0;
 eam(1,3) = (-ux*d21y + uy*d21x)/6.0;

 eam(2,1) = eam(1,1);
 eam(2,2) = eam(1,2);
 eam(2,3) = eam(1,3);

 eam(3,1) = eam(1,1);
 eam(3,2) = eam(1,2);
 eam(3,3) = eam(1,3);

%-----
% done
%-----

return;
```

TABLE 4.2.2   Function *eam3* evaluates the element advection matrix from the coordinates of the vertices of a three-node triangle.

PROBLEMS

4.2.1 *Gradient of the element interpolation functions*

Prove the geometrical interpretation of the gradients of the element interpolation functions shown in (4.2.32), as discussed in the text.

4.2.2 *The element diffusion matrix is singular*

Show that the determinant of the element diffusion matrix displayed in (4.2.33) is zero, which means that the matrix is singular.

4.2.3 *Gradient of a function*

Consider a function, $f(x, y)$, defined over a three-node triangle. The gradient of this function, $\nabla f = (\partial f/\partial x, \partial f/\partial y)$, can be computed in terms of ξ and η derivatives using the counterpart of (4.2.27),

$$
\begin{bmatrix} \dfrac{\partial x}{\partial \xi} & \dfrac{\partial y}{\partial \xi} \\[2mm] \dfrac{\partial x}{\partial \eta} & \dfrac{\partial y}{\partial \eta} \end{bmatrix} \cdot \nabla f = \begin{bmatrix} \dfrac{\partial f}{\partial \xi} \\[2mm] \dfrac{\partial f}{\partial \eta} \end{bmatrix}.
\tag{4.2.55}
$$

Solving this system by Cramer's rule, we obtain

$$
\frac{\partial f}{\partial x} = \frac{1}{h_S} \left(\frac{\partial f}{\partial \xi} \frac{\partial y}{\partial \eta} - \frac{\partial f}{\partial \eta} \frac{\partial y}{\partial \xi} \right), \qquad \frac{\partial f}{\partial y} = \frac{1}{h_S} \left(\frac{\partial f}{\partial \eta} \frac{\partial x}{\partial \xi} - \frac{\partial f}{\partial \xi} \frac{\partial x}{\partial \eta} \right).
\tag{4.2.56}
$$

Applying these equations first for $f = \xi$ and then for $f = \eta$, we obtain

$$
\frac{\partial \xi}{\partial x} = \frac{1}{h_S} \frac{\partial y}{\partial \eta}, \quad \frac{\partial \xi}{\partial y} = -\frac{1}{h_S} \frac{\partial x}{\partial \eta}, \quad \frac{\partial \eta}{\partial x} = -\frac{1}{h_S} \frac{\partial y}{\partial \xi}, \quad \frac{\partial \eta}{\partial y} = \frac{1}{h_S} \frac{\partial x}{\partial \xi}.
\tag{4.2.57}
$$

(a) Confirm that the matrix

$$
\mathbf{J}^{-1} \equiv \begin{bmatrix} \dfrac{\partial \xi}{\partial x} & \dfrac{\partial \xi}{\partial y} \\[2mm] \dfrac{\partial \eta}{\partial x} & \dfrac{\partial \eta}{\partial y} \end{bmatrix}
\tag{4.2.58}
$$

is the inverse of the Jacobian matrix,

$$
\mathbf{J} \equiv \begin{bmatrix} \dfrac{\partial x}{\partial \xi} & \dfrac{\partial x}{\partial \eta} \\[2mm] \dfrac{\partial y}{\partial \xi} & \dfrac{\partial y}{\partial \eta} \end{bmatrix}.
\tag{4.2.59}
$$

(b) Consider a vector function, $\mathbf{F} = (F_x, F_y)$. The divergence of this function is the scalar

$$
\nabla \cdot \mathbf{F} \equiv \frac{\partial F_x}{\partial m} + \frac{\partial F_y}{\partial y}.
\tag{4.2.60}
$$

Show that the divergence can be evaluated from the expression

$$
\nabla \cdot \mathbf{F} = \frac{1}{h_S} \left[\frac{\partial}{\partial \xi} \left(F_x \frac{\partial y}{\partial \eta} - F_y \frac{\partial x}{\partial \eta} \right) + \frac{\partial}{\partial \eta} \left(-F_x \frac{\partial y}{\partial \xi} + F_y \frac{\partial x}{\partial \xi} \right) \right].
\tag{4.2.61}
$$

(c) Show that the divergence of a vector function, $\mathbf{F}$, can be evaluated from the expression

$$
\nabla \cdot \mathbf{F} = \frac{1}{h_S} \left[\frac{\partial}{\partial \xi} \left(h_S \, \mathbf{F} \cdot \nabla \xi \right) + \frac{\partial}{\partial \eta} \left(h_S \, \mathbf{F} \cdot \nabla \eta \right) \right],
\tag{4.2.62}
$$

where $\nabla \xi = (\partial \xi / \partial x, \partial \xi / \partial y)$ and $\nabla \eta = (\partial \eta / \partial x, \partial \eta / \partial y)$.

4.3   Grid generation

To prepare the ground for the finite element expansion, we develop a computational module for domain discretization into the three-node triangles discussed in Section 4.2. First, we discuss a triangulation procedure based on the successive subdivision of a parental structure where a group of ancestral elements (root set) are hard-coded, and smaller descendant elements arise by subdivision. Second, we discuss a method of grid generation based on the Delaunay triangulation procedure implemented in a MATLAB function. Advanced methods of automatic domain triangulation and public domain codes are reviewed in Appendix E.

4.3.1   Successive subdivisions

FSELIB function *trgl3_disk*, listed in Table 4.3.1, triangulates the unit disk in the xy plane. The algorithm successively subdivides each of four ancestral triangles located at the four quadrants, as shown in the first frame of Figure 4.3.1, into four descendant triangles. The vertices of the descendant triangles lie at the vertices and edge mid-points of their parental elements. In the process of triangulation, boundary nodes are projected in the radial direction onto the unit circle.

The level of refinement is determined by an input flag, $ndiv$, as follows: $ndiv = 0$ generates the 4 ancestral elements; $ndiv = 1$ generates 16 first-generation elements; $ndiv = 2$ generates 64 second-generation elements; each time a subdivision is carried out, the number of elements is increased by a factor of 4. Triangulations for discretization levels $ndiv = 0, 1, 2, 3$ are shown in Figure 4.3.1. These shapes can be subsequently deformed or mapped by a transformation to yield other simply connected, disk-like shapes. Important features of the triangulation algorithm include the following:

- In the first part of the algorithm, only element nodes are generated. The unique global nodes and the connectivity matrix are defined in the second stage, after the triangulation has been carried out.

- The array efl is a boundary flag of the element nodes, as discussed in Section 4.1.3. The entry $efl(i,j)$ is set to zero at interior nodes or unity at boundary nodes.

- The flag efl of a new node is set according to the flags of the two vertex nodes defining the subdivided side. If both nodes are boundary nodes, the new node is also a boundary node.

- The global array $p(i,j)$ contains the x and y coordinates of the global nodes, where $i = 1, \ldots, N_G$ and $j = 1, 2$.

- The array $c(i,j)$ hosts the connectivity matrix, as discussed in Section 4.1.3.

- The array gfl is the boundary flag of the global nodes, as discussed in Section 4.1.3.

```
function [ne,ng,p,c,efl,gfl] = trgl3_disk(ndiv)

%===========================================================
% Triangulation of the unit disk into three-node elements
% by the successive subdivision of four
% ancestral elements
%
% ndiv: discretization level
%       0 yields the ancestral set
% ne:   number of elements
% ng:   number of global nodes
% p:    coordinates of global nodes
% c:    connectivity table
% efl:  element node boundary flag
% gfl:  global node boundary flag
%===========================================================

%----------------------------------------
% ancestral structure with four elements
%----------------------------------------

ne = 4;

x(1,1) = 0.0; y(1,1) = 0.0; efl(1,1) = 0;  % first element
x(1,2) = 1.0; y(1,2) = 0.0; efl(1,2) = 1;
x(1,3) = 0.0; y(1,3) = 1.0; efl(1,3) = 1;

x(2,1) = 0.0; y(2,1) = 0.0; efl(2,1) = 0;  % second element
x(2,2) = 0.0; y(2,2) = 1.0; efl(2,2) = 1;
x(2,3) =-1.0; y(2,3) = 0.0; efl(2,3) = 1;

x(3,1) = 0.0; y(3,1) = 0.0; efl(3,1) = 0;  % third element
x(3,2) =-1.0; y(3,2) = 0.0; efl(3,2) = 1;
x(3,3) = 0.0; y(3,3) =-1.0; efl(3,3) = 1;

x(4,1) = 0.0; y(4,1) = 0.0; efl(4,1) = 0;  % fourth element
x(4,2) = 0.0; y(4,2) =-1.0; efl(4,2) = 1;
x(4,3) = 1.0; y(4,3) = 0.0; efl(4,3) = 1;

if(ndiv > 0)      % refinement loop

for i=1:ndiv

  nm = 0; % count the new elements arising in each refinement;
          % four elements will be generated in each pass

  for j=1:ne      % loop over current elements

  % edge mid-nodes will become vertex nodes

    x(j,4) = 0.5*(x(j,1)+x(j,2)); y(j,4) = 0.5*(y(j,1)+y(j,2));
    x(j,5) = 0.5*(x(j,2)+x(j,3)); y(j,5) = 0.5*(y(j,2)+y(j,3));
    x(j,6) = 0.5*(x(j,3)+x(j,1)); y(j,6) = 0.5*(y(j,3)+y(j,1));
```

TABLE 4.3.1 Function *trgl3_disk* (Continuing →)

```
% set the flag of the mid-nodes; a mid-node is a boundary node
% only if the vertex lies on the boundary

   efl(j,4) = 0; efl(j,5) = 0; efl(j,6) = 0;

   if(efl(j,1)==1 & efl(j,2)==1) efl(j,4) = 1; end
   if(efl(j,2)==1 & efl(j,3)==1) efl(j,5) = 1; end
   if(efl(j,3)==1 & efl(j,1)==1) efl(j,6) = 1; end

% assign vertex nodes to sub-elements;
% these will be the "new" elements

   nm = nm+1; % first sub-element
   xn(nm,1) = x(j,1); yn(nm,1) = y(j,1); efln(nm,1) = efl(j,1);
   xn(nm,2) = x(j,4); yn(nm,2) = y(j,4); efln(nm,2) = efl(j,4);
   xn(nm,3) = x(j,6); yn(nm,3) = y(j,6); efln(nm,3) = efl(j,6);

   nm = nm+1; % second sub-element
   xn(nm,1) = x(j,4); yn(nm,1) = y(j,4); efln(nm,1) = efl(j,4);
   xn(nm,2) = x(j,2); yn(nm,2) = y(j,2); efln(nm,2) = efl(j,2);
   xn(nm,3) = x(j,5); yn(nm,3) = y(j,5); efln(nm,3) = efl(j,5);

   nm = nm+1; % third sub-element
   xn(nm,1) = x(j,6); yn(nm,1) = y(j,6); efln(nm,1) = efl(j,6);
   xn(nm,2) = x(j,5); yn(nm,2) = y(j,5); efln(nm,2) = efl(j,5);
   xn(nm,3) = x(j,3); yn(nm,3) = y(j,3); efln(nm,3) = efl(j,3);

   nm = nm+1; % fourth sub-element
   xn(nm,1) = x(j,4); yn(nm,1) = y(j,4); efln(nm,1) = efl(j,4);
   xn(nm,2) = x(j,5); yn(nm,2) = y(j,5); efln(nm,2) = efl(j,5);
   xn(nm,3) = x(j,6); yn(nm,3) = y(j,6); efln(nm,3) = efl(j,6);

end      % end of loop over current elements

ne = 4*ne;  % number of elements has increased
            % by a factor of four

 for k=1:ne     % relabel the new points
    for l=1:3   % and put them in the master list

     x(k,l) = xn(k,l); y(k,l) = yn(k,l); efl(k,l) = efln(k,l);

      % project the boundary nodes onto the unit circle:

      if(efl(k,l)==1)
         rad = sqrt(x(k,l)^2+y(k,l)^2);
         x(k,l) = x(k,l)/rad; y(k,l) = y(k,l)/rad;
      end

     end
  end

end % end do of refinement loop
end % end if of refinement loop
```

TABLE 4.3.1 Function *trgl3_disk* ($\to$ Continuing $\to$)

```
%--------------------
% plotting (optional)
%--------------------

for i=1:ne
  xp(1) = x(i,1); xp(2) = x(i,2);
  xp(3) = x(i,3); xp(4) = x(i,1);
  yp(1) = y(i,1); yp(2) = y(i,2);
  yp(3) = y(i,3); yp(4) = y(i,1);
  plot(xp, yp,'-o'); hold on;
  xlabel('x'); ylabel('y');
end

%-----------------------------------------------------
% define the global nodes and the connectivity table
%-----------------------------------------------------

% 3 nodes of the first element are entered manually

p(1,1) = x(1,1); p(1,2) = y(1,1); gfl(1) = efl(1,1);
p(2,1) = x(1,2); p(2,2) = y(1,2); gfl(2) = efl(1,2);
p(3,1) = x(1,3); p(3,2) = y(1,3); gfl(3) = efl(1,3);

c(1,1) = 1;  % first  node of first element is global node 1
c(1,2) = 2;  % second node of first element is global node 2
c(1,3) = 3;  % third  node of first element is global node 3

ng = 3;

% loop over further elements

eps = 0.000001;

% Iflag=0 will signal a new global node

for i=2:ne        % loop over elements
 for j=1:3            % loop over element nodes

  Iflag=0;        % initialize

  for k=1:ng

   if(abs(x(i,j)-p(k,1)) < eps)
    if(abs(y(i,j)-p(k,2)) < eps)
      Iflag = 1;    % the node has been recorded previously
      c(i,j) = k;   % the jth local node of element i
                    % is the kth global node
    end
   end

  end
```

TABLE 4.3.1 Function *trgl3_disk* ($\rightarrow$ Continuing $\rightarrow$)

```
if(Iflag==0)   % record the node
   ng = ng+1;

   p(ng,1) = x(i,j);
   p(ng,2) = y(i,j);

   gfl(ng) = efl(i,j);
   c(i,j) = ng;    % the jth local node of the element
                   % is the new global node
 end

 end % end of loop over element nodes
end  % end of loop over elements

%-----
% done
%-----

return;
```

TABLE 4.3.1 ($\rightarrow$ Continued) Function *trgl3_disk* triangulates the unit disk into three-node elements by successively subdividing four ancestral hard-coded elements.

Triangulation of other domains can be performed beginning with a different set of ancestral triangles in the spirit of the function *trgl3_disk*. In all cases, at least one element node in the ancestral set must be an interior node, otherwise the flag $efl(i, 1)$ will be erroneously set to unity for all generated element nodes.

FSELIB function *trgl3_sqr*, not listed in the text, triangulates the unit square using an algorithm that is similar to that implemented in the function *trgl3_disk* for a circular disk. In the case of the square, we begin by hard-coding eight ancestral triangles, two in each quadrant, as shown in Figure 4.3.2(*a*). Different levels of triangulation are then generated by element subdivision. In the end, the grid can be deformed to yield a desired shape containing four corners/ A deformed grid for discretization level $ndiv = 3$ is displayed in Figure 4.3.2(*b*).

4.3.2 Delaunay triangulation

In the inverse approach, the locations of the global nodes defining the triangle vertices are first specified, and the elements are generated by mapping global nodes to element nodes. Element definition from a given global set can be done by several methods, including the method of Delaunay triangulation. One limitation of this approach is that the standard algorithm triangulates the convex hull of the specified set of global nodes, which is defined as the area enclosed by a rubber band when it is stretched and then allowed to relax toward the global nodes.

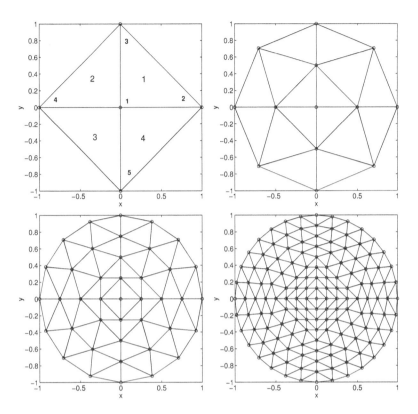

FIGURE 4.3.1   Discretization of the unit disk into three-node triangles generated by the FSELIB function *trgl3_disk* for discretization level $ndiv = 0$ (root set), 1, 2, and 3. The labels of the global nodes are printed in bold and the labels of the elements are printed in large font in the first frame.

Delaunay triangulation can be implemented in terms of the Dirichlet–Voronoi–Thiessen (DVT) tessellation. In this approach, the discretization domain is divided into polygons, subject to two conditions: (*a*) each polygon contains only one global node and (*b*) the distance of an arbitrary point inside a polygon from the native global node is smaller than the distance from any other node. It can be shown that the sides of the polygons thus generated are perpendicular bisectors of the straight segments connecting pairs of nodes. Once the DVT tessellation has been completed, the Delaunay triangulation emerges by connecting a node to all other nodes that share a polygon side. Direct methods of Delaunay triangulation are reviewed by Rebay [51].[1]

---

[1] A wealth of information, algorithms, and useful links can be found at the Internet site *http://www.voronoi.com*.

(a) (b)

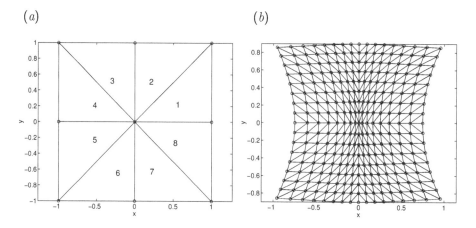

FIGURE 4.3.2 (a) Ancestral eight-element structure for triangulating a square and (b) discretization of a deformed square for $ndiv = 2$ generated by the FSELIB function *trgl3_sqr*.

MATLAB encapsulates the function *delaunay* that performs the Delaunay triangulation according to the function call:

$$\texttt{c = delaunay(x,y)}$$

where the input vectors, x and y, hold the coordinates of the specified global nodes. The familiar $N_E \times 3$ global connectivity matrix, c, holding the global labels of the N_E three-node triangles arising from the Delaunay triangulation, is generated in the output.

FSELIB function *trgl3_delaunay*, listed in Table 4.3.2, performs a triangulation based on a specified set of global nodes tabulated in the file *points.dat*. The first column in this file holds the x coordinates of the nodes, the second column holds the y coordinates of the nodes, and the third column holds the global boundary flag, gfl, in the format of the following sample file:

```
0.6375 0.2625 1
0.4950 0.3000 1
...
0.4725 0.4875 0
0.5475 0.5325 0
```

In this example, the first two nodes happen to be boundary nodes and the last two nodes happen to be interior nodes. A triangulation generated by this function is shown in Figure 4.3.3(a) along with the corresponding DVT tessellation generated by the MATLAB function *voronoi*, represented by the dotted lines.

```
function [ne,ng,p,c,efl,gfl] = trgl3_delaunay

%===========================================================
% Discretization of an arbitrary domain into three-node
% elements by the Delaunay triangulation
%===========================================================

%---------------------------------------------
% Read the global nodes and boundary flags
% from file: points.dat
% The three columns in this file are the
% x and y coordinates and the boundary
% flag of the global nodes:
%
%   x(1) y(1) gfl(1)
%   x(2) y(2) gfl(2)
%   ...   ...   ...
%   x(n) y(n) gfl(n)
%---------------------------------------------

file1 = fopen('points.dat');
 points = fscanf(file1,'%f');
fclose(file1);

%--------------------------
% number of global nodes, ng
%--------------------------

sp = size(points); ng = sp(1)/3;

%----------------------------------------------
% coordinates and boundary flag of global nodes
%----------------------------------------------

for i=1:ng
 p(i,1) = points(3*i-2);
 p(i,2) = points(3*i-1);
 gfl(i) = points(3*i);
 xdel(i) = p(i,1); ydel(i) = p(i,2);  % input to delaunay
end

%----------------------
% delaunay triangulation
%----------------------

c = delaunay (xdel,ydel);

%-----------------------------------
% extract the number of elements, ne
%-----------------------------------

sc = size(c); ne = sc(1,1);
```

TABLE 4.3.2 Function *trgl3_delaunay* (Continuing →)

```
%------------------------------------
% set the element-node boundary flags
%------------------------------------

for i=1:ne
  efl(i,1) = gfl(c(i,1));
  efl(i,2) = gfl(c(i,2));
  efl(i,3) = gfl(c(i,3));
end

%-----
% plot
%-----

for i=1:ne

   % plot the elements:

   xp(1) = p(c(i,1),1); yp(1) = p(c(i,1),2);
   xp(2) = p(c(i,2),1); yp(2) = p(c(i,2),2);
   xp(3) = p(c(i,3),1); yp(3) = p(c(i,3),2);
   xp(4) = p(c(i,1),1); yp(4) = p(c(i,1),2);

   plot(xp, yp);
   hold on

   % mark the boundary nodes:

   if(efl(i,1)==1)
      plot(xp(1), yp(1), 'o');
   elseif(efl(i,2)==1)
      plot(xp(2), yp(2), '+');
   elseif(efl(i,3)==1)
     plot(xp(3), yp(3), 'x');
   end

end

%---------------------
% Voronoi tessellation
%---------------------

figure;
voronoi(xdel,ydel)

%-----
% done
%-----

return
```

TABLE 4.3.2   ($\rightarrow$ Continued) Function *trgl3_delaunay* triangulates a domain with pre-assigned global nodes using the Delaunay triangulation.

(*a*)　　　　　　　　　　　　　　　　(*b*)

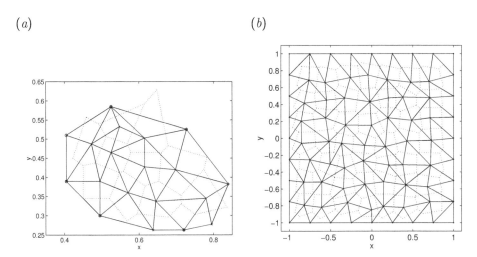

FIGURE 4.3.3　(*a*) Discretization of a domain into three-node triangles by Delaunay triangulation generated by the FSELIB function *trgl3_delaunay*. The boundary nodes are indicated by filled circles. (*b*) Discretization of a square with randomized interior nodes into three-node triangles by Delaunay triangulation generated by the FSELIB function *trgl3_delaunay_sqr*. The corresponding Voronoi tessellations are represented by the dotted lines.

Triangulation of a square

As a further application, we perform the triangulation of a square. FSELIB function *trgl3_delaunay_sqr*, listed in Table 4.3.3, generates a uniform $N \times M$ grid of global nodes, randomizes the coordinates of the interior nodes using the MATLAB random-number generator *rand*, performs the Delaunay triangulation, displays the triangulation using the MATLAB graphics function *trimesh*, and displays a graph of the Voronoi tessellation. A triangulation generated by this function is shown in Figure 4.3.3(*b*) along with the corresponding DVT tessellation.

4.3.3　Generalized connectivity matrices

Both discretization methods discussed in this section generate a connectivity matrix, $c(i, j)$, that relates the element nodes to the global nodes. The index i runs over the elements labels, $i = 1, \ldots, N_E$, and the index j runs over the three element node labels, $j = 1, 2, 3$. The components of the connectivity matrix, $c(i, j)$, take values in the range $1, \ldots, N_G$. We recall that $c(i, j)$ is the global label to the jth node of the ith triangle. Additional connectivity tables can be defined to facilitate various computations, as required.

```
function [ne,ng,p,c,efl,gfl] = trgl3_delaunay_sqr

%--------------------
% window and grid size
%--------------------

X1 = -1.0; X2 = 1.0;
Y1 = -1.0; Y2 = 1.0;

N = 8; M = 8;

%--------
% prepare
%--------

Dx = (X2-X1)/N;
Dy = (Y2-Y1)/M;

%---------------------------
% arrange points on a mesh grid
% set the boundary flag
% and count the nodes (ng)
%---------------------------

ng = 0;

for j=1:M+1

 for i=1:N+1
   ng = ng+1;
   p(ng,1) = X1+(i-1.0)*Dx;
   p(ng,2) = Y1+(j-1.0)*Dy;
   gfl(ng) = 0;
   if(i==1 | i==N+1 | j==1 | j==M+1) gfl(ng)=1; end;
 end

end

%---------------------------
% randomize the interior nodes
%---------------------------

Ic = N+2;

for j=2:M
 for i=2:N
   Ic=Ic+1;
   p(Ic,1) = p(Ic,1)+(rand-1.0)*0.5*Dx;
   p(Ic,2) = p(Ic,2)+(rand-1.0)*0.4*Dy;
 end
 Ic=Ic+2;
end
```

TABLE 4.3.3   Function *trgl3_delaunay_sqr* (Continuing →)

```
%----------------------
% Delaunay triangulation
%----------------------

for i=1:ng
 xdel(i) = p(i,1); ydel(i) = p(i,2);
end

c = delaunay (xdel,ydel);

%------------------------------
% extract the number of elements
%------------------------------

sc = size(c); ne = sc(1,1);

%-------------------------------------
% set the element-node boundary flags
%-------------------------------------

for i=1:ne
 efl(i,1) = gfl(c(i,1));
 efl(i,2) = gfl(c(i,2));
 efl(i,3) = gfl(c(i,3));
end

%------------------------
% display the triangulation
%------------------------

trimesh (c,p(:,1),p(:,2),zeros(ng,1));

%--------------------
% Voronoi tessellation
%--------------------

figure; voronoi (xdel,ydel);

%-----
% done
%-----

return
```

TABLE 4.3.3   ($\rightarrow$ Continued) Function *trgl3_delaunay_sqr* triangulates a square based on global nodes generated by a randomly perturbed square grid using Delaunay triangulation.

Labels of elements sharing a node

The element-node connectivity matrix $cen(i, j)$ is defined such that $cen(i, 1)$ is the number of elements sharing the ith global node for $i = 1, \ldots, N_G$, and $cen(i, j)$ for $j = 2, \ldots, cen(i, 1) + 1$ are the corresponding element labels. This matrix can be computed with the help of the connectivity matrix c by running over all global nodes and then over all element nodes according to the FSELIB function $cen3$, listed in Table 4.3.4. The matrix $cen(i, j)$ is useful for computing nodal values as averages of neighboring element values.

Element labels sharing an element edge

The connectivity matrix $ces(i, j)$ is defined such that $ces(i, j)$ is the label of the element sharing the jth side of the ith element, where $j = 1, 2, 3$ and $i = 1, \ldots, N_E$. By convention, side 1 is subtended between nodes 1 and 2, side 2 is subtended between nodes 2 and 3, and side 3 is subtended between nodes 3 and 1. If $ces(i, j) = 0$, the jth side of the ith element does not have a neighbor, which means that the node lies at a boundary. This matrix can be computed with the help of the connectivity matrix c by running over all global nodes and then over all element nodes according to the FSELIB function $ces3$, listed in Table 4.3.5. The matrix $ces(i, j)$ is useful for computing element side values as averages of neighboring element values. As an application, the FSELIB script see_elm3, listed in Table 4.3.6, scans the global nodes and draws neighboring elements.

4.3.4 Element and node labeling schemes

Given an element layout, there are many ways of labeling the elements, the element nodes, and the unique global nodes. For example, the labels of two unique global nodes can be interchanged, provided that appropriate modifications are made to the connectivity matrix connecting global nodes to element nodes. Although the element and node numbering schemes are inconsequential to the finite element solution, they do affect the structure of the global diffusion, mass, and advection matrices, and thus the efficiency of the finite element code.

Ideally, elements and nodes should be labeled so that the global matrices are as close to being diagonal as possible, that is, they have the smallest possible bandwidth. The objective is to economize storage and expedite the solution of the final system of algebraic or ordinary differential equations, as discussed in Appendix C. Optimal node labeling schemes are reviewed by Schwarz ([58], pp. 167–185).

PROBLEMS

4.3.1 *Triangulation of a square*

Execute the FSELIB function *trgl3_sqr* and display the grid layout for discretization levels $ndiv = 0$, 1, and 2.

```
function cen = cen3(ne,ng,c)

%================================================
% Evaluation of the element-to-node
% connectivity matrix "cen"
%
% cen(i,1) is the number of elements sharing
% the ith global node, where i=1, ..., ng
%
% cen(i,j), for j=2, 3, ..., ce(i,1)+1
% are the corresponding element labels
%
% ne: number of elements
% ng: number of global nodes
%================================================

for i=1:ng    % scan the global nodes

    cen(i,1) = 0;

    Icount = 1;

%---
    for j=1:ne    % scan the elements

      for k=1:3

        if(c(j,k)==i)              % the ith global node and the
            cen(i,1)=cen(i,1)+1;   % test element node are identical
            Icount = Icount +1;
            cen(i,Nodeplace) = j;
        end

      end

    end
%---

end

%-----
% done
%-----

return
```

TABLE 4.3.4   Function *cen3* generates the element-to-node connectivity matrix *cen* for three-node triangles.

```
function ces = ces3(ne,ng,c)

%=====================================
% Evaluation of the element-to-sides
% connectivity matrix "ces"
%=====================================

%--------------------
% wrap the first node
%--------------------

  for i=1:ne
    c(i,4) = c(i,1)
  end

%---------------------
% run over the elements
%---------------------

%---
  for i=1:ne
%---
   for j=1:3

    ces(i,j) = 0;

    for k=1:ne

      if(k ~= i) % skip the self-element
%----
        for l=1:3
          if( c(k,l)==c(i,j) & c(k,l+1)==c(i,j+1) )
            ces(i,j) = k;
          end
          if( c(k,l)==c(i,j+1) & c(k,l+1)==c(i,j) )
            ces(i,j) = k;
          end
        end
%----
      end

    end

   end
%---
  end
%---

%-----
% done
%-----

return
```

TABLE 4.3.5   Function *ces3* computes the element-to-side connectivity matrix *ces* for three-node triangles.

```
%----------------
% script see_elm3
%----------------

 ndiv=1;  % specify the discretization level

 [ne,ng,p,c,efl,gfl] = trgl3_disk (ndiv);  % triangulate the disk

 cen = cen3(ne,ng,c);   % element to node connectivity

 for node=1:ng   % run over nodes; or else specify "node"

  for i=2:cen(node,1)+1
   l = cen(node,i);     % element label
   j = c(l,1); xp(1) = p(j,1); yp(1) = p(j,2);
   j = c(l,2); xp(2) = p(j,1); yp(2) = p(j,2);
   j = c(l,3); xp(3) = p(j,1); yp(3) = p(j,2);
   j = c(l,1); xp(4) = p(j,1); yp(4) = p(j,2);
   cp(1) = 1.0; cp(2) = 0.0; cp(3) = 0.0; cp(4) = 1.0;  % color indices
   patch(xp,yp,cp); hold on;
  end

 end
```

TABLE 4.3.6 Script *see_elm3* scans the global nodes and displays the hosting neighboring elements.

4.3.2 *Delaunay triangulation*

Execute the FSELIB function *trgl3_delaunay* with a set of global nodes of your
choice and display the grid layout and associated Voronoi diagram.

4.3.3 *Element removal*

Write a function that removes a specified element from a triangulation generated
by the FSELIB function *trgl3_delaunay_sqr*, possibly also removing a global node.
Run the function until all elements have been removed one by one, and display
intermediate stages of the triangulation.

4.4 Laplace's equation with the Dirichlet boundary condition

In the first application of the finite element method in two dimensions, we consider steady heat conduction in the absence of a distributed source. The unknown
temperature field, $f(x, y)$, satisfies Laplace's equation,

$$\nabla^2 f = 0, \tag{4.4.1}$$

subject to the Dirichlet boundary condition around the entire contour of a simply
connected, disk like solution domain.

Carrying out the Galerkin projection, we find that the solution vector, $\mathbf{f}$, containing the N_G values of the function, f, at the global nodes, satisfies a homogeneous linear algebraic system,

$$\mathbf{D} \cdot \mathbf{f} = \mathbf{0}, \tag{4.4.2}$$

where $\mathbf{D}$ is the global diffusion matrix.

Implementing the Dirichlet boundary condition, as discussed in Section 4.1.5, we obtain an altered inhomogeneous system with a non-zero right-hand side,

$$\widehat{\mathbf{D}} \cdot \mathbf{f} = \mathbf{b}, \tag{4.4.3}$$

where $\widehat{\mathbf{D}}$ is a symmetric matrix.

The complete finite element code *lapl3_d* is listed in Table 4.4.1, where the suffix *d* stands for *Dirichlet*. The basic modules are as follows:

1. Specification of the input data.
2. Grid generation.
3. Implementation of the Dirichlet boundary condition at the global nodes.
4. Assembly of the global diffusion matrix from element diffusion matrices according to the algorithm shown in Table 1.2.2.
5. Implementation of the Dirichlet boundary condition by modifying the equations corresponding to the boundary nodes, as discussed in Section 4.1.5.
6. Solution of the linear system using a MATLAB linear solver.
7. Graphical display of the solution.

If $gfl(i) = 1$, the ith global node is a boundary node. The corresponding entry of the vector $bcd(i)$ hosts the prescribed boundary data. The element diffusion matrix is computed by the FSELIB function *edm3*, listed in Table 4.2.3.

For the lowest level of discretization, $ndiv = 0$, illustrated in the first frame of Figure 4.3.1, the 5×5 global diffusion matrix computed in the fourth step takes the form

$$\mathbf{D} = \begin{bmatrix} 4 & -1 & -1 & -1 & -1 \\ -1 & 1 & 0 & 0 & 0 \\ -1 & 0 & 1 & 0 & 0 \\ -1 & 0 & 0 & 1 & 0 \\ -1 & 0 & 0 & 0 & 1 \end{bmatrix}. \tag{4.4.4}$$

The modified global diffusion matrix incorporating the Dirichlet boundary conditions computed in the fifth step takes the diagonal form

$$\widehat{\mathbf{D}} = \begin{bmatrix} 4 & 0 & 0 & 0 & 0 \\ 0 & 1 & 0 & 0 & 0 \\ 0 & 0 & 1 & 0 & 0 \\ 0 & 0 & 0 & 1 & 0 \\ 0 & 0 & 0 & 0 & 1 \end{bmatrix}. \tag{4.4.5}$$

```
%=================================================
% Code lapl3_d
%
% Finite element code for Laplace's equation
% in a disk-like domain with the
% Dirichlet boundary condition
%=================================================

close all
clear all

%-----------
% input data
%-----------

ndiv = 3;    % level of triangulation

%------------
% triangulate
%------------

[ne,ng,p,c,efl,gfl] = trgl3_disk (ndiv);

%-------
% deform
%-------

defx = 0.6;                     % sample deformation

for i=1:ng
  p(i,1) = p(i,1)*(1.0-defx*p(i,2)^2);
end

%-----------------------------------------
% specify the Dirichlet boundary condition
%-----------------------------------------

for i=1:ng
 if(gfl(i)==1)
   bcd(i) = p(i,1)*sin(0.5*pi*p(i,2));
 end
end

%----------------------------------
% assemble the global diffusion matrix
%----------------------------------

gdm = zeros(ng,ng); % initialize
```

TABLE 4.4.1   Finite element code *lapl3_d* ($\rightarrow$ Continuing)

```
for l=1:ne    % loop over the elements

% compute the element diffusion matrix:

 j=c(l,1); x1=p(j,1); y1=p(j,2);
 j=c(l,2); x2=p(j,1); y2=p(j,2);
 j=c(l,3); x3=p(j,1); y3=p(j,2);

 edm_elm = edm3 (x1,y1,x2,y2,x3,y3);

   for i=1:3
      i1 = c(l,i);
      for j=1:3
         j1 = c(l,j);
         gdm(i1,j1) = gdm(i1,j1) + edm_elm(i,j);
      end
   end

end

%------------------------------
% compute the right-hand side
%------------------------------

for m=1:ng
  b(m) = 0.0;    % initialize
end

for m=1:ng

 if(gfl(m)==1)
    for i=1:ng
      b(i) = b(i) - gdm(i,m) * bcd(m);
      gdm(i,m) = 0; gdm(m,i) = 0;
    end
    gdm(m,m) = 1.0;
    b(m) = bcd(m);
 end

end

%------------------------
% solve the linear system
%------------------------

f = b/gdm';

%---------
% graphics
%---------

plot_3 (ne,ng,p,c,f); trimesh (c,p(:,1),p(:,2),f);
```

TABLE 4.4.1 (→ Continued) Finite element code *lapl3_d* for solving Laplace's equation with the Dirichlet boundary condition.

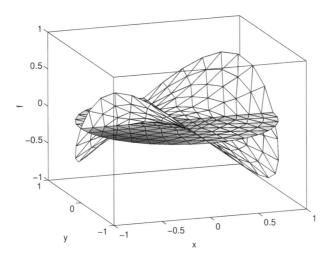

<small>FIGURE</small> 4.4.1   Finite element solution of Laplace's equation subject to the Dirichlet boundary condition, generated by the code *lapl3_d*.

In the adopted node labeling scheme, the first node is an interior node and the remaining nodes are boundary nodes. Consequently, all rows, except for the first row, implement the Dirichlet boundary condition.

For the next discretization level, $ndiv = 1$, illustrated in the second frame of Figure 4.3.1, the 13×13 global diffusion matrix takes the form shown in Table 4.4.2. The associated modified global diffusion matrix implementing the boundary conditions, truncated for convenience at the first decimal place, is shown in Table 4.4.3. The boundary nodes correspond to the 1.0 entries along the diagonal.

A color graph of the finite element solution is generated using (*a*) the MAT-LAB graphics function *patch* with color coding implemented in the FSELIB function *plot_3*, listed in Table 4.4.4, and (*b*) the MATLAB graphics function *trimesh*. Note that *trimesh* accepts as input the connectivity matrix, *c*, the *x* and *y* coordinates of the global nodes, and the solution vector. The graphics display of a computation with the sinusoidal boundary condition implemented in the code, listed in Table 4.4.1, is shown in Figure 4.4.1.

Computation of the boundary flux

After the finite element solution has been concluded, the boundary flux can be calculated based on the Galerkin projection of the global interpolation functions corresponding to the boundary nodes, expressed by the following redacted form of

$$
\begin{bmatrix}
4.0 & -1.0 & -1.0 & 0 & 0 & 0 & -1.0 \\
-1.0 & 3.9557 & -0.2050 & -1.4214 & -0.5621 & 0 & 0 \\
-1.0 & -0.2050 & 3.7610 & 0 & -0.8030 & -0.7448 & -0.2050 \\
0 & -1.4214 & 01.4143 & 0.0036 & 0 & 0 & \\
0 & -0.5621 & -0.8030 & 0.0036 & 1.5708 & -0.2092 & 0 \\
0 & 0 & -0.7448 & 0 & -0.2092 & 1.1633 & 0 \\
-1.0 & 0 & -0.2050 & 0 & 0 & 0 & 3.9557 \longrightarrow \\
0 & 0 & -0.8030 & 0 & 0 & -0.2092 & -0.5621 \\
0 & 0 & 0 & 0 & 0 & 0 & -1.4214 \\
-1.0 & -0.2050 & 0 & 0 & 0 & 0 & -0.2050 \\
0 & 0 & 0 & 0 & 0 & 0 & -0.5621 \\
0 & 0 & 0 & 0 & 0 & 0 & 0 \\
0 & -0.5621 & 0 & 0.0036 & 0 & 0 & 0
\end{bmatrix}
$$

$$
\begin{bmatrix}
0 & 0 & -1.0 & 0 & 0 & 0 \\
0 & 0 & -0.2050 & 0 & 0 & -0.5621 \\
-0.8030 & 0 & 0 & 0 & 0 & 0 \\
0 & 0 & 0 & 0 & 0 & 0.0036 \\
0 & 0 & 0 & 0 & 0 & 0 \\
-0.2092 & 0 & 0 & 0 & 0 & 0 \\
\longrightarrow \; -0.5621 & -1.4214 & -0.2050 & -0.5621 & 0 & 0 \\
1.5708 & 0.0036 & 0 & 0 & 0 & 0 \\
0.0036 & 1.4143 & 0 & 0.0036 & 0 & 0 \\
0 & 0 & 3.7610 & -0.8030 & -0.7448 & -0.8030 \\
0 & 0.0036 & -0.8030 & 1.5708 & -0.2092 & 0 \\
0 & 0 & -0.7448 & -0.2092 & 1.1633 & -0.2092 \\
0 & 0 & -0.8030 & 0 & -0.2092 & 1.5708
\end{bmatrix}
$$

TABLE 4.4.2   The global diffusion matrix, $\mathbf{D}$, for discretization level, $ndiv = 1$. The element layout is shown in the second frame of Figure 4.3.1.

$$
\begin{bmatrix}
4.0 & -1.0 & -1.0 & 0 & 0 & 0 & -1.0 & 0 & 0 & -1.0 & 0 & 0 & 0 \\
-1.0 & 4.0 & -0.2 & 0 & 0 & 0 & 0 & 0 & 0 & -0.2 & 0 & 0 & 0 \\
-1.0 & -0.2 & 3.8 & 0 & 0 & 0 & -0.2 & 0 & 0 & 0 & 0 & 0 & 0 \\
0 & 0 & 0 & 1.0 & 0 & 0 & 0 & 0 & 0 & 0 & 0 & 0 & 0 \\
0 & 0 & 0 & 0 & 1.0 & 0 & 0 & 0 & 0 & 0 & 0 & 0 & 0 \\
0 & 0 & 0 & 0 & 0 & 1.0 & 0 & 0 & 0 & 0 & 0 & 0 & 0 \\
-1.0 & 0 & -0.2 & 0 & 0 & 0 & 4.0 & 0 & 0 & -0.2 & 0 & 0 & 0 \\
0 & 0 & 0 & 0 & 0 & 0 & 0 & 1.0 & 0 & 0 & 0 & 0 & 0 \\
0 & 0 & 0 & 0 & 0 & 0 & 0 & 0 & 1.0 & 0 & 0 & 0 & 0 \\
-1.0 & -0.2 & 0 & 0 & 0 & 0 & -0.2 & 0 & 0 & 3.8 & 0 & 0 & 0 \\
0 & 0 & 0 & 0 & 0 & 0 & 0 & 0 & 0 & 0 & 1.0 & 0 & 0 \\
0 & 0 & 0 & 0 & 0 & 0 & 0 & 0 & 0 & 0 & 0 & 1.0 & 0 \\
0 & 0 & 0 & 0 & 0 & 0 & 0 & 0 & 0 & 0 & 0 & 0 & 1.0
\end{bmatrix}
$$

TABLE 4.4.3   The global diffusion matrix shown in Table 4.4.2 after implementing the Dirichlet boundary condition.

the general projection (4.1.19),

$$
\oint_C \phi_i \, q \, \mathrm{d}l = k \sum_{i=j}^{N_{\mathrm{G}}} D_{ij} f_j, \tag{4.4.6}
$$

with the understanding that the ith node is a boundary node. Note that (4.4.6) is the counterpart of the one-dimensional algebraic form (1.1.35) in the absence of a source. Applying (4.4.6) for the global interpolation functions of all boundary nodes, we obtain a system of integral equations for q whose solution can be found by standard numerical methods.

PROBLEMS

4.4.1 *Constant and linear boundary conditions*

Run the FSELIB code *lapl3_d* for a circular domain with boundary values that are (a) uniform (constant), (b) linear in x, and (c) linear in y. Discuss the numerical results in each case.

4.4.2 *Deformed domain*

Repeat Problem 4.4.1 with a deformed disk-like domain of your choice.

```
function plot_3 (ne,ng,p,c,f);

%=======================================================
% Color-mapped visualization of a function f
% in a domain discretized into three-node triangles
%=======================================================

% compute the maximum and minimum of the function f

fmax =-100.0;  % initialize
fmin = 100.0;  % initialize

for i=1:ng
 if(f(i) > fmax) fmax = f(i); end
 if(f(i) < fmin) fmin = f(i); end
end

range = 1.2*(fmax-fmin); shift = fmin;

%---------------------------------------------
% shift the color index in the range (0, 1)
% and plot 4-point patches
%---------------------------------------------

hold on

for l=1:ne

 j = c(1,1);
 xp(1)=p(j,1); yp(1)=p(j,2); cp(1)=(f(j)-shift)/range;

 j = c(1,2);
 xp(2)=p(j,1); yp(2)=p(j,2); cp(2)=(f(j)-shift)/range;

 j = c(1,3);
 xp(3)=p(j,1); yp(3)=p(j,2); cp(3)=(f(j)-shift)/range;

 j = c(1,1);
 xp(4)=p(j,1); yp(4)=p(j,2); cp(4)=(f(j)-shift)/range;

 patch (xp, yp,cp);

end

%-----
% done
%-----
```

TABLE 4.4.4 FSELIB function *plot_3* performs the color-mapped visualization of a function in a domain discretized into three-node triangles using the MATLAB function *patch*.

4.4.3 *Delaunay triangulation*

Replace the discretization function *trgl3_disk* with the Delaunay-based FSELIB function *trgl3_delaunay* discussed in Section 4.3.2. Run the code for a domain geometry and boundary conditions of your choice and discuss the results.

4.5   Eigenvalues of the Laplacian operator

An eigenfunction of the Laplacian operator, $f(x, y)$, satisfies the equation

$$\nabla^2 f + \lambda f = 0, \tag{4.5.1}$$

where ∇^2 is the Laplacian operator in the xy plane, λ is the corresponding eigenvalue, subject to the homogeneous Dirichlet boundary condition around the entire boundary of the solution domain, $f = 0$.

Applying the Galerkin finite element projection, we obtain a generalized eigenvalue problem expressed by a system of linear algebraic equations,

$$\mathbf{D} \cdot \mathbf{f} = \lambda \mathbf{M} \cdot \mathbf{f}, \tag{4.5.2}$$

where $\mathbf{D}$ is the global diffusion matrix, $\mathbf{M}$ is the global mass matrix, and the vector $\mathbf{f}$ encapsulates the N_{G} values of the eigenfunction at the global nodes.

Removing the rows and columns corresponding to the boundary nodes, we obtain a modified reduced system,

$$\widehat{\mathbf{D}} \cdot \mathbf{f} = \lambda \widehat{\mathbf{M}} \cdot \mathbf{f}, \tag{4.5.3}$$

whose eigenvalues are approximations of those of the Laplacian operator. The generalized eigenvalue problem can be solved using a function embedded in MATLAB.

The complete finite element code *lapl3_eig* incorporating the assembly of the global diffusion and mass matrices is listed in Table 4.5.1. The global diffusion matrix is assembled from element diffusion matrices computed using the FSELIB function *edm3* listed in Table 4.3.2. The global mass matrix is assembled from element mass matrices computed using the FSELIB function *emm3* listed in Table 4.5.2.

Circular disk

The first six eigenfunctions of the Laplacian operator on a circular disk of unit radius are shown in Figure 4.5.1 for discretization level $ndiv = 4$. The eigenvalue corresponding to the first eigenfunction, shown in Figure 4.5.1(a), is 5.8024. This numerical value is a good approximation of the exact eigenvalue, $\lambda_1 = \varsigma_0^2 = 5.7840$, where ς_0 is the first zero of the zeroth order Bessel function, J_0. The numerical error is $5.8024 - 5.7840 = 0.0758$.

Increasingly accurate numerical approximations of the eigenvalues can be obtained by selecting higher discretization levels. For the finer grid, $ndiv = 5$, the computed first eigenvalue is 5.7780, and the associated error is $5.7780 - 5.7840 = 0.0040$.

```
%==========================================
% Code lapl3_eig
%
% Eigenfunctions of Laplace's equation
% in a disk-like domain
% using three-node triangles
%==========================================

close all
clear all

%-----------
% input data
%-----------

ndiv = 3;  % discretization level

%------------
% triangulate
%------------

  [ne,ng,p,c,efl,gfl] = trgl3_disk (ndiv);

%-------
% deform
%-------

defx = 0.0;

for i=1:ng
 p(i,1) = p(i,1)*(1.0-defx*p(i,2)^2);
end

%----------------------------------------------------
% assemble the global diffusion and mass matrices
%----------------------------------------------------

gdm = zeros(ng,ng); % initialize
gmm = zeros(ng,ng); % initialize

for l=1:ne              % loop over the elements

% compute the element diffusion and mass matrices

j = c(l,1); x1 = p(j,1); y1 = p(j,2);
j = c(l,2); x2 = p(j,1); y2 = p(j,2);
j = c(l,3); x3 = p(j,1); y3 = p(j,2);

[edm_elm] = edm3 (x1,y1,x2,y2,x3,y3);
[emm_elm] = emm3 (x1,y1,x2,y2,x3,y3);
```

TABLE 4.5.1   Code *lapl3_eig* (Continuing →)

```
    for i=1:3
      i1 = c(1,i);
      for j=1:3
        j1 = c(1,j);
        gdm(i1,j1) = gdm(i1,j1) + edm_elm(i,j);
        gmm(i1,j1) = gmm(i1,j1) + emm_elm(i,j);
      end
    end

end

%--------------------
% reduce the matrices by removing equations
% corresponding to boundary nodes
%--------------------

Ic = 0;

for i=1:ng

 if(gfl(i,1)==0)
  Ic = Ic+1;
  map(Ic) = i;
  Jc = 0;

  for j=1:ng
   if(gfl(j,1)==0)
    Jc = Jc+1;
    A(Ic,Jc) = gdm(i,j);
    B(Ic,Jc) = gmm(i,j);
   end
  end

 end

end

ngred = Ic;

%------------------------
% compute the eigenvalues and eigenvectors
%------------------------

[V,D] = eig(A,B);

%-----
% map and plot
%-----

for order=1:6

 for i=1:ng
  f(i) = 0.0;
 end
```

TABLE 4.5.1   Code *lapl3_eig* ($\rightarrow$ Continuing $\rightarrow$)

```
for i=1:ngred
 f(map(i)) = V(i,order);
end

% one figure for each eigenfunction

figure(order)
plot_3 (ne,ng,p,c,f);
trimesh (c,p(:,1),p(:,2),f);
view([-19,18])
box on

end

%-----
% done
%-----
```

TABLE 4.5.1   ($\rightarrow$ Continued) Code *lapl3_eig* computes the eigenvalues and eigenfunctions of the Laplacian operator in a disk-like domain.

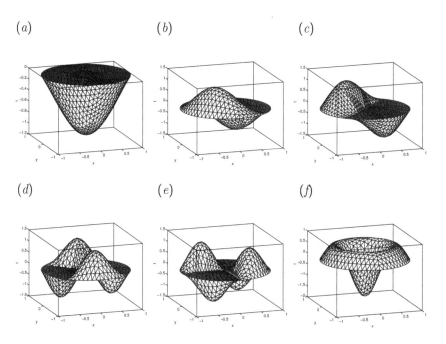

FIGURE 4.5.1   Finite element solutions for the first six eigenfunctions of the Laplacian operator on a circular disk.

```
function emm = emm3(x1,y1,x2,y2,x3,y3)

%========================================
% Evaluation of the element mass matrix
% from the coordinates of the vertices
% of a three-node triangle
%
% A: element area
%========================================

  d23x = x2-x3; d23y = y2-y3;
  d31x = x3-x1; d31y = y3-y1;
  d12x = x1-x2; d12y = y1-y2;

  A = 0.5*(d31x*d12y - d31y*d12x);

  fc = A/12.0; fs = 2.0*fc;

  emm(1,1) = fs; emm(1,2) = fc; emm(1,3) = fc;
  emm(2,1) = fc; emm(2,2) = fs; emm(2,3) = fc;
  enm(3,1) = fc; emm(3,2) = fc; emm(3,3) = fs;

%-----
% done
%-----

return;
```

TABLE 4.5.2   Function *emm3* evaluates the element mass matrix from the coordinates of the vertices of a three-node triangle.

Since $0.0758/16 \simeq 0.004$, and since increasing $ndiv$ by one unit quadruples the number of finite elements due to subdivision, the error scales with the inverse of the square of the number of elements.

PROBLEM

4.5.1 *Elliptical domain*

Compute and discuss the first eigenvalue of the Laplacian operator in an elliptical domain with aspect ratio 2:1.

4.6   Convection–diffusion with the Dirichlet boundary condition

As a third case study, we solve the steady linear convection–diffusion equation in two dimensions,

$$u_x \frac{\partial f}{\partial x} + u_y \frac{\partial f}{\partial y} = \kappa \nabla^2 f, \tag{4.6.1}$$

where u_x and u_y are specified constant velocities along the x and y axes, and κ is the medium diffusivity. The Dirichlet boundary condition is prescribed around the contiguous contour of a simply connected domain arising from the deformation of a circular disk.

The importance of convection relative to diffusion is expressed by the dimensionless Péclet number,

$$\mathrm{Pe} \equiv \frac{UR}{\kappa}, \tag{4.6.2}$$

where U is the maximum of $|u_x|$ and $|u_y|$, and R is the radius of the undeformed disk. When $\mathrm{Pe} = 0$, we recover the problem of steady diffusion governed by Laplace's equation, discussed in Section 4.4. As the Péclet number increases, convective transport becomes increasingly important everywhere except near the edges of the solution domain where thin boundary layers arise.

Finite element equations

Carrying out the Galerkin projection, we find that the solution vector, $\mathbf{f}$, containing the N_{G} nodal values of the requisite function at the global nodes, satisfies a homogeneous algebraic system,

$$(\kappa \mathbf{D} + \mathbf{N}) \cdot \mathbf{f} = \mathbf{0}, \tag{4.6.3}$$

where $\mathbf{D}$ is the global diffusion matrix and $\mathbf{N}$ is the global advection matrix.

Implementing the Dirichlet boundary condition, as discussed in Section 4.1.5, we derive an altered inhomogeneous system with non-zero right-hand side,

$$(\kappa \widehat{\mathbf{D}} + \widehat{\mathbf{N}}) \cdot \mathbf{f} = \mathbf{b}, \tag{4.6.4}$$

where the caret (hat) designates modified diffusion and advection matrices. The finite element implementation is a straightforward modification of that described in Section 4.4 for Laplace's equation. The complete finite element code *scd3_d* incorporating the assembly of the global diffusion and advection matrices from corresponding element matrices is listed in Table 4.6.1.

Numerical solutions

Finite element solutions for convection along the x axis, $u_x > 0$ and $u_y = 0$, are shown in Figure 4.6.1 for four values of the Péclet number, $\mathrm{Pe} = 0$, 2, 5, and 10. The boundary conditions specify that the temperature, f, varies linearly in the x coordinate around the boundary, as implemented in the code.

The numerical results confirm that, as the Péclet number increases, convection becomes increasingly important and steep gradients arise at the right half of the boundary, in agreement with physical intuition.

```
%=========================================================
% Code scd3_d
%
% Finite element code for steady convection--diffusion
% subject to the Dirichlet boundary condition
%
% gdm: global diffusion matrix
% gam: global advection matrix
% lsm: linear system matrix
%=========================================================

%-----------
% input data
%-----------

k = 1.0; rho = 1.0; cp = 1.0; ux = 5.0; uy = 0.0;

ndiv = 3;

%----------
% constants
%----------

kappa = k/(rho*cp);

%----------------------
% triangulate and deform
%----------------------

[ne,ng,p,c,efl,gfl] = trgl3_disk (ndiv);

for i=1:ng
 p(i,1) = p(i,1)*(1.0-0.5*p(i,2)^2 );
end

%-------------------------------------------
% specify the Dirichlet boundary condition
%-------------------------------------------

for i=1:ng
 if(gfl(i)==1)
   bcd(i) = p(i,1);  % example
 end
end

%---------------------------
% assemble the global diffusion
% and advection matrices
%---------------------------

gdm = zeros(ng,ng);   %  initialize
gam = zeros(ng,ng);   %  initialize
```

TABLE 4.6.1   Code *scd3_d* (Continuing →)

```
for l=1:ne          % loop over the elements

% compute the element diffusion
% and advection matrices

  j=c(l,1); x1=p(j,1); y1=p(j,2);
  j=c(l,2); x2=p(j,1); y2=p(j,2);
  j=c(l,3); x3=p(j,1); y3=p(j,2);

  [edm_elm] = edm3 (x1,y1,x2,y2,x3,y3);

  [eam_elm] = eam3 (ux,uy,x1,y1,x2,y2,x3,y3);

   for i=1:3
     i1 = c(l,i);
     for j=1:3
       j1 = c(l,j);
       gdm(i1,j1) = gdm(i1,j1) + edm_elm(i,j);
       gam(i1,j1) = gam(i1,j1) + eam_elm(i,j);
     end
   end

end

%--------------------------------
% compute the coefficient matrix
%--------------------------------

lsm = kappa*gdm-gam;

%----------------------------------------------
% compute the right-hand side and implement
% the Dirichlet boundary condition
%----------------------------------------------

for i=1:ng
 b(i) = 0.0;
end

for j=1:ng

 if(gfl(j)==1)
   for i=1:ng
     b(i) = b(i) - lsm(i,j) * bcd(j);
     lsm(i,j) = 0.0;
     lsm(j,i) = 0.0;
   end
   lsm(j,j) = 1.0;
   b(j) = bcd(j);
 end

end

%------------------------
% solve the linear system
%------------------------
```

TABLE 4.6.1   Code *scd3_d* ($\rightarrow$ Continuing $\rightarrow$)

```
f = b/lsm';

%------------------
% plot the solution
%------------------

plot_3 (ne,ng,p,c,f);

trimesh (c,p(:,1),p(:,2),f);

%-----
% done
%-----
```

TABLE 4.6.1   ($\rightarrow$ Continued) Finite element code *scd3_d* for the steady convection–diffusion equation with the Dirichlet boundary condition around the contour of a disk-like domain.

PROBLEMS

4.6.1 *Code scd3_d*

(*a*) Run the code *scd3_d* for a circular domain with parameter values and boundary conditions of your choice, and discuss the results.

(*b*) Repeat (*a*) for a deformed disk-like domain of your choice.

4.6.2 *Delaunay triangulation*

Replace the discretization function *trgl3_disk* implemented in the code *scd3_d* with the Delaunay-based FSELIB function *trgl3_delaunay* discussed in Section 4.3.2. Run the modified code for a domain geometry, convection velocity, and other parameter values of your choice and discuss the results.

4.6.3 *Convection-dominated transport*

Assess by numerical experimentation the performance of the numerical method implemented in the code *scd3_d* as the Péclet becomes higher for a domain geometry, parameter values, and boundary conditions of your choice.

4.6.4 *Convective transport*

Run the code *scd3_d* for convection-dominated transport corresponding to the limit $Pe \rightarrow \infty$. Discuss and explain on physical grounds the MATLAB response.

(a)   (b)

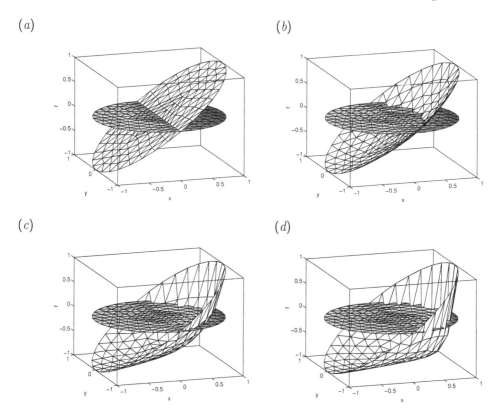

(c)   (d)

FIGURE 4.6.1   Finite element solution of the steady convection–diffusion equation
generated by the FSELIB code *scd3_d* with a Dirichlet boundary condition that
is linear in x. A uniform convection velocity, U, is imposed from left to right.
Solutions are shown for (a) Pe $= UR/\kappa = 0$, (b) 2, (c) 5, and (d) 10, where R
is the radius of the undeformed disk.

4.7   Helmholtz's equation with the Neumann boundary condition

Helmholtz's equation can be regarded as a special case of the steady diffusion equation, arising when the source term is proportional to the *a priori* unknown solution,
$s = \alpha f/k$, yielding the differential equation

$$\nabla^2 f + \alpha f = 0, \qquad (4.7.1)$$

where α is a specified constant with dimensions of inverse squared length. We assume that the Neumann boundary condition specifying the flux, q, is prescribed
around the boundary of a simply connected solution domain arising from the deformation of a circular disk.

Carrying out the Galerkin projection, we find that the preliminary Galerkin finite element system (4.1.23) simplifies into an algebraic system,

$$(\mathbf{D} - \alpha \, \mathbf{M}) \cdot \mathbf{f} = \mathbf{b}, \tag{4.7.2}$$

where the ith component of the vector $\mathbf{b}$ on the right-hand side is given by

$$b_i \equiv \frac{1}{k} \oint_C \phi_i \, q \, dl. \tag{4.7.3}$$

The flux integral on the right-hand side (4.7.3) is non-zero only if the ith node associated with the global interpolation function ϕ_i, corresponding to the ith Galerkin projection equation, is a boundary node. In the finite element implementation, this flux integral can be assembled in terms of *element edge integrals* in a process that is similar to that used to assemble the coefficient matrix, as follows:

1. We run over the elements and then loop along the three element vertex nodes. To facilitate the logistics, the element nodes are wrapped by introducing a ghost fourth node, which is identical to the first node.

2. If the jth element node is a boundary node for $j = 1, 2, 3$, we examine whether the next element node numbered $j + 1$ is also a boundary node.

3. If both element nodes are boundary nodes, the corresponding element edge lies at the boundary of the solution domain. In that case, the computation discussed in the next step is performed.

4. We evaluate the contour integral on the right-hand side of (4.7.3) along the element edge for the element interpolation functions corresponding to the first and second end-nodes, and add these integrals to the right-hand sides of the appropriate Galerkin equations identified with the help of a connectivity matrix.

An elementary computation shows that, consistent with the piecewise linear interpolation, the edge integrals in the fourth stage corresponding to the edge nodes are given by

$$I_j = \Big(\frac{1}{3} q_j + \frac{1}{6} q_{j+1} \Big) \Delta s, \qquad I_{j+1} = \Big(\frac{1}{6} q_j + \frac{1}{3} q_{j+1} \Big) \Delta s, \tag{4.7.4}$$

where Δs is the element edge length. The coefficients on the right-hand sides derive from the one-dimensional mass matrix with linear elements, as shown in (1.2.9).

The complete FSELIB code *hlm3_n* is listed in Table 4.7.1. The individual modules are:

1. Specification of input data.

2. Grid generation.

(a)   (b)

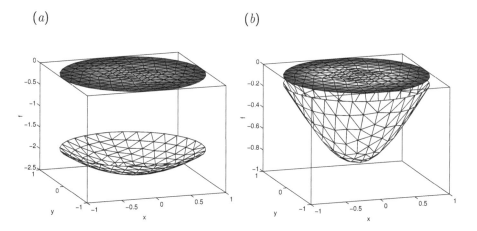

FIGURE 4.7.1   Solution of the Helmholtz equation with the Neumann boundary condition specifying a uniform boundary flux, $q = q_0$, generated by the FSELIB code
hlm3_n for (a) $\alpha R^2 = 1$ and (b) 5.

3. Assignment of the Neumann boundary conditions to the global nodes.

4. Assembly of the global diffusion and mass matrices from element diffusion
   matrices according to the algorithm shown in Table 1.2.2.

5. Assembly of the right-hand side of the linear system.

6. Solution of the linear system using an intrinsic MATLAB solver.

7. Plotting the solution using the FSELIB function *plot_3*, listed in Table 4.4.4,
   and the MATLAB function *trimesh*.

The element diffusion matrices in the fourth step are evaluated by the FSELIB function *edm3*, listed in Table 4.2.1. The element mass matrices are evaluated by the
FSELIB function *emm3*, listed in Table 4.5.2. The solution generated by the code
with the uniform flux boundary condition, $q = 1$, as implemented in the code, is
shown in Figure 4.7.1 for $\alpha = 1$ and 5.

As the coefficient α becomes smaller, the solution tends to become increasingly
uniform. A singular behavior is encountered in the limit $\alpha \to 0$ due to the compatibility condition of the Poisson equation,

$$\nabla^2 f + s = 0, \qquad\qquad\qquad (4.7.5)$$

where s is a source term. The compatibility condition requires that the areal integral
of the source function, s, over the solution domain, be equal to the line integral of the
flux across the boundary. Physically, the total rate of production of an entity must
be balanced by the total outward flow to prevent accumulation. If this condition is
not met, a solution cannot be found. If this condition is met, an infinite number of

```
%===========================================
% Code hlm3_n
%
% Code for Helmholtz's equation in a disk-like
% domain subject to the Neumann
% boundary condition
%===========================================

%-----
% input: conductivity,
%         Helmholtz coefficient,
%         discretization level
%-----

k = 1.0; alpha = 4.0; ndiv = 3;

%------------
% triangulate
%------------

[ne,ng,p,c,efl,gfl] = trgl3_disk (ndiv);

%-------
% deform
%-------

def = 0.20;

for i=1:ng
 p(i,1) = p(i,1)*(1.0-def*p(i,2)^2);
end

%---------------------------------------
% specify the Neumann boundary condition (bcn)
%---------------------------------------

for i=1:ng
 if(gfl(i)==1)
  bcn(i) = 1.0;     % example
 end
end

%-----------------------------------------------
% assemble the global diffusion and mass matrix
%-----------------------------------------------

gdm = zeros(ng,ng);  % initialize
gmm = zeros(ng,ng); % initialize

for l=1:ne     % loop over the elements

% compute the element matrices
```

TABLE 4.7.1   Code *hlm3_n* (Continuing $\rightarrow$)

```
j = c(1,1); x1 = p(j,1); y1 = p(j,2);
j = c(1,2); x2 = p(j,1); y2 = p(j,2);
j = c(1,3); x3 = p(j,1); y3 = p(j,2);

[edm_elm] = edm3(x1,y1,x2,y2,x3,y3);
[emm_elm] = emm3(x1,y1,x2,y2,x3,y3);

   for i=1:3
     i1 = c(1,i);
     for j=1:3
       j1 = c(1,j);
       gdm(i1,j1) = gdm(i1,j1) + edm_elm(i,j);
       gmm(i1,j1) = gmm(i1,j1) + emm_elm(i,j);
     end
   end

end

%----------------------------
% set up the right-hand side
%----------------------------

for i=1:ng       % initialize
 b(i) = 0.0;
end

%---------------------------------------------
% Neumann integral on the right-hand side
%---------------------------------------------

for i=1:ne       % loop over the elements

 efl(i,4) = efl(i,1);  % wrap around the element
 c(i,4) = c(i,1);

 for j=1:3                % run over the element edges

  j1 = c(i,j);
  j2 = c(i,j+1);

  if( gfl(j1)==1 & gfl(j2)==1 )
    xe1 = p(j1,1);
    ye1 = p(j1,2);
    xe2 = p(j2,1);
    ye2 = p(j2,2);
    edge = sqrt((xe2-xe1)^2+(ye2-ye1)^2);
    int1 = edge * (bcn(j1)/3 + bcn(j2)/6);
    int2 = edge * (bcn(j1)/6 + bcn(j2)/3);
    b(j1) = b(j1)+int1/k;
    b(j2) = b(j2)+int2/k;
  end

 end

end
```

TABLE 4.7.1   Code *hlm3_n* ($\to$ Continuing $\to$)

```
%-------------------
% coefficient matrix
%-------------------

lsm = gdm-alpha*gmm;

%-----------------------
% solve the linear system
%-----------------------

f = b/lsm';

%---------
% plotting
%---------

plot_3 (ne,ng,p,c,f);
trimesh (c,p(:,1),p(:,2),f);

%-----
% done
%-----
```

TABLE 4.7.1   ($\rightarrow$ Continued) Code *hlm3_n* for the Helmholtz equation in a disk-like domain subject to the Neumann boundary condition.

solutions differing by an arbitrary constant arise. In the case of Laplace's equation, $s = 0$, the compatibility condition requires that the line integral of the boundary flux is zero (Problem 4.7.2).

PROBLEMS

4.7.1 *Code hlm3_n*

Run the FSELIB code *hlm3_n* for a deformed shape, parameter values, and a Neumann boundary condition of your choice. Discuss the finite element solution.

4.7.2 *Compatibility condition for Poisson's equation*

Run the code *hlm3_d* with $\alpha = 0$, in which case Helmholtz's equation reduces to Laplace's equation, using parameter values and boundary conditions of your choice, and confirm that the computation fails. In particular, MATLAB will produce the warning "*Matrix is close to singular or badly scaled.*" Compute the determinant of the coefficient matrix using the MATLAB function *det* and explain the failure of the numerical method from a mathematical viewpoint.

4.8   Laplace's equation with arbitrary boundary conditions

In a more demanding application, we solve Laplace's equation,

$$\nabla^2 f = 0, \tag{4.8.1}$$

in a disk-like domain centered at the origin, subject to the Dirichlet boundary condition around the left part of the boundary, $x < 0$, and the Neumann boundary condition around the right part of the boundary, $x > 0$. Under these circumstances, the master Galerkin finite element system (4.1.23) simplifies into an algebraic system,

$$\mathbf{D} \cdot \mathbf{f} = \mathbf{b}, \tag{4.8.2}$$

where the ith component of the vector on the right-hand side, $\mathbf{b}$, corresponding to the ith global node, is given by

$$b_i \equiv \frac{1}{k} \oint_C \phi_i \, q \, \mathrm{d}l, \tag{4.8.3}$$

and k is the medium conductivity. The flux integral on the right-hand side of (4.8.3) is non-zero only if the ith node associated with the global interpolation function, ϕ_i, corresponding to the ith Galerkin projection, is a Neumann boundary node.

The complete FSELIB code *lapl3_dn* is listed in Table 4.8.1. Implicit in the algorithm is the splitting of each of two boundary nodes located at the junction of the Dirichlet and Neumann contours, $x = 0$, into a pair of companion nodes. Highlights of the algorithm include the following:

- After the solution domain, D, has been triangulated, element edges that lie on the boundary are identified and counted sequentially using an element edge counter, Ic. An element edge is a boundary edge only if both element nodes defining the edge are boundary nodes. The total number of boundary edges is denoted as *nbe* in the code.

- The variable $face1(i)$ is the label of the global node of the first point of the ith boundary edge, and the variable $face2(i)$ is the label of the global node of the second point of the ith boundary edge, where $i = 1, \ldots, nbe$.

- The flag $face3(i) = 1, 2$ indicates, respectively, that the Dirichlet or Neumann boundary condition is applied at the ith boundary edge, where $i = 1, \ldots, nbe$. The default value is $face3(i) = 1$. If $face3(i) = 2$, the vector $bcn1(i)$ holds the prescribed Neumann value at the first point of the ith edge and the vector $bcn2(i)$ holds the prescribed Neumann value at the second point of the ith edge.

- The flag $bcf(i) = 1$ indicates that the ith global node is a Dirichlet boundary node. This flag is set by running over the boundary edges and raising the flags

(a)   (b)

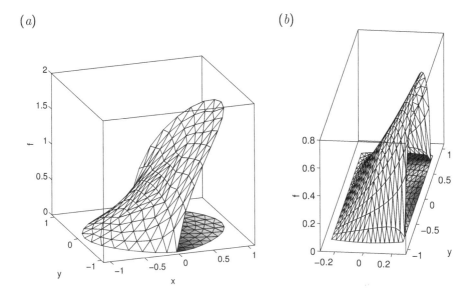

FIGURE 4.8.1   (a) Finite element solution of Laplace's equation subject to the Dirichlet boundary condition on the left half of the boundary, $x < 0$, $f = 0$, and
the uniform flux Neumann boundary condition on the right half of the boundary, $x > 0$, $q = q_0$, generated by the code *lapl3_dn*. (b) Finite element solution
of Laplace's equation in a deformed square resembling a slab generated by the
FSELIB code *lapl3_dn_sqr*. The homogeneous Dirichlet boundary condition is applied along three sides, $f = 0$, and the uniform Neumann boundary condition is
applied along the fourth side, $q = q_0$.

of the first and second edge nodes. If $bcf(i) = 1$, in which case the ith global
node is a Dirichlet boundary node, the entry $bcd(i)$ is assigned the prescribed
boundary value.

- The contour integral on the right-hand side of (4.8.3) is computed by running
over the *nbe* boundary element edges, while making additive contributions
to the equations corresponding to the first and second end-node, using the
formulas given in (4.7.4).

- The Dirichlet boundary condition is implemented after the linear system has
been compiled using a familiar method.

The finite element solution with the uniform flux boundary condition along the
Neumann portion of the boundary, $q = q_0$, and the homogeneous Dirichlet condition
on the Dirichlet portion of the boundary, $f = 0$, as implemented in the code, is
displayed in Figure 4.8.1(a).

```
%==================================================
% Code lapl3_dn
%
% Finite element solution of Laplace's equation
% in a disk-like domain, with Dirichlet
% and Neumann boundary conditions
%==================================================

clear all
close all

%-----------
% input data
%-----------

k = 1.0;        % conductivity
ndiv = 3;       % level of triangulation

%----------------------
% triangulate and deform
%----------------------

[ne,ng,p,c,efl,gfl] = trgl3_disk (ndiv);

for i=1:ng
 p(i,1) = p(i,1)*(1.0-0.30*p(i,2)^2);
end

%----------------------------------------
% find the element edges on the boundary
%----------------------------------------

Ic = 0;    % initialize the element edge counter

for i=1:ne

 efl(i,4) = efl(i,1); % wrap around the element
 c(i,4) = c(i,1);

 for j=1:3       % run over the element edges

  if(efl(i,j)==1 & efl(i,j+1)==1)   % new edge side
   Ic = Ic+1;
   face1(Ic) = c(i,j);
   face2(Ic) = c(i,j+1);
  end

 end

end

nbe = Ic;    % number of boundary edges
```

TABLE 4.8.1   Code *lapl3_dn* (Continuing →)

```
%--------------------------------------------
% specify the Neumann boundary condition on
% the right side of the solution domain
%--------------------------------------------

for i=1:nbe  % run along the boundary sides

 m = face1(i);
 l = face2(i);

 face3(i)=1;  % default Dirichlet flag

 if(p(m,1)>0.00001 | p(l,1)>0.00001)    % Neumann side
   face3(i)= 2;     % Neumann flag
   bcn1(i) = 1.0;   % Neumann condition at first  edge node
   bcn2(i) = 1.0;   % Neumann condition at second edge node
 end

end

%--------------------------------------------
% specify the Dirichlet boundary condition
% on the left side
%--------------------------------------------

for i=1:ng     % initialize node flag
 bcf(i) = 0;   % bcf(i)=1 will indicate a Dirichlet node
end

for i=1:nbe    % run over the boundary edges

 m = face1(i);
 l = face2(i);

 if(p(m,1)<-0.00001 | p(l,1)<-0.00001)    % Dirichlet edge
   bcf(m) = 1;     % Dirichlet flags
   bcf(l) = 1;
   bcd(m) = 0.0;   % Dirichlet condition at first  edge node
   bcd(l) = 0.0;   % Dirichlet condition at second edge node
 end

end

%--------------------------------------------
% assemble the global diffusion matrix
%--------------------------------------------

gdm = zeros(ng,ng); % initialize

for l=1:ne    % loop over the elements
               %to compute the global diffusion matrix
```

TABLE 4.8.1   Code *lapl3_dn* ($\rightarrow$ Continuing $\rightarrow$)

```
j=c(l,1); x1=p(j,1); y1=p(j,2);
j=c(l,2); x2=p(j,1); y2=p(j,2);
j=c(l,3); x3=p(j,1); y3=p(j,2);

[edm_elm] = edm3(x1,y1,x2,y2,x3,y3);

    for i=1:3
      i1 = c(l,i);
      for j=1:3
        j1 = c(l,j);
        gdm(i1,j1) = gdm(i1,j1) + edm_elm(i,j);
      end
    end

end

%-------------------------------
% initialize the right-hand side
%-------------------------------

for i=1:ng
 b(i) = 0.0;
end

%----------------------------------------
% Neumann integral on the right-hand side
%----------------------------------------

for i=1:nbe

 if(face3(i)==2)

  m = face1(i);
  l = face2(i);
  xe1 = p(m,1);
  ye1 = p(m,2);
  xe2 = p(l,1);
  ye2 = p(l,2);
  edge = sqrt((xe2-xe1)^2+(ye2-ye1)^2);
  int1 = edge*(bcn1(i)/3 + bcn2(i)/6);
  int2 = edge*(bcn1(i)/6 + bcn2(i)/3);
  b(m) = b(m)+int1/k;
  b(l) = b(l)+int2/k;

 end

end

%--------------------------------------------
% implement the Dirichlet boundary condition
%--------------------------------------------
```

TABLE 4.8.1   Code *lapl3_dn* ($\rightarrow$ Continuing $\rightarrow$)

```
for j=1:ng
 if(bcf(j)==1)
    for i=1:ng
       b(i) = b(i) - gdm(i,j) * bcd(j);
       gdm(i,j) = 0; gdm(j,i) = 0;
    end
    gdm(j,j) = 1.0; b(j) = bcd(j);
 end
end

%------------------------
% solve the linear system
%------------------------

f = b/gdm';

%--------
% plotting
%--------

plot_3 (ne,ng,p,c,f);
trimesh (c,p(:,1),p(:,2),f);

%-----
% done
%-----
```

TABLE 4.8.1   ($\longrightarrow$ Continued) Code *lapl3_dn* solves Laplace's equation in a disk-like domain. The Dirichlet boundary condition is prescribed on the right part of the boundary, $x < 0$, and the Neumann boundary condition is prescribed on the left part of the boundary, $x > 0$.

PROBLEMS

4.8.1 *Code lapl3_dn*

Run the FSELIB code *lapl3_dn* for a deformed disk-like domain of your choice and discuss the results.

4.8.2 *Modified code lapl3_dn*

Modify the FSELIB code *lapl3_dn* so that the Dirichlet boundary condition is specified all around the boundary of the undeformed disk, except at the first quadrant where the Neumann boundary condition is specified. Run the modified code for a deformed shape of your choice and discuss the results.

4.8.3 *Solution of Laplace's equation in a deformed square*

FSELIB code *lapl3_dn_sqr*, not listed in the text, solves Laplace's equation in a deformed square subject to the uniform Dirichlet boundary condition along the top, left, and bottom sides, $f = 0$, and the uniform Neumann boundary condition along the right side, $q = 1$. The implementation of the finite element method is similar to that described in the text for the corresponding code *lapl3_dn*. The numerical solution for a deformed rectangular shape resembling a slab is shown in Figure 4.8.1(b). Run the code *lapl3_dn_sqr* for a deformed shape of your choice and discuss the results.

4.9 Surface elements

The mathematical modeling of a class of problems in mathematical physics and engineering results in partial differential equations defined over three-dimensional, stationary or evolving surfaces. In physical applications, these surfaces can be material boundaries, interfaces, and propagating or moving fronts. An example is the transport equation governing the evolution of the surface concentration of an insoluble surfactant residing at the interface between two immiscible fluids (e.g., Pozrikidis [49]). The surfactant molecules are advected and diffuse tangentially over the interface, while their number density changes because of interfacial in-plane stretching and expansion associated with normal motion.

A key notion in the mathematical formulation is the *surface gradient*, which is a vector consisting of spatial derivatives in directions that are tangential to the surface at any instant. Specifically, the surface gradient operator is defined as

$$\nabla_s \equiv (\mathbf{I} - \mathbf{n} \otimes \mathbf{n}) \cdot \nabla, \tag{4.9.1}$$

where

$$\nabla = \left(\frac{\partial}{\partial x}, \frac{\partial}{\partial y}, \frac{\partial}{\partial z} \right) \tag{4.9.2}$$

is the usual three-dimensional gradient, $\mathbf{I}$ is the unit matrix, $\mathbf{n}$ is the unit vector normal to the surface, and the symbol $\otimes$ denotes the tensor product. The matrix

$$\mathbf{P} \equiv \mathbf{I} - \mathbf{n} \otimes \mathbf{n} \tag{4.9.3}$$

with components

$$P_{ij} = \delta_{ij} - n_i\, n_j \tag{4.9.4}$$

projects a vector onto the plane that is normal to $\mathbf{n}$, and therefore tangential to the surface by eliminating the normal component. Thus, if $\mathbf{v}$ is a three-dimensional vector defined over the surface, the vector

$$\mathbf{P} \cdot \mathbf{v} = (\mathbf{I} - \mathbf{n} \otimes \mathbf{n}) \cdot \mathbf{v} = \mathbf{v} - \mathbf{n}\,(\mathbf{v} \cdot \mathbf{v}) \tag{4.9.5}$$

is its tangential projection onto the surface. An identity allows us to write

$$\mathbf{P} \cdot \mathbf{v} = \mathbf{n} \times (\mathbf{v} \times \mathbf{n}) \tag{4.9.6}$$

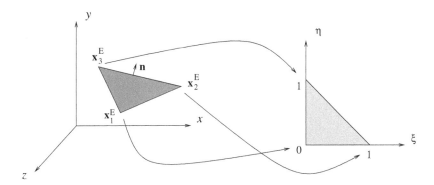

FIGURE 4.9.1 A three-node triangle in three-dimensional space is mapped to a right isosceles triangle in the $\xi\eta$ parametric plane.

(Problem 4.9.1). The right-hand side of (4.9.6) is the double rotation of the tangential component of **v**.

   The methodology described earlier in this chapter for planar three-node triangular elements can be extended in a straightforward fashion to surface elements in three-dimensional space, as illustrated in Figures 4.9.1. The mapping from the physical to the parameter space is described by (4.2.7) and the element interpolation functions are given in (4.2.2). The element diffusion, mass, and advection matrices are given by the right-hand sides of (4.2.21), (4.2.23), and (4.2.24).

   To compute the three-dimensional surface gradient of an element node interpolation functions, ψ_i, we complement the counterparts of the equations in (4.2.25),

$$\sum_{j=1}^{3} \frac{\partial \psi_j}{\partial \xi} \mathbf{x}_j^{\mathrm{E}} \cdot \nabla_s \psi_i = \frac{\partial \psi_i}{\partial \xi}, \qquad \sum_{j=1}^{3} \frac{\partial \psi_j}{\partial \eta} \mathbf{x}_j^{\mathrm{E}} \cdot \nabla_s \psi_i = \frac{\partial \psi_i}{\partial \eta}, \qquad (4.9.7)$$

with the condition

$$\mathbf{n} \cdot \nabla_s \psi_i = 0, \qquad (4.9.8)$$

where $\mathbf{n} = (n_x, n_y, n_z)$ is the unit vector normal to the element. The resulting 3×3 linear system takes the form

$$\boldsymbol{\Gamma} \cdot \nabla_s \psi_i = \begin{bmatrix} \partial \psi_i / \partial \xi \\ \partial \psi_i / \partial \eta \\ 0 \end{bmatrix}, \qquad (4.9.9)$$

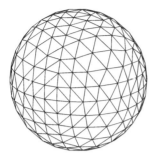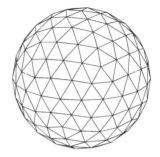

FIGURE 4.9.2   Discretization of the surface of a sphere into three-node triangles descending from the subdivision of an octahedron (left) or icosahedron (right).

involving the coefficient matrix

$$
\Gamma =
\begin{bmatrix}
\sum\limits_{j=1}^{3} \dfrac{\partial \psi_j}{\partial \xi}\, x_j^{\mathrm{E}} & \sum\limits_{j=1}^{3} \dfrac{\partial \psi_j}{\partial \xi}\, y_j^{\mathrm{E}} & \sum\limits_{j=1}^{3} \dfrac{\partial \psi_j}{\partial \xi}\, z_j^{\mathrm{E}} \\[3mm]
\sum\limits_{j=1}^{6} \dfrac{\partial \psi_j}{\partial \eta}\, x_j^{\mathrm{E}} & \sum\limits_{j=1}^{6} \dfrac{\partial \psi_j}{\partial \eta}\, y_j^{\mathrm{E}} & \sum\limits_{j=1}^{6} \dfrac{\partial \psi_j}{\partial \eta}\, z_j^{\mathrm{E}} \\[3mm]
n_x & n_y & n_z
\end{bmatrix}.
\tag{4.9.10}
$$

Making substitutions, we obtain

$$
\Gamma =
\begin{bmatrix}
x_2^{\mathrm{E}} - x_1^{\mathrm{E}} & y_2^{\mathrm{E}} - y_1^{\mathrm{E}} & z_2^{\mathrm{E}} - z_1^{\mathrm{E}} \\[1mm]
x_3^{\mathrm{E}} - x_1^{\mathrm{E}} & y_3^{\mathrm{E}} - y_1^{\mathrm{E}} & z_3^{\mathrm{E}} - z_1^{\mathrm{E}} \\[1mm]
n_x & n_y & n_z
\end{bmatrix}.
\tag{4.9.11}
$$

The solution of the linear system (4.9.9) can be found readily by elementary analytical or numerical methods.

Surface grids of three-node triangles can be generated by the successive subdivision of a hard-coded structure. FSELIB functions *trgl3_octa* and *trgl3_icos*, not listed in the text, discretize the surface of a sphere into three-node triangles descending from the subdivision of an octahedron or icosahedron (root sets), as shown in Figure 4.9.2.

PROBLEM

4.9.1 *Tangential projection*

To prove identity (4.9.6), we employ index notation and write

$$
(\delta_{ij} - n_i\, n_j)\, v_j = \epsilon_{ijk}\, n_j\, (\epsilon_{klm}\, v_l\, n_m),
\tag{4.9.12}
$$

where δ_{ij} is Kronecker's delta, ϵ_{ijk} is the alternating tensor, and summation is implied over the repeated indices j, k, l, and m. The alternating tensor is defined such that $\epsilon_{ijk} = 0$ if at least two of the indices take the same value, $\epsilon_{123} = \epsilon_{231} = \epsilon_{312} = 1$, and $\epsilon_{ijk} = -1$ otherwise, as discussed in Section A.3, Appendix A. The product of two alternating tensors with one shared index satisfies the identity

$$\epsilon_{kij} \, \epsilon_{klm} = \delta_{il} \, \delta_{jm} - \delta_{im} \, \delta_{jl}, \tag{4.9.13}$$

under the repeated index summation convention for the index k. Based on this identity, and using the fundamental property of Kronecker's delta, $\delta_{ij} v_j = v_i$, complete the proof of (4.9.6).

4.10 Bilinear quadrilateral elements

It is sometimes convenient to employ quadrilateral elements with four straight or curved sides. The finite element methodology for these elements is similar to that for the three-sided triangular element discussed previously in this chapter.

The simplest quadrilateral element has four straight edges defined by four vertices. To describe this element, we map it from the physical xy plane to a standard square in the parametric $\xi\eta$ plane confined between $-1 \le \xi \le 1$ and $-1 \le \eta \le 1$, as shown in Figure 4.10.1. The first node is mapped to the point $\xi = -1, \eta = -1$, the second to the point $\xi = 1, \eta = -1$, the third to the point $\xi = -1, \eta = 1$, and the fourth to the point $\xi = 1, \eta = 1$.

The mapping from the xy to the $\xi\eta$ plane is mediated by the function

$$\mathbf{x} = \sum_{i=1}^{4} \mathbf{x}_i^{\mathrm{E}} \, \psi_i(\xi, \eta), \tag{4.10.1}$$

where $\mathbf{x}_i^{\mathrm{E}}$ for $i = 1, 2, 3, 4$ are the element node positions in the xy plane.

The element-node cardinal interpolation functions, $\psi_i(\xi, \eta)$, are required to be bilinear. This means that, for a given ξ, $\psi_i(\xi, \eta)$ is linear in η; and for a given η, $\psi_i(\xi, \eta)$ is linear in ξ, so that

$$\psi_i(\xi, \eta) = (a\,\xi + b)\,(c\,\eta + d), \tag{4.10.2}$$

where a–d are appropriate expansion coefficients. Expanding the product on the right-hand side, we obtain an *incomplete* quadratic expansion displayed in the deliberate pictorial form

$$
\begin{aligned}
\psi_i(\xi, \eta) \quad = \quad & a_{00} \\
& + a_{10}\,\xi + a_{01}\,\eta \\
& + a_{11}\,\xi\eta,
\end{aligned}
\tag{4.10.3}
$$

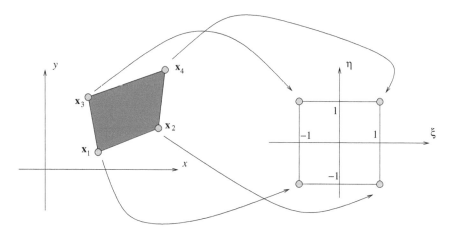

FIGURE 4.10.1   A four-node quadrilateral element in the xy plane is mapped to a square in the parametric $\xi\eta$ plane.

where $a_{00} = bd$, $a_{10} = ad$, $a_{01} = bc$, and $a_{11} = ac$. The interpolation functions are incomplete quadratics in (ξ, η), handicapped by the absence of the pure quadratic terms ξ^2 and η^2. Consequently, the element is spatially anisotropic.

To compute the four coefficients a–d, or the equivalent coefficients a_{00}, a_{10}, a_{01}, and a_{11}, we enforce the cardinal interpolation property demanding that $\psi_i(\xi, \eta) = 1$ at the ith node and $\psi_i(\xi, \eta) = 0$ at the other three nodes. The results are

$$\psi_1 = \frac{1}{4}(1 - \xi)(1 - \eta), \qquad \psi_2 = \frac{1}{4}(1 + \xi)(1 - \eta),$$

$$\psi_3 = \frac{1}{4}(1 - \xi)(1 + \eta), \qquad \psi_4 = \frac{1}{4}(1 + \xi)(1 + \eta). \tag{4.10.4}$$

A graph of the interpolation function ψ_1 corresponding to the southwestern vertex node, generated by the FSELIB function *psi_q4*, not listed in the text, is shown in Figure 4.10.2. The interpolation functions for the other three nodes have similar shapes.

In the isoparametric interpolation, the requisite solution over the element is expressed in a form that is analogous to that shown in (4.10.1),

$$f(x, y, t) = \sum_{i=1}^{4} f_i^{\mathrm{E}}(t)\,\psi_i(\xi, \eta), \tag{4.10.5}$$

where $f_i^{\mathrm{E}}(t)$ are the element nodal values. With this choice, the lth-element diffusion matrix is given by

$$A_{ij}^{(l)} = \iint_{E_l} \nabla\psi_i \cdot \nabla\psi_j \, \mathrm{d}x \, \mathrm{d}y = \int_{-1}^{1}\int_{-1}^{1} \nabla\psi_i \cdot \nabla\psi_j \, h_S \, \mathrm{d}\xi \, \mathrm{d}\eta \tag{4.10.6}$$

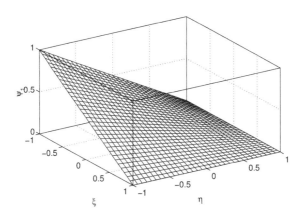

FIGURE 4.10.2   Graph of the bilinear interpolation function corresponding to the southwestern node of a quadrilateral element, labeled node 1.

for $i, j = 1$–4, where h_S is a *areal metric coefficient*, given by

$$h_S = \left| \frac{\partial \mathbf{x}}{\partial \xi} \times \frac{\partial \mathbf{x}}{\partial \eta} \right|. \tag{4.10.7}$$

Note that h_S is not constant, as in the case of the three-node triangle, but depends on the position inside the element. The element mass matrix is given by

$$B_{ij}^{(l)} = \iint_{E_l} \psi_i \psi_j \, dx \, dy = \int_{-1}^{1} \int_{-1}^{1} \psi_i \psi_j \, h_S \, d\xi \, d\eta \tag{4.10.8}$$

and the element advection matrix is given by

$$C_{ij}^{(l)} = \iint_{E_l} \psi_i \, \mathbf{u} \cdot \nabla \psi_j \, dx \, dy = \int_{-1}^{1} \int_{-1}^{1} \psi_i \, \mathbf{u} \cdot \nabla \psi_j \, h_S \, d\xi \, d\eta \tag{4.10.9}$$

for $i, j = 1$–4. To compute the interpolation function gradient, $\nabla \psi_i$, we work as in Section 4.2.1 for three-node triangles.

Since the integration limits in the $\xi \eta$ plane are fixed, the element diffusion, mass, and advection matrices can be computed by the dual application of a one-dimensional integration quadrature. The Lobatto quadrature shown in (3.4.14) provides us with the approximation

$$\int_{-1}^{1} \int_{-1}^{1} f(\xi, \eta) \, d\xi \, d\eta \simeq \sum_{p_1=1}^{k_1+1} \sum_{p_2=1}^{k_2+1} f(t_{p_1}, t_{p_2}) \, w_{p_1} \, w_{p_2} \tag{4.10.10}$$

for a regular function, $f(\xi, \eta)$, where k_1 and k_2 are specified quadrature orders.

PROBLEMS

4.10.1 *Surface metric coefficient*

Derive an expression for the surface metric coefficient of the four-node rectangular element, h_S, in terms of the node coordinates and derivatives of the element interpolation functions.

4.10.2 *Computation of an integral over a quadrilateral element*

Write a MATLAB function that evaluates the integral of a given function, $f(\xi, \eta)$, over a quadrilateral element,

$$\iint f(\xi, \eta) \, dx \, dy, \qquad (4.10.11)$$

using the double Lobatto quadrature. The input to this function should include the position of the four vertices, $\mathbf{x}_i$ for $i = 1, 2, 3, 4$. Run the function for $f(\xi, \eta) = 1$ and confirm that the answer is equal to the area of the quadrilateral element in the xy plane.

Quadratic and spectral elements in two dimensions

5

In Chapter 4, we discussed the implementation of the finite element method in two dimensions with linear triangular elements defined by three nodes. To improve the accuracy of the interpolation and also account for the boundary curvature, we may discretize the solution domain into triangular or quadrilateral elements with curved edges defined by a higher number of nodes. In addition, we may approximate the solution over the individual elements isoparametrically or superparametrically using quadratic or high-order polynomial expansions expressed in nodal or modal form. Spectral element methods arise by judiciously deploying the element interpolation nodes at positions corresponding to optimal interpolation sets that are specific to the selected element type.

In this chapter, we discuss the implementation of the finite element method with quadratic triangular elements, develop high-order and spectral element methods with nodal and modal basis functions, and review high-order and spectral methods for quadrilateral elements. The discussion of the spectral element method assumes some familiarity with the theory of orthogonal polynomials, summarized in Appendix B, and the theory of function interpolation, summarized in Appendix D.

Readers who are not presently interested in high-order and spectral element methods may skip the corresponding sections of this chapter and proceed without ramifications to Chapters 6 and 7 where applications of the finite element method in solid and fluid mechanics are discussed.

5.1 Six-node triangular elements

Planar six-node triangular elements with straight or curved edges are defined by three *vertex* nodes and three *edge* nodes, as illustrated in Figure 5.1.1. To describe an element in parametric form, we map it from the physical xy plane to the familiar right isosceles triangle in the $\xi\eta$ plane, as shown in Figure 5.1.1(a), so that:

- The first node is mapped to the origin of the $\xi\eta$ plane, $\xi = 0$, $\eta = 0$.
- The second node is mapped to the point $\xi = 1, \eta = 0$ on the ξ axis.
- The third node is mapped to the point $\xi = 0, \eta = 1$ on the η axis.
- The fourth node is mapped to the point $\xi - \alpha, \eta = 0$ on the ξ axis.

321

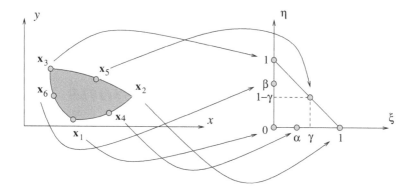

FIGURE 5.1.1 Mapping of a six-node triangle in the xy plane to a right isosceles triangle in the parametric $\xi\eta$ plane.

- The fifth node is mapped to the point $\xi = \gamma, \eta = 1 - \gamma$ on the hypotenuse.
- The sixth node is mapped to the point $\xi = 0, \eta = \beta$ on the η axis.

The dimensionless geometrical mapping coefficients, α, β, and γ are defined as

$$\alpha = \frac{1}{1 + \dfrac{|\mathbf{x}_4 - \mathbf{x}_2|}{|\mathbf{x}_4 - \mathbf{x}_1|}}, \qquad \beta = \frac{1}{1 + \dfrac{|\mathbf{x}_6 - \mathbf{x}_3|}{|\mathbf{x}_6 - \mathbf{x}_1|}}, \qquad \gamma = \frac{1}{1 + \dfrac{|\mathbf{x}_5 - \mathbf{x}_2|}{|\mathbf{x}_5 - \mathbf{x}_3|}}. \tag{5.1.1}$$

FSELIB function *elm6_abc*, listed in Table 5.1.1, evaluates these coefficients from the coordinates of the 6 vertices using the expressions in (5.1.1).

In finite element implementations, it is a common practice to set $\alpha = \beta = \gamma = 1/2$, which amounts to mapping the edge nodes numbered 4, 5, and 6, to the edge mid-points. However, we will see that the scaling embedded in (5.1.1) ensures the regularity of the mapping and prevents distortion that may occur when the edge nodes are located near the vertex nodes in the xy plane.

Element mapping

The mapping from the physical to the parametric space is mediated by the vector function

$$\mathbf{x} = \sum_{i=1}^{6} \mathbf{x}_i^{\mathrm{E}} \, \psi_i(\xi, \eta), \tag{5.1.2}$$

whose Cartesian components are

$$x = \sum_{i=1}^{6} x_i^{\mathrm{E}} \, \psi_i(\xi, \eta), \qquad y = \sum_{i=1}^{6} y_i^{\mathrm{E}} \, \psi_i(\xi, \eta). \tag{5.1.3}$$

```
function [al, be, ga] = elm6_abc ...
...
    (x1,y1, x2,y2, x3,y3, x4,y4, x5,y5, x6,y6)

%=====================================
% Computation of the (xi, eta) mapping
% coefficients alpha, beta, gamma
% for a six-node triangle
%=====================================

D42 = sqrt((x4-x2)^2 + (y4-y2)^2);
D41 = sqrt((x4-x1)^2 + (y4-y1)^2);
D63 = sqrt((x6-x3)^2 + (y6-y3)^2);
D61 = sqrt((x6-x1)^2 + (y6-y1)^2);
D52 = sqrt((x5-x2)^2 + (y5-y2)^2);
D53 = sqrt((x5-x3)^2 + (y5-y3)^2);

al = 1.0/(1.0+D42/D41);
be = 1.0/(1.0+D63/D61);
ga = 1.0/(1.0+D52/D53);

%-----
% done
%-----

return;
```

TABLE 5.1.1   FSELIB function *elm6_abc* evaluates the mapping coefficient α, β, and γ, for a six-node triangle.

The quadratic element interpolation functions, $\psi_i(\xi, \eta)$, are required to satisfy familiar cardinal interpolation properties requiring that $\psi_i = 1$ at the ith element node and $\psi_i = 0$ at the other five nodes. In terms of Kronecker's delta, $\delta_{i,j}$,

$$\psi_i(\xi_j, \eta_j) = \delta_{i,j} \tag{5.1.4}$$

for $i, j = 1, \ldots, 6$, where

$$(\xi_1, \eta_1) = (0, 0), \qquad (\xi_2, \eta_2) = (1, 0), \qquad (\xi_3, \eta_3) = (0, 1),$$
$$(\xi_4, \eta_4) = (\alpha, 0), \qquad (\xi_5, \eta_5) = (\gamma, 1 - \gamma), \qquad (\xi_6, \eta_6) = (0, \beta) \tag{5.1.5}$$

are the coordinates of the 6 nodes in the $\xi\eta$ plane.

Computation of the element interpolation functions

To derive the ith node interpolation function, we write

$$\psi_i(\xi, \eta) = a_i + b_i\xi + c_i\eta + d_i\xi^2 + e_i\xi\eta + f_i\eta^2, \tag{5.1.6}$$

and compute the 6 coefficients, a_i–f_i, to satisfy the aforementioned interpolation conditions. The results are shown in Table 5.1.2. As in the case of the three-node

$$\psi_2 = \frac{1}{1-\alpha}\, \xi \left(\xi - \alpha + \frac{\alpha - \gamma}{1 - \gamma}\, \eta \right)$$

$$\psi_3 = \frac{1}{1-\beta}\, \eta \left(\eta - \beta + \frac{\beta + \gamma - 1}{\gamma}\, \xi \right)$$

$$\psi_4 = \frac{1}{\alpha\,(1-\alpha)}\, \xi \zeta$$

$$\psi_5 = \frac{1}{\gamma\,(1-\gamma)}\, \xi \eta$$

$$\psi_6 = \frac{1}{\beta\,(1-\beta)}\, \eta \zeta$$

$$\psi_1 = 1 - \psi_2 - \psi_3 - \psi_4 - \psi_5 - \psi_6$$

TABLE 5.1.2   Element interpolation functions for a six-node triangle, where the variable $\zeta \equiv 1 - \xi - \eta$ is the third barycentric coordinate.

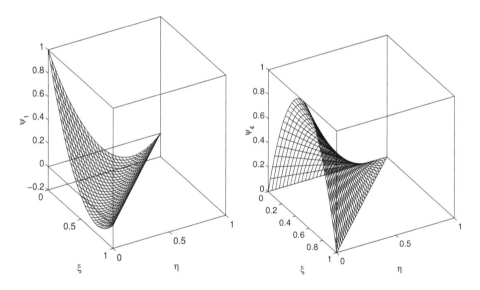

FIGURE 5.1.2   Graphs of the element node interpolation functions ψ_1 and ψ_4 associated with a vertex node (left) and an edge node (right).

triangles, $\zeta \equiv 1 - \xi - \eta$ is zero along the hypotenuse where $\eta = 1 - \xi$ and $\xi = 1 - \eta$. Graphs of the interpolation functions ψ_1 and ψ_4 associated with a vertex node and an edge node, generated by the FSELIB function *psi_6*, not listed in the text, are shown in Figure 5.1.2.

F<small>IGURE</small> 5.1.3   Two six-node adjacent elements join seamlessly to form a patch. The element edges are generated by the general mapping functions shown in Table 5.1.2 (solid lines) or by the simplified mapping functions (5.1.7) (dotted lines).

If the edge nodes are fortuitously or intentionally mapped to the edge midpoints in the $\xi\eta$ plane, $\alpha = 1/2$, $\beta = 1/2$, and $\gamma = 1/2$, then the node interpolation functions take the simpler forms

$$\psi_1 = \zeta\,(2\zeta - 1), \qquad \psi_2 = \xi\,(2\xi - 1), \qquad \psi_3 = \eta\,(2\eta - 1),$$
$$\psi_4 = 4\,\xi\zeta, \qquad\qquad \psi_5 = 4\,\xi\eta, \qquad\qquad \psi_6 = 4\,\eta\zeta. \tag{5.1.7}$$

In Section 5.8, these formulas will be recovered as special cases of more general expressions for high-order expansions.

Substituting the interpolation functions shown in Table 5.1.2 into the right-hand side of (5.1.2), we obtain a representation in terms of a *complete quadratic* function in ξ and η consisting of 6 terms, as shown in (5.1.6): a constant term, a term that is linear in ξ, a term that is linear in η, a term that is quadratic in ξ, a term that is quadratic in η, and a mixed quadratic term in $\xi\eta$, For a specified location in the triangle in the physical (x, y) plane, the corresponding location in the triangle in the parametric (ξ, η) plane can be found by solving the system of two quadratic equations (5.1.3), and retaining the physically meaningful solution.

Element edges

The interpolated geometry of the element edges in the physical xy plane is determined implicitly by the quadratic mapping function (5.1.2). An element has a straight interpolated edge only if the three nodes defining the edge are collinear in the physical xy plane. More generally, since each edge is a quadratic function of ξ or η with coefficients that depend on the location of the three corresponding nodes, shared edges of adjacent elements coincide. This means that two adjacent elements join seamlessly to form a two-element patch, as shown in Figure 5.1.3. The depiction in this figure was generated by the FSELIB script *edges6*, not listed in the text.

The solid lines in Figure 5.1.3 trace the element contours computed with the general mapping functions shown in Table 5.1.2. The broken lines trace the element contours computed with the simplified mapping functions (5.1.7). The results justify our earlier assertion that using the general mapping function helps ensure regularity by generating smooth element edges. Under extreme conditions, the simplified mapping functions may lead to edge crossing where a singular point is mapped to two pairs of barycentric coordinates, (ξ, η), resulting in loss of analyticity.

5.1.1 Integral over a triangle

The Jacobian matrix of the mapping from the physical xy plane to the parametric $\xi\eta$ plane was defined in (4.2.12) as

$$\mathbf{J} \equiv \begin{bmatrix} \dfrac{\partial x}{\partial \xi} & \dfrac{\partial x}{\partial \eta} \\[2mm] \dfrac{\partial y}{\partial \xi} & \dfrac{\partial y}{\partial \eta} \end{bmatrix}. \tag{5.1.8}$$

Substituting the expressions in (5.1.3), we find that the components of the Jacobian matrix are linear functions of ξ and η. Thus, unlike a three-node triangle discussed in Section 4.2, a six-node triangle has a position-dependent Jacobian. A singularity of the Jacobian matrix at a point signals the failure of the quadratic mapping function and the consequent loss of analyticity.

The determinant of the Jacobian matrix is the *surface metric coefficient* defined in (4.2.13) as

$$h_S \equiv \det[\mathbf{J}] = \left| \frac{\partial \mathbf{x}}{\partial \xi} \times \frac{\partial \mathbf{x}}{\partial \eta} \right|. \tag{5.1.9}$$

Substituting the expressions in (5.1.3), we find that h_S is a nonlinear function of ξ and η. Thus, unlike a three-node triangle, a six-node triangle has a position-dependent metric coefficient that cannot be extracted from an integral.

The integral of a function, $f(x, y)$, over the area of the triangle in the xy plane can be expressed as an integral over the area of the triangle in the parametric $\xi\eta$ plane as

$$\iint f(x, y) \, \mathrm{d}x \, \mathrm{d}y = \iint f(\xi, \eta) \, h_S \, \mathrm{d}\xi \, \mathrm{d}\eta. \tag{5.1.10}$$

In practice, the integral over the standard triangle on the right-hand side of (5.1.10) is computed by numerical methods.

5.1.2 Isoparametric interpolation and element matrices

In the isoparametric interpolation, a function of interest defined over the parametric triangle, $f(\xi, \eta)$, is expressed by the counterpart of the geometrical expansion (5.1.2) as

$$f(\xi, \eta) = \sum_{j=1}^{6} f^{E}(\xi_j, \eta_j)\, \psi_j(\xi, \eta), \tag{5.1.11}$$

where $f^{E}(\xi_j, \eta_j)$ are element nodal values, with the understanding that the field point $\mathbf{x} = (x, y)$ is mapped to the parametric point (ξ, η), and *vice versa*, through (5.1.2).

The lth-element diffusion matrix is given by

$$A_{ij}^{(l)} = \iint_{E_l} \nabla \psi_i \cdot \nabla \psi_j \, \mathrm{d}x \, \mathrm{d}y = \iint \nabla \psi_i \cdot \nabla \psi_j \, h_S \, \mathrm{d}\xi \, \mathrm{d}\eta, \tag{5.1.12}$$

the corresponding element mass matrix is given by

$$B_{ij}^{(l)} = \iint_{E_l} \psi_i \psi_j \, \mathrm{d}x \, \mathrm{d}y = \iint \psi_i \psi_j \, h_S \, \mathrm{d}\xi \, \mathrm{d}\eta, \tag{5.1.13}$$

and the corresponding element advection matrix is given by

$$C_{ij}^{(l)} = \iint_{E_l} \psi_i\, \mathbf{u} \cdot \nabla \psi_j \, \mathrm{d}x \, \mathrm{d}y = \iint \psi_i\, \mathbf{u} \cdot \nabla \psi_j \, h_S \, \mathrm{d}\xi \, \mathrm{d}\eta \tag{5.1.14}$$

for $i, j = 1, \ldots, 6$. The integrals on the right-hand sides are performed over the area of the right triangle in the parametric plane.

Gradient of the element interpolation functions

To compute the element diffusion and advection matrices, we require the gradients of the element interpolation functions, $\nabla \psi_i$. These can be found readily using the relations in (4.2.25), which amount to

$$\sum_{j=1}^{6} \frac{\partial \psi_j}{\partial \xi}\, \mathbf{x}_j^{E} \cdot \nabla \psi_i = \frac{\partial \psi_i}{\partial \xi}, \qquad \sum_{j=1}^{6} \frac{\partial \psi_j}{\partial \eta}\, \mathbf{x}_j^{E} \cdot \nabla \psi_i = \frac{\partial \psi_i}{\partial \eta} \tag{5.1.15}$$

for $i = 1, \ldots, 6$. The associated matrix form is

$$\mathbf{J}^T \cdot \nabla \psi_i = \begin{bmatrix} \dfrac{\partial \psi_i}{\partial \xi} \\[2mm] \dfrac{\partial \psi_i}{\partial \eta} \end{bmatrix}, \tag{5.1.16}$$

where $\mathbf{J}^T$ is the transpose of the Jacobian matrix,

$$\mathbf{J}^T = \begin{bmatrix} \displaystyle\sum_{j=1}^{6} \frac{\partial \psi_j}{\partial \xi}\, x_j^{E} & \displaystyle\sum_{j=1}^{6} \frac{\partial \psi_j}{\partial \xi}\, y_j^{E} \\[4mm] \displaystyle\sum_{j=1}^{6} \frac{\partial \psi_j}{\partial \eta}\, x_j^{E} & \displaystyle\sum_{j=1}^{6} \frac{\partial \psi_j}{\partial \eta}\, y_j^{E} \end{bmatrix}. \tag{5.1.17}$$

The determinant of $\mathbf{J}^T$ is equal to the determinant of $\mathbf{J}$, which is equal to the surface metric coefficient, h_S. Applying Cramer's rule, we find that the solution of (5.1.16) is given by

$$
\nabla \psi_i = \frac{1}{h_S}
\begin{bmatrix}
\sum\limits_{j=1}^{6} \dfrac{\partial \psi_j}{\partial \eta}\, y_j^{\mathrm{E}} & -\sum\limits_{j=1}^{6} \dfrac{\partial \psi_j}{\partial \xi}\, y_j^{\mathrm{E}} \\[2mm]
-\sum\limits_{j=1}^{6} \dfrac{\partial \psi_j}{\partial \eta}\, x_j^{\mathrm{E}} & \sum\limits_{j=1}^{6} \dfrac{\partial \psi_j}{\partial \xi}\, x_j^{\mathrm{E}}
\end{bmatrix}
\cdot
\begin{bmatrix}
\dfrac{\partial \psi_i}{\partial \xi} \\[2mm]
\dfrac{\partial \psi_i}{\partial \eta}
\end{bmatrix}.
\tag{5.1.18}
$$

FSELIB function *elm6_interp*, listed in Table 5.1.3, computes the node interpolation functions and their gradients and evaluates the surface metric coefficient, h_S, at a position corresponding to a specified pair of barycentric coordinates, (ξ, η). The input to *elm6_interp* includes the coordinates of the 6 element nodes and the mapping coefficients α, β, and γ. The latter are evaluated by the FSELIB function *elm6_abc*, listed in Table 5.1.1.

5.1.3 Element matrices and integration quadratures

Substituting the preceding expressions into (5.1.12), (5.1.13), and (5.1.14), we obtain lengthy expressions for the integrands of the element diffusion, mass, and advection matrices. In practice, the integrals over the parametric triangle in the $\xi\eta$ plane are computed most efficiently by numerical integration using a triangle integration quadrature (e.g., Pozrikidis [47], Chapter 7).

Triangle integration quadratures

Applying an integration quadrature amounts to approximating the integral of a function, $g(\xi, \eta)$, over the area of the triangle with a weighted sum,

$$
\iint g(\xi, \eta)\, \mathrm{d}\xi\, \mathrm{d}\eta \simeq \frac{1}{2} \sum_{i=1}^{N_Q} g(\xi_i, \eta_i)\, w_i,
\tag{5.1.19}
$$

where N_Q is a specified number of quadrature base points. The beginning and end of the FSELIB function *gauss_trgl* that assigns numerical values to the integration base points, (ξ_i, η_i), and corresponding weights, w_i, is listed in Table 5.1.4. The complete code is included in FSELIB.

Note that the sum of the weights is equal to unity for any number of integration base points, N_Q. Consequently, if $g(\xi, \eta) = 1$, the integral is equal to the area of the triangle in the $\xi\eta$ plane, which is equal to $1/2$.

FSELIB function edm6

FSELIB function *edm6*, listed in Table 5.1.5, evaluates the element diffusion matrix using the Gauss triangle quadrature. The code first calls the FSELIB function

```
function [psi, gpsi, hs] = elm6_interp ...
...
            (x1,y1, x2,y2, x3,y3, x4,y4, x5,y5, x6,y6 ...
            ,al,be,ga, xi,eta)
%============================================================
% Evaluation of the surface metric coefficient, hs, and
% computation of the interpolation functions and their
% gradients over a six-node triangle
%
% grad(psi_i) = [gpsi(i,1), gpsi(i,2)]
%============================================================

%--------
% prepare
%--------

alc = 1.0-al; bec = 1.0-be; gac = 1.0-ga;
alalc = al*alc; bebec = be*bec; gagac = ga*gac;

%----------------------------------
% compute the interpolation functions
%----------------------------------

psi(2) = xi*(xi-al+eta*(al-ga)/gac)/alc;
psi(3) = eta*(eta-be+xi*(be+ga-1.0)/ga)/bec;
psi(4) = xi*(1.0-xi-eta)/alalc;
psi(5) = xi*eta/gagac;
psi(6) = eta*(1.0-xi-eta)/bebec;
psi(1) = 1.0-psi(2)-psi(3)-psi(4)-psi(5)-psi(6);

%------------------------------------------------------------
% compute the xi derivatives of the interpolation functions
%------------------------------------------------------------

dps2 = (2.0*xi-al+eta*(al-ga)/gac)/alc;
dps3 = eta*(be+ga-1.0)/(ga*bec);
dps4 = (1.0-2.0*xi-eta)/alalc;
dps5 = eta/gagac; dps6 = -eta/bebec;
dps1 =-dps2-dps3-dps4-dps5-dps6;

%------------------------------------------------------------
% compute the eta derivatives of the interpolation functions
%------------------------------------------------------------

pps2 = xi*(al-ga)/(alc*gac);
pps3 = (2.0*eta-be+xi*(be+ga-1.0)/ga)/bec;
pps4 =-xi/alalc;
pps5 = xi/gagac;
pps6 = (1.0-xi-2.0*eta)/bebec;
pps1 =-pps2-pps3-pps4-pps5-pps6;
```

TABLE 5.1.3   Function *elm6_interp* (Continuing →)

```
%---------------------------------------
% compute the xi and eta derivatives of x
%---------------------------------------

DxDxi = x1*dps1 + x2*dps2 + x3*dps3 ...
        +x4*dps4 + x5*dps5 + x6*dps6;
DyDxi = y1*dps1 + y2*dps2 + y3*dps3 ...
        +y4*dps4 + y5*dps5 + y6*dps6;
DxDet = x1*pps1 + x2*pps2 + x3*pps3 ...
        +x4*pps4 + x5*pps5 + x6*pps6;
DyDet = y1*pps1 + y2*pps2 + y3*pps3 ...
        +y4*pps4 + y5*pps5 + y6*pps6;

%----------------------------
% compute the surface metric hs
%----------------------------

vnz = DxDxi * DyDet - DxDet * DyDxi;

hs = sqrt(vnz^2);

%-----------------------------------------------
% compute the gradient of the six interpolation
% functions by solving two linear equations:
%
% dx/dxi . grad = d psi/dxi
% dx/det . grad = d psi /det
%
% The solution is found by Cramer's rule
%-----------------------------------------------

A11 = DxDxi; A12 = DyDxi;
A21 = DxDet; A22 = DyDet;
Det = A11*A22-A21*A12;

%--- first

B1 = dps1; B2 = pps1;
Det1 =   B1*A22 - B2*A12;
Det2 = - B1*A21 + B2*A11;
gpsi(1,1) = Det1/Det;
gpsi(1,2) = Det2/Det;

%--- second

B1 = dps2; B2 = pps2;
Det1 =   B1*A22 - B2*A12;
Det2 = - B1*A21 + B2*A11;
gpsi(2,1) = Det1/Det;
gpsi(2,2) = Det2/Det;
```

TABLE 5.1.3   Function *elm6_interp* ($\rightarrow$ Continuing $\rightarrow$)

```
%--- third

B1 = dps3; B2 = pps3;
Det1 =   B1*A22 - B2*A12;
Det2 = - B1*A21 + B2*A11;
gpsi(3,1) = Det1/Det;
gpsi(3,2) = Det2/Det;

%--- fourth

B1 = dps4; B2 = pps4;
Det1 =   B1*A22 - B2*A12;
Det2 = - B1*A21 + B2*A11;
gpsi(4,1) = Det1/Det;
gpsi(4,2) = Det2/Det;

%--- fifth

B1 = dps5; B2 = pps5;
Det1 =   B1*A22 - B2*A12;
Det2 = - B1*A21 + B2*A11;
gpsi(5,1) = Det1/Det;
gpsi(5,2) = Det2/Det;

%--- sixth

B1 = dps6; B2 = pps6;
Det1 =   B1*A22 - B2*A12;
Det2 = - B1*A21 + B2*A11;
gpsi(6,1) = Det1/Det;
gpsi(6,2) = Det2/Det;

%-----
% done
%-----

return;
```

TABLE 5.1.3   ($\rightarrow$ Continued) Function *elm6_interp* evaluates the surface metric coefficient, h_S, and computes the node interpolation functions and their gradients over a six-node triangle. The input includes the coordinates of the 6 element nodes and the mapping coefficients α, β, and γ.

edm6_abc, listed in Table 5.1.1, to evaluate the mapping coefficients, and then calls the function *edm6_interp*, listed in Table 5.1.3, to evaluate the gradients of the interpolation functions at the integration base points. The quadrature is implemented in a single loop. The area of the triangle in physical space is also returned in the output.

```
function [xi, eta, w] = gauss_trgl(m)

%========================================================
% Abscissas (xi, eta) and weights (w) for Gaussian
% integration over a flat triangle in the xi-eta plane
%
%
% The integration is performed with respect
% to the triangle barycentric coordinates
%
% m: number of base points (NQ)
%    choose from 1,3,4,6,7,9,12,13; default is 7
%========================================================

%-----
% trap
%-----

if((m ~= 1) & (m ~= 3) & (m ~= 4) & (m ~= 6) & (m ~= 7) ...
 & (m ~= 9) & (m ~= 12) & (m ~= 13))
  disp('');disp(' Gauss_trgl: Chosen number of points');
           disp('  is not available; Will take m=7');
  m=7;
end

%-------
if(m==1)
%-------

xi(1) = 1.0/3.0; eta(1) = 1.0/3.0; w(1) = 1.0;

%-----------
elseif(m==3)
%-----------

xi(1) = 1.0/6.0; eta(1) = 1.0/6.0; w(1) = 1.0/3.0;
xi(2) = 2.0/3.0; eta(2) = 1.0/6.0; w(2) = w(1);
xi(3) = 1.0/6.0; eta(3) = 2.0/3.0; w(3) = w(1);

%-----------
elseif(m==4)
%-----------

xi(1) = 1.0/3.0; eta(1) = 1.0/3.0; w(1) = -27.0/48.0;
xi(2) = 1.0/5.0; eta(2) = 1.0/5.0; w(2) =  25.0/48.0;
xi(3) = 3.0/5.0; eta(3) = 1.0/5.0; w(3) =  25.0/48.0;
xi(4) = 1.0/5.0; eta(4) = 3.0/5.0; w(4) =  25.0/48.0;

%-----------
elseif(m==6)
%-----------

al = 0.816847572980459; be = 0.445948490915965;
ga = 0.108103018168070; de = 0.091576213509771;
o1 = 0.109951743655322; o2 = 0.223381589678011;
```

TABLE 5.1.4   Function *gauss_trgl* (Continuing →)

```
xi(1) = de; xi(2) = al; xi(3) = de; xi(4) = be;
xi(5) = ga; xi(6) = be;

eta(1) = de; eta(2) = de;eta(3) = al; eta(4) = be;
eta(5) = be; eta(6) = ga;

w(1) = o1; w(2) = o1; w(3) = o1; w(4) = o2;
w(5) = o2; w(6) = o2;

......

%--
end
%--

return;
```

TABLE 5.1.4   ($\rightarrow$ Continued) Function *gauss_trgl* defines base points and weights for integrating a function over a triangle in the $\xi\eta$ plane. The 6 dots at the end of the listing indicate additional code.

```
function [edm, arel] = edm6 ...
   ...
       (x1,y1, x2,y2, x3,y3, x4,y4, x5,y5, x6,y6, NQ)

%=======================================================
% Evaluation of the element diffusion matrix for a
% six-node triangle, using an integration quadrature
%=======================================================

%------------------------------------
% compute the mapping coefficients
%------------------------------------

[al, be, ga] = elm6_abc...
   ...
       (x1,y1, x2,y2, x3,y3, x4,y4, x5,y5, x6,y6)

%----------------------------
% read the triangle quadrature
%----------------------------

[xi, eta, w] = gauss_trgl(NQ);
```

TABLE 5.1.5   Function *edm6* (Continuing $\rightarrow$)

```
%---------------------------------------
% initialize the element diffusion matrix
%---------------------------------------

for k=1:6
 for l=1:6
  edm(k,l) = 0.0
 end
end

%----------------------
% perform the quadrature
%----------------------

arel = 0.0;  % element area

for i=1:NQ

 [psi, gpsi, hs] = elm6_interp ...
   ...
    (x1,y1, x2,y2, x3,y3, x4,y4, x5,y5, x6,y6 ...
    ,al,be,ga, xi(i),eta(i));

 cf = 0.5*hs*w(i);

 for k=1:6
  for l=1:6
   edm(k,l) = edm(k,l) + (gpsi(k,1)*gpsi(l,1) ...
                       +  gpsi(k,2)*gpsi(l,2))*cf;
  end
 end

 arel = arel + cf;

end

%-----
% done
%-----

return;
```

TABLE 5.1.5   ($\longrightarrow$ Continued) Function *edm6* evaluates the element diffusion matrix for a six-node triangle using a Gaussian integration quadrature.

FSELIB function edmm6

FSELIB function *edmm6*, listed in Table 5.1.6, evaluates the element diffusion and mass matrices using a similar method.

FSELIB function edam6

FSELIB function *edam6*, listed in Table 5.1.7, evaluates the diffusion and advection matrices by a similar method. Note that the velocity at the quadrature base points is computed by interpolation in terms of the element-node interpolation functions and the values of the velocity at the 6 element nodes, which are passed in the function call.

5.1.4 Elements with straight edges

When the element edges are straight and the element edge nodes are located at the mid-points, the element interpolation functions are given by the simplified expressions given in (5.1.7) and the transpose of the Jacobian matrix is given by

$$
\mathbf{J}^T = \begin{bmatrix} x_2^E - x_1^E & y_2^E - y_1^E \\ x_3^E - x_1^E & y_3^E - y_1^E \end{bmatrix}.
\tag{5.1.20}
$$

Accordingly, $h_S = 2A$ is a constant, where A is the area of the triangle in the physical xy plane. The gradients of the element interpolation functions are given by

$$
\nabla \psi_i = \frac{1}{2A} \, \mathbf{\Phi} \cdot \begin{bmatrix} \dfrac{\partial \psi_i}{\partial \xi} \\ \dfrac{\partial \psi_i}{\partial \eta} \end{bmatrix},
\tag{5.1.21}
$$

where

$$
\mathbf{\Phi} = \begin{bmatrix} y_3^E - y_1^E & -y_2^E + y_1^E \\ -x_3^E + x_1^E & x_2^E - x_1^E \end{bmatrix}.
\tag{5.1.22}
$$

Straightforward differentiation yields the expressions

$$
\begin{aligned}
&\frac{\partial \psi_1}{\partial \xi} = -4\,\zeta + 1, \quad &\frac{\partial \psi_1}{\partial \eta} = -4\,\zeta + 1, \quad &\frac{\partial \psi_2}{\partial \xi} = 4\,\xi - 1, \quad &\frac{\partial \psi_2}{\partial \eta} = 0, \\
&\frac{\partial \psi_3}{\partial \xi} = 0, \quad &\frac{\partial \psi_3}{\partial \eta} = 4\,\eta - 1, \quad &\frac{\partial \psi_4}{\partial \xi} = 4\,(\zeta - \xi), \quad &\frac{\partial \psi_4}{\partial \eta} = -4\,\xi, \\
&\frac{\partial \psi_5}{\partial \xi} = 4\,\eta, \quad &\frac{\partial \psi_5}{\partial \eta} = 4\,\xi, \quad &\frac{\partial \psi_6}{\partial \xi} = -4\,\eta, \quad &\frac{\partial \psi_6}{\partial \eta} = 4\,(\zeta - \eta).
\end{aligned}
\tag{5.1.23}
$$

The element diffusion, mass, and advection matrices can be computed using the integration formula (4.2.36), repeated below for convenience,

$$
\iint \zeta^p \, \xi^q \, \eta^r \, \mathrm{d}\xi \, \mathrm{d}\eta = \frac{p!\, q!\, r!}{(p + q + r + 2)!},
\tag{5.1.24}
$$

```
function [edm, emm, arel] = edmm6 ...
...
   (x1,y1, x2,y2, x3,y3, x4,y4, x5,y5, x6,y6, NQ)

%=================================================
% Evaluation of the element diffusion
% and mass matrix
% for a six-node triangle,
% using a Gauss-triangle integration quadrature
%=================================================

%---------------------------------
% compute the mapping coefficients
%---------------------------------

[al, be, ga] = elm6_abc ...
...
   (x1,y1, x2,y2, x3,y3, x4,y4, x5,y5, x6,y6);

%---------------------------
% read the triangle quadrature
%---------------------------

[xi, eta, w] = gauss_trgl(NQ);

%---------------------------------
% initialize the element diffusion
% and mass matrix
%---------------------------------

 for k=1:6
  for l=1:6
   edm(k,l) = 0.0;
   emm(k,l) = 0.0;
  end
 end

%----------------------
% perform the quadrature
%----------------------

arel = 0.0;  % element area (optional)

for i=1:NQ

 [psi, gpsi, hs] = elm6_interp ...
  ...
    (x1,y1, x2,y2, x3,y3, x4,y4, x5,y5, x6,y6 ...
    ,al,be,ga, xi(i),eta(i));

 cf = 0.5*hs*w(i);
```

TABLE 5.1.6   Function *edmm6* (Continuing →)

```
for k=1:6
  for l=1:6
    edm(k,l)  =  edm(k,l)  +  (gpsi(k,1)*gpsi(l,1)    ...
                           +   gpsi(k,2)*gpsi(l,2)  )*cf;
    emm(k,l)  =  emm(k,l)  +  psi(k)*psi(l)*cf;
  end
end

  arel = arel + cf;

end

%-----
% done
%-----

return;
```

TABLE 5.1.6   ($\longrightarrow$ Continued) Function *edmm6* evaluates the element diffusion and mass matrices for a six-node triangle using a Gaussian integration quadrature.

where p, q, and r are non-negative integers and an exclamation mark denotes the factorial, $m! = 1 \cdot 2 \cdots m$ for any integer, m.

Element diffusion matrix

The integrand of the element diffusion matrix can be expressed in the form

$$\nabla \psi_i \cdot \nabla \psi_j = \frac{1}{4A^2} \mathbf{O} : \mathbf{T}^{(i,j)}, \tag{5.1.25}$$

where the colon denotes the double matrix product product,[1] the matrix $\mathbf{O}$ is given by

$$\mathbf{O} \equiv \mathbf{\Phi}^I \cdot \mathbf{\Phi}, \tag{5.1.26}$$

the matrix $\mathbf{\Phi}$ is given in (5.1.22), and

$$\mathbf{T}^{(i,j)} = \begin{bmatrix} \dfrac{\partial \psi_i}{\partial \xi} \\[2mm] \dfrac{\partial \psi_i}{\partial \eta} \end{bmatrix} \otimes \begin{bmatrix} \dfrac{\partial \psi_j}{\partial \xi} & \dfrac{\partial \psi_j}{\partial \eta} \end{bmatrix} \tag{5.1.27}$$

is an outer tensor product.

---

[1]The double matrix product is a scalar defined as the sum of the products of the corresponding elements of two matrices.

```
function [edm, eam, arel] = edam6 ...
...
    (x1,y1, x2,y2, x3,y3, x4,y4, x5,y5, x6,y6  ...
    ,u1,v1, u2,v2, u3,v3, u4,v4, u5,v5, u6,v6, ...
    ,NQ)

%=====================================================
% Evaluation of the element diffusion and advection
% matrices and element area for a six-node triangle
%
% The element integrals are computed using
% the NQ-point Gauss-triangle quadrature
%
% u and v are the x and y velocity components
%
% The quadrature base points and weights are
% read from the function gauss_trgl
%=====================================================

%-----------------------------------
% compute the mapping coefficients
%-----------------------------------

[al, be, ga] = elm6_abc ...
      ...
        (x1,y1, x2,y2, x3,y3, x4,y4, x5,y5, x6,y6);

%----------------------------
% read the triangle quadrature
%----------------------------

[xi, eta, w] = gauss_trgl(NQ);

%-----------------------------------
% initialize the element diffusion
% and advection matrices
%-----------------------------------

for k=1:6
 for l=1:6
  edm(k,l) = 0.0;
  eam(k,l) = 0.0;
 end
end

%-----------------------
% perform the quadrature
%-----------------------

arel = 0.0; % element area  (optional)

for i=1:NQ

[psi, gpsi, hs] = elm6_interp ...
...
      (x1,y1, x2,y2, x3,y3, x4,y4, x5,y5, x6,y6 ...
      ,al,be,ga, xi(i),eta(i));
```

TABLE 5.1.7   Function *edam6* (Continuing →)

```
% interpolate the velocity at the base points:

u = u1*psi(1) + u2*psi(2) + u3*psi(3) ...
   + u4*psi(4) + u5*psi(5) + u6*psi(6)

v = v1*psi(1) + v2*psi(2) + v3*psi(3) ...
   + v4*psi(4) + v5*psi(5) + v6*psi(6)

cf = 0.5*hs*w(i);

for k=1:6
  for l=1:6
    edm(k,l) = edm(k,l) + (gpsi(k,1)*gpsi(l,1) ...
             +  gpsi(k,2)*gpsi(l,2))*hs*w(i);

    prj = u*gpsi(l,1)+v*gpsi(l,1);
    eam(k,l) = eam(k,l) + psi(k)*prj*cf;
  end
end

 arel = arel + cf;

end      % end of the quadrature loop

%-----
% done
%-----

return;
```

TABLE 5.1.7 ($\rightarrow$ Continued) Function *edam6* evaluates the element diffusion and advection matrices over a six-node triangle.

Substituting (5.1.21) into (5.1.26), we obtain

$$\mathbf{O} = \begin{bmatrix} y_3^E - y_1^E & -x_3^E + x_1^E \\ -y_2^F + y_1^F & m_2^E & x_1^E \end{bmatrix} \cdot \begin{bmatrix} y_3^E - y_1^F & -y_2^E + y_1^E \\ -x_0^E + x_1^E & x_2^E - x_1^E \end{bmatrix}, \tag{5.1.28}$$

yielding

$$\mathbf{O} = \begin{bmatrix} |\mathbf{d}_{31}|^2 & -\mathbf{d}_{31} \cdot \mathbf{d}_{21} \\ -\mathbf{d}_{31} \cdot \mathbf{d}_{21} & |\mathbf{d}_{21}|^2 \end{bmatrix}, \tag{5.1.29}$$

where

$$\mathbf{d}_{31} = \mathbf{x}_3^E - \mathbf{x}_1^E, \qquad \mathbf{d}_{21} = \mathbf{x}_2^E - \mathbf{x}_1^E. \tag{5.1.30}$$

The lth-element diffusion matrix is given by

$$A_{ij}^{(l)} = \iint_{E_l} \nabla \psi_i \cdot \nabla \psi_j \, \mathrm{d}x \, \mathrm{d}y = \frac{1}{2\Lambda} \mathbf{O} : \mathbf{\Psi}^{(i,j)}, \tag{5.1.31}$$

where

$$\mathbf{\Psi}^{(i,j)} \equiv \iint \mathbf{T}^{(i,j)} \, \mathrm{d}\xi \, \mathrm{d}\eta \qquad (5.1.32)$$

are ancillary matrices. Performing the integrations using the integration formula (5.1.24), we find the matrices shown in Table 5.1.8. These analytical results can be used to confirm the accuracy of numerical results obtained by an integration quadrature.

PROBLEM

5.1.1 *Numerical integration over a triangle*

Write a code that integrates a specified function, $g(\xi, \eta)$, over the standard triangle, using the integration quadrature discussed in the text. Run the code for $N_Q = 6$ and assess whether the quadrature is able to generate the exact answer when the integrand is (a) a linear function, $g(\xi) = \xi$ or $g(\eta) = \eta$, (b) a quadratic monomial product, $g(\xi, \eta) = \xi^p \eta^q$, where $p + q = 2$, (c) a cubic monomial product, $g(\xi, \eta) = \xi^p \eta^q$, where $p + q = 3$, and (d) a quartic monomial product, $g(\xi, \eta) = \xi^p \eta^q$, where $p + q = 4$.

5.2   Grid generation

To prepare the ground for the finite element expansion, we develop a computation module for domain discretization into the six-node triangles introduced in Section 5.1. Available discretization schemes are similar to those discussed in Section 4.3 for three-node triangles. Grid generation discussed in this section is based on the successive subdivision of a parental structure where a group of ancestral elements (root set) are hard-coded and smaller descendant elements arise by subdivision.

5.2.1   Circular disk

FSELIB function *trg6_disk*, listed in Table 5.2.1, discretizes the unit disk into six-node triangles, by successively subdividing an ancestral set of four triangles located in the four quadrants into four descendant elements. In the algorithm, the vertices of the emerging triangles are placed at the mid-points of the parental triangles, which are computed by linear interpolation. In the process of triangulation, new boundary nodes are projected in the radial direction onto the unit circle.

The refinement level is determined by an input flag *ndiv*, which is defined so that *ndiv* = 0 generates the 4 root elements, *ndiv* = 1 generates 16 first-generation elements, and *ndiv* = 2 generates 64 descendant elements. Each time a subdivision is carried out, the number of elements increases by a factor of 4. The methodology is similar to that discussed in Section 4.3.1 for three-node triangles.

$$\mathbf{\Psi}^{(1,1)} = \frac{1}{2} \begin{bmatrix} 1 & 1 \\ 1 & 1 \end{bmatrix}$$

$$\mathbf{\Psi}^{(2,2)} = \frac{1}{2} \begin{bmatrix} 1 & 0 \\ 0 & 0 \end{bmatrix}$$

$$\mathbf{\Psi}^{(3,3)} = \frac{1}{2} \begin{bmatrix} 0 & 0 \\ 0 & 1 \end{bmatrix}$$

$$\mathbf{\Psi}^{(4,4)} = \mathbf{\Psi}^{(5,5)} = \mathbf{\Psi}^{(6,6)} = \frac{2}{3} \begin{bmatrix} 2 & 1 \\ 1 & 2 \end{bmatrix}$$

$$\mathbf{\Psi}^{(1,2)} = \frac{1}{6} \begin{bmatrix} 1 & 1 \\ 0 & 0 \end{bmatrix}$$

$$\mathbf{\Psi}^{(1,3)} = \frac{1}{6} \begin{bmatrix} 0 & 0 \\ 1 & 1 \end{bmatrix}$$

$$\mathbf{\Psi}^{(1,4)} = \mathbf{\Psi}^{(2,4)} = -\frac{2}{3} \begin{bmatrix} 1 & 0 \\ 1 & 0 \end{bmatrix}$$

$$\mathbf{\Psi}^{(1,5)} = \mathbf{\Psi}^{(2,6)} = \mathbf{\Psi}^{(3,4)} = \begin{bmatrix} 0 & 0 \\ 0 & 0 \end{bmatrix}$$

$$\mathbf{\Psi}^{(1,6)} = \mathbf{\Psi}^{(3,6)} = -\frac{2}{3} \begin{bmatrix} 0 & 1 \\ 0 & 1 \end{bmatrix}$$

$$\mathbf{\Psi}^{(2,3)} = -\frac{1}{6} \begin{bmatrix} 0 & 1 \\ 0 & 0 \end{bmatrix}$$

$$\mathbf{\Psi}^{(2,5)} = \mathbf{\Psi}^{(3,5)} = \frac{2}{3} \begin{bmatrix} 0 & 1 \\ 0 & 0 \end{bmatrix}$$

$$\mathbf{\Psi}^{(4,5)} = -\frac{2}{3} \begin{bmatrix} 0 & 1 \\ 1 & 2 \end{bmatrix}$$

$$\mathbf{\Psi}^{(4,6)} = \frac{2}{3} \begin{bmatrix} 0 & 1 \\ 1 & 0 \end{bmatrix}$$

$$\mathbf{\Psi}^{(5,6)} = -\frac{2}{3} \begin{bmatrix} 2 & 1 \\ 1 & 0 \end{bmatrix}$$

TABLE 5.1.8   Ancillary matrices used for the calculation of the diffusion matrix for six-node elements with straight edges and mid-point edge nodes.

```
function [ne,ng,p,c,efl,gfl] = trgl6_disk(ndiv)

%=============================================================
% Triangulation of the unit disk into six-node elements by
% the successive subdivision of four ancestral elements
%=============================================================

act = 1.0;  % disk radius

%----------------------------------------
% ancestral structure with four elements
%----------------------------------------

ne = 4;

x(1,1) = 0.0; y(1,1) = 0.0; efl(1,1) = 0; % first element
x(1,2) = act; y(1,2) = 0.0; efl(1,2) = 1;
x(1,3) = 0.0; y(1,3) = act; efl(1,3) = 1;

x(2,1) = 0.0; y(2,1) = 0.0; efl(2,1) = 0; % second element
x(2,2) = 0.0; y(2,2) = act; efl(2,2) = 1;
x(2,3) =-act; y(2,3) = 0.0; efl(2,3) = 1;

x(3,1) = 0.0; y(3,1) = 0.0; efl(3,1) = 0; % third element
x(3,2) =-act; y(3,2) = 0.0; efl(3,2) = 1;
x(3,3) = 0.0; y(3,3) =-act; efl(3,3) = 1;

x(4,1) = 0.0; y(4,1) = 0.0; efl(4,1) = 0; % fourth element
x(4,2) = 0.0; y(4,2) =-act; efl(4,2) = 1;
x(4,3) = act; y(4,3) = 0.0; efl(4,3) = 1;

for i=1:ne
 efl(i,4) = 0;
 efl(i,5) = 1;
 efl(i,6) = 0;
end

%-------------------
% element mid-points
%-------------------

for i=1:ne
 x(i,4) = 0.5*(x(i,1)+x(i,2)); y(i,4) = 0.5*(y(i,1)+y(i,2));
 x(i,5) = 0.5*(x(i,2)+x(i,3)); y(i,5) = 0.5*(y(i,2)+y(i,3));
 x(i,6) = 0.5*(x(i,3)+x(i,1)); y(i,6) = 0.5*(y(i,3)+y(i,1));
end

% project boundary nodes onto a circle

for k=1:ne
    rad = sqrt(x(k,5)^2+y(k,5)^2);
    fc = act/rad;
    x(k,5) = fc*x(k,5);
    y(k,5) = fc*y(k,5);
end
```

TABLE 5.2.1 Function *trgl6_disk* (Continuing →)

```
if(ndiv > 0)

for i=1:ndiv

 nm = 0; % count the new elements arising by each refinement loop
         % four elements will be generated in each pass

 for j=1:ne    % loop over current elements

  % assign vertex nodes to sub-elements;
  % these will become the "new" elements

   nm = nm+1;                    %  first sub-element

   xn(nm,1)=x(j,1); yn(nm,1)=y(j,1); efln(nm,1)=efl(j,1);
   xn(nm,2)=x(j,4); yn(nm,2)=y(j,4); efln(nm,2)=efl(j,4);
   xn(nm,3)=x(j,6); yn(nm,3)=y(j,6); efln(nm,3)=efl(j,6);

   xn(nm,4) = 0.5*(xn(nm,1)+xn(nm,2));
   yn(nm,4) = 0.5*(yn(nm,1)+yn(nm,2));
   xn(nm,5) = 0.5*(xn(nm,2)+xn(nm,3));
   yn(nm,5) = 0.5*(yn(nm,2)+yn(nm,3));
   xn(nm,6) = 0.5*(xn(nm,3)+xn(nm,1));
   yn(nm,6) = 0.5*(yn(nm,3)+yn(nm,1));

   efln(nm,4) = 0; efln(nm,5) = 0; efln(nm,6) = 0;

   if(efln(nm,1)==1 & efln(nm,2)==1) efln(nm,4) = 1; end
   if(efln(nm,2)==1 & efln(nm,3)==1) efln(nm,5) = 1; end
   if(efln(nm,3)==1 & efln(nm,1)==1) efln(nm,6) = 1; end

   nm = nm+1;                    %  second sub-element

   xn(nm,1)=x(j,4); yn(nm,1)=y(j,4); efln(nm,1)=efl(j,4);
   xn(nm,2)=x(j,2); yn(nm,2)=y(j,2); efln(nm,2)=efl(j,2);
   xn(nm,3)=x(j,5); yn(nm,3)=y(j,5); efln(nm,3)=efl(j,5);

   xn(nm,4) = 0.5*(xn(nm,1)+xn(nm,2));
   yn(nm,4) = 0.5*(yn(nm,1)+yn(nm,2));
   xn(nm,5) = 0.5*(xn(nm,2)+xn(nm,3));
   yn(nm,5) = 0.5*(yn(nm,2)+yn(nm,3));
   xn(nm,6) = 0.5*(xn(nm,3)+xn(nm,1));
   yn(nm,6) = 0.5*(yn(nm,3)+yn(nm,1));

   efln(nm,4) = 0; efln(nm,5) = 0; efln(nm,6) = 0;

   if(efln(nm,1)==1 & efln(nm,2)==1) efln(nm,4) = 1; end
   if(efln(nm,2)==1 & efln(nm,3)==1) efln(nm,5) = 1; end
   if(efln(nm,3)==1 & efln(nm,1)==1) efln(nm,6) = 1; end
```

TABLE 5.2.1 Function *trgl6_disk* ($\rightarrow$ Continuing $\rightarrow$)

```
nm = nm+1; %  third sub-element

xn(nm,1)=x(j,6); yn(nm,1)=y(j,6); efln(nm,1)=efl(j,6);
xn(nm,2)=x(j,5); yn(nm,2)=y(j,5); efln(nm,2)=efl(j,5);
xn(nm,3)=x(j,3); yn(nm,3)=y(j,3); efln(nm,3)=efl(j,3);

xn(nm,4) = 0.5*(xn(nm,1)+xn(nm,2));
yn(nm,4) = 0.5*(yn(nm,1)+yn(nm,2));
xn(nm,5) = 0.5*(xn(nm,2)+xn(nm,3));
yn(nm,5) = 0.5*(yn(nm,2)+yn(nm,3));
xn(nm,6) = 0.5*(xn(nm,3)+xn(nm,1));
yn(nm,6) = 0.5*(yn(nm,3)+yn(nm,1));

efln(nm,4) = 0; efln(nm,5) = 0; efln(nm,6) = 0;
if(efln(nm,1)==1 & efln(nm,2)==1) efln(nm,4) = 1; end
if(efln(nm,2)==1 & efln(nm,3)==1) efln(nm,5) = 1; end
if(efln(nm,3)==1 & efln(nm,1)==1) efln(nm,6) = 1; end

nm = nm+1; %  fourth sub-element

xn(nm,1)=x(j,4); yn(nm,1)=y(j,4); efln(nm,1)=efl(j,4);
xn(nm,2)=x(j,5); yn(nm,2)=y(j,5); efln(nm,2)=efl(j,5);
xn(nm,3)=x(j,6); yn(nm,3)=y(j,6); efln(nm,3)=efl(j,6);

xn(nm,4) = 0.5*(xn(nm,1)+xn(nm,2));
yn(nm,4) = 0.5*(yn(nm,1)+yn(nm,2));
xn(nm,5) = 0.5*(xn(nm,2)+xn(nm,3));
yn(nm,5) = 0.5*(yn(nm,2)+yn(nm,3));
xn(nm,6) = 0.5*(xn(nm,3)+xn(nm,1));
yn(nm,6) = 0.5*(yn(nm,3)+yn(nm,1));

efln(nm,4) = 0; efln(nm,5) = 0; efln(nm,6) = 0;
if(efln(nm,1)==1 & efln(nm,2)==1) efln(nm,4) = 1; end
if(efln(nm,2)==1 & efln(nm,3)==1) efln(nm,5) = 1; end
if(efln(nm,3)==1 & efln(nm,1)==1) efln(nm,6) = 1; end

end % end of loop over current elements

ne = 4*ne;  % number of elements has increased
            % by a factor of four

for k=1:ne      % relabel the new points
                % and put them in the master list
                % project boundary nodes onto the circle
   for l=1:6
    x(k,l)=xn(k,l); y(k,l)=yn(k,l); efl(k,l)=efln(k,l);
     if(efl(k,l) == 1)
       rad = sqrt(x(k,l)^2+y(k,l)^2);
       x(k,l) = x(k,l)/rad;
       y(k,l) = y(k,l)/rad;
     end
   end

end
```

TABLE 5.2.1 Function *trgl6_disk* ($\rightarrow$ Continuing $\rightarrow$)

```
 Irepo = 1;

%---
 if(Irepo==1)

 for k=1:ne      % reposition the mid-nodes

     if(efl(k,4)  == 0)
        x(k,4) = 0.5*(x(k,1)+x(k,2));
        y(k,4) = 0.5*(y(k,1)+y(k,2)); end

     if(efl(k,5)  == 0)
        x(k,5) = 0.5*(x(k,2)+x(k,3));
        y(k,5) = 0.5*(y(k,2)+y(k,3)); end

     if(efl(k,6)  == 0)
        x(k,6) = 0.5*(x(k,3)+x(k,1));
        y(k,6) = 0.5*(y(k,3)+y(k,1)); end

  end
  end
%---

 end % end do of refinement loop
 end % end if of refinement loop

% --------
% define the global nodes and the connectivity table
% --------

% 6 nodes of the first element are entered manually

p(1,1) = x(1,1); p(1,2) = y(1,1); gfl(1) = efl(1,1);
p(2,1) = x(1,2); p(2,2) = y(1,2); gfl(2) = efl(1,2);
p(3,1) = x(1,3); p(3,2) = y(1,3); gfl(3) = efl(1,3);

p(4,1) = x(1,4); p(4,2) = y(1,4); gfl(4) = efl(1,4);
p(5,1) = x(1,5); p(5,2) = y(1,5); gfl(5) = efl(1,5);
p(6,1) = x(1,6); p(6,2) = y(1,6); gfl(6) = efl(1,6);

c(1,1) = 1;  % first  node of first element is
             % global node 1; similarly:
c(1,2) = 2,
c(1,3) = 3; c(1,4) = 4;
c(1,5) = 5; c(1,6) = 6;

% loop over further elements;
% Iflag = 0 will signal a new global node

eps = 0.000001;
```

TABLE 5.2.1 Function *trgl6_disk* ($\rightarrow$ Continuing $\rightarrow$)

```
for i=2:ne          % loop over elements

 for j=1:6          % loop over element nodes

 Iflag = 0;

  for k=1:ng

   if(abs(x(i,j)-p(k,1)) < eps)
    if(abs(y(i,j)-p(k,2)) < eps)

      Iflag = 1;    % the node has been recorded previously
      c(i,j) = k;   % the jth local node of element i
                    % is the kth global node
    end
   end

  end

  if(Iflag==0)  % record the node

   ng = ng+1;
   p(ng,1) = x(i,j);
   p(ng,2) = y(i,j);
   gfl(ng) = efl(i,j);

   c(i,j) = ng;   % the jth local node of this element
                  % is the new global node
  end

 end

end   % end of loop over elements
%-----
% done
%-----

return;
```

TABLE 5.2.1 (→ Continued) Function *trgl6_disk* triangulates the unit disk into six-node elements by successively subdividing four ancestral elements located in the four quadrants. Each time a subdivision is carried out, the number of elements increases by a factor of 4. Triangulations generated by this function are shown in Figure 5.2.1(a).

We recall that the array efl is the boundary flag of the element nodes, as discussed in Section 4.1.3. Specifically, $efl(i,j)$ is set to zero at the interior nodes or to unity at the boundary nodes. The flag efl of a new node is set according to the flags of the two nodes defining the subdivided edge. If both nodes are boundary nodes, the new node is also a boundary node; otherwise, the new node is an interior node.

Triangulations generated by the function *trgl6_disk* are shown in Figure 5.2.1(*a*) for triangulation levels $ndiv = 0$ (ancestral), 1, 2, and 3. These shapes can be subsequently deformed or mapped by a transformation to produce other isomorphic simply connected domains.

5.2.2   Square

FSELIB function *trgl6_sqr* triangulates a square confined by $-1 \le x \le 1$ and $-1 \le y \le 1$ into six-node triangles using a similar method. In the case of the square, we begin by defining eight ancestral triangles, two elements in each quadrant. Different levels of triangulation are then generated by element subdivision, as shown in Figure 5.2.1(*b*). The beginning of the triangular code defining the ancestral triangulation (root set) is listed in Table 5.2.2. After the triangulation has been completed, the grid can be deformed to yield another desirable isomorphic shape that exhibits four corners.

5.2.3   L-shaped domain

FSELIB function *trgl6_L* triangulates an L-shaped domain in six-node triangles by the successive subdivision of six ancestral triangles residing in the first, third, and fourth quadrants, as shown in Figure 5.2.2. The beginning of the triangular code defining the ancestral triangulation is listed in Table 5.2.3.

5.2.4   Square with a square or circular hole

FSELIB function *trgl6_ss* triangulates the unit square pierced by a square hole with half side length equal to $a < 1$, as shown in Figure 5.2.3(*a*). The companion function *trgl6_sc* triangulates the unit square pierced by a circular hole of radius $a < 1$, as shown in Figure 5.2.3(*b*). The beginning of each triangulation code showing the ancestral triangular structure of 12 elements is listed in Tables 5.2.4 and 5.2.5.

One new feature of the discretization functions *trgl6_ss* and *trgl6_sc* is that, because the solution domain is doubly connected, the element-node flags $efl(i,1)$ and corresponding global node flags $gfl(i)$ may take one of three values, 0, 1, or 2. The value 0 indicates an interior node, the value 1 indicates a node that lies at the outer boundary, and the value 2 indicates a node that lies at the inner boundary. The flag of a new node generated during element subdivision is set accordingly depending on whether the new node lies on a segment defined by two end-nodes with identical flags.

(a)

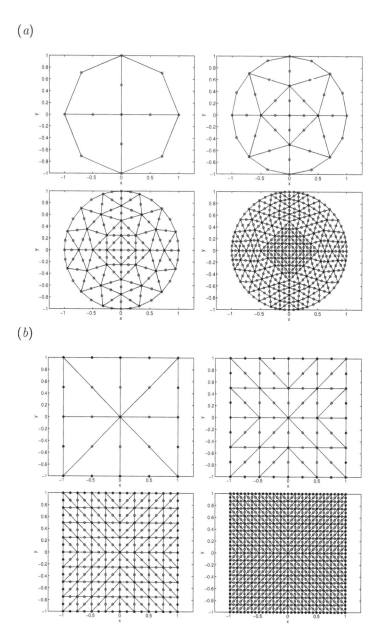

(b)

FIGURE 5.2.1   Triangulation of (a) a disk generated by the FSELIB function *trgl6_disk* and (b) a square generated by the FSELIB function *trgl6_sqr* for discretization levels $ndiv = 0$ (ancestral), 1, 2, and 3. Each triangle is defined by six nodes, including three vertex nodes and three edge nodes.

```
function [ne,ng,p,c,efl,gfl] = trgl6_sqr(ndiv)

%==========================================================
% triangulation of a square into six-node triangles by
% the successive subdivision of eight ancestral elements
%==========================================================

%------------------------------------------
% ancestral structure with eight elements
%------------------------------------------

ne = 8;

x(1,1) = 0.0; y(1,1) = 0.0; efl(1,1) = 0; % first element
x(1,2) = 1.0; y(1,2) = 0.0; efl(1,2) = 1;
x(1,3) = 1.0; y(1,3) = 1.0; efl(1,3) = 1;
x(1,4) = 0.5*(x(1,1)+x(1,2)); y(1,4) = 0.5*(y(1,1)+y(1,2)); efl(1,4) = 0;
x(1,5) = 0.5*(x(1,2)+x(1,3)); y(1,5) = 0.5*(y(1,2)+y(1,3)); efl(1,5) = 1;
x(1,6) = 0.5*(x(1,3)+x(1,1)); y(1,6) = 0.5*(y(1,3)+y(1,1)); efl(1,6) = 0;

x(2,1) = 0.0; y(2,1) = 0.0; efl(2,1)=0;  % second element
x(2,2) = 1.0; y(2,2) = 1.0; efl(2,2)=1;
x(2,3) = 0.0; y(2,3) = 1.0; efl(2,3)=1;
x(2,4) = 0.5*(x(2,1)+x(2,2)); y(2,4) = 0.5*(y(2,1)+y(2,2)); efl(2,4) = 0;
x(2,5) = 0.5*(x(2,2)+x(2,3)); y(2,5) = 0.5*(y(2,2)+y(2,3)); efl(2,5) = 1;
x(2,6) = 0.5*(x(2,3)+x(2,1)); y(2,6) = 0.5*(y(2,3)+y(2,1)); efl(2,6) = 0;

x(3,1) = 0.0; y(3,1) = 0.0; efl(3,1)=0;  % third element
x(3,2) = 0.0; y(3,2) = 1.0; efl(3,2)=1;
x(3,3) =-1.0; y(3,3) = 1.0; efl(3,3)=1;
x(3,4) = 0.5*(x(3,1)+x(3,2)); y(3,4) = 0.5*(y(3,1)+y(3,2)); efl(3,4) = 0;
x(3,5) = 0.5*(x(3,2)+x(3,3)); y(3,5) = 0.5*(y(3,2)+y(3,3)); efl(3,5) = 1;
x(3,6) = 0.5*(x(3,3)+x(3,1)); y(3,6) = 0.5*(y(3,3)+y(3,1)); efl(3,6) = 0;

x(4,1) = 0.0; y(4,1) = 0.0; efl(4,1)=0;  % fourth element
x(4,2) =-1.0; y(4,2) = 1.0; efl(4,2)=1;
x(4,3) =-1.0; y(4,3) = 0.0; efl(4,3)=1;
x(4,4) = 0.5*(x(4,1)+x(4,2)); y(4,4) = 0.5*(y(4,1)+y(4,2)); efl(4,4) = 0;
x(4,5) = 0.5*(x(4,2)+x(4,3)); y(4,5) = 0.5*(y(4,2)+y(4,3)); efl(4,5) = 1;
x(4,6) = 0.5*(x(4,3)+x(4,1)); y(4,6) = 0.5*(y(4,3)+y(4,1)); efl(4,6) = 0;

x(5,1) = 0.0; y(5,1) = 0.0; efl(5,1)=0;  % fifth element
x(5,2) =-1.0; y(5,2) = 0.0; efl(5,2)=1;
x(5,3) = 1.0; y(5,3) = 1.0; ofl(5,3)=1;
x(5,4) = 0.5*(x(5,1)+x(5,2)); y(5,4) = 0.5*(y(5,1)+y(5,2)); efl(5,4) = 0;
x(5,5) = 0.5*(x(5,2)+x(5,3)); y(5,5) = 0.5*(y(5,2)+y(5,3)); efl(5,5) = 1;
x(5,6) = 0.5*(x(5,3)+x(5,1)); y(5,6) = 0.5*(y(5,3)+y(5,1)); efl(5,6) = 0;

x(6,1) = 0.0; y(6,1) = 0.0; efl(6,1)=0;  % sixth element
x(6,2) =-1.0; y(6,2) =-1.0; efl(6,2)=1;
x(6,3) = 0.0; y(6,3) =-1.0; efl(6,3)=1;
```

TABLE 5.2.2   Function *trgl6_sqr* (Continuing →)

```
x(6,4) = 0.5*(x(6,1)+x(6,2)); y(6,4) = 0.5*(y(6,1)+y(6,2)); efl(6,4) = 0;
x(6,5) = 0.5*(x(6,2)+x(6,3)); y(6,5) = 0.5*(y(6,2)+y(6,3)); efl(6,5) = 1;
x(6,6) = 0.5*(x(6,3)+x(6,1)); y(6,6) = 0.5*(y(6,3)+y(6,1)); efl(6,6) = 0;

x(7,1) = 0.0; y(7,1) = 0.0; efl(7,1)=0;  % seventh element
x(7,2) = 0.0; y(7,2) =-1.0; efl(7,2)=1;
x(7,3) = 1.0; y(7,3) =-1.0; efl(7,3)=1;
x(7,4) = 0.5*(x(7,1)+x(7,2)); y(7,4) = 0.5*(y(7,1)+y(7,2)); efl(7,4) = 0;
x(7,5) = 0.5*(x(7,2)+x(7,3)); y(7,5) = 0.5*(y(7,2)+y(7,3)); efl(7,5) = 1;
x(7,6) = 0.5*(x(7,3)+x(7,1)); y(7,6) = 0.5*(y(7,3)+y(7,1)); efl(7,6) = 0;

x(8,1) = 0.0; y(8,1) = 0.0; efl(8,1)=0;  % eighth element
x(8,2) = 1.0; y(8,2) =-1.0; efl(8,2)=1;
x(8,3) = 1.0; y(8,3) = 0.0; efl(8,3)=1;
x(8,4) = 0.5*(x(8,1)+x(8,2)); y(8,4) = 0.5*(y(8,1)+y(8,2)); efl(8,4) = 0;
x(8,5) = 0.5*(x(8,2)+x(8,3)); y(8,5) = 0.5*(y(8,2)+y(8,3)); efl(8,5) = 1;
x(8,6) = 0.5*(x(8,3)+x(8,1)); y(8,6) = 0.5*(y(8,3)+y(8,1)); efl(8,6) = 0;

%-----------------
% refinement loop
%-----------------

......
```

TABLE 5.2.2 ($\rightarrow$ Continued) Function *trgl6_sqr* triangulates a square into six-node elements generated by the successive subdivision of eight hard-coded ancestral elements. The six dots at the end of the table indicate additional code implementing grid refinement, included in FSELIB. Triangulations generated by this function are shown in Figure 5.2.1(b).

5.2.5 A rectangle with a circular hole

FSELIB function *trgl6_rc*, not listed in the text, triangulates a rectangular domain pierced by a circular hole based on a root set with 20 triangles, as shown in Figure 5.2.4(a). Different levels of triangulation are generated by element subdivision, as shown in Figure 5.2.4(b, c). The element node flag $efl(i,1)$ and the corresponding global node flag $gfl(i)$ take the values 0, 1, or 2, respectively, for interior, exterior boundary, and interior boundary nodes, as discussed in Section 5.2.4.

PROBLEMS

5.2.1 *Elliptical hole*

Modify the FSELIB function *trgl6_sc* to generate an elliptical hole with a specified aspect ratio.

```
function [ne,ng,p,c,efl,gfl] = trgl6_L(ndiv)

%================================================================
% Triangulation of an L-shaped domain into six-node elements
% by the successive subdivision of six ancestral elements
%================================================================

%----------------------------------------
% ancestral structure with six elements
%----------------------------------------

ne = 6;

x(1,1) = 0.0; y(1,1) = 0.0; efl(1,1) = 1;  % first element
x(1,2) = 1.0; y(1,2) = 0.0; efl(1,2) = 1;
x(1,3) = 0.0; y(1,3) = 1.0; efl(1,3) = 1;
x(1,4) = 0.5*(x(1,1)+x(1,2)); y(1,4) = 0.5*(y(1,1)+y(1,2)); efl(1,4) = 0;
x(1,5) = 0.5*(x(1,2)+x(1,3)); y(1,5) = 0.5*(y(1,2)+y(1,3)); efl(1,5) = 0;
x(1,6) = 0.5*(x(1,3)+x(1,1)); y(1,6) = 0.5*(y(1,3)+y(1,1)); efl(1,6) = 1;


x(2,1) = 1.0; y(2,1) = 0.0; efl(2,1) = 1;  % second element
x(2,2) = 1.0; y(2,2) = 1.0; efl(2,2) = 1;
x(2,3) = 0.0; y(2,3) = 1.0; efl(2,3) = 1;
x(2,4) = 0.5*(x(2,1)+x(2,2)); y(2,4) = 0.5*(y(2,1)+y(2,2)); efl(2,4)=1;
x(2,5) = 0.5*(x(2,2)+x(2,3)); y(2,5) = 0.5*(y(2,2)+y(2,3)); efl(2,5)=1;
x(2,6) = 0.5*(x(2,3)+x(2,1)); y(2,6) = 0.5*(y(2,3)+y(2,1)); efl(2,6)=0;

x(3,1) = 0.0; y(3,1) = 0.0; efl(3,1)=1;  % third element
x(3,2) =-1.0; y(3,2) = 0.0; efl(3,2)=1;
x(3,3) = 0.0; y(3,3) =-1.0; efl(3,3)=1;
x(3,4) = 0.5*(x(3,1)+x(3,2)); y(3,4) = 0.5*(y(3,1)+y(3,2)); efl(3,4)=1;
x(3,5) = 0.5*(x(3,2)+x(3,3)); y(3,5) = 0.5*(y(3,2)+y(3,3)); efl(3,5)=0;
x(3,6) = 0.5*(x(3,3)+x(3,1)); y(3,6) = 0.5*(y(3,3)+y(3,1)); efl(3,6)=0;

x(4,1) =-1.0; y(4,1) = 0.0; efl(4,1)=1;  % fourth element
x(4,2) =-1.0; y(4,2) =-1.0; efl(4,2)=1;
x(4,3) = 0.0; y(4,3) =-1.0; efl(4,3)=1;
x(4,4) = 0.5*(x(4,1)+x(4,2)); y(4,4) = 0.5*(y(4,1)+y(4,2)); efl(4,4)=1;
x(4,5) = 0.5*(x(4,2)+x(4,3)); y(4,5) = 0.5*(y(4,2)+y(4,3)); efl(4,5)=1;
x(4,6) = 0.5*(x(4,3)+x(4,1)); y(4,6) = 0.5*(y(4,3)+y(4,1)); efl(4,6)=0;

x(5,1) = 0.0; y(5,1) = 0.0; efl(5,1)=0;  % fifth element
x(5,2) = 0.0; y(5,2) =-1.0; efl(5,2)=1;
x(5,3) = 1.0; y(5,3) = 0.0; efl(5,3)=1;
x(5,4) = 0.5*(x(5,1)+x(5,2)); y(5,4) = 0.5*(y(5,1)+y(5,2)); efl(5,4)=0;
x(5,5) = 0.5*(x(5,2)+x(5,3)); y(5,5) = 0.5*(y(5,2)+y(5,3)); efl(5,5)=0;
x(5,6) = 0.5*(x(5,3)+x(5,1)); y(5,6) = 0.5*(y(5,3)+y(5,1)); efl(5,6)=0;

x(6,1) = 1.0; y(6,1) = 0.0; efl(6,1)=1;  % sixth element
x(6,2) = 0.0; y(6,2) =-1.0; efl(6,2)=1;
x(6,3) = 1.0; y(6,3) =-1.0; efl(6,3)=1;
x(6,4) = 0.5*(x(6,1)+x(6,2)); y(6,4) = 0.5*(y(6,1)+y(6,2)); efl(6,4)=0;
x(6,5) = 0.5*(x(6,2)+x(6,3)); y(6,5) = 0.5*(y(6,2)+y(6,3)); efl(6,5)=1;
x(6,6) = 0.5*(x(6,3)+x(6,1)); y(6,6) = 0.5*(y(6,3)+y(6,1)); efl(6,6)=1;
```

TABLE 5.2.3   Function *trgl6_L* (Continuing →)

```
%----------------
% refinement loop
%----------------

. . . . . .
```

TABLE 5.2.3 ($\rightarrow$ Continued) Function *trgl6_L* triangulates an L-shaped domain into six-node elements generated by the successive subdivision of six hard-coded ancestral elements. The six dots at the end of the table indicate additional code included in FSELIB.

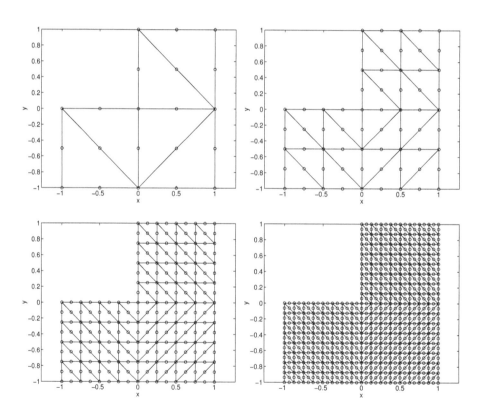

FIGURE 5.2.2   Triangulations generated by the FSELIB function *trgl6_L* for discretization levels $ndiv = 0$ (root set), 1, 2, and 3. Each triangle is defined by six nodes, including three vertex nodes and three edge nodes.

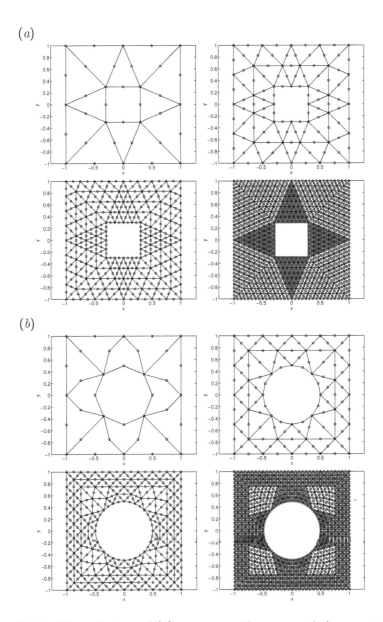

FIGURE 5.2.3   Triangulations of (*a*) a square with a square hole generated by the FSELIB function *trgl6_ss* and (*b*) a square with a circular hole generated by the FSELIB function *trgl6_sc* for discretization levels $ndiv = 0$ (root set), 1, 2, and 3. Each triangle is defined by six nodes, including three vertex nodes and three edge nodes.

```
function [ne,ng,p,c,efl,gfl] = trgl6_ss(a, ndiv)

%==========================================================
% Triangulation of a unit square with a square hole of
% radius "a" into six-node elements by the successive
% subdivision of 12 ancestral elements (root set)
%==========================================================

%-------------------------------------
% ancestral structure with 12 elements
%-------------------------------------

ne = 12;

x(1,1) =   a;y(1,1) =  -a; efl(1,1) = 2;  % first element
x(1,2) = 1.0;y(1,2) =-1.0; efl(1,2) = 1;
x(1,3) = 1.0;y(1,3) = 0.0; efl(1,3) = 1;

x(1,4) = 0.5*(x(1,1)+x(1,2));
y(1,4) = 0.5*(y(1,1)+y(1,2)); efl(1,4)=0;
x(1,5) = 0.5*(x(1,2)+x(1,3));
y(1,5) = 0.5*(y(1,2)+y(1,3)); efl(1,5)=1;
x(1,6) = 0.5*(x(1,3)+x(1,1));
y(1,6) = 0.5*(y(1,3)+y(1,1)); efl(1,6) = 0;

x(2,1) =   a;y(2,1) =  -a; efl(2,1) = 2;  % second element
x(2,2) = 1.0;y(2,2) = 0.0; efl(2,2) = 1;
x(2,3) =   a;y(2,3) =   a; efl(2,3) = 2;

x(2,4) = 0.5*(x(2,1)+x(2,2));
y(2,4) = 0.5*(y(2,1)+y(2,2)); efl(2,4) = 0;
x(2,5) = 0.5*(x(2,2)+x(2,3));
y(2,5) = 0.5*(y(2,2)+y(2,3)); efl(2,5) = 0;
x(2,6) = 0.5*(x(2,3)+x(2,1));
y(2,6) = 0.5*(y(2,3)+y(2,1)); efl(2,6) = 2;

x(3,1) =   a;y(3,1) =   a; efl(3,1)=2;  % third element
x(3,2) = 1.0;y(3,2) = 0.0; efl(3,2)=1;
x(3,3) = 1.0;y(3,3) = 1.0; efl(3,3)=1;

x(3,4) = 0.5*(x(3,1)+x(3,2));
y(3,4) = 0.5*(y(3,1)+y(3,2)); efl(3,4)=0;
x(3,5) = 0.5*(x(3,2)+x(3,3));
y(3,5) = 0.5*(y(3,2)+y(3,3)); efl(3,5)=1;
x(3,6) = 0.5*(x(3,3)+x(3,1));
y(3,6) = 0.5*(y(3,3)+y(3,1)); efl(3,6)=0;

x(4,1) =   a; y(4,1) =   a; efl(4,1) = 2;  % fourth element
x(4,2) = 1.0; y(4,2) = 1.0; efl(4,2) = 1;
x(4,3) = 0.0; y(4,3) = 1.0; efl(4,3) = 1;

x(4,4) = 0.5*(x(4,1)+x(4,2));
y(4,4) = 0.5*(y(4,1)+y(4,2)); efl(4,4)=0;
x(4,5) = 0.5*(x(4,2)+x(4,3));
y(4,5) = 0.5*(y(4,2)+y(4,3)); efl(4,5)=1;
```

TABLE 5.2.4   Function *trgl6_ss* (Continuing →)

```
x(4,6) = 0.5*(x(4,3)+x(4,1));
y(4,6) = 0.5*(y(4,3)+y(4,1)); efl(4,6)=0;

x(5,1) = 0.0;y(5,1) = 1.0; efl(5,1) = 1;  % fifth element
x(5,2) =   -a;y(5,2) =   a; efl(5,2) = 2;
x(5,3) =    a;y(5,3) =   a; efl(5,3) = 2;

x(5,4) = 0.5*(x(5,1)+x(5,2));
y(5,4) = 0.5*(y(5,1)+y(5,2)); efl(5,4)=0;
x(5,5) = 0.5*(x(5,2)+x(5,3));
y(5,5) = 0.5*(y(5,2)+y(5,3)); efl(5,5)=2;
x(5,6) = 0.5*(x(5,3)+x(5,1));
y(5,6) = 0.5*(y(5,3)+y(5,1)); efl(5,6)=0;

x(6,1) =   -a;y(6,1) =   a; efl(6,1) = 2;  % sixth element
x(6,2) = 0.0;y(6,2) = 1.0; efl(6,2) = 1;
x(6,3) =-1.0;y(6,3) = 1.0; efl(6,3) = 1;

x(6,4) = 0.5*(x(6,1)+x(6,2));
y(6,4) = 0.5*(y(6,1)+y(6,2)); efl(6,4) = 0;
x(6,5) = 0.5*(x(6,2)+x(6,3));
y(6,5) = 0.5*(y(6,2)+y(6,3)); efl(6,5) = 1;
x(6,6) = 0.5*(x(6,3)+x(6,1));
y(6,6) = 0.5*(y(6,3)+y(6,1)); efl(6,6) = 0;

%-----------------------------------
% rest of the elements by reflection
%-----------------------------------

for i=1:6
 for j=1:6

  x(6+i,j) =-x(i,j);
  y(6+i,j) =-y(i,j);
  efl(6+i,j) = efl(i,j);

 end
end

%----------------
% refinement loop
%----------------

. . . . . .
```

TABLE 5.2.4 ($\rightarrow$ Continued) Function *trgl6_ss* triangulates the unit square pierced by a square hole into six-node elements generated by the successive subdivision of an ancestral structure with 12 elements (root set). The six dots at the end of the listing indicate unprinted lines of code.

```
function [ne,ng,p,c,efl,gfl] = trgl6_sc(a, ndiv)

%===========================================================
% Triangulation of a unit square with a circular hole of
% radius "a" into six-node elements by the successive
% subdivision of 12 ancestral elements (root set)
%===========================================================

%------------------------------------
% ancestral structure with 12 elements
%------------------------------------

ne = 12;

x(1,1) =    a; y(1,1) =  -a; efl(1,1) = 2;  % first element
x(1,2) = 1.0; y(1,2) =-1.0; efl(1,2) = 1;
x(1,3) = 1.0; y(1,3) = 0.0; efl(1,3) = 1;

x(1,4) = 0.5*(x(1,1)+x(1,2));
y(1,4) = 0.5*(y(1,1)+y(1,2)); efl(1,4)=0;
x(1,5) = 0.5*(x(1,2)+x(1,3));
y(1,5) = 0.5*(y(1,2)+y(1,3)); efl(1,5)=1;
x(1,6) = 0.5*(x(1,3)+x(1,1));
y(1,6) = 0.5*(y(1,3)+y(1,1)); efl(1,6)=0;

x(2,1) =    a; y(2,1) =  -a; efl(2,1)=2;  % second element
x(2,2) = 1.0; y(2,2) = 0.0; efl(2,2)=1;
x(2,3) =    a; y(2,3) =    a; efl(2,3)=2;
x(2,4) = 0.5*(x(2,1)+x(2,2));

y(2,4) = 0.5*(y(2,1)+y(2,2)); efl(2,4)=0;
x(2,5) = 0.5*(x(2,2)+x(2,3));
y(2,5) = 0.5*(y(2,2)+y(2,3)); efl(2,5)=0;
x(2,6) = 0.5*(x(2,3)+x(2,1));
y(2,6) = 0.5*(y(2,3)+y(2,1)); efl(2,6)=2;

x(3,1) =    a; y(3,1) =    a; efl(3,1)=2;  % third element
x(3,2) = 1.0; y(3,2) = 0.0; efl(3,2)=1;
x(3,3) = 1.0; y(3,3) = 1.0; efl(3,3)=1;

x(3,4) = 0.5*(x(3,1)+x(3,2));
y(3,4) = 0.5*(y(3,1)+y(3,2)); efl(3,4)=0;
x(3,5) = 0.5*(x(3,2)+x(3,3));
y(3,5) = 0.5*(y(3,2)+y(3,3)); efl(3,5)=1;
x(3,6) = 0.5*(x(3,3)+x(3,1));
y(3,6) = 0.5*(y(3,3)+y(3,1)); efl(3,6)=0;

x(4,1) =    a;y(4,1) =    a; efl(4,1)=2;  % fourth element
x(4,2) = 1.0;y(4,2) = 1.0; efl(4,2)=1;
x(4,3) = 0.0;y(4,3) = 1.0; efl(4,3)=1;
x(4,4) = 0.5*(x(4,1)+x(4,2));
y(4,4) = 0.5*(y(4,1)+y(4,2)); efl(4,4)=0;
x(4,5) = 0.5*(x(4,2)+x(4,3));
y(4,5) = 0.5*(y(4,2)+y(4,3)); efl(4,5) = 1;
x(4,6) = 0.5*(x(4,3)+x(4,1));
y(4,6) = 0.5*(y(4,3)+y(4,1)); efl(4,6) = 0;
```

TABLE 5.2.5   Function *trgl6_sc* (Continuing →)

```
x(5,1) = 0.0; y(5,1) = 1.0; efl(5,1) = 1;  % fifth element
x(5,2) =   -a; y(5,2) =    a; efl(5,2) = 2;
x(5,3) =    a; y(5,3) =    a; efl(5,3) = 2;
x(5,4) = 0.5*(x(5,1)+x(5,2));

y(5,4) = 0.5*(y(5,1)+y(5,2)); efl(5,4) = 0;
x(5,5) = 0.5*(x(5,2)+x(5,3));
y(5,5) = 0.5*(y(5,2)+y(5,3)); efl(5,5) = 2;
x(5,6) = 0.5*(x(5,3)+x(5,1));
y(5,6) = 0.5*(y(5,3)+y(5,1)); efl(5,6) = 0;

x(6,1) =   -a; y(6,1) =    a; efl(6,1) = 2;  % sixth element
x(6,2) = 0.0; y(6,2) = 1.0; efl(6,2) = 1;
x(6,3) =-1.0; y(6,3) = 1.0; efl(6,3) = 1;

x(6,4) = 0.5*(x(6,1)+x(6,2));
y(6,4) = 0.5*(y(6,1)+y(6,2)); efl(6,4) = 0;
x(6,5) = 0.5*(x(6,2)+x(6,3));
y(6,5) = 0.5*(y(6,2)+y(6,3)); efl(6,5) = 1;
x(6,6) = 0.5*(x(6,3)+x(6,1));
y(6,6) = 0.5*(y(6,3)+y(6,1)); efl(6,6) = 0;

%------------------------------------
% rest of the elements by reflection
%------------------------------------

for i=1:6
 for j=1:6
  x(6+i,j) =-x(i,j); y(6+i,j) =-y(i,j); efl(6+i,j) = efl(i,j);
 end
end

%---------------------------------------------------------------------
% project the interior-boundary nodes onto a circle of radius "a"
%---------------------------------------------------------------------

for i=1:12
  for j=1:6
    if (efl(i,j)==2)
      rad = a/sqrt(x(i,j)^2+y(i,j)^2);
      x(i,j) = x(i,j)*rad;y(i,j) = y(i,j)*rad;
    end
  end
end

%----------------
% refinement loop
%----------------

. . . . . .
```

TABLE 5.2.5   ($\rightarrow$ Continued) Function *trgl6_sc* triangulates a unit square pierced by a circular hole into six-node elements by the successive subdivision of an ancestral structure with 12 elements. The six dots at the end of the listing indicate additional lines of unprinted code.

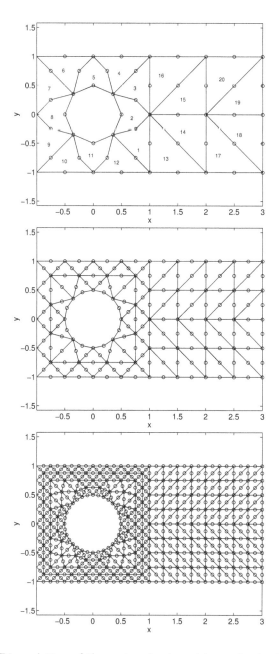

FIGURE 5.2.4 Triangulation of the rectangle pierced by a circular hole generated by the FSELIB function *trgl6_rc* for discretization levels $ndiv = 0$ (root set), 1, 2, and 3. Each triangle is defined by six nodes, including three vertex nodes and three edge nodes.

5.2.2 *T-shaped domain*

Write a function named *trgl6_T* that triangulates a T-shaped domain based on the subdivision of an ancestral hard-coded structure.

5.3   Laplace and Poisson equations

Finite element methods with six-node triangular elements arise as straightforward extensions of those for three-node triangular elements discussed in Chapter 4. The formulation provides us with a system of linear equations for the values of the solution at the global nodes, which can be element vertex or edge nodes.

To display the computed solution, we may use the FSELIB function *plot_6*, listed in Table 5.3.1, which is based on the MATLAB graphics function *patch*. An alternative graphics module is based on the MATLAB function *trimesh*. The input to *trimesh* includes an extended connectivity matrix pertaining to three-node sub-triangles that arise by dividing each six-node triangle into four sub-elements.

5.3.1   Laplace equation

Consider Laplace's equation for a function, $f(x, y)$,

$$\nabla^2 f = 0, \tag{5.3.1}$$

subject to the Dirichlet boundary condition specifying the values of f around the entire boundary of the solution domain. The input-output (I/O) and computational modules of the finite element code are as follows:

1. Specify the input data.
2. Perform grid generation.
3. Specify the Dirichlet boundary condition at the boundary nodes.
4. Assemble the global diffusion matrix from the element diffusion matrices aided by the connectivity matrix.
5. Assemble the right-hand side of the linear system.
6. Implement the Dirichlet boundary condition using the algorithm shown in Table 4.1.2.
7. Solve the linear system.
8. Display the solution.

We recall that the Dirichlet boundary condition in Step 6 is implemented in a way that preserves the symmetry of the compiled linear system.

Disk

FSELIB code *lapl6_d*, listed in Table 5.3.2, solves Laplace's equation in a disk-like domain. The triangulation is performed by the function *trgl6_disk*, listed in Table 5.2.1. The solution for discretization level $ndiv = 2$ subject to the boundary condition implemented in the code is displayed in Figure 5.3.1(*a*).

```
function plot_6 (ne,ng,p,c,f);

%====================================================
% Color mapped visualization of a function f in a
% domain discretized into six-node triangles
% using the patch graphics function
%====================================================

figure(1)
hold on
xlabel('x');
ylabel('y');

%----------------------------------------------------
% compute the maximum and minimum of the function f
%----------------------------------------------------

fmax = -100.0;  fmin= 100; % trial values

for i=1:ng
 if(f(i) > fmax) fmax = f(i); end
 if(f(i) < fmin) fmin = f(i); end
end

range = 1.2*(fmax-fmin); shift = fmin;

%--------------------------------------------
% shift the color index in the range (0, 1)
% and paint seven-point patches
%--------------------------------------------

for l=1:ne

 j = c(1,1); xp(1) = p(j,1); yp(1) = p(j,2); cp(1) = (f(j)-shift)/range;
 j = c(1,4); xp(2) = p(j,1); yp(2) = p(j,2); cp(2) = (f(j)-shift)/range;
 j = c(1,2); xp(3) = p(j,1); yp(3) = p(j,2); cp(3) = (f(j)-shift)/range;
 j = c(1,5); xp(4) = p(j,1); yp(4) = p(j,2); cp(4) = (f(j)-shift)/range;
 j = c(1,3); xp(5) = p(j,1); yp(5) = p(j,2); cp(5) = (f(j)-shift)/range;
 j = c(1,6); xp(6) = p(j,1); yp(6) = p(j,2); cp(6) = (f(j)-shift)/range;
 j = c(1,1); xp(7) = p(j,1); yp(7) = p(j,2); cp(7) = (f(j)-shift)/range;

 patch(xp, yp,cp);

end

%-----
% done
%-----

return;
```

TABLE 5.3.1   Function *plot_6* performs the color mapped visualization of a function in a domain discretized into six-node triangles.

```
%================================================
% Code lapl6_d
%
%
% Finite element code for Laplace's equation
% in a disk-like domain with the Dirichlet
% boundary condition, using six-node triangles
%================================================

%-----------
% input data
%-----------

ndiv = 1;   % discretization level

NQ = 6;     % Gauss-triangle quadrature

%------------
% triangulate
%------------

[ne,ng,p,c,efl,gfl] = trgl6_disk (ndiv);

%-------
% deform
%-------

for i=1:ng
  p(i,1) = p(i,1)*(1.0-0.60*p(i,2)^2);
end

%----------------------------------------
% specify the Dirichlet boundary condition
%----------------------------------------

for i=1:ng

 if(gfl(i)==1)
   bcd(i) = p(i,1)*sin(0.5*pi*p(i,2));
 end

end

%------------------------------------------------
% assemble the global diffusion matrix
% and compute the domain surface area (optional)
%------------------------------------------------

gdm = zeros(ng,ng);  % initialize
area = 0.0;           % initialize
```

TABLE 5.3.2   Code *lapl6_d* (Continuing →)

```
for l=1:ne               % loop over the elements
                         % compute the element diffusion matrix
 j=c(l,1); x1=p(j,1); y1=p(j,2);
 j=c(l,2); x2=p(j,1); y2=p(j,2);
 j=c(l,3); x3=p(j,1); y3=p(j,2);
 j=c(l,4); x4=p(j,1); y4=p(j,2);
 j=c(l,5); x5=p(j,1); y5=p(j,2);
 j=c(l,6); x6=p(j,1); y6=p(j,2);

  [edm_elm, arel] = edm6 ...
  ...
    (x1,y1, x2,y2, x3,y3, x4,y4, x5,y5, x6,y6 ,NQ);

  area = area + arel;

   for i=1:6
     i1 = c(l,i);
     for j=1:6
       j1 = c(l,j);
       gdm(i1,j1) = gdm(i1,j1) + edm_elm(i,j);
     end
   end

end

%---------------------------------------------------
% compute the right-hand side
% and implement the Dirichlet boundary condition
%---------------------------------------------------

for i=1:ng
 b(i) = 0.0;
end

for j=1:ng

 if(gfl(j)==1)

    for i=1:ng
      b(i) = b(i) - gdm(i,j) * bcd(j);
      gdm(i,j) = 0; gdm(j,i) = 0;
    end

    gdm(j,j) = 1.0; b(j) = bcd(j);

 end

end

%------------------------
% solve the linear system
%------------------------

  f = b/gdm';
```

TABLE 5.3.2   Code *lapl6_d* ($\rightarrow$ Continuing $\rightarrow$)

```
%-----
% plot
%-----

plot_6 (ne,ng,p,c,f);

%-----
% connectivity matrix for three-node sub-triangles
%-----

Ic = 0;

for i=1:ne
  Ic = Ic+1;
  c3(Ic,1) = c(i,1); c3(Ic,2) = c(i,4); c3(Ic,3) = c(i,6);
  Ic = Ic+1;
  c3(Ic,1) = c(i,4); c3(Ic,2) = c(i,2); c3(Ic,3) = c(i,5);
  Ic = Ic+1;
  c3(Ic,1) = c(i,5); c3(Ic,2) = c(i,3); c3(Ic,3) = c(i,6);
  Ic = Ic+1;
  c3(Ic,1) = c(i,4); c3(Ic,2) = c(i,5); c3(Ic,3) = c(i,6);
end

%-----
% plot
%-----

trimesh (c3,p(:,1),p(:,2),f);

%-----
% done
%-----
```

TABLE 5.3.2 ($\rightarrow$ Continued) Finite element code *lapl6_d* solves Laplace's equation in a disk-like domain subject to the Dirichlet boundary condition using six-node triangles.

L-shaped domain

FSELIB code *lapl6_d_L*, not listed in the text, solves Laplace's equation in the L-shaped domain discussed in Section 5.2, subject to the following Dirichlet boundary condition around the polygonal boundary:

$$f = xy\,(1-x)(y+1). \tag{5.3.2}$$

The solution for discretization level $ndiv = 2$ is displayed in Figure 5.3.1(*b*).

Square with a square or circular hole

FSELIB codes *lapl6_d_ss* and *lapl6_d_sc*, not listed in the text, implement the finite element method for Laplace's equation in a square with a square or circular hole.

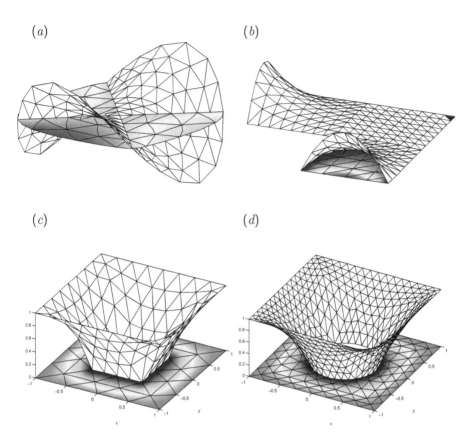

FIGURE 5.3.1   (*a*) Finite element solution of Laplace's equation subject to the Dirichlet boundary condition, generated by the FSELIB code *lapl6_d* for discretization
level $ndiv = 2$. (*b*) Solution in an L-shaped domain generated by the FSELIB code
lapl6_d_L for discretization level $ndiv = 2$. (*c*) Solution in a square domain with
a square hole generated by the FSELIB code *lapl6_d_ss* for discretization level
$ndiv = 1$. (*d*) Solution in a square domain with a circular hole generated by the
FSELIB code *lapl6_d_sc* for discretization level $ndiv = 2$.

The Dirichlet boundary condition specifies that $f = 1$ around the exterior boundary
and $f = 0$ around the interior boundary. The triangulation is performed by the
functions *trgl6_ss* and *trgl6_sc* discussed in Section 5.2. Physically, the solution
domain can be identified with the cross-section of a chimney wall. Comparing
the solutions displayed in Figure 5.3.1(*c, d*) indicates that the shape of the inner
boundary–square versus circular–is significant only in its proximity.

5.3.2   Eigenvalues of the Laplacian operator

In Section 4.5, we discussed a finite element code for computing the eigenvalues of the Laplacian operator on a circular disk using three-node triangles. An analogous code for six-node triangles, entitled *lapl6_eig*, is shown in Table 5.3.3.

The smallest computed eigenvalue for discretization level $ndiv = 2$ is 5.7850. This numerical value is in excellent agreement with the known exact value, 5.7840. We recall from Section 4.5 that the smallest computed eigenvalue for discretization level $ndiv = 4$ using three-node triangles is 5.8024, which is noticeably less accurate. This comparison demonstrates that six-node elements are more powerful than their three-node counterparts.

Other domain shapes can be accommodated by replacing the triangulation function in the code with another suitable function. As an example, the first six eigenfunctions of the Laplacian operator on a square pierced by a circular hole are shown in Figure 5.3.2 for discretization level $ndiv = 2$.

5.3.3   Poisson equation

The finite element code for Laplace's equation can be readily modified to solve the Poisson equation,

$$\nabla^2 f + g = 0, \tag{5.3.3}$$

where g is a specified distributed source. The finite element method is implemented in the FSELIB code *pois6_d*, listed in Table 5.3.4. The code is able to handle six computational domain geometries selected by the integer flag *ishape*. Other domains can be included by straightforward modifications. The source function, g, named *source* in the code, is given an arbitrary constant value equal to -1. The finite element formulation provides us with a linear system whose right-hand side, named b in the code, incorporates the effect of the distributed source through the global mass matrix. The Dirichlet boundary condition is implemented in a way that preserves the symmetry of the coefficient matrix.

The numerical solution in an L-shaped domain or a rectangle with a circular hole is shown in Figure 5.3.3(a, b) for a uniform distributed source, g. Physically, the solution represents the velocity distribution over the cross-section of a straight L-shaped channel or rectangular channel in the presence of an inner circular tube in pressure- or gravity-driven unidirectional fluid flow (e.g., Pozrikidis [46]).

PROBLEMS

5.3.1 *Laplace's equation in an L-shaped domain*

Execute the FSELIB code *lapl6_d_L* with boundary conditions of your choice and discuss the numerical solution.

```
%=======================================
% Code lapl6_eig
%
% Eigenfunctions of Laplace's equation
% in a disk-like domain
% using six-node triangles
%=======================================

%-----------
% input data
%-----------

ndiv = 2;   % discretization level

NQ = 12;    % Gauss-triangle quadrature

%------------
% triangulate
%------------

  [ne,ng,p,c,efl,gfl] = trgl6_disk (ndiv);

%-------
% deform
%-------

defx = 0.0;

for i=1:ng
 p(i,1) = p(i,1)*(1.0-defx*p(i,2)^2);
end

%--------------------------------------
% assemble the global diffusion and mass matrices
%--------------------------------------

gdm = zeros(ng,ng); % initialize
gmm = zeros(ng,ng); % initialize

for l=1:ne              % loop over the elements

 % compute the element diffusion and mass matrices

 j = c(l,1); x1 = p(j,1); y1 = p(j,2);
 j = c(l,2); x2 = p(j,1); y2 = p(j,2);
 j = c(l,3); x3 = p(j,1); y3 = p(j,2);
 j = c(l,4); x4 = p(j,1); y4 = p(j,2);
 j = c(l,5); x5 = p(j,1); y5 = p(j,2);
 j = c(l,6); x6 = p(j,1); y6 = p(j,2);

 [edm_elm, emm_elm, arel] = edmm6 ...
  ...
    (x1,y1, x2,y2, x3,y3, x4,y4, x5,y5, x6,y6, NQ);
```

TABLE 5.3.3   Code *lapl6_eig* (Continuing →)

```
    for i=1:6
      i1 = c(l,i);
      for j=1:6
        j1 = c(l,j);
        gdm(i1,j1) = gdm(i1,j1) + edm_elm(i,j);
        gmm(i1,j1) = gmm(i1,j1) + emm_elm(i,j);
      end
    end
end

%------------------------------------------
% reduce the matrices by removing equations
% corresponding to boundary nodes
%------------------------------------------

Ic = 0;

for i=1:ng

 if(gfl(i)==0)
   Ic = Ic+1;
   map(Ic) = i;

   Jc = 0;
   for j=1:ng
    if(gfl(j)==0)
     Jc = Jc+1;
     A(Ic,Jc) = gdm(i,j);
     B(Ic,Jc) = gmm(i,j);
    end
   end

 end

end

ngred = Ic;

%------------------------
% compute the eigenvalues
%------------------------

[V,D] = eig(A,D),

%-----
% map and plot
%-----

Ic = 0;
```

TABLE 5.3.3   Code *lapl6_eig* ($\rightarrow$ Continuing $\rightarrow$)

```
for i=1:ne

 Ic = Ic+1;

 c3(Ic,1) = c(i,1); c3(Ic,2) = c(i,4);
 c3(Ic,3) = c(i,6);

 Ic = Ic+1;

 c3(Ic,1) = c(i,4); c3(Ic,2) = c(i,2);
 c3(Ic,3) = c(i,5);

 Ic = Ic+1;

 c3(Ic,1) = c(i,5); c3(Ic,2) = c(i,3);
 c3(Ic,3) = c(i,6);

 Ic = Ic+1;
 c3(Ic,1) = c(i,4); c3(Ic,2) = c(i,5);
 c3(Ic,3) = c(i,6);

end

%---
% run over six eigensolutions
%---

for order=1:6

 for i=1:ng
  f(i) = 0.0;
 end

 for i=1:ngred
  f(map(i)) = V(i,order);
 end

 %---
 % plot by two methods
 %---

 figure(order)
 plot_6 (ne,ng,p,c,f);

 trimesh (c3,p(:,1),p(:,2),f);

end

%-----
% done
%-----
```

TABLE 5.3.3   ($\rightarrow$ Continued) Finite element code *lapl6_eig* computes the eigenvalues of the Laplacian operator and displays the eigenfunctions.

```
%=============
% Code pois6_d
%
% FSELIB
%
% Finite element code for solving Poisson's equation
% in a disk-like domain subject to
% the Dirichlet boundary condition,
% using six-node triangles.
%=============

close all
clear all

%-----------
% input data
%-----------

ndiv = 3;  % triangulation level

NQ = 6;    % Gauss-triangle quadrature

ishape = 1;

%------------
% triangulate
%------------

if(ishape==1)
 ndiv = 3;
 [ne,ng,p,c,efl,gfl] = trgl6_disk(ndiv);
elseif(ishape==2)
 ndiv = 3;
 [ne,ng,p,c,efl,gfl] = trgl6_sqr(ndiv);
elseif(ishape==3)
 ndiv = 3;
 [ne,ng,p,c,efl,gfl] = trgl6_L(ndiv);
elseif(ishape==4)
 ndiv = 2;
 a = 0.5;
 [ne,ng,p,c,efl,gfl] = trgl6_sc(a, ndiv);
elseif(ishape==5)
 ndiv = 2;
 a = 0.5;
 [ne,ng,p,c,efl,gfl] = trgl6_ss(a,ndiv);
elseif(ishape==6)
 ndiv = 2;
 a = 0.5;
 [ne,ng,p,c,efl,gfl] = trgl6_rc (a,ndiv)
end

%-------------------------------------------
% specify the Dirichlet boundary condition
%-------------------------------------------
```

TABLE 5.3.4   Code *pois6_d* (Continuing →)

```
for i=1:ng
 if(gfl(i)==1)
   bcd(i) = p(i,1)*sin(0.5*pi*p(i,2));   % example
   bcd(i) = 0.0;              % another example
 end
 if(gfl(i)==2)
   bcd(i) = 0.0;              % another example
 end
end

%-------------------
% specify the source
%-------------------

for i=1:ng
  source(i) =-1.0;
end

%-------------------------------------
% assemble the global diffusion matrix
% and global mass matrix
% and compute the domain surface area (optional)
%-------------------------------------

gdm = zeros(ng,ng); % initialize
gmm = zeros(ng,ng); % initialize

for l=1:ne           % loop over the elements

% compute the element diffusion and mass matrices

j=c(l,1); x1=p(j,1); y1=p(j,2);
j=c(l,2); x2=p(j,1); y2=p(j,2);
j=c(l,3); x3=p(j,1); y3=p(j,2);
j=c(l,4); x4=p(j,1); y4=p(j,2);
j=c(l,5); x5=p(j,1); y5=p(j,2);
j=c(l,6); x6=p(j,1); y6=p(j,2);

[edm_elm, emm_elm, arel] = edmm6 ...
...
     (x1,y1, x2,y2, x3,y3, x4,y4, x5,y5, x6,y6 ...
     ,NQ);

   for i=1:6
     i1 = c(l,i);
     for j=1:6
       j1 = c(l,j);
       gdm(i1,j1) = gdm(i1,j1) + edm_elm(i,j);
       gmm(i1,j1) = gmm(i1,j1) + emm_elm(i,j);
     end
   end

end
```

TABLE 5.3.4   Code *pois6_d* ($\rightarrow$ Continuing $\rightarrow$)

```
%------------------------
% set the right-hand side
%------------------------

for i=1:ng
 b(i) = 0.0;
 for j=1:ng
  b(i) = b(i)-gmm(i,j)*source(j);
 end
end

%---------------------------------------------
% implement the Dirichlet boundary condition
%---------------------------------------------

for j=1:ng
 if(gfl(j)==1|gfl(j)==2)
   for i=1:ng
    b(i) = b(i) - gdm(i,j) * bcd(j);
    gdm(i,j) = 0; gdm(j,i) = 0;
   end
   gdm(j,j) = 1.0; b(j) = bcd(j);
 end
end

%------------------------
% solve the linear system
%------------------------

  f = b/gdm';

%---------
% plotting
%---------

figure(1)
hold on
view([-19,18])
%axis equal
box on
xlabel('x','fontsize',14);
ylabel('y','fontsize',14);
zlabel('f','fontsize',14);
set(gca,'fontsize',14)

%---
% done
%---
```

TABLE 5.3.4 (→ Continued) Finite element code *pois6_d* for solving the Poisson equation subject to the Dirichlet boundary condition using six-node triangles.

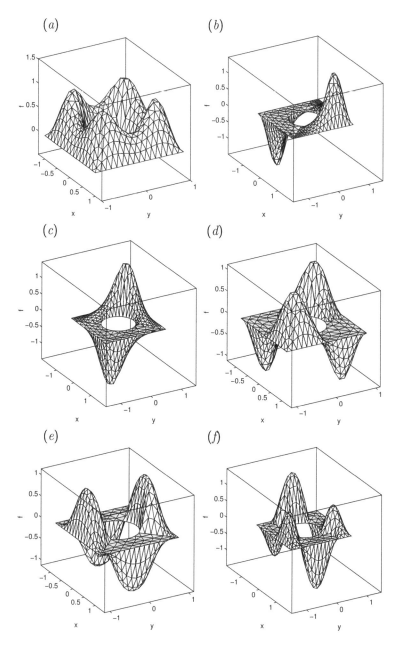

FIGURE 5.3.2   Eigenfunctions corresponding to the first six eigenvalues of the Laplacian operator on a square with a circular hole.

(a)

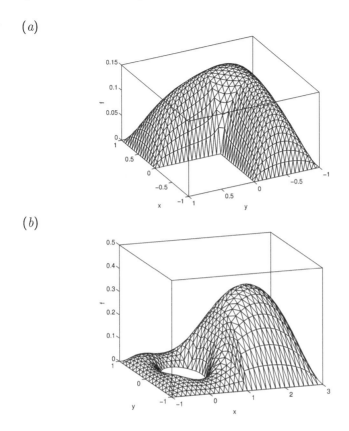

(b)

FIGURE 5.3.3   Finite element solution of the Poisson equation with a uniform distributed source in (a) an L-shaped domain and (b) a rectangle with a circular
hole.

5.3.2 *Laplace's equation in a square with a square hole*

Execute the FSELIB code *lapl6_d_ss* with boundary conditions of your choice and
discuss the numerical solution.

5.3.3 *Laplace's equation in a square with a circular hole*

Execute the FSELIB code *lapl6_d_sc* with boundary conditions of your choice and
discuss the numerical solution.

5.3.4 *Eigenvalues of the Laplacian operator in an L-shaped domain*

Compute the eigenvalues and display the first six eigenfunctions of the Laplacian
operator in an L-shaped domain.

5.4 Convection–diffusion with the Dirichlet boundary condition

FSELIB code *scd6_d_rc*, listed in Table 5.4.1, solves the steady convection–diffusion equation,

$$u_x \frac{\partial f}{\partial x} + u_y \frac{\partial f}{\partial y} = \kappa \nabla^2 f, \tag{5.4.1}$$

where $u_x(x,y)$ and $u_y(x,y)$ are the x and y components of the advection velocity, and κ is the medium diffusivity. The solution domain is a rectangle with a circular hole of radius a, physically representing the contour of a circular cylinder embedded in streaming flow inside a channel, as shown in Figure 5.4.1. The Dirichlet boundary condition specifies the distribution of the function f around the outer rectangular boundary and along the inner circular boundary. The triangulation is performed by the FSELIB function *trgl6_rc* discussed in Section 5.2.5.

The advection velocity prescribed in the code corresponds to streaming potential flow with velocity U past a circular cylinder, given by

$$u_x = U\left(1 + \frac{a^2}{r^2}\left(1 - 2\frac{x^2}{r^2}\right)\right), \qquad u_y = -2Ua^2\frac{xy}{r^4}, \tag{5.4.2}$$

where r is the distance from the cylinder center (e.g., Pozrikidis [46]). The importance of convection relative to diffusion is determined by the Péclet number, $\text{Pe} = Ua/\kappa$.

Finite element code

A new module of the finite element code is the interpolation of the x and y velocity components at the element integration base points from corresponding values at the six element nodes. This computation is performed inside the function *edam6* that evaluates the global diffusion and advection matrices, listed in Table 5.1.7. The nodal velocities themselves are evaluated by the FSELIB function *scd6_vel*, listed in Table 5.4.2.

Numerical solutions

The finite element solution subject to the uniform Dirichlet boundary condition requiring that $f = 1$ around the outer rectangular boundary and $f = 0$ around the inner circular boundary is displayed in Figure 5.4.1 for $\text{Pe} = 0.0$ and 2.5. Physically, the results illustrate the temperature field established around an isothermal cold cylinder in a channel confined between two isothermal parallel plane walls in the presence of a convective stream.

As Pe tends to zero, the convective contribution diminishes and heat transport occurs due to conduction alone. As Pe increases, convective transport becomes increasingly important and the temperature field extends in the downstream direction, in agreement with physical intuition.

(a)

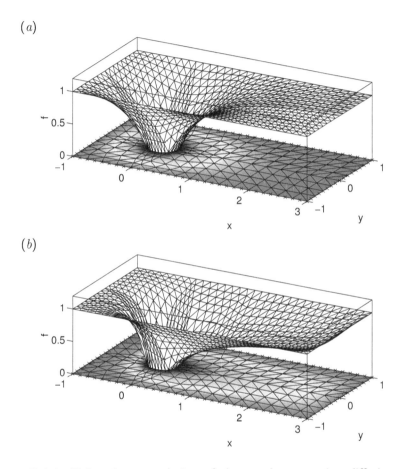

(b)

FIGURE 5.4.1   Finite element solution of the steady convection–diffusion equation
subject to the Dirichlet boundary condition, generated by the FSELIB code
scd6_d_rc for discretization level $ndiv = 2$ and (a) $\mathrm{Pe} = 0.0$ (pure diffusion)
and (b) 2.5 (diffusion combined with significant convection).

```
%======================================================
% Code scd6_d_rc
%
% Code for the steady convection--diffusion equation
% with the Dirichlet boundary condition in a
% rectangular domain with a circular hole
%======================================================

%-----------
% input data
%-----------

k = 1.0; rho = 1.0; cp = 1.0;
U = 0.5;
NQ = 6;
ndiv = 2;
a = 0.25;   % cylinder radius

%------------
% diffusivity
%------------

kappa = k/(rho*cp);

%------------
% triangulate
%------------

[ne,ng,p,c,efl,gfl] = trgl6_rc (a, ndiv);

%----------------------------------------
% specify the Dirichlet boundary condition
%----------------------------------------

for i=1:ng

 if(gfl(i)==1)
   bcd(i) = 1.0;
 end

 if(gfl(i)==2)
   bcd(i) = 0.0;
 end

end

%-------------------------------
% assemble the global diffusion
% and advection matrices
%-------------------------------

gdm = zeros(ng,ng); % initialize
gam = zeros(ng,ng); % initialize
```

TABLE 5.4.1 Code *scd6_d_rc* (Continuing →)

```
for l=1:ne

% compute the element diffusion
% and advection matrices

   j = c(l,1); x1 = p(j,1); y1 = p(j,2); % node positions
   j = c(l,2); x2 = p(j,1); y2 = p(j,2);
   j = c(l,3); x3 = p(j,1); y3 = p(j,2);
   j = c(l,4); x4 = p(j,1); y4 = p(j,2);
   j = c(l,5); x5 = p(j,1); y5 = p(j,2);
   j = c(l,6); x6 = p(j,1); y6 = p(j,2);

% define the nodal velocities (ux=u, uy=v)

   [u1, v1] = scd6_vel(U,a,x1,y1);
   [u2, v2] = scd6_vel(U,a,x2,y2);
   [u3, v3] = scd6_vel(U,a,x3,y3);
   [u4, v4] = scd6_vel(U,a,x4,y4);
   [u5, v5] = scd6_vel(U,a,x5,y5);
   [u6, v6] = scd6_vel(U,a,x6,y6);

% element matrices

[edm_elm, eam_elm, arel] = edam6 ...
...
   (x1,y1, x2,y2, x3,y3, x4,y4, x5,y5, x6,y6 ...
   ,u1,v1, u2,v2, u3,v3, u4,v4, u5,v5, u6,v6 ...
   ,NQ);

   for i=1:6
     i1 = c(l,i);
     for j=1:6
       j1 = c(l,j);
       gdm(i1,j1) = gdm(i1,j1) + edm_elm(i,j);
       gam(i1,j1) = gam(i1,j1) + eam_elm(i,j);
     end
   end
end

%-------------------
% coefficient matrix
%-------------------

lsm = kappa*gdm+gam;

%------------------------------------
% compute the right-hand side and
% implement the boundary conditions
%------------------------------------

for i=1:ng
 b(i) = 0.0;
end
```

TABLE 5.4.1 Code *scd6_d_rc* ($\rightarrow$ Continuing $\rightarrow$)

```
for j=1:ng

 if(gfl(j)==1 | gfl(j)==2)  % outer and inner
                            % boundary nodes
    for i=1:ng
      b(i) = b(i) - lsm(i,j)*bcd(j);
      lsm(i,j) = 0;
      lsm(j,i) = 0;
    end

    lsm(j,j) = 1.0;
    b(j) = bcd(j);

 end

end

%------------------------
% solve the linear system
%------------------------

f = b/lsm';

%---------
% plotting
%---------

plot_6 (ne,ng,p,c,f);

%-----------------------------
% extended connectivity matrix
% for three-node sub-triangles
%-----------------------------

Ic = 0;

for i=1:ne

 Ic = Ic+1;
 c3(Ic,1) = c(i,1); c3(Ic,2) = c(i,4); c3(Ic,3) = c(i,6);
 Ic = Ic+1;
 c3(Ic,1) = c(i,4); c3(Ic,2) = c(i,2); c3(Ic,3) = c(i,5);
 Ic = Ic+1;
 c3(Ic,1) = c(i,5); c3(Ic,2) = c(i,3); c3(Ic,3) = c(i,6);
 Ic = Ic+1;
 c3(Ic,1) = c(i,4); c3(Ic,2) = c(i,5); c3(Ic,3) = c(i,6);

end
```

TABLE 5.4.1 Code *scd6_d_rc* ($\rightarrow$ Continuing $\rightarrow$)

```
%---------
% plotting
%---------

trimesh (c3,p(:,1),p(:,2),f);

%-----
% done
%-----
```

TABLE 5.4.1 (⟶ Continued) Code *scd6_d_rc* implements the finite element solution of the steady convection–diffusion with the Dirichlet boundary condition in a rectangular domain with a circular hole.

```
function [u, v] = scd6_vel (U,a,x,y)

%====================================
% velocity evaluation corresponding
% to streaming potential flow past
% a cylinder of radius "a"
%====================================

  as = a^2;
  rs = x^2+y^2;
  rq = rs^2;

  u = U*(1.0+as/rs - 2.0*as*x^2/rq);
  v = U*( -2.0*as*x*y/rq);

%-----
% done
%-----

return;
```

TABLE 5.4.2 Function *scd6_vel* evaluates the nodal velocities required by code *scd6_d_rc*.

PROBLEM

5.4.1 *Convection–diffusion equation*

Execute the FSELIB code *scd6_d_rc* for Pe = 0, 1.0, 5.0, and 10.0. Discuss the effect of convection on the temperature distribution.

5.5 High-order triangle expansions

Equation (4.2.5) for the ith node interpolation function on a three-node triangle expresses a complete linear expansion over the area of the standard triangle in the parametric $\xi\eta$ plane,

$$\psi_i(\xi,\eta) = a_i + b_i\xi + c_i\eta, \tag{5.5.1}$$

where a_i–c_i are appropriate coefficients.

Equation (5.1.6) for the ith node interpolation function on six-node triangle expresses a complete quadratic expansion over the area of the standard triangle in the parametric $\xi\eta$ plane,

$$\psi_i(\xi,\eta) = a_i + b_i\xi + c_i\eta + d_i\xi^2 + e_i\xi\eta + f_i\eta^2, \tag{5.5.2}$$

where a_i–f_i are appropriate coefficients.

A complete mth-order polynomial expansion of any suitable function, $f(\xi,\eta)$, takes the triangular form

$$
\begin{aligned}
f(\xi,\eta) = \quad & a_{00} \\
& + a_{10}\xi + a_{01}\eta \\
& + a_{20}\xi^2 + a_{11}\xi\,\eta + a_{02}\eta^2 \\
& + a_{30}\xi^3 + a_{21}\xi^2\,\eta + a_{12}\,\xi\,\eta^2 + a_{03}\eta^3 \\
& \cdots \quad \cdots \quad \cdots \quad \cdots \quad \cdots \quad \cdots \quad \cdots \quad \cdots \\
& + a_{m,0}\,\xi^m + a_{m-1,1}\,\xi^{m-1}\eta + \ldots + a_{1,m-1}\,\xi\eta^{m-1} + a_{0,m}\,\eta^m.
\end{aligned}
\tag{5.5.3}
$$

Note that the sum of the indices of the coefficients a_{ij} across each row is constant. The total number of coefficients is

$$N = 1+2+3+\cdots+m+(m+1) = \frac{1}{2}(m+1)(m+2) = \binom{m+2}{2}, \tag{5.5.4}$$

where

$$\binom{l}{k} = \frac{l!}{k!\,(l-k)!} \tag{5.5.5}$$

is the combinatorial expressing the number of possible ways by which k objects can be chosen from a set of l identical objects, leaving $l-k$ objects behind, where $k \le l$. The exclamation mark denotes the factorial, $q! = 1 \cdot 2 \cdots q$, with the understanding that $0! = 1$. When $k = l-1$, we obtain

$$\binom{l}{l-1} = \frac{l!}{(l-1)!\,1!} = l, \tag{5.5.6}$$

which is correct, since we have l choices for leaving one object behind. As expected, the total number of coefficients summed in (5.5.4) is an integer irrespective of whether the expansion order, m, is even or odd.

   An essential property of the complete mth-order expansion, as compared to an incomplete expansion where some of the terms are missing from the right-hand side of (5.5.3), is that the function, $f(\xi, \eta)$, is an mth-degree polynomial with respect to distance along any straight line in the $\xi\eta$ plane. More important, the function is a complete mth-degree polynomial of ξ or η along the edges of the triangle. Consequently, C^0 continuity across each element edge is guaranteed by the presence of $m + 1$ shared element nodes, including vertex and edge nodes.

Pascal triangle

The triangular structure displayed in (5.5.3) can be compared to Pascal's triangle,

$$
\begin{array}{ccccccccccc}
 & & & & & 1 & & & & & \\
 & & & & 1 & & 1 & & & & \\
 & & & 1 & & 2 & & 1 & & & \\
 & & 1 & & 3 & & 3 & & 1 & & \\
 & 1 & & 4 & & 6 & & 4 & & 1 & \\
1 & & 5 & & 10 & & 10 & & 5 & & 1 \\
\cdots & & \cdots & & \cdots & & \cdots & & \cdots & & \cdots
\end{array}
\qquad (5.5.7)
$$

Each entry is the sum of the two entries immediately above it, while all outermost entries are equal to unity. The entries of the pth row provide us with the coefficients of the binomial expansion of the power

$$(\xi + \eta)^{p-1}, \qquad (5.5.8)$$

where p is an integer. For example, when $p = 3$, the third row provides us with the coefficients of the quadratic expansion,

$$(\xi + \eta)^2 = \mathbf{1}\,\xi^2 + \mathbf{2}\,\xi\eta + \mathbf{1}\,\eta^2.$$

More generally,

$$(\xi + \eta)^m = \sum_{k=0}^{m} \binom{m}{k} \xi^{m-k}\,\eta^k, \qquad (5.5.9)$$

where the tall parentheses designate the combinatorial.

Leibniz high-order product differentiation rule

Let $f(x)$ and $g(x)$ be two functions of an independent variable, x. Applying the rules of product function differentiation, we derive the Leibniz high-order product differentiation rule,

$$\frac{\mathrm{d}^m(fg)}{\mathrm{d}x^m} = \sum_{k=0}^{m} \binom{m}{k} \frac{\mathrm{d}^{m-k} f}{\mathrm{d}x^{m-k}} \frac{\mathrm{d}^k g}{\mathrm{d}x^k}. \qquad (5.5.10)$$

Note that the sum of the orders of the derivatives on the right-hand side is m, independent of the running index, k.

Polynomial forms and interpolation functions

A complete mth-degree polynomial can be expressed in the form

$$
\begin{aligned}
f(\xi, \eta) \;=\; & d_1 \eta^m \\
+\; & (d_2 + d_3 \xi)\eta^{m-1} \\
+\; & (d_4 + d_5 \xi + d_6 \xi^2)\,\eta^{m-2} \\
+\; & (d_7 + d_8 \xi + d_9 \xi^2 + d_{10}\xi^3)\eta^{m-3} \\
+\; & \cdots \\
+\; & (d_{N-m} + d_{N-m+1}\xi + \cdots + d_{N-1}\xi^{m-1} + d_N \xi^m),
\end{aligned}
\tag{5.5.11}
$$

where d_1–d_N is a new set of coefficients. This expression demonstrates that the coefficient multiplying the monomial η^p is an $(m - p)$-degree polynomial in ξ. For example, when $p = m$, the coefficient is a zeroth-degree polynomial identified with a constant, d_1.

Rearranging, we obtain the complementary form

$$
\begin{aligned}
f(\xi, \eta) \;=\; & e_1 \xi^m \\
+\; & (e_2 + e_3\, \eta)\xi^{m-1} \\
+\; & (e_4 + e_5 \eta + e_6\, \eta^2)\xi^{m-2} \\
+\; & (e_7 + e_8 \eta + e_9 \eta^2 + e_{10}\eta^3)\xi^{m-3} \\
+\; & \cdots \\
+\; & (e_{N-m} + e_{N-m+1}\eta + \cdots + e_{N-1}\eta^{m-1} + e_N \eta^m),
\end{aligned}
\tag{5.5.12}
$$

where e_1–e_N is a new set of coefficients. This expression demonstrates that the coefficient multiplying the monomial ξ^p is an $(m - p)$-degree polynomial in η.

To formalize the mth-order expansion, we denote the monomial products on the right-hand side of (5.5.3) as

$$
\mathcal{M}_{ij}(\xi, \eta) \equiv \xi^i \eta^j,
\tag{5.5.13}
$$

and write

$$
f(\xi, \eta) = \sum_{j=0}^{m} \left[\sum_{i=0}^{m-j} a_{ij}\, \mathcal{M}_{ij}(\xi, \eta) \right],
\tag{5.5.14}
$$

corresponding to (5.5.11), or

$$
f(\xi, \eta) = \sum_{i=0}^{m} \left[\sum_{j=0}^{m-i} a_{ij}\, \mathcal{M}_{ij}(\xi, \eta) \right],
\tag{5.5.15}
$$

corresponding to (5.5.12).

Element interpolation nodes

To implement the finite element method, we specify the polynomial order, m, and introduce N interpolation nodes over the area and along the edges of each triangle, (ξ_i, η_i), where $i = 1, \ldots, N$. The number of nodes, N, is related to the polynomial order, m, by (5.5.4).

By definition, the cardinal interpolation function for the ith node, $\psi_i(\xi, \eta)$, is a complete mth-degree polynomial in ξ and η, required to satisfy N interpolation conditions,

$$\psi_i(\xi_j, \eta_j) = \delta_{i,j} \tag{5.5.16}$$

for $j = 1, \ldots, N$, where $\delta_{i,j}$ is Kronecker's delta. The nodal expansion over the parametric triangle takes the familiar form

$$f(\xi, \eta) = \sum_{j=1}^{N} f(\xi_j, \eta_j)\, \psi_i(\xi_j, \eta_j), \tag{5.5.17}$$

involving the specified or *a priori* unknown nodal values, $f(\xi_j, \eta_j)$.

5.5.1   Computation of the node interpolation functions

In certain special cases, the node interpolation functions, $\psi_i(\xi, \eta)$, can be deduced by inspection aided by geometrical intuition based on a simple rule. Consider the interpolation function of the ith node and assume that m straight lines can be found passing through all nodes, except for the ith node. If these lines are described by the linear functions

$$\Phi_j(\xi, \eta) = A_j\, \xi + B_j\, \eta + C_j = 0 \tag{5.5.18}$$

for $j = 1, \ldots, m$, where A_j, B_j, and C_j are constant coefficients, then the ith-node interpolation function is given by

$$\psi_i(\xi, \eta) = \prod_{j=1}^{m} \frac{\Phi_j(\xi, \eta)}{\Phi_j(\xi_i, \eta_i)}. \tag{5.5.19}$$

A specific example will be given in Section 5.8.2. It is a simple matter to show that (5.5.19) satisfies the cardinal interpolation condition (5.5.16).

Polynomial expansions

More generally, the cardinal interpolation function corresponding to a node can be expressed in the polynomial form shown in (5.5.15) or (5.5.14), and the N coefficients, a_{ij}, can be computed by solving a system of linear equations originating from the cardinal interpolation condition expressed by equation (5.5.16).

In the most general approach, the interpolation functions are expressed as linear combinations of a set of N independent polynomials that form a complete base of the mth-order expansion in the $\xi\eta$ plane, $\phi_j(\xi, \eta)$ for $j = 1, \ldots, N$,

$$\psi_i(\xi, \eta) = c_N \, \phi_1(\xi, \eta) + c_{N-1} \, \phi_2(\xi, \eta) \\ + \cdots + c_2 \, \phi_{N-1}(\xi, \eta) + c_1 \, \phi_N(\xi, \eta), \qquad (5.5.20)$$

where c_j comprise a set of $N + 1$ expansion coefficients for the ith node.

For example, identifying the basis function with the monomial products shown in (5.5.13), we obtain

$$\phi_1 = \mathcal{M}_{00} = 1, \qquad \phi_2 = \mathcal{M}_{10} = \xi, \qquad \phi_3 = \mathcal{M}_{01} = \eta, \qquad \ldots, \\ \phi_{N-1} = \mathcal{M}_{1,m-1} = \xi \, \eta^{m-1}, \qquad \phi_N = \mathcal{M}_{0,m} = \eta^m. \qquad (5.5.21)$$

Another possible polynomial base is provided by the nodal interpolation functions corresponding to a nodal set that is different from the set under consideration. Better choices are provided by the Appell and Proriol families of semi-orthogonal and orthogonal polynomials discussed in Sections 5.6 and 5.7.

Enforcing the cardinal interpolation condition (5.5.16) for any choice of basis functions, we obtain a linear system,

$$\mathbf{V}_\phi^T \cdot \mathbf{c} = \mathbf{e}_i, \qquad (5.5.22)$$

where the superscript T denotes the matrix transpose,

$$\mathbf{V}_\phi \equiv \begin{bmatrix} \phi_1(\xi_1, \eta_1) & \phi_1(\xi_2, \eta_2) & \cdots & \phi_1(\xi_N, \eta_N) \\ \phi_2(\xi_1, \eta_1) & \phi_2(\xi_2, \eta_2) & \cdots & \phi_2(\xi_N, \eta_N) \\ \cdots & \cdots & \cdots & \cdots \\ \phi_{N-1}(\xi_1, \eta_1) & \phi_{N-1}(\xi_2, \eta_2) & \cdots & \phi_{N-1}(\xi_N, \eta_N) \\ \phi_N(\xi_1, \eta_1) & \phi_N(\xi_2, \eta_2) & \cdots & \phi_N(\xi_N, \eta_N) \end{bmatrix} \qquad (5.5.23)$$

is the $N \times N$ *generalized Vandermonde matrix* with components

$$\left(V_\phi\right)_{ij} = \phi_i(\xi_j, \eta_j), \qquad (5.5.24)$$

$\mathbf{c}$ is the basis functions coefficient vector,

$$\mathbf{c} \equiv \begin{bmatrix} c_N, & c_{N-1}, & \ldots & c_2, & c_1 \end{bmatrix}^T, \qquad (5.5.25)$$

and

$$\mathbf{e}_i \equiv \begin{bmatrix} 0, & \ldots, & 0, & 1, & \ldots, & 0 \end{bmatrix}^T, \qquad (5.5.26)$$

is the unit vector of the N-dimensional space associated with the ith node. The unity on the right-hand side of (5.5.26) appears in the ith entry.

Inverting the linear system (5.5.22), we obtain

$$\mathbf{c} = \mathbf{V}_\phi^{T^{-1}} \cdot \mathbf{e}_i. \tag{5.5.27}$$

The ith nodal interpolation function is given by

$$\psi_i(\xi, \eta) = \boldsymbol{\phi}(\xi, \eta) \cdot \mathbf{V}_\phi^{T^{-1}} \cdot \mathbf{e}_i, \tag{5.5.28}$$

where

$$\boldsymbol{\phi}(\xi, \eta) \equiv \left[\begin{array}{cccc} \phi_1, & \phi_2, & \cdots & \phi_{N-1}, & \phi_N \end{array} \right]^T \tag{5.5.29}$$

is the vector of basis functions.

Compiling the expressions for all nodes, we obtain the nodal function interpolation vector

$$\boldsymbol{\psi}(\xi, \eta) = \mathbf{V}_\phi^{-1} \cdot \boldsymbol{\phi}(\xi, \eta), \tag{5.5.30}$$

where

$$\boldsymbol{\psi}(\xi, \eta) \equiv \left[\begin{array}{cccc} \psi_1, & \psi_2 & \cdots, & \psi_{N-1}, & \phi_N \end{array} \right]^T. \tag{5.5.31}$$

Rearranging, we derive a linear system,

$$\mathbf{V}_\phi \cdot \boldsymbol{\psi}(\xi, \eta) = \boldsymbol{\phi}(\xi, \eta). \tag{5.5.32}$$

Alternatively, the linear system (5.5.32) can be derived immediately, following the discussion of Section 3.2. Using the Lagrange interpolation formula discussed in Section D.2, Appendix D, we write the exact representation

$$\phi_i(\xi, \eta) = \sum_{j=1}^{N} \phi_i(\xi_j, \eta_j) \, \psi_j(\xi, \eta) \tag{5.5.33}$$

for $i = 1, \ldots, N$, which is equivalent to (5.5.32).

Computing the solution of the linear system by Cramer's rule, we find that the ith-node element interpolation function is given by

$$\psi_i(\xi, \eta) = \frac{\det[\mathbf{V}_\phi(\xi_1, \eta_1, \xi_2, \eta_2, \ldots, \xi_{i-1}, \eta_{i-1}, \xi, \eta, \xi_{i+1}, \eta_{i+1}, \ldots \xi_N, \eta_N)]}{\det[\mathbf{V}_\phi(\xi_1, \eta_1, \xi_2, \eta_2, \ldots, \xi_N, \eta_N)]} \tag{5.5.34}$$

for $i = 1, \ldots, N$. To confirm this representation, we make three key observations:

1. Since the basis functions $\phi_i(\xi, \eta)$ involved in the generalized Vandermonde matrix form a complete basis for the mth-order expansion in (ξ, η), the numerator in (5.5.34) is an mth-degree polynomial in (ξ, η). This can be demonstrated readily by considering the Laplace expansion of the determinant with respect to the column involving ξ and η.

2. The numerator in (5.5.34) is zero when $\xi = \xi_j$ and $\eta = \eta_j$, where $j \neq i$, since two columns of the generalized Vandermonde matrix become identical.

3. The right-hand side of (5.5.34) is equal to unity when $\xi = \xi_i$ and $\eta = \eta_i$.

For low polynomial orders, m, the linear system can be solved analytically by elementary methods.

As an example, we consider the linear expansion corresponding to $m = 1$ and $N = 3$, and choose

$$\phi_1 = 1, \qquad \phi_2 = \xi, \qquad \phi_3 = \eta. \tag{5.5.35}$$

The nodal set is comprised of the three vertex nodes with barycentric coordinates $\xi_1 = 0, \eta_1 = 0$ for the first node, $\xi_2 = 1, \eta_2 = 0$ for the second node, and $\xi_3 = 0, \eta_3 = 1$ for the third node, yielding the Vandermonde matrix

$$\mathbf{V}_\phi \equiv \begin{bmatrix} 1 & 1 & 1 \\ 0 & 1 & 0 \\ 0 & 0 & 1 \end{bmatrix}, \tag{5.5.36}$$

whose inverse is

$$\mathbf{V}_\phi^{-1} \equiv \begin{bmatrix} 1 & -1 & -1 \\ 0 & 1 & 0 \\ 0 & 0 & 1 \end{bmatrix}. \tag{5.5.37}$$

The nodal interpolation function vector is thus given by

$$\psi = \begin{bmatrix} 1 & -1 & -1 \\ 0 & 1 & 0 \\ 0 & 0 & 1 \end{bmatrix} \cdot \begin{bmatrix} 1 \\ \xi \\ \eta \end{bmatrix}, \tag{5.5.38}$$

in agreement with previously obtained expressions shown in (4.2.7).

For higher polynomial orders, the solution of the linear system for the nodal interpolation functions can be found by numerical methods. To ensure that the coefficient matrix is well-conditioned, it is beneficial to employ basis functions that are partially or entirely orthogonal, such as those provided by the Appell and Proriol polynomials discussed in Sections 5.6 and 5.7. The linear system may then be solved without difficulty using standard methods, such as the method of Gauss elimination discussed in Appendix C.

The nodal set that maximizes the magnitude of the determinant of the generalized Vandermonde matrix within the confines of the triangle is the highly desirable Fekete set. Because any two polynomial bases are linearly dependent, the Fekete set is independent of the working base chosen to carry out the optimization.

5.5.2   The Lebesgue constant

A complete nodal set is comprised on N interpolation nodes, where N is related to a specified polynomial order, m, through (5.5.4). To assess the relative merits of two candidate nodal sets, we require a measure of success quantified by an objective function.

A heuristic generalization of the theory of one-dimensional interpolation discussed in Appendix D leads us to the Lebesgue function in two dimensions,

$$\mathcal{L}_N(\xi,\eta) \equiv \sum_{i=1}^{N} \Big|\psi_i(\xi,\eta)\Big|, \tag{5.5.39}$$

and associated Lebesgue constant

$$\Lambda_N \equiv \max\big[\mathcal{L}(\xi,\eta)\big], \tag{5.5.40}$$

where $\max[\bullet]$ denotes the maximum of the bracketed function over the area of the triangle. The theory of interpolation suggests that, the lower the value of Λ_N, the more desirable the nodal set. Other objective functions can be defined by exercising common sense or physical intuition, as will be discussed in Section 5.8.

5.5.3   Node condensation

In a typical nodal set, three nodes are placed at the triangle vertices, a number of nodes are placed along the triangle edges, and other nodes are placed in the triangle interior. When node condensation is applied, the unknowns associated with the interior nodes are eliminated from the finite element equations arising from the Galerkin projection of the global interpolation functions, and the final system is condensed into a system of smaller size involving only the shared vertex and edge nodes. The process of elimination is similar to that for one-dimensional problems discussed in Sections 1.5.6 and 3.5.3 (Wilson [71]). Condensation schemes and relevant algorithms are reviewed by Schwarz ([58], pp. 185–201).

PROBLEM

5.5.1 *Linear interpolation functions*

Consider the linear expansion corresponding to $m = 1$ and $N = 3$, and choose $\phi_1 = 1$, $\phi_2 = \xi + \eta$, and $\phi_3 = \xi - \eta$. Compute the corresponding generalized Vandermonde matrix and its inverse. Confirm that the results are consistent with the expressions shown in (4.2.7).

5.6   Appell polynomial base

A function of interest defined over the standard triangle in the $\xi\eta$ plane can be approximated with a complete mth-degree polynomial in ξ and η. The approximating polynomial can be expressed in the form

$$f(\xi, \eta) = \sum_{k=0}^{m} \left(\sum_{l=0}^{m-k} \hat{a}_{kl} \, \mathcal{A}_{kl}(\xi, \eta) \right), \tag{5.6.1}$$

where $\hat{a}_{kl}$ are expansion coefficients, $\mathcal{A}_{kl}$ are the Appell polynomials defined as

$$\mathcal{A}_{kl}(\xi, \eta) \equiv \frac{\partial^{k+l}}{\partial \xi^k \partial \eta^l} \left(\xi^k \eta^l \zeta^{k+l} \right), \tag{5.6.2}$$

and $\zeta = 1 - \xi - \eta$ is the third barycentric coordinate (Appell [2]). In the literature of orthogonal polynomials, the definition (5.6.2) is sometimes called the *Rodrigues formula*.

Carrying out the differentiation with respect to ξ and η with the help of Leibniz's rule discussed in Section 5.5, we find that $\mathcal{A}_{kl}$ is a $(k+l)$-degree polynomial in ξ and η, and may thus be expressed in the form

$$\mathcal{A}_{kl}(\xi, \eta) = \sum_{i=0}^{k+l} \left(\sum_{j=0}^{k+l-i} a_{kl}^{ij} \, \xi^i \eta^j \right), \tag{5.6.3}$$

where a_{kl}^{ij} are appropriate coefficients. In this notation, ij is *not* an exponent.

To derive an explicit form of the Appell polynomials, we apply Leibniz's rule (5.5.10) to compute the η derivatives in (5.6.2), finding

$$\mathcal{A}_{kl}(\xi, \eta) = \frac{\partial^k}{\partial \xi^k} \left[\xi^k \sum_{j=0}^{l} \binom{l}{j} \frac{\mathrm{d}^{l-j} \eta^l}{\mathrm{d}\eta^{l-j}} \frac{\partial^j \zeta^{k+l}}{\partial \eta^j} \right]. \tag{5.6.4}$$

Carrying out the differentiations with respect to η, we obtain

$$\mathcal{A}_{kl}(\xi, \eta) = \frac{\partial^k}{\partial \xi^k} \left[\xi^k \sum_{j=0}^{l} \binom{l}{j} \, \frac{l!}{j!} \, \eta^j \times (-1)^j \, \frac{(k+l)!}{(k+l-j)!} \, \zeta^{k+l-j} \right], \tag{5.6.5}$$

which can be recast into the form

$$\mathcal{A}_{kl}(\xi, \eta) = \sum_{j=0}^{l} \eta^j \, (-1)^j \, \frac{(l!)^2 \, (k+l)!}{(j!)^2 \, (l-j)! \, (k+l-j)!} \, \frac{\partial^k}{\partial \xi^k} \left(\xi^k \, \zeta^{k+l-j} \right). \tag{5.6.6}$$

Next, we perform the differentiation with respect to ξ using once again Leibniz's rule, and simplify to obtain the explicit formula

$$\mathcal{A}_{kl}(\xi, \eta) = k! \, l! \sum_{i=0}^{k} \sum_{j=0}^{l} (-1)^{i+j} \, \frac{(k)_i \, (l)_j \, (k+l)_{i+j}}{i!^2 \, j!^2} \, \xi^i \, \eta^j \, \zeta^{k+l-i-j}, \tag{5.6.7}$$

$$\mathcal{A}_{00} = 1$$
$$\mathcal{A}_{10} = \zeta - \xi = 1 - 2\xi - \eta$$
$$\mathcal{A}_{01} = \zeta - \eta = 1 - 2\eta - \xi$$
$$\mathcal{A}_{20} = 2\zeta(\zeta - 4\xi) + 2\xi^2$$
$$\mathcal{A}_{11} = \zeta(\zeta - 2\xi - 2\eta) + 2\xi\eta$$
$$\mathcal{A}_{02} = 2\zeta(\zeta - 4\eta) + 2\eta^2$$

TABLE 5.6.1   Explicit expressions for the first few Appell polynomials. Note that $\mathcal{A}_{kl}$ is a complete $(k+l)$-degree polynomial in ξ and η.

subject to the definitions $(p)_0 \equiv 1$ and

$$(p)_q \equiv (-p)(-p+1)(-p+2)\cdots(-p+q-1) \tag{5.6.8}$$

for $q \geq 1$, for an arbitrary pair of integers, p and q (Suetin [64], p. 65). For example, $(-1)_q = q!$.

Further manipulation shows that the coefficients a_{kl}^{ij} defined in (5.6.3) are given by

$$a_{kl}^{ij} = (-1)^{i+j}\frac{(k+j)!\,(l+i)!}{i!^2\,j!^2}\frac{(k+l)!}{(k+l-i-j)!}. \tag{5.6.9}$$

Note that $a_{kl}^{ij} = a_{lk}^{ji}$, as required by symmetry.

Applying these formulas, or else using the Rodrigues formula, we derive the first few Appell polynomials shown in Table 5.6.1.

Governing differential equation

It can be shown that the Appell polynomials satisfy the homogeneous partial differential equation

$$\xi(\xi-1)\mathcal{A}_{\xi\xi} + 2\xi\eta\mathcal{A}_{\xi\eta} + \eta(\eta-1)\mathcal{A}_{\eta\eta} \tag{5.6.10}$$
$$+(3\xi-1)\mathcal{A}_{\xi} + (3\eta-1)\mathcal{A}_{\eta} = (k+l)(k+l+2)\mathcal{A},$$

where a subscript denotes a partial derivative with respect to the corresponding variable; for simplicity, we have denoted $\mathcal{A} = \mathcal{A}_{kl}$ (Suetin [64], p. 75). Equation (5.6.10) has $k+l+1$ independent eigensolutions, which are the Appell polynomials of the same total degree, $k+l$. For example, when $k+l=1$, the two eigensolutions are the polynomials $\mathcal{A}_{10}$ and $\mathcal{A}_{01}$ shown in Table 5.6.1.

5.6.1 Incomplete biorthogonality

The Appell polynomials are orthogonal against a lower class of monomials, $\mathcal{M}_{pq} = \xi^p \eta^q$, that is,

$$\iint \mathcal{A}_{kl}(\xi, \eta)\, \xi^p \eta^q \, \mathrm{d}\xi \, \mathrm{d}\eta = 0 \tag{5.6.11}$$

for

$$p + q < k + l \equiv m, \tag{5.6.12}$$

and the integration is performed over the area of the right isosceles triangle in the parametric plane. Consequently,

$$\iint \mathcal{A}_{kl}(\xi, \eta) \, \mathrm{d}\xi \, \mathrm{d}\eta = 0 \tag{5.6.13}$$

for $k > 0$ or $l > 0$, which states that the mean value of all polynomials, except for the zeroth-degree polynomial, is zero over the area of the parametric triangle.

To demonstrate the biorthogonality property (5.6.11), we may integrate in ξ and η, finding

$$\iint \mathcal{A}_{kl}(\xi, \eta)\, \xi^p \eta^q \, \mathrm{d}\xi \, \mathrm{d}\eta = \int_0^1 \left(\int_0^{1-\eta} \frac{\partial^{k+l}}{\partial \xi^k \partial \eta^l} \left(\xi^k \eta^l \zeta^{k+l} \right) \xi^p \, \mathrm{d}\xi \right) \eta^q \, \mathrm{d}\eta. \tag{5.6.14}$$

Focusing on the inner integral on the right-hand side, we integrate by parts and find that

$$\int_0^{1-\eta} \frac{\partial^{k+l}}{\partial \xi^k \partial \eta^l} \left(\xi^k \eta^l \zeta^{k+l} \right) \xi^p \, \mathrm{d}\xi = \left[\frac{\partial^{k+l-1}}{\partial \xi^{k-1} \partial \eta^l} \left(\xi^k \eta^l \zeta^{k+l} \right) \xi^p \right]_{\xi=0}^{\xi=1-\eta}$$

$$- \int_0^{1-\eta} \frac{\partial^{k+l-1}}{\partial \xi^{k-1} \partial \eta^l} \left(\xi^k \eta^l \zeta^{k+l} \right) \frac{\partial \xi^p}{\partial \xi} \, \mathrm{d}\xi. \tag{5.6.15}$$

Recalling the definition $\zeta = 1 - \xi - \eta$, we find that the first term on the right-hand side is zero. Repeating the integration by parts, we obtain

$$\int_0^{1-\eta} \frac{\partial^{k+l}}{\partial \xi^k \partial \eta^l} \left(\xi^k \eta^l \zeta^{k+l} \right) \xi^p \, \mathrm{d}\xi = (-1)^k \int_0^{1-\eta} \frac{\partial^l}{\partial \eta^l} \left(\xi^k \eta^l \zeta^{k+l} \right) \frac{\partial^k \xi^p}{\partial \xi^k} \, \mathrm{d}\xi, \tag{5.6.16}$$

which is zero if $p < k$. This conclusion may also be reached independently using the explicit form shown in (5.6.6). Working in a similar fashion, we write

$$\iint \mathcal{A}_{kl}(\xi, \eta)\, \xi^p \eta^q \, \mathrm{d}\xi \, \mathrm{d}\eta = \int_0^1 \left(\int_0^{1-\xi} \frac{\partial^{k+l}}{\partial \xi^k \partial \eta^l} \left(\xi^k \eta^l \zeta^{k+l} \right) \eta^q \, \mathrm{d}\eta \right) \xi^p \, \mathrm{d}\xi, \tag{5.6.17}$$

and find that this integral is zero when $q < l$, which completes the proof.

A detailed calculation shows that

$$\iint \mathcal{A}_{kl}(\xi, \eta)\, \xi^p \eta^q \, \mathrm{d}\xi \, \mathrm{d}\eta = (-1)^{k+l} \frac{(p!)^2 \, (q!)^2 \, (k+l)!}{(p-k)! \, (q-l)! \, (k+l+p+q+2)!} \tag{5.6.18}$$

for $p \geq k$ and $q \geq l$.

5.6.2 Incomplete orthogonality

Biorthogonality implies that the polynomial $\mathcal{A}_{kl}(\xi, \eta)$ is orthogonal against any polynomial of lower total degree in ξ and η, that is, against any polynomial involving monomials $\xi^p \eta^q$ with $p + q < k + l$. A consequence is the incomplete self-orthogonality property[2]

$$\iint \mathcal{A}_{kl}(\xi, \eta)\, \mathcal{A}_{pq}(\xi, \eta)\, \mathrm{d}\xi\, \mathrm{d}\eta = 0 \tag{5.6.19}$$

for $p + q \neq k + l$. For example, the Appell polynomial $\mathcal{A}_{10}$ is orthogonal against any other Appell polynomial, except for the same-row polynomial $\mathcal{A}_{01}$.

5.6.3 Generalized Appell polynomials

The Appell polynomials defined in (5.6.2) can be generalized into a broader family of polynomials parametrized by the three exponents, $\alpha > -1$, $\beta > -1$, and $\gamma > -1$. The generalized polynomials are defined as

$$\mathcal{A}_{kl}^{(\alpha, \beta, \gamma)}(\xi, \eta) \equiv \frac{1}{\xi^\alpha\, \eta^\eta\, \zeta^\gamma} \frac{\partial^{k+l}}{\partial \xi^k \partial \eta^l} \left(\xi^{k+\alpha} \eta^{l+\beta} \zeta^{k+l+\gamma} \right), \tag{5.6.20}$$

where $\zeta = 1 - \xi - \eta$. The standard Appell polynomials shown in (5.6.2) correspond to $\alpha = 0$, $\beta = 0$, and $\gamma = 0$.

Repeating the preceding analysis, we find that the generalized Appell polynomials satisfy the weighted orthogonality properties

$$\iint \mathcal{A}_{kl}(\xi, \eta)\, \xi^p\, \eta^q\, w(\xi, \eta)\, \mathrm{d}\xi\, \mathrm{d}\eta = 0 \tag{5.6.21}$$

for $p + q < k + l \equiv m$, and

$$\iint \mathcal{A}_{kl}(\xi, \eta)\, \mathcal{A}_{pq}(\xi, \eta)\, w(\xi, \eta)\, \mathrm{d}\xi\, \mathrm{d}\eta = 0 \tag{5.6.22}$$

for $p + q \neq k + l$, with weighting function

$$w(\xi, \eta) = \xi^\alpha \eta^\beta \zeta^\gamma. \tag{5.6.23}$$

When $p \geq k$ and $q \geq l$, we find that

$$\iint \mathcal{A}_{kl}(\xi, \eta)\, \xi^p\, \eta^q\, w(\xi, \eta)\, \mathrm{d}\xi\, \mathrm{d}\eta \tag{5.6.24}$$

$$= (-1)^{k+l} \frac{p!\, q!}{(p-k)!\, (q-l)!} \frac{\Gamma(p+\alpha+1)\, \Gamma(q+\beta+1)\, \Gamma(k+l+\gamma+1)}{\Gamma(p+q+k+l+\alpha+\beta+\gamma+3)},$$

---

[2]Complete orthogonality would require that the projection expressed by the integral on the left-hand side of (5.6.19) is zero for $k \neq p$ and $l \neq q$, which is not true.

where Γ is the Gamma function; if m is an integer, then $\Gamma(m+1) = m!$.

PROBLEM

5.6.1 *Appell polynomials*

Verify property (5.6.13) for the polynomials $\mathcal{A}_{10}$ and $\mathcal{A}_{01}$ shown in Table 5.6.1.
Hint: Use the integration formula (5.1.24).

5.7   Proriol polynomial base

The most desirable polynomial base over a triangle with *straight edges* is provided
by the Proriol polynomials, which are entirely orthogonal over the triangle area
(Proriol [50]). To introduce these polynomials, we map the standard triangle from
the $\xi\eta$ parameter plane to the standard square in the $\xi'\eta'$ parametric plane, as shown
in Figure 5.7.1, using the transformations (4.2.42), repeated below for convenience,

$$\xi' = \frac{2\xi}{1-\eta} - 1, \qquad \eta' = 2\eta - 1. \tag{5.7.1}$$

The Proriol polynomials are given by

$$\mathcal{P}_{kl} = \mathrm{L}_k(\xi') \left(\frac{1-\eta'}{2}\right)^k \mathcal{J}_l^{(2k+1,0)}(\eta') \tag{5.7.2}$$

or

$$\mathcal{P}_{kl} = \mathrm{L}_k(\xi')(1-\eta)^k \mathcal{J}_l^{(2k+1,0)}(\eta'), \tag{5.7.3}$$

where L_k is a Legendre polynomial and $\mathcal{J}_l^{(2k+1,0)}$ is a Jacobi polynomial, as discussed in Section B.8, Appendix B, (e.g., Dubiner [19], Koornwinder [34]).

The factor $(1-\eta)^k$ in (5.7.3) cancels the denominator of the Legendre polynomial
$\mathrm{L}_k(\xi')$ arising from the fraction on the right-hand side of the transformation rule for
ξ' shown in (5.7.1). Consequently, the Proriol polynomial, $\mathcal{P}_{kl}$, involves products,
$\xi^p\eta^{(k-p+q)}$, with combined order $k+q$, where $p = 1, \ldots, k$ and $q = 1, \ldots, l$.

Explicit expressions for the first few Proriol polynomials can be derived by using
the following expressions for the first few Jacobi polynomials:

$$\mathcal{J}_0^{(\alpha,0)}(t) = 1,$$
$$\mathcal{J}_1^{(\alpha,0)}(t) = \frac{1}{2}\left[(\alpha+2)t + \alpha\right], \tag{5.7.4}$$
$$\mathcal{J}_2^{(\alpha,0)}(t) = \frac{1}{8}(\alpha+3)(\alpha+4)t^2 + \frac{1}{4}\alpha(\alpha+3)t + \frac{1}{8}(\alpha^2 - \alpha - 4).$$

Using these formulas together with corresponding formulas for the Legendre polynomials, we find the constant polynomial

$$\mathcal{P}_{00} = \mathrm{L}_0(\xi')\,\mathcal{J}_0^{(1,0)}(\eta') = 1, \tag{5.7.5}$$

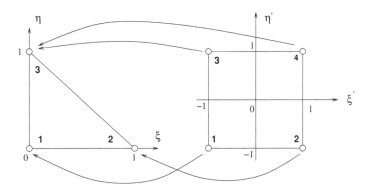

FIGURE 5.7.1   Mapping of the standard triangle to the standard square using Duffy's transformation. The bold-faced numbers are the element node labels.

the linear polynomials

$$\mathcal{P}_{10} = L_1(\xi')\,(1-\eta)\,\mathcal{J}_0^{(3,0)}(\eta') = \xi'\,(1-\eta) = 2\xi + \eta - 1 = \xi - \zeta,$$

$$\mathcal{P}_{01} = L_0(\xi')\,\mathcal{J}_1^{(1,0)}(\eta') = \frac{1}{2}\,(3\,\eta'+1) = 3\,\eta - 1, \qquad (5.7.6)$$

and the quadratic polynomials

$$\mathcal{P}_{20} = L_2(\xi')\,(1-\eta)^2\,\mathcal{J}_0^{(5,0)}(\eta')$$
$$= \frac{1}{2}\,(3\,\xi'^2 - 1)\,(1-\eta)^2 = 6\,\xi^2 + 6\,\xi\eta + \eta^2 - 6\,\xi - 2\,\eta + 1,$$

$$\mathcal{P}_{11} = L_1(\xi')\,(1-\eta)\,\mathcal{J}_1^{(3,0)}(\eta') = \xi'\,(1-\eta)\,\frac{1}{2}\,(5\,\eta'+3)$$
$$= (2\,\xi + \eta - 1)\,(5\,\eta - 1), \qquad (5.7.7)$$

$$\mathcal{P}_{02} = L_0(\xi')\,\mathcal{J}_2^{(1,0)}(\eta') = \frac{5}{2}\,\eta'^2 + \eta' - \frac{1}{2} = 10\,\eta^2 - 8\,\eta + 1,$$

Note that $\mathcal{P}_{01}, \mathcal{P}_{02}, \mathcal{P}_{03}, \ldots$ are pure polynomials in η, whereas $\mathcal{P}_{10}, \mathcal{P}_{20}, \mathcal{P}_{30}, \ldots$ are polynomials in both ξ and η. A similar calculation yields

$$\mathcal{P}_{22} = 216\,y^2x^2 - 96\,y^2x + 6\,y^2 + 216\,yx^3 - 312\,yx^2$$
$$+ 102\,yx - 6\,y + 36\,x^4 - 88\,x^3 + 69\,x^2 - 18\,x + 1. \qquad (5.7.8)$$

FSELIB function *proriol*, not listed in the text, evaluates the Proriol polynomials based on the FSELIB function *jacobi*, not listed in the text. Graphs of two quadratic polynomials generated by the FSELIB function *proriol_graph*, not listed in the text, are shown in Figure 5.7.2.

(a)   (b)

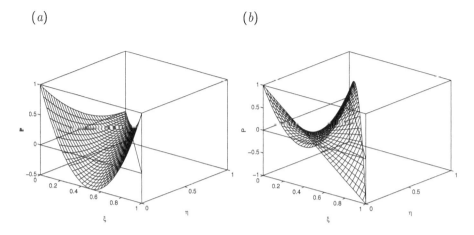

FIGURE 5.7.2   Graphs of the first two quadratic Proriol polynomials, (a) $\mathcal{P}_{20}$ and (b) $\mathcal{P}_{11}$. The third quadratic polynomial, $\mathcal{P}_{02}$, is only a function of η.

5.7.1   Orthogonality

Using the integration formula (4.2.45), we find that the integral of the product of the ij and kl Proriol polynomials over the area of the standard triangle is given by

$$\mathcal{B}_{ij,kl} \equiv \iint \mathcal{P}_{ij}\,\mathcal{P}_{kl}\,\mathrm{d}\xi\,\mathrm{d}\eta = \frac{1}{8}\int_{-1}^{1}\int_{-1}^{1}\mathcal{P}_{ij}\,\mathcal{P}_{kl}\,(1-\eta')\,\mathrm{d}\xi'\,\mathrm{d}\eta' \qquad (5.7.9)$$

or

$$\mathcal{B}_{ij,kl} = \frac{1}{4}\int_{-1}^{1}\int_{-1}^{1}\mathrm{L}_i(\xi')\,\mathrm{L}_k(\xi')\left(\frac{1-\eta'}{2}\right)^{i+k+1}$$
$$\times \mathcal{J}_j^{(2i+1,0)}(\eta')\,\mathcal{J}_l^{(2k+1,0)}(\eta')\,\mathrm{d}\xi'\,\mathrm{d}\eta', \qquad (5.7.10)$$

which can be recast into the form

$$\mathcal{B}_{ij,kl} = \frac{1}{4}\int_{-1}^{1}\left(\int_{-1}^{1}\mathrm{L}_i(\xi')\,\mathrm{L}_k(\xi')\,\mathrm{d}\xi'\right) \qquad (5.7.11)$$
$$\times \left(\frac{1-\eta'}{2}\right)^{i+k+1}\mathcal{J}_j^{(2i+1,0)}(\eta')\,\mathcal{J}_l^{(2k+1,0)}(\eta')\,\mathrm{d}\eta'.$$

Using the orthogonality properties of the Legendre and Jacobi polynomials discussed in Appendix B, we find that $\mathcal{B}_{ij,kl} = 0$ if $i \neq k$ or $j \neq l$.

The self-projection integral is given by

$$\mathcal{G}_{ij} \equiv \mathcal{B}_{ij,ij} = \frac{1}{(2i+1)(2i+2j+2)}. \qquad (5.7.12)$$

For example, $\mathcal{G}_{00} = 1/2$, which is equal to the area of the standard triangle in the $\xi\eta$ plane. FSELIB scripts *proriol_ortho* and *proriol_ortho1*, not listed in the text, confirm the orthogonality of the Proriol polynomials by direct integration using a Gaussian quadrature or integration rule.

5.7.2 Orthogonal expansion

Proriol's polynomials provide us with a complete orthogonal basis over a triangle. Any function defined over the standard triangle in the $\xi\eta$ plane, $f(\xi, \eta)$, can be approximated with a complete mth-degree polynomial in ξ and η, expressed in the form

$$f(\xi, \eta) \simeq \sum_{k=0}^{m} \left(\sum_{l=0}^{m-k} a_{kl}\, \mathcal{P}_{kl}(\xi, \eta) \right). \tag{5.7.13}$$

Multiplying (5.7.13) by $\mathcal{P}_{ij}$, integrating the product over the surface of the triangle, and using the orthogonality property, we derive the expansion coefficients

$$a_{kl} = \frac{1}{\mathcal{G}_{kl}} \iint f(\xi, \eta)\, \mathcal{P}_{kl}(\xi, \eta)\, \mathrm{d}\xi\, \mathrm{d}\eta, \tag{5.7.14}$$

where $\mathcal{G}_{kl}$ is defined in (5.7.12).

To convert a monomial product series into an equivalent Proriol series, we may use the expressions

$$1 = \mathcal{P}_{00},$$
$$\xi = \frac{1}{6} \left(3\mathcal{P}_{10} - \mathcal{P}_{01} + 2\,\mathcal{P}_{00} \right),$$
$$\eta = \frac{1}{3} \left(\mathcal{P}_{01} + \mathcal{P}_{00} \right),$$
$$\xi^2 = \frac{1}{30} \left(5\mathcal{P}_{20} - 3\mathcal{P}_{11} + 3\mathcal{P}_{02} - 4\mathcal{P}_{01} + 12\mathcal{P}_{10} + c\mathcal{P}_{00} \right), \tag{5.7.15}$$
$$\xi\eta = \frac{1}{60} \left(3\mathcal{P}_{02} + 6\mathcal{P}_{11} + 6\mathcal{P}_{10} + 2\mathcal{P}_{01} + 5\,\mathcal{P}_{00} \right),$$
$$\eta^2 = \frac{1}{30} \left(3\mathcal{P}_{02} + 8\mathcal{P}_{01} + 5\,\mathcal{P}_{00} \right).$$

Expressions for higher-order monomials can be obtained by symbolic algebraic manipulation.

PROBLEMS

5.7.1 *Orthogonality of the Proriol polynomials*

(*a*) Confirm the orthogonality of the first three Proriol polynomials shown in (5.7.5) and (5.7.6).

(*b*) Confirm formula (5.7.12) for the first three Proriol polynomials shown in (5.7.5) and (5.7.6).

5.7.2 *Orthogonal expansion*

(*a*) Expand each of the three linear interpolation functions shown in (4.2.7) in a sum of Proriol polynomials.

(*b*) Repeat (*a*) for the quadratic functions shown in (5.1.7).

5.8   High-order node distributions

In Section 5.5, we discussed the computation of element node interpolation functions but evaded the important issue of how the nodes should be distributed over the area of the triangle in the physical xy or parametric $\xi\eta$ plane. Two desirable features of a node distribution are: (*a*) the corresponding global matrices should be well-conditioned, and (*b*) the accuracy of the interpolation should be as high as possible for a given polynomial degree, m.

With these considerations in mind, we proceed to establish guidelines for determining node distributions over a triangle working in two stages. First, we develop a general policy of node distribution that ensures that the number of available interpolation nodes is exactly equal to that required by the complete mth-degree polynomial expansion in ξ and η. Second, we optimize the node distribution guided by the theory of orthogonal polynomials discussed in Appendix B and polynomial interpolation discussed in Appendix D.

5.8.1   *Node distribution based on a one-dimensional master grid*

One way of ensuring that the number of interpolation nodes, N, is equal to the number of coefficients in the complete mth-order expansion,

$$N = \frac{1}{2}(m+1)(m+2),\qquad(5.8.1)$$

is to distribute the nodes as shown in Figure 5.8.1.

To generate these distributions, we introduce a one-dimensional *master grid* defined by a set of $m+1$ points, v_i for $i = 1,\ldots,m+1$, subject to the end-node constraints

$$v_1 = 0,\qquad v_{m+1} = 1.\qquad(5.8.2)$$

If the master grid is symmetric with respect to the mid-point, $v = 0.5$, we have

$$v_{m+2-j} = 1 - v_j\qquad(5.8.3)$$

for $j = 1,\ldots,m+1$. Nodes are then distributed along the ξ and η axes at positions

$$\xi_i = v_i,\qquad \eta_j = 1 - v_{m+2-j}\qquad(5.8.4)$$

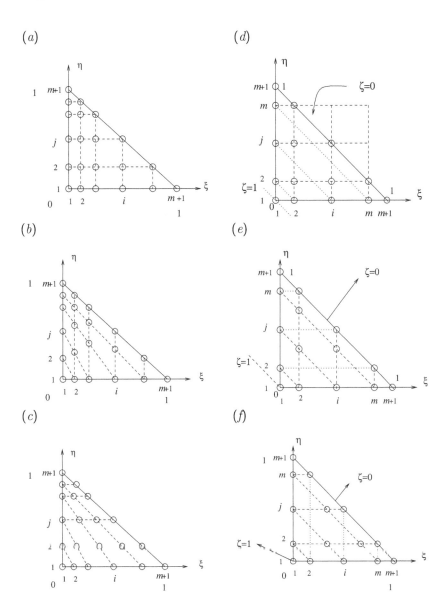

FIGURE 5.8.1   (*a–c*) Three possible distributions of interpolation nodes corresponding to a complete *m*th-degree polynomial expansion over the standard triangle,
generated by an arbitrary one-dimensional master grid. In the illustrations shown,
$m = 5$ and $N = 21$. (*d–f*) Distribution of interpolation nodes corresponding to a
complete *m*th-degree polynomial expansion over the standard triangle generated
by a symmetric one-dimensional master grid. In the illustrations shown, $m = 4$
and $N = 15$.

for $i, j = 1, \ldots, m + 1$. Nodes along the hypotenuse of the triangle are identified by moving either vertically upward from the nodes along the ξ axis, or horizontally to the right from the nodes along the η axis.

The drawings in Figure 5.8.1(a–c) illustrate three possible ways of locating the interior nodes for an arbitrary one-dimensional master grid. The interior nodes in Figure 5.8.1(a) are located at intersections of vertical and horizontal grid lines, yielding pairs (ξ_i, η_j), where $i = 1, \ldots, m + 1$. For each value of the index i, the index j takes values in the range $j = 1, \ldots, m + 2 - i$. The interior nodes in Figure 5.8.1(b) are located at intersections of vertical and diagonal lines, while the interior nodes in Figure 5.8.1(c) are located at intersections of horizontal and diagonal lines. The drawings in Figure 5.8.1(d–f) correspond to those shown in Figure 5.8.1(a–c) for a symmetric master grid.

When $m = 1$, we obtain a three-node triangle with 3 vertex nodes, as discussed in Section 4.2. When $m = 2$, we obtain a six-node triangle with three vertex nodes and three edge nodes, as discussed in Section 5.1. When $m = 3$, we obtain a 10-node triangle with three vertex nodes, six edge nodes, and one interior node. When $m = 9$, we obtain a 55-node triangle with 3 vertex nodes, 24 edge nodes, and 28 interior nodes.

The nodes can be labeled sequentially moving horizontally along the ξ axis and then vertically or diagonally, or *vice versa*, within the confines of the triangle. In all cases, to enforce C^0 continuity of an interpolated function over the entire solution domain consisting of the union of the elements, we require that the nodal values of the function at the vertex and edge nodes are shared by neighboring elements at corresponding positions.

5.8.2 Uniform grid

In the case of a uniform one-dimensional master grid corresponding to evenly spaced vertical and horizontal lines,

$$v_i = \frac{i - 1}{m} \tag{5.8.5}$$

for $i = 1, \ldots, m + 1$, the three node distributions shown in Figure 5.8.1 coincide and the interior nodes lie along diagonal lines, drawn as dotted lines in Figure 5.8.2(a).

The diagonal lines correspond to constant values of the third barycentric coordinate $\zeta = 1 - \xi - \eta$, ranging from $\zeta = 0$ along the hypotenuse to $\zeta = 1$ at the first vertex node labeled 1. Each diagonal line is identified by the index

$$k = m + 3 - i - j, \tag{5.8.6}$$

which increases from the value of 1 along the hypotenuse where $i + j = m + 2$, to the value of $m + 1$ at the origin where $i + j = 2$.

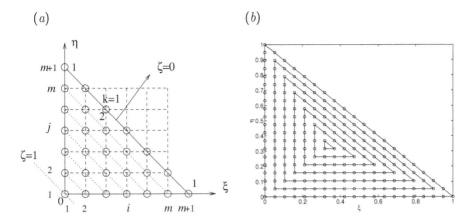

FIGURE 5.8.2   (*a*) Illustration of a uniform grid over the triangle. (*b*) The interpolation
nodes fall on a nested sequence of concentric triangles.

Bos [5] observed that the nodes are distributed around the perimeters of

$$n_T = \left[\frac{m-1}{3}\right] + 1 \tag{5.8.7}$$

nested triangles, where the square brackets denote the integral part of the enclosed
fraction. An example for $m = 19$ is shown in Figure 5.8.2(*b*). The nested triangles
in this illustration were generated by the FSELIB script *bos*, not listed in the text.

Node interpolation functions

The cardinal node interpolation functions for the uniform grid can be constructed
using formula (5.5.19). The validity of this formula hinges on the observation that
for each node, all other nodes lie along preceding horizontal lines, preceding vertical
lines, or subsequent diagonal lines. Specifically, the interpolation function corresponding to the (i, j) node can be expressed as the product of three polynomials,

$$\psi_{ij}(\xi, \eta) = \Xi_i^{(i-1)}(\xi) \cdot H_j^{(j-1)}(\eta) \cdot Z_k^{(k-1)}(\zeta), \tag{5.8.8}$$

where $\Xi_i^{(i-1)}(\xi)$ is an $(i - 1)$-degree polynomial, defined such that

$$\Xi_1^0(\xi) = 1, \qquad \Xi_i^{(i-1)}(\xi) = \frac{(\xi - v_1)(\xi - v_2) \cdots (\xi - v_{i-2})(\xi - v_{i-1})}{(v_i - v_1)(v_i - v_2) \cdots (v_i - v_{i-2})(v_i - v_{i-1})} \tag{5.8.9}$$

for $i = 2, \ldots, m + 1$, $H_j^{(j-1)}(\eta)$ is a $(j - 1)$-degree polynomial, defined such that

$$H_1^0(\eta) = 1, \qquad H_j^{(j-1)}(\eta) = \frac{(\eta - v_1)(\eta - v_2) \cdots (\eta - v_{j-2})(\eta - v_{j-1})}{(v_j - v_1)(v_j - v_2) \cdots (v_j - v_{j-2})(v_j - v_{j-1})} \tag{5.8.10}$$

for $j = 2, \ldots, m+1$, and $Z_k^{(j-1)}(\zeta)$ is a $(k-1)$-degree polynomial, defined such that

$$Z_1^0(\zeta) = 1, \qquad Z_k^{(k-1)}(\zeta) = \frac{(\zeta - v_1)(\zeta - v_2)\ldots(\zeta - v_{k-2})(\zeta - v_{k-1})}{(v_k - v_1)(v_k - v_2)\ldots(v_k - v_{k-2})(v_k - v_{k-1})}$$

(5.8.11)

for $k = 2, \ldots, m + 1$. It is a straightforward exercise to verify that $\psi_{ij}(\xi, \eta)$ is a polynomial of degree

$$(i - 1) + (j - 1) + (k - 1) = i + j + k - 3 = m \tag{5.8.12}$$

with respect to ξ and η, as required by (5.8.6), and also confirm that all cardinal interpolation conditions are met.

Quadratic expansion

As an example, we consider the familiar six-node triangle with straight edges, corresponding to $m = 2$, $v_1 = 0$, $v_2 = 0.5$, and $v_3 = 1$. According to (5.8.8), the interpolation function of the first vertex node corresponding to $i = 1$, $j = 1$, $k = 3$, is

$$\psi_{11}(\xi, \eta) = \Xi_1^{(0)}(\xi) \cdot H_1^{(0)}(\eta) \cdot Z_3^{(2)}(\zeta), \tag{5.8.13}$$

yielding

$$\psi_{11}(\xi, \eta) = 1 \cdot 1 \cdot \frac{(\zeta - v_1)(\zeta - v_2)}{(v_3 - v_1)(v_3 - v_2)} = \zeta\,(2\zeta - 1), \tag{5.8.14}$$

which is consistent with the expression for ψ_1 given in (5.1.7). Working in a similar fashion with the mid-edge node $i = 2$, $j = 1$, $k = 2$, we obtain

$$\psi_{21}(\xi, \eta) = \Xi_2^{(1)}(\xi) \cdot H_1^{(0)}(\eta) \cdot Z_2^{(1)}(\zeta), \tag{5.8.15}$$

yielding

$$\psi_{21}(\xi, \eta) = \frac{\xi - v_1}{v_2 - v_1} \cdot 1 \cdot \frac{\zeta - v_1}{v_2 - v_1} = 4\,\xi\,\zeta, \tag{5.8.16}$$

which is consistent with the expression for ψ_4 given in (5.1.7).

Standard equilateral triangle

The standard right triangle in the $\xi\eta$ plane can be mapped to an equilateral triangle in the $\hat{\xi}\hat{\eta}$ plane using the transformation rules

$$\hat{\xi} = \xi + \frac{1}{2}\,\eta, \qquad \hat{\eta} = \frac{\sqrt{3}}{2}\,\eta. \tag{5.8.17}$$

The inverse transformation rules are

$$\xi = \hat{\xi} - \frac{1}{\sqrt{3}}\,\hat{\eta}, \qquad \eta = \frac{2}{\sqrt{3}}\,\hat{\eta}. \tag{5.8.18}$$

(a)   (b)

(c)   (d)

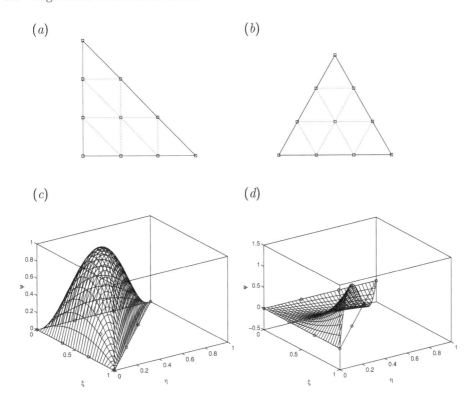

FIGURE 5.8.3   (a, b) Uniform node distributions generated by the FSELIB script *nodes*
for $m = 3$ (a) on the standard right triangle in the $\xi\eta$ plane and (b) on the standard equilateral triangle in the $\hat{\xi}\hat{\eta}$ plane. (c, d) Cardinal interpolation functions
corresponding to (c) the interior (bubble) mode $i = 2$ and $j = 2$, and (b) the
edge node $i = 3$ and $j = 2$.

Node distributions in the $\xi\eta$ plane and the corresponding distributions in the $\hat{\xi}\hat{\eta}$
plane are shown in Figure 5.8.3(a, b) for polynomial expansion degree $m = 3$. The
distributions shown were generated by the FSELIB script *nodes*, not listed in the
text.

Bubble and edge modes

Graphs of node interpolation functions generated by the FSELIB function *psi_uni*,
not listed in the text, are shown in Figure 5.8.3(c, d). The interpolation function
corresponding to the interior node, $i = 2$ and $j = 2$ is shown in Figure 5.8.3(c).
For obvious reasons, this function is described as a *bubble mode*. The interpolation
function corresponding to the edge node, $i = 3$ and $j = 2$ is shown in Figure 5.8.3(d).
Note that this function is zero along the entire length of the two perpendicular edges
described by $\xi = 0$ and $\eta = 0$.

Accuracy and convergence

Unfortunately, as the expansion order m increases, the accuracy of the interpolation does not necessarily improve uniformly due to the two-dimensional version of the Runge effect discussed in Section 3.1.2, manifested by growing oscillations near the triangle edges. Numerical computations show that the Lebesgue constant increases rapidly with the polynomial order, m, or size of the nodal set, N (Bos [5]). Thus, the uniform node distribution is recommended only for low-order polynomial expansions, typically $m \leq 3$.

To circumvent this difficulty, Bos [5] modified the uniform node distribution by adjusting the sizes of the individual nested triangles illustrated in Figure 5.8.2(b), while maintaining the uniformity of the point distribution around the triangle perimeters. Although the optimized distributions considerably improve the accuracy and convergence properties of the interpolation, they do not provide us with an optimal set.

5.8.3 Lobatto grid on the triangle

In Section 3.3, we saw that the Lobatto nodal set is optimal for one-dimensional interpolation, subject to the constraint that one interpolation node is placed at each element end-node. Motivated by this discovery, we employ a one-dimensional master grid with $v_1 = 0$, $v_{m+1} = 1$, and interior nodes, v_i for $i = 2, \ldots, m$, positioned at the scaled zeros of the $(m-1)$-degree Lobatto polynomial. The resulting Lobatto master grid is defined as

$$v_1 = 0, \quad v_2 = \frac{1}{2}(1 + t_1), \quad v_3 = \frac{1}{2}(1 + t_2), \quad \ldots,$$
$$v_m = \frac{1}{2}(1 + t_{m-1}), \quad v_{m+1} = 1, \tag{5.8.19}$$

where t_i for $i = 1, \ldots, m-1$ are the zeros of the $(m-1)$-degree Lobatto polynomial, $\mathrm{Lo}_{m-1}(t)$, distributed in the interval $(-1, 1)$.

Asymmetric distribution

Node distributions can be generated by the intersection of horizontal and vertical grid lines, as shown in Figure 5.8.1(d). The corresponding distribution for $m = 7$, generated by the FSELIB script *nodes*, not listed in the text, is shown in Figure 5.8.4(a). The nodes are identified by the coordinates

$$(\xi_i, \eta_j) = (v_i, v_j) \tag{5.8.20}$$

for $i = 1, \ldots, m+1$ and $j = 1, \ldots, m+2-i$. The corresponding node distribution on the associated equilateral triangle in the $\hat{\xi}\hat{\eta}$ plane is shown in Figure 5.8.4(b).

The asymmetry of the distribution with respect to the three vertices is due to the choice of horizontal and vertical pairs of lines for identifying the interior

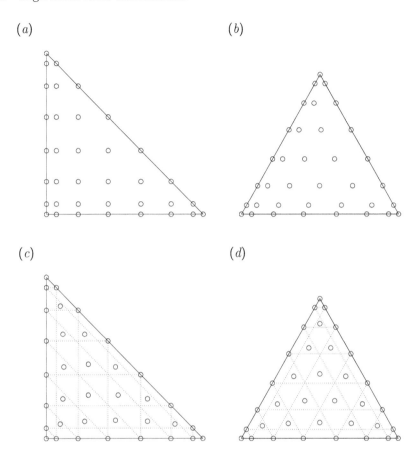

FIGURE 5.8.4   (a) Node distribution for $m = 7$ generated by the intersection of horizontal and vertical Lobatto grid lines over the right triangle in the $\xi\eta$ plane; (b)
node distribution on the corresponding equilateral triangle in the $\hat{\xi}\hat{\eta}$ plane showing the lack of rotational symmetry. (c) Improved distributions with rotational
symmetry over the right triangle and (d) corresponding equilateral triangle.

nodes. Alternatives are pairs of horizontal and diagonal lines, and pairs of vertical
and diagonal lines, as shown in Figure 5.8.1(e, f). Because the three vertices in
the physical xy plane have been labeled arbitrarily, 1, 2, and 3, subject to the
counterclockwise convention, the lack of threefold symmetry is unsatisfactory.

Symmetric distribution

To eliminate the asymmetry, we compute all three node distributions shown Figure
5.8.1(d–f), and then average the coordinates of the interior nodes over the three
realizations. The result is an improved node distribution that respects the desirable

(a)   (b)

(c)   (d)

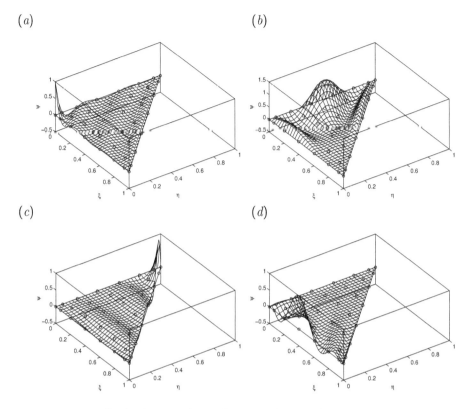

FIGURE 5.8.5 Cardinal interpolation functions for sample nodal sets corresponding to the symmetric node distribution shown in Figure 5.8.4(c, d).

threefold symmetry. The nodal coordinates are given by

$$\xi_{ij} = \frac{1}{3}\left(1 + 2\,v_i - v_j - v_k\right), \qquad \eta_{ij} = \frac{1}{3}\left(1 - v_i + 2\,v_j - v_k\right), \qquad (5.8.21)$$

where

$$k = m + 3 - i - j \qquad (5.8.22)$$

for $i = 1, \ldots, m+1$ and $j = 1, \ldots, m+2-i$.

The symmetric node distribution for $m = 7$, generated by the FSELIB script *nodes*, is shown in Figure 5.8.4(c, d) over the right isosceles triangle and corresponding equilateral triangle. Drawing a mesh of dotted lines connecting the edge nodes on the equilateral triangle reveals that the interior nodes are situated at the centroids of smaller inner triangles surrounding them.

An assortment of node interpolation functions generated by the FSELIB function *psi_lob*, not listed in the text, is shown in Figure 5.8.5. The interpolation functions were computed using the Vandermonde matrix method discussed in Section 5.3.1, where the basis functions, ϕ_i, are identified with the Proriol polynomials.

Numerical investigation has shown that the Lobatto triangle distribution with threefold symmetry described in this section enjoys interpolation convergence properties that compare favorably with those of the best-known distributions comprised of the Fekete set (Blyth & Pozrikidis [4]).

5.8.4 The Fekete set

An alternative method of distributing the interpolation nodes over a triangle is based on the notion of the Fekete points (Bos [5]). To introduce these points, we apply the polynomial expansion (5.5.3) at $N = \frac{1}{2}(m+1)(m+2)$ nodes over the area of the triangle, (ξ_i, η_i), and consider the $N \times N$ generalized Vandermonde matrix defined in (5.5.23),

$$\mathbf{V}(\xi_1, \eta_1, \quad \xi_2, \eta_2, \quad \ldots, \quad \xi_N, \eta_N), \tag{5.8.23}$$

regarded as a function of all nodal positions. By definition, the Fekete set maximizes the magnitude of the determinant of this matrix within the confines of the triangle. Since the determinant of a matrix is multiplied by a constant factor when a multiple of a row is added to another row to alter the polynomial base, the Fekete set is independent of the working polynomial base.

Invoking the expression of the node interpolation functions shown in (5.5.34), we find that, by construction of the Fekete points,

$$|\psi_i(\xi, \eta)| \leq 1. \tag{5.8.24}$$

The equality holds only when $\xi = \xi_i$ and $\eta = \eta_i$. This desirable property ensures rapid interpolation convergence with respect to polynomial order, as discussed in Appendix D.

The Fekete set over the triangle includes three vertex nodes and a group of $m - 1$ nodes distributed along each edge at positions corresponding to the scaled zeros of the $(m-1)$-degree Lobatto polynomial. The edge nodes are optimal for one-dimensional interpolation over a finite interval with fixed end-nodes, as discussed in Section D.5, Appendix D. The remaining nodes are deployed inside the triangular element at non-obvious positions.

As an example, when $m = 3$, the Fekete set includes ten nodes, including three vertex nodes, two nodes along each edge corresponding to the zeros of the Lo_2 Lobatto polynomial, and one interior node located at the element centroid. The Fekete point coordinates and associated Lebesgue constants for higher-order expansions are available in tabular form (Chen & Babuška [8], Taylor, Wingate, & Vincent [67]). The node coordinates are generated by the FSELIB function *fekete*, not listed in the text.

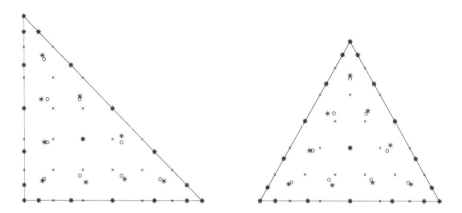

FIGURE 5.8.6   Comparison of the Fekete nodes (asterisks) with the symmetric Lobatto
triangle nodes (circles) and with the uniform nodes ($\times$) for polynomial order
$m = 6$ over the right and equilateral triangle.

The Fekete distribution, the uniform distribution, and the symmetric Lobatto
distribution discussed in Section 5.8.3 are compared in Figure 5.8.6. The graphs in
this figure were generated by the FSELIB script *node_fekete*, not listed in the text.
It is interesting that the Lobatto interior nodes are displaced only slightly inward
with respect to the corresponding Fekete nodes.

5.8.5   *Further nodal distributions*

Chen and Babuška [8] computed node distributions by minimizing the $\mathcal{L}_2$ norm

$$\mathcal{L}_2 \equiv \Big(\iint \sum_{i=1}^{N} |\psi_i(\xi, \eta)|^2 \, \mathrm{d}\xi \, \mathrm{d}\eta \Big)^{1/2}, \qquad (5.8.25)$$

and found that the properties of the resulting sets compare favorably with those of
the Fekete set. Note that the Lebesgue function defined in (5.5.39) is the $\mathcal{L}_\infty$ norm.
Unlike the Fekete sets, the sets corresponding to the L_2 norm do not include edge
nodes corresponding to the zeros of Lobatto polynomials.

Hesthaven [28] performed a similar calculation by minimizing a properly defined
energy function that depends on the relative position of all pairs of nodes. Like
the Fekete sets, the energy sets include edge nodes corresponding to the zeros of
the Lobatto polynomials. Other point distributions are computed and discussed
critically by Briani, Sommariva, and Vianello [7]. In the absence of a rigorous theory
of two-dimensional polynomial interpolation, the relative merits of a candidate set
computed based on different objective functions can be assessed only by numerical
experimentation.

PROBLEMS

5.8.1 *Quadratic expansion*

Verify that formula (5.8.8) reproduces the quadratic element node interpolation functions given in (5.1.7).

5.8.2 *Cardinal functions with uniform node distribution*

Run a properly modified version of the FSELIB code *psi_uni* to generate graphs of the three vertex node interpolation functions corresponding to a uniform node distribution for $m = 4$.

5.9 Modal expansions in a triangle

A function of interest over the standard triangle in the $\xi\eta$ plane, $f(\xi, \eta)$, can be approximated with an mth-degree polynomial, according to Dubiner's modal expansion

$$f(\xi, \eta) \simeq F_v + F_e + F_i, \tag{5.9.1}$$

where

$$F_v(\xi, \eta) = f_1 \, \zeta_1^v(\xi, \eta) + f_2 \, \zeta_2^v(\xi, \eta) + f_3 \, \zeta_3^v(\xi, \eta) \tag{5.9.2}$$

is the *vertex part*,

$$F_e(\xi, \eta) = \sum_{i=1}^{m-1} c_i^{12} \zeta_i^{12}(\xi, \eta) + \sum_{i=1}^{m-1} c_i^{13} \zeta_i^{13}(\xi, \eta) + \sum_{i=1}^{m-1} c_i^{23} \zeta_i^{23}(\xi, \eta) \tag{5.9.3}$$

is the *edge part*, arising when $m \geq 2$, and

$$F_i(\xi, \eta) = \sum_{i=1}^{m-2} \left(\sum_{j=1}^{m-i-1} c_{ij} \, \zeta_{ij}(\xi, \eta) \right) \tag{5.9.4}$$

is the *interior part*, arising when $m \geq 3$ (Dubiner [19]). Each part consists of a linear combination of modes multiplied by respective coefficients, sometimes also called *degrees of freedom*.

The modal expansion includes $N_v = 3$ vertex modes, $N_e = 3\,(m-1)$ edge modes, and

$$N_i = (m-2) + (m-1) + \cdots + 1 = \frac{1}{2}\,(m-1)(m-2) \tag{5.9.5}$$

interior modes. The total number of modes is

$$N = N_v + N_e + N_i = 3 + 3\,(m-1) + \frac{1}{2}\,(m-1)(m-2), \tag{5.9.6}$$

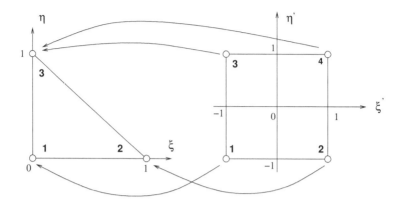

FIGURE 5.9.1   Mapping of the standard triangle to the standard square using Duffy's transformation. The bold-faced numbers are the element node labels.

adding up to

$$N = \frac{1}{2}(m+1)(m+2), \tag{5.9.7}$$

which is precisely equal to the number of terms in the complete mth-degree polynomial expansion in ξ and η.

The modes are designed so that the modal expansion expressed by (5.9.1) is a complete mth-degree polynomial in ξ and η. For example, when $m = 1$, we obtain the familiar three-term linear expansion involving only the vertex modes. When $m = 2$, we obtain the familiar six-term quadratic expansion involving three vertex modes and three edge modes. Interior modes arise for higher-order expansions.

Mapping the triangle to the standard square

To develop expressions for the three families of modes, we map the standard triangle in the $\xi\eta$ plane to the standard square in the $\xi'\eta'$ parametric plane, as shown in Figure 5.9.1, where $-1 \leq \xi' \leq 1$ and $-1 \leq \eta' \leq 1$. The barycentric coordinates, ξ and η, are related to the Cartesian coordinates, ξ' and η', and *vice versa*, by the transformation rules (4.2.39), (4.2.40), and (4.2.42), repeated below for convenience,

$$\xi = \frac{1+\xi'}{2}\frac{1-\eta'}{2}, \qquad \eta = \frac{1+\eta'}{2}, \tag{5.9.8}$$

yielding

$$\zeta \equiv 1 - \xi - \eta = \frac{1-\xi'}{2}\frac{1-\eta'}{2}, \qquad 1 - \eta = 1 - \frac{1+\eta'}{2} = \frac{1-\eta'}{2}. \tag{5.9.9}$$

The inverse transformations are

$$\xi' = \frac{2\xi}{1-\eta} - 1, \qquad \eta' = 2\eta - 1. \tag{5.9.10}$$

Vertex modes

The coefficients f_1, f_2, and f_3 in (5.9.2) are the values of the function $f(\xi, \eta)$ at the three vertex nodes labeled 1, 2, and 3. To ensure C^0 continuity of the finite element expansion, these values must be shared by the neighboring elements sharing nodes. The corresponding cardinal interpolation functions, ζ_1^v, ζ_2^v, and ζ_3^v, are identical to those given in (4.2.7) for the linear triangle, given by

$$\zeta_1^v = \zeta = \frac{1-\xi'}{2}\frac{1-\eta'}{2}, \qquad \zeta_2^v = \xi = \frac{1+\xi'}{2}\frac{1-\eta'}{2}, \qquad \zeta_3^v = \eta = \frac{1+\eta'}{2}.$$
(5.9.11)

Edge modes

The coefficients c_i^{12}, c_i^{13}, and c_i^{23} in (5.9.3) are associated with three corresponding families of edge interpolation functions ζ_i^{12}, ζ_i^{13}, and ζ_i^{23}, where $i = 1, \ldots, m - 1$. To ensure C^0 continuity of the finite element expansion, these coefficients must be shared by the neighboring elements sharing edges. The interpolation functions ζ_i^{12} are zero along the two edges 13 and 23, corresponding to $\xi = 0$ and $\zeta = 0$. To satisfy this requirement, we write

$$\zeta_i^{12} = \xi\zeta\,\Psi_i^{12}(\xi', \eta'),$$
(5.9.12)

where $\Psi_i^{12}(\xi', \eta')$ is a new set of functions. Moreover, we set

$$\Psi_i^{12}(\xi', \eta') = \mathrm{Lo}_{i-1}(\xi')\,\Phi_i^{12}(\eta')$$
(5.9.13)

for $i = 1, \ldots, m - 1$, where $\mathrm{Lo}_{i-1}(\xi')$ is a Lobatto polynomial, and $\Phi_i^{12}(\eta')$ is a new set of functions. The choice of Lobatto polynomials is motivated by the partial diagonalization of the element diffusion matrix, as discussed in Section 3.7. In summary, the edge modes are given by

$$\zeta_i^{12} = \xi\zeta\,\mathrm{Lo}_{i-1}(\xi')\,\Phi_i^{12}(\eta').$$
(5.9.14)

Next, we require that ζ_i^{12} is a polynomial in ξ and η. Noting the denominator of the fraction in the transformation rule for ξ', shown in the first equation of (5.9.10), we set

$$\Phi_i^{12}(\eta') = (1 - \eta)^{i-1} = \left(\frac{1-\eta'}{2}\right)^{i-1},$$
(5.9.15)

and derive the *hierarchical*[3] modal expansion

$$\zeta_i^{12} = \xi\,\zeta\,\mathrm{Lo}_{i-1}(\xi')\left(\frac{1-\eta'}{2}\right)^{i-1}$$
(5.9.16)

or

$$\zeta_i^{12} = \frac{1-\xi'}{2}\frac{1+\xi'}{2}\left(\frac{1-\eta'}{2}\right)^{i+1}\mathrm{Lo}_{i-1}(\xi')$$
(5.9.17)

for $i = 1, \ldots, m - 1$.

---

[3]The terminology *hierarchical* implies that the members of this function set can be arranged consecutively with respect to the polynomial order.

Working in a similar fashion with the other two families of edge modes, we find that

$$\zeta_i^{13} = \eta \, \zeta \, \mathrm{Lo}_{i-1}(\eta') \tag{5.9.18}$$

or

$$\zeta_i^{13} = \frac{1+\eta'}{2} \frac{1-\xi'}{2} \frac{1-\eta'}{2} \, \mathrm{Lo}_{i-1}(\eta') \tag{5.9.19}$$

for $i = 1, \ldots, m-1$, and

$$\zeta_i^{23} = \xi \, \eta \, \mathrm{Lo}_{i-1}(\eta') \tag{5.9.20}$$

or

$$\zeta_i^{23} = \frac{1+\xi'}{2} \frac{1+\eta'}{2} \frac{1-\eta'}{2} \, \mathrm{Lo}_{i-1}(\eta') \tag{5.9.21}$$

for $i = 1, \ldots, m-1$.

For example, when $i = 1$, which is possible only when $m \geq 2$, we substitute $\mathrm{Lo}_0 = 1$, and obtain the familiar quadratic edge modes

$$\zeta_1^{12} = \xi\zeta, \qquad \zeta_1^{13} = \eta\zeta, \qquad \zeta_1^{23} = \xi\eta. \tag{5.9.22}$$

Graphs of the six edge modes for $m = 3$ generated by the FSELIB script *psi_modes*, not listed in the text, are shown in Figure 5.9.2.

Interior modes

The coefficients c_{ij} in (5.9.4) correspond to the interior or bubble modes described by the interpolation functions ζ_{ij}. To ensure that these functions are zero all around the perimeter of the triangle, we express them in the form

$$\zeta_{ij} = \xi \, \eta \, \zeta \, X_{ij}(\xi', \eta') = \frac{1+\xi'}{2} \frac{1-\xi'}{2} \left(\frac{1-\eta'}{2}\right)^2 \frac{1+\eta'}{2} \, X_{ij}(\xi', \eta'), \tag{5.9.23}$$

where $X_{ij}(\xi', \eta')$ is a new set of functions. Working as previously for the edge modes, we set

$$X_{ij}(\xi', \eta') = \mathrm{Lo}_{i-1}(\xi') \left(\frac{1-\eta'}{2}\right)^{i-1} \mathcal{J}_{j-1}^{(2i+1,1)}(\eta'), \tag{5.9.24}$$

where $\mathcal{J}_{j-1}^{(2i+1,1)}$ are the Jacobi polynomials discussed in Section B.8, Appendix B. Substituting this expression into (5.9.23), we obtain the final form

$$\zeta_{ij} = \xi \, \eta \, \zeta \, \mathrm{Lo}_{i-1}(\xi') \left(\frac{1-\eta'}{2}\right)^{i-1} \mathcal{J}_{j-1}^{(2i+1,1)}(\eta') \tag{5.9.25}$$

or

$$\zeta_{ij} = \frac{1+\xi'}{2} \frac{1-\xi'}{2} \frac{1+\eta'}{2} \left(\frac{1-\eta'}{2}\right)^{i+1} \mathrm{Lo}_{i-1}(\xi') \, \mathcal{J}_{j-1}^{(2i+1,1)}(\eta') \tag{5.9.26}$$

for $i = 1, \ldots, m-2$, $j = 1, \ldots, m-i-1$, and $m \geq 3$.

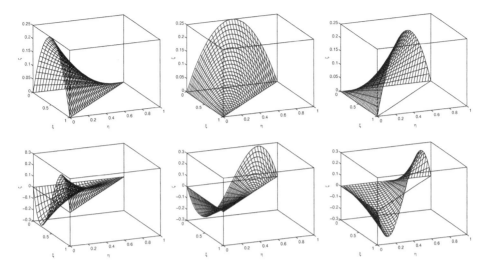

FIGURE 5.9.2 Edge modes generated by the FSELIB script *psi_modes* for $m = 3$. The functions ζ_1^{12}, ζ_1^{13}, and ζ_1^{23} are displayed in the first row, and the functions ζ_2^{12}, ζ_2^{13}, and ζ_2^{23} are displayed in the second row.

For example, when $m = 3$, we obtain a single bubble mode corresponding to $i = 1$ and $j = 1$, given by

$$\zeta_{11} = \xi \, \eta \, \zeta. \tag{5.9.27}$$

Note that this mode is identical to the interior-node interpolation function plotted in Figure 5.8.3(a) for $m = 3$, corresponding to $\xi = 1/3$ and $\eta = 1/3$, except that it is scaled by the factor $3^3 = 27$.

Partial orthogonality of the bubble modes

Applying the integration formula (4.2.45), we find that the integral of the product of the ij and kl interior modes over the triangle is given by the following element of the modal mass matrix:

$$\mathcal{B}_{ij,kl} = \frac{1}{8} \int_{-1}^{1} \int_{-1}^{1} \zeta_{ij} \, \zeta_{kl} \, (1 - \eta') \, \mathrm{d}\xi' \, \mathrm{d}\eta'. \tag{5.9.28}$$

Explicitly,

$$\mathcal{B}_{ij,kl} = \frac{1}{8} \int_{-1}^{1} \int_{-1}^{1} \Big(\frac{1 - \xi'^2}{4} \frac{1 + \eta'}{2} \Big)^2 \Big(\frac{1 - \eta'}{2} \Big)^{i+1} \Big(\frac{1 - \eta'}{2} \Big)^{k+1} \tag{5.9.29}$$

$$\times \mathrm{Lo}_{i-1}(\xi') \, \mathrm{Lo}_{k-1}(\xi') \, \mathcal{J}_{j-1}^{(2i+1,1)}(\eta') \, \mathcal{J}_{l-1}^{(2k+1,1)}(\eta') \, (1 - \eta') \, \mathrm{d}\xi' \, \mathrm{d}\eta',$$

which can be rearranged into

$$\mathcal{B}_{ij,kl} = \frac{1}{4} \int_{-1}^{1} \left(\frac{1-\xi'^2}{4}\right)^2 \mathrm{Lo}_{i-1}(\xi')\, \mathrm{Lo}_{k-1}(\xi')\, \mathrm{d}\xi' \tag{5.9.30}$$

$$\times \int_{-1}^{1} \left(\frac{1-\eta'}{2}\right)^{i+k+3} \left(\frac{1+\eta'}{2}\right)^2 \mathcal{J}_{j-1}^{(2i+1,1)}(\eta')\, \mathcal{J}_{l-1}^{(2k+1,1)}(\eta')\, \mathrm{d}\eta'.$$

The orthogonality of the Lobatto polynomials requires that

$$\int_{-1}^{1} (1-\xi'^2)\, \mathcal{P}_q(\xi')\, \mathrm{Lo}_r(\xi')\, \mathrm{d}\xi' = 0, \tag{5.9.31}$$

where $\mathcal{P}_q(\xi')$ is a q-degree polynomial with $q < r$. Consequently, the first integral on the right-hand side of (5.9.30) is zero when either $2+(i-1) < k-1$ or $2+(k-1) < i-1$, that is, when

$$i+2 < k < i-2, \tag{5.9.32}$$

which shows that the mass matrix has a finite bandwidth (Sherwin & Karniadakis [59]).

It is interesting to observe that, if the Lobatto polynomials, Lo_{i-1}, were replaced by the Jacobi polynomials, $\mathcal{J}_{i-1}^{(2,2)}$, discussed in Section B.8, Appendix B, the first integral on the right-hand side of (5.9.30) would be zero for $i \neq k$. For similar reasons, if $\mathcal{J}_{j-1}^{(2i+1,1)}$ were replaced by $\mathcal{J}_{j-1}^{(2i+3,2)}$, the second integral on the right-hand side of (5.9.30) would be zero for $i = k$. However, these modifications promote the coupling of the interior with the vertex and edge modes (Karniadakis & Sherwin [32], p. 91).

5.9.1 Implementation of the modal expansion

The Galerkin finite element method for the modal expansion is implemented as discussed previously in this chapter for the nodal expansion. The only new feature is that the element-node interpolation functions are replaced by the non-cardinal modal interpolation functions and the interior nodal values are replaced by the corresponding expansion coefficients, also called degrees of freedom. The numerical computation of the element mass and diffusion matrices is discussed in detail by Sherwin and Karniadakis [32, 59].

5.9.2 Properties of the modal expansion

Two important features of the modal expansion can be called to attention. First, the modal decomposition (5.9.1) is *not* entirely orthogonal, which means that the corresponding mass matrix is only *nearly* diagonal. The lack of complete orthogonality can be traced to the presence of vertex and edge modes, which must be included to ensure the C^0 continuity of the finite element expansion. In the absence of this constraint, it is beneficial to use a modal expansion in terms of the entirely orthogonal family of Proriol polynomials.

Second, the aforementioned partial orthogonality of the modal expansion applies only to the super-parametric interpolation over the three-node triangle in the physical xy plane. The reason is that this triangle has a constant surface metric coefficient, h_S, which can be extracted from the integrals defining the various element matrices in the $\xi\eta$ parametric plane. In contrast, because the surface metric coefficient of the six-node triangle generally depends on position, that is, it is a function of ξ and η, bubble-mode orthogonality with a constant weighting function discussed in this section does not produce a sparse mass matrix.

PROBLEMS

5.9.1 *Cubic modal expansion*

When $m = 3$, we obtain the 10-mode cubic expansion involving three vertex modes, six edge modes, and one bubble mode.

(a) Write the explicit form of the six edge modes as functions of ξ and η.

(b) Prepare a graph of the bubble mode.

5.9.2 *Vandermonde matrix*

Discuss the structure of the Vandermonde matrix, $V_{ij} = \zeta_i(\xi_j, \eta_j)$, for the Lobatto triangle and the Fekete sets discussed in Section 5.8.4, where the modal functions are used as a polynomial base, ζ_i, and the nodes are ordered in the following sequence: vertex nodes, edge nodes, and interior nodes. Explain why the determinant of the Vandermonde matrix is the product of three determinants defined with respect to the vertex nodes and modes, edge nodes and modes, and interior nodes and modes.

5.10   Surface elements

The methodology described in Section 4.9 for three-node triangular surface elements can be extended in a straightforward fashion to six-node triangular surface elements, as illustrated in Figure 5.10.1. The mapping from the physical space to the parametric $\xi\eta$ plane is described by the function (5.1.2), where the element interpolation functions are given in Table 5.1.2.

FSELIB script *edges6_3d*, not listed in the text, generates the edges of two adjacent triangles in terms of the coordinates of the vertex and edge nodes. A typical configuration is shown in Figure 5.10.2. The requisite mapping coefficients are computed by the FSELIB function *elm6_3d_abc*, not listed in the text. Note that the triangles join seamlessly along a common quadratic edge.

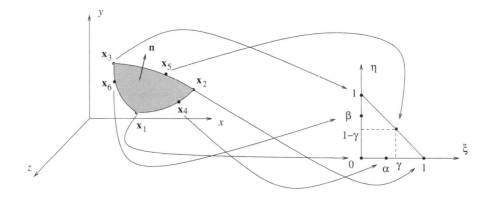

FIGURE 5.10.1   A curved six-node triangle in three-dimensional space is mapped to a flat right isosceles triangle in the $\xi\eta$ parametric plane.

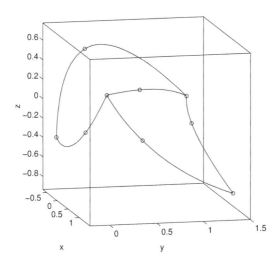

FIGURE 5.10.2   Reconstruction of two adjacent surface triangles by the FSELIB script *edges6_3d*.

5.10.1 Surface gradient

The element diffusion and mass matrices are given by the right-hand sides of (5.1.12) and (5.1.13). To compute the three-dimensional surface gradient of the element interpolation functions, ψ_i, we complement the counterparts of the equations in (5.1.15),

$$\sum_{j=1}^{6} \frac{\partial \psi_j}{\partial \xi} \, \mathbf{x}_j^{\mathrm{E}} \cdot \nabla_s \psi_i = \frac{\partial \psi_i}{\partial \xi}, \qquad \sum_{j=1}^{6} \frac{\partial \psi_j}{\partial \eta} \, \mathbf{x}_j^{\mathrm{E}} \cdot \nabla_s \psi_i = \frac{\partial \psi_i}{\partial \eta}, \qquad (5.10.1)$$

with the condition

$$\mathbf{n} \cdot \nabla_s \psi_i = 0, \qquad (5.10.2)$$

where $\mathbf{n} = (n_x, n_y, n_z)$ is the unit vector normal to the element. The resulting 3×3 linear system takes the form

$$\boldsymbol{\Gamma} \cdot \nabla_s \psi_i = \begin{bmatrix} \partial \psi_i / \partial \xi \\ \partial \psi_i / \partial \eta \\ 0 \end{bmatrix}, \qquad (5.10.3)$$

where

$$\boldsymbol{\Gamma} = \begin{bmatrix} \sum_{j=1}^{6} \frac{\partial \psi_j}{\partial \xi} x_j^{\mathrm{E}} & \sum_{j=1}^{6} \frac{\partial \psi_j}{\partial \xi} y_j^{\mathrm{E}} & \sum_{j=1}^{6} \frac{\partial \psi_j}{\partial \xi} z_j^{\mathrm{E}} \\ \sum_{j=1}^{6} \frac{\partial \psi_j}{\partial \eta} x_j^{\mathrm{E}} & \sum_{j=1}^{6} \frac{\partial \psi_j}{\partial \eta} y_j^{\mathrm{E}} & \sum_{j=1}^{6} \frac{\partial \psi_j}{\partial \eta} z_j^{\mathrm{E}} \\ n_x & n_y & n_z \end{bmatrix}. \qquad (5.10.4)$$

The solution of this system can be found readily by elementary analytical or numerical methods. The unit normal vector itself can be computed from the expression

$$\mathbf{n} = \frac{1}{h_S} \frac{\partial \mathbf{x}}{\partial \xi} \times \frac{\partial \mathbf{x}}{\partial \eta}, \qquad (5.10.5)$$

where h_S is the surface metric coefficient.

5.10.2   Grid generation

Surface grids of six-node triangles can be generated by successively subdividing a hard-coded root structure. FSELIB functions *trgl6_octa* and *trgl6_icos*, not listed in the text, discretize the surface of a sphere into six-node triangles descending from the subdivision of an octahedron icosahedron, as shown in Figure 5.10.3.

PROBLEM

5.10.1 *Adjacent triangles*

Add one mode adjacent triangle of your choice to the two-element assembly shown in Figure 5.10.2.

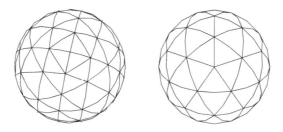

FIGURE 5.10.3   Discretization of the surface of a sphere into six-node triangles descending from the subdivision of an octahedron (left) or icosahedron (right).

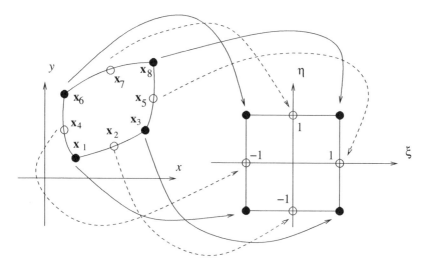

FIGURE 5.11.1   Mapping of an eight-node quadrilateral in the physical xy plane to the standard square in the parametric $\xi\eta$ plane.

5.11   High-order quadrilateral elements

It is sometimes desirable to use quadrilateral elements with four straight or curved edges, instead of their triangular counterparts. The finite and spectral element methodology for these elements is similar to that for the triangular elements discussed in Chapter 4 and previously in this chapter.

To describe a quadrilateral element, we map it from the physical xy plane to the standard square in the parametric $\xi\eta$ plane confined within $-1 \le \xi \le 1$ and $-1 \le \eta \le 1$, as shown in Figure 5.11.1. The precise form of the mapping depends on the selected number of element nodes and chosen polynomial expansion.

A complete mth-order polynomial expansion of a function over the standard square in the $\xi\eta$ plane, $f(\xi,\eta)$, has the flat-base triangular form (5.5.3), repeated below for convenience,

$$
\begin{aligned}
f(\xi,\eta) = \quad & a_{00} \\
& + a_{10}\,\xi + a_{01}\,\eta \\
& + a_{20}\,\xi^2 + a_{11}\,\xi\,\eta + a_{02}\,\eta^2 \\
& + a_{30}\,\xi^3 + a_{21}\,\xi^2\,\eta + a_{13}\,\xi\,\eta^2 + a_{03}\,\eta^3 \\
& \cdots \quad \cdots \quad \cdots \quad \cdots \quad \cdots \quad \cdots \quad \cdots \quad \cdots \quad \cdots \\
& + a_{m0}\,\xi^m + a_{m-1,1}\,\xi^{m-1}\,\eta + \cdots + a_{1,m-1}\,\xi\,\eta^{m-1} + a_{0m}\,\eta^m,
\end{aligned}
\tag{5.11.1}
$$

where a_{ij} are expansion coefficients. Quadrilateral elements with various numbers of interpolation nodes can accommodate different truncations of this triangular form, typically corresponding to incomplete polynomial expansions. One disadvantage of the incomplete expansion is that the solution is not necessarily invariant with respect to the rotation of the global Cartesian axes.

The simplest quadrilateral element is the four-node bilinear element discussed in Section 4.10. In the remainder of this section, we discuss selected higher-order and spectral elements described by a higher number of vertex, edge, and interior interpolation nodes.

5.11.1 *Eight-node serendipity elements*

An eight-node quadrilateral element is defined by four vertex nodes and four edge nodes, as shown in Figure 5.11.1. To describe the element in parametric form, we map it from the xy plane to the standard square in the $\xi\eta$ plane so that:

- The first node is mapped to the point $\xi = -1, \eta = -1$.
- The second node is mapped to the point $\xi = 0, \eta = -1$.
- The third node is mapped to the point $\xi = 1, \eta = -1$.
- The fourth node is mapped to the point $\xi = -1, \eta = 0$.
- The fifth node is mapped to the point $\xi = 1, \eta = 0$.
- The sixth node is mapped to the point $\xi = -1, \eta = 1$.
- The seventh node is mapped to the point $\xi = 0, \eta = 1$.
- The eighth node is mapped to the point $\xi = 1, \eta = 1$.

The mapping from the physical to the parametric space is mediated by the function

$$
\mathbf{x} = \sum_{i=1}^{8} \mathbf{x}_i^{E}\,\psi_i(\xi,\eta),
\tag{5.11.2}
$$

where $\mathbf{x}_i^{E}$ are the element node positions in the xy plane

Element node interpolation functions

The element-node cardinal interpolation functions, $\psi_i(\xi, \eta)$, are biquadratic of the serendipity class,[4] given by

$$
\begin{aligned}
\psi_i(\xi, \eta) \;=\; & (a + b\,\xi + c\,\xi^2) \\
& +(d + e\,\xi + f\,\xi^2)\,\eta \\
& +(g + h\,\xi)\,\eta^2,
\end{aligned}
\qquad (5.11.3)
$$

where a–h are expansion coefficients. Rearranging, we obtain an *incomplete* cubic expansion written in the deliberate form

$$
\begin{aligned}
\psi_i(\xi, \eta) = \quad & a_{00} \\
& +a_{10}\,\xi + a_{01}\,\eta, \\
& +a_{20}\,\xi^2 + a_{11}\,\xi\eta + a_{02}\,\eta^2 \\
& +a_{21}\,\xi^2\eta + a_{12}\,\xi\eta^2,
\end{aligned}
\qquad (5.11.4)
$$

where the new coefficients, a_{ij}, are reincarnations of the previous coefficients a–h. The following observations are significant:

- For a given ξ, $\psi_i(\xi, \eta)$ is quadratic in η.

- For a given η, $\psi_i(\xi, \eta)$ is quadratic in ξ.

- Only terms $\xi^m \eta^n$ with $m \le 2$ and $n \le 2$ appear.

The expansion shown in (5.11.4) is a Pascal triangle with a chopped flat base.

To compute the polynomial coefficients of a node interpolation function, we require the cardinal interpolation property demanding that $\psi_i(\xi, \eta) = 1$ at the ith node, and $\psi_i(\xi, \eta) = 0$ at all other nodes. A detailed calculation yields the formulas shown in Table 5.11.1. Graphs of the interpolation functions ψ_1 and ψ_2 corresponding to the southwestern vertex node and adjacent edge node, generated by the FSELIB function *psi_q8*, not listed in the text, are shown in Figure 5.11.2.

In the isoparametric interpolation, a function of interest over the element is expressed in a form that is analogous to that shown in (5.11.2), as

$$
f(x, y) = \sum_{i=1}^{8} f_i^{\mathrm{E}}\,\psi_i(\xi, \eta).
\qquad (5.11.5)
$$

The element diffusion, mass, and advection matrices are computed as discussed in Section 4.10 for the four-node bilinear quadrilateral element.

---

[4]Named after Horace Walpole's novel *The Three Princes of Serendip.*

$$\psi_1 = -\frac{1}{4}\left(1-\xi\right)\left(1-\eta\right)\left(\xi+\eta+1\right)$$

$$\psi_2 = \frac{1}{2}\left(1-\xi^2\right)\left(1-\eta\right)$$

$$\psi_3 = \frac{1}{4}\left(1+\xi\right)\left(1-\eta\right)\left(\xi-\eta-1\right)$$

$$\psi_4 = \frac{1}{2}\left(1-\xi\right)\left(1-\eta^2\right)$$

$$\psi_5 = \frac{1}{2}\left(1+\xi\right)\left(1-\eta^2\right)$$

$$\psi_6 = \frac{1}{4}\left(1-\xi\right)\left(1+\eta\right)\left(\eta-\xi-1\right)$$

$$\psi_7 = \frac{1}{2}\left(1-\xi^2\right)\left(1+\eta\right)$$

$$\psi_8 = \frac{1}{4}\left(1+\xi\right)\left(1+\eta\right)\left(\xi+\eta-1\right)$$

TABLE 5.11.1 Element interpolation functions for the eight-node quadrilateral element depicted in Figure 5.11.1.

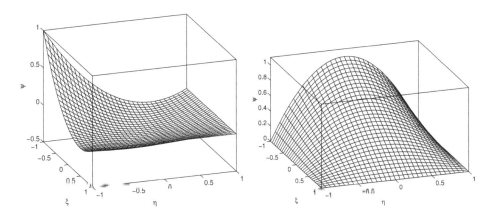

FIGURE 5.11.2 Graphs of the interpolation functions corresponding to a vertex node (left) and a mid-node (right) of an eight-node quadrilateral element.

5.11.2   12-node serendipity elements

The four-node element discussed in Section 4.10 and the eight-node element discussed in 5.11.1 belong to a broader class of incomplete serendipity expansions. The next member in this family is the quartic serendipity element whose node

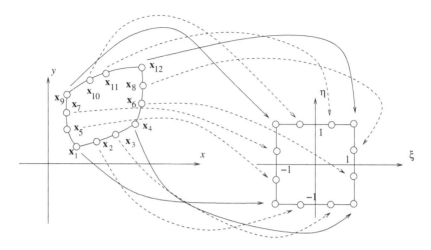

FIGURE 5.11.3   Mapping of a 12-node quadrilateral element in the xy plane to the standard square in the parametric $\xi\eta$ plane.

interpolation functions have the general form

$$
\begin{aligned}
\psi_i(\xi, \eta) = \quad & a_{00} \\
& + a_{10}\,\xi + a_{01}\,\eta, \\
& + a_{20}\,\xi^2 + a_{11}\,\xi\eta + a_{12}\,\eta^2 \\
& + a_{30}\,\xi^3 + a_{21}\,\xi^2\eta + a_{12}\,\xi\eta^2 + a_{03}\,\eta^3 \\
& + a_{31}\,\xi^3\eta \qquad\quad + a_{13}\,\xi\eta^3,
\end{aligned}
\tag{5.11.6}
$$

involving 12 coefficients. Four interpolation nodes are placed at the vertices and four pairs of interpolation nodes are placed along the four edges, as shown in Figure 5.11.3.

To compute the polynomial coefficients, we require that $\psi_i(\xi, \eta) = 1$ at the ith node and $\psi_i(\xi, \eta) = 0$ at the other 11 nodes for $i = 1, \ldots, 12$, and derive the expressions shown in Table 5.11.2. Graphs of the interpolation functions ψ_1 and ψ_4, corresponding to the southwestern vertex node and adjacent edge node, generated by the FSELIB function psi_q12, not listed in the text, are shown in Figure 5.11.4.

In the isoparametric interpolation, a function of interest over the element is expressed in the form

$$
f(x, y) = \sum_{i=1}^{12} f_i^{\mathrm{E}}\,\psi_i(\xi, \eta).
\tag{5.11.7}
$$

The element diffusion, mass, and advection matrices are computed as discussed in Section 4.10 for four-node quadrilateral elements.

$$\psi_1 = -\frac{1}{32}\,(1-\xi)\,(1-\eta)\,(10-9\,\xi^2-9\,\eta^2)$$

$$\psi_2 = \frac{9}{32}\,(1-\xi^2)\,(1-\eta)\,(1-3\,\xi)$$

$$\psi_3 = \frac{9}{32}\,(1-\xi^2)\,(1-\eta)\,(1+3\,\xi)$$

$$\psi_4 = -\frac{1}{32}\,(1+\xi)\,(1-\eta)\,(10-9\,\xi^2-9\,\eta^2)$$

$$\psi_5 = \frac{9}{32}\,(1-\xi)\,(1-\eta^2)\,(1-3\,\eta)$$

$$\psi_6 = \frac{9}{32}\,(1+\xi)\,(1-\eta^2)\,(1-3\,\eta)$$

$$\psi_7 = \frac{9}{32}\,(1-\xi)\,(1-\eta^2)\,(1+3\,\eta)$$

$$\psi_8 = \frac{9}{32}\,(1+\xi)\,(1-\eta^2)\,(1+3\,\eta)$$

$$\psi_9 = -\frac{1}{32}\,(1-\xi)\,(1+\eta)\,(10-9\,\xi^2-9\,\eta^2)$$

$$\psi_{10} = \frac{9}{32}\,(1-\xi^2)\,(1-3\,\xi)\,(1+\eta)$$

$$\psi_{11} = \frac{9}{32}\,(1-\xi^2)\,(1+3\,\xi)\,(1+\eta)$$

$$\psi_{12} = -\frac{1}{32}\,(1+\xi)\,(1+\eta)\,(10-9\,\xi^2-9\,\eta^2)$$

TABLE 5.11.2   Element interpolation functions for the 12-node quadrilateral element depicted in Figure 5.11.3.

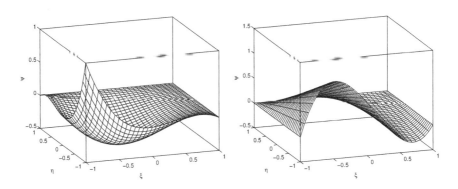

FIGURE 5.11.4   Graphs of the interpolation functions corresponding to a vertex node (left) and adjacent edge node (right) of a 12-node quadrilateral element.

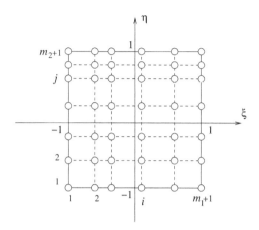

FIGURE 5.11.5   Distribution of interpolation nodes corresponding to a tensor-product expansion over the standard square.

5.11.3   Grid nodes via tensor-product expansions

In this class of quadrilateral elements, the interpolation nodes are distributed on a $(m_1 + 1) \times (m_2 + 1)$ Cartesian grid covering the standard square in the $\xi \eta$ plane, as shown in Figure 5.11.5. The nodes are identified by a coordinate doublet, (ξ_i, η_j), where $i = 1, \ldots, m_1 + 1$ and $j = 1, \ldots, m_2 + 1$. To ensure that shared nodes are present at the element vertices and along the edges, and thereby enforce C^0 continuity condition of the finite element expansion, we require that

$$\xi_1 = -1, \qquad \xi_{m_1+1} = 1, \qquad \eta_1 = -1, \qquad \eta_{m_2+1} = 1. \qquad (5.11.8)$$

The cardinal interpolation function of the (i, j) node is given by the tensor product

$$\psi_{ij}(\xi, \eta) = \mathcal{L}_i(\xi) \, \mathcal{M}_j(\eta), \qquad (5.11.9)$$

where $\mathcal{L}_i(\xi)$ is an m_1-degree Lagrange interpolating polynomial defined with respect to the ξ grid lines, given by

$$\mathcal{L}_i(\xi) = \frac{(\xi - \xi_1)(\xi - \xi_2) \cdots (\xi - \xi_{i-1})(\xi - \xi_{i+1}) \cdots (\xi - \xi_{m_1+1})}{(\xi_i - \xi_1)(\xi_i - \xi_2) \cdots (\xi_i - \xi_{i-1})(\xi_i - \xi_{i+1}) \cdots (\xi_i - \xi_{m_1+1})}, \qquad (5.11.10)$$

and $\mathcal{M}_j(\eta)$ is an m_2-degree Lagrange interpolating polynomial defined with respect to the η grid lines, given by

$$\mathcal{M}_j(\eta) = \frac{(\eta - \eta_1)(\eta - \eta_2) \cdots (\eta - \eta_{j-1})(\eta - \eta_{j+1}) \cdots (\eta - \eta_{m_2+1})}{(\eta_j - \eta_1)(\eta_j - \eta_2) \cdots (\eta_j - \eta_{j-1})(\eta_j - \eta_{j+1}) \cdots (\eta_j - \eta_{m_2+1})}. \qquad (5.11.11)$$

Substituting these expressions into (5.11.9), we obtain a polynomial expansion in (ξ, η) involving $(m_1 + 1) \times (m_2 + 1)$ coefficients.

As an example, when $m_1 = 3$ and $m_2 = 1$, the element interpolation functions have the polynomial form

$$
\begin{aligned}
\psi_{ij}(\xi, \eta) = \quad & a_{00} \\
+ & a_{10}\,\xi + a_{01}\,\eta \\
+ a_{20}\,\xi^2 + & a_{11}\,\xi\eta \\
+ a_{30}\,\xi^3 + a_{21}\,\xi^2\,\eta & \\
+ a_{31}\,\xi^3\,\eta, \quad &
\end{aligned}
\tag{5.11.12}
$$

involving eight coefficients.

When $m_1 = m_2$, the expansion takes the following symmetric diamond-like form:

which can be regarded as a truncated Pascal triangle.

In the case of a uniform grid with $m_1 = 2$ and $m_2 = 2$, we obtain the vertical grid lines

$$
\xi_1 = -1, \qquad \xi_2 = 0, \qquad \xi_3 = 1, \tag{5.11.13}
$$

and the horizontal grid lines

$$
\eta_1 = -1, \qquad \eta_2 = 0, \qquad \eta_3 = 1. \tag{5.11.14}
$$

Graphs of several node interpolation functions generated by the FSELIB function *psi_q2x2*, not listed in the text, are shown in Figure 5.11.6.

Spectral node distributions

For best interpolation accuracy, the vertical interior grid lines, ξ_i for $i = 2, \ldots, m_1$, should be positioned at the zeros of the (m_1-1)-degree Lobatto polynomial, and the horizontal interior grid lines, η_j for $j = 2, \ldots, m_2$, should be positioned at the zeros of the $(m_2 - 1)$-degree Lobatto polynomial. When this is done, the interpolation error behaves spectrally, that is, it decreases faster than algebraically with respect to the polynomial order m_1 or m_2. The spectral properties of the tensor product expansion in two dimensions derive from those of its one-dimensional constituents, as discussed in Chapter 3. Use of the dual Lobatto grid is further motivated by the discovery that the resulting nodal set is identical to the Fekete nodal set for the square (Bos, Taylor, & Wingate [6]). By definition, the Fekete set maximizes the magnitude of the determinant of the generalized Vandermonde matrix within the confines of the square.

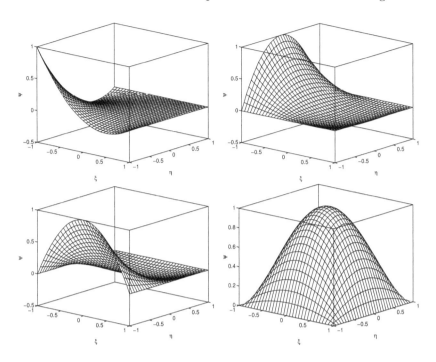

Figure 5.11.6 Graphs of node interpolation functions for a uniform node distribution with $m_1 = 2$ and $m_2 = 2$.

5.11.4 Modal expansion

We consider the standard square in the parametric $\xi\eta$ plane and number the vertices as shown in Figure 4.10.1, repeated for convenience in Figure 5.11.7. In the modal expansion, a function of interest defined over the standard square, $f(\xi, \eta)$, is approximated with the sum of three components comprised of expansion modes multiplied by corresponding coefficients, as follows:

$$f(\xi, \eta) \simeq F_v + F_e + F_i, \tag{5.11.15}$$

where

$$F_v(\xi, \eta) = f_1 \, \zeta_1^v(\xi, \eta) + f_2 \, \zeta_2^v(\xi, \eta) + f_3 \, \zeta_3^v(\xi, \eta) + f_4 \, \zeta_4^v(\xi, \eta) \tag{5.11.16}$$

is the *vertex part*,

$$F_e(\xi, \eta) = \sum_{i=1}^{m_1-1} c_i^{12} \zeta_i^{12}(\xi, \eta) + \sum_{i=1}^{m_2-1} c_i^{34} \zeta_i^{34}(\xi, \eta)$$

$$+ \sum_{i=1}^{m_1-1} c_i^{13} \zeta_i^{13}(\xi, \eta) + \sum_{i=1}^{m_2-1} c_i^{24} \zeta_i^{24}(\xi, \eta) \tag{5.11.17}$$

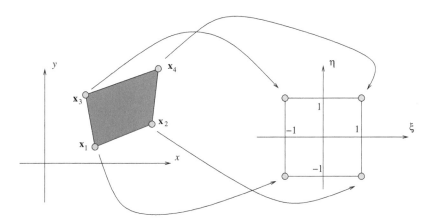

FIGURE 5.11.7   A four-node quadrilateral element in the xy plane is mapped to a square element in the parametric $\xi\eta$ plane.

is the *edge part*, present when $m_1 > 1$ or $m_2 > 1$, and

$$F_i(\xi, \eta) = \sum_{i=1}^{m_1-1} \left(\sum_{j=1}^{m_2-1} c_{ij}\, \zeta_{ij}(\xi, \eta) \right), \qquad (5.11.18)$$

is the *interior part*, present when $m_1 > 1$ or $m_2 > 1$. The truncation limits, m_1 and m_2, are specified polynomial orders.

Adding the $N_v = 4$ vertex modes, the $N_e = 2 \times (m_1 - 1) + 2 \times (m_2 - 1)$ edge modes, and the $N_i = (m_1 - 1)(m_2 - 1)$ interior modes, we obtain

$$N_t = 4 + 2\,(m_1 - 1) + 2\,(m_2 - 1) + (m_1 - 1)(m_2 - 1) \qquad (5.11.19)$$

or

$$N_t = (m_1 + 1)(m_2 + 1) \qquad (5.11.20)$$

total element modes, which is precisely equal to that involved in the tensor-product nodal expansion discussed in Section 5.11.3.

Vertex modes

The coefficients f_1, f_2, f_3, and f_4 in (5.11.16) are the values of the function $f(\xi, \eta)$ at the four vertices of the quadrilateral. To ensure C^0 continuity, these values must be shared by neighboring elements at the common nodes. The corresponding cardinal interpolation functions, ζ_1, ζ_2, ζ_3, and ζ_4, are identical to the bilinear node interpolation functions for the four-node quadrilateral element given in (4.10.4), repeated below for convenient reference,

$$\zeta_1 = \frac{1-\xi}{2}\frac{1-\eta}{2}, \qquad \zeta_2 = \frac{1+\xi}{2}\frac{1-\eta}{2},$$

$$\zeta_3 = \frac{1-\xi}{2}\frac{1+\eta}{2}, \qquad \zeta_4 = \frac{1+\xi}{2}\frac{1+\eta}{2}. \tag{5.11.21}$$

Edge modes

The coefficients c_i^{12}, c_i^{34}, c_i^{13}, and c_i^{24} in (5.11.17) are associated with the horizontal edge interpolation functions ζ_i^{12} and ζ_i^{34} for $i = 1, \ldots, m_1 - 1$, and with the vertical edge interpolation functions ζ_i^{13} and ζ_i^{24} for $i = 1, \ldots, m_2 - 1$. To ensure C^0 continuity, these coefficients must be shared by neighboring elements with common edges.

The interpolation function, ζ_i^{12}, is zero along the 13, 34, and 23 edges of the ith triangle. To satisfy this requirement, we write

$$\zeta_i^{12} = \frac{1-\xi}{2}\frac{1+\xi}{2}\frac{1-\eta}{2}\,\mathrm{Lo}_{i-1}(\xi) \tag{5.11.22}$$

for $i = 1, \ldots, m_1 - 1$. The choice of Lobatto polynomials on the right-hand side ensures the orthogonality of the $\zeta_i^{12}(\xi)$ set,

$$\int_{-1}^{1} \zeta_i^{12}(\xi,\eta)\,\zeta_j^{12}(\xi,\eta)\,\mathrm{d}\xi = 0 \tag{5.11.23}$$

if $i \neq j$.

Working in a similar fashion, we derive the remaining edge modes,

$$\zeta_i^{34} = \frac{1-\xi}{2}\frac{1+\xi}{2}\frac{1+\eta}{2}\,\mathrm{Lo}_{i-1}(\xi) \tag{5.11.24}$$

for $i = 1, \ldots, m_1 - 1$,

$$\zeta_i^{13} = \frac{1-\xi}{2}\frac{1-\eta}{2}\frac{1+\eta}{2}\,\mathrm{Lo}_{i-1}(\eta) \tag{5.11.25}$$

for $i = 1, \ldots, m_2 - 1$, and

$$\zeta_i^{24} = \frac{1+\xi}{2}\frac{1-\eta}{2}\frac{1+\eta}{2}\,\mathrm{Lo}_{i-1}(\eta) \tag{5.11.26}$$

for $i = 1, \ldots, m_2 - 1$, which enjoy similar orthogonality properties.

As an example, we consider the modal expansion for $m_1 = m_2 = 2$, recall that the zeroth-degree Lobatto polynomial is constant, $\mathrm{Lo}_0(\xi) = 1$, and obtain the cubic edge interpolation functions

$$\zeta_1^{12} = \frac{1-\xi}{2}\frac{1+\xi}{2}\frac{1-\eta}{2}, \qquad \zeta_1^{34} = \frac{1-\xi}{2}\frac{1+\xi}{2}\frac{1+\eta}{2},$$

$$\zeta_1^{13} = \frac{1-\eta}{2}\frac{1+\eta}{2}\frac{1-\xi}{2}, \qquad \zeta_1^{24} = \frac{1-\eta}{2}\frac{1+\eta}{2}\frac{1+\xi}{2}. \tag{5.11.27}$$

Note that these modes are proportional, respectively, to the mid-point edge interpolation functions of the eight-node element $\zeta_2, \zeta_7, \zeta_4$, and ζ_5, shown in Table 5.11.1.

Interior modes

The coefficients c_{ij} in (5.11.18) correspond to the interior modes, also called bubble modes, described by the interpolation functions, ζ_{ij}. To ensure that these functions are zero all around the perimeter of the standard square, we write

$$\zeta_{ij} = \frac{1-\xi}{2}\frac{1+\xi}{2}\,\mathrm{Lo}_{i-1}(\xi)\,\frac{1-\eta}{2}\frac{1+\eta}{2}\,\mathrm{Lo}_{j-1}(\eta) \tag{5.11.28}$$

for $i = 1, \ldots, m_1 - 1$ and $j = 1, \ldots, m_2 - 1$. The choice of the Lobatto polynomial on the right-hand side is motivated by the orthogonality of the ζ_{ij} set,

$$\int_{-1}^{1}\int_{-1}^{1} \zeta_{ij}(\xi,\eta)\,\zeta_{kl}(\xi,\eta)\,\mathrm{d}\xi\,\mathrm{d}\eta \neq 0 \tag{5.11.29}$$

only if $i = k$ and $j = l$. The proof follows immediately from the properties of the Lobatto polynomials discussed in Section B.6, Appendix B.

As an example, we consider the quadratic modal expansion for $m_1 = m_2 = 2$, recall that the zeroth-degree Lobatto polynomial is constant, $\mathrm{Lo}_0(\xi) = 1$ and $\mathrm{Lo}_0(\eta) = 1$, and obtain a solitary quartic interior mode,

$$\zeta_{11} = \frac{1-\xi}{2}\frac{1+\xi}{2}\frac{1-\eta}{2}\frac{1+\eta}{2}. \tag{5.11.30}$$

PROBLEMS

5.11.1 *Element node interpolation functions*

Prepare graphs of the element node interpolation functions corresponding to the Lobatto node distribution with $m_1 = 3$ and $m_2 = 2$.

5.11.2 *Interior modes*

Prepare graphs and discuss the structure of the interior modes for $m_1 = 2$ and $m_2 = 2$.

Applications in mechanics 6

The finite element method finds important practical applications in the field of computational mechanics, including solid mechanics and structural statics and dynamics, where it is used to contact a *finite element analysis.* Several problems involving the bending and buckling of beams governed by fourth-order ordinary differential equations were discussed in Sections 2.4–2.6.

The general methodology and implementation of the finite element method in computational mechanics is similar to that discussed in previous chapters for the diffusion equation and similar scalar equations. Two new features are that the mathematical modeling typically results in a system of partial differential equations instead of a single partial differential equation, and the order of the individual differential equations may vary across the unknown functions.

In this chapter, finite element formulations and solution procedures for selected problems are discussed and integrated finite element codes, complete with grid generation and data visualization modules, are developed. To set up the theoretical framework, basic concepts from the theory of three-dimensional elasticity are reviewed in Section 6.1. The equations governing plane stress and plane strain analysis and the corresponding finite element formulations and implementations are discussed in Sections 6.2 and 6.3. Advanced formulations for plate bending, buckling, and wrinkling are discussed in Sections 6.4–6.9.

6.1 Elements of elasticity theory

To describe the forces developing in a deformed three-dimensional solid, we introduce the Cauchy stress tensor,

$$
\boldsymbol{\sigma} = \begin{bmatrix} \sigma_{xx} & \sigma_{xy} & \sigma_{xz} \\ \sigma_{yx} & \sigma_{yy} & \sigma_{yz} \\ \sigma_{zx} & \sigma_{zy} & \sigma_{zz} \end{bmatrix}. \tag{6.1.1}
$$

A tensor is a matrix whose components are physical entities endowed with special properties that allow us to compute values corresponding to one coordinate system from values corresponding to another coordinate system, using simple geometrical transformation rules (e.g., Pozrikidis [46]).

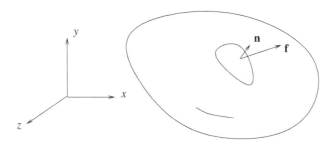

FIGURE 6.1.1   Illustration of the traction, **f**, exerted on a small surface inside or at
the boundary of a deformed elastic material. The unit vector **n** is normal to the
small surface.

The force surface density is defined as the force exerted on an infinitesimal
surface element drawn inside or at the boundary of a solid material divided by the
element surface area, as shown in Figure 6.1.1. The components of the Cauchy
stress tensor are defined such that the force surface density *in the deformed state* is
given by the *traction vector*

$$\mathbf{f} = \mathbf{n} \cdot \boldsymbol{\sigma}, \tag{6.1.2}$$

where $\mathbf{n} = (n_x, n_y, n_z)$ is the unit vector normal to the infinitesimal surface. By
convention, the traction is exerted on the side of the material that lies behind the
normal vector, **n**. Newton's third law requires that the traction exerted on the other
side pulls with equal strength in the opposite direction. Explicitly, the Cartesian
components of the traction vector are given by

$$
\begin{aligned}
f_x &= n_x\,\sigma_{xx} + n_y\,\sigma_{yx} + n_z\,\sigma_{zx}, \\
f_y &= n_x\,\sigma_{xy} + n_y\,\sigma_{yy} + n_z\,\sigma_{zy}, \\
f_z &= n_x\,\sigma_{xz} + n_y\,\sigma_{yz} + n_z\,\sigma_{zz}.
\end{aligned}
\tag{6.1.3}
$$

Since the unit normal vector is dimensionless, the traction has units of stress.

In the absence of an external torque field, moment equilibrium requires that the
stress tensor is symmetric,

$$\sigma_{xy} = \sigma_{yx}, \qquad \sigma_{xz} = \sigma_{zx}, \qquad \sigma_{yz} = \sigma_{zy}. \tag{6.1.4}$$

A torque field arises when a material made of small magnetized particles is subjected
to an electrical field.

Performing a force balance over an infinitesimal parallelepiped whose sides are
parallel to the x, y, or z axes, we derive the equilibrium equation

$$\frac{\partial \sigma_{xx}}{\partial x} + \frac{\partial \sigma_{yx}}{\partial y} + \frac{\partial \sigma_{zx}}{\partial z} + b_x = 0 \tag{6.1.5}$$

for the x direction,

$$\frac{\partial \sigma_{xy}}{\partial x} + \frac{\partial \sigma_{yy}}{\partial y} + \frac{\partial \sigma_{zy}}{\partial z} + b_y = 0 \qquad (6.1.6)$$

for the y direction, and

$$\frac{\partial \sigma_{xz}}{\partial x} + \frac{\partial \sigma_{yz}}{\partial y} + \frac{\partial \sigma_{zz}}{\partial z} + b_z = 0, \qquad (6.1.7)$$

for the z direction, where $\mathbf{b} = (b_x, b_y, b_z)$ is the body force exerted on the material, defined as the force divided by the volume of the material. In the case of the gravitational force, $\mathbf{b} = \rho \mathbf{g}$, where ρ is the material density and $\mathbf{g}$ is the acceleration of gravity.

In compact vector notation, the three scalar equilibrium equations (6.1.5)–(6.1.7) combine into the vectorial equation

$$\nabla \cdot \boldsymbol{\sigma} + \mathbf{b} = \mathbf{0}, \qquad (6.1.8)$$

where $\nabla = (\partial/\partial x, \partial/\partial y, \partial/\partial z)$ is the three-dimensional gradient and $\nabla \cdot \boldsymbol{\sigma}$ is the divergence of the stress tensor.

6.1.1 Deformation and constitutive equations

Consider a point particle of the elastic material in the undeformed configuration and then in the deformed configuration. The displacement of the point particle is encapsulated in the vector

$$\mathbf{v} = \begin{bmatrix} v_x \\ v_y \\ v_z \end{bmatrix}, \qquad (6.1.9)$$

where v_x, v_y, and v_z are the scalar displacements along the x, y, and z axes. To simplify the notation, we denote $x_1 = x$, $x_2 = y$, $x_3 = z$ and $v_1 = v_x$, $v_2 = v_y$, $v_3 = v_z$.

Physical arguments dictate that the components of the stress tensor developing in the material due to the deformation at the position of a point particle are functions of the components of the strain tensor with components

$$\epsilon_{kl} = \frac{1}{2}\left(\frac{\partial v_k}{\partial x_l} + \frac{\partial v_l}{\partial x_k}\right) \qquad (6.1.10)$$

for $k, l = 1, 2, 3$. This functional dependence is mediated by a constitutive equation that reflects the physical structure and mechanical properties of the material. Explicitly, the diagonal components of the strain tensor are given by

$$\epsilon_{11} \equiv \epsilon_{xx} = \frac{\partial v_x}{\partial x}, \qquad \epsilon_{22} \equiv \epsilon_{yy} = \frac{\partial v_y}{\partial y}, \qquad \epsilon_{33} \equiv \epsilon_{zz} = \frac{\partial v_z}{\partial z}. \qquad (6.1.11)$$

The off-diagonal components are given by

$$\epsilon_{12} = \epsilon_{21} \equiv \epsilon_{xy} = \epsilon_{yx} = \frac{1}{2}\left(\frac{\partial v_x}{\partial y} + \frac{\partial v_y}{\partial x}\right),$$

$$\epsilon_{13} = \epsilon_{31} \equiv \epsilon_{xz} = \epsilon_{zx} = \frac{1}{2}\left(\frac{\partial v_x}{\partial z} + \frac{\partial v_z}{\partial x}\right), \qquad (6.1.12)$$

$$\epsilon_{23} = \epsilon_{32} \equiv \epsilon_{yz} = \epsilon_{zx} = \frac{1}{2}\left(\frac{\partial v_y}{\partial z} + \frac{\partial v_z}{\partial y}\right).$$

These definitions require the compatibility conditions

$$\frac{\partial^2 \epsilon_{xx}}{\partial y^2} + \frac{\partial^2 \epsilon_{yy}}{\partial x^2} = 2\frac{\partial^2 \epsilon_{xy}}{\partial x \partial y}, \qquad \frac{\partial^2 \epsilon_{xx}}{\partial z^2} + \frac{\partial^2 \epsilon_{zz}}{\partial x^2} = 2\frac{\partial^2 \epsilon_{xz}}{\partial x \partial z},$$

$$\frac{\partial^2 \epsilon_{yy}}{\partial z^2} + \frac{\partial^2 \epsilon_{zz}}{\partial y^2} = 2\frac{\partial^2 \epsilon_{yz}}{\partial y \partial z}, \qquad (6.1.13)$$

which prevent us from specifying arbitrarily all components of the strain tensor.

6.1.2 Linear elasticity

If a material behaves like a linearly elastic medium, the constitutive equation relating stress to deformation takes the form

$$\sigma_{ij} = \alpha \lambda \delta_{ij} + 2\mu \epsilon_{ij}, \qquad (6.1.14)$$

where δ_{ij} is the Kronecker delta, λ and μ are the Lamé constants, and

$$\alpha \equiv \epsilon_{xx} + \epsilon_{yy} + \epsilon_{zz} = \frac{\partial v_k}{\partial x_k} = \nabla \cdot \mathbf{v} \qquad (6.1.15)$$

is the dilatation due to the deformation expressed by the divergence of the displacement vector field, which is equal to the trace of the strain tensor. Summation of the repeated index k over $1, 2, 3$ or x, y, z is implied in (6.1.15). Explicitly, the linear constitutive equation reads

$$\sigma_{ij} = \frac{\partial v_k}{\partial x_k}\lambda \delta_{ij} + \mu\left(\frac{\partial v_i}{\partial x_j} + \frac{\partial v_j}{\partial x_i}\right), \qquad (6.1.16)$$

where summation is implied over the repeated index k. Note that the stress tensor is symmetric, $\sigma_{ij} = \sigma_{ji}$, in agreement with our earlier discussion.

Alternatively, the components of the strain tensor can be related to the components of the stress tensor by a generalized Hooke's law for an isotropic medium,

$$\begin{bmatrix} \epsilon_{xx} \\ \epsilon_{yy} \\ \epsilon_{zz} \\ \epsilon_{xy} \\ \epsilon_{xz} \\ \epsilon_{yz} \end{bmatrix} = \frac{1}{E}\begin{bmatrix} 1 & -\nu & -\nu & 0 & 0 & 0 \\ -\nu & 1 & -\nu & 0 & 0 & 0 \\ -\nu & -\nu & 1 & 0 & 0 & 0 \\ 0 & 0 & 0 & 1+\nu & 0 & 0 \\ 0 & 0 & 0 & 0 & 1+\nu & 0 \\ 0 & 0 & 0 & 0 & 0 & 1+\nu \end{bmatrix} \cdot \begin{bmatrix} \sigma_{xx} \\ \sigma_{yy} \\ \sigma_{zz} \\ \sigma_{xy} \\ \sigma_{xz} \\ \sigma_{yz} \end{bmatrix}, \qquad (6.1.17)$$

where E is the Young modulus of elasticity and ν is the Poisson ratio taking values in the range $[-1, 0.5]$. The extreme value $\nu = 0.5$ corresponds to an idealized incompressible solid. While negative values of the Poisson ratio are theoretically possible and have been observed for wrinkled materials that unfold under deformation, most common materials have a positive Poisson ratio in the range $(0, 0.5)$.

As an application, we consider the third scalar component of (6.1.17),

$$\epsilon_{zz} = \frac{1}{E} \left[\sigma_{zz} - \nu \left(\sigma_{xx} + \sigma_{yy} \right) \right], \tag{6.1.18}$$

and note the coupling of the directional stresses and strain in three dimensions for any non-zero Poisson ratio.

In fact, the constitutive equations (6.1.14) and (6.1.17) are identical, provided that the Lamé constants are related to the modulus of elasticity, E, and Poisson's ratio, ν, by

$$\mu = \frac{E}{2\,(1+\nu)}, \qquad \lambda = \frac{E\nu}{(1+\nu)(1-2\nu)}. \tag{6.1.19}$$

Substituting the constitutive equation (6.1.14) into the equilibrium equation (6.1.8), we derive Navier's equation,

$$\mu \nabla^2 \mathbf{v} + (\mu + \lambda) \nabla \alpha + \mathbf{b} = \mathbf{0}. \tag{6.1.20}$$

Using the relations in (6.1.19), we obtain

$$\mu + \lambda = \frac{E}{2\,(1+\nu)(1-2\,\nu)} = \frac{\mu}{1-2\,\nu}, \tag{6.1.21}$$

and derive the alternative form

$$\mu \nabla^2 \mathbf{v} + \frac{\mu}{1-2\,\nu} \nabla \alpha + \mathbf{b} = \mathbf{0}, \tag{6.1.22}$$

where ∇^2 is the three-dimensional Laplacian operator. Explicitly, the three scalar components of (6.1.22) read

$$\mu \left(\frac{\partial^2 v_x}{\partial x^2} + \frac{\partial^2 v_x}{\partial y^2} + \frac{\partial^2 v_x}{\partial z^2} \right) + \frac{\mu}{1-2\,\nu} \frac{\partial \alpha}{\partial x} + b_x = 0,$$

$$\mu \left(\frac{\partial^2 v_y}{\partial x^2} + \frac{\partial^2 v_y}{\partial y^2} + \frac{\partial^2 v_y}{\partial z^2} \right) + \frac{\mu}{1-2\,\nu} \frac{\partial \alpha}{\partial y} + b_y = 0, \tag{6.1.23}$$

$$\mu \left(\frac{\partial^2 v_z}{\partial x^2} + \frac{\partial^2 v_z}{\partial y^2} + \frac{\partial^2 v_z}{\partial z^2} \right) + \frac{\mu}{1-2\,\nu} \frac{\partial \alpha}{\partial z} + b_z = 0.$$

PROBLEM

6.1.1 *Linear constitutive equations*

Confirm that the constitutive equations (6.1.14) and (6.1.17) are identical, provided that the Lamé constants are related to the modulus of elasticity, E, and Poisson's ratio, ν, by the equations in (6.1.19).

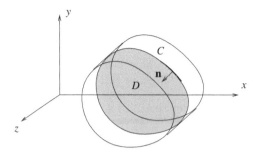

FIGURE 6.2.1   Illustration of a solid material confined by two parallel surfaces used to
perform the two-dimensional (plane) stress or strain analysis.

6.2   Plane stress and plane strain analysis

Consider an elastic material deforming under the action of an external surface load
or body force. The plane stress and plane strain analysis problems are distinguished
by the assumption that material point particles are deflected so that their displacement in the xy plane is independent of the transverse coordinate, z, as illustrated
in Figure 6.2.1. Consequently, all components of the stress tensor and all components of the strain tensor are independent of the z coordinate (e.g., Novoshilov [41],
Chapter 7).

Physically, plane stress and plane strain occur when an in-plane force that is
parallel to the xy plane is exerted around the edges, at the upper surface, or at the
lower surface of an elastic plate, or else when an in-plane distributed body force,
such as gravity, acts over the volume of the plate.

The complementary situation where the plate exhibits transverse deformation
normal to the xy plane under the action of a distributed or localized transverse
load or body force directed along the z axis will be discussed in Section 6.4 in the
context of plate bending.

6.2.1   Plane stress analysis

The plane stress problem is distinguished further by the stress conditions

$$\sigma_{iz} = \sigma_{zi} = 0 \qquad\qquad (6.2.1)$$

for $i = x, y, z$. Physically, the two parallel surfaces of the material normal to the z
axis are free to deform in the z direction, but are unable to support traction. This
situation occurs when a flat wrench, modeled as a plate, is pushed on its handle to
loosen a bolt. Under these assumptions, the general constitutive equation (6.1.17)
simplifies to

$$
\begin{bmatrix}
\epsilon_{xx} \\
\epsilon_{yy} \\
\epsilon_{xy} \\
\epsilon_{zz}
\end{bmatrix}
=
\frac{1}{E}
\begin{bmatrix}
1 & -\nu & 0 \\
-\nu & 1 & 0 \\
0 & 0 & 1+\nu \\
-\nu & -\nu & 0
\end{bmatrix}
\cdot
\begin{bmatrix}
\sigma_{xx} \\
\sigma_{yy} \\
\sigma_{xy}
\end{bmatrix},
\tag{6.2.2}
$$

complemented by the conditions

$$
\epsilon_{xz} = 0, \qquad\qquad \epsilon_{yz} = 0.
\tag{6.2.3}
$$

Note that the right-hand side of (6.2.2) involves a 4×3 matrix.

Solving the first three equations in (6.2.2) for σ_{xx}, σ_{yy}, and σ_{xy} in terms of ϵ_{xx}, ϵ_{yy}, and ϵ_{xy}, we obtain

$$
\begin{bmatrix}
\sigma_{xx} \\
\sigma_{yy} \\
\sigma_{xy}
\end{bmatrix}
=
\frac{E}{1-\nu^2}
\begin{bmatrix}
1 & \nu & 0 \\
\nu & 1 & 0 \\
0 & 0 & 1-\nu
\end{bmatrix}
\cdot
\begin{bmatrix}
\epsilon_{xx} \\
\epsilon_{yy} \\
\epsilon_{xy}
\end{bmatrix}.
\tag{6.2.4}
$$

The fourth equation in (6.2.2) is a simplification of equation (6.1.18),

$$
\epsilon_{zz} = -\frac{\nu}{E} \left(\sigma_{xx} + \sigma_{yy} \right).
\tag{6.2.5}
$$

This relation reveals that a material with a Poisson ratio equal to zero, such as a bottle cork, will not expand laterally when subjected to an in-plane load. When the Poisson ratio is negative, stretching of the material in the xy plane, $\sigma_{xx} + \sigma_{yy} > 0$, causes expansion in the lateral direction, $\epsilon_{zz} > 0$, which is counterintuitive.

Equilibrium equations

In plane stress analysis, the equilibrium equations (6.1.5) and (6.1.6) simplify to

$$
\frac{\partial \sigma_{xx}}{\partial x} + \frac{\partial \sigma_{yx}}{\partial y} + b_x = 0, \qquad\qquad \frac{\partial \sigma_{xy}}{\partial x} + \frac{\partial \sigma_{yy}}{\partial y} + b_y = 0.
\tag{6.2.6}
$$

Taking the partial derivative of the first equation with respect to x and the partial derivative of the second equation with respect to y, adding the resulting expressions and rearranging, we obtain

$$
\frac{\partial^2 \sigma_{xx}}{\partial x^2} + \frac{\partial^2 \sigma_{yy}}{\partial y^2} + 2 \frac{\partial^2 \sigma_{xy}}{\partial x \, \partial y} = - \left(\frac{\partial b_x}{\partial x} + \frac{\partial b_y}{\partial y} \right).
\tag{6.2.7}
$$

Combining the first of the compatibility conditions in (6.1.13) with the equations in (6.2.2), we obtain

$$
\frac{\partial^2 \epsilon_{xx}}{\partial y^2} + \frac{\partial^2 \epsilon_{yy}}{\partial x^2} = \frac{1}{E} \left(\frac{\partial^2 (\sigma_{xx} - \nu\sigma_{yy})}{\partial y^2} + \frac{\partial^2 (\sigma_{yy} - \nu\sigma_{xx})}{\partial x^2} \right) = 2 \frac{\partial^2 \epsilon_{xy}}{\partial x \partial y}
\tag{6.2.8}
$$

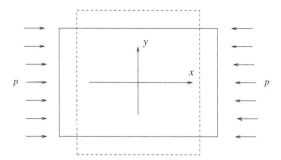

<small>FIGURE</small> 6.2.2 Compression of a rectangular plate along the x axis with two free
bottom and top sides.

or

$$\frac{\partial^2 (\sigma_{xx} - \nu \sigma_{yy})}{\partial y^2} + \frac{\partial^2 (\sigma_{yy} - \nu \sigma_{xx})}{\partial x^2} = 2 (1 + \nu) \frac{\partial^2 \sigma_{xy}}{\partial x \partial y}, \tag{6.2.9}$$

which can be rearranged into

$$2 \frac{\partial^2 \sigma_{xy}}{\partial x \, \partial y} = \frac{1}{1 + \nu} \left(\frac{\partial^2 (\sigma_{xx} - \nu \sigma_{yy})}{\partial y^2} + \frac{\partial^2 (\sigma_{yy} - \nu \sigma_{xx})}{\partial x^2} \right). \tag{6.2.10}$$

Combining (6.2.7) and (6.2.10) to eliminate the derivative $\partial^2 \sigma_{xy} / (\partial x \partial y)$, we derive
a Poisson equation for the sum of the normal stresses, which is equal to the trace
of the stress tensor in the xy plane,

$$\left(\frac{\partial^2}{\partial x^2} + \frac{\partial^2}{\partial y^2} \right) (\sigma_{xx} + \sigma_{yy}) = -(1 + \nu) \left(\frac{\partial b_x}{\partial x} + \frac{\partial b_y}{\partial y} \right). \tag{6.2.11}$$

In compact notation,

$$\nabla^2 (\sigma_{xx} + \sigma_{yy}) = -(1 + \nu) \, \nabla \cdot \mathbf{b}. \tag{6.2.12}$$

The solution is subject to appropriate boundary conditions.

Compression of a rectangular plate

As an application, we consider the compression of a rectangular plate of thickness
h with two sides parallel to the x axis and two sides parallel to the y axis. When
the plate is compressed normal to the two edges that are parallel to the y axes with
a uniform force with linear density p, while the other two edges are free, as shown
in Figure 6.2.2, the displacement field is given by

$$v_x = -\frac{p}{Eh} x, \qquad v_y = \frac{p}{Eh} \nu y. \tag{6.2.13}$$

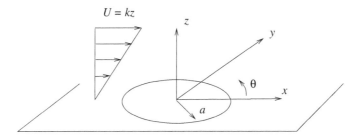

FIGURE 6.2.3   Illustration of shear flow past a discoidal membrane clamped around a circular rim on a plane wall.

When the Poisson ratio is positive, compression induces expansion in the lateral direction, in agreement with physical intuition. The associated stress field is

$$\sigma_{xx} = -\frac{p}{h}, \qquad \sigma_{yy} = -\frac{p}{(1+\nu)\,h}, \qquad \sigma_{xy} = 0. \qquad (6.2.14)$$

Shear stresses arise when the plate is compressed along arbitrary axes.

Shear flow over a circular membrane

In another application, we consider shear flow of a viscous fluid along the x axis with velocity $U = kz$, over a discoidal membrane of radius a that is clamped around a circular rim on a rigid plane wall, as shown in Figure 6.2.3, where k is the shear rate. The upper surface of the membrane is exposed to the hydrodynamic shear stress, $\tau = \eta\,k$, which amounts to an in-plane body force with components $b_x = \tau/h$ and $b_y = 0$, where η is the fluid viscosity and h is the membrane thickness.

By inspection, we find that the displacement field is given by

$$v_x = \frac{\tau}{Eh}\frac{1-\nu^2}{3-\nu}(a^2 - x^2 - y^2), \qquad v_y = 0. \qquad (6.2.15)$$

The associated stress field, satisfying the equilibrium equations in (6.2.6), can be derived from (6.2.4) and is given by

$$\sigma_{xx} = \frac{E}{1-\nu^2}\frac{\partial v_x}{\partial x} = -\frac{2}{3-\nu}\frac{\tau}{h}x,$$

$$\sigma_{xy} = \frac{E}{1-\nu^2}\frac{\partial v_x}{\partial y} = -\frac{1-\nu}{3-\nu}\frac{\tau}{h}y, \qquad (6.2.16)$$

$$\sigma_{yy} = \nu\,\sigma_{xx}.$$

Since $\sigma_{xx} > 0$ when $x < 0$, the membrane is stretched on the upstream side and compressed on the downstream side, in agreement with physical intuition.

Shear flow over an elliptical membrane

In the case of shear flow past an elliptical membrane with principal axes in the x and y directions, the membrane rim is described by the equation $(x/a_x)^2 + (y/a_y)^2 = 1$, where a_x and a_y are the ellipse semi-axes. By inspection, we find that the displacement field is given by

$$v_x = V\left(1 - \frac{x^2}{a_x^2} - \frac{y^2}{a_y^2}\right), \qquad v_y = 0, \qquad (6.2.17)$$

where

$$V = \frac{\tau}{Eh}(1 - \nu^2)\frac{a_x^2\, a_y^2}{(1-\nu)\, a_x^2 + 2\, a_y^2} \qquad (6.2.18)$$

is a constant with dimensions of length. The associated membrane stress field is given by

$$\sigma_{xx} = \frac{E}{1-\nu^2}\frac{\partial v_x}{\partial x} = -2\frac{EV}{(1-\nu^2)\,a_x^2}\,x = -\frac{\tau}{h}\frac{2\,a_y^2}{(1-\nu)\,a_x^2 + 2\,a_y^2}\,x,$$

$$\sigma_{xy} = \frac{E}{2(1+\nu)}\frac{\partial v_x}{\partial y} = -\frac{EV}{(1+\nu)\,a_y^2}\,y = -\frac{\tau}{h}\frac{(1-\nu)\,a_x^2}{(1-\nu)\,a_x^2 + 2\,a_y^2}\,y, \qquad (6.2.19)$$

$$\sigma_{yy} = \nu\,\sigma_{xx}.$$

When $a_x = a_y$, for a circular membrane, we recover the expressions in (6.2.16).

Biharmonic equation for the Airy stress function

The Cartesian components of a conservative body force, b_x and b_y, derive from a scalar potential function, $\mathcal{V}$, as

$$b_x = \frac{\partial \mathcal{V}}{\partial x}, \qquad b_y = \frac{\partial \mathcal{V}}{\partial y}. \qquad (6.2.20)$$

In vector notation,

$$\mathbf{b} = \nabla\mathcal{V}. \qquad (6.2.21)$$

For example, if the body force is uniform in space,

$$\mathcal{V} = b_x x + b_y y = \mathbf{b}\cdot\mathbf{x}. \qquad (6.2.22)$$

Under these conditions, the equilibrium equations in (6.2.6) become

$$\frac{\partial(\sigma_{xx} + \mathcal{V})}{\partial x} + \frac{\partial\sigma_{yx}}{\partial y} = 0, \qquad \frac{\partial\sigma_{xy}}{\partial x} + \frac{\partial(\sigma_{yy} + \mathcal{V})}{\partial y} = 0, \qquad (6.2.23)$$

which are satisfied if the stress components derive from the Airy stress function, ϕ, as

$$\sigma_{xx} = \frac{\partial^2 \phi}{\partial y^2} - \mathcal{V}, \qquad \sigma_{yy} = \frac{\partial^2 \phi}{\partial x^2} - \mathcal{V}, \qquad \sigma_{xy} = \sigma_{yx} = -\frac{\partial^2 \phi}{\partial x \partial y}. \qquad (6.2.24)$$

The trace of the stress tensor is given by

$$\sigma_{xx} + \sigma_{yy} = \nabla^2 \phi - 2\mathcal{V}. \qquad (6.2.25)$$

Substituting this expression along with expressions (6.2.20) into (6.2.12), we find that

$$\nabla^2 (\nabla^2 \phi - 2\mathcal{V}) = -(1+\nu)\nabla \cdot \mathbf{b} = -(1+\nu)\nabla^2 \mathcal{V}. \qquad (6.2.26)$$

Rearranging, we find that the Airy stress function satisfies a forced biharmonic equation,

$$\nabla^4 \phi \equiv \left(\frac{\partial^2}{\partial x^2} + \frac{\partial^2}{\partial y^2} \right)^2 \phi = \frac{\partial^4 \phi}{\partial x^4} + 2\frac{\partial^4 \phi}{\partial x^2 \partial y^2} + \frac{\partial^4 \phi}{\partial y^4} = (1-\nu)\nabla^2 \mathcal{V}. \quad (6.2.27)$$

If the potential $\mathcal{V}$ is harmonic, satisfying Laplace's equation, $\nabla\mathcal{V}^2 = 0$, which is true when the components of the body force, b_x and b_y, are constant in space and accordingly $\mathcal{V}$ is linear in x, y, or both, the right-hand side of (6.2.27) is zero, yielding the homogeneous biharmonic equation

$$\nabla^4 \phi = 0. \qquad (6.2.28)$$

This equation is a common point of departure for computing solutions of the plane stress problem by finite difference methods.

The finite element formulation of the biharmonic equation will be discussed in Section 6.9 in the context of plate bending.

6.2.2 *Plane strain analysis*

In the plane strain analysis problem, all material points are assumed to undergo the same displacement along the z axis so that

$$\epsilon_{iz} = \epsilon_{zi} = 0 \qquad (6.2.29)$$

for $i = x, y, z$. Physically, the planar boundaries of the two-dimensional material normal to the z axis are anchored as though they were sandwiched between two impenetrable walls that only allow sliding displacement. In practice, the plane-strain problem occurs when a water dam is immobilized by side-walls or when a slab of earth is held fixed between two hard rocks.

The general constitutive equation (6.1.17) for the plane strain problem simplifies to

$$
\begin{bmatrix}
\epsilon_{xx} \\
\epsilon_{yy} \\
0 \\
\epsilon_{xy} \\
0 \\
0
\end{bmatrix}
= \frac{1}{E}
\begin{bmatrix}
1 & -\nu & -\nu & 0 & 0 & 0 \\
-\nu & 1 & -\nu & 0 & 0 & 0 \\
-\nu & -\nu & 1 & 0 & 0 & 0 \\
0 & 0 & 0 & 1+\nu & 0 & 0 \\
0 & 0 & 0 & 0 & 1+\nu & 0 \\
0 & 0 & 0 & 0 & 0 & 1+\nu
\end{bmatrix}
\cdot
\begin{bmatrix}
\sigma_{xx} \\
\sigma_{yy} \\
\sigma_{zz} \\
\sigma_{xy} \\
\sigma_{xz} \\
\sigma_{yz}
\end{bmatrix}.
\qquad (6.2.30)
$$

The last two equations require that $\sigma_{xz} = 0$ and $\sigma_{yz} = 0$. Consequently, the equilibrium equations (6.1.5) and (6.1.6) reduce to those shown in (6.2.6).

Taking the partial derivative of the first equation in (6.2.30) with respect to x and the partial derivative of the second equation with respect to y, adding the resulting expressions and rearranging, we derive the second-order equation (6.2.7). The fourth equation in (6.2.30) requires that

$$
\sigma_{zz} = \nu \, (\sigma_{xx} + \sigma_{yy}). \qquad (6.2.31)
$$

Using this constraint to eliminate σ_{zz}, we obtain a simplified constitutive equation,

$$
\begin{bmatrix}
\epsilon_{xx} \\
\epsilon_{yy} \\
\epsilon_{xy}
\end{bmatrix}
= \frac{1+\nu}{E}
\begin{bmatrix}
1-\nu & -\nu & 0 \\
-\nu & 1-\nu & 0 \\
0 & 0 & 1
\end{bmatrix}
\cdot
\begin{bmatrix}
\sigma_{xx} \\
\sigma_{yy} \\
\sigma_{xy}
\end{bmatrix}.
\qquad (6.2.32)
$$

Solving for the stresses in terms of the strains, we finally arrive at the targeted constitutive equation

$$
\begin{bmatrix}
\sigma_{xx} \\
\sigma_{yy} \\
\sigma_{xy}
\end{bmatrix}
= \frac{E}{(1+\nu)(1-2\nu)}
\begin{bmatrix}
1-\nu & \nu & 0 \\
\nu & 1-\nu & 0 \\
0 & 0 & 1-2\nu
\end{bmatrix}
\cdot
\begin{bmatrix}
\epsilon_{xx} \\
\epsilon_{yy} \\
\epsilon_{xy}
\end{bmatrix}.
\qquad (6.2.33)
$$

Alternatively, the equations in (6.2.33) can be derived directly from (6.1.14) subject to the relations in (6.1.19) (Problem 6.2.2).

A singular behavior is encountered when $\nu = 0.5$, corresponding to an incompressible material. When $\nu = 0$, and only then, the constitutive equations for plane strain reduce to those for plane stress discussed in Section 6.2.1.

6.2.3 *Finite element formulation*

To develop the finite element method for plane stress or plane strain analysis, we discretize solution domain in the xy plane into finite elements, as discussed in Chapters 4 and 5 for the Laplace, Helmholtz, and convection–diffusion equations. In the second step, we express the x and y components of the displacement field in terms of global interpolation functions, ϕ_j, in the familiar form

$$
\mathbf{v}(x,y) = \sum_{j=1}^{N_G} \mathbf{v}_j \, \phi_j(x,y), \qquad (6.2.34)
$$

where N_G is the number of global interpolation nodes and

$$\mathbf{v}_j = \begin{bmatrix} v_{x_j} \\ v_{y_j} \end{bmatrix} \tag{6.2.35}$$

is the vectorial displacement at the position of the jth node. The scalar components, v_{x_j} and v_{x_j}, are the x and y displacements of a point particle at the position of the jth global node.

Galerkin projection

Next, we carry out the Galerkin projection of the vectorial equilibrium equation (6.1.8) over a solution domain, D, as shown in Figure 6.2.1,

$$\iint_D \phi_i(x, y) \left(\nabla \cdot \boldsymbol{\sigma} + \mathbf{b} \right) dx \, dy = \mathbf{0}. \tag{6.2.36}$$

Using the Gauss divergence theorem, we remove the divergence of the stress tensor working in two stages. First, we write

$$\iint_D \phi_i \nabla \cdot \boldsymbol{\sigma} \, dx \, dy = \iint_D \nabla \cdot (\phi_i \boldsymbol{\sigma}) \, dx \, dy - \iint_D \nabla \phi_i \cdot \boldsymbol{\sigma} \, dx \, dy. \tag{6.2.37}$$

Second, we use the divergence theorem to convert the first integral on the right-hand side to a contour integral, obtaining

$$\iint_D \phi_i \nabla \cdot \boldsymbol{\sigma} \, dx \, dy = \oint_C \phi_i \, \mathbf{n} \cdot \boldsymbol{\sigma} \, dl - \iint_D \nabla \phi_i \cdot \boldsymbol{\sigma} \, dx \, dy, \tag{6.2.38}$$

where $\mathbf{n}$ is the unit vector normal to the boundary of the solution domain, C, pointing *inward*, and l is the arc length along C. Substituting this expression into (6.2.36) and rearranging, we derive the final form

$$\iint_D \nabla \phi_i \cdot \boldsymbol{\sigma} \, dx \, dy = \oint_C \phi_i \, \mathbf{f} \, dl + \iint_D \phi_i \, \mathbf{b} \, dx \, dy, \tag{6.2.39}$$

where $\mathbf{f} = \mathbf{n} \cdot \boldsymbol{\sigma}$ is the boundary traction.

Explicitly, the x and y components of the ith vectorial Galerkin equation (6.2.39) read

$$\iint_D \left(\frac{\partial \phi_i}{\partial x} \sigma_{xx} + \frac{\partial \phi_i}{\partial y} \sigma_{yx} \right) dx \, dy = F_{x_i} + B_{x_i} \tag{6.2.40}$$

and

$$\iint_D \left(\frac{\partial \phi_i}{\partial x} \sigma_{xy} + \frac{\partial \phi_i}{\partial y} \sigma_{yy} \right) dx \, dy = F_{y_i} + B_{y_i}, \tag{6.2.41}$$

where

$$F_{x_i} = \oint_C \phi_i \, f_x \, dl, \qquad F_{y_i} \equiv \oint_C \phi_i \, f_y \, dl \tag{6.2.42}$$

are the *projected nodal edge forces* and

$$B_{x_i} \equiv \iint_D \phi_i \, b_x \, dx \, dy, \qquad B_{y_i} \equiv \iint_D \phi_i \, b_y \, dx \, dy \qquad (6.2.43)$$

are the *projected nodal body forces*. Note that, since ϕ_i is zero around the boundary for an interior node, the projected nodal edge forces are non-zero only if the ith node is a boundary node.

Implementation

The implementation of the finite element method for plane-stress or plane-strain analysis involves the following steps:

First step

Substitute the finite element expansion for the nodal displacement (6.2.34) into the definition of the strain tensor (6.1.10) to obtain the components of the strain tensor in terms of the node displacements,

$$\epsilon_{xx} = \sum_{j=1}^{N_G} v_{x_j} \frac{\partial \phi_j}{\partial x}, \quad \epsilon_{yy} = \sum_{j=1}^{N_G} v_{y_j} \frac{\partial \phi_j}{\partial y}, \quad \epsilon_{xy} = \frac{1}{2} \sum_{j=1}^{N_G} \left(v_{x_j} \frac{\partial \phi_j}{\partial y} + v_{y_j} \frac{\partial \phi_j}{\partial x} \right).$$

$$(6.2.44)$$

Second step

Substitute the strain components into (6.2.4) for the plane stress problem, or into (6.2.33) for the plane strain problem, to obtain the stresses in terms of the nodal displacements and their corresponding interpolation functions.

Third step

Substitute the stresses computed in the previous step into equations (6.2.40) and (6.2.41) to derive a linear system of algebraic equations for the x and y components of the nodal displacements.

Fourth step

Implement the boundary conditions in a way that preserves the symmetry of the coefficient matrix.

Fifth step

Solve a linear system for the nodal displacements.

Sixth step

Having recovered the displacements, compute the stresses using (6.2.4) for the plane stress problem or (6.2.33) for the plane strain problem.

Seventh step

Visualize the displacement and stress fields.

The particulars of the finite element code will be discussed in Section 6.3 for the plane stress problem.

PROBLEMS

6.2.1 *Shear flow past a circular or elliptical membrane*

(*a*) Confirm the solutions given in (6.2.15) and (6.2.16).

(*b*) Repeat (*a*) for (6.2.17) and (6.2.19).

6.2.2 *Plane strain analysis*

Derive the equations in (6.2.33) directly from (6.1.14), subject to the relations in (6.1.19).

6.3   Finite element plane stress analysis

In the plane stress analysis problem, the stresses field derives from the displacement field introduced in (6.2.34) as

$$
\sigma_{xx} = \frac{E}{1-\nu^2} \sum_{j=1}^{N_{\mathrm{G}}} \left(v_{x_j} \frac{\partial \phi_j}{\partial x} + \nu\, v_{y_j} \frac{\partial \phi_j}{\partial y} \right),
$$

$$
\sigma_{yy} = \frac{E}{1-\nu^2} \sum_{j=1}^{N_{\mathrm{G}}} \left(\nu\, v_{x_j} \frac{\partial \phi_j}{\partial x} + v_{y_j} \frac{\partial \phi_j}{\partial y} \right), \tag{6.3.1}
$$

$$
\sigma_{xy} = \frac{E}{2\left(1+\nu\right)} \sum_{j=1}^{N_{\mathrm{G}}} \left(v_{x_j} \frac{\partial \phi_j}{\partial y} + v_{y_j} \frac{\partial \phi_j}{\partial x} \right),
$$

where $\mathbf{v}_j$ are the nodal displacements. Substituting these expressions into the projection equations (6.2.40) and (6.2.41), we derive the equilibrium equations

$$
E \sum_{j=1}^{N_{\mathrm{G}}} \left(R_{ij}^{xx}\, v_{x_j} + R_{ij}^{xy}\, v_{y_j} \right) = F_{x_i} + B_{x_i} \tag{6.3.2}
$$

and

$$
E \sum_{j=1}^{N_{\mathrm{G}}} \left(R_{ij}^{yx}\, v_{x_j} + R_{ij}^{yy}\, v_{y_j} \right) = F_{y_i} + B_{y_i}, \tag{6.3.3}
$$

where

$$
R_{ij}^{xx} = \frac{1}{1-\nu^2} K_{ij}^{xx} + \frac{1}{2(1+\nu)} K_{ij}^{yy}, \qquad
R_{ij}^{xy} = \frac{\nu}{1-\nu^2} K_{ij}^{xy} + \frac{1}{2(1+\nu)} K_{ij}^{yx},
$$

$$
R_{ij}^{yx} = \frac{\nu}{1-\nu^2} K_{ij}^{yx} + \frac{1}{2(1+\nu)} K_{ij}^{xy}, \qquad
R_{ij}^{yy} = \frac{1}{1-\nu^2} K_{ij}^{yy} + \frac{1}{2(1+\nu)} K_{ij}^{xx}.
$$

$$\tag{6.3.4}$$

We have introduced four stiffness matrices with components

$$K_{ij}^{xx} \equiv \iint_D \frac{\partial \phi_i}{\partial x} \frac{\partial \phi_j}{\partial x} \, dx \, dy, \qquad K_{ij}^{xy} \equiv \iint_D \frac{\partial \phi_i}{\partial x} \frac{\partial \phi_j}{\partial y} \, dx \, dy,$$

$$K_{ij}^{yx} \equiv \iint_D \frac{\partial \phi_i}{\partial y} \frac{\partial \phi_j}{\partial x} \, dx \, dy, \qquad K_{ij}^{yy} \equiv \iint_D \frac{\partial \phi_i}{\partial y} \frac{\partial \phi_j}{\partial y} \, dx \, dy,$$

(6.3.5)

where i and j run over an appropriate number of nodes. Note the symmetry properties

$$K_{ij}^{xy} = K_{ji}^{yx}, \qquad\qquad R_{ij}^{xy} = R_{ji}^{yx}.$$

(6.3.6)

Equations (6.3.2) and (6.3.3) are collected into a linear system,

$$\mathbf{R} \cdot \begin{bmatrix} v_{x_1} \\ v_{x_2} \\ \vdots \\ v_{x_{N_G}} \\ --- \\ v_{y_1} \\ v_{y_2} \\ \vdots \\ v_{y_{N_G}} \end{bmatrix} = \begin{bmatrix} F_{x_1} \\ F_{x_2} \\ \vdots \\ F_{x_{N_G}} \\ --- \\ F_{y_1} \\ F_{y_2} \\ \vdots \\ F_{y_{N_G}} \end{bmatrix} + \begin{bmatrix} B_{x_1} \\ B_{x_2} \\ \vdots \\ B_{x_{N_G}} \\ --- \\ B_{y_1} \\ B_{y_2} \\ \vdots \\ B_{y_{N_G}} \end{bmatrix},$$

(6.3.7)

where

$$\mathbf{R} = E \left[\begin{array}{c|c} \mathbf{R}^{xx} & \mathbf{R}^{xy} \\ \hline \mathbf{R}^{yx} & \mathbf{R}^{yy} \end{array} \right]$$

(6.3.8)

is the *global block stiffness matrix*. Because of the symmetry properties (6.3.6), the matrix $\mathbf{R}$ is symmetric. The first vector on the right-hand side of (6.3.7) represents the effect of the edge force and the second vector represents the effect of the body force.

Projected nodal body force

To compute the nodal body forces expressed by the last term on the right-hand side of (6.3.7), we introduce the finite element expansion

$$\mathbf{b}(x, y) = \sum_{j=1}^{N_G} \mathbf{b}_j \, \phi_j(x, y).$$

(6.3.9)

Substituting this expression into (6.2.43), we obtain

$$B_{x_i} = \sum_{j=1}^{N_G} \left(\iint_D \phi_i \phi_j \, dx \, dy \right) \times b_{x_j}$$

(6.3.10)

and

$$B_{y_i} = \sum_{j=1}^{N_G} \left(\iint_D \phi_i \, \phi_j \, \mathrm{d}x \, \mathrm{d}y \right) \times b_{y_j}. \tag{6.3.11}$$

Consequently, we may write

$$\begin{bmatrix} B_{x_1} \\ B_{x_2} \\ \vdots \\ B_{x_{N_G}} \end{bmatrix} = \mathbf{M} \cdot \begin{bmatrix} b_{x_1} \\ b_{x_2} \\ \vdots \\ b_{x_{N_G}} \end{bmatrix}, \qquad \begin{bmatrix} B_{y_1} \\ B_{y_2} \\ \vdots \\ B_{y_{N_G}} \end{bmatrix} = \mathbf{M} \cdot \begin{bmatrix} b_{y_1} \\ b_{y_2} \\ \vdots \\ b_{y_{N_G}} \end{bmatrix}, \qquad (6.3.12)$$

where $\mathbf{M}$ is the global mass matrix.

Element stiffness matrix

FSELIB function *esmm6*, listed in Table 6.3.1, evaluates the element stiffness and mass matrices for a six-node triangle using a triangle Gaussian quadrature. The input includes the coordinates of the six element nodes in the xy plane and the order of the quadrature. The output includes the components of the element stiffness matrix, the mass matrix, and the element area.

6.3.1   Deformation due to an edge force

FSELIB code *psa6*, listed in Table 6.3.2, solves the equations of plane stress analysis in a square domain, possibly pierced by a circular hole, in the absence of a body force.

In the case of an unpierced square, the domain is discretized into six-node triangles using the FSELIB function *trgl6_sqr* discussed in Section 5.2.2. In the case of a pierced square, the domain is discretized into six-node triangles using the FSELIB function *trgl6_sc* discussed in Section 5.2.4. Other shapes can be implemented by straightforward substitutions.

The layout of the finite element code is similar to that discussed in Section 5.3 for Laplace's equation, involving the assembly of the stiffness matrices from corresponding element matrices evaluated using the FSELIB function *esmm6*, listed in Table 6.3.1.

Projected boundary traction

One new feature concerns the computation of the projected nodal values of the boundary traction components F_{x_i} and F_{y_i} defined in (6.2.42). In the main code *psa6*, the corresponding integrals are assembled from integrals around the element edges. Specifically, if the traction boundary condition is imposed at the element vertex nodes labeled 1 and 2, then the element edge subtended between these nodes contributes to the integrals shown in (6.2.42).

```
function [esm_xx, esm_xy, esm_yy, emass, arel] = esmm6 ...
...
     (x1,y1, x2,y2, x3,y3, x4,y4, x5,y5, x6,y6, NQ)

%=========================================================
% Evaluation of the element stiffness and mass matrices
% and element area (arel) for plane stress analysis
%=========================================================

%---------------------------------
% compute the mapping coefficients
%---------------------------------

[al, be, ga] = elm6_abc ...
...
  (x1,y1, x2,y2, x3,y3, x4,y4, x5,y5, x6,y6);

%----------------------------
% read the triangle quadrature
%----------------------------

[xi, eta, w] = gauss_trgl(NQ);

%----------------------------------------------
% initialize the stiffness and mass matrices
%----------------------------------------------

 for k=1:6

  emass(k) = 0.0;

  for l=1:6
   esm_xx(k,l) = 0.0;
   esm_xy(k,l) = 0.0;
   esm_yy(k,l) = 0.0;
  end

 end

%-----------------------
% perform the quadrature
%-----------------------

arel = 0.0;     % element area

for i=1:NQ

[psi, gpsi, hs] = elm6_interp ...
...
     (x1,y1, x2,y2, x3,y3, x4,y4, x5,y5, x6,y6 ...
     ,al,be,ga, xi(i),eta(i));

 cf = 0.5*hs*w(i);
```

TABLE 6.3.1   Function *esmm6* (Continuing →)

```
 for k=1:6
   emass(k) = emass(k) + psi(k)*cf;
   for l=1:6
     esm_xx(k,l) = esm_xx(k,l) + gpsi(k,1)*gpsi(l,1)*cf;
     esm_xy(k,l) = esm_xy(k,l) + gpsi(k,1)*gpsi(l,2)*cf;
     esm_yy(k,l) = esm_yy(k,l) + gpsi(k,2)*gpsi(l,2)*cf;
   end
 end

 arel = arel + cf;

end

%-----
% done
%-----

return;
```

TABLE 6.3.1 (→ Continued) Function *esmm6* evaluates the element stiffness and
mass matrices for plane stress analysis.

Assume that the mid-side node labeled 4 is located midway between vertex
nodes labeled 1 and 2. Consideration of the element node interpolation functions
shows that the contribution to a global node labeled i, corresponding to the element
vertex nodes labeled 1 or 2, is given by a consistent lumping approximation that
arises by applying the product integration rule, yielding

$$\Delta F_{x_i} = \Delta l \int_0^1 \psi_1(\xi) f_x \, d\xi \simeq \Delta l \frac{1}{6} f_{x_i} \tag{6.3.13}$$

and

$$\Delta F_{y_i} = \Delta l \int_0^1 \psi_1(\xi) f_y \, d\xi \sim \Delta l \frac{1}{6} f_{y_i}, \tag{6.3.14}$$

where Δl is the side length.

The contributions to the global node labeled i corresponding to the mid-side
element node labeled 4 is

$$\Delta F_{x_i} = \Delta l \int_0^1 \psi_4(\xi) f_x \, d\xi \simeq \Delta l \frac{2}{3} f_{x_i} \tag{6.3.15}$$

and

$$\Delta F_{y_i} = \Delta l \int_0^1 \psi_1(\xi) f_y \, d\xi \simeq \Delta l \frac{2}{3} f_{y_i}. \tag{6.3.16}$$

```
%===========
% Code psa6
%
% Plane stress analysis in a square
% or a square with a circular hole
%
% dependencies: elm6_abc elm_6_interp esmm6 gauss_trgl psa6_stress
%=============

%-----------
% input data
%-----------

E = 1.0;   % modulus of elasticity
nu = 0.5; % Poisson ratio
NQ = 6;    % Gauss-triangle quadrature
ndiv = 2; % discretization level

%------------
% triangulate
%------------

[ne,ng,p,c,efl,gfl] = trgl6_sqr(ndiv);   % square

% a = 0.5;   % hole radius
% [ne,ng,p,c,efl,gfl] = trgl6_sc(a, ndiv); % square with a hole

disp('Number of elements:'); ne

%--------
% prepare
%--------

ng2 = 2*ng;
nus = nu^2; % square of the Poisson ratio

%-------------------------
% count the interior nodes (optional)
%-------------------------

Ic = 0;

for j=1:ng
 if(gfl(j,1)==0) Ic = Ic+1; end
end

disp('Number of interior nodes:'); ic

%---------------------------------------
% specify the boundary condition at the
% top, bottom, and left sides,
% and the traction at the right side
%---------------------------------------
```

TABLE 6.3.2   Code *psa6* (Continuing →)

```
for i=1:ng

 gfl(i,2) = 0;    % initialize

 if(gfl(i,1)==1)     % boundary node

% default:  zero traction boundary condition

  gfl(i,2) = 1;    gfl(i,3) = 0.0;    % fx = 0
                   gfl(i,4) = 0.0;    % fy = 0

% apply the load at the right edge:

  if(p(i,1) > 0.99)

    gfl(i,3) = 0.0;                      % fx
    gfl(i,4) = 0.01*(1.0-p(i,2)^2); % fy

  end

% zero displacement at the left edge

  if(p(i,1) < -0.99)
    gfl(i,2) = 2;    % displacement boundary condition
    gfl(i,3) = 0.0;  % vx: x-displacement
    gfl(i,4) = 0.0;  % vy: y-displacement
  end

 end
end

%----------------------
% deform to a rectangle
%----------------------

defx = 1.0;    % x deformation factors
defy = 1.0;    % y deformation factors

for i=1:ng
  p(i,1) = p(i,1)*defx;
  p(i,2) = p(i,2)*defy;
end

%----------------------
% plot the element nodes
%----------------------

for i=1:ne
 i1=c(i,1); i2=c(i,2); i3=c(i,3); i4=c(i,4); i5=c(i,5); i6=c(i,6);
 xp(1)=p(i1,1); xp(2)=p(i4,1); xp(3)=p(i2,1); xp(4)=p(i5,1);
             xp(5)=p(i3,1); xp(6)=p(i6,1); xp(7)=p(i1,1);
 yp(1)=p(i1,2); yp(2)=p(i4,2); yp(3)=p(i2,2); yp(4)=p(i5,2);
             yp(5)=p(i3,2); yp(6)=p(i6,2); yp(7)=p(i1,2);
```

TABLE 6.3.2   Code *psa6* ($\rightarrow$ Continuing $\rightarrow$)

```
 plot(xp, yp,'+');
end

%-----------------------------------------
% assemble the global stiffness matrices
%-----------------------------------------

gsm_xx = zeros(ng,ng);    % initialize
gsm_xy = zeros(ng,ng);
gsm_yy = zeros(ng,ng);

% compute the element stiffness matrices

j=c(1,1); x1=p(j,1); y1=p(j,2);
j=c(1,2); x2=p(j,1); y2=p(j,2);
j=c(1,3); x3=p(j,1); y3=p(j,2);
j=c(1,4); x4=p(j,1); y4=p(j,2);
j=c(1,5); x5=p(j,1); y5=p(j,2);
j=c(1,6); x6=p(j,1); y6=p(j,2);

[esm_xx, esm_xy, esm_yy, arel] = esm6 ...
...
     (x1,y1, x2,y2, x3,y3, x4,y4, x5,y5, x6,y6 ,NQ);

  for i=1:6

    i1 = c(1,i);

    for j=1:6
      j1 = c(1,j);
      gsm_xx(i1,j1) = gsm_xx(i1,j1) + esm_xx(i,j);
      gsm_xy(i1,j1) = gsm_xy(i1,j1) + esm_xy(i,j);
      gsm_yy(i1,j1) = gsm_yy(i1,j1) + esm_yy(i,j);
    end

  end

end

%------------------------
% form the transpose block
%------------------------

gsm_yx = gsm_xy';   % a prime denotes the transpose

%-------------------------
% define the block matrices
%-------------------------

Rxu =    gsm_xx/(1-nus) + 0.5*gsm_yy/(1+nu);
Rxv = nu*gsm_xy/(1-nus) + 0.5*gsm_yx/(1+nu);
Ryu = Rxv';
Ryv =    gsm_yy/(1-nus) + 0.5*gsm_xx/(1+nu);
```

TABLE 6.3.2   Code *psa6* ($\rightarrow$ Continuing $\rightarrow$)

```
%------------------------
% compile the grand matrix
%------------------------

for i=1:ng
 for j=1:ng
  Gm(i,    j)    = E*Rxu(i,j);
  Gm(i,    j+ng) = E*Rxv(i,j);
  Gm(ng+i,j)     = E*Ryu(i,j);
  Gm(ng+i,j+ng)  = E*Ryv(i,j);
 end
end

%---------------------------
% assemble the right-hand side
% of the linear system
%---------------------------

for i=1:ng2
 b(i) = 0.0;
end

for i=1:ne

  % first side

  l1=c(i,1); l2=c(i,2); l3=c(i,4);

  if( gfl(l1,2)==1 & gfl(l2,2)==1)
    side = sqrt((p(l2,1)-p(l1,1))^2+(p(l2,2)-p(l1,2))^2);
    b(l1) = b(l1) + gfl(l1,3)*side*1.0/6.0;
    b(l2) = b(l2) + gfl(l2,3)*side*1.0/6.0;
    b(l3) = b(l3) + gfl(l3,3)*side*2.0/3.0;
    k1 = l1+ng;   k2 = l2+ng;   k3 = l3+ng;
    b(k1) = b(k1) + gfl(k1,4)*side*1.0/6.0;
    b(k2) = b(k2) + gfl(k2,4)*side*1.0/6.0;
    b(k3) = b(k3) + gfl(k3,4)*side*2.0/3.0;
  end

  % second side

  l1=c(i,2); l2=c(i,3); l3=c(i,5);

  if(gfl(l1,2)==1 & gfl(l2,2)==1)
    side = sqrt((p(l2,1)-p(l1,1))^2+(p(l2,2)-p(l1,2))^2);
    b(l1) = b(l1) + gfl(l1,3)*side*1.0/6.0;
    b(l2) = b(l2) + gfl(l2,3)*side*1.0/6.0;
    b(l3) = b(l3) + gfl(l3,3)*side*2.0/3.0;
    k1 = l1+ng; k2 = l2+ng; k3 = l3+ng;
    b(k1) = b(k1) + gfl(l1,4)*side*1.0/6.0;
    b(k2) = b(k2) + gfl(l2,4)*side*1.0/6.0;
    b(k3) = b(k3) + gfl(l3,4)*side*2.0/3.0;
  end
```

TABLE 6.3.2   Code *psa6* ($\rightarrow$ Continuing $\rightarrow$)

```
% third side

l1=c(i,3); l2=c(i,1); l3=c(i,6);

if(gfl(l1,2)==1 & gfl(l2,2)==1)
 side = sqrt((p(l2,1)-p(l1,1))^2+(p(l2,2)-p(l1,2))^2);
 b(l1) = b(l1) + gfl(l1,3)*side*1.0/6.0;
 b(l2) = b(l2) + gfl(l2,3)*side*1.0/6.0;
 b(l3) = b(l3) + gfl(l3,3)*side*2.0/3.0;
 k1 = l1+ng; k2 = l2+ng; k3 = l3+ng;
 b(k1) = b(k1) + gfl(k1,4)*side*1.0/6.0;
 b(k2) = b(k2) + gfl(k2,4)*side*1.0/6.0;
 b(k3) = b(k3) + gfl(k3,4)*side*2.0/3.0;
 end

end

%-----------------------------------------------
% implement the Dirichlet boundary condition
%-----------------------------------------------

for j=1:ng  % run over global nodes

%---
 if(gfl(j,2)==2)  % node with displacement boundary condition

   for i=1:ng2
    b(i) = b(i) - Gm(i,j) * gfl(j,3) - Gm(i,ng+j) * gfl(j,4);
    Gm(i,j) = 0; Gm(i,ng+j) = 0;
    Gm(j,i) = 0; Gm(ng+j,i) = 0;
   end

   Gm(j,j) = 1.0;
   Gm(j+ng,j+ng) = 1.0;
   b(j) = gfl(j,3);
   b(j+ng) = gfl(j,4);

 end
%---

end

%-----------------------
% solve the linear system
%-----------------------

f = b/Gm';

%-----------------------
% assign the displacements
%-----------------------
```

TABLE 6.3.2   Code *psa6* ($\rightarrow$ Continuing $\rightarrow$)

```
for i=1:ng
 vx(i) = f(i);
 vy(i) = f(ng+i);
end

%-------------------------
% plot the deformed elements
%-------------------------

for i=1:ng
  p(i,1) = p(i,1)+vx(i);
  p(i,2) = p(i,2)+vy(i);
end

for i=1:ne

 i1=c(i,1); i2=c(i,2); i3=c(i,3); i4=c(i,4); i5=c(i,5); i6=c(i,6);

 xp(1)=p(i1,1); xp(2)=p(i4,1); xp(3)=p(i2,1); xp(4)=p(i5,1);
               xp(5)=p(i3,1); xp(6)=p(i6,1); xp(7)=p(i1,1);

 yp(1)=p(i1,2); yp(2)=p(i4,2); yp(3)=p(i2,2); yp(4)=p(i5,2);
               yp(5)=p(i3,2); yp(6)=p(i6,2); yp(7)=p(i1,2);

 plot(xp, yp,'k-o');

end

axis([-1.2 1.2 -1.2 1.2]);
xlabel('x'); ylabel('y');

%---
% reset
%---

for i=1:ng
  p(i,1) = p(i,1)-vx(i);
  p(i,2) = p(i,2)-vy(i);
end

%-------------------------------------------------
% average the element nodal stresses to compute
% the stresses at the global nodes
%-------------------------------------------------

%---
% initialize the global node stresses:
%---

for j=1:ng
 gsig_xx(j) = 0.0; gsig_xy(j) = 0.0; gsig_yy(j) = 0.0;
 gitally(j) = 0;
end
```

TABLE 6.3.2   Code *psa6* ($\rightarrow$ Continuing $\rightarrow$)

```
%---
% loop over the element nodes and compute
% the element nodal stresses
%---

for l=1:ne

 j=c(l,1); xe(1)=p(j,1); ye(1)=p(j,2); vxe(1)=vx(j); vye(1)=vy(j);
 j=c(l,2); xe(2)=p(j,1); ye(2)=p(j,2); vxe(2)=vx(j); vye(2)=vy(j);
 j=c(l,3); xe(3)=p(j,1); ye(3)=p(j,2); vxe(3)=vx(j); vye(3)=vy(j);
 j=c(l,4); xe(4)=p(j,1); ye(4)=p(j,2); vxe(4)=vx(j); vye(4)=vy(j);
 j=c(l,5); xe(5)=p(j,1); ye(5)=p(j,2); vxe(5)=vx(j); vye(5)=vy(j);
 j=c(l,6); xe(6)=p(j,1); ye(6)=p(j,2); vxe(6)=vx(j); vye(6)=vy(j);

 [sig_xx, sig_xy, sig_yy] = psa6_stress ...
 ...
     (xe,ye,vxe,vye,E,nu);

%---
% add element node to the global node and increase tally by one
%---

 for k=1:6
   j=c(l,k);
   gitally(j) = gitally(j) + 1;
   gsig_xx(j) = gsig_xx(j) + sig_xx(k);
   gsig_xy(j) = gsig_xy(j) + sig_xy(k);
   gsig_yy(j) = gsig_yy(j) + sig_yy(k);
 end

end

for j=1:ng
 gsig_xx(j) = gsig_xx(j)/gitally(j);
 gsig_xy(j) = gsig_xy(j)/gitally(j);
 gsig_yy(j) = gsig_yy(j)/gitally(j);
end

%-----------------------------------
% generalize the connectivity matrix
% for three-node sub-triangles
%-----------------------------------

Ic = 0;

for i=1:ne

 Ic = Ic+1;
 c3(Ic,1) = c(i,1); c3(Ic,2) = c(i,4); c3(Ic,3) = c(i,6);
 Ic = Ic+1;
 c3(Ic,1) = c(i,4); c3(Ic,2) = c(i,2); c3(Ic,3) = c(i,5);
 Ic = Ic+1;
 c3(Ic,1) = c(i,5); c3(Ic,2) = c(i,3); c3(Ic,3) = c(i,6);
 Ic = Ic+1;
 c3(Ic,1) = c(i,4); c3(Ic,2) = c(i,5); c3(Ic,3) = c(i,6);

end
```

TABLE 6.3.2   Code *psa6* ($\rightarrow$ Continuing $\rightarrow$)

```
%----------------------------
% plot the stress field using
% the matlab function trimesh
%----------------------------

figure
%trimesh (c3,p(:,1),p(:,2),gsig_xx);
%trimesh (c3,p(:,1),p(:,2),gsig_yy);
 trimesh (c3,p(:,1),p(:,2),gsig_xy);
hold on;

%----------------------------------
% plot the displacement vector field
%----------------------------------

quiver (p(:,1)',p(:,2)',vx,vy);
xlabel('x'); ylabel('y');

%-----
% done
%-----
```

TABLE 6.3.2 (→ Continued) Code *psa6* for plane stress analysis in a square block possibly pierced by a circular hole. Other shapes can be implemented by straight-forward substitutions.

Similar contributions are made by element edges subtended by the element vertex node pairs $(1, 2)$ and $(3, 1)$.

Stress recovery module

Once the finite solution has been computed, post-processing is performed to extract various quantities of interest deriving from the nodal displacements. In the *stress recovery* module, the element node stresses are evaluated directly, as implemented in the FSELIB function *psa6_stress*, listed in Table 6.3.3. In some finite element implementations with quadrilateral elements, the stresses are computed at Gaussian integration points and then extrapolated to the element interpolation nodes.

Because the solution for the nodal displacements is only C^0 continuous, the strains are discontinuous across the element edges, and so are the stresses. To construct continuous fields, we may evaluate the stresses at the global nodes as unweighted averages of the corresponding element nodal values, as shown in the finite element code *psa6*. Alternatively, we may evaluate the global nodal stresses as weighted averages of the element nodal values, where the weights are adjusted according to the apertures of the corners subtended by the element edges. Finally, the components of the stress field are visualized based on the values at the global nodes, as shown in code *psa6*.

```
function [sig_xx, sig_xy, sig_yy] = psa6_stress ...
 ...
      (x,y,u,v,E,nu);

%=========================================
% Evaluation of the node element stresses
% in plane stress analysis
%=========================================

%----------------------------------------
% compute the mapping coefficients
%----------------------------------------

[al, be, ga] = elm6_abc ...
 ...
 (x(1),y(1), x(2),y(2), x(3),y(3), x(4),y(4), x(5),y(5), x(6),y(6));

%----------------------------
% nodal (xi, eta) coordinates
%----------------------------

xi(1) = 0.0; eta(1) = 0.0;
xi(2) = 1.0; eta(2) = 0.0;
xi(3) = 0.0; eta(3) = 1.0;
xi(4) = al;  eta(4) = 0.0;
xi(5) = be ; eta(5) = 1-be;
xi(6) = 0.0; eta(6) = ga;

%-------------------------
% compute the node stresses
%-------------------------

nus = nu^2;

for l=1:6  % run over the nodes

 sig_xx(l) = 0.0; sig_xy(l) = 0.0; sig_yy(l) = 0.0;

  [psi, gpsi, hs] = elm6_interp ...
   ...
   (x(1),y(1), x(2),y(2), x(3),y(3) ...
   ,x(4),y(4), x(5),y(5), x(6),y(6) ...
   ,al,be,ga, xi(l),eta(l));

  for k=1:6

    sig_xx(l) = sig_xx(l)...
        +(u(k)*gpsi(k,1)+nu*v(k)*gpsi(k,2))/(1-nus);

    sig_xy(l) = sig_xy(l)...
        +0.5*(u(k)*gpsi(k,2)+v(k)*gpsi(k,1))/(1+nu);
```

TABLE 6.3.3 Function *psa6_stress* (Continuing →)

```
    sig_yy(1) = sig_yy(1)...
          +(nu*u(k)*gpsi(k,1)+v(k)*gpsi(k,2))/(1-nus);
  end

  sig_xx(1) = E*sig_xx(1);
  sig_xy(1) = E*sig_xy(1);
  sig_yy(1) = E*sig_yy(1);

end

%-----
% done
%-----

return;
```

TABLE 6.3.3 ($\rightarrow$ Continued) Function *psa6_stress* evaluates of the element node stresses in plane stress analysis at the conclusion of the finite element solution.

Deformation of a square plate

As an example, we consider the deformation of a square plate subject to the following boundary conditions implemented in the code *psa6*, listed in Table 6.3.2:

- Zero displacement boundary condition on the left side (fixed side).

- Zero traction boundary condition at the bottom and top sides (free sides).

- Zero traction boundary condition for the x component and a quadratically varying boundary condition for the y component of the traction on the right side (loaded side), as implemented in the code.

The finite element solution generated by the code is shown in Figure 6.3.1. The element shapes and nodal positions before and after deformation are marked with crosses and circles in Figure 6.3.1(a).

The components of the stress tensor visualized by the MATLAB function *trimesh* are shown in Figure 6.3.1(b–d) along with an overlapping *quiver* (arrow) plot of the nodal displacements. It is reassuring that the computed boundary stress distribution is consistent with the specified boundary conditions. In particular, σ_{xx} and σ_{yy} are zero along the right edge of the plate, whereas σ_{xy} displays the specified parabolic dependence on y, as implemented in the code.

Similar results are presented in Figure 6.3.2 for a square pierced by a circular hole. In this application, the zero stress condition is imposed around the contour of the hole (free boundary). The circular hole deform to obtain an elliptical shape under the action of the edge force applied at the right side of the plate.

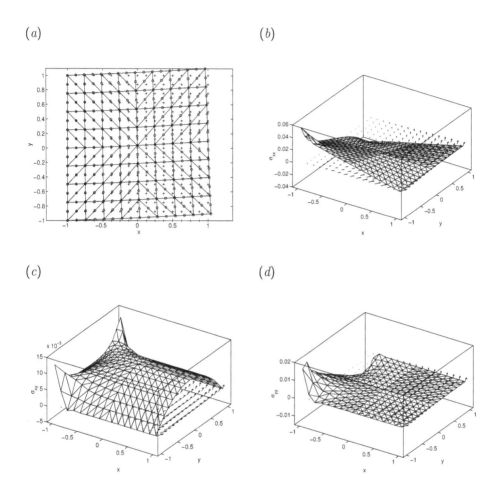

FIGURE 6.3.1   Plane stress analysis for a square plate. The zero displacement boundary condition is applied at the left edge (fixed edge), the zero traction boundary
condition is applied at the bottom and top edges (free edges), and a vertical upward traction boundary condition is applied on the right side (loaded edge). (a)
Element shapes and nodal positions before (crosses) and after (circles) deformation. (b–d) Graphs of the xx, xy, and yy components of the stress tensor. The
vector fields in (b–d) illustrate the nodal displacements.

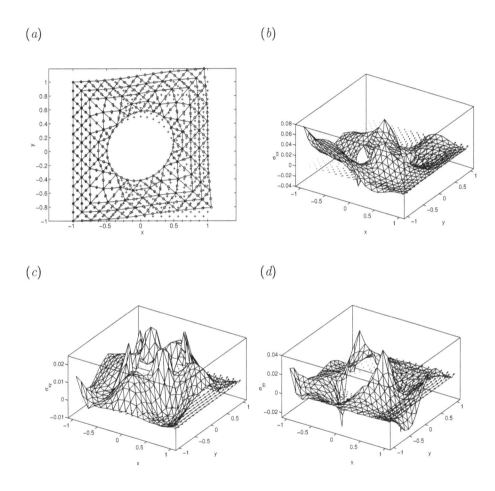

FIGURE 6.3.2 Same as Figure 6.3.1, except that the square plate is pierced by a
circular hole where the zero-traction boundary condition is imposed.

6.3.2 Deformation due to a body force

FSELIB code *psa6M*, listed in Table 6.3.4, solves the equations of plane stress analysis in the presence of a uniform body force along the x axis, subject to the zero traction and zero displacement boundary condition around the edges.

In the physical circumstances implemented in the code, we have stipulated that b_x is constant and $b_y = 0$. Physically, the elastic plate is a membrane attached to a rigid rim, undergoing deformation under the action of gravity or shear stress induced by an overpassing shear flow. Other body force distributions can be implemented by straightforward modifications. Exact solutions for a circular or elliptical membrane were discussed in Section 6.2.2.

In the finite element implementation, the equations in (6.3.12) are used to compute the x and y components of the projected nodal body force in terms of the global mass matrix. Following standard practice, the zero displacement boundary condition is implemented after a linear system is compiled for all global nodes to preserve the symmetry of the coefficient matrix.

The finite element solutions generated by the code for a discoidal and a square membrane are shown in Figures 6.3.3 and 6.3.4. The element shapes and nodal positions before and after deformation are shown as crosses and circles in Figures 6.3.3(a) and 6.3.4(a). The components of the stress tensor generated by the MATLAB function *trimesh* and an overlapping *quiver* (arrow) plot of the nodal displacements, **u**, are shown in Figures 6.3.3(b–d) and 6.3.4(b–d). It is reassuring that the computed displacement field is consistent with the specified boundary conditions.

The numerical results for the deformation of the discoidal membrane shown in Figure 6.3.3(b–d) are in excellent agreement with the exact solution discussed in Section 6.2.2.

PROBLEMS

6.3.1 *Plane stress analysis*

(a) Run the code *psa6* for a rectangular plate with aspect ratio 1:4 or 4:1. Compare and discuss the finite element solutions.

(b) Repeat (a) for a rectangular plate with a circular hole.

(c) Modify the code *psa6* so that the boundary conditions specify zero displacement on the left and right edges (fixed sides), zero traction at the bottom (free side), and a specified non-zero boundary condition for the x and y components of the traction on the upper edge (loaded side). Run the modified code for a rectangular plate and an upper-side traction boundary condition of your choice and discuss the numerical results.

```
%===========
% Code  psa6M
%
% In-plane deformation of a membrane patch under
% a constant body force in plane stress analysis
%=============

clear all
close all

%-----------
% input data
%-----------

E = 1.0;   % modulus of elasticity
nu = 0.5;  % Poisson ratio
NQ = 6;    % Gauss-triangle quadrature
bx = 1.00; % body force
by = 0.00; % body force

ndiv = 2; % discretization level

ishape = 3;

%------------
% triangulate
%------------

if(ishape==1)
  [ne,ng,p,c,efl,gfl] = trgl6_sqr(ndiv);    % square
elseif(ishape==2)
  a = 0.5;   % hole radius
  [ne,ng,p,c,efl,gfl] = trgl6_sc(a, ndiv); % square with a hole
elseif(ishape==3)
  [ne,ng,p,c,efl,gfl] = trgl6_disk(ndiv);
end

disp('Number of elements:'); ne

%--------
% prepare
%--------

ng2 = 2*ng;
nus = nu^2;  % square of the Poisson ratio

%-------------------------
% count the interior nodes (optional)
%-------------------------

Ic = 0;
```

TABLE 6.3.4   Code *psa6M* (Continuing →)

```
for j=1:ng
 if(gfl(j,1)==0) Ic = Ic+1; end
end

disp('Number of interior nodes:'); Ic

%-------------------------------------------------
% specify the zero displacement boundary condition
%-------------------------------------------------

for i=1:ng
 if(gfl(i,1)==1)     % boundary node
   gfl(i,2) = 2;     % displacement BC
   gfl(i,3) = 0.0;   % u x-displacement
   gfl(i,4) = 0.0;   % v y-displacement
 end
end

%-------
% deform
%-------

def = 0.0;    % deformation factor

for i=1:ng
  p(i,1) = p(i,1)*(1.0 + def);
  p(i,2) = p(i,2)*(1.0 - def);
end

%------------------
% plot the elements
%------------------

for i=1:ne

 i1=c(i,1); i2=c(i,2); i3=c(i,3); i4=c(i,4); i5=c(i,5); i6=c(i,6);

 xp(1)=p(i1,1); xp(2)=p(i4,1); xp(3)=p(i2,1); xp(4)=p(i5,1);
               xp(5)=p(i3,1); xp(6)=p(i6,1); xp(7)=p(i1,1);
 yp(1)=p(i1,2); yp(2)=p(i4,2); yp(3)=p(i2,2); yp(4)=p(i5,2);
               yp(5)=p(i3,2); yp(6)=p(i6,2); yp(7)=p(i1,2);

 plot (xp, yp,'k+','markersize',5)

end

%----------------------------
% assemble the global stiffness
% and mass matrices
%----------------------------
```

TABLE 6.3.4   Code *psa6M* ($\rightarrow$ Continuing $\rightarrow$)

```
gsm_xx = zeros(ng,ng); % initialize
gsm_xy = zeros(ng,ng); % initialize
gsm_yy = zeros(ng,ng); % initialize

for i=1:ng
  gmass(i) = 0.0;
end
for l=1:ne            % loop over the elements

% compute the element stiffness matrices and RHS

j=c(l,1); x1=p(j,1); y1=p(j,2);
j=c(l,2); x2=p(j,1); y2=p(j,2);
j=c(l,3); x3=p(j,1); y3=p(j,2);
j=c(l,4); x4=p(j,1); y4=p(j,2);
j=c(l,5); x5=p(j,1); y5=p(j,2);
j=c(l,6); x6=p(j,1); y6=p(j,2);

[esm_xx, esm_xy, esm_yy, emass, arel] = esmm6 ...
...
     (x1,y1, x2,y2, x3,y3, x4,y4, x5,y5, x6,y6, NQ);

   for i=1:6
     i1 = c(l,i);
     gmass(i1) = gmass(i1) + emass(i);
     for j=1:6
       j1 = c(l,j);
       gsm_xx(i1,j1) = gsm_xx(i1,j1) + esm_xx(i,j);
       gsm_xy(i1,j1) = gsm_xy(i1,j1) + esm_xy(i,j);
       gsm_yy(i1,j1) = gsm_yy(i1,j1) + esm_yy(i,j);
     end
   end

end

%-------------------------
% form the transpose block
%-------------------------

gsm_yx = gsm_xy';      % a prime denotes the transpose

%-------------------------
% form the transpose block
%-------------------------

gsm_yx = gsm_xy';      % a prime denotes the transpose

%-------------------------
% define the block matrices
%-------------------------
```

TABLE 6.3.4   Code *psa6M* ($\rightarrow$ Continuing $\rightarrow$)

```
Rxu =    gsm_xx/(1-nus) + 0.5*gsm_yy/(1+nu);
Rxv = nu*gsm_xy/(1-nus) + 0.5*gsm_yx/(1+nu);
Ryu = Rxv';
Ryv =    gsm_yy/(1-nus) + 0.5*gsm_xx/(1+nu);

%------------------------
% compile the grand matrix
%------------------------

for i=1:ng
 for j=1:ng
  Gm(i,   j) = E*Rxu(i,j); Gm(i,   j+ng) = E*Rxv(i,j);
  Gm(ng+i,j) = E*Ryu(i,j); Gm(ng+i,j+ng) = E*Ryv(i,j);
 end
end

%------------------------
% set the right-hand side of the linear system
% and implement the Dirichlet BC
%------------------------

for i=1:ng2
 b(i) = 0.0;
end

for j=1:ng  % run over nodes

b(j)    = gmass(j)*bx;
b(j+ng) = gmass(j)*by;

%---
 if(gfl(j,1)==1 & gfl(j,2)==2)  % boundary node with displacement BC

   for i=1:ng2
    b(i) = b(i) - Gm(i,j) * gfl(j,3) - Gm(i,ng+j) * gfl(j,4);
    Gm(i,j) = 0; Gm(i,ng+j) = 0;
    Gm(j,i) = 0; Gm(ng+j,i) = 0;
   end

  Gm(j,   j)     = 1.0;
  Gm(j+ng,j+ng) = 1.0;
  b(j)     = gfl(j,3);
  b(j+ng) = gfl(j,4);

 end
%---

end
```

TABLE 6.3.4   Code *psa6M* ($\rightarrow$ Continuing $\rightarrow$)

```
%------------------------
% solve the linear system
%------------------------

f = b/Gm';

%------------------------
% assign the displacements
%------------------------

for i=1:ng
 vx(i) = f(i);
 vy(i) = f(ng+i);
end

%--------------------------
% plot the deformed elements
%--------------------------

for i=1:ng
  p(i,1) = p(i,1)+vx(i);
  p(i,2) = p(i,2)+vy(i);
end

for i=1:ne

 i1=c(i,1); i2=c(i,2); i3=c(i,3);i4=c(i,4);i5=c(i,5);i6=c(i,6);

 xp(1)=p(i1,1); xp(2)=p(i4,1); xp(3)=p(i2,1); xp(4)=p(i5,1);
                xp(5)=p(i3,1); xp(6)=p(i6,1); xp(7)=p(i1,1);
 yp(1)=p(i1,2); yp(2)=p(i4,2); yp(3)=p(i2,2); yp(4)=p(i5,2);
                yp(5)=p(i3,2); yp(6)=p(i6,2); yp(7)=p(i1,2);
 plot(xp, yp,'-ko','markersize',5)
end

%---
% reset
%---

for i=1:ng
  p(i,1) = p(i,1)-vx(i);
  p(i,2) = p(i,2)-vy(i);
end

%-------------------------------
% average the element nodal stresses
% to compute the stresses
% at the global nodes
%-------------------------------

for j=1:ng                    % initialize the global node stresses
 gsig_xx(j) = 0.0;
 gsig_xy(j) = 0.0;
```

TABLE 6.3.4   Code *psa6M* ($\rightarrow$ Continuing $\rightarrow$)

```
 gsig_yy(j) = 0.0;
 gitally(j) = 0;
end

%---
% loop over the element nodes and compute
% the element nodal stresses
%---

for l=1:ne

j=c(l,1); xe(1)=p(j,1); ye(1)=p(j,2); ue(1)=u(j); ve(1)=v(j);
j=c(l,2); xe(2)=p(j,1); ye(2)=p(j,2); ue(2)=u(j); ve(2)=v(j);
j=c(l,3); xe(3)=p(j,1); ye(3)=p(j,2); ue(3)=u(j); ve(3)=v(j);
j=c(l,4); xe(4)=p(j,1); ye(4)=p(j,2); ue(4)=u(j); ve(4)=v(j);
j=c(l,5); xe(5)=p(j,1); ye(5)=p(j,2); ue(5)=u(j); ve(5)=v(j);
j=c(l,6); xe(6)=p(j,1); ye(6)=p(j,2); ue(6)=u(j); ve(6)=v(j);

[sig_xx, sig_xy, sig_yy] = psa6_stress ...
...
     (xe,ye,ue,ve,E,nu);

 for k=1:6
   j=c(l,k);
   gitally(j) = gitally(j) + 1;
   gsig_xx(j) = gsig_xx(j) + sig_xx(k);
   gsig_xy(j) = gsig_xy(j) + sig_xy(k);
   gsig_yy(j) = gsig_yy(j) + sig_yy(k);
 end

end

for j=1:ng
 gsig_xx(j) = gsig_xx(j)/gitally(j);
 gsig_xy(j) = gsig_xy(j)/gitally(j);
 gsig_yy(j) = gsig_yy(j)/gitally(j);
end

%-----
% extend the connectivity matrix
% for three-node sub-triangles
%-----
```

TABLE 6.3.4   Code *psa6M* ($\rightarrow$ Continuing $\rightarrow$)

```
Ic = 0;

for i=1:ne
  Ic = Ic+1;
  c3(Ic,1) = c(i,1); c3(Ic,2) = c(i,4); c3(Ic,3) = c(i,6);
  Ic = Ic+1;
  c3(Ic,1) = c(i,4); c3(Ic,2) = c(i,2); c3(Ic,3) = c(i,5);
  Ic = Ic+1;
  c3(Ic,1) = c(i,5); c3(Ic,2) = c(i,3); c3(Ic,3) = c(i,6);
  Ic = Ic+1;
  c3(Ic,1) = c(i,4); c3(Ic,2) = c(i,5); c3(Ic,3) = c(i,6);
end

%---------------------
% plot the stress field
% using the matlab trimesh function
%---------------------

figure
hold on
trimesh (c3,p(:,1),p(:,2),gsig_xx);
%trimesh (c3,p(:,1),p(:,2),gsig_yy);
%trimesh (c3,p(:,1),p(:,2),gsig_xy);
quiver (p(:,1)',p(:,2)',u,v);
xlabel('x','fontsize',14)
ylabel('y','fontsize',14)
zlabel('\sigma xx','fontsize',14)

%-----
% done
%-----
```

TABLE 6.3.4   (→ Continued) Code *psa6M* for plane stress analysis of a clamped block subject to a distributed body force.

6.3.2 *Shear flow past a membrane*

Run the code *psa6M* for an elliptical membrane with aspect ratio 2:1. Confirm that the numerical results are consistent with the exact solution discussed in Section 6.2.2.

6.3.3 *Plane strain analysis*

(*a*) Write the counterparts of the equations in (6.3.1) for plane strain analysis.

(*b*) Modify the FSELIB code *psa6* into a code named *pstrna6* that performs plane strain analysis. Run the modified code for a rectangular plate and boundary conditions of your choice and discuss the finite element solution.

(a)   (b)

(c)   (d)

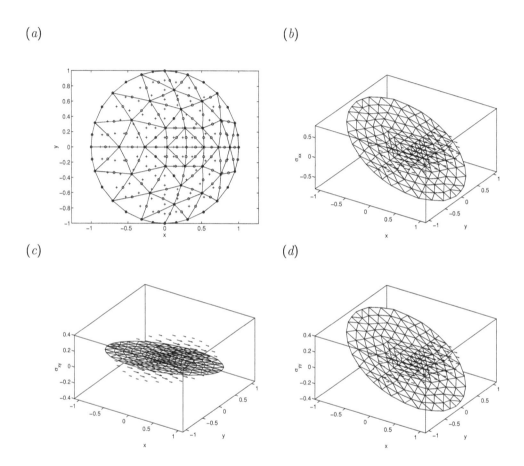

FIGURE 6.3.3   Plane stress analysis of a clamped discoidal plate under the action of
a uniform body force along the x axis generated by the FSELIB code *psa6M*. The
zero displacement boundary condition is applied along the circular contour. (a) Element shapes and nodal positions before (crosses) and after (circles) deformation.
(b–d) Graphs of the xx, xy, and yy components of the stress tensor for Poisson
ratio $\nu = 0.5$. The vector fields in (b–d) illustrate the nodal displacements.

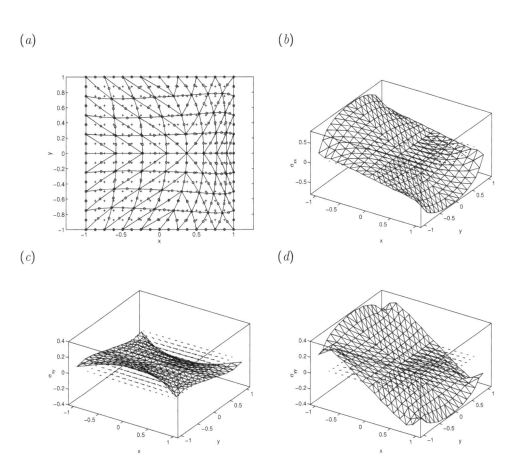

FIGURE 6.3.4 Plane stress analysis of a clamped square plate under the action of a uniform body force along the x axis generated by the FSELIB code *psa6M*. The zero displacement boundary condition is applied along the four edges. (a) Element shapes and nodal positions before (crosses) and after (circles) deformation. (b–d) Graphs of the xx, xy, and yy components of the stress tensor for Poisson ratio $\nu = 0.5$. The vector fields in (b–d) illustrate the nodal displacements.

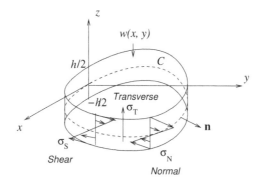

FIGURE 6.4.1 Illustration of an elastic flat plate deforming under the influence of a transverse load with surface density $w(x, y)$ parallel to the z axis. The contour, C, corresponds to the plate mid-surface. The tangential unit vector, $\mathbf{n}$, is normal to C.

6.4 Plate bending

Consider a flat elastic plate with small thickness, h, positioned parallel to the xy plane, as shown in Figure 6.4.1. We are interested in computing the transverse displacement of the plate along the z axis under the influence of gravity or a mechanical load applied normal to the upper or lower surface. The complementary plane stress analysis problem where the plate deforms predominantly in its plane under the influence of an in-plane edge or body force in the xy plane was discussed in Sections 6.1–6.3. The bending of a compressed plate under the influence of a combined transverse and in-plane load is discussed in Section 6.9.

Transverse, shear, and normal stresses

Because of the predominantly transverse deformation along the z axis, stresses develop over a cross-section of the plate, as shown in Figure 6.4.1. In the classical theory of plates, the cross-sectional stress is resolved into the following three components:

- A transverse shear stress pointing along the z axis, denoted as σ_T.

- An in-plane normal stress that is tangential to the plate and perpendicular to the cross-sectional contour, C, denoted as σ_N.

- An in-plane shear stress that is tangential to the plate and also tangential to the cross-sectional contour, C, denoted as σ_S.

To develop a theory for plate bending, we integrate the transverse, normal, and shear stresses over the plate cross-section to obtain corresponding tensions and bending moments, which we then use to formulate force and torque equilibrium equations and subsequently shape-governing equations.

Tensions and bending moments

Integrating the transverse shear stress over a cross-section of the plate, we obtain the transverse shear tension,

$$q \equiv \int_{-h/2}^{h/2} \sigma_T \, dz, \tag{6.4.1}$$

where h is the plate thickness. Physically, the transverse shear tension is the force per unit length exerted around the contour of the plate cross-section in the mid-plane, C, as shown in Figure 6.4.1.

Likewise integrating the in-plane stresses over a cross-section of the plate, we derive expressions for the in-plane normal and shear tensions,

$$T_N \equiv \int_{-h/2}^{h/2} \sigma_N \, dz, \qquad T_S \equiv \int_{-h/2}^{h/2} \sigma_S \, dz. \tag{6.4.2}$$

Variations of the in-plane stresses over the plate cross-section are responsible for normal and tangential bending moments computed with respect to the undeformed mid-surface,

$$M_N \equiv \int_{-h/2}^{h/2} \sigma_N \, z \, dz, \qquad M_S \equiv \int_{-h/2}^{h/2} \sigma_S \, z \, dz. \tag{6.4.3}$$

In the classical theory of plates and shells, equilibrium equations are written with respect to the tensions and bending moments instead of the stresses. One important step in completing the theory and achieving closure is the derivation of constitutive equations relating tensions and bending moments to mid-plane deformation. This challenging task must be accomplished in a way that neither introduces mathematical incompatibilities nor compromises the fundamental physical laws.

Tension and bending moment tensors

To derive force and moment equilibrium equations, we consider two intersecting cross-sections of the plate that are normal to the x and y axes, as illustrated in Figure 6.4.2, and introduce transverse shear tensions and in-plane stresses as follows:

- The transverse shear tensions, q_x and q_y, act on a cross-section that is normal to the x or y axis.

- The in-plane normal and shear stresses, σ_{xx} and σ_{xy}, act on a cross-section that is normal to the x axis. The corresponding tensions are denoted by τ_{xx} and τ_{xy}.

- The in-plane normal and shear stresses, σ_{yy} and σ_{yx}, act on a cross-section that is normal to the y axis. The corresponding tensions are denoted by τ_{yy} and τ_{yx}.

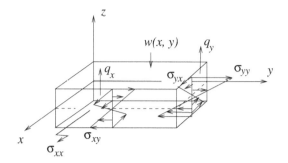

FIGURE 6.4.2   Illustration of transverse shear tensions and in-plane stresses acting on a cross-section of the plate that is normal to the x or y axis.

It is expeditious to collect the transverse shear tensions to collect into a tangential vector field,

$$\mathbf{q} = (\, q_x, \, q_y \,),\qquad(6.4.4)$$

defined so that the transverse shear tension exerted on a cross-section of the plate that is normal to the tangential unit vector, $\mathbf{n}$, is given by

$$q = \mathbf{n} \cdot \mathbf{q},\qquad(6.4.5)$$

as illustrated in Figure 6.4.1. For example, setting $\mathbf{n}$ parallel to the x axis, $\mathbf{n} = (1,0)$, we obtain the expected result, $q = q_x$.

The in-plane stresses and corresponding tensions can be collected into tangential stress and tension fields,

$$\mathbf{\Sigma} = \begin{bmatrix} \sigma_{xx} & \sigma_{xy} \\ \sigma_{yx} & \sigma_{yy} \end{bmatrix}, \qquad \mathbf{\tau} = \begin{bmatrix} \tau_{xx} & \tau_{xy} \\ \tau_{yx} & \tau_{yy} \end{bmatrix}.\qquad(6.4.6)$$

The tension field is defined such that the in-plane tension exerted on a cross-section of the plate that is normal to the tangential unit vector, $\mathbf{n}$, is given by

$$\mathbf{f} = \mathbf{n} \cdot \mathbf{\tau}.\qquad(6.4.7)$$

For example, setting $\mathbf{n}$ parallel to the y axis, $\mathbf{n} = (0,1)$, we obtain the expected result, $f_x = \tau_{yx}$ and $f_y = \tau_{yy}$.

The tangential bending moments computed with respect to the undeformed mid-surface are defined as

$$M_{ij} \equiv \int_{-h/2}^{h/2} \sigma_{ij}\, z\, \mathrm{d}z\qquad(6.4.8)$$

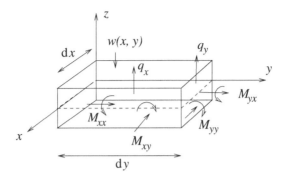

FIGURE 6.4.3   Illustration of bending moments exerted at the cross-section of a plate that is normal to the x or y axis.

for $i, j = x, y$, as shown in Figure 6.4.3, where $z = 0$ marks the location of the undeformed mid-surface. The four scalar bending moments can be collected into a Cartesian bending moments tensor,

$$\mathbf{M} = \begin{bmatrix} M_{xx} & M_{xy} \\ M_{yx} & M_{yy} \end{bmatrix}. \tag{6.4.9}$$

For obvious reasons, the off-diagonal components of $\mathbf{M}$ are called the *twisting moments*.

The bending moments tensor, $\mathbf{M}$, has been defined such that the vector of bending moments acting on a cross-section of the plate that is normal to the tangential unit vector, $\mathbf{n}$, is given by

$$\mathbf{m} = \mathbf{e}_z \times (\mathbf{n} \cdot \mathbf{M}), \tag{6.4.10}$$

where $\mathbf{e}_z$ is the unit vector along the z axis and $\times$ denotes the outer vector product, as illustrated in Figure 6.4.1. For example, setting $\mathbf{n}$ parallel to the x axis, $\mathbf{n} = (1, 0)$, we find that

$$\mathbf{m} = \mathbf{e}_z \times \begin{bmatrix} M_{xx} \\ M_{xy} \end{bmatrix} = \begin{bmatrix} -M_{xy} \\ M_{xx} \end{bmatrix}, \tag{6.4.11}$$

which is consistent with the vector drawings on the front side of the rectangular section depicted in Figure 6.4.3. Similarly, setting $\mathbf{n}$ parallel to the y axis, $\mathbf{n} = (0, 1)$, we find that

$$\mathbf{m} = \mathbf{e}_z \times \begin{bmatrix} M_{yx} \\ M_{yy} \end{bmatrix} = \begin{bmatrix} -M_{yy} \\ M_{yx} \end{bmatrix}, \tag{6.4.12}$$

which is consistent with the vector drawings on the right side of the rectangular section depicted in Figure 6.4.3.

The vector of bending moments can be resolved into a normal twisting component, denoted by m_{twist}, and a remaining tangential pure bending component, denoted by $\mathbf{m}_{\text{bend}}$,

$$\mathbf{m} = m_{\text{twist}}\,\mathbf{n} + \mathbf{m}_{\text{bend}}, \tag{6.4.13}$$

where

$$m_{\text{twist}} \equiv \mathbf{m} \cdot \mathbf{n} = \big[\mathbf{e}_z \times (\mathbf{n} \cdot \mathbf{M})\big] \cdot \mathbf{n} \tag{6.4.14}$$

and

$$\mathbf{m}_{\text{bend}} = \mathbf{m} - m_{\text{twist}}\,\mathbf{n} = \mathbf{m} - (\mathbf{m} \cdot \mathbf{n})\,\mathbf{n} = (\mathbf{I} - \mathbf{n} \otimes \mathbf{n}) \cdot \mathbf{m}, \tag{6.4.15}$$

$\otimes$ denotes the tensor product and $\mathbf{I}$ is the identity matrix. The matrix $\mathbf{I} - \mathbf{n} \otimes \mathbf{n}$ projects the in-plane vector $\mathbf{m}$ tangentially to the boundary contour C illustrated in Figure 6.4.1.

In index notation, the ith component of the pure bending moment is given by

$$\big(\mathbf{m}_{\text{bend}}\big)_i = (\delta_{ij} - n_i n_j)\,m_j, \tag{6.4.16}$$

where δ_{ij} is Kronecker's delta representing the identity matrix and summation is implied over the repeated index, j. When $\mathbf{n} = (1,0)$, we find that

$$\mathbf{m}_{\text{bend}} = \big(0, \quad M_{xx}\big). \tag{6.4.17}$$

When $\mathbf{n} = (0,1)$, we find that

$$\mathbf{m}_{\text{bend}} = \big(-M_{yy}, \quad 0\big). \tag{6.4.18}$$

Both expressions are consistent with the vector drawings in Figure 6.4.3.

6.4.1 Equilibrium equations

To derive equilibrium equations, we perform a force balance in the z direction around the four edges of a rectangular section of the plate with infinitesimal dimensions dx and dy, as shown in Figure 6.4.2. Balancing the elastic forces exerted around the edges with the vertical load exerted on the rectangular surface area, $dx\,dy$, we obtain

$$\big[q_x(x + dx) - q_x(x)\big]\,dy + \big[q_y(y + dy) - q_y(y)\big]\,dx - w\,dx\,dy = 0. \tag{6.4.19}$$

Dividing each side by $dx\,dy$, taking the limit as dx and dy tend to zero, and rearranging, we derive the differential equation

$$\frac{\partial q_x}{\partial x} + \frac{\partial q_y}{\partial y} = w. \tag{6.4.20}$$

In vector notation,

$$\nabla \cdot \mathbf{q} = w, \tag{6.4.21}$$

where $\mathbf{q} = (q_x, q_y)$ has been defined as the vector of transverse shear tensions.

Next, we perform a moment balance in the y direction about the center-point of the rectangular section, finding

$$\left[M_{xx}(x + dx) - M_{xx}(x) \right] dy$$
$$+ \left[M_{yx}(y + dy) - M_{yx}(y) \right] dx - (q_x \, dy) \, dx = 0. \tag{6.4.22}$$

Dividing both sides by $dx \, dy$, taking the limit as dx and dy tend to zero, and rearranging, we derive the differential relation

$$q_x = \frac{\partial M_{xx}}{\partial x} + \frac{\partial M_{yx}}{\partial y}. \tag{6.4.23}$$

Performing a similar moment balance in the x direction, we derive the companion relation

$$q_y = \frac{\partial M_{xy}}{\partial x} + \frac{\partial M_{yy}}{\partial y}. \tag{6.4.24}$$

In vector notation, equations (6.4.23) and (6.4.24) take the compact form

$$\mathbf{q} = \nabla \cdot \mathbf{M}, \tag{6.4.25}$$

where $\mathbf{M}$ is the Cartesian tensor of bending moments introduced in (6.4.9).

Substituting expressions (6.4.23) and (6.4.24) into (6.4.20), we derive an equilibrium equation involving the bending moments and the vertical load alone,

$$\nabla \cdot (\nabla \cdot \mathbf{M}) = \frac{\partial^2 M_{xx}}{\partial x^2} + \frac{\partial^2 (M_{xy} + M_{yx})}{\partial x \, \partial y} + \frac{\partial^2 M_{yy}}{\partial y^2} = w. \tag{6.4.26}$$

Once this equation has been solved, subject to appropriate boundary conditions, the transverse shear tensions can be computed from (6.4.25).

6.4.2   Boundary conditions

At the edge of the plate represented by the contour C, we impose different types of boundary conditions according to the prevailing physical circumstances, as shown in Figure 6.4.4. In all cases, two scalar boundary conditions are required along each edge, which is consistent with our impending discovery that plate deflection is governed by a fourth-order differential equation for a function describing the position of the plate mid-surface after deformation,

$$z = f(x, y), \tag{6.4.27}$$

where $f(x, y) = 0$ describes the undeformed state.

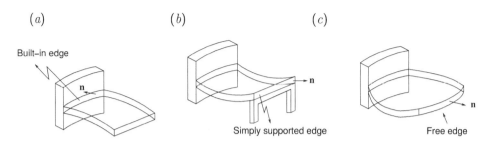

FIGURE 6.4.4 Illustration of a plate with (a) a clamped edge, also called a built-in or fixed edge, (b) a simply supported edge, and (c) a free edge.

In the case of a *clamped, built-in,* or *fixed edge,* the deflection and its derivative normal to the edge are required to be zero,

$$f = 0, \qquad \mathbf{n} \cdot \nabla f \equiv \frac{\partial f}{\partial l_n} = 0, \qquad (6.4.28)$$

where l_n is the arc length measured in the direction of the normal vector, $\mathbf{n}$, as illustrated in Figure 6.4.4(a).

In the case of a *simply supported* edge, the deflection and the pure bending moment are required to be zero,

$$f = 0, \qquad |\mathbf{m}_{\text{bend}}| = 0, \qquad (6.4.29)$$

as illustrated in Figure 6.4.4(b). Physically, the plate rests on a straight beam or curved bar.

In the case of a *free edge,* the transverse shear tension and the pure bending moment are required to be zero,

$$q = 0, \qquad |\mathbf{m}_{\text{bend}}| = 0, \qquad (6.4.30)$$

as illustrated in Figure 6.4.4(c). Using (6.4.4) and (6.4.25), we restate the first condition as

$$q \equiv \mathbf{n} \cdot \mathbf{q} = \mathbf{n} \cdot \nabla \cdot \mathbf{M} \qquad (6.4.31)$$

or

$$q = n_x \left(\frac{\partial M_{xx}}{\partial x} + \frac{\partial M_{yx}}{\partial y} \right) + n_y \left(\frac{\partial M_{xy}}{\partial x} + \frac{\partial M_{yy}}{\partial y} \right) = 0. \qquad (6.4.32)$$

Physically, the edge can be identified with the edge of a balcony or with the far side of an aircraft wing.

6.4.3   *Constitutive and governing equations*

To make further progress, we must relate the bending moments to the deformation of the plate mid-surface described by the deflection function, $f(x, y)$. This can be done by adopting constitutive equations for the in-plane stresses, multiplying these expressions by the distance from the plate mid-surface, and integrating the product over the plate cross-section.

Kirchhoff theory of plates

In the classical Kirchhoff theory of plates, material lines that are normal to the plate mid-surface before deformation are assumed to remain normal after deformation. Physically, this assumption is justified when (a) the plate is structurally and geometrically symmetric with respect to the mid-surface, (b) the thickness of the plate is either constant or varies slowly over the mid-surface, and (c) the mid-surface of the plate remains virtually undeformed in the lateral plane after a transverse load has been applied.

The displacement of material points along the x or y axis, denoted by v_x and v_y, are given by

$$v_x = -z \frac{\partial f}{\partial x}, \qquad v_y = -z \frac{\partial f}{\partial y}. \tag{6.4.33}$$

Note that, at this level of approximation, material points at the mid-plane, located at $z = 0$, remain stationary at the undeformed position in the xy plane, as required. Using these equations, we find that the in-plane strain components are given by

$$\epsilon_{xx} \equiv \frac{\partial v_x}{\partial x} = -z \frac{\partial^2 f}{\partial x^2}, \qquad \epsilon_{xy} \equiv \frac{1}{2}\left(\frac{\partial v_x}{\partial y} + \frac{\partial v_y}{\partial x}\right) = -z \frac{\partial^2 f}{\partial x \partial y},$$

$$\epsilon_{yy} \equiv \frac{\partial v_y}{\partial y} = -z \frac{\partial^2 f}{\partial y^2}. \tag{6.4.34}$$

Substituting these expressions into the plane-stress constitutive equations given in (6.2.4), we find that

$$\begin{bmatrix} \sigma_{xx} \\ \sigma_{yy} \\ \sigma_{xy} \end{bmatrix} = -z \frac{E}{1-\nu^2} \begin{bmatrix} 1 & \nu & 0 \\ \nu & 1 & 0 \\ 0 & 0 & 1-\nu \end{bmatrix} \cdot \begin{bmatrix} \frac{\partial^2 f}{\partial x^2} \\ \frac{\partial^2 f}{\partial y^2} \\ \frac{\partial^2 f}{\partial x \partial y} \end{bmatrix}. \tag{6.4.35}$$

Substituting these expressions into the equations in (6.4.8) and performing the integration with respect to z, we find that the bending moments are given by

$$M_{xx} = -E_B \left(\frac{\partial^2 f}{\partial x^2} + \nu \frac{\partial^2 f}{\partial y^2}\right), \qquad M_{yy} = -E_B \left(\frac{\partial^2 f}{\partial y^2} + \nu \frac{\partial^2 f}{\partial x^2}\right),$$

$$M_{xy} = M_{yx} = -E_B (1-\nu) \frac{\partial^2 f}{\partial x \partial y}, \tag{6.4.36}$$

involving the plate modulus of bending,

$$E_B \equiv \frac{Eh^3}{12(1-\nu^2)}. \tag{6.4.37}$$

It is convenient to collect these equations into the vector form

$$\begin{bmatrix} M_{xx} \\ M_{yy} \\ M_{xy} \end{bmatrix} = -E_B \begin{bmatrix} 1 & \nu & 0 \\ \nu & 1 & 0 \\ 0 & 0 & 1-\nu \end{bmatrix} \cdot \begin{bmatrix} \dfrac{\partial^2 f}{\partial x^2} \\ \dfrac{\partial^2 f}{\partial y^2} \\ \dfrac{\partial^2 f}{\partial x\,\partial y} \end{bmatrix}. \tag{6.4.38}$$

The trace of the bending moments tensor is given by

$$M \equiv M_{xx} + M_{yy} = -E_B(1+\nu)\nabla^2 f, \tag{6.4.39}$$

where

$$\nabla^2 f = \frac{\partial^2 f}{\partial x^2} + \frac{\partial^2 f}{\partial y^2} \tag{6.4.40}$$

is the Laplacian of the normal displacement. Rearranging (6.4.39), we obtain a Poisson equation for the displacement function,

$$\nabla^2 f = -\frac{M}{E_B(1+\nu)}, \tag{6.4.41}$$

where M is an *a priori* unknown distribution.

To derive expressions for the transverse shear tensions in terms of the displacement, we substitute expressions (6.4.36) into (6.4.23) and (6.4.24), finding

$$q_x = -E_B \frac{\partial \nabla^2 f}{\partial x} = \frac{1}{1+\nu}\frac{\partial M}{\partial x}, \qquad q_y = -E_B \frac{\partial \nabla^2 f}{\partial y} = \frac{1}{1+\nu}\frac{\partial M}{\partial y}. \tag{6.4.42}$$

In vector notation, these equations combine into

$$\mathbf{q} = -E_B \nabla(\nabla^2 f) = \frac{1}{1+\nu}\nabla M. \tag{6.4.43}$$

Finally, we substitute expressions (6.4.36) into (6.4.26) and rearrange to derive Lagrange's inhomogeneous biharmonic equation for the mid-surface deflection,

$$\nabla^4 f \equiv \nabla^2\nabla^2 f = \frac{\partial^4 f}{\partial x^4} + 2\frac{\partial^4 f}{\partial x^2\,\partial y^2} + \frac{\partial^4 f}{\partial y^4} = -\frac{1}{E_B}w(x,y). \tag{6.4.44}$$

This fourth-order partial differential equation is the counterpart of the fourth-order ordinary differential equation (2.4.13) governing the bending of a prismatic beam.

Because (6.4.44) is a fourth-order differential equation, the finite element solution must be C^1 continuous, which means that the transverse deflection, $f(x, y)$, and its gradient, consisting of the first derivatives with respect to x and y, must be continuous at shared element nodes and along element edges. To ensure C^1 continuity, special Hermitian bending elements, also called plate elements, must be employed, as discussed in Section 2.4 for the corresponding problem in one dimension.

The requirement of C^1 continuity can be relaxed by converting the fourth-order equation into a system of two second-order equations. For example, substituting (6.4.41) into (6.4.44), we derive a Poisson equation,

$$\nabla^2 M = (1 + \nu)\, w(x, y), \tag{6.4.45}$$

which is to be solved together with the companion Poisson equation (6.4.41) subject to appropriate boundary conditions. Other methods of deriving Poisson-like equations are discussed in Section 6.8.

The limitations of the Kirchhoff theory of plates become evident by setting the displacement of a material point particle normal to the plate equal to the deflection of the mid-surface,

$$v_z = f(x, y). \tag{6.4.46}$$

The transverse strain components turn out to be zero,

$$\epsilon_{zz} \equiv \frac{\partial v_z}{\partial z} = 0, \qquad \epsilon_{xz} \equiv \frac{1}{2}\left(\frac{\partial v_x}{\partial z} + \frac{\partial v_z}{\partial x}\right) = 0, \qquad \epsilon_{yz} \equiv \frac{1}{2}\left(\frac{\partial v_y}{\partial z} + \frac{\partial v_z}{\partial y}\right) = 0. \tag{6.4.47}$$

The vanishing of ϵ_{zz} implies that the plate is in a state of *plane strain* instead of the assumed state of *plane stress*, which cannot be true if the plate locally supports a transverse load. Moreover, the vanishing of the transverse shear strains, ϵ_{xz} and ϵ_{yz}, implies the absence of corresponding transverse shear stresses, σ_{xz} and σ_{yz}, at least for some types of materials, and this contradicts the underlying theoretical framework. Notwithstanding these limitations, the Kirchhoff theory is an acceptable idealization when the in-plane stresses are larger than their transverse shear counterparts.

Advanced theories of plates

Improvements to the Kirchhoff theory of plates are built into the geometrically nonlinear von Kármán model, which reduces to the Kirchhoff model for small deflections, and into the Reissner–Mindlin model, which takes into consideration the transverse shear tensions. More sophisticated models based on the theory of three-dimensional elasticity are available.

6.4.4   Circular plate

As an application of Kirchhoff's theory of plates, we consider the deflection of a circular plate under a normal load. It is convenient to work in plane polar coordinates, (r, θ), related to the Cartesian coordinates by

$$x = r \cos \theta, \qquad y = r \sin \theta, \tag{6.4.48}$$

where r is the distance from the center of the plate and θ is the associated polar angle measured in the counterclockwise direction. Using standard coordinate transformation rules, we derive the following relations between the first partial derivatives of a function, f, with respect to Cartesian and plane polar coordinates,

$$\frac{\partial f}{\partial x} = \cos \theta \, \frac{\partial f}{\partial r} - \frac{\sin \theta}{r} \frac{\partial f}{\partial \theta}, \qquad \frac{\partial f}{\partial y} = \sin \theta \, \frac{\partial f}{\partial r} + \frac{\cos \theta}{r} \frac{\partial f}{\partial \theta}. \tag{6.4.49}$$

For the second partial derivatives, we find that

$$\frac{\partial^2 f}{\partial x^2} = \cos^2 \theta \, \frac{\partial^2 f}{\partial r^2} - \frac{\sin 2\theta}{r} \frac{\partial^2 f}{\partial r \partial \theta} + \frac{\sin^2 \theta}{r} \frac{\partial f}{\partial r} + \frac{\sin 2\theta}{r^2} \frac{\partial f}{\partial \theta} + \frac{\sin^2 \theta}{r^2} \frac{\partial^2 f}{\partial \theta^2},$$

$$\frac{\partial^2 f}{\partial y^2} = \sin^2 \theta \, \frac{\partial^2 f}{\partial r^2} + \frac{\sin 2\theta}{r} \frac{\partial^2 f}{\partial r \partial \theta} + \frac{\cos^2 \theta}{r} \frac{\partial f}{\partial r} - \frac{\sin 2\theta}{r^2} \frac{\partial f}{\partial \theta} + \frac{\cos^2 \theta}{r^2} \frac{\partial^2 f}{\partial \theta^2},$$

$$\frac{\partial^2 f}{\partial x \partial y} = \frac{1}{2} \Big(\sin 2\theta \, \frac{\partial^2 f}{\partial r^2} + 2 \frac{\cos 2\theta}{r} \frac{\partial^2 f}{\partial r \partial \theta} - \frac{\sin 2\theta}{r} \frac{\partial f}{\partial r} \tag{6.4.50}$$

$$- 2 \frac{\cos 2\theta}{r^2} \frac{\partial f}{\partial \theta} - \frac{\sin 2\theta}{r^2} \frac{\partial^2 f}{\partial \theta^2} \Big).$$

Using these relations, we recover a well-known expression for the Laplacian of a function, f, in plane polar coordinates,

$$\nabla^2 f = \frac{1}{r} \frac{\partial}{\partial r} \Big(r \, \frac{\partial f}{\partial r} \Big) + \frac{1}{r^2} \frac{\partial^2 f}{\partial \theta^2}. \tag{6.4.51}$$

The biharmonic operator acting on f yields

$$\nabla^4 f = \mathcal{L}(f) + \frac{2}{r^2} \frac{\partial^4 f}{\partial r^2 \partial \theta^2} - \frac{2}{r^3} \frac{\partial^3 f}{\partial r \partial \theta^2} + \frac{4}{r^4} \frac{\partial^2 f}{\partial \theta^2} + \frac{1}{r^4} \frac{\partial^4 f}{\partial \theta^4}, \tag{6.4.52}$$

where

$$\mathcal{L}(f) \equiv \frac{1}{r} \frac{\partial}{\partial r} \Big[r \, \frac{\partial}{\partial r} \Big\{ \frac{1}{r} \frac{\partial}{\partial r} \Big(r \, \frac{\partial f}{\partial r} \Big) \Big\} \Big] = \frac{\partial^4 f}{\partial r^4} + \frac{2}{r} \frac{\partial^3 f}{\partial r^3} - \frac{1}{r^2} \frac{\partial^2 f}{\partial r^2} + \frac{1}{r^3} \frac{\partial f}{\partial r}. \tag{6.4.53}$$

The general solution of the plate bending equation for the transverse displacement can be expressed as a Fourier series with respect to the polar angle, θ,

$$f(r, \theta) = \frac{1}{2} p_0(r) + \sum_{n=1}^{\infty} \big[p_n(r) \cos n\theta + q_n(r) \sin n\theta \big], \tag{6.4.54}$$

where $p_n(r)$ and $q_n(r)$ are *a priori* unknown functions (e.g., McFarland, Smith, & Bernhart [39]). The equivalent complex form is

$$f(r,\theta) = \sum_{n=-\infty}^{\infty} \mathcal{F}_n(r) \exp(-\mathrm{i} n\theta), \tag{6.4.55}$$

where i is the imaginary unit. We have employed Euler's formula for the complex exponential,

$$\exp(-\mathrm{i} n\theta) = \cos(n\theta) - \mathrm{i}\sin(n\theta), \tag{6.4.56}$$

and we have introduced the complex functions $\mathcal{F}_n(r)$, defined such that

$$\mathcal{F}_n(r) \equiv \frac{1}{2}\left[p_n(r) + \mathrm{i}\, q_n(r)\right] \tag{6.4.57}$$

for $n \geq 0$ and

$$\mathcal{F}_n(r) \equiv \frac{1}{2}\left[p_n(r) - \mathrm{i}\, q_n(r)\right] \tag{6.4.58}$$

for $n < 0$. Note that $\mathcal{F}_n(r) = \mathcal{F}_{-n}^*(r)$, which guarantees that the right-hand side of (6.4.55) is real, where an asterisk denotes the complex conjugate.

Substituting the Fourier series into the Kirchhoff plate bending governing equation, $\nabla^4 f = -w/E_B$, we obtain

$$\nabla^4 f = \sum_{n=-\infty}^{\infty} \Psi_n \exp(-\mathrm{i} n\theta) = -\frac{w}{E_B}, \tag{6.4.59}$$

where

$$\Psi_n \equiv \mathcal{F}_n'''' + \frac{2}{r}\mathcal{F}_n''' - \frac{1+2n^2}{r^2}\mathcal{F}_n'' + \frac{1+2n^2}{r^3}\mathcal{F}_n' + n^2\frac{n^2-4}{r^4}\mathcal{F}_n. \tag{6.4.60}$$

The general solution consists of a particular solution, f^P, that absorbs the forcing term on the right-hand side of (6.4.59), and a homogeneous solution, f^H, corresponding to $w = 0$. The Fourier coefficients of the homogeneous solution are found by setting $\Psi_n = 0$ to obtain a system of equidimensional linear ordinary differential equations. The solution is given by

$$\mathcal{F}_0 = A_0\, r^2 + B_0\, r^2 \ln\frac{r}{a} + C_0 \ln\frac{r}{a} + D_0,$$
$$\mathcal{F}_1 = A_1\, r^3 + B_1\, r \ln\frac{r}{a} + C_1\, r + D_1\,\frac{1}{r}, \tag{6.4.61}$$
$$\mathcal{F}_n = A_n\, r^{n+2} + B_n\, r^n + C_n\, r^{-n+2} + D_n\, r^{-n}$$

for $n \geq 2$, where a is the plate radius and the constants, A_i, B_i, C_i, and D_i, are determined by the boundary conditions.

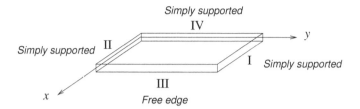

FIGURE 6.4.5   Illustration of a rectangular plate with sides parallel to the x or y axes where different boundary conditions are imposed.

When the load density w is uniform over the entire area of the plate, a suitable particular solution is

$$f^P = -\frac{w}{64E_B}\, r^4,\tag{6.4.62}$$

and the plate deflection function is given by

$$f = -\frac{w}{64E_B}(a^2 - r^2)^2.\tag{6.4.63}$$

The constant and quadratic terms in r correspond to the homogeneous solution (Problem 6.4.2).

PROBLEMS

6.4.1 *Bending of a rectangular plate*

Consider a rectangular plate with two sides parallel to the x and y axes, as shown in Figure 6.4.5. The plate is simply supported along the two edges that are parallel to the x axis and along one edge that is parallel to the y axis, and has a free fourth edge. Write the boundary conditions to be applied along each side.

6.4.2 *Bending of a clamped circular plate due to a uniform load*

Confirm that the deflection of a clamped circular plate of radius a under the action of a spatially uniform load, w, is given in (6.4.63).

6.5   Hermite triangles

Pursuant to our discussion of the plate bending problem formulated in Section 6.4, we require C^1 continuity of the finite element solution and use polynomial expansions that are defined not only with respect to the nodal values, but also with respect to their first and possibly higher-order derivatives with respect to x and y.

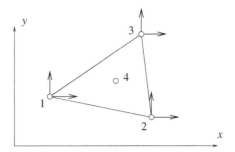

FIGURE 6.5.1   A Hermite triangle with three vertex nodes where the displacement
and its gradient are defined, and one interior node where the displacement alone
is defined.

A complete cubic expansion in x and y over an element takes the form

$$
\begin{aligned}
f(x,y) = \quad & a_{00} \\
& +a_{10}\,x + a_{01}\,y \\
& +a_{20}\,x^2 + a_{11}\,xy + a_{02}\,y^2 \\
& +a_{30}\,x^3 + a_{21}\,x^2 y + a_{12}\,xy^2 + a_{03}\,y^3,
\end{aligned}
\tag{6.5.1}
$$

involving 10 coefficients, a_{ij}. To ensure an equal number of unknowns, we introduce a four-node triangle defined by three vertex nodes and one interior node, as
illustrated in Figure 6.5.1.

At the vertex nodes, we define the values of the function and its gradient consisting of x and y derivatives, denoted by

$$
f_i, \qquad f_i^x \equiv \left(\frac{\partial f}{\partial x}\right)_i, \qquad f_i^y \equiv \left(\frac{\partial f}{\partial y}\right)_i
\tag{6.5.2}
$$

for $i = 1, 2, 3$. At the interior node labeled 4, we define only the value of the
function, denoted by f_4. The unknown values are collected into a 10-dimensional
vector,

$$
\mathbf{h} \equiv \begin{bmatrix} f_1, & f_1^x, & f_1^y, & f_2, & f_2^x, & f_2^y, & f_3, & f_3^x, & f_3^y, & f_4 \end{bmatrix}^T,
\tag{6.5.3}
$$

encapsulating 10 *degrees of freedom* (dof). The requirement of C^1 continuity demands that the function values and its derivatives at the vertex nodes are shared
by neighboring elements at corresponding positions.

To compute the coefficients of the cubic expansion, a_{ij}, we require the interpolation conditions

$$
f(x_i, y_i) = f_i
\tag{6.5.4}
$$

for $i = 1, 2, 3, 4$, and obtain four equations,

$$a_{00} + a_{10} x_i + a_{01} y_i + a_{20} x_i^2 + a_{11} x_i y_i + a_{02} y_i^2$$
$$+ a_{30} x_i^3 + a_{21} x_i^2 y_i + a_{12} x_i y_i^2 + a_{03} y_i^3 = f_i, \qquad (6.5.5)$$

which can be expressed in the form of a vector product,

$$\begin{bmatrix} 1 & x_i & y_i & x_i^2 & x_i y_i & y_i^2 & x_i^3 & x_i^2 y_i & x_i y_i^2 & y_i^3 \end{bmatrix} \cdot \mathbf{a} = f_i, \qquad (6.5.6)$$

where

$$\mathbf{a} \equiv \begin{bmatrix} a_{00}, & a_{10}, & a_{01}, & a_{20}, & a_{11}, & a_{02}, & a_{30}, & a_{03} \end{bmatrix}^T \qquad (6.5.7)$$

is the expansion coefficient vector. Requiring further interpolation conditions at the vertex nodes,

$$\left(\frac{\partial f}{\partial x} \right)_{(x_i, y_i)} = f_i^x, \qquad \left(\frac{\partial f}{\partial y} \right)_{(x_i, y_i)} = f_i^y \qquad (6.5.8)$$

for $i = 1, 2, 3$, we obtain six additional equations,

$$\begin{bmatrix} 0, & 1, & 0, & 2x_i, & y_i, & 0, & 3x_i^2, & 2x_i y_i, & y_i^2 & 0 \end{bmatrix} \cdot \mathbf{a} = f_i^x,$$

$$\begin{bmatrix} 0, & 0, & 1, & 0, & x_i, & 2y_i, & 0, & x_i^2, & 2x_i y_i, & 3y_i^2 \end{bmatrix} \cdot \mathbf{a} = f_i^y. \qquad (6.5.9)$$

Collecting equations (6.5.6) and (6.5.9), we formulate a system of 10 linear equations for the coefficient vector, $\mathbf{a}$,

$$\mathbf{D} \cdot \mathbf{a} = \mathbf{h}, \qquad (6.5.10)$$

where

$$\mathbf{D} = \begin{bmatrix} 1 & x_1 & y_1 & x_1^2 & x_1 y_1 & y_1^2 & x_1^3 & x_1^2 y_1 & x_1 y_1^2 & y_1^3 \\ 0 & 1 & 0 & 2x_1 & y_1 & 0 & 3x_1^2 & 2x_1 y_1 & y_1^2 & 0 \\ 0 & 0 & 1 & 0 & x_1 & 2y_1 & 0 & x_1^2 & 2x_1 y_1 & 3y_1^2 \\ 1 & x_2 & y_2 & x_2^2 & x_2 y_2 & y_2^2 & x_2^3 & x_2^2 y_2 & x_2 y_2^2 & y_2^3 \\ 0 & 1 & 0 & 2x_2 & y_2 & 0 & 3x_2^2 & 2x_2 y_2 & y_2^2 & 0 \\ 0 & 0 & 1 & 0 & x_2 & 2y_2 & 0 & x_1^2 & 2x_1 y_1 & 3y_1^2 \\ 1 & x_3 & y_3 & x_3^2 & x_3 y_3 & y_3^2 & x_3^3 & x_3^2 y_3 & x_3 y_3^2 & y_3^3 \\ 0 & 1 & 0 & 2x_3 & y_3 & 0 & 3x_3^2 & 2x_3 y_3 & y_3^2 & 0 \\ 0 & 0 & 1 & 0 & x_3 & 2y_3 & 0 & x_3^2 & 2x_3 y_3 & 3y_1^2 \\ 1 & x_4 & y_4 & x_4^2 & x_4 y_2 & y_4^2 & x_4^3 & x_4^2 y_4 & x_4 y_4^2 & y_4^3 \end{bmatrix} \qquad (6.5.11)$$

is a grand coefficient matrix. Solving for $\mathbf{a}$ in terms of the inverse matrix, $\mathbf{D}^{-1}$, we obtain

$$\mathbf{a} = \mathbf{D}^{-1} \cdot \mathbf{h}. \qquad (6.5.12)$$

The grand coefficient matrix $\mathbf{D}$ and its inverse are determined completely by the position of the four nodes.

```
function Dinv = herm10_Dinv (x1,y1, x2,y2, x3,y3, x4,y4)

%================================================================
% Evaluation of the inverse of the grand coefficient matrix
% for a four-node, ten-dof Hermite triangle
%================================================================

%-------------------------------
% define the coefficient matrix
%-------------------------------

D = [ ...
1  x1  y1  x1^2  x1*y1  y1^2  x1^3  x1^2*y1  x1*y1^2  y1^3  ; ...
0  1   0   2*x1  y1     0     3*x1^2  2*x1*y1  y1^2     0     ; ...
0  0   1   0     x1     2*y1  0       x1^2     2*x1*y1  3*y1^2 ; ...
1  x2  y2  x2^2  x2*y2  y2^2  x2^3  x2^2*y2  x2*y2^2  y2^3  ; ...
0  1   0   2*x2  y2     0     3*x2^2  2*x2*y2  y2^2     0     ; ...
0  0   1   0     x2     2*y2  0       x2^2     2*x2*y2  3*y2^2 ; ...
1  x3  y3  x3^2  x3*y3  y3^2  x3^3  x3^2*y3  x3*y3^2  y3^3  ; ...
0  1   0   2*x3  y3     0     3*x3^2  2*x3*y3  y3^2     0     ; ...
0  0   1   0     x3     2*y3  0       x3^2     2*x3*y3  3*y3^2 ; ...
1  x4  y4  x4^2  x4*y4  y4^2  x4^3  x4^2*y4  x4*y4^2  y4^3  ; ...
];

%--------------------
% compute the inverse
%--------------------

Dinv = inv(D);

%-----
% done
%-----

return
```

TABLE 6.5.1 Function *herm10_Dinv* computes the inverse of the grand coefficient matrix, **D**, for a four-node Hermite triangle with ten degrees of freedom (dof).

FSELIB function *herm10_Dinv*, listed in Table 6.5.1, evaluates the matrix **D** and computes its inverse using the MATLAB function *inv*. Other numerical methods for computing the inverse of an arbitrary matrix based on Gauss elimination and related algorithms are available (e.g., Pozrikidis [47], see also Appendix C).

The complete cubic expansion (6.5.1) can be expressed as the inner product of two vectors,

$$f(x,y) = \varphi \cdot \mathbf{a}, \tag{6.5.13}$$

where the vector

$$\varphi \equiv \begin{bmatrix} 1, & x, & y, & x^2, & xy & y^2, & x^3, & x^2y, & xy^2, & y^3 \end{bmatrix}^T \tag{6.5.14}$$

encapsulates linear and quadratic monomial products. Substituting the coefficient
vector **a** from (6.5.12), we obtain

$$f(x, y) = \varphi \cdot \mathbf{D}^{-1} \cdot \mathbf{h} \equiv \psi \cdot \mathbf{h}, \tag{6.5.15}$$

where the 10-dimensional vector function

$$\psi(x, y) \equiv \varphi \cdot \mathbf{D}^{-1}, \tag{6.5.16}$$

encapsulates the Hermite interpolation functions, also called *expansion modes*, associated with the 10 degrees of freedom composing the vector **h**. Specifically, the
ith interpolation function is given by

$$\psi_i(x, y) = \sum_{j=1}^{10} \varphi_j(x, y) \, D_{ji}^{-1} \tag{6.5.17}$$

for $i = 1, \ldots, 10$.

Alternatively, the individual functions, ψ_i, arise by solving a linear system for
the individual coefficient vector **a**,

$$\mathbf{D} \cdot \mathbf{a} = \mathbf{e}_i, \tag{6.5.18}$$

where the right-hand side is a unit vector,

$$\mathbf{e}_i = \begin{bmatrix} 0, & \cdots, & 0, & 1, & 0, & \cdots, & 0 \end{bmatrix}^T, \tag{6.5.19}$$

and the unity resides in the ith entry.

Although the triangle can be mapped from the physical xy plane to the standard
triangle in the $\xi\eta$ plane, and the independent variables x and y can be expressed in
terms of ξ and η using the relations in (4.2.11), the endeavor is both cumbersome
and unnecessary.

FSELIB code *herm10_psi*, listed in Table 6.5.2, generates graphs of the interpolation functions contained in the vector ψ. Mesh lines are generated in the parametric
$\xi\eta$ plane and then transferred to the physical xy plane using the transformation rules
given in (4.2.11).

Graphs of four selected interpolation functions are shown in Figure 6.5.2 for
an arrangement where the first vertex is located at $x_1 = 0$, $y_1 = 0$, the second
at $x_2 = 1$, $y_2 = 0$, the third at $x_3 = 0$, $y_3 = 1$, and the fourth at the element
centroid. The interpolation functions ψ_1, ψ_2, and ψ_3, associated with the first node,
corresponding to f_1, f_1^x, and f_1^y, are shown in Figure 6.5.2(a-c) . The interpolation
function associated with the interior node, ψ_{10}, corresponding to f_4, is shown in
Figure 6.5.2(d).

The function ψ_1 displayed in Figure 6.5.2(a) is zero along the entire triangle edge
2–3. The reason is that, because ψ_1 is a cubic polynomial with respect to arc length

```
%============================================================
% herm10_psi
%
% prepare a graph of the element interpolation functions
% for a four-node, ten-dof Hermite triangle
%============================================================

%-----------------------------------
% define the vertices arbitrarily
%-----------------------------------

x1 = 0.0; y1 = 0.0;
x2 = 1.0; y2 = 0.0;
x3 = 0.0; y3 = 1.0;

%-------------------------------------------------------
% put the interior node at the centroid (arbitrary)
%-------------------------------------------------------

x4 = (x1+x2+x3)/3.0;
y4 = (y1+y2+y3)/3.0;

%-----------------------------------
% inquire on the mode to be graphed
%-----------------------------------

mode = input('Please enter the mode to plot:')

%--------------------------
% compute the inverse matrix
%--------------------------

Dinv = psi_10_Dinv (x1,y1, x2,y2, x3,y3, x4,y4);

%-------------------------------------------------
% define the plotting grid in the xi-eta plane
%-------------------------------------------------

N = 32; M = 32;      % arbitrary

Dxi = 1.0/N; Deta = 1.0/M;

for i=1:N+1
  xi(i) = Dxi*(i-1.0);
end

for j=1:M+1
  eta(j) = Deta*(j-1.0);
end

%----------------------------------------
% draw horizontal and vertical mesh lines
%----------------------------------------
```

TABLE 6.5.2   Code *herm10_psi* (Continuing →)

```
%---------------------
% horizontal mesh lines
%---------------------

for i=1:N
 for j=1:M+2-i
  x = x1 + (x2-x1)*xi(i) + (x3-x1)*eta(j);
  y = y1 + (y2-y1)*xi(i) + (y3-y1)*eta(j);
  monvec = [1  x y  x^2 x*y y^2  x^3 x^2*y x*y^2 y^3];
  psi = monvec*Dinv;
  xp(j) = x; yp(j)=y; zp(j) = psi(mode);
 end
 plot3 (xp,yp,zp);
end

%---------------------
% vertical mesh lines
%---------------------

for j=1:M
 for i=1:N+2-j
  x = x1 + (x2-x1)*xi(i) + (x3-x1)*eta(j);
  y = y1 + (y2-y1)*xi(i) + (y3-y1)*eta(j);
  monvec = [1  x y  x^2 x*y y^2  x^3 x^2*y x*y^2 y^3];
  psi = monvec*Dinv;
  xp(i) = x; yp(i)=y; zp(i) = psi(mode);
 end
 plot3 (xp,yp,zp);
end

%-----------------------------------
% plot the perimeter of the triangle
%-----------------------------------

xx(1) = x1; yy(1) = y1; zz(1) = 0.0;
xx(2) = x2; yy(2) = y2; zz(2) = 0.0;
xx(3) = x3; yy(3) = y3; zz(3) = 0.0;
xx(4) = x1; yy(4) = y1; zz(4) = 0.0;

plot3 (xx,yy,zz)

%----------------------
% plot the interior node
%----------------------

plot3 (x4,y4,0.0,'o'); xlabel('x'); ylabel('y')

%-----
% done
%-----
```

TABLE 6.5.2   ($\rightarrow$ Continued) Code *herm10_psi* generates graphs of the Hermite interpolation functions for a four-node, ten-dof Hermite triangle.

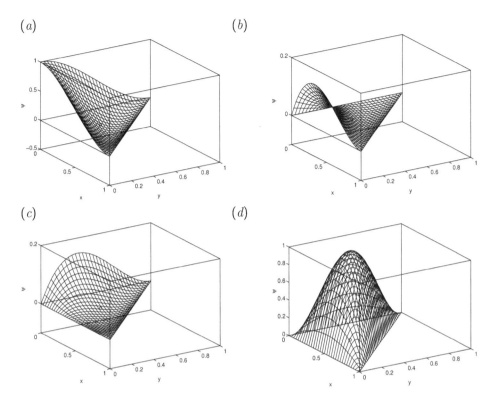

FIGURE 6.5.2   Element-node interpolation functions for a four-node, ten-dof Hermite
triangle: (*a*) ψ_1 corresponding to f_1, (*b*) ψ_2 corresponding to f_1^x, (*c*) ψ_3 corresponding to f_1^y, and (*d*) ψ_{10} corresponding to f_4.

along this edge with a double root at each vertex, the cubic must be identically zero.
The distribution of ψ_1 along the 1–2 edge depends on the position of the first and
second node and is independent of the position of the third and fourth node. To
explain this, we introduce the arc length along the 1–2 edge measured from point
1, denoted by s, and require that

$$\psi_1(0) = 1, \qquad (\mathrm{d}\psi_1/\mathrm{d}s)_0 = 0, \qquad \psi_1(a) = 0, \qquad (\mathrm{d}\psi_1/\mathrm{d}s)_a = 0, \qquad (6.5.20)$$

where a is the length of the 1–2 edge. These conditions completely determine the
cubic function $\psi_1(s)$ as

$$\psi_1(s) = \frac{2\ell + 1}{(1 - \ell)^2}, \tag{6.5.21}$$

where $\ell \equiv s/a$. Similar arguments can be made to show that the distribution of
ψ_1 along the 1–3 edge depends on the positions of the first and third nodes and is
independent of the positions of the second and fourth nodes.

The function ψ_2 displayed in Figure 6.5.2(b) is zero along the entire 2–3 and 1–3 edges. The reason is that, because ψ_2 is a cubic polynomial with respect to arc length along each edge with a double root at each vertex, the cubic must be identically zero. The distribution of ψ_2 along the 1–2 edge depends on the position of the first and second node and is independent of the position of the third and fourth node. To explain this property, we introduce the arc length along the 1–2 edge measured from point 1, denoted by s, and require that

$$\psi_2(0) = 0, \quad (d\psi_2/ds)_0 = 1, \quad \psi_2(a) = 0, \quad (d\psi_2/ds)_a = 0, \quad (6.5.22)$$

where a is the length of the $1 - 2$ edge. These conditions completely determine the cubic function $\psi_2(s)$ as

$$\psi_2(s) = \ell \, (1 - \ell)^2, \quad (6.5.23)$$

where $\ell \equiv s/a$. A similar behavior pertains to the function ψ_3 shown in Figure 6.5.2(c).

Finally, we observe that the function ψ_{10} shown in Figure 6.5.2(d) is zero around all three edges. The reason is that, because ψ_{10} is a cubic polynomial with respect to arc length along each triangle edge with a double root at each vertex, all three cubics must be identically zero. It is not surprising that this interpolation function describes a bubble mode.

Global interpolation functions

The union of selected element interpolation functions provides us with global interpolation functions, ϕ_i, corresponding to the deflection and its gradient at the global nodes. The index i takes values over an appropriate number of *global modes* associated with global degrees of freedom, which is typically far greater than the number of global nodes.

As an example, we consider the two-element arrangement shown in Figure 6.5.3(a) where the labels of the global nodes are printed in bold. The finite element assembly involves four global vertex nodes and two interior element nodes. The finite element expansion involves

$$N_{G_m} = 4 \times 3 + 2 \times 1 = 14 \quad (6.5.24)$$

global modes, three for each global vertex node and one for each interior node. For example, if $\psi_i^{(1)}$ are the Hermite interpolation functions of the first element and $\psi_i^{(2)}$ are the Hermite interpolation functions of the second element, then:

- ϕ_1 is identical to $\psi_1^{(1)}$, ϕ_2 is identical to $\psi_2^{(1)}$, and ϕ_3 is identical to $\psi_3^{(1)}$. All of these functions are associated with the first global node.

- ϕ_4 is the union of $\psi_4^{(1)}$ and $\psi_1^{(2)}$, ϕ_5 is the union of $\psi_5^{(1)}$ and $\psi_2^{(2)}$, and ϕ_6 is the union of $\psi_6^{(1)}$ and $\psi_3^{(2)}$. All of these functions are associated with the second global node.

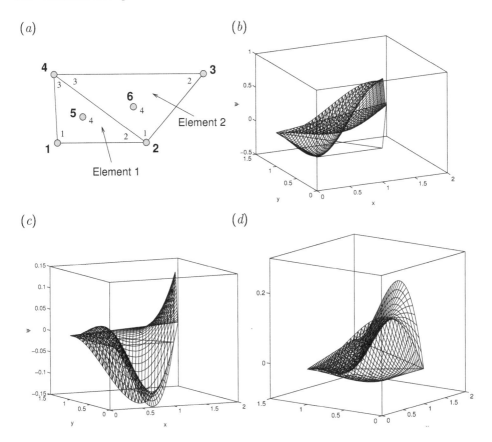

FIGURE 6.5.3 (*a*) A two-element assembly serves as a prototype for illustrating the relation between the element nodes, the global nodes (bold faced), and the corresponding interpolation functions. (*b–d*) Graphs of three global interpolation functions, ϕ_4, ϕ_5, and ϕ_6, associated with the second global node.

Graphs of the functions ϕ_4, ϕ_5, and ϕ_6 over the two-element assembly associated with the second global node are shown in Figure 6.5.3(*b–d*).

For a general element arrangement, the total number of global interpolation functions is

$$N_{G_m} = 3N_V + N_I,\tag{6.5.25}$$

where N_V is the number of global vertex nodes and N_I is the number of element interior nodes. Any suitable function admits the expansion

$$f(x,y) = \sum_{j=1}^{N_{G_m}} h_j^G \, \phi_j(x,y),\tag{6.5.26}$$

where the coefficients h_j^G represent the values of the function or the Cartesian components of its gradient at appropriate global nodes. The Galerkin finite element equations are derived based on this expansion, as discussed in previous chapters.

Edge incompatibility

Cursory inspection of the element interpolation functions used to assemble the global interpolation functions reveals that the global expansion (6.5.26) represents a continuous function whose derivatives are continuous at the element nodes, but are not necessarily continuous across the element edges. Elements with this undesirable property are called *incompatible* or *non-conforming*.

Some, but not all, non-conforming elements generate numerical solutions that do not converge to the exact solution of the plate-bending governing equations as the element size becomes smaller. Unfortunately, the four-node, ten-dof element discussed in this section falls in this undesirable category in the context of the biharmonic equation. The reason can be traced to the lack of polynomial invariance associated with the interior node (e.g., Ciarlet [11], p. 373).

PROBLEMS

6.5.1 *Element interpolation functions and global modes*

With reference to Figure 6.5.3, complete the definition of the global modes discussed in the text.

6.5.2 *A Hermite, four-node, twelve-mode rectangle*

Consider a four-node quadrilateral element and introduce the value of a function, $f(x, y)$, and the components of its gradient at each node,

$$f_i, \qquad f_i^x \equiv \left(\frac{\partial f}{\partial x}\right)_i, \qquad f_i^y \equiv \left(\frac{\partial f}{\partial y}\right)_i \qquad (6.5.27)$$

for $i = 1, 2, 3, 4$, providing us with 12 degrees of freedom. In the finite element literature, it is a standard practice to approximate the function $f(x, y)$ over the element with an incomplete cubic, as

$$
\begin{aligned}
f(x, y) \simeq \quad & a_{00} \\
& + a_{10}\, x + a_{01}\, y \\
& + a_{20}\, x^2 + a_{11}\, x\, y + a_{02}\, y^2 \\
& + a_{30}\, x^3 + a_{21}\, x^2\, y + a_{12}\, x\, y^2 + a_{03}\, y^3 \\
+ a_{31}\, x^3\, y \quad & \qquad\qquad\qquad\qquad + a_{13}\, x\, y^3,
\end{aligned}
\qquad (6.5.28)
$$

involving 12 coefficients, a_{ij}. Write a program that computes the interpolation functions (modes) corresponding to the 12 degrees of freedom. Prepare graphs of the three modes associated with a vertex of your choice and assess whether this element is conforming.

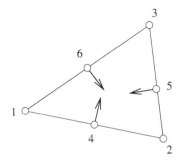

FIGURE 6.6.1 Illustration of the Morley element described by three vertex nodes where the value of the solution is defined, and three mid-side nodes where the normal inward derivative is defined.

6.6 Morley's triangle

Morley [40] introduced a simple non-conforming triangular element defined by six nodes, including three vertex nodes and three mid-side nodes, as illustrated in Figure 6.6.1. At the vertex nodes, the value of the function, f_i for $i = 1, 2, 3$, is defined. At the mid-side nodes, the normal derivative is defined,

$$f_i^n \equiv \mathbf{n}_i \cdot (\nabla f)_{x_i, y_i} \tag{6.6.1}$$

for $i = 4, 5, 6$, where $\mathbf{n}_i$ is the inward unit vector normal to the edge hosting the ith node. Using elementary geometry, we find that

$$\mathbf{n}_4 = \frac{1}{|\mathbf{x}_2 - \mathbf{x}_1|} \begin{bmatrix} -(y_2 - y_1) \\ (x_2 - x_1) \end{bmatrix}, \qquad \mathbf{n}_5 = \frac{1}{|\mathbf{x}_3 - \mathbf{x}_2|} \begin{bmatrix} -(y_3 - y_2) \\ (x_3 - x_2) \end{bmatrix},$$

$$\mathbf{n}_6 = \frac{1}{|\mathbf{x}_1 - \mathbf{x}_3|} \begin{bmatrix} -(y_1 - y_3) \\ (x_1 - x_3) \end{bmatrix}. \tag{6.6.2}$$

The unknowns can be collected into a six-dimensional vector,

$$\mathbf{h} \equiv \begin{bmatrix} f_1, & f_2, & f_3, & f_4^n, & f_5^n, & f_6^n \end{bmatrix}^T, \tag{6.6.3}$$

encapsulating six degrees of freedom.

The requirement of C^0 continuity demands that the vertex values are shared by neighboring elements at corresponding positions. The additional requirement of C^1 continuity demands that the normal derivatives at the mid-side nodes are shared by neighboring elements at corresponding positions, provided that allowance is made for the opposite orientation of the normal vector.

Having six degrees of freedom available, we may introduce a complete quadratic expansion,

$$f(x, y) = \quad a_{00}$$
$$+ a_{10} \, x + a_{01} \, y$$
$$+ a_{20} \, x^2 + a_{11} \, x \, y + a_{02} \, y^2, \qquad (6.6.4)$$

involving six coefficients, a_{ij}. To compute these coefficients, we require the interpolation conditions $f(x_i, y_i) = f_i$, for $i = 1, 2, 3$, and obtain three equations,

$$a_{00} + a_{10} \, x_i + a_{01} \, y_i + a_{20} \, x_i^2 + a_{11} \, x_i \, y_i + a_{02} \, y_i^2 = f_i, \qquad (6.6.5)$$

which can be recast as a vector product,

$$\begin{bmatrix} 1, & x_i, & y_i, & x_i^2, & x_i \, y_i, & y_i^2 \end{bmatrix} \cdot \mathbf{a} = f_i, \qquad (6.6.6)$$

where

$$\mathbf{a} \equiv \begin{bmatrix} a_{00}, & a_{10}, & a_{01}, & a_{20}, & a_{11}, & a_{02} \end{bmatrix} \qquad (6.6.7)$$

is the expansion coefficient vector.

Requiring further the derivative interpolation conditions (6.6.1) at the mid-side nodes, we obtain three additional equations,

$$\begin{bmatrix} 0, & n_{x_i}, & n_{y_i}, & 2 \, x_i \, n_{x_i}, & y_i \, n_{x_i}, + x_i \, n_{y_i}, & 2 \, y_i \, n_{y_i} \end{bmatrix} \cdot \mathbf{a} = f_i^n \qquad (6.6.8)$$

for $i = 4, 5, 6$. Collecting equations (6.6.6) and (6.6.8), we formulate a system of six linear equations,

$$\mathbf{D} \cdot \mathbf{a} = \mathbf{h}, \qquad (6.6.9)$$

where

$$\mathbf{D} = \begin{bmatrix} 1 & x_1 & y_1 & x_1^2 & x_1 \, y_1 & y_1^2 \\ 1 & x_2 & y_2 & x_2^2 & x_2 \, y_2 & y_2^2 \\ 1 & x_3 & y_3 & x_3^2 & x_3 \, y_3 & y_3^2 \\ 0 & n_{x_4} & n_{y_4} & 2 \, x_4 \, n_{x_4} & y_4 \, n_{x_4} + x_4 \, n_{y_4} & 2 \, y_4 \, n_{y_4} \\ 0 & n_{x_5} & n_{y_5} & 2 \, x_5 \, n_{x_5} & y_5 \, n_{x_5} + x_5 \, n_{y_5} & 2 \, y_5 \, n_{y_5} \\ 0 & n_{x_6} & n_{y_6} & 2 \, x_6 \, n_{x_6} & y_6 \, n_{x_6} + x_6 \, n_{y_6} & 2 \, y_6 \, n_{y_6} \end{bmatrix} \qquad (6.6.10)$$

is a grand coefficient matrix. Solving for the coefficient vector $\mathbf{a}$ in terms of the inverse matrix, $\mathbf{D}^{-1}$, we obtain $\mathbf{a} = \mathbf{D}^{-1} \cdot \mathbf{h}$.

FSELIB function *morley_Dinv*, listed in Table 6.6.1, compiles the matrix $\mathbf{D}$ and computes its inverse using the MATLAB function *inv*.

The complete quadratic expansion (6.6.4) can be expressed as the inner product of two vectors,

$$f(x, y) = \varphi \cdot \mathbf{a}, \qquad (6.6.11)$$

```
function Dinv = morley_Dinv (x1,y1, x2,y2, x3,y3)

%==============================================
% Computation of the inverse of the matrix D
% for the Morley element.
%==============================================

%----------------------------------
% mid-side nodes and normal vectors
%----------------------------------

x4 = 0.5*(x1+x2); y4 = 0.5*(y1+y2);
x5 = 0.5*(x2+x3); y5 = 0.5*(y2+y3);
x6 = 0.5*(x3+x1); y6 = 0.5*(y3+y1);

d12 = sqrt((x2-x1)^2+(y2-y1)^2);

  nx4 = -(y2-y1)/d12; ny4 = (x2-x1)/d12;

d23 = sqrt((x3-x2)^2+(y3-y2)^2);

  nx5 = -(y3-y2)/d23; ny5 = (x3-x2)/d23;

d31 = sqrt((x1-x3)^2+(y1-y3)^2);

  nx6 = -(y1-y3)/d31; ny6 = (x1-x3)/d31;

%---------------------------
% define the coefficient matrix
%---------------------------

D = [ ...
1  x1  y1  x1^2  x1*y1  y1^2 ; ...
1  x2  y2  x2^2  x2*y2  y2^2 ; ...
1  x3  y3  x3^2  x3*y3  y3^2 ; ...
0 nx4 ny4 2*x4*nx4 y4*nx4+x4*ny4 2*y4*ny4 ; ...
0 nx5 ny5 2*x5*nx5 y5*nx5+x5*ny5 2*y5*ny5 ; ...
0 nx6 ny6 2*x6*nx6 y6*nx6+x6*ny6 2*y6*ny6 ...
];

%--------------------
% compute the inverse
%--------------------

Dinv = inv(D);

%-----
% done
%-----

return
```

TABLE 6.6.1 Function *morley_Dinv* computes the inverse of the grand coefficient matrix for the Morley element.

where

$$\varphi \equiv \begin{bmatrix} 1, & x, & y, & x^2, & xy, & y^2 \end{bmatrix} \qquad (6.6.12)$$

is the vector of linear and quadratic monomial products. Substituting the expression for the coefficient vector, $\mathbf{a}$, we obtain

$$f(x,y) = \phi \cdot \mathbf{D}^{-1} \cdot \mathbf{h} \equiv \psi \cdot \mathbf{h}. \qquad (6.6.13)$$

The six-dimensional vector function

$$\psi(x,y) \equiv \phi \cdot \mathbf{D}^{-1} \qquad (6.6.14)$$

hosts the Hermite interpolation functions associated with the components of the vector $\mathbf{h}$, which are regarded as *expansion modes* associated with the available degrees of freedom. Specifically, the ith modal function is given by

$$\psi_i(x,y) = \sum_{j=1}^{6} \varphi_j(x,y)\, D_{ji}^{-1} \qquad (6.6.15)$$

for $i = 1, \dots, 6$.

FSELIB script *morley_psi*, not listed in the text, generates graphs of the interpolation functions contained in the vector ψ. The script is similar to the script *herm10_psi*, listed in Section 6.5.2, for the 10-dof Hermitian triangle.

Graphs of interpolation functions for an arrangement where the first vertex is located at $x_1 = 0$, $y_1 = 0$, the second at $x_2 = 1$, $y_2 = 0$, and the third at $x_3 = 0$, $y_3 = 1$ are shown in Figure 6.6.2. The interpolation function ψ_1 corresponding to f_1 is shown in Figure 6.6.2(a) and the interpolation function ψ_4 corresponding to f_4^n is shown in Figure 6.6.2(b).

Like the four-node, ten-dof element discussed in Section 6.5, the Morley element discussed in this section does not guarantee C^1 continuity of the finite element expansion. In fact, the Morley element does not even guarantee C^0 continuity across the entire length of the element edges. In spite of this deficiency, the Morley element remarkably allows for convergent solutions of the plate bending problem for reasons that are not entirely clear (e.g., Ciarlet [11], pp. 374–376).

PROBLEM

6.6.1 *Global interpolation functions*
Prepare the counterpart of Figure 6.5.3 for the Morley element.

6.7 Conforming triangles

To achieve C^1 continuity across the element edges, we employ triangles with additional degrees of freedom. The advantage of using these more complicated elements

(a)   (b)

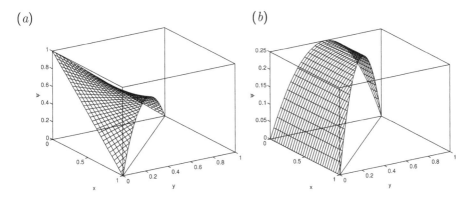

FIGURE 6.6.2 Element node interpolation functions for the Morley element. (*a*) Interpolation function ψ_1 corresponding to f_1 and (*b*) interpolation function ψ_4
corresponding to f_4^n.

is that the corresponding global expansion provides us with a continuous function whose derivatives are continuous across the entire length of the element edges.
To describe this property, we say that the elements are *compatible* or *conforming*.
Conformity guarantees that the numerical solution of the governing plate bending equations converges to the exact solution as the finite element grid becomes
increasingly fine (e.g., Ciarlet [11], Chapter 6).

6.7.1   Six-node, 21-dof triangle

A complete quintic expansion in x and y takes the form

$$
\begin{aligned}
f(x,y) = \quad & a_{00} \\
& + a_{10}\,x + a_{01}\,y \\
& + a_{20}\,x^2 + a_{11}\,x\,y + a_{02}\,y^2 \\
& + a_{30}\,x^3 + a_{21}\,x^2\,y + a_{12}\,x\,y^2 + a_{03}\,y^3 \\
& + a_{40}\,x^4 + a_{31}\,x^3\,y + a_{22}\,x^2\,y^2 + a_{13}\,x\,y^3 + a_{04}\,x\,y^3 \\
& + a_{50}\,x^5 + a_{41}\,x^4\,y + a_{32}\,x^3\,y^2 + a_{23}\,x^2\,y^3 + a_{14}\,x\,y^4 + a_{05}\,y^5,
\end{aligned}
\tag{6.7.1}
$$

involving 21 coefficients, a_{ij}. To ensure an equal number of unknowns, we introduce
a six-node triangle with three vertex nodes and three edge nodes, as illustrated in
Figure 6.7.1. At the vertex nodes, the value of the function, its first and second
derivatives are defined,

$$
f_i, \qquad f_i^x \equiv \left(\frac{\partial f}{\partial x}\right)_i, \qquad f_i^y \equiv \left(\frac{\partial f}{\partial y}\right)_i,
$$

$$
f_i^{xx} \equiv \left(\frac{\partial^2 f}{\partial x^2}\right)_i, \qquad f_i^{xy} \equiv \left(\frac{\partial^2 f}{\partial x \partial y}\right)_i, \qquad f_i^{yy} \equiv \left(\frac{\partial^2 f}{\partial y^2}\right)_i
\tag{6.7.2}
$$

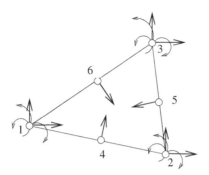

FIGURE 6.7.1　A Hermite triangle with six nodes and 21 degrees of freedom. The straight arrows indicate degrees of freedom associated with first derivatives, and the curved arrows indicate degrees of freedom associated with second derivatives.

for $i = 1, 2, 3$. At the mid-side nodes, labeled 4, 5, and 6, the normal derivative is defined,

$$f_i^n \equiv \left(\mathbf{n} \cdot \nabla f\right)_{(x_i, y_i)} \tag{6.7.3}$$

for $i = 4, 5, 6$, where $\mathbf{n}$ is the unit vector normal to the corresponding edge. The associated *modes* are computed as discussed previously for the Morley element. To ensure C^1 continuity, the nodal values of the function and its derivatives are shared by neighboring elements at corresponding locations.

6.7.2　The Hsieh–Clough–Tocher (HCT) element

To avoid introducing second derivatives as degrees of freedom, Clough and Felippa [12] proposed subdividing a triangular element into three sub-elements and using individual cubic expansions over each sub-element, as illustrated in Figure 6.7.2 where the sub-element labels are printed in bold. The resulting element is known as the Hsieh–Clough–Tocher (HCT) triangle. The parental element is sometimes called a macro-element.

Consider a function, $f(x, y)$, defined over the triangle. A complete cubic expansion over the qth sub-element takes the form

$$\begin{aligned}
f^{(q)}(x, y) = \quad & a_{00}^{(q)} \\
+ & a_{10}^{(q)}\, x + a_{01}^{(q)}\, y \\
+ & a_{20}^{(q)}\, x^2 + a_{11}^{(q)}\, x\, y + a_{02}^{(q)}\, y^2 \\
+ & a_{30}^{(q)}\, x^3 + a_{21}^{(q)}\, x^2\, y + a_{12}^{(q)}\, x\, y^2 + a_{03}^{(q)}\, y^3
\end{aligned} \tag{6.7.4}$$

for $q = 1, 2, 3$, involving 30 coefficients, $a_{ij}^{(q)}$. To ensure a sufficient number of degrees of freedom, we introduce the values of the function and its first derivatives

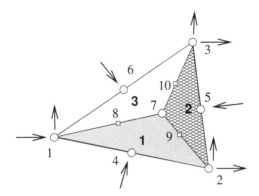

FIGURE 6.7.2 Illustration of the HCT element defined by three vertex nodes, three
mid-side nodes, and several interior nodes defined with reference to three sub-
triangles whose labels are printed in bold. The arrows indicate degrees of freedom
associated with the x, y, and normal derivatives.

at the vertex nodes,

$$f_i, \qquad f_i^x \equiv \left(\frac{\partial f}{\partial x}\right)_i, \qquad f_i^y \equiv \left(\frac{\partial f}{\partial y}\right)_i \qquad (6.7.5)$$

for $i = 1, 2, 3$, and the values of the normal derivatives at the mid-side nodes,

$$f_i^n \equiv (\mathbf{n} \cdot \nabla f)_i \qquad (6.7.6)$$

for $i = 4, 5, 6$. These 12 values are collected into a vector,

$$\mathbf{h} \equiv \left[\ f_1,\ f_1^x,\ f_1^y,\ \ f_2,\ f_2^x,\ f_2^y,\ f_3,\ \ f_3^x,\ f_3^y,\ f_4^n,\ \ f_5^n,\ \ f_6^n\ \right]^T, \qquad (6.7.7)$$

encapsulating 12 degrees of freedom.

To compute the 30 coefficients of the three cubic expansions, $a_{ij}^{(q)}$, we impose
the vertex node interpolation conditions

$$f^{(q)}(x_i, y_i) = f_i, \qquad \left(\frac{\partial f^{(q)}}{\partial y}\right)_{(x_i, y_i)} = f_i^y, \qquad \left(\frac{\partial f^{(q)}}{\partial x}\right)_{(x_i, y_i)} = f_i^x, \qquad (6.7.8)$$

where $i = 1, 2$ for $q = 1$, $i = 2, 3$ for $q = 2$, and $i = 3, 1$ for $q = 3$, which provide
us with $3 \times (2 \times 3) = 18$ equations. In addition, we require the mid-side node
interpolation conditions

$$\mathbf{n}_4 \cdot \left(\nabla f^{(1)}\right)_{(x_4, y_4)} = f_4^n, \qquad \mathbf{n}_5 \cdot \left(\nabla f^{(2)}\right)_{(x_5, y_5)} = f_5^n,$$

$$\mathbf{n}_6 \cdot \left(\nabla f^{(3)}\right)_{(x_6, y_6)} = f_6^n, \qquad (6.7.9)$$

which provide us with three additional equations. The remaining nine equations express (*a*) continuity of the three polynomial expansions and their gradients at the interior node labeled 7 and (*b*) continuity of the normal derivative at the interior mid-side nodes labeled 8, 9, and 10, indicated by the small square symbols in Figure 6.7.2.

Compiling the aforementioned $18 + 3 + 9 = 30$ conditions, we formulate a system of linear equations,

$$\mathbf{D} \cdot \begin{bmatrix} \mathbf{a}^{(1)} \\ \mathbf{a}^{(2)} \\ \mathbf{a}^{(3)} \end{bmatrix} = \mathbf{b}, \tag{6.7.10}$$

where $\mathbf{D}$ is a grand coefficient matrix and $\mathbf{b}$ is a suitable right-hand side defined in terms of the available degrees of freedom. The last nine entries of the vector $\mathbf{b}$ are zero, reflecting the interior node continuity conditions.

FSELIB function *HCT_sys*, listed in Table 6.7.1, generates and solves the linear system for the polynomial coefficients. The 12-dimensional vector $\mathbf{h}$, defined in (6.7.7), is contained in the input vector *dof*. The element interpolation functions can be generated by setting the value of a selected entry of this vector equal to unity, while holding all other values to zero, and then running over the available degrees of freedom.

FSELIB code *HCT_psi*, not listed in the text, generates graphs of the modes associated with the available degrees of freedom. Graphs of four modes for an arrangement where the first vertex is located at $x_1 = 0$, $y_1 = 0$, the second at $x_2 = 1$, $y_2 = 0$, and the third at $x_3 = 0$, $y_3 = 1$, are shown in Figure 6.7.3.

PROBLEM

6.7.1 *Interpolation over the HCT element*

Prepare a graph of the interpolated function $f(x, y) = \exp(-x^2 - y^2)$ over the HCT element shown in Figure 6.7.3.

6.8 Finite element methods for plate bending

The elements discussed in Sections 6.6 and 6.7 belong to an extensive library of triangular and quadrilateral elements designed specifically for applications involving plate bending. A variety of other non-conforming and conforming elements are reviewed and tabulated by Hrabok and Hrudey [30]. While the use of conforming elements guarantees that the finite element solution converges to the exact solution as the finite element grid becomes finer, some non-conforming elements have also been used with success, as discussed by Ciarlet [11].

```
function [a] = HCT_sys (x1,y1, x2,y2, x3,y3, dof)

%======================================================
% Generate the modal (dof) polynomial coefficients
% for the HCT element
%
% dof: degrees of freedom
%======================================================

%-----------------------------------
% mid-side nodes and normal vectors
%-----------------------------------

x4 = 0.5*(x1+x2); y4 = 0.5*(y1+y2);
d4 = sqrt((x2-x1)^2+(y2-y1)^2);
nx4 =-(y2-y1)/d4; ny4 = (x2-x1)/d4;

x5 = 0.5*(x2+x3); y5 = 0.5*(y2+y3);
d5 = sqrt((x3-x2)^2+(y3-y2)^2);
nx5 = -(y3-y2)/d5; ny5 = (x3-x2)/d5;
x6 = 0.5*(x3+x1); y6 = 0.5*(y3+y1);
d6 = sqrt((x1-x3)^2+(y1-y3)^2);
nx6 = -(y1-y3)/d6; ny6 = (x1-x3)/d6;

%--------------
% interior node
%--------------

x7 = (x1+x2+x3)/3.0; y7 = (y1+y2+y3)/3.0;

%-----------------------------------------
% interior mid-side nodes and normal vectors
%-----------------------------------------

x8 = 0.5*(x1+x7); y8=0.5*(y1+y7);
d8 = sqrt((x7-x1)^2+(y7-y1)^2);
nx8 =-(y7-y1)/d8; ny8=(x7-x1)/d8;

x9 = 0.5*(x2+x7); y9=0.5*(y2+y7);
d9 = sqrt((x7-x2)^2+(y7-y2)^2);
nx9 =-(y7-y2)/d9; ny9=(x7-x2)/d9;

x10 = 0.5*(x3+x7); y10=0.5*(y3+y7);
d10 = sqrt((x7-x3)^2+(y7-y3)^2);
nx10 =-(y7-y3)/d10; ny10=(x7-x3)/d10;

%---------------------------
% define the coefficient matrix
%---------------------------

D = [ ...
   % first sub-element: (6 eqns)
...
```

TABLE 6.7.1 Function *HCT_sys* (Continuing →)

```
1  x1  y1  x1^2  x1*y1  y1^2  x1^3  x1^2*y1  x1*y1^2  y1^3 ...
0 0 0 0 0 0 0 0 0 0 0 0 0 0 0 0 0 0 0 0 0 ; ...
...
0  1  0  2*x1  y1  0  3*x1^2  2*x1*y1  y1^2  0   ...
0 0 0 0 0 0 0 0 0 0 0 0 0 0 0 0 0 0 0 0 0 ; ...
...
0  0  1  0  x1  2*y1  0  x1^2  2*x1*y1  3*y1^2 ...
0 0 0 0 0 0 0 0 0 0 0 0 0 0 0 0 0 0 0 0 ; ...
...
1  x2  y2  x2^2  x2*y2  y2^2  x2^3  x2^2*y2  x2*y2^2  y2^3 ...
0 0 0 0 0 0 0 0 0 0 0 0 0 0 0 0 0 0 0 0 0 ; ...
...
0  1  0  2*x2  y2  0  3*x2^2  2*x2*y2  y2^2  0   ...
0 0 0 0 0 0 0 0 0 0 0 0 0 0 0 0 0 0 0 0 0 ; ...
...
0  0  1  0  x2  2*y2  0  x2^2  2*x2*y2  3*y2^2 ...
0 0 0 0 0 0 0 0 0 0 0 0 0 0 0 0 0 0 0 0 0 ; ...
...
  % second sub-element: (6 eqns)
...
0 0 0 0 0 0 0 0 0 ...
1  x2  y2  x2^2  x2*y2  y2^2  x2^3  x2^2*y2  x2*y2^2  y2^3 ...
0 0 0 0 0 0 0 0 0 0 ; ...
0 0 0 0 0 0 0 0 0 0 ...
0  1  0  2*x2  y2  0  3*x2^2  2*x2*y2  y2^2  0   ...
0 0 0 0 0 0 0 0 0 0 ; ...
0 0 0 0 0 0 0 0 0 0 ...
0  0  1  0  x2  2*y2  0  x2^2  2*x2*y2  3*y2^2 ...
0 0 0 0 0 0 0 0 0 0 ; ...
...
0 0 0 0 0 0 0 0 0 0 ...
1  x3  y3  x3^2  x3*y3  y3^2  x3^3  x3^2*y3  x3*y3^2  y3^3   ...
0 0 0 0 0 0 0 0 0 0 ; ...
0 0 0 0 0 0 0 0 0 0 ...
0 1 0 2*x3 y3 0  3*x3^2  2*x3*y3  y3^2   0   ...
0 0 0 0 0 0 0 0 0 0 ; ...
0 0 0 0 0 0 0 0 0 0 ...
0  0  1  0  x3  2*y3  0  x3^2  2*x3*y3  3*y3^2 ...
0 0 0 0 0 0 0 0 0 0 ; ...
...
  % third sub-element: (6 eqns)
...
0 0 0 0 0 0 0 0 0 0 0 0 0 0 0 0 0 0 0 0 ...
1  x3  y3  x3^2  x3*y3  y3^2  x3^3  x3^2*y3  x3*y3^2  y3^3 ; ...
0 0 0 0 0 0 0 0 0 0 0 0 0 0 0 0 0 0 0 0 ...
0  1  0  2*x3  y3    0  3*x3^2  2*x3*y3   y3^2  0 ; ...
0 0 0 0 0 0 0 0 0 0 0 0 0 0 0 0 0 0 0 0 ...
0  0  1  0  x3  2*y3  0  x3^2  2*x3*y3  3*y3^2 ; ...
...
0 0 0 0 0 0 0 0 0 0 0 0 0 0 0 0 0 0 0 0 ...
1  x1  y1  x1^2  x1*y1  y1^2  x1^3  x1^2*y1  x1*y1^2  y1^3 ; ...
0 0 0 0 0 0 0 0 0 0 0 0 0 0 0 0 0 0 0 0 ...
0  1  0  2*x1   y1  0  3*x1^2  2*x1*y1  y1^2   0 ; ...
0 0 0 0 0 0 0 0 0 0 0 0 0 0 0 0 0 0 0 0 ...
0  0  1  0     x1  2*y1  0  x1^2  2*x1*y1  3*y1^2 ; ...
```

TABLE 6.7.1 Function *HCT_sys* ($\rightarrow$ Continuing $\rightarrow$)

```
 . . .
  % mid-side node 4: (1 eqn)
 . . .
0  nx4  ny4  2*x4*nx4  y4*nx4+x4*ny4  2*y4*ny4  ...
  3*x4^2*nx4  x4*(2*y4*nx4+x4*ny4)  y4*(y4*nx4+2*x4*ny4)  3*y4^2*ny4  ...
0 0 0 0 0 0 0 0 0 0 0 0 0 0 0 0 0 0 0 0 ;
 . . .
  % mid-side node 5: (1 eqn)
 . . .
0 0 0 0 0 0 0 0 0 ...
0  nx5  ny5  2*x5*nx5  y5*nx5+x5*ny5  2*y5*ny5  ...
  3*x5^2*nx5  x5*(2*y5*nx5+x5*ny5)  y5*(y5*nx5+2*x5*ny5)  3*y5^2*ny5  ...
0 0 0 0 0 0 0 0 0 ; ...
 . . .
  % mid-side node 6: (1 eqn)
 . . .
0 0 0 0 0 0 0 0 0 0 0 0 0 0 0 0 0 0 0 0 0 ...
0  nx6  ny6  2*x6*nx6  y6*nx6+x6*ny6  2*y6*ny6  ...
  3*x6^2*nx6  x6*(2*y6*nx6+x6*ny6)  y6*(y6*nx6+2*x6*ny6)  3*y6^2*ny6; ...
 . . .
 . . .
  % continuity at interior node 7 for sub-elements 1 and 2:  (3 eqns)
 . . .
 1  x7  y7  x7^2  x7*y7  y7^2  x7^3  x7^2*y7  x7*y7^2  y7^3  ...
-1 -x7 -y7 -x7^2 -x7*y7 -y7^2 -x7^3 -x7^2*y7 -x7*y7^2 -y7^3  ...
0 0 0 0 0 0 0 0 0 ; ...
 . . .
0  1  0  2*x7  y7  0  3*x7^2  2*x7*y7  y7^2  0  ...
0 -1  0 -2*x7 -y7  0 -3*x7^2 -2*x7*y7 -y7^2  0  ...
0 0 0 0 0 0 0 0 0 ; ...
 . . .
0  0  1  0  x7  2*y7  0  x7^2  2*x7*y7  3*y7^2 ...
0  0 -1  0 -x7 -2*y7  0 -x7^2 -2*x7*y7 -3*y7^2 ...
0 0 0 0 0 0 0 0 0 ; ...
 . . .
  % continuity at interior node 7 for sub-elements 2 and 3:  (3 eqns)
 . . .
0 0 0 0 0 0 0 0 0 ...
 1  x7  y7  x7^2  x7*y7  y7^2  x7^3  x7^2*y7  x7*y7^2  y7^3   ...
-1 -x7 -y7 -x7^2 -x7*y7 -y7^2 -x7^3 -x7^2*y7 -x7*y7^2 -y7^3 ;  ...
 . . .
0 0 0 0 0 0 0 0 0 ...
0  1  0  2*x7  y7  0  3*x7^2  2*x7*y7  y7^2  0  ...
0 -1  0 -2*x7 -y7  0 -3*x7^2 -2*x7*y7 -y7^2  0 ;
 . . .
0 0 0 0 0 0 0 0 0 ...
0  0  1  0  x7  2*y7 0  x7^2  2*x7*y7  3*y7^2 ...
0  0 -1  0 -x7 -2*y7 0 -x7^2 -2*x7*y7 -3*y7^2 ;
 . . .
  % continuity of normal derivative at mid-side node 8: (1 eqn)
 . . .
0  nx8  ny8  2*x8*nx8  y8*nx8+x8*ny8  2*y8*ny8  ...
   3*x8^2*nx8  x8*(2*y8*nx8+x8*ny8)  y8*(y8*nx8+2*x8*ny8) ...
```

TABLE 6.7.1 Function *HCT_sys* (→ Continuing →)

```
    3*y8^2*ny8   ...
0 0 0 0 0 0 0 0 0 0 ...
0 -nx8 -ny8 -2*x8*nx8 -y8*nx8-x8*ny8 -2*y8*ny8 ...
   -3*x8^2*nx8 -x8*(2*y8*nx8+x8*ny8) -y8*(y8*nx8+2*x8*ny8) ...
   -3*y8^2*ny8 ;...
...
    % Continuity of normal derivative at mid-side node 9: (1 eqn)
...
0  nx9  ny9  2*x9*nx9  y9*nx9+x9*ny9  2*y9*ny9  ...
   3*x9^2*nx9  x9*(2*y9*nx9+x9*ny9)  y9*(y9*nx9+2*x9*ny9) ...
   3*y9^2*ny9 ...
0 -nx9 -ny9 -2*x9*nx9 -y9*nx9-x9*ny9 -2*y9*ny9  ...
   -3*x9^2*nx9 -x9*(2*y9*nx9+x9*ny9) -y9*(y9*nx9+2*x9*ny9) ...
   -3*y9^2*ny9 ...
0 0 0 0 0 0 0 0 0 0 ; ...
...
    % Continuity of normal derivative at mid-side node 10: (1 eqn)
...
0 0 0 0 0 0 0 0 0 0 ...
0  nx10  ny10  2*x10*nx10  y10*nx10+x10*ny10  2*y10*ny10  ...
   3*x10^2*nx10  x10*(2*y10*nx10+x10*ny10)  y10*(y10*nx10+2*x10*ny10) ...
   3*y10^2*ny10 ...
0 -nx10 -ny10 -2*x10*nx10 -y10*nx10-x10*ny10 -2*y10*ny10  ...
   -3*x10^2*nx10 -x10*(2*y10*nx10+x10*ny10) -y10*(y10*nx10+2*x10*ny10) ...
   -3*y10^2*ny10  ;
];

%--------------------------
% define the right-hand side
%--------------------------

b=zeros(1,30);

b(1)  = dof(1);   b(2)  = dof(2);   b(3)  = dof(3);
b(4)  = dof(4);   b(5)  = dof(5);   b(6)  = dof(6);
b(7)  = dof(4);   b(8)  = dof(5);   b(9)  = dof(6);
b(10) = dof(7);   b(11) = dof(8);   b(12) = dof(9);
b(13) = dof(7);   b(14) = dof(8);   b(15) = dof(9);
b(16) = dof(1);   b(17) = dof(2);   b(18) = dof(3);
b(19) = dof(10);  b(20) = dof(11);  b(21) = dof(12);

%--------------------------
% solve the linear system
%--------------------------

a = b/D';

return
```

TABLE 6.7.1 ($\rightarrow$ Continued) Function *HCT_sys* computes the cubic polynomial coefficients for the HCT element.

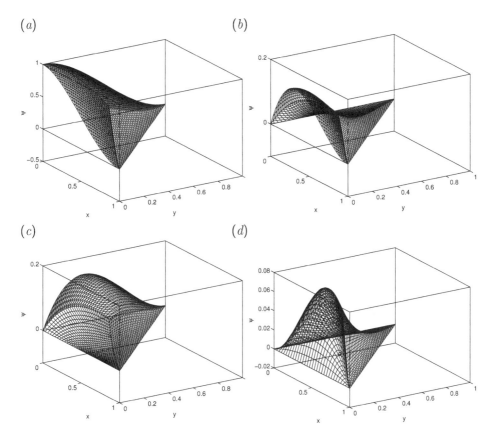

FIGURE 6.7.3 Element interpolation functions (modes) for the HCT element. Graphs of (a) ψ_1 corresponding to f_1, (b) ψ_2 corresponding to f_1^x, (c) ψ_3 corresponding to f_1^y, and (d) ψ_{10} corresponding to f_4^n.

Finite element formulations

A variety of finite element formulations are available for computing plate deflection under a normal load. Direct methods provide us with numerical solutions of the governing differential equations, whereas indirect methods provide us with numerical solutions based on functional minimization associated with an appropriate variational formulation. Indirect methods are discussed extensively, and sometimes exclusively, in the literature on finite element methods in structural and solid mechanics (e.g., Gallagher [23], Kwon & Bang [37]).

In the remainder of this section, we discuss a direct formulation based on the biharmonic equation and an indirect formulation based on force and moment equilibrium equations.

6.8.1   *Formulation as a biharmonic equation*

The formulation based on the biharmonic equation (6.4.44), repeated below for convenience,

$$\nabla^4 f = -\frac{1}{E_B}\, w(x, y), \tag{6.8.1}$$

is the counterpart of the beam bending formulation discussed in Section 2.4, culminating in a fourth-order ordinary differential equation.

   To develop the Galerkin finite element equations, we multiply the governing equation (6.8.1) by each one of the global interpolation functions associated with the available degrees of freedom, ϕ_i, and integrate the product over the solution domain, D, to obtain

$$\iint_D \phi_i(x, y)\, \nabla^4 f \, dx \, dy = -\frac{1}{E_B} \iint_D \phi_i(x, y)\, w(x, y)\, dx \, dy. \tag{6.8.2}$$

Further manipulation involves the reduction of the order of the fourth derivatives involved in the biharmonic operator on the left-hand side.

   For convenience, we denote the left-hand side as

$$\mathcal{J}_i \equiv \iint_D \phi_i\, \nabla^4 f \, dx \, dy = \iint_D \phi_i\, \nabla \cdot \left(\nabla \left(\nabla^2 f \right) \right) dx \, dy, \tag{6.8.3}$$

and write

$$\mathcal{J}_i = \iint_D \nabla \cdot \left(\phi_i\, \nabla \left(\nabla^2 f \right) \right) dx \, dy - \iint_D \nabla \phi_i \cdot \nabla (\nabla^2 f)\, dx \, dy. \tag{6.8.4}$$

Now using the divergence theorem, we obtain

$$\mathcal{J}_i = -\oint_C \phi_i\, \mathbf{n} \cdot \nabla \left(\nabla^2 f \right) dl - \iint_D \nabla \phi_i \cdot \nabla \left(\nabla^2 f \right) dx \, dy, \tag{6.8.5}$$

where C is the boundary of D, $\mathbf{n}$ is the *inward* unit vector normal to C, and l is the arc length along C.

   Also integrating by parts the last integral in (6.8.5), we write

$$\iint_D \nabla \phi_i \cdot \nabla \left(\nabla^2 f \right) dx \, dy = \iint_D \nabla \cdot \left(\nabla \phi_i \left(\nabla^2 f \right) \right) dx \, dy - \iint_D \nabla^2 \phi_i\, \nabla^2 f \, dx \, dy$$

$$\tag{6.8.6}$$

and then

$$\iint_D \nabla \phi_i \cdot \nabla \left(\nabla^2 f \right) dx \, dy = -\oint_C (\mathbf{n} \cdot \nabla \phi_i)\, \nabla^2 f \, dl - \iint_D \nabla^2 \phi_i\, \nabla^2 f \, dx \, dy. \tag{6.8.7}$$

Combining (6.8.5) and (6.8.7), we obtain

$$\iint_D \phi_i \nabla^4 f \, dx \, dy = - \oint_C \phi_i \, \mathbf{n} \cdot \nabla(\nabla^2 f) \, dl$$

$$+ \oint_C (\mathbf{n} \cdot \nabla \phi_i) \, \nabla^2 f \, dl + \iint_D \nabla^2 \phi_i \, \nabla^2 f \, dx \, dy. \tag{6.8.8}$$

Finally, we substitute this expression into the left-hand side of (6.8.2) and rearrange to obtain the Galerkin equation

$$\iint_D \nabla^2 \phi_i \, \nabla^2 f \, dx \, dy = \oint_C \phi_i \, \mathbf{n} \cdot \nabla(\nabla^2 f) \, dl$$

$$- \oint_C (\mathbf{n} \cdot \nabla \phi_i) \, \nabla^2 f \, dl - \frac{1}{E_B} \iint_D \phi_i \, w \, dx \, dy. \tag{6.8.9}$$

In terms of the trace of the bending moments tensor given in (6.4.39),

$$M \equiv M_{xx} + M_{yy} = -E_B(1 + \nu) \nabla^2 f, \tag{6.8.10}$$

and the transverse shear tension vector given in (6.4.43),

$$\mathbf{q} = -E_B \nabla(\nabla^2 f), \tag{6.8.11}$$

equation (6.8.9) becomes

$$E_B \iint_D \nabla^2 \phi_i \, \nabla^2 f \, dx \, dy = - \oint_C \phi_i \, \mathbf{n} \cdot \mathbf{q} \, dl$$

$$+ \frac{1}{1 + \nu} \oint_C M \, \mathbf{n} \cdot \nabla \phi_i \, dl - \iint_D \phi_i \, w \, dx \, dy, \tag{6.8.12}$$

where ν is the Poisson ratio. The Galerkin equation (6.8.12) is the counterpart of the beam equation (2.5.16).

The contour integrals on the right-hand sides of (6.8.9) and (6.8.12) are non-zero only if ϕ_i is a mode associated with a boundary node. If ϕ_i is a mode associated with an interior node, these integrals are zero, yielding the simplified equation

$$\iint_D \nabla^2 \phi_i \, \nabla^2 f \, dx \, dy = - \frac{1}{E_B} \iint_D \phi_i \, w \, dx \, dy. \tag{6.8.13}$$

In the case of a clamped plate, the transverse displacement and its gradient are specified along the boundary, C. Consequently, the corresponding projections are excluded from the finite element formulation and the simplified formulation applies.

Derivation of a linear algebraic system

Assuming that the finite element expansion involves N_{G_m} global modes associated with all available degrees of freedom, we introduce the global expansion

$$f(x, y) = \sum_{j=1}^{N_{G_m}} h_j^G \, \phi_j(x, y), \tag{6.8.14}$$

where the coefficients h_j^G represent either the value of the solution or the value of a spatial derivative of the solution at a global node. Inserting this expression into (6.8.13), we derive a linear system,

$$K_{ij} \, h_j^G = -\frac{1}{E_B} \, b_i, \tag{6.8.15}$$

where summation over j is implied on the left-hand side,

$$K_{ij} \equiv \iint_D \nabla^2 \phi_i \, \nabla^2 \phi_j \, dx \, dy \tag{6.8.16}$$

is the *global bending stiffness matrix*, and

$$b_i \equiv \iint_{E_l} \phi_i \, w \, dx \, dy \tag{6.8.17}$$

is the vector on the right-hand side. The global stiffness matrix and right-hand side can be compiled in the usual way from corresponding element stiffness matrices,

$$A_{ij}^{(l)} \equiv \iint_{E_l} \nabla^2 \psi_i \, \nabla^2 \psi_j \, dx \, dy \tag{6.8.18}$$

and

$$b_i^{(l)} \equiv \iint_{E_l} \psi_i \, w \, dx \, dy, \tag{6.8.19}$$

where E_l denotes the lth element. Like the element diffusion matrix, the element stiffness matrix is singular (Problem 6.8.1).

Evaluation of the element bending stiffness matrices

FSELIB function *HCT_ebsm*, listed in Table 6.8.1, evaluates the element stiffness matrix, $A_{ij}^{(l)}$, and element integral, $b_i^{(l)}$ corresponding to a uniform load, $w = w_0$, for the HCT element discussed in Section 6.7.2.

To compute the element integrals, we map each sub-triangle of the HCT triangle from the physical xy plane to the parametric $\xi\eta$ plane using the mapping functions in (4.2.11). The integrals over the sub-triangles are then computed using a Gaussian triangle quadrature.

```
function [ebsm, rhs, arel] = HCT_ebsm (x1,y1, x2,y2, x3,y3, NQ, w0)

%====================================================
% Evaluation of the element bending stiffness matrix (ebsm)
% and right-hand side for the HCT triangle,
% using a Gauss integration quadrature.
%
% NQ:   number of quadrature base points
% arel: element surface area
% rhs:  right-hand side
%====================================================

%---------------------------------
% compute the area of the triangle
%---------------------------------

d12x = x1-x2; d12y = y1-y2;
d31x = x3-x1; d31y = y3-y1;

arel = 0.5*(d31x*d12y - d31y*d12x);

%--------------
% interior node
%--------------

x7 = (x1+x2+x3)/3.0;
y7 = (y1+y2+y3)/3.0;

%---------------------------
% read the triangle quadrature
%---------------------------

[xi, eta, w] = gauss_trgl(NQ);

%--------------------------------------------
% initialize the element bending stiffness matrix
% and right-hand side
%--------------------------------------------

for k=1:12
 for l=1:12
  ebsm(k,l) = 0.0;
 end
 rhs(k) = 0.0;
end

%----------------------------------
% compute the 30 cubic coefficients
% and put them in the vector "am"
%----------------------------------
```

TABLE 6.8.1   Function *HCT_ebsm* (Continuing →)

```
for mode=1:12

  dof = zeros(1,12);
  dof(mode) = 1.0;
  [a] = HCT_sys (x1,y1, x2,y2, x3,y3, dof);

  for i=1:30
    am(i,mode) = a(i);
  end

end

%-----------------------
% perform the quadrature
%-----------------------

for pass=1:3  % run over the 3 sub-triangles

 for i=1:NQ    % quadrature

%----
% modes and laplacian:
%---

 for mode=1:12

   for j=1:30
     c(j) = am(j,mode);
   end

   if(pass==1)        % triangle 1-2-7

     x = x1 + (x2-x1)*xi(i) + (x7-x1)*eta(i);
     y = y1 + (y2-y1)*xi(i) + (y7-y1)*eta(i);

     psi(mode) = c(1) + c(2)*x  +c(3)*y ...
                   + c(4)*x^2 +c(5)*x*y + c(6)*y^2 ...
                   + c(7)*x^3 +c(8)*x^2*y+ c(9)*x*y^2+c(10)*y^3;

     lpsi(mode) = 2.0*c(4)+2.0*c(6) ...
                   +6.0*c(7)*x + 2.0*c(8)*y ...
                   +2.0*c(9)*x + 6.0*c(10)*y;

    elseif(pass==2)  % triangle 2-3-7

     x = x2 + (x3-x2)*xi(i) + (x7-x2)*eta(i);
     y = y2 + (y3-y2)*xi(i) + (y7-y2)*eta(i);

     psi(mode) = c(11)+c(12)*x+c(13)*y...
                 +c(14)*x^2+c(15)*x*y+c(16)*y^2 ...
                 +c(17)*x^3+c(18)*x^2*y...
                 +c(19)*x*y^2+c(20)*y^3;
```

TABLE 6.8.1   Function *HCT_ebsm* ($\rightarrow$ Continuing $\rightarrow$)

```
        lpsi(mode) = 2.0*c(14)+2.0*c(16) + 6.0*c(17)*x+2.0*c(18)*y...
                     +2.0*c(19)*x+6.0*c(20)*y;

     else                % triangle 3-1-7

     x = x3 + (x1-x3)*xi(i) + (x7-x3)*eta(i);
     y = y3 + (y1-y3)*xi(i) + (y7-y3)*eta(i);

     psi(mode) = c(21)+c(22)*x+c(23)*y ...
                     +c(24)*x^2 + c(25)*x*y...
                     +c(26)*y^2 + c(27)*x^3+c(28)*x^2*y...
                     +c(29)*x*y^2+c(30)*y^3;

     lpsi(mode) = 2.0*c(24)+2.0*c(26) ...
                     +6.0*c(27)*x+2.0*c(28)*y...
                     +2.0*c(29)*x+6.0*c(30)*y;

     end
   end    % end of loop over modes
%--

   cf = arel*w(i)/3.0;

   for k=1:12
    for l=k:12
     ebsm(k,l) = ebsm(k,l) + lpsi(k)*lpsi(l)*cf;
    end
    wload = w0;                     % uniform load
    rhs(k) = rhs(k) + wload*psi(k)*cf;
   end

 end    % end of quadrature
end % end of run over sub-triangles

%--------------------------------
% fill in lower diagonal of ebsm
%--------------------------------

 for k=1:12
  for l=1:k-1
   ebsm(k,l) = ebsm(l,k);
  end
 end

%-----
% done
%-----

return;
```

TABLE 6.8.1   (→ Continued) Function *HCT_ebsm* evaluates the element bending stiffness matrix.

Other types of load can be incorporated by straightforward modifications. For example, in the case of a unidirectional load along the x axis that is linear in x and y, such that

$$w = w_x(x - x_0) + w_y(y - y_0), \tag{6.8.20}$$

where w_x, w_y, x_0, and y_0 are specified constants, the line:

```
wload = w0;
```

17 lines before the end of the function should be replaced with the line

```
wload = wx*(x-x0)+wy*(y-y0);    .
```

Physically, a linear load pushes against a vertical foundation wall of a building due to the back filled soil or percolating rain water. The amount of water retained around the foundation depends on the soil constitution. Sand retains the least amount of water, heavy clay retains the most amount of water.

Bending of a clamped plate

FSELIB code *HCT_bend*, listed in Table 6.8.2, solves the inhomogeneous biharmonic equation governing the deflection of a plate with an entirely clamped contour based on the HCT element. The layout of the code is similar to that of the codes for steady diffusion discussed in Chapters 4 and 5.

One important new feature is that, because the normal derivative at the mid-side nodes of an element points into each element by convention, the sign must be switched in assembling the corresponding entries of the global linear system after the first contribution has been made. This is done by introducing a flag named *Idid* whose value is switched from 1 to -1 after the first contribution has been made.

Results for a circular or square plate subjected to a uniform vertical load corresponding to a constant load density function, w, are shown in Figure 6.8.1. In the case of the circular plate, the exact analytical solution predicts that the maximum deflection, occurring at the center of the plate, is given by

$$f_{\max} = \frac{wa^4}{64E_B} \simeq 0.0156 \frac{wa^4}{E_B}, \tag{6.8.21}$$

where a is the plate radius (e.g., Timoshenko & Woinowsky-Krieger [69], pp. 54–55). The finite element solution for discretization level $ndiv = 3$ predicts $f_{\max} = 0.0154\, wa^4/E_B$, which is in good agreement with the exact value. In the case of the square plate, an approximate series solution predicts the maximum deflection

$$f_{\max} \simeq 0.0202 \frac{wa^4}{E_B} \tag{6.8.22}$$

occurring at the center-point, where a is the plate edge half-width (e.g., Timoshenko & Woinowsky-Krieger [69], p. 202). The finite element solution for discretization level $ndiv = 3$ agrees with this prediction up to the fourth decimal place.

```
%=============
% Code bend_HCT
%
% Solution of the biharmonic equation
% for plate bending using the HCT element.
%=============

%-----------
% input data
%-----------

E_B = 1.0;  % bending modulus
NQ = 4;     % quadrature order
w0 = 4;     % load
ndiv = 2;   % discretization level

%-----------
% triangulate
%-----------

 [ne,ng,p,c,efl,gfl] = trgl6_sqr(ndiv);
% [ne,ng,p,c,efl,gfl] = trgl6_disk(ndiv);

%--------------------------------------------------
% make sure the mid-nodes are along straight edges
%--------------------------------------------------

for i=1:ne
 for j=1:2
  p(c(i,4),j) = 0.5*( p(c(i,1),j)+p(c(i,2),j) );
  p(c(i,5),j) = 0.5*( p(c(i,2),j)+p(c(i,3),j) );
  p(c(i,6),j) = 0.5*( p(c(i,3),j)+p(c(i,1),j) );
 end
end

%---------------------------------------
% mark the global vertex and edge nodes
%---------------------------------------

for i=1:ne
 gfln(c(i,1)) = 0;      % vertex nodes
    gfln(c(i,2)) = 0;
       gfln(c(i,3)) = 0;
 gfln(c(i,4)) = 1;      % edge nodes
    gfln(c(i,5)) = 1;
       gfln(c(i,6)) = 1;
end

%---------------------------------------
% count the number of global edge nodes
%---------------------------------------

nge = 0;  % number of edge nodes
```

TABLE 6.8.2   Code *bend_HCT* (Continuing →)

```
for i=1:ng
 if(gfln(i) == 1)
  nge = nge+1;
 end
end

%-----------------------------
% total number of global modes
%-----------------------------

ngv = ng-nge;     % vertex nodes

ngm = 3*ngv+nge;  % number of global modes

%-----------------------------------------
% mark the position of the rows and columns
% of the modes in the global system
%-----------------------------------------

for i=1:ng

  s(i) = 1;

  for j=1:i-1
   if(gfln(j) == 1)    % edge node, 1 dof
     s(i) = s(i)+1;
   else                % vertex node, 3 dofs
     s(i) = s(i)+3;
   end
  end

end

%-----
% initialize the orientation index of the normal derivative
% at the mid-side nodes
%-----

for i=1:ng
 Idid(i) = 1.0;
end

%-------------------------------------------------
% assemble the global bending stiffness matrix
% and right-hand side
%-------------------------------------------------

gbsm = zeros(ngm,ngm); % initialize
b = zeros(1,ngm);      % right-hand side
```

TABLE 6.8.2   Code *bend_HCT* ($\rightarrow$ Continuing $\rightarrow$)

```
for l=1:ne              % loop over the elements

% compute the element bending stiffness matrix and rhs
% and right-hand side

j1 = c(l,1); j2 = c(l,2); j3 = c(l,3);   % global element labels
j4 = c(l,4); j5 = c(l,5); j6 = c(l,6);

x1 = p(j1,1); y1 = p(j1,2);              % three vertices
x2 = p(j2,1); y2 = p(j2,2);
x3 = p(j3,1); y3 = p(j3,2);

[ebsm_elm, rhs, arel] = HCT_ebsm (x1,y1, x2,y2, x3,y3, NQ, w0);

e(1) = s(j1);  e(2) = s(j1)+1; e(3) = s(j1)+2;  % position in
e(4) = s(j2);  e(5) = s(j2)+1; e(6) = s(j2)+2;  % global matrix
e(7) = s(j3);  e(8) = s(j3)+1; e(9) = s(j3)+2;
e(10)= s(j4);  e(11)= s(j5);   e(12)= s(j6);

   for i=1:12

     fci = 1.0;
         if(i==10) fci = Idid(j4);
     elseif(i==11) fci = Idid(j5);
     elseif(i==12) fci = Idid(j6);
     end

     for j=1:12
       fcj = 1.0;
           if(j==10) fcj = Idid(j4);
       elseif(j==11) fcj = Idid(j5);
       elseif(j==12) fcj = Idid(j6);
       end
         gbsm(e(i),e(j)) = gbsm(e(i),e(j)) + fci*fcj*ebsm_elm(i,j);
     end

     b(e(i)) = b(e(i)) - fci*rhs(i);

   end

 Idid(j4) = -1;
 Idid(j5) = -1;
 Idid(j6) = -1;

end

%-------------------------------------------------------------
% implement the homogeneous Dirichlet boundary condition
% no need to modify the right-hand side
%-------------------------------------------------------------
```

TABLE 6.8.2   Code *bend_HCT* ($\rightarrow$ Continuing $\rightarrow$)

```
for j=1:ng

 if(gfl(j,1)==1)    % boundary node

    j1 = s(j);
    j2 = s(j)+1;
    j3 = s(j)+2;

    for i=1:ngm

      gbsm(i,j1)=0.0; gbsm(j1,i)=0.0;

      if(gfln(j) == 0)    % vertex node
        gbsm(i,j2) = 0.0; gbsm(j2,i) = 0.0;
        gbsm(i,j3) = 0.0; gbsm(j3,i) = 0.0;
      end
    end

    gbsm(j1,j1) = 1.0; b(j1) = 0.0;

    if(gfln(j) == 0)    % vertex node
      gbsm(j2,j2) = 1.0; b(j2) = 0.0;
      gbsm(j3,j3) = 1.0; b(j3) = 0.0;
    end

 end
end

%-----------------------
% solve the linear system
%-----------------------

sol = b/gbsm';

%--------------------------------------
% extract the deflection at the vertices
%--------------------------------------

for i=1:ng
  if(gfln(i) == 0)
    f(i) = sol(s(i));
  end
end

%-------------------------------
% interpolate the mid-side values
%-------------------------------

for l=1:ne

  j1 = c(l,1); j2 = c(l,2); j3 = c(l,3);    % global element labels
  j4 = c(l,4); j5 = c(l,5); j6 = c(l,6);

  e(1) = s(j1); e(2) = s(j1)+1; e(3) = s(j1)+2; % position in the
  e(4) = s(j2); e(5) = s(j2)+1; e(6) = s(j2)+2; % global matrix
```

TABLE 6.8.2   Code *bend_HCT* ($\rightarrow$ Continuing $\rightarrow$)

```
e(7) = s(j3); e(8) = s(j3)+1; e(9) = s(j3)+2;
e(10)= s(j4); e(11)= s(j5);   e(12)= s(j6);

for j=1:12
  dof(j) = sol(e(j));
end

x1 = p(j1,1); y1 = p(j1,2);   % three vertices
x2 = p(j2,1); y2 = p(j2,2);
x3 = p(j3,1); y3 = p(j3,2);

[a] = HCT_sys (x1,y1, x2,y2, x3,y3, dof);

x=p(j4,1); y=p(j4,2);

f(j4) = a(1)+a(2)*x+a(3)*y+a(4)*x^2+a(5)*x*y+a(6)*y^2 ...
        +a(7)*x^3+a(8)*x^2*y+a(9)*x*y^2+a(10)*y^3;

x=p(j5,1); y=p(j5,2);

f(j5) = a(11)+a(12)*x+a(13)*y+a(14)*x^2+a(15)*x*y+a(16)*y^2 ...
        +a(17)*x^3+a(18)*x^2*y+a(19)*x*y^2+a(20)*y^3;

x=p(j6,1); y=p(j6,2);

f(j6) = a(21)+a(22)*x+a(23)*y+a(24)*x^2+a(25)*x*y+a(26)*y^2 ...
        +a(27)*x^3+a(28)*x^2*y+a(29)*x*y^2+a(30)*y^3;
end

%----------------------------
% plot using an fselib function
%----------------------------

plot_6 (ne,ng,p,c,f);

%----------------------------
% extended connectivity matrix
% for three-node sub-triangles
%----------------------------

Ic = 0;

for i=1:ne
  Ic = Ic+1;
  c3(Ic,1) = c(i,1); c3(Ic,2) = c(i,4); c3(Ic,3) = c(i,6);
  Ic = Ic+1;
  c3(Ic,1) = c(i,4); c3(Ic,2) = c(i,2); c3(Ic,3) = c(i,5);
  Ic = Ic+1;
  c3(Ic,1) = c(i,5); c3(Ic,2) = c(i,3); c3(Ic,3) = c(i,6);
  Ic = Ic+1;
  c3(Ic,1) = c(i,4); c3(Ic,2) = c(i,5); c3(Ic,3) = c(i,6);
end
```

TABLE 6.8.2 Code *bend_HCT* ($\rightarrow$ Continuing $\rightarrow$)

```
%----------------------------
% plot using a matlab function
%----------------------------

trimesh (c3,p(:,1),p(:,2),f); xlabel('x'); ylabel('y');

%-----
% done
%-----
```

TABLE 6.8.2 ($\rightarrow$ Continued) Code *bend_HCT* computes the finite element solution of the forced biharmonic equation describing the bending of a clamped plate.

(a) (b)

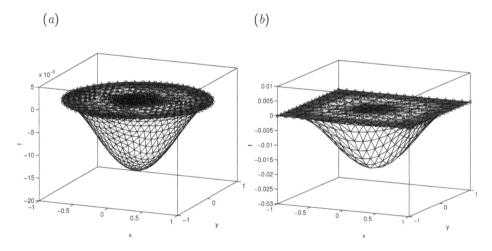

FIGURE 6.8.1 Bending of (a) an entirely clamped circular plate or (b) and entirely clamped square plate, subject to a uniform vertical load representing, for example, the gravitational force.

6.8.2 *Formulation as a system of Poisson equations*

In an alternative approach, the constitutive equations in (6.4.36) and the equilibrium equation (6.4.26) are restated as a system of four Poisson-like equations, as shown in Table 6.8.3(a). The four unknown functions include the deflection, $f(x, y)$, and the bending and twisting moments, M_{xx}, M_{yy}, and M_{xy}.

In the finite element implementation, each of these functions is expanded in terms of the global modes and associated degrees of freedom, as shown in (6.5.26),

(a)

$$\frac{\partial^2 f}{\partial x^2} + \nu \frac{\partial^2 f}{\partial y^2} = -\frac{1}{E_B} M_{xx} \qquad\qquad \frac{\partial^2 f}{\partial y^2} + \nu \frac{\partial^2 f}{\partial x^2} = -\frac{1}{E_B} M_{yy}$$

$$\frac{\partial^2 f}{\partial x\,\partial y} = -\frac{1}{E_B\,(1-\nu)} M_{xy} \qquad\qquad \frac{\partial^2 M_{xx}}{\partial x^2} + 2\frac{\partial^2 M_{xy}}{\partial x\,\partial y} + \frac{\partial^2 M_{yy}}{\partial y^2} = w$$

(b)

$$K_{ij}^{xx}\, f_j^{\mathrm{G}} + \nu\, K_{ij}^{yy}\, f_j^{\mathrm{G}} - M_{ij}\, M_{xx_j}^{\mathrm{G}} = \mathcal{K}_i$$

$$\nu\, K_{ij}^{xx}\, f_j^{\mathrm{G}} + K_{ij}^{yy}\, f_j^{\mathrm{G}} - M_{ij}\, M_{yy_j}^{\mathrm{G}} = \mathcal{L}_i$$

$$K_{ij}^{yx}\, f_j^{\mathrm{G}} + K_{ij}^{xy}\, f_j^{\mathrm{G}} - \frac{2}{1-\nu}\, M_{ij}\, M_{xy_j}^{\mathrm{G}} = \mathcal{M}_i$$

$$K_{ij}^{xx}\, M_{xx_j}^{\mathrm{G}} + K_{ij}^{yy}\, M_{yy_j}^{\mathrm{G}} + K_{ij}^{yx}\, M_{xy_j}^{\mathrm{G}} + K_{ij}^{xy}\, M_{xy_j}^{\mathrm{G}} = \mathcal{N}_i$$

TABLE 6.8.3 (a) Four Poisson-like equations governing plate bending and (b) the corresponding finite element equations. Summation over the repeated index j is implied on the left-hand sides.

$$\begin{bmatrix} M_{xx} \\ M_{yy} \\ M_{xy} \\ f \end{bmatrix} = \sum_{j=1}^{N_{\mathrm{G}m}} \begin{bmatrix} M_{xx_j}^{\mathrm{G}} \\ M_{yy_j}^{\mathrm{G}} \\ M_{xy_j}^{\mathrm{G}} \\ f_j^{\mathrm{G}} \end{bmatrix} \phi_j(x, y), \tag{6.8.23}$$

where $N_{\mathrm{G}m}$ is the number of global modes, and the vector on the right-hand side contains the respective coefficients of the global nodes. To simplify the notation, we introduce a vector field, $\mathbf{F}$,

$$\mathbf{F}(x, y) \equiv \begin{bmatrix} M_{xx} \\ M_{yy} \\ M_{xy} \\ f \end{bmatrix} (x, y), \tag{6.8.24}$$

and associated modal coefficient vector,

$$\mathbf{F}_j^{\mathrm{G}} \equiv \begin{bmatrix} M_{xx_j}^{\mathrm{G}} \\ M_{yy_j}^{\mathrm{G}} \\ M_{xy_j}^{\mathrm{G}} \\ f_j^{\mathrm{G}} \end{bmatrix} \tag{6.8.25}$$

for $j = 1, \ldots N_{Gm}$, and express the expansion (6.8.23) in the compact form

$$\mathbf{F}(x,y) = \sum_{j=1}^{N_{Gm}} \mathbf{F}_j^G \, \phi_j(x,y).$$ (6.8.26)

For a plate with simply supported edges, $\mathbf{F}_j^G = \mathbf{0}$ at the boundary nodes.

Galerkin projection

Next, we carry out the Galerkin projection by multiplying each equation shown in Table 6.8.3(a) by ϕ_i and then integrating by parts to eliminate the second derivatives on the left-hand sides.

The Galerkin projection of the first equation reads

$$\iint_D \phi_i \left(\frac{\partial^2 f}{\partial x^2} + \nu \frac{\partial^2 f}{\partial y^2} \right) dx \, dy = -\frac{1}{E_B} \iint_D \phi_i \, M_{xx} \, dx \, dy.$$ (6.8.27)

Integrating by parts on the left-hand side, we obtain

$$\iint_D \phi_i \left(\frac{\partial^2 f}{\partial x^2} + \nu \frac{\partial^2 f}{\partial y^2} \right) dx \, dy$$

$$= \iint_D \left[\frac{\partial}{\partial x}\left(\phi_i \frac{\partial f}{\partial x} \right) + \nu \frac{\partial}{\partial y}\left(\phi_i \frac{\partial f}{\partial y} \right) - \frac{\partial \phi_i}{\partial x} \frac{\partial f}{\partial x} - \nu \frac{\partial \phi_i}{\partial y} \frac{\partial f}{\partial y} \right] dx \, dy$$

$$= -\oint_C \phi_i \left(\frac{\partial f}{\partial x} n_x + \nu \frac{\partial f}{\partial y} n_y \right) dl - \iint_D \left(\frac{\partial \phi_i}{\partial x} \frac{\partial f}{\partial x} + \nu \frac{\partial \phi_i}{\partial y} \frac{\partial f}{\partial y} \right) dx \, dy,$$ (6.8.28)

where $\mathbf{n}$ is the *inward* unit normal vector. Substituting the last expression into the left-hand side of (6.8.27) and rearranging, we derive the final result

$$\iint_D \left(\frac{\partial \phi_i}{\partial x} \frac{\partial f}{\partial x} + \nu \frac{\partial \phi_i}{\partial y} \frac{\partial f}{\partial y} \right) dx \, dy - \frac{1}{E_B} \iint_D \phi_i M_{xx} \, dx \, dy$$

$$= -\oint_C \phi_i \left(\frac{\partial f}{\partial x} n_x + \nu \frac{\partial f}{\partial y} n_y \right) dl.$$ (6.8.29)

Working in a similar fashion with the second and third equations in Table 6.8.3(a), we obtain

$$\iint_D \left(\nu \frac{\partial \phi_i}{\partial x} \frac{\partial f}{\partial x} + \frac{\partial \phi_i}{\partial y} \frac{\partial f}{\partial y} \right) dx \, dy - \frac{1}{E_B} \iint_D \phi_i M_{yy} \, dx \, dy$$

$$= -\oint_C \phi_i \left(\nu \frac{\partial f}{\partial x} n_x + \frac{\partial f}{\partial y} n_y \right) dl$$ (6.8.30)

and

$$
\iint_D \left(\frac{\partial \phi_i}{\partial x} \frac{\partial f}{\partial y} + \frac{\partial \phi_i}{\partial y} \frac{\partial f}{\partial x} \right) dx\, dy - \frac{2}{E_B (1 - \nu)} \iint_D \phi_i M_{xy}\, dx\, dy
$$

$$
= - \oint_C \phi_i \left(\frac{\partial f}{\partial x} n_y + \frac{\partial f}{\partial y} n_x \right) dl.
$$

(6.8.31)

Combining the fourth equation in Table 6.8.3(a) with the equilibrium equation (6.4.25), $\mathbf{q} = \nabla \cdot \mathbf{M}$, and working in a similar fashion, we obtain

$$
\iint_D \left(\frac{\partial \phi_i}{\partial x} \frac{\partial M_{xx}}{\partial x} + \frac{\partial \phi_i}{\partial y} \frac{\partial M_{yy}}{\partial y} + \frac{\partial \phi_i}{\partial y} \frac{\partial M_{xy}}{\partial x} + \frac{\partial \phi_i}{\partial x} \frac{\partial M_{xy}}{\partial y} \right) dx\, dy
$$

$$
= - \oint_C \phi_i \left[\left(\frac{\partial M_{xx}}{\partial x} + \frac{\partial M_{xy}}{\partial y} \right) n_x + \left(\frac{\partial M_{xy}}{\partial x} + \frac{\partial M_{yy}}{\partial y} \right) n_y \right] dl
$$

(6.8.32)

$$
- \iint_D \phi_i w\, dx\, dy = - \oint_C \phi_i \mathbf{n} \cdot \mathbf{q}\, dl - \iint_D \phi_i w\, dx\, dy.
$$

Diffusion and mass matrices

Next we introduce expansion (6.8.23) and recast equations (6.8.29)–(6.8.32) into the form shown in Table 6.8.3(b), where summation of the repeated index j is implied on the left-hand sides in the range $j = 1, \ldots, N_{G_m}$. We have defined the familiar mass matrix,

$$
M_{ij} = \frac{1}{E_B} \iint_D \phi_i \phi_j\, dx\, dy
$$

(6.8.33)

and the directional diffusion-like matrices

$$
K_{ij}^{xx} = \iint_D \frac{\partial \phi_i}{\partial x} \frac{\partial \phi_j}{\partial x}\, dx\, dy, \qquad\qquad K_{ij}^{xy} = \iint_D \frac{\partial \phi_i}{\partial x} \frac{\partial \phi_j}{\partial y}\, dx\, dy,
$$

$$
K_{ij}^{yx} = K_{ji}^{xy} = \iint_D \frac{\partial \phi_i}{\partial y} \frac{\partial \phi_j}{\partial x}\, dx\, dy, \qquad\qquad K_{ij}^{yy} = \iint_D \frac{\partial \phi_i}{\partial y} \frac{\partial \phi_j}{\partial y}\, dx\, dy.
$$

(6.8.34)

The right-hand sides of the equations shown in Table 6.8.3(b) are given by

$$
\mathcal{K}_i = - \oint_C \phi_i \left(\frac{\partial f}{\partial x} n_x + \nu \frac{\partial f}{\partial y} n_y \right) dl,
$$

$$
\mathcal{L}_i = - \oint_C \phi_i \left(\nu \frac{\partial f}{\partial x} n_x + \frac{\partial f}{\partial y} n_y \right) dl,
$$

$$
\mathcal{M}_i = - \oint_C \phi_i \left(\frac{\partial f}{\partial x} n_y + \frac{\partial f}{\partial y} n_x \right) dl,
$$

(6.8.35)

$$
\mathcal{N}_i = - \oint_C \phi_i \mathbf{n} \cdot \mathbf{q}\, dl - \iint_D \phi_i w\, dx\, dy.
$$

The equations in Table 6.8.3(*b*) can be collected into a matrix form,

$$
\sum_{j=1}^{N_{Gm}}
\begin{bmatrix}
-M_{ij} & 0 & 0 & K_{ij}^{xx} + \nu K_{ij}^{yy} \\
0 & -M_{ij} & 0 & \nu K_{ij}^{xx} + K_{ij}^{yy} \\
0 & 0 & -\dfrac{2}{1-\nu} M_{ij} & K_{ij}^{xy} + K_{ij}^{yx} \\
K_{ij}^{xx} & K_{ij}^{yy} & K_{ij}^{xy} + K_{ij}^{yx} & 0
\end{bmatrix}
\cdot \mathbf{F}_j^G =
\begin{bmatrix}
\mathcal{K}_i \\
\mathcal{L}_i \\
\mathcal{M}_i \\
\mathcal{N}_i
\end{bmatrix},
$$

$$(6.8.36)$$

where the modal coefficient vector $\mathbf{F}_j^G$ is defined in (6.8.26).

The coefficient matrix on the right-hand side of (6.8.36) can be rendered symmetric by subtracting from the first equation the second equation multiplied by ν, and from the second equation the first equation multiplied by ν, and then dividing the resulting equations by $1 - \nu^2$ to obtain

$$
\sum_{j=1}^{N_{Gm}} \mathbf{\Phi}_{ij} \cdot \mathbf{F}_j^G =
\begin{bmatrix}
\dfrac{1}{1-\nu^2} (\mathcal{K}_i - \nu \mathcal{L}_i) \\
\dfrac{1}{1-\nu^2} (\mathcal{L}_i - \nu \mathcal{K}_i) \\
\mathcal{M}_i \\
\mathcal{N}_i
\end{bmatrix}
=
\begin{bmatrix}
-\oint_C \phi_i \frac{\partial f}{\partial x}\, dl \\
-\oint_C \phi_i \frac{\partial f}{\partial y}\, dl \\
\mathcal{M}_i \\
\mathcal{N}_i
\end{bmatrix},
$$

$$(6.8.37)$$

where

$$
\mathbf{\Phi}_{ij} =
\begin{bmatrix}
-M'_{ij} & \nu M'_{ij} & 0 & K_{ij}^{xx} \\
\nu M'_{ij} & -M'_{ij} & 0 & K_{ij}^{yy} \\
0 & 0 & -\frac{2}{1+\nu} M_{ij} & K_{ij}^{xy} + K_{ij}^{yx} \\
K_{ij}^{xx} & K_{ij}^{yy} & K_{ij}^{xy} + K_{ij}^{yx} & 0
\end{bmatrix}
$$

$$(6.8.38)$$

and

$$
M'_{ij} \equiv \frac{1}{1-\nu^2} M_{ij} = \frac{12}{E\,h^3} \iint_D \phi_i\, \phi_j\, dx\, dy.
$$

$$(6.8.39)$$

In practice, the domain integrals are assembled from corresponding element integrals involving the element interpolation functions.

One-element solution

For illustration, we consider a solution domain consisting of a single four-node, ten-mode element, as discussed in Section 6.6. The linear system (6.8.37) reduces to

$$
\sum_{j=1}^{10} \mathbf{\Phi}_{ij} \cdot \mathbf{F}_j^G = -
\begin{bmatrix}
\oint_C \phi_i \frac{\partial f}{\partial x}\, dl \\
\oint_C \phi_i \frac{\partial f}{\partial y}\, dl \\
\oint_C \phi_i \left(\frac{\partial f}{\partial x} n_y + \frac{\partial f}{\partial y} n_x \right) dl \\
\oint_C \phi_i\, \mathbf{n} \cdot \mathbf{q}\, dl + \iint_E \phi_i\, w\, dx\, dy
\end{bmatrix}
$$

$$(6.8.40)$$

for $i = 1$–10, where E denotes the element surface and C denotes the element contour. For a plate with simply supported edges, the boundary conditions require that the deflection is zero, $f = 0$, and the bending moments vanish at the three vertices. Consequently, the nine modal coefficients associated with the vertices are zero, $\mathbf{F}_j^G = \mathbf{0}$ for $j = 1, \ldots, 9$, and the left-hand side of (6.8.40) involves a single term, $\mathbf{\Phi}_{i,10} \cdot \mathbf{F}_{10}^G$.

Applying the equations in (6.8.40) for the tenth global mode corresponding to the interior element node, $i = 10$, and recalling that ϕ_{10} is zero around the perimeter of the triangle, we obtain a system of four linear equations for the bending and twisting moments and the displacement at the interior node,

$$
\mathbf{\Phi}_{10,10} \cdot
\begin{bmatrix}
M_{xx} \\
M_{yy} \\
M_{xy} \\
f
\end{bmatrix}_{10}^{G}
= -
\begin{bmatrix}
0 \\
0 \\
0 \\
\iint_E \phi_{10}\, w\, \mathrm{d}x\, \mathrm{d}y
\end{bmatrix}.
\tag{6.8.41}
$$

PROBLEMS

6.8.1 *Bending of a rectangular plate*

Run the code *bend_HCT* to compute the bending of a rectangular plate with aspect ratio 2:1 subject to a uniform load. Compare your results with the series solution predicting the maximum deflection

$$
f_{\max} \simeq 0.0406 \, \frac{wa^4}{E_B},
\tag{6.8.42}
$$

where a is the small plate edge half-width (e.g., Timoshenko & Woinowsky-Krieger [69], p. 202).

6.8.2 *Element stiffness matrix for the biharmonic equation*

Verify by numerical computation that the element stiffness matrix for the HCT triangle is singular.

6.8.3 *One-element bending problem*

Write a code that formulates and solves system (6.8.41). Run the code for a triangle with a shape of your choice and discuss the results.

6.9   Buckling and wrinkling

In the analysis of Section 6.8, we have assumed that the plate is not subjected to an in-plane stress other than that developing due to the deformation, that is, the plate is tangentially (laterally) unstressed.

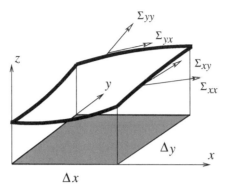

FIGURE 6.9.1 Schematic illustration of deformed plate used to derive equilibrium
equations governing the vertical deflection of a compressed plate.

An applied in-plane edge or body force generates an independent in-plane stress
field, denoted by $\boldsymbol{\Sigma}$, as discussed in Section 6.4. After the plate has buckled, the
in-plane stresses are oriented in tangential directions, thereby generating a net force
component normal to the undeformed plate with a *downward* force density given
by

$$w(x,y) = -h\left[\frac{\partial}{\partial x}\left(\Sigma_{xx}\frac{\partial f}{\partial x}\right) + \frac{\partial}{\partial y}\left(\Sigma_{xy}\frac{\partial f}{\partial x}\right)\right.$$
$$\left. + \frac{\partial}{\partial x}\left(\Sigma_{xy}\frac{\partial f}{\partial y}\right) + \frac{\partial}{\partial x}\left(\Sigma_{yy}\frac{\partial f}{\partial y}\right)\right], \qquad (6.9.1)$$

where h is the plate thickness (e.g., Timoshenko & Gere [68], p. 305).

To derive expression (6.9.1), we have projected the tangential forces exerted
around the edges of a rectangular section onto the z axis, as shown in Figure 6.9.1.
The downward contribution of Σ_{xx} is

$$-\left(\Sigma_{xx}\frac{\partial f}{\partial x}h\,\Delta y\right)_{x+\Delta x} + \left(\Sigma_{xx}\frac{\partial f}{\partial x}h\,\Delta y\right)_{x}$$
$$\simeq -\Delta x\Delta y\,h\,\frac{\left(\Sigma_{xx}\frac{\partial f}{\partial x}\right)_{x+\Delta x} - \left(\Sigma_{xx}\frac{\partial f}{\partial x}\right)_{x}}{\Delta x}. \qquad (6.9.2)$$

Taking the limit as Δx and Δy tend to zero, we derive the first term on the right-
hand side of (6.9.1). The remaining terms are derived working in a similar fashion
(Problem 6.9.2).

Now expanding the derivatives on the right-hand side of (6.9.1) and using the
equilibrium equations for $\boldsymbol{\Sigma}$ stated in (6.2.6), we obtain

$$w(x,y) = -h\left(\Sigma_{xx}\frac{\partial^2 f}{\partial x^2} + 2\Sigma_{xy}\frac{\partial^2 f}{\partial x\partial y} + \Sigma_{yy}\frac{\partial^2 f}{\partial y^2} - b_x\frac{\partial f}{\partial x} - b_y\frac{\partial f}{\partial y}\right), \qquad (6.9.3)$$

where **b** is the in-plane body force. In the absence of a body force, the last two terms on the right-hand side of (6.9.3) do not appear.

Linearized von Kármán equation

Substituting expression (6.9.3) into the equilibrium equation (6.4.44), we obtain the Saint Venant or linearized von Kármán equation

$$\nabla^4 f \equiv \nabla^2 \nabla^2 f = \frac{\partial^4 f}{\partial x^4} + 2 \frac{\partial^4 f}{\partial x^2 \, \partial y^2} + \frac{\partial^4 f}{\partial y^4} = -\frac{h}{E_B} \omega, \tag{6.9.4}$$

where ∇ is the two-dimensional gradient in the xy plane and

$$\omega = -\Sigma_{xx} \frac{\partial^2 f}{\partial x^2} - 2 \Sigma_{xy} \frac{\partial^2 f}{\partial x \partial y} - \Sigma_{yy} \frac{\partial^2 f}{\partial y^2} + b_x \frac{\partial f}{\partial x} + b_y \frac{\partial f}{\partial y} \tag{6.9.5}$$

is a distributed source. Setting $\mathbf{b} = -\nabla \cdot \boldsymbol{\Sigma}$, we obtain the compact form

$$\omega = -\nabla \cdot (\boldsymbol{\Sigma} \cdot \nabla f). \tag{6.9.6}$$

In the absence of a body force, $b_x = 0$ and $b_y = 0$, we obtain the counterpart of the beam equation (2.6.6).

The solution procedure involves determining the in-plane stresses $\boldsymbol{\Sigma}$ by analytical methods neglecting the transverse deformation, and then solving equation (6.9.4) for the deflection, f. A trivial solution is $f = 0$. Nontrivial solutions are possible when the in-plane stresses reach a sequence of thresholds, physically representing critical in-plane loads inducing different modes of plate buckling. These critical loads and the corresponding modes of deformation are found by solving an algebraic eigenvalue problem arising from the finite element formulation.

For example, in the case of a circular plate subjected to a uniform tangential body force along the x axis with force density τ due, for example, to gravity, we substitute into (6.9.4) the stress field given in (6.2.15) and set $b_x = \tau/h$ to obtain the specific form

$$\nabla^4 f = -\frac{2\tau}{E_B(3-\nu)} \left(x \frac{\partial^2 f}{\partial x^2} + (1-\nu) y \frac{\partial^2 f}{\partial x \partial y} + \nu x \frac{\partial^2 f}{\partial y^2} + \frac{3-\nu}{2} \frac{\partial f}{\partial x} \right). \tag{6.9.7}$$

A similar equation can be obtained for an elliptical plate using the stress field given in (6.2.19).

Generalized eigenvalue problem

The finite element formulation provides us with a generalized algebraic eigenvalue problem expressed by the equation

$$\mathbf{A} \cdot \mathbf{f} = \tau \, \mathbf{B} \cdot \mathbf{f}, \tag{6.9.8}$$

where **A** and **B** are two suitable matrices and τ is an eigenvalue.

Conservative body force

When the body force is conservative, we may set $\mathbf{b} = \nabla \mathcal{V}$ and express the in-plane stresses, $\mathbf{\Sigma}$, in terms of the Airy stress function ϕ defined in (6.2.24), to obtain an alternative statement of the linearized von Kármán equation,

$$\nabla^4 f = -\frac{h}{E_B}\left[-\frac{\partial^2 f}{\partial x^2}\frac{\partial^2 \phi}{\partial y^2} - \frac{\partial^2 f}{\partial y^2}\frac{\partial^2 \phi}{\partial x^2} + 2\frac{\partial^2 f}{\partial x \partial y}\frac{\partial^2 \phi}{\partial x \partial y} + \nabla \cdot (\mathcal{V} \nabla f)\right], \quad (6.9.9)$$

which is a common point of departure in finite element formulations (e.g., Dossou & Pierre [18]).

Deformation-dependent body force

A subtlety arises when the body force, $\mathbf{b}$, is not constant but changes due to deformation. In the case of shear flow past a membrane,

$$\omega = -\Sigma_{xx}\frac{\partial^2 f}{\partial x^2} - 2\,\Sigma_{xy}\frac{\partial^2 f}{\partial x \partial y} - \Sigma_{yy}\frac{\partial^2 f}{\partial y^2} + p_n, \quad (6.9.10)$$

where p_n is the normal load in the deformed configuration, linearized with respect to f. When the body force is independent of the deformation,

$$p_n \equiv b_x\,\frac{\partial f}{\partial x} + b_y\,\frac{\partial f}{\partial y}, \quad (6.9.11)$$

where b_x and b_y are two functions of x and y.

Element matrices

FSELIB function *HCT_buckle*, listed in Table 6.9.1, evaluates the HCT element matrices corresponding to the eigenvalue problem expressed by (6.9.8). The input includes the components of the in-plane stress Σ_{xx}, Σ_{xy}, and Σ_{yy}, at the three vertices of the triangle, named *txx,txy*, and *tyy*, the two components of the body force, named *bfx* and *bfy*, the order of the quadrature, named *NQ*, and the Poisson ratio, named *nu*.

Finite element code

FSELIB code *buckle_HCT*, listed in Table 6.9.2, solves the generalized eigenvalue problem for a circular or square compressed plate. Other shapes and physical circumstances can be implemented by straightforward substitutions. After the eigenvalues have been computed, the eigenfunctions are plotted using the FSELIB function *buckle_wrinkle*, listed in Table 6.9.3. The input includes a selected eigenvalue in the first argument. Typical results are shown in Figure 6.9.2.

PROBLEM

6.9.1 *Buckling of a plate under compression*

Derive the second and third terms on the right-hand side of expression (6.9.1).

```
function [ebsm, rhsm, arel] = HCT_buckle ...
...
   (x1,y1, x2,y2, x3,y3 ...
   ,txx1,txx2,txx3 ...
   ,txy1,txy2,txy3 ...
   ,tyy1,tyy2,tyy3 ...
   ,bfx,bfy, ...
   ,NQ,nu)

%==================================================
% Evaluation of the element matrices
% and right-hand side matrix
% for the HCT triangle,
% using a Gauss integration quadrature.
%
% arel: element area
%==================================================

%----------------------------------
% compute the area of the triangle
%----------------------------------

d12x = x1-x2; d12y = y1-y2;
d31x = x3-x1; d31y = y3-y1;

arel = 0.5*(d31x*d12y - d31y*d12x);

%----
% interior node
%-----

x7 = (x1+x2+x3)/3.0; y7 = (y1+y2+y3)/3.0;

txx7 = (txx1+txx2+txx3)/3.0;
txy7 = (txy1+txy2+txy3)/3.0;
tyy7 = (tyy1+tyy2+tyy3)/3.0;

%----------------------------
% read the triangle quadrature
%----------------------------

[xi, eta, w] = gauss_trgl(NQ);

%---------
% initialize the element bending stiffness matrix
% and right-hand side matrix
%---------

for k=1:12
 for l=1:12
  ebsm(k,l) = 0.0;
  rhsm(k,l) = 0.0;
 end
end
```

TABLE 6.9.1   Function *HCT_buckle* (Continuing →)

```
%-------------------------------
% compute the 30 cubic coefficients
% and put them in a vector: am
%-------------------------------

for mode=1:12

   dof = zeros(1,12);
   dof(mode) = 1.0;

   [a] = HCT_sys (x1,y1, x2,y2, x3,y3, dof);

   for i=1:30
      am(i,mode) = a(i);
   end

end

%----------------------
% perform the quadrature
%----------------------

for pass=1:3  % run over the 3 sub-triangles

 sum = 0.0;

 for i=1:NQ     % quadrature

%----
% modes and laplacian:
%---

%--
 for mode=1:12

    for j=1:30
      c(j) = am(j,mode);
    end

    %---
    if(pass==1)          % triangle 1-2-7
    %---

       x = x1 + (x2-x1)*xi(i) + (x7-x1)*eta(i);
       y = y1 + (y2-y1)*xi(i) + (y7-y1)*eta(i);

       tau_xx = txx1+(txx2-txx1)*xi(i) + (txx7-txx1)*eta(i);
       tau_xy = txy1+(txy2-txy1)*xi(i) + (txy7-txy1)*eta(i);
       tau_yy = tyy1+(tyy2-tyy1)*xi(i) + (tyy7-tyy1)*eta(i);

       psi(mode) = c(1) + c(2)*x    +c(3)*y ...
                        + c(4)*x^2 +c(5)*x*y   + c(6)*y^2 ...
                        + c(7)*x^3 +c(8)*x^2*y + c(9)*x*y^2+c(10)*y^3;
```

TABLE 6.9.1   Function *HCT_buckle* ($\rightarrow$ Continuing $\rightarrow$)

```
    psi_x(mode) = c(2)+2.0*c(4)*x+c(5)*y+3.0*c(7)*x^2+2.0*c(8)*x*y ...
                  + c(9)*y^2;

    lpsi(mode) = 2.0*c(4)+2.0*c(6) ...
                 +6.0*c(7)*x + 2.0*c(8)*y + 2.0*c(9)*x + 6.0*c(10)*y;

    psi_xx(mode) = 2.0*c(4)+6.0*c(7)*x + 2.0*c(8)*y;
    psi_xy(mode) = c(5)+2.0*c(8)*x+2.0*c(9)*y;
    psi_yy(mode) = 2.0*c(6)+2.0*c(9)*x + 6.0*c(10)*y;

%---
elseif(pass==2)  % triangle 2-3-7
%---

    x = x2 + (x3-x2)*xi(i) + (x7-x2)*eta(i);
    y = y2 + (y3-y2)*xi(i) + (y7-y2)*eta(i);

    tau_xx = txx2+(txx3-txx2)*xi(i) + (txx7-txx2)*eta(i);
    tau_xy = txy2+(txy3-txy2)*xi(i) + (txy7-txy2)*eta(i);
    tau_yy = tyy2+(tyy3-tyy2)*xi(i) + (tyy7-tyy2)*eta(i);

    psi(mode) = c(11)+c(12)*x+c(13)*y+c(14)*x^2+c(15)*x*y+c(16)*y^2 ...
                +c(17)*x^3+c(18)*x^2*y+c(19)*x*y^2+c(20)*y^3;

    psi_x(mode) = c(12)+2.0*c(14)*x+c(15)*y+3.0*c(17)*x^2 ...
                  +2.0*c(18)*x*y+c(19)*y^2;

    lpsi(mode) = 2.0*c(14)+2.0*c(16) ...
                 +6.0*c(17)*x+2.0*c(18)*y+2.0*c(19)*x+6.0*c(20)*y;

    psi_xx(mode) = 2.0*c(14)+6.0*c(17)*x + 2.0*c(18)*y;
    psi_xy(mode) = c(15)+2.0*c(18)*x+2.0*c(19)*y;
    psi_yy(mode) = 2.0*c(16)+2.0*c(19)*x + 6.0*c(20)*y;

%---
elseif(pass==3)  % triangle 3-1-7
%---

    x = x3 + (x1-x3)*xi(i) + (x7-x3)*eta(i);
    y = y3 + (y1-y3)*xi(i) + (y7-y3)*eta(i);

    tau_xx = txx3+(txx1-txx3)*xi(i) + (txx7-txx3)*eta(i);
    tau_xy = txy3+(txy1-txy3)*xi(i) + (txy7-txy3)*eta(i);
    tau_yy = tyy3+(tyy1-tyy3)*xi(i) + (tyy7-tyy3)*eta(i);

    psi(mode) = c(21)+c(22)*x+c(23)*y+c(24)*x^2+c(25)*x*y+c(26)*y^2 ...
                +c(27)*x^3+c(28)*x^2*y+c(29)*x*y^2+c(30)*y^3;

    psi_x(mode) = c(22)+2.0*c(24)*x+c(25)*y+3.0*c(27)*x^2 ...
                  +2.0*c(28)*x*y +c(29)*y^2;

    lpsi(mode) = 2.0*c(24)+2.0*c(26) ...
                 +6.0*c(27)*x+2.0*c(28)*y+2.0*c(29)*x+6.0*c(30)*y;
```

TABLE 6.9.1 Function *HCT_buckle* ($\rightarrow$ Continuing $\rightarrow$)

```
        psi_xx(mode) = 2.0*c(24)+6.0*c(27)*x + 2.0*c(28)*y;
        psi_xy(mode) = c(25)+2.0*c(28)*x+2.0*c(29)*y;
        psi_yy(mode) = 2.0*c(26)+2.0*c(29)*x + 6.0*c(30)*y;

      end

    end   % end of loop over modes
%--

    sum = sum + w(i);

    cf = arel*w(i)/3.0;

    for k=1:12
     for l=1:12

      ebsm(k,l) = ebsm(k,l) + lpsi(k)*lpsi(l)*cf;

      rhsm(k,l) = rhsm(k,l) + psi(k) * ( ...
                    +      tau_xx * psi_xx(l) ...
                    + 2.0*tau_xy * psi_xy(l) ...
                    +      tau_yy * psi_yy(l) ...
                    - bfx*psi_x(l) ...
                    )*cf;
     end
    end

  end   % end of quadrature

%--
end % end of run over sub-triangles
%--

%-----
% done
%-----

return;
```

TABLE 6.9.1 ($\rightarrow$ Continued) Function *HCT_buckle* compute the element matrices for the buckling problem.

```
%==============
% Code buckle_HCT
%
% buckling of a plate using the HCT element
%
% depends on HCT_buckle, HCT_wrinkle
%==============

%-----
% input data
%-----

E_B = 1.0; % bending modulus
nu = 0.5; % Poisson ratio
ndiv = 3;  % discretization level
NQ = 7;       % quadrature order

shape = 1;

%-----------------
% elliptical shape
%-----------------

if(shape==1)

 [ne,ng,p,c,efl,gfl] = trgl6_disk(ndiv);

 asp = 2.0;
 eps = (asp-1.0)/(asp+1.0);
 ax = 1+eps;
 ay = 1-eps;
 scale = sqrt(ax*ay)
 ax = ax/scale;
 ay = ay/scale;

  for i=1:ng
   p(i,1) = ax*p(i,1);
   p(i,2) = ay*p(i,2);
   end

end

%-------
% square
%-------

if(shape==2)

 [ne,ng,p,c,efl,gfl] = trgl6_sqr(ndiv);

end
```

TABLE 6.9.2   Code *buckle_HCT* (Continuing →)

```
%-----------------------------------------------------
% make sure that the mid-nodes lie along the edges
%-----------------------------------------------------

for i=1:ne
 for j=1:2
  p(c(i,4),j) = 0.5*(p(c(i,1),j)+p(c(i,2),j));
  p(c(i,5),j) = 0.5*(p(c(i,2),j)+p(c(i,3),j));
  p(c(i,6),j) = 0.5*(p(c(i,3),j)+p(c(i,1),j));
 end
end

%---------------------------------------
% mark the global vertex and edge nodes
%---------------------------------------

for i=1:ne
  gfln(c(i,1)) = 0; gfln(c(i,2)) = 0; gfln(c(i,3)) = 0;  % vertex nodes
  gfln(c(i,4)) = 1; gfln(c(i,5)) = 1; gfln(c(i,6)) = 1;  % edge nodes
end

%--------
% count the number of global edge nodes
%-------

nge=0;  % number of edge nodes

for i=1:ng
 if(gfln(i) == 1)
   nge = nge+1;
   xpp(nge) = p(i,1);
   ypp(nge) = p(i,2);
 end
end

Ic = 0;
for i=1:ng
 if(gfln(i) == 0)
   Ic = Ic+1;
   xppp(Ic) = p(i,1);
   yppp(Ic) = p(i,2);
 end
end

%---------------------------
% total number of global modes
%---------------------------

ngv = ng-nge;     % vertex nodes

ngm = 3*ngv+nge;  % number of global modes
```

TABLE 6.9.2   Code *buckle_HCT* ($\to$ Continuing $\to$)

```
%----------------------------------------------
% mark the position of the rows and columns
% of the modes corresponding to the global nodes
%----------------------------------------------

for i=1:ng
  s(i) = 1;
  for j=1:i-1
   if(gfln(j) == 1)     % edge node, 1 dof
     s(i) = s(i)+1;
   else                 % vertex node, 3 dofs
     s(i) = s(i)+3;
   end
  end
end

%-----
% initialize the orientation index of the normal derivative
% at the mid-side nodes
%-----

for i=1:ng
 Idid(i) = 1;
end

%-----------------------------------------------------
% assemble the global bending stiffness matrix
% and rhs
%-----------------------------------------------------

gbsm = zeros(ngm,ngm); % initialize
grhs = zeros(ngm,ngm);    % right-hand side

area = 0;

for l=1:ne              % loop over the elements

% compute the element bending stiffness matrix and rhs

j1 = c(l,1); j2 = c(l,2); j3 = c(l,3);   % global element labels
j4 = c(l,4); j5 = c(l,5); j6 = c(l,6);

x1 = p(j1,1); y1 = p(j1,2);   % three vertices
x2 = p(j2,1); y2 = p(j2,2);
x3 = p(j3,1); y3 = p(j3,2);

%----- tensions for an elliptical patch in shear flow:

if(shape==1)
 DN = (1-nu)*ax^2+2.0*ay^2;V1 =-2*ay^2/DN;V2 = -(1-nu)*ax^2/DN;
 txx1 = V1*x1; txx2 = V1*x2; txx3 = V1*x3;
 txy1 = V2*y1; txy2 = V2*y2; txy3 = V2*y3;
```

TABLE 6.9.2   Code *buckle_HCT* ($\rightarrow$ Continuing $\rightarrow$)

```
 tyy1 = nu*txx1; tyy2 = nu*txx2; tyy3 = nu*txx3;
 bfx = 1.0; bfy = 0.0;
end

%----- TENSIONS for a uniformly compressed square patch:

if(shape==2)
 txx1 =-1.0;txx2 =-1.0;txx3 =-1.0;
 txy1 = 0.0;txy2 = 0.0;txy3 = 0.0;
 tyy1 =-1.0;tyy2 =-1.0;tyy3 =-1.0;
 bfx = 0.0;bfy = 0.0;
end

%-----

[ebsm_elm, rhs_elm, arel] = HCT_buckle ...
  ...
    (x1,y1, x2,y2, x3,y3, ...
    txx1,txx2,txx3,...
    txy1,txy2,txy3,...
    tyy1,tyy2,tyy3,...
    bfx, bfy, ...
    NQ, nu);

e(1) = s(j1); e(2) = s(j1)+1; e(3) = s(j1)+2;  % position in the
e(4) = s(j2); e(5) = s(j2)+1; e(6) = s(j2)+2;  % global matrix
e(7) = s(j3); e(8) = s(j3)+1; e(9) = s(j3)+2;
e(10)= s(j4); e(11)= s(j5);   e(12)= s(j6);

    for i=1:12

      fci = 1.0;
          if(i==10) fci = Idid(j4);
      elseif(i==11) fci = Idid(j5);
      elseif(i==12) fci = Idid(j6); end

      for j=1:12
        fcj = 1.0;
            if(j==10) fcj = Idid(j4);
        elseif(j==11) fcj = Idid(j5);
        elseif(j==12) fcj = Idid(j6); end
          gbsm(e(i),e(j)) = gbsm(e(i),e(j)) + fci*fcj*ebsm_elm(i,j);
          grhs(e(i),e(j)) = grhs(e(i),e(j)) - fci*fcj* rhs_elm(i,j);
        end
    end

 Idid(j4) = -1; Idid(j5) = -1; Idid(j6) = -1;

area = area+arel;

end
```

TABLE 6.9.2   Code *buckle_HCT* ($\rightarrow$ Continuing $\rightarrow$)

```
%-------------------------------------------------------------
% remove the equations corresponding to the boundary nodes
%-------------------------------------------------------------

Ic = 0;

for i=1:ng

 if(gfl(i,1) == 0)   % pick up the interior nodes

    Ic = Ic+1;
    for j=1:ngm
      gbsm(Ic,j) = gbsm(s(i),j);
      grhs(Ic,j) = grhs(s(i),j);
    end

    if(gfln(i) == 0)  % vertex node with 3 modes
    Ic = Ic+1;
    for j=1:ngm
        gbsm(Ic,j) = gbsm(s(i)+1,j);
        grhs(Ic,j) = grhs(s(i)+1,j);
    end
    Ic = Ic+1;
    for j=1:ngm
      gbsm(Ic,j) = gbsm(s(i)+2,j);
      grhs(Ic,j) = grhs(s(i)+2,j);
    end
    end

 end
end

ngm1 = Ic;

%---
% repeat for the columns:
%---

Ic = 0;

for i=1:ng

 if(gfl(i,1) == 0) % interior node

    Ic = Ic+1;
    for j=1:ngm1
      gbsm(j,Ic) = gbsm(j,s(i));
      grhs(j,Ic) = grhs(j,s(i));
    end
```

TABLE 6.9.2   Code *buckle_HCT* ($\rightarrow$ Continuing $\rightarrow$)

```
   if(gfln(i) == 0)   % vertex node
     Ic = Ic+1;
     for j=1:ngm1
       gbsm(j,Ic) = gbsm(j,s(i)+1);
       grhs(j,Ic) = grhs(j,s(i)+1);
     end

     Ic = Ic+1;
     for j=1:ngm1
       gbsm(j,Ic) = gbsm(j,s(i)+2);
       grhs(j,Ic) = grhs(j,s(i)+2);
     end
   end
 end
end

%----
% formulate the compact matrices
%---

for i=1:ngm1
 for j=1:ngm1
  gbsm1(i,j) = gbsm(i,j);
  grhs1(i,j) = grhs(i,j);
 end
end

%----
% solve the generalized eigenvalue problem
%----

gbsm1_inv = inv(gbsm1);
mat = gbsm1_inv*grhs1;
charval = eig(mat);

%---
% display the first six eigenfunctions
%---

for mode=1:6

 figure(mode)
 hold on
 box on
 eigenvalue = charval(mode)
 buckle_wrinkles (eigenvalue,ne,ng,c,p,gfl,gfln,ngm1,mat)

end

%---
% done
%---
```

TABLE 6.9.2   ($\rightarrow$ Continued) Code *buckle_HCT* solves the plate buckling problem.

```
function buckle_wrinkle (eigenvalue,ne,ng,c,p,gfl,gfln,ngm1,mat)

%=====================================
% compute and plot the eigenfunctions
%=====================================

mat1 = mat - eigenvalue*eye(ngm1);

for i=1:ngm1-1
 rhs2(i) = - mat1(i,ngm1);
end

%----
% formulate the eigenvector system
%---

for i=1:ngm1-1
 for j=1:ngm1-1
  mat2(i,j) = mat1(i,j);
 end
end

eigenvector = rhs2/mat2';
eigenvector(ngm1) = 1.0;

%---
% extract transverse displacement
% at the vertex nodes
%---

Ic = 0;

for i=1:ng

 if(gfl(i,1) == 0) % interior node
   Ic = Ic+1;
   f(i) = eigenvector(Ic);
   if(gfln(i) == 0)  % vertex node
     Ic = Ic+2;
   end
 else
   f(i) = 0;
 end

end

%-------------------------------
% interpolate the mid-side values
%-------------------------------

for i=1:ne
  f(c(i,4)) = 0.5*(f(c(i,1))+f(c(i,2)));
  f(c(i,5)) = 0.5*(f(c(i,2))+f(c(i,3)));
  f(c(i,6)) = 0.5*(f(c(i,3))+f(c(i,1)));
end
```

TABLE 6.9.3   Function *buckle_wrinkle* (Continuing →)

```
%-----
% plot
%-----

%-----
% generate the extended connectivity matrix
% for three-node sub-triangles
%-----

Ic = 0;

for i=1:ne
 Ic = Ic+1;
 c3(Ic,1) = c(i,1); c3(Ic,2) = c(i,4); c3(Ic,3) = c(i,6);
 Ic = Ic+1;
 c3(Ic,1) = c(i,4); c3(Ic,2) = c(i,2); c3(Ic,3) = c(i,5);
 Ic = Ic+1;
 c3(Ic,1) = c(i,5); c3(Ic,2) = c(i,3); c3(Ic,3) = c(i,6);
 Ic = Ic+1;
 c3(Ic,1) = c(i,4); c3(Ic,2) = c(i,5); c3(Ic,3) = c(i,6);
end

trimesh (c3,p(:,1),p(:,2),f);

%-----
% done
%-----

return
```

TABLE 6.9.3 ($\rightarrow$ Continued) Function *buckle_wrinkle* computes and displays the eigenfunctions of the buckling problem.

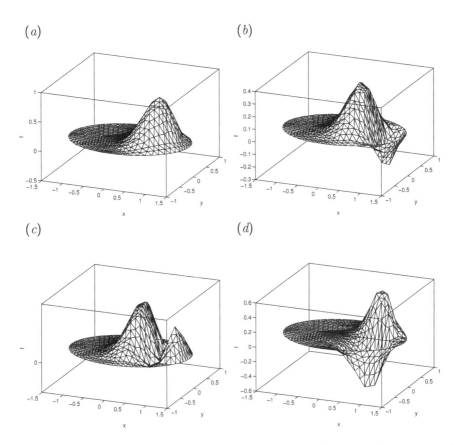

FIGURE 6.9.2   Buckling modes of a clamped circular plate due to a uniform tangential force.

7

Viscous flow

The finite element method finds important applications in the field of computational fluid dynamics (CFD), where it is used to solve the equations of fluid flow consisting of Cauchy's equation of motion and the continuity equation. The former expresses Newton's second law of motion for a small fluid parcel idealized as an infinitesimal point particle, and the latter ensures mass conservation. The primary unknowns are the velocity and pressure established in response to boundary motion, a body force, or a prescribed pressure drop.

The general methodology and implementation of the finite element method in fluid mechanics is similar to that discussed in Chapter 6 in solid mechanics. The formulation of the Galerkin finite element will be discussed in this chapter along with pertinent finite element codes.

7.1 Governing equations

Cauchy's equation of motion relates the acceleration of a point particle in a flow, $\mathbf{a}$, to the local divergence of the Cauchy stress tensor, $\boldsymbol{\sigma}$, introduced in Section 6.1.1, and to the body force, $\mathbf{b}$,

$$\rho\,\mathbf{a} = \nabla \cdot \boldsymbol{\sigma} + \mathbf{b}, \qquad (7.1.1)$$

where ρ is the fluid density (e.g., Pozrikidis [46]). In the case of the gravitational force, $\mathbf{b} = \rho\mathbf{g}$, where $\mathbf{g} = (g_x, g_y, g_z)$ is the acceleration of gravity. The Cartesian components of equation (7.1.1) with the gravitational body force are

$$\rho\,a_x = \frac{\partial \sigma_{xx}}{\partial x} + \frac{\partial \sigma_{yx}}{\partial y} + \frac{\partial \sigma_{zx}}{\partial z} + \rho g_x,$$

$$\rho\,a_y = \frac{\partial \sigma_{xy}}{\partial x} + \frac{\partial \sigma_{yy}}{\partial y} + \frac{\partial \sigma_{zy}}{\partial z} + \rho g_y, \qquad (7.1.2)$$

$$\rho\,a_z = \frac{\partial \sigma_{xz}}{\partial x} + \frac{\partial \sigma_{yz}}{\partial y} + \frac{\partial \sigma_{zz}}{\partial z} + \rho g_z.$$

These equations govern the motion of any arbitrary homogeneous fluid that can be regarded as a continuum.

To evaluate the point particle acceleration, $\mathbf{a}$, we introduce the material derivative, D/Dt, expressing the rate of change of a property following the motion of a material point particle. By definition, the point particle acceleration is the material derivative of the fluid velocity, $\mathbf{u} = (u_x, u_y, u_z)$,

$$\mathbf{a} = \frac{D\mathbf{u}}{Dt} \equiv \frac{\partial \mathbf{u}}{\partial t} + \mathbf{u} \cdot \nabla \mathbf{u}, \tag{7.1.3}$$

where $\nabla \mathbf{u}$ is the velocity gradient tensor. In index notation,

$$a_i = \frac{\partial u_i}{\partial t} + u_x \frac{\partial u_i}{\partial x} + u_y \frac{\partial u_i}{\partial y} + u_z \frac{\partial u_i}{\partial z} \tag{7.1.4}$$

for $i = x, y, z$.

Mass conservation for an incompressible fluid is enforced by the continuity equation,

$$\nabla \cdot \mathbf{u} = 0 \tag{7.1.5}$$

(e.g., Pozrikidis [46]). Explicitly, the continuity equation reads

$$\frac{\partial u_j}{\partial x_j} = \frac{\partial u_x}{\partial x} + \frac{\partial u_y}{\partial y} + \frac{\partial u_z}{\partial z} = 0, \tag{7.1.6}$$

where summation of the repeated index, j, is implied on the left-hand side over $j = 1, 2, 3$ or x, y, z. To simplify the notation, we have denoted $x_1 = x$, $x_2 = y$, $x_3 = z$, and $u_1 = u_x$, $u_2 = u_y$, $u_3 = u_z$.

Newtonian fluids

The stress field in an incompressible *Newtonian* fluid is related to the pressure, p, and to the rate-of-deformation tensor,

$$\mathbf{E} \equiv \frac{1}{2} \left[\nabla \mathbf{u} + (\nabla \mathbf{u})^T \right], \tag{7.1.7}$$

by the linear constitutive equation

$$\boldsymbol{\sigma} = -p\mathbf{I} + 2\mu \mathbf{E}, \tag{7.1.8}$$

where $\mathbf{I}$ is the identity matrix, μ is the fluid viscosity, and the superscript T denotes the matrix transpose. In index notation, we have

$$\sigma_{ij} = -p\delta_{ij} + \mu \left(\frac{\partial u_i}{\partial x_j} + \frac{\partial u_j}{\partial x_i} \right), \tag{7.1.9}$$

where δ_{ij} is Kronecker's delta.

Navier–Stokes equation

Substituting expressions (7.1.3) and (7.1.8) into the equation of motion (7.1.1) and simplifying, taking into account the continuity equation, we derive the Navier–Stokes equation,

$$\rho\left(\frac{\partial \mathbf{u}}{\partial t} + \mathbf{u}\cdot\nabla\mathbf{u}\right) = -\nabla p + \mu\nabla^2\mathbf{u} + \rho\mathbf{g} \tag{7.1.10}$$

(Problem 7.1.1). In index notation, the ith component of the Navier–Stokes equation reads

$$\rho\left(\frac{\partial u_i}{\partial t} + u_j\frac{\partial u_i}{\partial x_j}\right) = -\frac{\partial p}{\partial x_i} + \mu\,\nabla^2 u_i + \rho g_i \tag{7.1.11}$$

for $i = 1, 2, 3$ or x, y, z, where summation of the repeated index, j, is implied on the left-hand side.

When the fluid density is uniform throughout the domain of flow, the effect of gravity can be incorporated into a hydrodynamic pressure indicated by a tilde, defined as

$$\tilde{p} \equiv p - \rho\,(g_x\,x + g_y\,y + g_z\,z) = p - \rho\mathbf{g}\cdot\mathbf{x}. \tag{7.1.12}$$

Written in terms of the hydrodynamic pressure, the gravitational term on the right-hand side of the Navier–Stokes equation does not appear. For simplicity, the tilde will be omitted under the convection that the pressure has been modified to incorporate the body force. A distinction between the physical and the hydrodynamic pressure arises only when boundary conditions involving the physical pressure are prescribed.

In the case of two-dimensional flow in the xy plane, the x and y components of the Navier–Stokes equation read

$$\rho\left(\frac{\partial u_x}{\partial t} + u_x\frac{\partial u_x}{\partial x} + u_y\frac{\partial u_x}{\partial y}\right) = -\frac{\partial p}{\partial x} + \mu\left(\frac{\partial^2 u_x}{\partial x^2} + \frac{\partial^2 u_x}{\partial y^2}\right),$$
$$\tag{7.1.13}$$
$$\rho\left(\frac{\partial u_y}{\partial t} + u_x\frac{\partial u_y}{\partial x} + u_y\frac{\partial u_y}{\partial y}\right) = -\frac{\partial p}{\partial y} + \mu\left(\frac{\partial^2 u_y}{\partial x^2} + \frac{\partial^2 u_y}{\partial y^2}\right),$$

and the continuity equation reads

$$\frac{\partial u_x}{\partial x} + \frac{\partial u_y}{\partial y} = 0, \tag{7.1.14}$$

where p is the hydrodynamic pressure.

Boundary conditions

To solve the Navier–Stokes equation, we require boundary conditions at the physical or computational boundaries of the flow. Typically, we specify the velocity, $\mathbf{u}$, at a solid surface, the traction, $\mathbf{f} \equiv \mathbf{n} \cdot \boldsymbol{\sigma}$, at a free surface, and the jump in the traction across a fluid interface, where $\mathbf{n}$ is the unit normal vector. A velocity boundary condition is a Dirichlet boundary condition, and a traction boundary condition is a Neumann boundary condition.

PROBLEM

7.1.1 *Navier–Stokes equation from the Cauchy equation of motion*

Derive the Navier–Stokes equation from the Cauchy equation of motion, as discussed in the text.

7.2 Finite element formulation

To formulate the finite element method for fluid flow, we discretize the domain of flow into finite elements, as discussed in Chapters 3 and 4, and approximate the velocity and pressure fields using two generally *different* sets of global interpolation functions, as

$$\mathbf{u} = \sum_{j=1}^{N_{\mathrm{G}}^u} \mathbf{u}_j(t)\, \phi_j^u, \qquad p = \sum_{j=1}^{N_{\mathrm{G}}^p} p_j(t)\, \phi_j^p, \qquad (7.2.1)$$

where:

- N_{G}^u is the number of velocity global interpolation nodes and $\mathbf{u}_j(t)$ are the corresponding nodal values of the velocity vector for $j = 1, \ldots, N_{\mathrm{G}}^u$.

- N_{G}^p is the number of pressure global interpolation nodes and $p_j(t)$ are the corresponding nodal values of the pressure for $j = 1, \ldots, N_{\mathrm{G}}^p$.

Specific examples of these global interpolation functions will be given in Section 7.3.

7.2.1 Galerkin projections

Consider a flow in a two-dimensional domain, D, that is bounded by a single contour or a collection of contours whose union is denoted by C. To derive the Galerkin finite element equations, we work in two stages:

- We project each component of the Navier–Stokes equation (7.1.10) onto the velocity global interpolation functions, ϕ_j^u, and thus obtain the vector equation

$$\iint_D \phi_i^u\, \rho \left(\frac{\partial \mathbf{u}}{\partial t} + \mathbf{u} \cdot \nabla \mathbf{u} \right) \mathrm{d}x\,\mathrm{d}y = \iint_D \phi_i^u \left(-\nabla p + \mu \nabla^2 \mathbf{u} \right) \mathrm{d}x\,\mathrm{d}y \quad (7.2.2)$$

for $i = 1, \ldots, N_{\mathrm{G}}^u$.

- We project the continuity equation (7.1.5) onto the pressure global interpolation functions, ϕ_j^p, and thus obtain the scalar equation

$$\iint_D \phi_i^p \left(\nabla \cdot \mathbf{u} \right) \mathrm{d}x \, \mathrm{d}y = 0 \qquad (7.2.3)$$

for $i = 1, \dots, N_G^p$.

Next, we reduce the order of differentiation determined by the Laplacian of the velocity on the right-hand side of (7.2.2). This is accomplished by restating the Laplacian operator as the divergence of the gradient, writing

$$\iint_D \phi_i^u \left(\nabla^2 \mathbf{u} \right) \mathrm{d}x \, \mathrm{d}y = \iint_D \phi_i^u \left(\nabla \cdot \nabla \mathbf{u} \right) \mathrm{d}x \, \mathrm{d}y$$
$$= \iint_D \nabla \cdot \left(\phi_i^u \, \nabla \mathbf{u} \right) \mathrm{d}x \, \mathrm{d}y - \iint_D \nabla \phi_i^u \cdot \nabla \mathbf{u} \, \mathrm{d}x \, \mathrm{d}y. \qquad (7.2.4)$$

Using the Gauss divergence theorem to convert the first integral on the right-hand side to a contour integral along the boundary, C, we obtain

$$\iint_D \phi_i^u \left(\nabla^2 \mathbf{u} \right) \mathrm{d}x \, \mathrm{d}y = - \oint_C \phi_i^u \, \mathbf{n} \cdot \nabla \mathbf{u} \, \mathrm{d}l - \iint_D \nabla \phi_i^u \cdot \nabla \mathbf{u} \, \mathrm{d}x \, \mathrm{d}y, \qquad (7.2.5)$$

where $\mathbf{n}$ is the unit vector normal to C pointing *into* the domain of flow, and l is the arc length along the boundary contour, C.

Working in a similar fashion with the pressure term on the right-hand side of (7.2.2), we obtain

$$\iint_D \phi_i^u \, \nabla p \, \mathrm{d}x \, \mathrm{d}y = \iint_D \nabla(\phi_i^u \, p) \, \mathrm{d}x \, \mathrm{d}y - \iint_D \nabla \phi_i^u \, p \, \mathrm{d}x \, \mathrm{d}y$$
$$= - \oint_C \phi_i^u \, p \mathbf{n} \, \mathrm{d}l - \iint_D \nabla \phi_i^u \, p \, \mathrm{d}x \, \mathrm{d}y. \qquad (7.2.6)$$

Substituting expression (7.2.5) and (7.2.6) into (7.2.2), we derive the final result,

$$\iint_D \phi_i^u \rho \left(\frac{\partial \mathbf{u}}{\partial t} + \mathbf{u} \cdot \nabla \mathbf{u} \right) \mathrm{d}x \, \mathrm{d}y = \oint_C \phi_i^u \, p \mathbf{n} \, \mathrm{d}l - \mu \oint_C \phi_i^u \, \mathbf{n} \cdot \nabla \mathbf{u} \, \mathrm{d}l$$
$$+ \iint_D \nabla \phi_i^u \, p \, \mathrm{d}x \, \mathrm{d}y - \mu \iint_D \nabla \phi_i^u \cdot \nabla \mathbf{u} \, \mathrm{d}x \, \mathrm{d}y \qquad (7.2.7)$$

for $i = 1, \dots N_G^u$.

The x and y components of the vectorial equation (7.2.7) read

$$\iint_D \phi_i^u \rho \left(\frac{\partial \mathbf{u}}{\partial t} + \mathbf{u} \cdot \nabla \mathbf{u} \right)_x \mathrm{d}x \, \mathrm{d}y = \oint_C \phi_i^u p n_x \, \mathrm{d}l - \mu \oint_C \phi_i^u \, \mathbf{n} \cdot \nabla u_x \, \mathrm{d}l$$
$$+ \iint_D \frac{\partial \phi_i^u}{\partial x} p \, \mathrm{d}x \, \mathrm{d}y - \mu \iint_D \nabla \phi_i^u \cdot \nabla u_x \, \mathrm{d}x \, \mathrm{d}y \qquad (7.2.8)$$

and

$$
\iint_D \phi_i^u \rho \left(\frac{\partial \mathbf{u}}{\partial t} + \mathbf{u} \cdot \nabla \mathbf{u} \right)_y dx\, dy = \oint_C \phi_i^u p\, n_y\, dl - \mu \oint_C \phi_i^u\, \mathbf{n} \cdot \nabla u_y\, dl
$$
$$
+ \iint_D \frac{\partial \phi_i^u}{\partial y} p\, dx\, dy - \mu \iint_D \nabla \phi_i^u \cdot \nabla u_y\, dx\, dy, \qquad (7.2.9)
$$

where the x and y subscripts on the left-hand sides denote the corresponding Cartesian components of the enclosed quantity.

To simplify the notation, we can multiply the x component (7.2.8) by an arbitrary constant, c_x, and the y component (7.2.9) by another arbitrary constant, c_y, and add the resulting equations to obtain

$$
\iint_D \rho\, \mathbf{a} \cdot \left(\frac{\partial \mathbf{u}}{\partial t} + \mathbf{u} \cdot \nabla \mathbf{u} \right) dx\, dy = \oint_C \mathbf{n} \cdot \mathbf{a}\, p\, dl - \mu \oint_C \mathbf{n} \cdot (\nabla \mathbf{u}) \cdot \mathbf{a}\, dl
$$
$$
+ \iint_D p\, \nabla \cdot \mathbf{a}\, dx\, dy - \mu \iint_D \nabla \mathbf{a} : \nabla \mathbf{u}\, dx\, dy, \qquad (7.2.10)
$$

where

$$
\mathbf{a} \equiv \phi_i^u \begin{bmatrix} c_x \\ c_y \end{bmatrix} \qquad (7.2.11)
$$

is a modulated interpolation function. The double dot product in the last integral of (7.2.10) denotes the sum of the products of corresponding elements of the matrices on either side of the colon. For every global interpolation function, we can introduce two independent vectors $\mathbf{a}$ corresponding to different choices of the parameters c_x and c_y, which can be combined to yield the x and y components shown in (7.2.8) and (7.2.9).

7.2.2 Discrete equations

To conclude the formulation of the finite element method, we substitute the velocity and pressure finite element expansions (7.2.1) into (7.2.7) and (7.2.3), and thus derive the following equations:

- A system of ordinary differential equations (ODEs) for the nodal velocities, $\mathbf{u}_i(t)$, also involving the nodal pressures, $p_i(t)$. If the flow is steady, we obtain instead a system of algebraic equations.

- A complementary linear system of linear algebraic equations for the nodal velocities associated with the continuity equation playing the role of a mathematical constraint.

The structure and properties of this system will be discussed in Section 7.3 for linearized flow.

PROBLEM

7.2.1 *Galerkin projection of Cauchy's equation of motion*

Perform the Galerkin projection of Cauchy's equation of motion with the velocity interpolation function, ϕ_i^u, to derive the Galerkin equation

$$\iint_D \phi_i^u \rho \left(\frac{\partial \mathbf{u}}{\partial t} + \mathbf{u} \cdot \nabla \mathbf{u} \right) \mathrm{d}x\, \mathrm{d}y = -\oint_C \phi_i^u \, \mathbf{f} \, \mathrm{d}l - \iint_D \nabla \phi_i^u \cdot \boldsymbol{\sigma} \, \mathrm{d}x\, \mathrm{d}y \qquad (7.2.12)$$

for $i = 1, \ldots N_G^u$, where $\mathbf{f} \equiv \mathbf{n} \cdot \boldsymbol{\sigma}$ is the boundary traction and $\mathbf{n}$ is the unit normal vector pointing into the domain of flow. This equation is a point of departure for computing free-surface flows where a condition for the boundary traction is specified.

7.3 Stokes flow

Consider a fluid flow in a domain with overall size L, where the magnitude of the velocity field is comparable to a characteristic scale, U. In the case of flow through a circular tube, the length L can be identified with the tube diameter and the velocity U can be identified with the fluid velocity at the tube centerline. In the case of flow due to a settling particle, the length L can be identified with the particle diameter and the velocity U can be identified with the particle settling velocity.

Scaling arguments can be made to show that, when the Reynolds number, $\mathrm{Re} = \rho U L/\mu$, is sufficiently small, typically less than unity, the left-hand side of the Navier–Stokes equation expressing the effect of fluid inertia is small compared to the right-hand side expressing contributions from pressure, viscous, and body forces, and may be neglected without significant error (e.g., Pozrikidis [46]).

One consequence of discarding the time derivative $\partial \mathbf{u}/\partial t$ in the case of unsteady flow is that the evolution becomes *quasi-steady*, which means that time-dependence is due exclusively to changes in the boundary geometry. In the absence of inertia, if the boundaries are stationary, the flow is steady.

7.3.1 *Governing equations*

Setting the left-hand side of the Navier–Stokes equation equal to zero, we obtain the Stokes equation describing creeping flow. In terms of the hydrodynamic pressure incorporating the effect of gravity, the Stokes equation reads

$$-\nabla p + \mu \nabla^2 \mathbf{u} = \mathbf{0}. \qquad (7.3.1)$$

The numerical task is to solve this equation together with the continuity equation for an incompressible fluid, $\nabla \cdot \mathbf{u} = 0$, subject to appropriate conditions at the physical or computational boundaries of the flow.

Variational formulation

The rate of viscous dissipation in an incompressible Newtonian fluid is given by

$$\dot{\mathcal{E}} = 2\mu \iiint \mathbf{E} : \mathbf{E}\, dV, \tag{7.3.2}$$

where $\mathbf{E}$ is the rate-of-deformation tensor given in (7.1.7) and the integration is performed over the volume of flow (e.g., Pozrikidis [46]). We recall that the double dot product denotes the sum of the products of corresponding elements of the matrices on either side of the colon.

It can be shown that, given the boundary velocity, $\mathbf{u} = \mathbf{0}$, a flow that minimizes the rate of energy dissipation must satisfy the equations of Stokes flow. To demonstrate this, we compute the variation

$$\delta\dot{\mathcal{E}} = 4\mu \iiint \mathbf{E} : \delta\mathbf{E}\, dV = 2 \iiint (\boldsymbol{\sigma} + p\mathbf{I}) : \delta\mathbf{E}\, dV. \tag{7.3.3}$$

The continuity equation requires that the trace of $\delta\mathbf{E}$ is zero, yielding

$$\delta\dot{\mathcal{E}} = 2 \iiint \boldsymbol{\sigma} : \delta\mathbf{E}\, dV = 4 \iiint \boldsymbol{\sigma} : \left[\nabla\delta\mathbf{u} + (\nabla\delta\mathbf{u})^T \right] dV. \tag{7.3.4}$$

Now using the Stokes equation and applying the divergence theorem, we obtain

$$\delta\dot{\mathcal{E}} = 4 \iiint \nabla \cdot (\boldsymbol{\sigma} \cdot \delta\mathbf{u})\, dV = -4 \iint \mathbf{n} \cdot \boldsymbol{\sigma} \cdot \delta\mathbf{u}\, dS, \tag{7.3.5}$$

where the surface integral is performed over the boundaries of the flow, and $\mathbf{n}$ is the unit normal vector pointing into the flow. The variation $\delta\dot{\mathcal{E}}$ is zero in light of the homogeneous boundary condition $\delta\mathbf{u} = \mathbf{0}$. A similar proof is conducted for two-dimensional flow.

7.3.2 Galerkin finite element equations

In the case of Stokes flow, the Galerkin finite element projection of the equation of motion shown in (7.2.7) takes the simpler form

$$\oint_C \phi_i^u\, p\,\mathbf{n}\, dl - \mu \oint_C \phi_i^u\, \mathbf{n} \cdot \nabla\mathbf{u}\, dl + \iint_D \nabla\phi_i^u\, p\, dx\, dy$$
$$-\mu \iint_D \nabla\phi_i^u \cdot \nabla\mathbf{u}\, dx\, dy = \mathbf{0} \tag{7.3.6}$$

for $i = 1, \ldots, N_G^u$, where $\mathbf{n}$ is the unit vector normal to the boundary, C, pointing into the flow.

The x and y components of the vectorial equation (7.3.6) read

$$\oint_C \phi_i^u\, p\, n_x\, dl - \mu \oint_C \phi_i^u\, \mathbf{n} \cdot \nabla u_x\, dl + \iint_D \frac{\partial \phi_i^u}{\partial x}\, p\, dx\, dy$$
$$-\mu \iint_D \nabla\phi_i^u \cdot \nabla u_x\, dx\, dy = 0 \tag{7.3.7}$$

and

$$\oint_C \phi_i^u \, p \, n_y \, dl - \mu \oint_C \phi_i^u \, \mathbf{n} \cdot \nabla u_y \, dl + \iint_D \frac{\partial \phi_i^u}{\partial y} \, p \, dx \, dy$$

$$-\mu \iint_D \nabla \phi_i^u \cdot \nabla u_y \, dx \, dy = 0. \qquad (7.3.8)$$

Substituting the velocity and pressure finite element expansions (7.2.1) into (7.3.6), and setting

$$\nabla u_x = \sum_{j=1}^{N_G^u} (\nabla \phi_j^u) \, u_{x_j}, \qquad \nabla u_y = \sum_{j=1}^{N_G^u} (\nabla \phi_j^u) \, u_{y_j}, \qquad (7.3.9)$$

and thus

$$\nabla \mathbf{u} = \sum_{j=1}^{N_G^u} (\nabla \phi_j^u) \, \mathbf{u}_j, \qquad (7.3.10)$$

we obtain

$$\sum_{j=1}^{N_G^p} p_j \oint_C \phi_i^u \, \phi_j^p \, \mathbf{n} \, dl - \mu \sum_{j=1}^{N_G^u} \mathbf{u}_j \oint_C \phi_i^u \, \mathbf{n} \cdot \nabla \phi_j^u \, dl \qquad (7.3.11)$$

$$+ \sum_{j=1}^{N_G^p} p_j \iint_D (\nabla \phi_i^u) \, \phi_j^p \, dx \, dy - \mu \sum_{j=1}^{N_G^u} \mathbf{u}_j \iint_D \nabla \phi_i^u \cdot \nabla \phi_j^u \, dx \, dy = \mathbf{0}.$$

In terms of the x and y components of the velocity at the jth node encapsulated in the nodal vector, $\mathbf{u}_j = (u_{x_j}, u_{y_j})$, the x and y components of the vector equation (7.3.11) read

$$\sum_{j=1}^{N_G^p} p_j \oint_C \phi_i^u \, \phi_j^p \, n_x \, dl - \mu \sum_{j=1}^{N_G^u} u_{x_j} \oint_C \phi_i^u \, \mathbf{n} \cdot \nabla \phi_j^u \, dl \qquad (7.3.12)$$

$$+ \sum_{j=1}^{N_G^p} p_j \iint_D \frac{\partial \phi_i^u}{\partial x} \, \phi_j^p \, dx \, dy - \mu \sum_{j=1}^{N_G^u} u_{x_j} \iint_D \nabla \phi_i^u \cdot \nabla \phi_j^u \, dx \, dy = 0$$

and

$$\sum_{j=1}^{N_G^p} p_j \oint_C \phi_i^u \, \phi_j^p \, n_y \, dl - \mu \sum_{j=1}^{N_G^u} u_{y_j} \oint_C \phi_i^u \, \mathbf{n} \cdot \nabla \phi_j^u \, dl \qquad (7.3.13)$$

$$+ \sum_{j=1}^{N_G^p} p_j \iint_D \frac{\partial \phi_i^u}{\partial y} \, \phi_j^p \, dx \, dy - \mu \sum_{j=1}^{N_G^u} u_{y_j} \iint_D \nabla \phi_i^u \cdot \nabla \phi_j^u \, dx \, dy = 0.$$

The last integral on the right-hand sides is recognized as the global diffusion matrix for the velocity interpolation functions, denoted by

$$D_{ij}^u \equiv \iint_D \nabla \phi_i^u \cdot \nabla \phi_j^u \, dx \, dy. \tag{7.3.14}$$

Substituting the velocity finite element expansion into the Galerkin projection of the continuity equation (7.2.3), we obtain

$$\sum_{j=1}^{N_G^u} \mathbf{u}_j \cdot \iint_D \phi_i^p \, \nabla \phi_j^u \, dx \, dy = 0 \tag{7.3.15}$$

for $i = 1, \ldots, N_G^p$.

In compact notation, equations (7.3.11) and (7.3.15) take the form

$$\sum_{j=1}^{N_G^p} \mathbf{Q}_{ij} \, p_j - \mu \sum_{j=1}^{N_G^u} R_{ij} \, \mathbf{u}_j + \sum_{j=1}^{N_G^p} \mathbf{D}_{ji} \, p_j - \mu \sum_{j=1}^{N_G^u} D_{ij}^u \, \mathbf{u}_j = \mathbf{0} \tag{7.3.16}$$

and

$$\sum_{j=1}^{N_G^u} \mathbf{D}_{ij} \cdot \mathbf{u}_j = 0, \tag{7.3.17}$$

where

$$\mathbf{Q}_{ij} \equiv \oint_C \phi_i^u \, \phi_j^p \, \mathbf{n} \, dl, \qquad R_{ij} \equiv \oint_C \phi_i^u \, \mathbf{n} \cdot \nabla \phi_j^u \, dl, \tag{7.3.18}$$

and

$$\mathbf{D}_{ij} \equiv \iint_D \phi_i^p \, \nabla \phi_j^u \, dx \, dy. \tag{7.3.19}$$

Specifically,

$$\mathbf{Q}_{ij} = (Q_{x_{ij}}, Q_{y_{ij}}), \tag{7.3.20}$$

where

$$Q_{x_{ij}} = \oint_C \phi_i^u \, \phi_j^p \, n_x \, dl, \qquad Q_{y_{ij}} = \oint_C \phi_i^u \, \phi_j^p \, n_y \, dl, \tag{7.3.21}$$

and

$$\mathbf{D}_{ij} = (D_{x_{ij}}, D_{y_{ij}}), \tag{7.3.22}$$

where

$$D_{x_{ij}} = \iint_D \phi_i^p \, \frac{\partial \phi_j^u}{\partial x} \, dl, \qquad D_{y_{ij}} = \iint_D \phi_i^p \, \frac{\partial \phi_j^u}{\partial y} \, dl. \tag{7.3.23}$$

Subject to these definitions, the x and y components of the finite element equation (7.3.16) read

$$\sum_{j=1}^{N_G^p} Q_{x_{ij}} p_j - \mu \sum_{j=1}^{N_G^u} R_{ij}\, u_{x_j} + \sum_{j=1}^{N_G^p} D_{x_{ji}}\, p_j - \mu \sum_{j=1}^{N_G^u} D_{ij}^u\, u_{x_j} = 0,$$

$$(7.3.24)$$

$$\sum_{j=1}^{N_G^p} Q_{y_{ij}} p_j - \mu \sum_{j=1}^{N_G^u} R_{ij}\, u_{y_j} + \sum_{j=1}^{N_G^p} D_{y_{ji}}\, p_j - \mu \sum_{j=1}^{N_G^u} D_{ij}^u\, u_{y_j} = 0.$$

If the velocity is provided as a boundary condition around the entire boundary contour, C, the projection of the Stokes equation onto ϕ_i^u is excluded for those nodes that lie at the boundary. For all other nodes, because $\phi_i^u = 0$ on C, the matrices R_{ij} and $\mathbf{Q}_{ij}$ are identically zero. Thus, the Galerkin projection for the interior velocity nodes simplifies to

$$\mu \sum_{j=1}^{N_G^u} D_{ij}^u\, \mathbf{u}_j - \sum_{j=1}^{N_G^p} \mathbf{D}_{ji}\, p_j = \mathbf{0}. \qquad (7.3.25)$$

The x and y components of this equation are

$$\mu \sum_{j=1}^{N_G^u} D_{ij}^u u_{x_j} - \sum_{j=1}^{N_G^p} D_{x_{ji}} p_j = 0, \qquad \mu \sum_{j=1}^{N_G^u} D_{ij}^u u_{y_j} - \sum_{j=1}^{N_G^p} D_{y_{ji}} p_j = 0. \qquad (7.3.26)$$

Writing these equations for all velocity nodes, bearing in mind that additional terms must be included for the boundary nodes, and appending to the resulting system the continuity equation (7.3.17), we derive a grand linear system,

$$\begin{bmatrix} \mu\,\mathbf{D}^u & \mathbf{0} & -\mathbf{D}_x^T \\ \mathbf{0} & \mu\,\mathbf{D}^u & -\mathbf{D}_y^T \\ -\mathbf{D}_x & -\mathbf{D}_y & \mathbf{0} \end{bmatrix} \cdot \boldsymbol{\psi} = \mathbf{0}, \qquad (7.3.27)$$

where $\mathbf{D}^u$ is the $N_G^u \times N_G^u$ global diffusion matrix associated with the velocity nodes, $\mathbf{D}_x$ and $\mathbf{D}_y$ are rectangular $N_G^p \times N_G^u$ matrices, the superscript T denotes the matrix transpose, and

$$\boldsymbol{\psi} = \begin{bmatrix} u_{x_1}, u_{x_2}, \ldots, u_{x_{N_G^u}}, & u_{y_1}, u_{y_2}, \ldots, u_{y_{N_G^u}}, & p_1, p_2, \ldots, p_{N_G^p} \end{bmatrix}^T \quad (7.3.28)$$

is the vector of all unknowns. Note the symmetry of the block coefficient matrix on the left-hand side of the grand linear system (7.3.27).

Constant-pressure elements

In the simplest implementation of the finite element method, the pressure is approximated with a constant function over each element. To implement this approximation, we place pressure interpolation nodes at designated element centers, the exact location being immaterial, and then set $\phi_j^p = 1$ over the jth element and $\phi_j^p = 0$ outside the element. With this choice, we obtain the simplified expressions

$$D_{x_{ij}} = \iint_{E_i} \frac{\partial \phi_j^u}{\partial x} \, dx \, dy, \qquad D_{y_{ij}} = \iint_{E_i} \frac{\partial \phi_j^u}{\partial y} \, dx \, dy, \qquad (7.3.29)$$

where E_j denotes the jth element. The integrals in (7.3.29) are non-zero only if the jth global velocity node is an ith-element node.

7.3.3 Triangularization

It is convenient to introduce the $2N_G^u \times 2N_G^u$ square Laplacian matrix

$$\mathbf{L} \equiv \begin{bmatrix} \mathbf{D}^u & \mathbf{0} \\ \mathbf{0} & \mathbf{D}^u \end{bmatrix}, \qquad (7.3.30)$$

and the $N_G^p \times 2N_G^u$ rectangular gradient matrix

$$\mathbf{D} \equiv \begin{bmatrix} \mathbf{D}_x & \mathbf{D}_y \end{bmatrix}. \qquad (7.3.31)$$

Subject to these definitions, the system (7.3.27) takes the form

$$\begin{bmatrix} \mu \mathbf{L} & -\mathbf{D}^T \\ -\mathbf{D} & \mathbf{0} \end{bmatrix} \cdot \begin{bmatrix} \mathbf{U} \\ \mathbf{P} \end{bmatrix} = \mathbf{0}, \qquad (7.3.32)$$

where

$$\mathbf{U} \equiv \begin{bmatrix} u_{x1} \\ u_{x2} \\ \vdots \\ u_{xN_G^u} \\ - \\ u_{y1} \\ u_{y2} \\ \vdots \\ u_{yN_G^u} \end{bmatrix}, \qquad \mathbf{P} \equiv \begin{bmatrix} p_1 \\ p_2 \\ \vdots \\ p_{N_G^p} \end{bmatrix} \qquad (7.3.33)$$

are nodal velocity and pressure vectors.

After implementing the boundary conditions, the homogeneous system (7.3.32) reduces into an inhomogeneous system,

$$\begin{bmatrix} \mu \widehat{\mathbf{L}} & -\widehat{\mathbf{D}}^T \\ -\widehat{\mathbf{D}} & \mathbf{0} \end{bmatrix} \cdot \begin{bmatrix} \mathbf{U} \\ \mathbf{P} \end{bmatrix} = \begin{bmatrix} \mathbf{b}_U \\ \mathbf{b}_P \end{bmatrix}, \qquad (7.3.34)$$

where the caret designates modified matrices, and $\mathbf{b}_U$ and $\mathbf{b}_P$ are properly constructed right-hand sides.

To compute the solution of the modified linear system, we may apply the method of Gauss elimination discussed in Section C.1, Appendix C. Solving the first block equation in (7.3.34) for the velocity vector $\mathbf{U}$, we obtain

$$\mathbf{U} = \frac{1}{\mu} \widehat{\mathbf{L}}^{-1} \cdot (\widehat{\mathbf{D}}^T \cdot \mathbf{P} + \mathbf{b}_U). \tag{7.3.35}$$

The inverse matrix, $\widehat{\mathbf{L}}^{-1}$, consists of two diagonal blocks, each containing the inverse of the modified diffusion matrix, $\widehat{\mathbf{D}}^u$. Substituting (7.3.35) into the second block equation of (7.3.34), we obtain an equation for the pressure,

$$-\widehat{\mathbf{D}} \cdot \widehat{\mathbf{L}}^{-1} \cdot (\widehat{\mathbf{D}}^T \cdot \mathbf{P} + \mathbf{b}_U) = \mu \, \mathbf{b}_P. \tag{7.3.36}$$

Replacing the second block equation in (7.3.34) with (7.3.36) and rearranging, we derive an upper-triangular system,

$$\begin{bmatrix} \mu \widehat{\mathbf{L}} & -\widehat{\mathbf{D}}^T \\ 0 & \widehat{\mathbf{D}} \cdot \widehat{\mathbf{L}}^{-1} \cdot \widehat{\mathbf{D}}^T \end{bmatrix} \cdot \begin{bmatrix} \mathbf{U} \\ \mathbf{P} \end{bmatrix} = \begin{bmatrix} \mathbf{b}_U \\ -\mu \mathbf{b}_P - \widehat{\mathbf{D}} \cdot \widehat{\mathbf{L}}^{-1} \cdot \mathbf{b}_U \end{bmatrix}, \tag{7.3.37}$$

which can be solved by backward substitution according to Uzawa's algorithm: solve (7.3.36) for $\mathbf{P}$, and then recover $\mathbf{U}$ from (7.3.35). The matrix

$$\mathbf{S} \equiv \widehat{\mathbf{D}} \cdot \widehat{\mathbf{L}}^{-1} \cdot \widehat{\mathbf{D}}^T, \tag{7.3.38}$$

occupying the southeastern block of (7.3.37), is known as the pressure matrix. Since $\widehat{\mathbf{L}}$ is symmetric, $\mathbf{S}$ is also symmetric. In practice, the pressure equation (7.3.36) is solved using iterative methods, such as the method of conjugate gradients discussed in Section C.3, Appendix C.

PROBLEM

7.3.1 *Three-dimensional flow*

Write the counterpart of the linear system (7.3.27) for three-dimensional flow.

7.4 Stokes flow in a rectangular cavity

As an application, we consider Stokes flow in a rectangular cavity that is filled with a liquid. The motion of the fluid is driven by the translation of the cavity lid parallel to itself with constant velocity, V, as illustrated in Figure 7.4.1. The non-slip and no-penetration boundary conditions require that the x and y velocity components, u_x and u_y, are zero over the bottom and side walls, whereas $u_x = V$ and $u_y = 0$ at the cavity lid.

FSELIB code *cvt6*, listed in Table 7.4.1, assembles and solves the linear system (7.3.27) for the nodal velocities and pressures. Results of a computation are shown in Figure 7.4.1. The layout of the code is similar to that of code *lapl6_d* for Laplace's

(a)

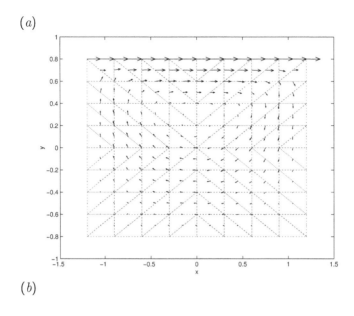

(b)

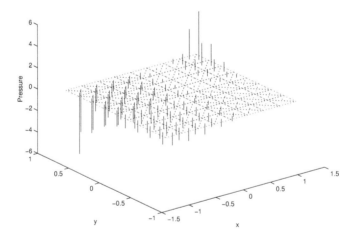

FIGURE 7.4.1   (a) Velocity vector field and (b) pressure field indicated by vertical
    bars passing through the element centroids for Stokes flow in a rectangular cavity
    driven by a translating lid, generated by the FSELIB code *cvt6*.

equation subject to the Dirichlet boundary condition, as discussed in Section 5.3.1,
with some parts being virtually identical. The flow domain is discretized into six-
node triangular elements using the FSELIB function *trgl6_sqr* discussed in Section
5.2.2. Noteworthy new features are discussed in the remainder of this section.

```
%==================================
% Code cvt6
%
% Stokes flow in a cavity computed
% with six-node triangular elements
%==================================

%-----------
% input data
%-----------

visc = 1.0; % viscosity
V = 1.0;    % lid velocity
NQ = 6;     % Gauss-triangle quadrature
ndiv = 1;   % discretization level

%-----------
% triangulate
%-----------

[ne,ng,p,c,efl,gfl] = trgl6_sqr (ndiv);

%-----------------------------
% specify the boundary velocity
%-----------------------------

for i=1:ng

 if(gfl(i,1)==1)  % boundary node

   gfl(i,2) = 0.0; % x velocity
   gfl(i,3) = 0.0; % y velocity

   if(p(i,2) > 0.99)  % point is on the lid
    gfl(i,2) = V;     % x velocity of the lid
   end

 end

end

%----------------------------------
% deform the square to a rectangle
%----------------------------------

def = 0.2; % example

for i=1:ng
 p(i,1) = p(i,1)*(1.0 + def);
 p(i,2) = p(i,2)*(1.0 - def);
end
```

TABLE 7.4.1   Code *cvt6* (Continuing →)

```
%----------------------------------------
% assemble: the global diffusion matrix
%           the Dx and Dy matrices
%----------------------------------------

gdm = zeros(ng,ng); % initialize

gDx = zeros(ne,ng); % initialize

gDy = zeros(ne,ng); % initialize

for l=1:ne           % loop over the elements

% compute the element diffusion matrix

 j = c(l,1); x1 = p(j,1); y1 = p(j,2);
 j = c(l,2); x2 = p(j,1); y2 = p(j,2);
 j = c(l,3); x3 = p(j,1); y3 = p(j,2);
 j = c(l,4); x4 = p(j,1); y4 = p(j,2);
 j = c(l,5); x5 = p(j,1); y5 = p(j,2);
 j = c(l,6); x6 = p(j,1); y6 = p(j,2);

 [edm_elm, arel] = edm6 ...
 ...
     (x1,y1, x2,y2, x3,y3, x4,y4, x5,y5, x6,y6, NQ);

% compute the element matrices Dx, Dy

 [Dx, Dy] = cvt6_D ...
 ...
     (x1,y1, x2,y2, x3,y3, x4,y4, x5,y5, x6,y6, NQ);

   for i=1:6

     i1 = c(l,i);
     for j=1:6
       j1 = c(l,j);
       gdm(i1,j1) = gdm(i1,j1) + edm_elm(i,j);
     end
     gDx(l,i1) = Dx(i);
     gDy(l,i1) = Dy(i);

   end

end    % end of loop over the elements

%------------------------
% compile the grand matrix
%------------------------

for i=1:ng       % first block

 ngi = ng+i;
```

TABLE 7.4.1   Code *cvt6* ($\rightarrow$ Continuing $\rightarrow$)

```
for j=1:ng
 ngj = ng+j;
 Gm(i,j)   = visc*gdm(i,j);  Gm(i,ngj) = 0.0;
 Gm(ngi,j) = 0.0;            Gm(ngi,ngj) = Gm(i,j);
end

for j=1:ne
 Gm(i,  2*ng+j) = -gDx(j,i);
 Gm(ngi,2*ng+j) = -gDy(j,i);
end

end

for i=1:ne     % second block

 for j=1:ng
 Gm(2*ng+i,j)    = -gDx(i,j);
 Gm(2*ng+i,ng+j) = -gDy(i,j);
end

 for j=1:ne
 Gm(2*ng+i,2*ng+j) = 0;
 end

end

%------------
% system size
%------------

nsys = 2*ng+ne;

%-------------------------------------------------
% compute the right-hand side and implement the
% velocity (Dirichlet) boundary condition
%-------------------------------------------------

for i=1:nsys
 b(i) = 0.0;   % initialize
end

for j=1:ng
 if(gfl(j,1)==1)
   for i=1:nsys
     b(i) = b(i) - Gm(i,j) * gfl(j,2) - Gm(i,ng+j) * gfl(j,3);
     Gm(i,j) = 0; Gm(i,ng+j) = 0;
     Gm(j,i) = 0; Gm(ng+j,i) = 0;
   end

   Gm(j,j) = 1.0; Gm(ng+j,ng+j) = 1.0;
   b(j) = gfl(j,2); b(ng+j) = gfl(j,3);
 end
end
```

TABLE 7.4.1   Code *cvt6* ($\rightarrow$ Continuing $\rightarrow$)

```
%------------------------
% solve the linear system
%------------------------

Gm(:,nsys) = [];  % remove the last (nsys) column

Gm(nsys,:) = [];  % remove the last (nsys) row

b(:,nsys)  = [];  % remove the last (nsys) element

f = b/Gm';

%---------------------
% recover the velocity
%---------------------

for i=1:ng
 ux(i) = f(i);
 uy(i) = f(ng+i);
end

%---------------------
% recover the pressure
%---------------------

for i=1:ne-1
 press(i) = f(2*ng+i);
end

press(ne) = 0.0;   % arbitrary reference level

%-----------------------
% plot the element edges
%-----------------------

for i=1:ne

 i1 = c(i,1); i2 = c(i,2); i3 = c(i,3);
 i4 = c(i,4); i5 = c(i,5); i6 = c(i,6);

 xp(1) = p(i1,1); xp(2) = p(i4,1); xp(3) = p(i2,1); xp(4) = p(i5,1);
                  xp(5) = p(i3,1); xp(6) = p(i6,1); xp(7) = p(i1,1);

 yp(1) = p(i1,2); yp(2) = p(i4,2); yp(3) = p(i2,2); yp(4) = p(i5,2);
                  yp(5) = p(i3,2); yp(6) = p(i6,2); yp(7) = p(i1,2);

 plot(xp, yp,'-'); hold on;  % plot(xp, yp,'-o');

end

%-----------------------------
% plot the velocity vector field
%-----------------------------
```

TABLE 7.4.1   Code *cvt6* ($\rightarrow$ Continuing $\rightarrow$)

```
plot_6 (ne,ng,p,c,ux);

quiver (p(:,1)',p(:,2)',ux,uy);

xlabel('x'); ylabel('y');

%-----------------------
% plot the pressure field
%-----------------------

figure;    % new graphic window

for i=1:ne

  i1=c(i,1); i2=c(i,2); i3=c(i,3);

  % compute the element centroid:

  xplt(1) = (p(i1,1)+p(i2,1)+p(i3,1))/3.0;
  yplt(1) = (p(i1,2)+p(i2,2)+p(i3,2))/3.0;
  zplt(1) = press(i);

  xplt(2) = xplt(1); yplt(2) = yplt(1);
  zplt(2) = 0.0;

  plot3(xplt,yplt,zplt,'-'); hold on;

end

xlabel('x'); ylabel('y'); zlabel('Pressure');

%-----------------------
% plot the element edges
%-----------------------

for i=1:ne

  i1 = c(i,1); i2 = c(i,2); i3 = c(i,3);
  i4 = c(i,4); i5 = c(i,5); i6 = c(i,6);

  xp(1) = p(i1,1); xp(2) = p(i4,1); xp(3) = p(i2,1); xp(4) = p(i5,1);
            xp(5) = p(i3,1); xp(6) = p(i6,1); xp(7) = p(i1,1);

  yp(1) = p(i1,2); yp(2) = p(i4,2); yp(3) = p(i2,2); yp(4) = p(i5,2);
            yp(5) = p(i3,2); yp(6) = p(i6,2); yp(7) = p(i1,2);

  plot(xp, yp, ':'); hold on;

end

%-----
% done
%-----
```

TABLE 7.4.1   ($\rightarrow$ Continued) Code *cvt6* for Stokes flow in a rectangular cavity.

Mixed expansion

The x and y velocity components are interpolated isoparametrically, that is, they are approximated with a quadratic function over each element. Accordingly, the number of velocity global nodes is equal to the number of unique geometrical global nodes,

$$N_G^u = N_G. \tag{7.4.1}$$

The pressure is approximated with a constant function over each element. Accordingly, the number of pressure global nodes is equal to the number of elements,

$$N_G^p = N_E. \tag{7.4.2}$$

Because the velocity nodes are different than the pressure nodes, the finite element expansion is recognized as *mixed.*

Computation of the diffusion matrices

The computation of the global diffusion matrix, $\mathbf{D}^u$, named *gdm* in the code, follows the procedure described in Section 5.3.1 for code *lapl6_d*. To compile the matrices $\mathbf{D}_x$ and $\mathbf{D}_y$ given in (7.3.29), we run over the elements, regarded as pressure nodes, compute the matrix entries with the help of the function *cvt6_D*, listed in Table 7.4.2, and make contributions to the finite element equations.

Velocity boundary condition

To implement the velocity boundary condition, we employ a straightforward generalization of the algorithm listed in Table 4.2.2.

Nonuniqueness of the pressure

Because the pressure gradient, but not the pressure itself, appears in the Stokes or Navier–Stokes equation, the solution for the pressure is defined only up to an arbitrary constant. Accordingly, the pressure of an arbitrary element can be assigned an arbitrary yet inconsequential value. In the code, the pressure at the last pressure node is arbitrarily set to zero, $p_{N_G^p} = 0$ by discarding the last equation of the grand system and the last column of the grand matrix.

Visualization

The velocity vector field is visualized using the MATLAB function *quiver.* The pressure field is visualized using low-level graphics functions by drawing vertical bars passing through the element centroids.

Solvability of the finite element equations

It is important to note that, after the boundary conditions have been implemented, the number of equations involving the nodal values of the pressure is reduced from

```
function [Dx, Dy] = cvt6_D ...
   ...
    (x1,y1, x2,y2, x3,y3, x4,y4, x5,y5, x6,y6, NQ)

%================================================
% Computation of the gradient matrices Dx and Dy
% required by cvt6 for viscous flow
%================================================

%-----------------------------------
% compute the mapping coefficients:
%-----------------------------------

[al, be, ga] = elm6_abc ...
   ...
    (x1,y1, x2,y2, x3,y3, x4,y4, x5,y5, x6,y6);

%-----------------------------
% read the triangle quadrature
%-----------------------------

[xi, eta, w] = gauss_trgl(NQ);

%----------------------------------
% initialize the matrices Dx and Dy
%----------------------------------

for k=1:6
   Dx(k) = 0.0;
   Dy(k) = 0.0;
end

%----------------------
% perform the quadrature
%----------------------

for i=1:NQ

  [psi, gpsi, hs] = elm6_interp ...
    ...
      (x1,y1, x2,y2, x3,y3, x4,y4, x5,y5, x6,y6 ...
      ,al,be,ga,
      ,xi(i),eta(i));

  cf = 0.5*hs*w(i)
  for k=1:6
    Dx(k) = Dx(k) + gpsi(k,1)*cf;
    Dy(k) = Dy(k) + gpsi(k,2)*cf;
  end

end

%-----
% done
%-----

return;
```

TABLE 7.4.2   Function *cvt6_D* computes the matrices $\mathbf{D}^x$ and $\mathbf{D}^y$ required by the finite element code *cvt6* for viscous flow in a cavity.

$2N_{\mathrm{G}}^u$, corresponding to the first two row blocks of (7.3.27), to $2\,(N_{\mathrm{G}}^u)_{\mathrm{int}}$, where $(N_{\mathrm{G}}^u)_{\mathrm{int}}$ is the number of *interior* global nodes. For a solution to exist, the inequality

$$\frac{(N_{\mathrm{G}}^u)_{\mathrm{int}}}{N_{\mathrm{G}}^p} > \frac{1}{2} \tag{7.4.3}$$

must be fulfilled, otherwise the computations will fail. In practice, the fraction on the left-hand side of (7.4.3) must be greater than two or three to ensure numerical stability. In the case of the cavity flow, if we had used three-node triangular elements with an isoparametric velocity interpolation and constant element pressures, criterion (7.4.3) would fail for any discretization level.

PROBLEM

7.4.1 *Stokes flow in a cavity*

Run the code *cvt6* for flow in a slender cavity with width to depth ratio of 5:1. Discuss the pressure distribution and the structure of the velocity field far from the side-walls.

7.5   Navier–Stokes flow

To develop a finite element formulation for flow at non-zero Reynolds numbers, we substitute the finite element expansion for the velocity into the left-hand side of the Navier–Stokes equation (7.1.10), denoted by $\mathbf{H}$, and obtain

$$\mathbf{H} \equiv \rho \sum_{j=1}^{N_{\mathrm{G}}^u} \frac{\mathrm{d}\mathbf{u}_j}{\mathrm{d}t}\, \phi_j^u + \rho \sum_{j=1}^{N_{\mathrm{G}}^u} \phi_j^u\, \mathbf{u}_j \cdot \sum_{k=1}^{N_{\mathrm{G}}^u} \nabla\phi_k^u\, \mathbf{u}_k$$

$$= \rho \sum_{j=1}^{N_{\mathrm{G}}^u} \frac{\mathrm{d}\mathbf{u}_j}{\mathrm{d}t}\, \phi_j^u + \rho \sum_{j=1}^{N_{\mathrm{G}}^u} \sum_{k=1}^{N_{\mathrm{G}}^u} \phi_j^u\, (\mathbf{u}_j \cdot \nabla\phi_k^u)\, \mathbf{u}_k. \tag{7.5.1}$$

The x and y components of this equation are

$$H_x = \rho \sum_{j=1}^{N_{\mathrm{G}}^u} \frac{\mathrm{d}u_{x_j}}{\mathrm{d}t}\, \phi_j^u + \rho \sum_{j=1}^{N_{\mathrm{G}}^u} \sum_{k=1}^{N_{\mathrm{G}}^u} \phi_j^u\, (\mathbf{u}_j \cdot \nabla\phi_k^u)\, u_{x_k},$$

$$\tag{7.5.2}$$

$$H_y = \rho \sum_{j=1}^{N_{\mathrm{G}}^u} \frac{\mathrm{d}u_{x_j}}{\mathrm{d}t}\, \phi_j^u + \rho \sum_{j=1}^{N_{\mathrm{G}}^u} \sum_{k=1}^{N_{\mathrm{G}}^u} \phi_j^u\, (\mathbf{u}_j \cdot \nabla\phi_k^u)\, u_{y_k}.$$

To carry out the Galerkin projection, we multiply each expression by the ith velocity global interpolation function, ϕ_i^u, and integrate the product over the solution domain to obtain

$$\iint \phi_i^u \mathbf{H}\, \mathrm{d}x\, \mathrm{d}y = \rho \sum_{j=1}^{N_{\mathrm{G}}^u} M_{ij}^u\, \frac{\mathrm{d}\mathbf{u}_j}{\mathrm{d}t} + \rho \sum_{j=1}^{N_{\mathrm{G}}^u} \sum_{k=1}^{N_{\mathrm{G}}^u} (\mathbf{u}_j \cdot \mathbf{N}_{jik})\, \mathbf{u}_k, \tag{7.5.3}$$

where

$$M_{ij}^u = \iint_D \phi_i^u \, \phi_j^u \, \mathrm{d}x \, \mathrm{d}y \tag{7.5.4}$$

is the familiar mass matrix and

$$\mathbf{N}_{jik} = \iint_D \phi_j^u \phi_i^u \, (\nabla \phi_k^u) \, \mathrm{d}x \, \mathrm{d}y \tag{7.5.5}$$

is the advection matrix.

Setting the left-hand side of (7.5.3) equal to the right-hand side of the projected Navier–Stokes equation, as given in (7.3.16), we obtain

$$\rho \sum_{j=1}^{N_G^u} M_{ij}^u \frac{\mathrm{d}\mathbf{u}_j}{\mathrm{d}t} + \rho \sum_{j=1}^{N_G^u} \sum_{k=1}^{N_G^u} (\mathbf{u}_j \cdot \mathbf{N}_{jik}) \, \mathbf{u}_k$$
$$= \sum_{j=1}^{N_G^p} \mathbf{Q}_{ij} \, p_j - \mu \sum_{j=1}^{N_G^u} R_{ij} \, \mathbf{u}_j + \sum_{j=1}^{N_G^p} \mathbf{D}_{ji} \, p_j - \mu \sum_{j=1}^{N_G^u} D_{ij}^u \, \mathbf{u}_j, \tag{7.5.6}$$

which is to be solved together with the Galerkin projection of the continuity equation (7.3.17), repeated below for convenience,

$$\sum_{j=1}^{N_G^u} \mathbf{D}_{ij} \cdot \mathbf{u}_j = 0. \tag{7.5.7}$$

Explicitly, the x and y components of the vector equation (7.5.6) read

$$\rho \sum_{j=1}^{N_G^u} M_{ij}^u \frac{\mathrm{d}u_{x_j}}{\mathrm{d}t} + \rho \sum_{j=1}^{N_G^u} \sum_{k=1}^{N_G^u} (\mathbf{u}_j \cdot \mathbf{N}_{jik}) \, u_{x_k} \tag{7.5.8}$$
$$= \sum_{j=1}^{N_G^p} Q_{x_{ij}} \, p_j - \mu \sum_{j=1}^{N_G^u} R_{ij} \, u_{x_j} + \sum_{j=1}^{N_G^p} D_{x_{ji}} \, p_j - \mu \sum_{j=1}^{N_G^u} D_{ij}^u \, u_{x_j} = 0$$

and

$$\rho \sum_{j=1}^{N_G^u} M_{ij}^u \frac{\mathrm{d}u_{y_j}}{\mathrm{d}t} + \rho \sum_{j=1}^{N_G^u} \sum_{k=1}^{N_G^u} (\mathbf{u}_j \cdot \mathbf{N}_{jik}) \, u_{y_k} \tag{7.5.9}$$
$$= \sum_{j=1}^{N_G^p} Q_{y_{ij}} \, p_j - \mu \sum_{j=1}^{N_G^u} R_{ij} \, u_{y_j} + \sum_{j=1}^{N_G^p} D_{y_{ji}} \, p_j - \mu \sum_{j=1}^{N_G^u} D_{ij}^u \, u_{y_j}.$$

Setting $\rho = 0$ provides us with the equations of Stokes flow.

7.5.1 Steady state

At steady steady, $d\mathbf{u}_j/dt = \mathbf{0}$, we obtain a nonlinear quadratic system of algebraic equations that must be solved by iteration. Two options are general-purpose Newton or quasi-Newton methods (e.g., Pozrikidis [17]), and iterative methods based on functional minimization involving the solution of a sequence of discrete Stokes flow problems (Glowinski [24]).

7.5.2 Time integration

One method of integrating (7.5.6) forward in time from a specified state at time t involves the following steps:

1. Select a time step, Δt.

2. Apply the differential equation at time $t + \frac{1}{2}\Delta t$.

3. Approximate the time derivative on the left-hand side with a centered difference and implement the Crank–Nicolson discretization for the viscous term to obtain

$$\frac{\rho}{\Delta t} \sum_{j=1}^{N_G^u} M_{ij}^u \left(\mathbf{u}_j^{n+1} - \mathbf{u}_j^n \right) + \rho \sum_{j=1}^{N_G^u} \sum_{k=1}^{N_G^u} \left[\left(\mathbf{u}_j \cdot \mathbf{N}_{jik} \right) \mathbf{u}_k \right]^{n+1/2}$$

$$= \sum_{j=1}^{N_G^p} \mathbf{Q}_{ij}\, p_j^{n+1/2} - \mu \frac{1}{2} \sum_{j=1}^{N_G^u} R_{ij} \left(\mathbf{u}_j^n + \mathbf{u}_j^{n+1} \right) \qquad (7.5.10)$$

$$+ \sum_{j=1}^{N_G^p} \mathbf{D}_{ji}\, p_j^{n+1/2} - \mu \frac{1}{2} \sum_{j=1}^{N_G^u} D_{ij}^u \left(\mathbf{u}_j^n + \mathbf{u}_j^{n+1} \right),$$

where the superscripts $n, n + \frac{1}{2}$, and $n + 1$ denote evaluation, respectively, at times $t, t + \frac{1}{2}\Delta t$, and $t + \Delta t$.

The second term on the left-hand side of (7.5.10), expressing the nonlinear effect of convection, is assumed to be available by extrapolation from previous steps. To simplify the algorithm, this term can be evaluated at the initial instant at the first step.

The continuity equation requires that

$$\sum_{j=1}^{N_G^u} \mathbf{D}_{ij} \cdot \mathbf{u}_j^{n+1} = 0. \qquad (7.5.11)$$

The unknowns in equations (7.5.10) and (7.5.11) are the nodal velocities and pressures, $p_j^{n+1/2}$ and $\mathbf{u}_j^{n+1}$. Rearranging (7.5.10) to bring the unknowns to the left-hand side, we obtain

$$
\mu \sum_{j=1}^{N_G^u} (\frac{1}{\nu \, \Delta t} M_{ij}^u + \frac{1}{2} D_{ij}^u) \, \mathbf{u}_j^{n+1} - \sum_{j=1}^{N_G^p} \mathbf{Q}_{ij} \, p_j^{n+1/2} + \mu \frac{1}{2} \sum_{j=1}^{N_G^u} R_{ij} \, \mathbf{u}_j^{n+1} \tag{7.5.12}
$$

$$
- \sum_{j=1}^{N_G^p} \mathbf{D}_{ji} \, p_j^{n+1/2} = \mu \sum_{j=1}^{N_G^u} (\frac{1}{\nu \, \Delta t} M_{ij}^u - \frac{1}{2} D_{ij}^u) \, \mathbf{u}_j^n - \mu \frac{1}{2} \sum_{j=1}^{N_G^u} R_{ij} \, \mathbf{u}_j^n + \mathbf{F}_i,
$$

where $\nu \equiv \mu/\rho$ is the kinematic viscosity. The vector $\mathbf{F}_i$ on the right-hand side encapsulates the nonlinear convection term,

$$
\mathbf{F}_i \equiv -\rho \sum_{j=1}^{N_G^u} \sum_{k=1}^{N_G^u} \Big[(\mathbf{u}_j \cdot \mathbf{N}_{jik}) \, \mathbf{u}_k \Big]^{n+1/2}. \tag{7.5.13}
$$

If the velocity is specified as a boundary condition around the entire boundary, C, the projection of the Navier–Stokes equation on ϕ_i^u is excluded for nodes that lie at the boundary. For all other nodes, because $\phi_i^u = 0$ on C, the matrices R_{ij} and $\mathbf{Q}_{ij}$ are identically zero.

Writing equation (7.5.13) for all velocity nodes, bearing in mind that additional terms should have been included for the boundary nodes, and appending to the resulting system the continuity equation (7.5.11), we obtain a grand system,

$$
\begin{bmatrix} \mu \, (\frac{1}{\nu \, \Delta t} \mathbf{M}^u + \frac{1}{2} \mathbf{D}^u) & \mathbf{0} & -\mathbf{D}_x^T \\ \mathbf{0} & \mu \, (\frac{1}{\nu \, \Delta t} \mathbf{M}^u + \frac{1}{2} \mathbf{D}^u) & -\mathbf{D}_y^T \\ -\mathbf{D}_x & -\mathbf{D}_y & \mathbf{0} \end{bmatrix} \cdot \mathbf{X}^{n+1} = \mathbf{RHS}, \tag{7.5.14}
$$

involving a symmetric coefficient matrix, where the right-hand side is given by

$$
\mathbf{RHS} = \begin{bmatrix} \mu \, (\frac{1}{\nu \, \Delta t} \mathbf{M}^u - \frac{1}{2} \mathbf{D}^u) & \mathbf{0} \\ \mathbf{0} & \mu \, (\frac{1}{\nu \, \Delta t} \mathbf{M}^u - \frac{1}{2} \mathbf{D}^u) \\ \mathbf{0} & \mathbf{0} \end{bmatrix} \cdot \mathbf{U}^n - \mathbf{b}. \tag{7.5.15}
$$

The unknown vector, $\mathbf{X}^{n+1}$, velocity vector, $\mathbf{U}^n$, and nonlinear term, $\mathbf{b}$, are displayed in Table 7.5.1. In the case of unsteady Stokes flow, the nonlinear term, $\mathbf{b}$, is zero (Problem 7.5.1). The solution of the linear system can be found as discussed in Section 7.4 for Stokes flow.

$$
\mathbf{X}^{n+1} \equiv
\begin{bmatrix}
u_{x_1}^{n+1} \\
u_{x_2}^{n+1} \\
\vdots \\
u_{x_{N_G^u}}^{n+1} \\
- \\
u_{y_1}^{n+1} \\
u_{y_2}^{n+1} \\
\vdots \\
u_{y_{N_G^u}}^{n+1} \\
- \\
p_1^{n+1/2} \\
p_2^{n+1/2} \\
\vdots \\
p_{N_G^p}^{n+1/2}
\end{bmatrix}
\qquad
\mathbf{U}^n \equiv
\begin{bmatrix}
u_{x_1}^n \\
u_{x_2}^n \\
\vdots \\
u_{x_{N_G^u}}^n \\
- \\
u_{y_1}^n \\
u_{y_2}^n \\
\vdots \\
u_{y_{N_G^u}}^n
\end{bmatrix}
\qquad
\mathbf{b} \equiv
\begin{bmatrix}
F_{x_1} \\
F_{x_2} \\
\vdots \\
F_{x_{N_G^u}} \\
- \\
F_{y_1}^{n+1} \\
F_{y_2}^{n+1} \\
\vdots \\
F_{y_{N_G^u}} \\
- \\
0 \\
0 \\
\vdots \\
0
\end{bmatrix}
$$

TABLE 7.5.1 Unknown solution vector, velocity vector, and nonlinear term on the right-hand side of system (7.5.14).

7.5.3 *Formulation based on the pressure Poisson equation*

In the formulation discussed in this previously in this chapter, the equation of motion was solved simultaneously with the continuity equation by the finite element method. In an alternative approach, the continuity equation is replaced by a Poisson equation that emerges by taking the divergence of the Navier–Stokes equation and then enforcing the incompressibility condition, finding

$$
\nabla^2 p = -\rho \, \nabla \cdot \left(\mathbf{u} \cdot \nabla \mathbf{u} \right). \tag{7.5.16}
$$

A variety of methods are available for integrating the equation of motion in space or time together with the pressure Poisson equation, as discussed in texts of computational fluid dynamics (CFD) (e.g., Karniadakis & Sherwin [32], Pozrikidis [47]).

PROBLEM

7.5.1 *Unsteady Stokes flow*

Unsteady Stokes flow is governed by the unsteady Stokes equation,

$$
\rho \frac{\partial \mathbf{u}}{\partial t} = -\nabla p + \mu \nabla^2 \mathbf{u}, \tag{7.5.17}
$$

subject to the continuity equation expressed by (7.1.5). Modify the FSELIB code *cvt6* into a code named *cvt6_u_cn* that computes the evolution of a flow from a specified initial state using the Crank–Nicolson method. Run the code to describe the onset of steady lid-driven cavity flow from a quiescent initial condition, subject to impulsively started lid translation where the lid velocity suddenly jumps to a constant value at the origin of computational time.

Finite and spectral element methods in three dimensions

8

The development and implementation of the Galerkin finite and spectral element method in three dimensions are straightforward generalizations of those in two dimensions discussed in Chapters 4–7. However, not surprisingly, the bookkeeping is more demanding and the implementation presents us with new challenges related to the efficient computation of various quantities involved in the formulation and the affordable solution of the final system of algebraic or ordinary differential equations. Main issues and basic procedures are discussed in this chapter with reference to the convection–diffusion equation, and complete finite element codes are provided.

8.1 Convection–diffusion in three dimensions

Consider unsteady heat transport in a three-dimensional domain in the presence of a distributed source, as illustrated in Figure 8.1.1. The evolution of the temperature field, $f(x, y, z, t)$, is governed by the convection–diffusion equation

$$\rho\, c_p \left(\frac{\partial f}{\partial t} + u_x \frac{\partial f}{\partial x} + u_y \frac{\partial f}{\partial y} + u_z \frac{\partial f}{\partial z} \right) = k \left(\frac{\partial^2 f}{\partial x^2} + \frac{\partial^2 f}{\partial y^2} + \frac{\partial^2 f}{\partial z^2} \right) + s, \qquad (8.1.1)$$

where ρ, c_p, and k are the medium density, heat capacity, and thermal conductivity,

$$u_x(x, y, z, f, t), \qquad u_y(x, y, z, f, t), \qquad u_z(x, y, z, f, t) \qquad (8.1.2)$$

are the $x, y,$ and z components of the convection velocity, and $s(x, y, z, f, t)$ is a distributed source. Equation (8.1.1) is the three-dimensional counterpart of the one-dimensional rod equation (2.3.1) and of the two-dimensional plate equation (4.1.1).

For simplicity, we assume that the convection velocity and source term depend on position and time explicitly, but not implicitly through f, that is, $u_x(x, y, z, t)$, $u_y(x, y, z, t)$, $u_z(x, y, z, t)$, and $s(x, y, z, t)$.

In vector notation, the convection–diffusion equation (8.1.1) takes the compact form

$$\frac{\partial f}{\partial t} + \mathbf{u} \cdot \nabla f = \kappa \nabla^2 f + \frac{s}{\rho\, c_p}, \qquad (8.1.3)$$

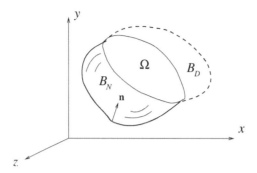

FIGURE 8.1.1   Illustration of heat conduction in a three-dimensional domain with arbitrary geometry, Ω, bounded by a closed surface, B.

where $\kappa \equiv k/(\rho\, c_p)$ is the thermal diffusivity,

$$\nabla f = \left(\frac{\partial f}{\partial x}, \; \frac{\partial f}{\partial y}, \; \frac{\partial f}{\partial z} \right) \tag{8.1.4}$$

is the three-dimensional gradient, and

$$\nabla^2 f \equiv \nabla \cdot \nabla f = \frac{\partial^2 f}{\partial x^2} + \frac{\partial^2 f}{\partial y^2} + \frac{\partial^2 f}{\partial z^2} \tag{8.1.5}$$

is the three-dimensional Laplacian operator. Equation (8.1.3) is to be solved in a domain, Ω, that is enclosed by a surface, B, subject to the *Neumann* or *Dirichlet* boundary condition.

8.1.1   Boundary conditions

The Neumann boundary condition prescribes the heat flux along the Neumann portion of B, denoted as B_N, drawn with the solid line in Figure 8.1.1,

$$q(\mathbf{x}) \equiv -k\,\mathbf{n} \cdot \nabla f \equiv -k\, \frac{\partial f}{\partial l_n}, \tag{8.1.6}$$

where $\mathbf{n}$ is the unit vector normal to B_N pointing into the solution domain, as shown in Figure 8.1.1, l_n is the arc length in the normal direction, and $q(\mathbf{x})$ is a given function of position over B_N. If $q(\mathbf{x}) > 0$, in which case $\mathbf{n} \cdot \nabla f < 0$, heat enters the solution domain Ω by diffusion; whereas if $q(\mathbf{x}) < 0$, in which case $\mathbf{n} \cdot \nabla f > 0$, heat escapes from the solution domain by diffusion.

The Dirichlet boundary condition prescribes the temperature distribution over the complementary portion of the boundary, denoted as B_D, drawn with the broken line in Figure 8.1.1,

$$f(\mathbf{x}) = g(\mathbf{x}), \tag{8.1.7}$$

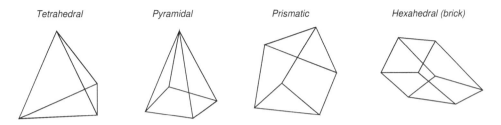

Tetrahedral   Pyramidal   Prismatic   Hexahedral (brick)

FIGURE 8.1.2   An assortment of elements are available for discretizing a three-dimensional domain.

where $B_D = B - B_N$ is the complement of B_N and $g(\mathbf{x})$ is a given function of position over B_D.

8.1.2   Domain discretization

In the first step of the finite or spectral element method, the solution domain is discretized into three-dimensional elements defined by a small group of geometrical element nodes. The union of these element nodes comprises a larger set of unique global nodes. For simplicity, in the remainder of this section we discuss the isoparametric representation where the interpolation element nodes coincide with the geometrical element nodes.

Several element types are available, as shown in Figure 8.1.2. Choices include tetrahedral and hexahedral (brick) elements, discussed in this chapter, and five-faced pyramidal or prismatic elements. Being the counterparts of the two-dimensional triangular and quadrilateral elements, the tetrahedral and hexahedral elements are favored in general-purpose finite and spectral element implementations.

8.1.3   Galerkin projection

To develop the finite element equations, we carry out the Galerkin projection of the governing differential equation (8.1.1) using as weighting functions the global cardinal interpolation functions associated with the unique global nodes, $\phi_i(x, y, z)$. By definition, the function $\phi_i(x, y, z)$ takes the value of unity at the ith global interpolation node and the value of zero at all other global interpolation nodes.

Multiplying first the right-hand side of (8.1.1) by $\phi_i(x, y, z)$, and integrating the product over the solution domain, Ω, we obtain

$$\mathcal{R} \equiv \iiint_\Omega \phi_i \left(k\,\nabla^2 f + s \right) \mathrm{d}x\,\mathrm{d}y\,\mathrm{d}z = \iiint_\Omega \left(\phi_i\,k\,(\nabla \cdot \nabla f) + \phi_i\,s \right) \mathrm{d}x\,\mathrm{d}y\,\mathrm{d}z.$$

$$(8.1.8)$$

Assuming that ϕ_i is continuous throughout Ω, we write

$$\mathcal{R} = k \iiint_\Omega \nabla \cdot (\phi_i \, \nabla f) \, \mathrm{d}x \, \mathrm{d}y \, \mathrm{d}z + \iiint_\Omega \left(-k \, \nabla \phi_i \cdot \nabla f + \phi_i \, s \right) \mathrm{d}x \, \mathrm{d}y \, \mathrm{d}z \quad (8.1.9)$$

and use the Gauss divergence theorem to obtain

$$\mathcal{R} = -k \iint_B \phi_i \, \mathbf{n} \cdot \nabla f \, \mathrm{d}S + \iiint_\Omega \left(-k \, \nabla \phi_i \cdot \nabla f + \phi_i \, s \right) \mathrm{d}x \, \mathrm{d}y \, \mathrm{d}z. \quad (8.1.10)$$

Substituting the boundary flux definition in the boundary integral, $q \equiv -k \, \mathbf{n} \cdot \nabla f$, recalling the definition of $\mathcal{R}$, and rearranging, we obtain

$$\iiint_\Omega \phi_i \left(k \, \nabla^2 f + s \right) \mathrm{d}x \, \mathrm{d}y \, \mathrm{d}z = -k \iiint_\Omega \nabla f \cdot \nabla \phi_i \, \mathrm{d}x \, \mathrm{d}y \, \mathrm{d}z + Q_i + S_i, \quad (8.1.11)$$

where

$$Q_i \equiv \iint_B \phi_i \, q \, \mathrm{d}S, \qquad S_i \equiv \iiint_\Omega \phi_i \, s \, \mathrm{d}x \, \mathrm{d}y \, \mathrm{d}z \quad (8.1.12)$$

are boundary and volume integrals of the surface flux and distributed source, both weighted by the ith global interpolation functions. If the ith node is an interior node, ϕ_i is zero over the entire boundary, B, and $Q_i = 0$.

Next, we multiply the left-hand side of (8.1.1) by ϕ_i, integrate the product over the solution domain, Ω, set the result equal to the right-hand side of (8.1.11), divide each term by $\rho \, c_p$, and thus derive the targeted Galerkin equation

$$\iiint_\Omega \phi_i \frac{\partial f}{\partial t} \, \mathrm{d}x \, \mathrm{d}y \, \mathrm{d}z + \iiint_\Omega \phi_i \, \mathbf{u} \cdot \nabla f \, \mathrm{d}x \, \mathrm{d}y \, \mathrm{d}z$$
$$= -\kappa \iiint_\Omega \nabla \phi_i \cdot \nabla f \, \mathrm{d}x \, \mathrm{d}y \, \mathrm{d}z + \frac{1}{\rho \, c_p} \left(Q_i + S_i \right), \quad (8.1.13)$$

which provides us with a basis for the Galerkin finite element method. The first term on the left-hand side involving the time derivative is zero at steady state.

8.1.4 *Galerkin finite element equations*

In the isoparametric interpolation, the requisite solution, f, is expressed in terms of *a priori* unknown values at the N_G global interpolation nodes, $f_j(t)$, and associated global cardinal interpolation functions, ϕ_j, as

$$f(x, y, z, t) = \sum_{j=1}^{N_\mathrm{G}} f_j(t) \, \phi_j(x, y, z). \quad (8.1.14)$$

Inserting this and a similar expansion for the source term, s, into (8.1.13), and recalling the definition of the boundary and source integrals Q_i and S_i in (8.1.12), we obtain

$$\sum_{j=1}^{N_G} M_{ij} \frac{\mathrm{d}f_j}{\mathrm{d}t} + \sum_{j=1}^{N_G} N_{ij}\, f_j = -\kappa \sum_{j=1}^{N_G} D_{ij}\, f_j + \frac{1}{\rho\, c_p} \iint_B \phi_i\, q \,\mathrm{d}S + \frac{1}{\rho\, c_p} \sum_{j=1}^{N_G} M_{ij}\, s_j,$$

$$(8.1.15)$$

where

$$D_{ij} \equiv \iiint_\Omega \nabla\phi_i \cdot \nabla\phi_j \,\mathrm{d}x\,\mathrm{d}y\,\mathrm{d}z \qquad (8.1.16)$$

is the global diffusion matrix,

$$M_{ij} \equiv \iiint_\Omega \phi_i\, \phi_j \,\mathrm{d}x\,\mathrm{d}y\,\mathrm{d}z \qquad (8.1.17)$$

is the global mass matrix, and

$$N_{ij} \equiv \iiint_\Omega \phi_i\, \mathbf{u} \cdot \nabla\phi_j \,\mathrm{d}x\,\mathrm{d}y\,\mathrm{d}z \qquad (8.1.18)$$

is the global advection matrix.

Ordinary differential equations

Applying (8.1.15) for the global interpolation functions associated with the interpolation nodes where the Dirichlet boundary condition is *not* prescribed, we obtain a system of ordinary differential equations (ODEs) for the unknown nodal values,

$$\mathbf{M} \cdot \frac{\mathrm{d}\mathbf{f}}{\mathrm{d}t} + \mathbf{N} \cdot \mathbf{f} = \kappa\,(-\mathbf{D} \cdot \mathbf{f} + \mathbf{b}), \qquad (8.1.19)$$

where the vector $\mathbf{b}$ on the right-hand side, with components

$$b_i \equiv \frac{1}{k}\left(\iint_B \phi_i\, q \,\mathrm{d}S + \sum_{j=1}^{N_G} M_{ij}\, s_j \right), \qquad (8.1.20)$$

incorporates the given source term and implements the prescribed Neumann boundary conditions. The first integral on the right-hand side of (8.1.20) is non-zero only if the ith node is a boundary node.

8.1.5   Element matrices

In practice, the global matrices are compiled from corresponding element matrices, which are defined in terms of the element interpolation functions, $\psi_i^{(l)}$, using a connectivity table, as discussed in previous chapters. The lth-element diffusion matrix is given by

$$A_{ij}^{(l)} \equiv \iiint_{E_l} \nabla\psi_i^{(l)} \cdot \nabla\psi_j^{(l)} \,\mathrm{d}x\,\mathrm{d}y\,\mathrm{d}z, \qquad (8.1.21)$$

the corresponding mass matrix is given by

$$B_{ij}^{(l)} \equiv \iiint_{E_l} \psi_i^{(l)}\, \psi_j^{(l)}\, \mathrm{d}x\, \mathrm{d}y\, \mathrm{d}z, \tag{8.1.22}$$

and the corresponding advection matrix is given by

$$C_{ij}^{(l)} \equiv \iiint_{E_l} \psi_i^{(l)}\, \mathbf{u} \cdot \nabla\psi_j^{(l)}\, \mathrm{d}x\, \mathrm{d}y\, \mathrm{d}z, \tag{8.1.23}$$

where $\mathbf{u}$ is the velocity field, E_l stands for the lth element, and the integration is performed over the element volume.

The assembly methodology hinges on the observation that these integrals are non-zero only if nodes i and j lie inside, at the faces, or at the vertices of the lth element. The assembly algorithm is identical to that shown in Table 4.1.1 for the corresponding problem in two dimensions.

8.1.6   Implementation of the Dirichlet boundary condition

The Dirichlet boundary condition is implemented using the algorithm shown in Table 4.1.2. subject to the convention that the entry $bcd(m)$ contains a prescribed value corresponding to the Dirichlet boundary condition at the position of the mth global node.

PROBLEM

8.1.1 *Flux across a spherical surface*

Departing from equation (8.1.6), show that the inward flux across a spherical boundary is given by $q(\mathbf{x}) = k\, \partial f/\partial r$, where r is the distance from the center. Explain on physical grounds why the flux is positive when $\partial f/\partial r > 0$, and negative otherwise.

8.2   Tetrahedral elements

Tetrahedral elements have four planar or curved *faces* intersecting at six straight or curved *edges* defined by the four element vertices.[1]   These elements are the counterparts of triangular elements with straight or curved edges in two dimensions discussed in Chapters 4–8.

---

[1] The word *tetrahedron* derives from the Greek word τετραεδρον, which consists of the words τεσσερα, meaning *four*, and εδρα, meaning *base* or *side*. Another composite Greek word is φιλοτιμο, which consists of the words φιλος, which means *friend*, and τιμη, which means *honor*. Linguists have discovered that *filotimo* is impossible to translate and can be understood only if honor is interpreted as a psychologically internalized yardstick of generosity, goodness, and integrity.

(a)   (b)

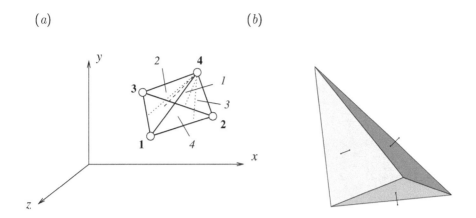

FIGURE 8.2.1   (*a*) A four-node tetrahedral element with six straight edges in physical
space. The node labels are printed in bold, and the face labels are printed in
italic. (*b*) Visualization of a tetrahedral element using the FSELIB script *tetra4*.
The short segments are perpendicular to the corresponding sides.

Four-node tetrahedra

The simplest tetrahedron has four planar faces and six straight edges defined by
four geometrical vertex nodes,

$$\mathbf{x}_i^{\mathrm{E}} = \left(x_i^{\mathrm{E}}, y_i^{\mathrm{E}}, z_i^{\mathrm{E}} \right) \tag{8.2.1}$$

for $i = 1, 2, 3, 4$, as shown in Figure 8.2.1(*a*). The superscript E emphasizes that
these are local or element nodes, which can be mapped to unique global nodes
through a connectivity table.

Face 1 is defined by nodes 2, 3, and 4, face 2 is defined by nodes 3, 1, and 4,
face 3 is defined by nodes 1, 2, and 4, and face 4 is defined by nodes 1, 2, and
3. The vertices 1, 2, and 3 are labeled such that the contour of face 4 defined by
these vertices is traced in the counterclockwise direction when the tetrahedron is
observed from the inside.

The first altitude of the tetrahedron is the minimum distance of vertex 4 from
the opposite face defined by vertices 1, 2, and 3, drawn with the dashed line in
Figure 8.2.1(*a*). The corresponding face altitudes are the minimum distances of
vertex 4 from the three edges defined by vertices 1, 2, and 3, drawn with the dotted
lines in Figure 8.2.1(*a*). Similar definitions apply to the other three vertices. If a
tetrahedron is regular, all altitudes and face altitudes are the same.

FSELIB script *tetra4*, listed in Table 8.2.1, defines the nodes of an arbitrary
tetrahedron and visualizes the element by painting the faces, as shown in Figure
8.2.1(*b*).

```
% vertex coordinates

x(1) = 0.0;  y(1) = 0.0;  z(1) = 0.0;
x(2) = 0.9;  y(2) = 0.2;  z(2) = 0.02;
x(3) = 0.1;  y(3) = 1.1;  z(3) =-0.05;
x(4) =-0.1;  y(4) =-0.1;  z(4) = 1.2;

% face node labels

s(1,1) = 2;  s(1,2) = 3;  s(1,3) = 4;  % first  face
s(2,1) = 3;  s(2,2) = 1;  s(2,3) = 4;  % second face
s(3,1) = 1;  s(3,2) = 2;  s(3,3) = 4;  % third  face
s(4,1) = 1;  s(4,2) = 2;  s(4,3) = 3;  % fourth face

% paint the faces

for i=1:4

  for j=1:3
   xp(j) = x(s(i,j));  yp(j) = y(s(i,j));  zp(j) = z(s(i,j));
  end

  if(i==1) patch(xp,yp,zp,'r'); end
  if(i==2) patch(xp,yp,zp,'b'); end
  if(i==3) patch(xp,yp,zp,'y'); end
  if(i==4) patch(xp,yp,zp,'g'); end

end
```

TABLE 8.2.1 Abbreviation of the FSELIB script *tetra4* for visualizing a four-node tetrahedron.

8.2.1 *Parametric representation*

To describe a four-node tetrahedral element in parametric form, we map it from
the physical xyz space to an orthogonal tetrahedron with three perpendicular faces
of equal size in the $\xi\eta\zeta$ space, as shown in Figure 8.2.2(a). The volume of the
parametric tetrahedron is equal to $1/6$. The mapping from the physical to the
parametric space is mediated by the function

$$\mathbf{x} = \mathbf{x}_1^E\,\psi_1(\xi,\eta,\zeta) + \mathbf{x}_2^E\,\psi_2(\xi,\eta,\zeta) + \mathbf{x}_3^E\,\psi_3(\xi,\eta,\zeta) + \mathbf{x}_4^E\,\psi_4(\xi,\eta,\zeta), \qquad (8.2.2)$$

where $\psi_i(\xi,\eta,\zeta)$ for $i = 1$–4 are element node interpolation functions satisfying
familiar cardinal properties: $\psi_i = 1$ at the ith element node, and $\psi_i = 0$ at the
other three element nodes.

Working as in Section 4.2 for three-node triangular elements in the plane, we
find that the tetrahedral element interpolation functions are given by

$$\begin{aligned} \psi_1(\xi,\eta,\zeta) &= \omega, & \psi_2(\xi,\eta,\zeta) &= \xi, \\ \psi_3(\xi,\eta,\zeta) &= \eta, & \psi_4(\xi,\eta,\zeta) &= \zeta, \end{aligned} \qquad (8.2.3)$$

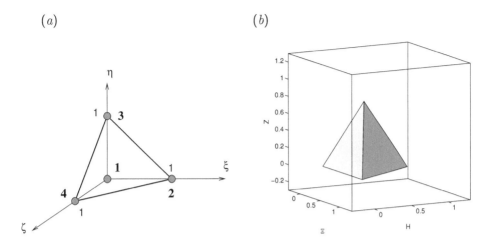

FIGURE 8.2.2   (*a*) The universal image of a four-node tetrahedral element with six
straight edges in parametric space. The node labels are printed in bold. (*b*)
Illustration of the corresponding standard regular tetrahedron.

where

$$w \equiv 1 - \xi - \eta - \zeta. \tag{8.2.4}$$

Physically, the coordinate ξ is the ratio of the volume of a tetrahedron having one
vertex at the field point $\mathbf{x}$ and three vertices at the tetrahedral vertices $(4, 3, 1)$, to
the volume of the entire tetrahedron,

$$\xi = \frac{V_2}{V}, \tag{8.2.5}$$

as shown in Figure 8.2.3. Similar interpretations apply to the rest of the tetrahedral
barycentric coordinates, η, ζ, and ω.

Substituting the element interpolation functions into (8.2.2), we derive a mapping function that is a *complete* linear function of ξ, η, and ζ, consisting of a
constant term, a term that is linear in ξ, a term that is linear in η, and a term that
is linear in ζ,

$$\mathbf{x} = \mathbf{x}_1^{E} + (\mathbf{x}_2^{E} - \mathbf{x}_1^{E})\,\xi + (\mathbf{x}_3^{E} - \mathbf{x}_1^{E})\,\eta + (\mathbf{x}_4^{E} - \mathbf{x}_1^{E})\,\zeta. \tag{8.2.6}$$

Explicitly, the x, y, and z components of the mapping are given by

$$\begin{aligned}
x &= x_1^{E} + (x_2^{E} - x_1^{E})\,\xi + (x_3^{E} - x_1^{E})\,\eta + (x_4^{E} - x_1^{E})\,\zeta, \\
y &= y_1^{E} + (y_2^{E} - y_1^{E})\,\xi + (y_3^{E} - y_1^{E})\,\eta + (y_4^{E} - y_1^{E})\,\zeta, \\
z &= z_1^{E} + (z_2^{F} - z_1^{E})\,\xi + (z_3^{E} - z_1^{E})\,\eta + (z_4^{E} - z_1^{E})\,\zeta.
\end{aligned} \tag{8.2.7}$$

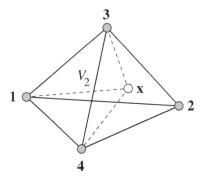

FIGURE 8.2.3   The barycentric coordinates are defined with respect to the volumes of
four sub-tetrahedra defined by a field point and each face.

For a specified location in physical space, (x, y, z), the corresponding barycentric
coordinates, (ξ, η, ζ), can be found in terms of the vertex coordinates by solving a
system of three linear equations originating from (8.2.7).

Mapping to a regular tetrahedron

The standard orthogonal tetrahedron in the $\xi \eta \zeta$ space can be mapped to a regular
tetrahedron with edges of unit length in the $\Xi H Z$ space, as shown in Figure 8.2.2(b).
The pertinent transformation rules are

$$\Xi = \xi + \frac{1}{2}(\xi + \zeta), \qquad H = \frac{\sqrt{3}}{2}\left(\eta + \frac{1}{3}\zeta\right), \qquad Z = \left(\frac{2}{3}\right)^{1/2}\zeta. \qquad (8.2.8)$$

The third rule shows that all altitudes of the regular octahedron are equal to
$(2/3)^{1/2}$.

8.2.2   *Integral over the volume of a tetrahedron*

The Jacobian matrix of the mapping from the physical xyz space to the parametric
$\xi \eta \zeta$ space is defined as

$$\mathbf{J} \equiv \begin{bmatrix} \dfrac{\partial x}{\partial \xi} & \dfrac{\partial x}{\partial \eta} & \dfrac{\partial x}{\partial \zeta} \\[2mm] \dfrac{\partial y}{\partial \xi} & \dfrac{\partial y}{\partial \eta} & \dfrac{\partial y}{\partial \zeta} \\[2mm] \dfrac{\partial z}{\partial \xi} & \dfrac{\partial z}{\partial \eta} & \dfrac{\partial z}{\partial \zeta} \end{bmatrix}. \qquad (8.2.9)$$

The determinant of the Jacobian matrix is the *volume metric coefficient*,

$$h_V \equiv \det(\mathbf{J}) = \left(\frac{\partial \mathbf{x}}{\partial \xi} \times \frac{\partial \mathbf{x}}{\partial \eta}\right) \cdot \frac{\partial \mathbf{x}}{\partial \zeta}. \qquad (8.2.10)$$

Substituting the linear mapping functions shown in (8.2.7), we find that

$$\frac{\partial \mathbf{x}}{\partial \xi} = \mathbf{x}_2^E - \mathbf{x}_1^E, \qquad \frac{\partial \mathbf{x}}{\partial \eta} = \mathbf{x}_3^E - \mathbf{x}_1^E, \qquad \frac{\partial \mathbf{x}}{\partial \zeta} = \mathbf{x}_4^E - \mathbf{x}_1^E, \qquad (8.2.11)$$

and calculate

$$\mathbf{J} = \begin{bmatrix} x_2^E - x_1^E & x_3^E - x_1^E & x_4^E - x_1^E \\ y_2^E - y_1^E & y_3^E - y_1^E & y_4^E - y_1^E \\ z_2^E - z_1^E & z_3^E - z_1^E & z_4^E - z_1^E \end{bmatrix}. \qquad (8.2.12)$$

Invoking the geometrical interpretation of the outer vector product in (8.2.10), as discussed in Appendix A, we find that h_V is equal to the volume of a parallelepiped with three sides along the vectors $\mathbf{x}_2^E - \mathbf{x}_1^E$, $\mathbf{x}_3^E - \mathbf{x}_1^E$, and $\mathbf{x}_4^E - \mathbf{x}_1^E$, which is equal to six times the volume of the tetrahedron in the physical xyz space. Thus, the volume metric coefficient of the tetrahedron is

$$h_V = 6V, \qquad (8.2.13)$$

where V is the volume of the tetrahedron in the physical xyz space.

Using elementary calculus, we find that the integral of a function $f(x, y, z)$ over the volume of the physical tetrahedron in the xyz space can be expressed as an integral over the parametric tetrahedron in the $\xi\eta\zeta$ space as

$$\iiint f(x, y, z) \, dx \, dy \, dz = \iiint f(\xi, \eta, \zeta) \, h_V \, d\xi \, d\eta \, d\zeta, \qquad (8.2.14)$$

yielding

$$\iiint f(x, y, z) \, dx \, dy \, dz = 6V \iiint f(\xi, \eta, \zeta) \, d\xi \, d\eta \, d\zeta \qquad (8.2.15)$$

or

$$\iiint f(x, y, z) \, dx \, dy \, dz = 6V \int_0^1 \left[\int_0^{1-\zeta} \left(\int_0^{1-\zeta-\xi} f(\xi, \eta, \zeta) \, d\eta \right) d\xi \right] d\zeta. \qquad (8.2.16)$$

The integral in parametric space can be computed analytically when the function f is a polynomial, or numerically under more general circumstances, as discussed in Section 8.6.5.

8.2.3 Element subdivision into eight tetrahedra

A tetrahedron can be subdivided into eight descendant tetrahedra defined with respect to the edge mid-points, as shown in Figure 8.2.4(a).

The subdivision algorithm is implemented in the FSELIB function *tetra4_sub8*, listed in Table 8.2.2. The input to this function includes the coordinates of the

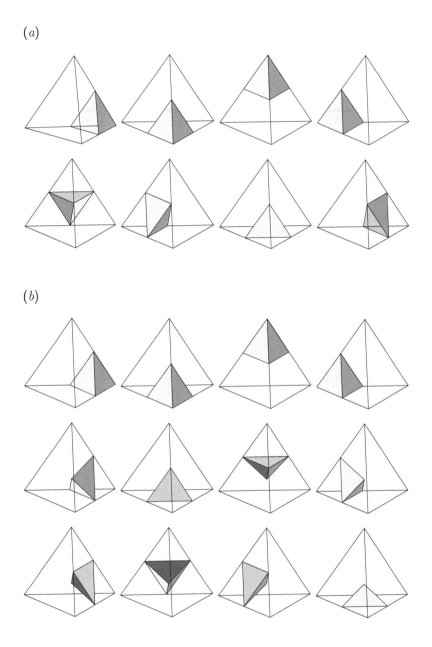

FIGURE 8.2.4 Subdivision of a regular tetrahedron into (*a*) 8 or (*b*) 12 descendant tetrahedra performed by the FSELIB function *tetra4_sub8* or *tetra4_sub12*.

```
function [x1,y1,z1, x2,y2,z2, x3,y3,z3, x4,y4,z4 ...
         ,x5,y5,z5, x6,y6,z6, x7,y7,z7, x8,y8,z8 ]...
         ...
         = tetra4_sub8 (x,y,z)

%========================
% tetra4_sub8
%
% subdivide a tetrahedron
% into eight sub-elements
%========================

% mid-edge nodes

xe12 = 0.5*(x(1)+x(2)); ye12 = 0.5*(y(1)+y(2)); ze12 = 0.5*(z(1)+z(2));
xe13 = 0.5*(x(1)+x(3)); ye13 = 0.5*(y(1)+y(3)); ze13 = 0.5*(z(1)+z(3));
xe14 = 0.5*(x(1)+x(4)); ye14 = 0.5*(y(1)+y(4)); ze14 = 0.5*(z(1)+z(4));
xe23 = 0.5*(x(2)+x(3)); ye23 = 0.5*(y(2)+y(3)); ze23 = 0.5*(z(2)+z(3));
xe24 = 0.5*(x(2)+x(4)); ye24 = 0.5*(y(2)+y(4)); ze24 = 0.5*(z(2)+z(4));
xe34 = 0.5*(x(3)+x(4)); ye34 = 0.5*(y(3)+y(4)); ze34 = 0.5*(z(3)+z(4));

%------------------
% first tetrahedron
%------------------

x1(1) = xe13; y1(1) = ye13; z1(1) = ze13;
x1(2) = xe23; y1(2) = ye23; z1(2) = ze23;
x1(3) = x(3); y1(3) = y(3); z1(3) = z(3);
x1(4) = xe34; y1(4) = ye34; z1(4) = ze34;

%------------------
% second tetrahedron
%------------------

x2(1) = xe12; y2(1) = ye12; z2(1) = ze12;
x2(2) = x(2); y2(2) = y(2); z2(2) = z(2);
x2(3) = xe23; y2(3) = ye23; z2(3) = ze23;
x2(4) = xe24; y2(4) = ye24; z2(4) = ze24;

%------------------
% third tetrahedron
%------------------

x3(1) = xe14; y3(1) = ye14; z3(1) = ze14;
x3(2) = xe24; y3(2) = ye24; z3(2) = ze24;
x3(3) = xe34; y3(3) = ye34; z3(3) = ze34;
x3(4) = x(4); y3(4) = y(4); z3(4) = z(4);

%------------------
% fourth tetrahedron
%------------------

x4(1) = x(1); y4(1) = y(1); z4(1) = z(1);
x4(2) = xe12; y4(2) = ye12; z4(2) = ze12;
```

TABLE 8.2.2   Function *tetra4_sub8* (Continuing →)

```
x4(3) = xe13; y4(3) = ye13; z4(3) = ze13;
x4(4) = xe14; y4(4) = ye14; z4(4) = ze14;

%-------------------
% fifth tetrahedron
%-------------------

x5(1) = xe34; y5(1) = ye34; z5(1) = ze34;
x5(2) = xe24; y5(2) = ye24; z5(2) = ze24;
x5(3) = xe14; y5(3) = ye14; z5(3) = ze14;
x5(4) = xe13; y5(4) = ye13; z5(4) = ze13;

%-------------------
% sixth tetrahedron
%-------------------

x6(1) = xe14; y6(1) = ye14; z6(1) = ze14;
x6(2) = xe12; y6(2) = ye12; z6(2) = ze12;
x6(3) = xe13; y6(3) = ye13; z6(3) = ze13;
x6(4) = xe24; y6(4) = ye24; z6(4) = ze24;

%-------------------
% seventh tetrahedron
%-------------------

x7(1) = xe12; y7(1) = ye12; z7(1) = ze12;
x7(2) = xe23; y7(2) = ye23; z7(2) = ze23;
x7(3) = xe13; y7(3) = ye13; z7(3) = ze13;
x7(4) = xe24; y7(4) = ye24; z7(4) = ze24;

%-------------------
% eighth tetrahedron
%-------------------

x8(1) = xe13; y8(1) = ye13; z8(1) = ze13;
x8(2) = xe24; y8(2) = ye24; z8(2) = ze24;
x8(3) = xe23; y8(3) = ye23; z8(3) = ze23;
x8(4) = xe34; y8(4) = ye34; z8(4) = ze34;

%-----
% done
%-----

return
```

TABLE 8.2.2 ($\rightarrow$ Continued) Function *tetra4_sub8* subdivides a four-node tetrahedron into eight descendant elements.

nodes of the original subdivided tetrahedron. The output includes the coordinates
of the nodes of the eight tetrahedra arising from the subdivision. The volumes of
all eight descendant tetrahedra are equal.

FSELIB function *tetra4_sub8_dr*, listed in Table 8.2.3, drives the subdivision
function and visualizes the elements using the FSELIB function *tetra4_vis*, listed in
Table 8.2.4.

8.2.4   Element subdivision into 12 tetrahedra

A tetrahedron may also be divided into 12 descendant tetrahedra defined with
respect to the edge mid-points and the element centroid, as shown in Figure 8.2.4(*b*).
The sub-elements in the first row of this figure have two faces on the faces of the
parental tetrahedron; the sub-elements in the second row have no faces on the faces
of the parental tetrahedron; the sub-elements in the third row have one face on the
faces of the parental tetrahedron. The volumes of the 12 descendant tetrahedra are
not all equal.

The subdivision algorithm is implemented in the FSELIB function *tetra4_sub12*,
listed in Table 8.2.5. The input to this function includes the coordinates of the
nodes of the original subdivided tetrahedron. The output includes the coordinates of the nodes of the 12 tetrahedra arising from the subdivision. FSELIB code
tetra4_sub12_dr, not listed in the text, drives the subdivision and visualizes the elements using the FSELIB function *tetra4_vis*, listed in Table 8.2.4. The layout of
the code *tetra4_sub12_dr* is similar to that of the code *tetra4_sub8_dr*, listed in Table
8.2.3.

8.2.5   Isoparametric interpolation

In the isoparametric interpolation, a function of interest defined over the tetrahedron is expressed in a form that is analogous to that shown in (8.2.2) for the position
vector,

$$f(x, y, z, t) = \sum_{i=1}^{4} f_i^{\mathrm{E}}(t)\, \psi_i(\xi, \eta, \zeta), \qquad (8.2.17)$$

with the understanding that the point $\mathbf{x} = (x, y, z)$ is mapped to the parametric
point (ξ, η, ζ), and *vice versa*, through (8.2.2).

8.2.6   Element diffusion matrix

The *l*th-element diffusion matrix for a tetrahedron is given by

$$A_{ij}^{(l)} \equiv \iiint_{E_l} \nabla \psi_i \cdot \nabla \psi_j \, \mathrm{d}x \, \mathrm{d}y \, \mathrm{d}z = 6\, V \iiint \nabla \psi_i \cdot \nabla \psi_j \, \mathrm{d}\xi \, \mathrm{d}\eta \, \mathrm{d}\zeta \qquad (8.2.18)$$

```
%===============================
% tetra4_sub8_dr
%
% driver for subdividing
% a tetrahedron into 8 elements
%===============================

% constants

wait = 0.5;

iopt = 1;
iopt = 2;

%---
if(iopt==1) % canonical vertex coordinates:
%---

  x(1) = 0.0; y(1) = 0.0; z(1)= 0.0;
  x(2) = 1.0; y(2) = 0.0; z(2)= 0.0;
  x(3) = 0.0; y(3) = 1.0; z(3)= 0.0;
  x(4) = 0.0; y(4) = 0.0; z(4)= 1.0;

  % canonical regular tetrahedron:

  for j=1:4
    x(j) = x(j)+0.5*(y(j)+z(j));
    y(j) = 0.5*sqrt(3)*(y(j)+z(j)/3.0);
    z(j) = sqrt(2/3)*z(j);
  end

%---
elseif(iopt==2) % arbitrary vertex coordinates:
%---

  x(1) = 0.0; y(1) = 0.1; z(1) = 0.0;
  x(2) = 0.9; y(2) = 0.2; z(2) = 0.5;
  x(3) = 0.1; y(3) = 1.1; z(3) =-0.05;
  x(4) =-0.1; y(4) =-0.1; z(4) = 1.2;

end

%---
% subdivide
%---


  [x1,y1,z1, x2,y2,z2, x3,y3,z3, x4,y4,z4 ...
  ,x5,y5,z5, x6,y6,z6, x7,y7,z7, x8,y8,z8 ] ...
  ...
  = tetra4_sub8(x,y,z);
```

TABLE 8.2.3   Function *tetra4_sub8_dr* (Continuing →)

```
%----------
% visualize
%----------

 f=x; f1=x; f2=x; f3=x; f4=x; f5=x; f6=x; f7=x; f8=x;

%----------
% plot
%----------

 figure(88)
 hold on
 axis equal

% display the parental element

 option = 2; % wireframe
 volume = tetra4_vis (x,y,z,f,option);

% display the eight sub-elements

 option = 1; % in color

 for i=1:8
   vol(i) = 0.0;
 end

 vol(8) = tetra4_vis (x8, y8, z8, f8, option);pause(wait);
 vol(7) = tetra4_vis (x7, y7, z7, f7, option);pause(wait);
 vol(6) = tetra4_vis (x6, y6, z6, f6, option);pause(wait);
 vol(5) = tetra4_vis (x5, y5, z5, f5, option);pause(wait);
 vol(4) = tetra4_vis (x4, y4, z4, f4, option);pause(wait);
 vol(3) = tetra4_vis (x3, y3, z3, f3, option);pause(wait);
 vol(2) = tetra4_vis (x2, y2, z2, f2, option);pause(wait);
 vol(1) = tetra4_vis (x1, y1, z1, f1, option);pause(wait);

%---
 end
%---

%--------
% volumes
%--------

 vlm = 0.0;
 for i=1:8
  vlm = vlm+vol(i);
 end

[vlm volume] % should be the same
```

TABLE 8.2.3   ($\rightarrow$ Continued) Code *tetra4_sub8_dr* drives the function *tetra4_sub8* to subdivide a four-node tetrahedron into eight descendant elements.

```
function volume = tetra4_vis (x,y,z,f,option)

%======================================
% visualize a tetrahedron defined by
% four vertices hosted in  x, y, z
%
% option = 1,2,3,4,5
%======================================

%---
% face node labels
%---

s(1,1) = 2; s(1,2) = 3; s(1,3) = 4;    % face 1
s(2,1) = 3; s(2,2) = 1; s(2,3) = 4;    % face 2
s(3,1) = 1; s(3,2) = 2; s(3,3) = 4;    % face 3
s(4,1) = 1; s(4,2) = 2; s(4,3) = 3;    % face 4

%---
% paint the faces
%---

%---
 for i=1:4  % closes at arkouda
%---

   for j=1:3
    xp(j) = x(s(i,j));
    yp(j) = y(s(i,j));
    zp(j) = z(s(i,j));
    fp(j) = f(s(i,j));
   end

   xp(4) = xp(1); yp(4) = yp(1); zp(4) = zp(1); fp(4) = fp(1);

   %---
    if(option==1)   % solid
   %---

     if(i==1) patch (xp,yp,zp,'r'); end
     if(i==2) patch (xp,yp,zp,'b'); end
     if(i==3) patch (xp,yp,zp,'y'); end
     if(i==4) patch (xp,yp,zp,'g'); end

   %---
    elseif(option==2)    % wireframe
   %---

     plot3 (xp,yp,zp,'k');

   %---
    elseif(option==3)  % graded
   %---
```

TABLE 8.2.4   Function *tetra4_vis* (Continuing →)

```
      cp(1) = 0.5+0.3*xp(1);
      cp(2) = 0.5+0.3*xp(2);
      cp(3) = 0.5+0.3*xp(3);
      cp(4) = 0.5+0.3*xp(4);

      cp(1) = 0.5+0.3*zp(1);
      cp(2) = 0.5+0.3*zp(2);
      cp(3) = 0.5+0.3*zp(3);
      cp(4) = 0.5+0.3*zp(4);

      patch(xp,yp,zp,cp);

   %---
    elseif(option==4)  % graded
   %---

      patch (xp,yp,zp,fp);

   %---
    elseif(option==5)   % solid white
   %---

      patch (xp,yp,zp,'w');

   %---
    end
   %---

%---
 end   % of arkouda
%---

%-------------------
% compute the volume
%-------------------

matrix = [1 1 1 1;
          x(1) x(2) x(3) x(4); ...
          y(1) y(2) y(3) y(4); ...
          z(1) z(2) z(3) z(4)];

volume = det(matrix)/6.0;

%-----
% done
%-----

return
```

TABLE 8.2.4   ($\rightarrow$ Continued) Function *tetra4_vis* visualizes a four-node tetrahedron.

```
function [x1,y1,z1, x2,y2,z2, x3,y3,z3, x4,y4,z4 ...
         ,x5,y5,z5, x6,y6,z6, x7,y7,z7, x8,y8,z8 ...
         ,x9,y9,z9, x10,y10,z10, x11,y11,z11, x12,y12,z12] ...
          ...
         = tetra4_sub12 (x,y,z)

%=========================
% tetra4_sub12
%
% subdivide a tetrahedron
% into 12 sub-elements
%=========================

% edge mid-nodes

xe12 = 0.5*(x(1)+x(2)); ye12 = 0.5*(y(1)+y(2)); ze12 = 0.5*(z(1)+z(2));
xe13 = 0.5*(x(1)+x(3)); ye13 = 0.5*(y(1)+y(3)); ze13 = 0.5*(z(1)+z(3));
xe14 = 0.5*(x(1)+x(4)); ye14 = 0.5*(y(1)+y(4)); ze14 = 0.5*(z(1)+z(4));
xe23 = 0.5*(x(2)+x(3)); ye23 = 0.5*(y(2)+y(3)); ze23 = 0.5*(z(2)+z(3));
xe24 = 0.5*(x(2)+x(4)); ye24 = 0.5*(y(2)+y(4)); ze24 = 0.5*(z(2)+z(4));
xe34 = 0.5*(x(3)+x(4)); ye34 = 0.5*(y(3)+y(4)); ze34 = 0.5*(z(3)+z(4));

% element centroid

xc = 0.25*(x(1)+x(2)+x(3)+x(4));
yc = 0.25*(y(1)+y(2)+y(3)+y(4));
zc = 0.25*(z(1)+z(2)+z(3)+z(4));

%------------------
% first tetrahedron
%------------------

x1(1) = xe13; y1(1) = ye13; z1(1) = ze13;
x1(2) = xe23; y1(2) = ye23; z1(2) = ze23;
x1(3) = x(3); y1(3) = y(3); z1(3) = z(3);
x1(4) = xe34; y1(4) = ye34; z1(4) = ze34;

%------------------
% second tetrahedron
%------------------

x2(1) = xe12; y2(1) = ye12; z2(1) = ze12;
x2(2) = x(2); y2(2) = y(2); z2(2) = z(2);
x2(3) = xe23; y2(3) = ye23; z2(3) = ze23;
x2(4) = xe24; y2(4) = ye24; z2(4) = ze24;

%------------------
% third tetrahedron
%------------------

x3(1) = xe14; y3(1) = ye14; z3(1) = ze14;
x3(2) = xe24; y3(2) = ye24; z3(2) = ze24;
x3(3) = xe34; y3(3) = ye34; z3(3) = ze34;
x3(4) = x(4); y3(4) = y(4); z3(4) = z(4);
```

TABLE 8.2.5   Function *tetra4_sub12* (Continuing →)

```
%------------------
% fourth tetrahedron
%------------------

x4(1) = x(1);  y4(1) = y(1);  z4(1) = z(1);
x4(2) = xe12;  y4(2) = ye12;  z4(2) = ze12;
x4(3) = xe13;  y4(3) = ye13;  z4(3) = ze13;
x4(4) = xe14;  y4(4) = ye14;  z4(4) = ze14;

%------------------
% fifth tetrahedron
%------------------

x5(1) = xe13;  y5(1) = ye13;  z5(1) = ze13;
x5(2) = xe23;  y5(2) = ye23;  z5(2) = ze23;
x5(3) = xe34;  y5(3) = ye34;  z5(3) = ze34;
x5(4) = xc;    y5(4) = yc;    z5(4) = zc;

%------------------
% sixth tetrahedron
%------------------

x6(1) = xe12;  y6(1) = ye12;  z6(1) = ze12;
x6(2) = xe24;  y6(2) = ye24;  z6(2) = ze24;
x6(3) = xe23;  y6(3) = ye23;  z6(3) = ze23;
x6(4) = xc;    y6(4) = yc;    z6(4) = zc;

%-------------------
% seventh tetrahedron
%-------------------

x7(1) = xe14;  y7(1) = ye14;  z7(1) = ze14;
x7(2) = xe34;  y7(2) = ye34;  z7(2) = ze34;
x7(3) = xe24;  y7(3) = ye24;  z7(3) = ze24;
x7(4) = xc;    y7(4) = yc;    z7(4) = zc;

%-------------------
% eighth tetrahedron
%-------------------

x8(1) = xe14;  y8(1) = ye14;  z8(1) = ze14;
x8(2) = xe12;  y8(2) = ye12;  z8(2) = ze12;
x8(3) = xe13;  y8(3) = ye13;  z8(3) = ze13;
x8(4) = xc;    y8(4) = yc;    z8(4) = zc;

%-------------------
% ninth tetrahedron
%------------------

x9(1) = xe24;  y9(1) = ye24;  z9(1) = ze24;
x9(2) = xe34;  y9(2) = ye34;  z9(2) = ze34;
x9(3) = xe23;  y9(3) = ye23;  z9(3) = ze23;
x9(4) = xc;    y9(4) = yc;    z9(4) = zc;
```

TABLE 8.2.5   Function *tetra4_sub12* ($\rightarrow$ Continuing $\rightarrow$)

```
%-------------------
% tenth tetrahedron
%-------------------

x10(1) = xe14; y10(1) = ye14; z10(1) = ze14;
x10(2) = xe13; y10(2) = ye13; z10(2) = ze13;
x10(3) = xe34; y10(3) = ye34; z10(3) = ze34;
x10(4) = xc;   y10(4) = yc;   z10(4) = zc;

%-------------------
% eleventh tetrahedron
%-------------------

x11(1) = xe14; y11(1) = ye14; z11(1) = ze14;
x11(2) = xe24; y11(2) = ye24; z11(2) = ze24;
x11(3) = xe12; y11(3) = ye12; z11(3) = ze12;
x11(4) = xc;   y11(4) = yc;   z11(4) = zc;

%-------------------
% twelfth tetrahedron
%-------------------

x12(1) = xe12; y12(1) = ye12; z12(1) = ze12;
x12(2) = xe23; y12(2) = ye23; z12(2) = ze23;
x12(3) = xe13; y12(3) = ye13; z12(3) = ze13;
x12(4) = xc;   y12(4) = yc;   z12(4) = zc;

%-----
% done
%-----

return
```

TABLE 8.2.5 (Continued $\rightarrow$) Function *tetra4_sub12* subdivides a four-node tetrahedron into 12 descendant elements.

for $i, j = 1$–4, where

$$\nabla \psi_i = \left(\frac{\partial \psi_i}{\partial x}, \quad \frac{\partial \psi_i}{\partial y}, \quad \frac{\partial \psi_i}{\partial z} \right) \tag{8.2.19}$$

is the three-dimensional gradient.

Computation of the gradient

The gradient of an element node interpolation function, $\nabla \psi_i$ for $i = 1, 2, 3, 4$, can be computed readily using the relations

$$\frac{\partial \mathbf{x}}{\partial \xi} \cdot \nabla \psi_i = \frac{\partial \psi_i}{\partial \xi}, \qquad \frac{\partial \mathbf{x}}{\partial \eta} \cdot \nabla \psi_i = \frac{\partial \psi_i}{\partial \eta}, \qquad \frac{\partial \mathbf{x}}{\partial \zeta} \cdot \nabla \psi_i = \frac{\partial \psi_i}{\partial \zeta}. \tag{8.2.20}$$

The first relation states that the projection of the gradient vector on the ξ-line vector, $\partial \mathbf{x}/\partial \xi$, is the partial derivative with respect to ξ. Similar statements can be made for η and ζ. Explicitly, the equations in (8.2.20) read

$$\frac{\partial x}{\partial \xi}\frac{\partial \psi_i}{\partial x} + \frac{\partial y}{\partial \xi}\frac{\partial \psi_i}{\partial y} + \frac{\partial z}{\partial \xi}\frac{\partial \psi_i}{\partial z} = \frac{\partial \psi_i}{\partial \xi},$$

$$\frac{\partial x}{\partial \eta}\frac{\partial \psi_i}{\partial x} + \frac{\partial y}{\partial \eta}\frac{\partial \psi_i}{\partial y} + \frac{\partial z}{\partial \eta}\frac{\partial \psi_i}{\partial z} = \frac{\partial \psi_i}{\partial \eta}, \qquad (8.2.21)$$

$$\frac{\partial x}{\partial \zeta}\frac{\partial \psi_i}{\partial x} + \frac{\partial y}{\partial \zeta}\frac{\partial \psi_i}{\partial y} + \frac{\partial z}{\partial \zeta}\frac{\partial \psi_i}{\partial z} = \frac{\partial \psi_i}{\partial \zeta}.$$

Collecting these equations into a linear system, we obtain

$$\mathbf{J}^T \cdot \nabla \psi_i = \left[\frac{\partial \psi_i}{\partial \xi}, \frac{\partial \psi_i}{\partial \eta}, \frac{\partial \psi_i}{\partial \zeta} \right]^T, \qquad (8.2.22)$$

where

$$\mathbf{J}^T \equiv \begin{bmatrix} \dfrac{\partial x}{\partial \xi} & \dfrac{\partial y}{\partial \xi} & \dfrac{\partial z}{\partial \xi} \\[6pt] \dfrac{\partial x}{\partial \eta} & \dfrac{\partial y}{\partial \eta} & \dfrac{\partial z}{\partial \eta} \\[6pt] \dfrac{\partial x}{\partial \zeta} & \dfrac{\partial y}{\partial \zeta} & \dfrac{\partial z}{\partial \zeta} \end{bmatrix} \qquad (8.2.23)$$

is the transpose of the Jacobian matrix defined in (8.2.9).

The determinant of $\mathbf{J}^T$ is equal to the determinant of $\mathbf{J}$, which is equal to the volume metric coefficient,

$$\det(\mathbf{J}) = \det(\mathbf{J}^T) = h_V = 6V. \qquad (8.2.24)$$

Using the expressions in (8.2.11), we find that the equations in (8.2.20) take the specific form

$$(\mathbf{x}_2^E - \mathbf{x}_1^E) \cdot \nabla \psi_i = \frac{\partial \psi_i}{\partial \xi}, \qquad (\mathbf{x}_3^E - \mathbf{x}_1^E) \cdot \nabla \psi_i = \frac{\partial \psi_i}{\partial \eta},$$

$$(\mathbf{x}_4^E - \mathbf{x}_1^E) \cdot \nabla \psi_i = \frac{\partial \psi_i}{\partial \zeta}. \qquad (8.2.25)$$

Accordingly,

$$\mathbf{J}^T = \begin{bmatrix} x_2^E - x_1^E & y_2^E - y_1^E & z_2^E - z_1^E \\ x_3^E - x_1^E & y_3^E - y_1^E & z_3^E - z_1^E \\ x_4^E - x_1^E & y_4^E - y_1^E & z_4^E - z_1^E \end{bmatrix}, \qquad (8.2.26)$$

in agreement with (8.2.9). Substituting the specific expressions of the cardinal interpolation functions given in (8.2.3), we derive the linear systems

$$\mathbf{J}^T \cdot \nabla \psi_1 = -\begin{bmatrix} 1 \\ 1 \\ 1 \end{bmatrix}, \qquad \mathbf{J}^T \cdot \nabla \psi_2 = \begin{bmatrix} 1 \\ 0 \\ 0 \end{bmatrix} \qquad (8.2.27)$$

for the first and second element nodes, and

$$\mathbf{J}^T \cdot \nabla \psi_3 = \begin{bmatrix} 0 \\ 1 \\ 0 \end{bmatrix}, \qquad \mathbf{J}^T \cdot \nabla \psi_4 = \begin{bmatrix} 0 \\ 0 \\ 1 \end{bmatrix} \qquad (8.2.28)$$

for the third and fourth element nodes. Applying Cramer's rule, we obtain the solutions

$$\nabla \psi_1 = \frac{1}{6V} \begin{bmatrix} -y_{32}\, z_{42} + z_{32}\, y_{42} \\ x_{32}\, z_{42} - z_{32}\, x_{42} \\ -x_{32}\, y_{42} + y_{32}\, x_{42} \end{bmatrix}, \qquad \nabla \psi_2 = \frac{1}{6V} \begin{bmatrix} y_{31}\, z_{41} - z_{31}\, y_{41} \\ -x_{31}\, z_{41} + z_{31}\, x_{41} \\ x_{31}\, y_{41} - y_{31}\, x_{41} \end{bmatrix}$$

$$(8.2.29)$$

for the first and second element nodes, and

$$\nabla \psi_3 = \frac{1}{6V} \begin{bmatrix} -y_{21}\, z_{41} + z_{21}\, y_{41} \\ x_{21}\, z_{41} - z_{21}\, x_{41} \\ -x_{21}\, y_{41} + z_{21}\, y_{41} \end{bmatrix}, \qquad \nabla \psi_4 = \frac{1}{6V} \begin{bmatrix} y_{21}\, z_{31} - z_{21}\, y_{31} \\ -x_{21}\, z_{31} + z_{21}\, x_{31} \\ x_{21}\, y_{31} - y_{21}\, x_{31} \end{bmatrix},$$

$$(8.2.30)$$

for the third and fourth element nodes, where

$$\mathbf{x}_{21} = \mathbf{x}_2^{\mathrm{E}} - \mathbf{x}_1^{\mathrm{E}}, \qquad \mathbf{x}_{31} = \mathbf{x}_3^{\mathrm{E}} - \mathbf{x}_1^{\mathrm{E}}, \qquad \mathbf{x}_{41} = \mathbf{x}_4^{\mathrm{E}} - \mathbf{x}_1^{\mathrm{E}}, \qquad (8.2.31)$$

and

$$\mathbf{x}_{32} = \mathbf{x}_3^{\mathrm{E}} - \mathbf{x}_2^{\mathrm{E}}, \qquad \mathbf{x}_{42} = \mathbf{x}_4^{\mathrm{E}} - \mathbf{x}_2^{\mathrm{E}}. \qquad (8.2.32)$$

The gradient $\nabla \psi_1$ is perpendicular to the face 234, the gradient $\nabla \psi_2$ is perpendicular to the face 314, the gradient $\nabla \psi_3$ is perpendicular to the face 124, and the gradient $\nabla \psi_4$ is perpendicular to the face 132.

The expressions in (8.2.29) and (8.2.30) are used to evaluate the element diffusion and advection matrices.

Computation of the element diffusion matrix

Because the gradient of the element interpolation functions is a vectorial constant independent of position inside the element, the element diffusion matrix defined in (8.2.18) simplifies to

$$A_{ij}^{(l)} = V \, \nabla \psi_i \cdot \nabla \psi_j \qquad (8.2.33)$$

for $i, j = 1$–4, where V is the element volume in physical space. Consequently, the element diffusion matrix can be computed by direct evaluation using the expressions in (8.2.29) and (8.2.30). The computation is performed by the FSELIB function *edm_t4*, listed in Table 8.2.6.

```
function [edm, vol] = edm_t4 (x1,y1,z1,x2,y2,z2 ...
                             ,x3,y3,z3,x4,y4,z4)

%==========================================
% Evaluation of the element diffusion matrix
% and element volume
% for a four-node tetrahedron
%==========================================

%--------
% prepare
%--------

x21 = x2-x1; y21 = y2-y1; z21 = z2-z1;
x31 = x3-x1; y31 = y3-y1; z31 = z3-z1;
x41 = x4-x1; y41 = y4-y1; z41 = z4-z1;

x32 = x3-x2; y32 = y3-y2; z32 = z3-z2;
x42 = x4-x2; y42 = y4-y2; z42 = z4-z2;

%----------------
% Jacobian matrix
%----------------

Jac(1,1) = x21; Jac(1,2) = x31; Jac(1,3) = x41;
Jac(2,1) = y21; Jac(2,2) = y31; Jac(2,3) = y41;
Jac(3,1) = z21; Jac(3,2) = z31; Jac(3,3) = z41;

%-------
% volume
%-------

vol = det(Jac)/6.0;

%-------
% psi gradients
%-------

V6 = 6.0*vol;

% first gradient

gx(1) = (-y32*z42 + z32*y42)/V6;
gy(1) = ( x32*z42 - z32*x42)/V6;
gz(1) = (-x32*y42 + y32*x42)/V6;

% second gradient

gx(2) = ( y31*z41 - z31*y41)/V6;
gy(2) = (-x31*z41 + z31*x41)/V6;
gz(2) = ( x31*y41 - y31*x41)/V6;

% third gradient

gx(3) = (-y21*z41 + z21*y41)/V6;
gy(3) = ( x21*z41 - z21*x41)/V6;
gz(3) = (-x21*y41 + y21*x41)/V6;
```

TABLE 8.2.6   Function *edm_t4* (Continuing →)

```
% fourth gradient

gx(4) = ( y21*z31 - z21*y31)/V6;
gy(4) = (-x21*z31 + z21*x31)/V6;
gz(4) = ( x21*y31 - y21*x31)/V6;

% element diffusion matrix

for i=1:4
 for j=1:4
  edm(i,j) = V*( gx(i)*gx(j)+gy(i)*gy(j)+gz(i)*gz(j) );
 end
end

%-----
% done
%-----

return;
```

TABLE 8.2.6   (Continued →) Function *edm_t4* evaluates the element diffusion matrix for a four-node tetrahedron.

8.2.7   *Element mass matrix*

The lth element mass matrix is given by

$$B_{ij}^{(l)} = \iiint_{E_l} \psi_i \psi_j \, dx \, dy \, dz = 6V \iiint \psi_i \psi_j \, d\xi \, d\eta \, d\zeta. \tag{8.2.34}$$

Substituting the expressions in (8.2.3), we obtain

$$\mathbf{B}^{(l)} = 6V \iiint \begin{bmatrix} \omega^2 & \omega\xi & \omega\eta & \omega\zeta \\ \xi\omega & \xi^2 & \xi\eta & \xi\zeta \\ \eta\omega & \eta\xi & \eta^2 & \eta\zeta \\ \zeta\omega & \zeta\xi & \zeta\eta & \zeta^2 \end{bmatrix} d\xi \, d\eta \, d\zeta. \tag{8.2.35}$$

To compute these integrals, we use the integration formula

$$\iiint \omega^p \xi^q \eta^r \zeta^s \, d\xi \, d\eta \, d\zeta = \frac{p! \, q! \, r! \, s!}{(p+q+r+s+3)!}, \tag{8.2.36}$$

proved in Section 8.2.8, where p, q, r and s are non-negative integers and an exclamation mark denotes the factorial of an integer, $m! = 1 \cdot 2 \cdots m$. A straightforward calculation yields

$$\mathbf{B}^{(l)} = V \frac{1}{20} \begin{bmatrix} 2 & 1 & 1 & 1 \\ 1 & 2 & 1 & 1 \\ 1 & 1 & 2 & 1 \\ 1 & 1 & 1 & 2 \end{bmatrix}. \tag{8.2.37}$$

FSELIB function *emm_t4*, listed in Table 8.2.7, evaluates the element mass matrix based in these expressions.

In the case of the four-node tetrahedron presently discussed, but not more generally, the element mass matrix depends only on the element volume, V, and is independent of the element shape. The sum of all entries of the element mass matrix is equal to the volume of the tetrahedron, V.

The mass-lumped, diagonal element mass matrix, denoted by a hat, is given by

$$\widehat{\mathbf{B}}^{(l)} = V \frac{1}{4} \begin{bmatrix} 1 & 0 & 0 & 0 \\ 0 & 1 & 0 & 0 \\ 0 & 0 & 1 & 0 \\ 0 & 0 & 0 & 1 \end{bmatrix}. \tag{8.2.38}$$

The trace of the lumped element mass matrix is equal to the volume of the tetrahedron.

8.2.8 *Proof of the integration formula (8.2.36)*

To prove the integration formula (8.2.36), we define, for convenience,

$$\mathcal{I}_{pqrs} \equiv \iiint \omega^p \, \xi^q \, \eta^r \, \zeta^s \, \mathrm{d}\xi \, \mathrm{d}\eta \, \mathrm{d}\zeta. \tag{8.2.39}$$

Using (8.2.16), we obtain

$$\mathcal{I}_{pqrs} = \int_0^1 \left[\int_0^{1-\zeta} \left(\int_0^{1-\zeta-\xi} (1 - \zeta - \xi - \eta)^p \, \xi^q \, \eta^r \, \mathrm{d}\eta \right) \mathrm{d}\xi \right] \zeta^s \, \mathrm{d}\zeta \tag{8.2.40}$$

and then

$$\mathcal{I}_{pqrs} = \int_0^1 \left[\int_0^1 \left(\int_0^{1-\hat{\xi}} (1 - \hat{\xi} - \hat{\eta})^p \, \hat{\xi}^q \, \hat{\eta}^r \, \mathrm{d}\hat{\eta} \right) \mathrm{d}\hat{\xi} \right] \zeta^s \, (1-\zeta)^{p+q+r+2} \, \mathrm{d}\zeta,$$

where

$$\hat{\xi} \equiv \frac{\xi}{1-\zeta}, \qquad \hat{\eta} \equiv \frac{\eta}{1-\zeta}. \tag{8.2.41}$$

Next, we compute the double integral enclosed by the square brackets in the last expression using the integration formula (4.2.36), finding

$$\mathcal{I}_{pqrs} = \frac{p! \, q! \, r!}{(p+q+r+2)!} \int_0^1 \zeta^s \, (1-\zeta)^{p+q+r} \, \mathrm{d}\zeta, \tag{8.2.42}$$

then

$$\mathcal{I}_{pqrs} = \frac{p! \, q! \, r!}{(p+q+r+2)!} \, \mathrm{B}(s+1, p+q+r+3), \tag{8.2.43}$$

```
function emm = emm_t4 (x1,y1,z1,x2,y2,z2 ...
                      ,x3,y3,z3,x4,y4,z4);

%==========================================
% Evaluation of the element mass matrix
% for a four-node tetrahedron
%==========================================

%--------
% prepare
%--------

x21 = x2-x1; y21 = y2-y1; z21 = z2-z1;
x31 = x3-x1; y31 = y3-y1; z31 = z3-z1;
x41 = x4-x1; y41 = y4-y1; z41 = z4-z1;
x32 = x3-x2; y32 = y3-y2; z32 = z3-z2;
x42 = x4-x2; y42 = y4-y2; z42 = z4-z2;

%----------------
% Jacobian matrix
%----------------

Jac(1,1) = x21; Jac(1,2) = x31; Jac(1,3) = x41;
Jac(2,1) = y21; Jac(2,2) = y31; Jac(2,3) = y41;
Jac(3,1) = z21; Jac(3,2) = z31; Jac(3,3) = z41;

%-------
% volume
%-------

V = det(Jac)/6.0;

%--------------------
% element mass matrix
%--------------------

fc = V/20;
fo = 2.0*fc;

emm(1,1) = fo; emm(1,2) = fc; emm(1,3) = fc; emm(1,4) = fc;
emm(2,1) = fc; emm(2,2) = fo; emm(2,3) = fc; emm(2,4) = fc;
emm(3,1) = fc; emm(3,2) = fc; emm(3,3) = fo; emm(3,4) = fc;
emm(4,1) = fc; emm(4,2) = fc; emm(4,3) = fc; emm(4,4) = fo;

%-----
% done
%-----

return;
```

TABLE 8.2.7 Function *emm_t4* evaluates the element mass matrix for a four-node tetrahedron.

and finally

$$\mathcal{I}_{pqrs} = \frac{p!\,q!\,r!}{(p+q+r+2)!}\,\frac{s!\,(p+q+r+2)!}{(p+q+r+s+3)!},\tag{8.2.44}$$

where $B(k,l)$ is the beta function. Simplifying, we derive the integration formula (8.2.36).

8.2.9   Element advection matrix

The lth-element advection matrix is given by

$$C_{ij}^{(l)} = \iiint_{E_l} \psi_i\,\mathbf{u}\cdot\nabla\psi_j \,\mathrm{d}x\,\mathrm{d}y\,\mathrm{d}z = 6V \iiint \psi_i\,\mathbf{u}\cdot\nabla\psi_j \,\mathrm{d}\xi\,\mathrm{d}\eta\,\mathrm{d}\zeta,\tag{8.2.45}$$

where $\mathbf{u} = (u_x, u_y, u_z)$ is a specified advection velocity.

When the advection velocity is constant, equal to $\mathbf{U} = (U_x, U_y, U_z)$, the element advection matrix simplifies to

$$C_{ij}^{(l)} = 6V\,\mathbf{U}\cdot\nabla\psi_j \iiint \psi_i\,\mathrm{d}\xi\,\mathrm{d}\eta\,\mathrm{d}\zeta = 6V\,\alpha\,\mathbf{U}\cdot\nabla\psi_j,\tag{8.2.46}$$

where α is a numerical coefficient given by

$$\alpha = \iiint \psi_i\,\mathrm{d}\xi\,\mathrm{d}\eta\,\mathrm{d}\zeta = \frac{1}{24}.\tag{8.2.47}$$

The element advection matrix may thus be computed by direct evaluation using the expressions given in (8.2.29) and (8.2.30).

PROBLEMS

8.2.1 *Volume of a tetrahedron*

Show that the volume of a tetrahedron is given by

$$V = \frac{1}{6}\det\left(\begin{bmatrix} 1 & 1 & 1 & 1 \\ x_1^E & x_2^E & x_3^E & x_4^E \\ y_1^E & y_2^E & y_3^E & y_4^E \\ z_1^E & z_2^E & z_3^E & z_4^E \end{bmatrix}\right).\tag{8.2.48}$$

Hint: The determinant of a matrix remains unchanged when one column is subtracted from another column.

8.2.2 *Gradient of the element node interpolation functions*

Prove the geometrical interpretation of the gradients of the element interpolation functions shown in (8.2.29) and (8.2.30), as discussed in the text.

8.3   Domain discretization into four-node tetrahedra

A three-dimensional finite element grid consisting of four-node tetrahedra can be generated by subdividing each element of an ancestral hard-coded structure (root set) into 8 or 12 elements, as discussed in Section 8.2. Sample discretizations discussed in this section are shown in Figure 8.3.1.

tetra4_sphere8

FSELIB function *tetra4_sphere8*, listed in Table 8.3.1, discretizes the unit sphere into four-node tetrahedra by successively subdividing each one of the 8 hard-coded ancestral elements (root set) shown in the first frame of Figure 8.3.1(*a*) into 8 descendant tetrahedra. In the course of the discretization, boundary nodes are projected in the radial direction onto the unit sphere. Each time a subdivision is carried out, the number of elements increases by a factor of 8.

The element structure generated by the FSELIB driver code *tetra4_sphere8_dr*, not listed in the text, is shown in Figure 8.3.1(*a*) for four discretization levels. For clarity, only elements in the left hemisphere, $x > 0$, are displayed.

tetra4_sphere12

FSELIB function *tetra4_sphere12*, not listed in the text, performs a similar discretization of the sphere by subdividing each one of 12 ancestral tetrahedra into 8 descendant triangles.

tetra4_cube8

FSELIB function *tetra4_cube8*, not listed in the text, performs a similar discretization for a cube. The algorithm successively subdivides each one of the 12 ancestral hard-coded tetrahedra shown in the first frame of Figure 8.3.1(*b*) into 8 descendant tetrahedra.

The element structure generated by the FSELIB function *tetra4_cube_dr*, not listed in the text, is shown in Figure 8.3.1(*b*) for four discretization levels.

tetra4_cube12

FSELIB function *tetra4_cube12*, not listed in the text, performs a similar discretization by subdividing each one of 12 ancestral tetrahedra into 8 descendant triangles.

8.3.1   Delaunay tessellation

In the method of Delaunay tessellation, the locations of the global nodes defining the tetrahedral vertices are specified, and the elements are generated by mapping global nodes to element nodes.

(*a*)

(*b*)

(*c*)

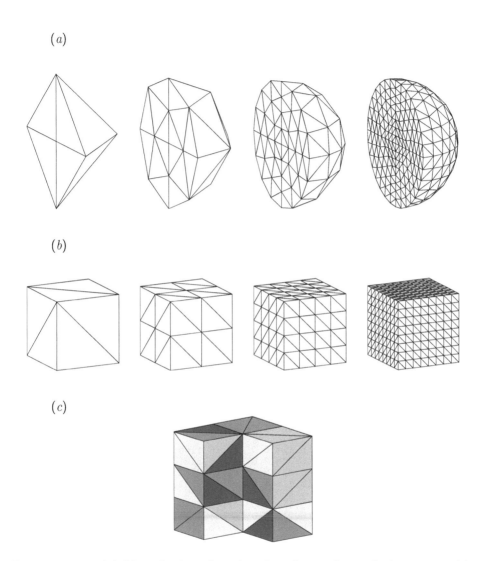

FIGURE 8.3.1 (*a*) Discretization of a sphere into four-node tetrahedra generated by the FSELIB function *tetra4_sphere8* for $ndiv = 0$ (root set), 1, 2, and 3. For clarity, only half of the sphere is shown. (*b*) Discretization of a cube into four-node tetrahedra generated by the FSELIB function *tetra4_cube8* for $ndiv = 0$ (root set), 1, 2, and 3. (*c*) Discretization of a cube generated by the MATLAB function *delaunay3*.

```
function [ne,ng,p,c,efl,gfl] = tetra4_sphere8 (ndiv)

%=========================================================
% Discretize a sphere into four-node tetrahedral elements
% by successive subdivisions of an 8-element
% parental structure into 8 descendant elements
%
% LEGEND:
%
% efl: element flag
% gfl: global node flag
%=========================================================

%----------------------------------------
% parental structure with eight elements
%----------------------------------------

ne=8;

x(1,1)= 0.0; y(1,1)= 0.0; z(1,1)= 0.0; efl(1,1)=0;  % first element
x(1,2)= 1.0; y(1,2)= 0.0; z(1,2)= 0.0; efl(1,2)=1;
x(1,3)= 0.0; y(1,3)= 1.0; z(1,3)= 0.0; efl(1,3)=1;
x(1,4)= 0.0; y(1,4)= 0.0; z(1,4)= 1.0; efl(1,4)=1;

x(2,1)= 0.0; y(2,1)= 0.0; z(2,1)= 0.0; efl(2,1)=0;  % second element
x(2,2)= 0.0; y(2,2)=-1.0; z(2,2)= 0.0; efl(2,2)=1;
x(2,3)= 1.0; y(2,3)= 0.0; z(2,3)= 0.0; efl(2,3)=1;
x(2,4)= 0.0; y(2,4)= 0.0; z(2,4)= 1.0; efl(2,4)=1;

x(3,1)= 0.0; y(3,1)= 0.0; z(3,1)= 0.0; efl(3,1)=0;  % third element
x(3,2)= 0.0; y(3,2)= 1.0; z(3,2)= 0.0; efl(3,2)=1;
x(3,3)=-1.0; y(3,3)= 0.0; z(3,3)= 0.0; efl(3,3)=1;
x(3,4)= 0.0; y(3,4)= 0.0; z(3,4)= 1.0; efl(3,4)=1;

x(4,1)= 0.0; y(4,1)= 0.0; z(4,1)= 0.0; efl(4,1)=0;  % fourth element
x(4,2)=-1.0; y(4,2)= 0.0; z(4,2)= 0.0; efl(4,2)=1;
x(4,3)= 0.0; y(4,3)=-1.0; z(4,3)= 0.0; efl(4,3)=1;
x(4,4)= 0.0; y(4,4)= 0.0; z(4,4)= 1.0; efl(4,4)=1;

x(5,1)= 0.0; y(5,1)= 0.0; z(5,1)= 0.0; efl(5,1)=0;  % fifth element
x(5,2)= 0.0; y(5,2)= 1.0; z(5,2)= 0.0; efl(5,3)=1;
x(5,3)= 1.0; y(5,3)= 0.0; z(5,3)= 0.0; efl(5,2)=1;
x(5,4)= 0.0; y(5,4)= 0.0; z(5,4)=-1.0; efl(5,4)=1;

x(6,1)= 0.0; y(6,1)= 0.0; z(6,1)= 0.0; efl(6,1)=0;  % sixth element
x(6,2)= 1.0; y(6,3)= 0.0; z(6,2)= 0.0; efl(6,2)=1;
x(6,3)= 0.0; y(6,3)=-1.0; z(6,3)= 0.0; efl(6,3)=1;
x(6,4)= 0.0; y(6,4)= 0.0; z(6,4)=-1.0; efl(6,4)=1;

x(7,1)= 0.0; y(7,1)= 0.0; z(7,1)= 0.0; efl(7,1)=0;  % seventh element
x(7,2)=-1.0; y(7,2)= 0.0; z(7,2)= 0.0; efl(7,2)=1;
x(7,3)= 0.0; y(7,3)= 1.0; z(7,3)= 0.0; efl(7,3)=1;
x(7,4)= 0.0; y(7,4)= 0.0; z(7,4)=-1.0; efl(7,4)=1;
```

TABLE 8.3.1 Function *tetra4_sphere8* (Continuing →)

```
x(8,1)= 0.0; y(8,1)= 0.0; z(8,1)= 0.0; efl(8,1)=0;   % eighth element
x(8,2)= 0.0; y(8,2)=-1.0; z(8,2)= 0.0; efl(8,2)=1;
x(8,3)=-1.0; y(8,3)= 0.0; z(8,3)= 0.0; efl(8,3)=1;
x(8,4)= 0.0; y(8,4)= 0.0; z(8,4)=-1.0; efl(8,4)=1;

%----------------
% refinement loop
%----------------

if(ndiv > 0)

for i=1:ndiv

  nm = 0; % count the new elements arising by each refinement loop
          % eight elements will be generated in each pass

  for j=1:ne    % loop over current elements

    % edge mid-nodes

    x12 = 0.5*(x(j,1)+x(j,2));
    y12 = 0.5*(y(j,1)+y(j,2));
    z12 = 0.5*(z(j,1)+z(j,2));
    efl12 = 0; if(efl(j,1)==1 & efl(j,2)==1) efl12 = 1; end

    x13 = 0.5*(x(j,1)+x(j,3));
    y13 = 0.5*(y(j,1)+y(j,3));
    z13 = 0.5*(z(j,1)+z(j,3));
    efl13 = 0; if(efl(j,1)==1 & efl(j,3)==1) efl13 = 1; end

    x14 = 0.5*(x(j,1)+x(j,4));
    y14 = 0.5*(y(j,1)+y(j,4));
    z14 = 0.5*(z(j,1)+z(j,4));
    efl14 = 0; if(efl(j,1)==1 & efl(j,4)==1) efl14 = 1; end

    x23 = 0.5*(x(j,2)+x(j,3));
    y23 = 0.5*(y(j,2)+y(j,3));
    z23 = 0.5*(z(j,2)+z(j,3));
    efl23 = 0; if(efl(j,2)==1 & efl(j,3)==1) efl23 = 1; end

    x24 = 0.5*(x(j,2)+x(j,4));
    y24 = 0.5*(y(j,2)+y(j,4));
    z24 = 0.5*(z(j,2)+z(j,4));
    efl24 = 0; if(efl(j,2)==1 & efl(j,4)==1) efl24 = 1; end

    x34 = 0.5*(x(j,3)+x(j,4));
    y34 = 0.5*(y(j,3)+y(j,4));
    z34 = 0.5*(z(j,3)+z(j,4));
    efl34 = 0; if(efl(j,3)==1 & efl(j,4)==1) efl34 = 1; end
```

TABLE 8.3.1 Function *tetra4_sphere8* ($\rightarrow$ Continuing $\rightarrow$)

```
% assign vertex nodes to sub-elements
% these will become the "new" elements

  nm = nm+1;  % first sub-element

  xn(nm,1)=x13;     yn(nm,1)=y13;     zn(nm,1)=z13;     efln(nm,1)=efl13;
  xn(nm,2)=x23;     yn(nm,2)=y23;     zn(nm,2)=z23;     efln(nm,2)=efl23;
  xn(nm,3)=x(j,3);  yn(nm,3)=y(j,3);  zn(nm,3)=z(j,3);  efln(nm,3)=efl(j,3);
  xn(nm,4)=x34;     yn(nm,4)=y34;     zn(nm,4)=z34;     efln(nm,4)=efl34;

  nm = nm+1;  % second sub-element

  xn(nm,1)=x12;     yn(nm,1)=y12;     zn(nm,1)=z12;     efln(nm,1)=efl12;
  xn(nm,2)=x(j,2);  yn(nm,2)=y(j,2);  zn(nm,2)=z(j,2);  efln(nm,2)=efl(j,2);
  xn(nm,3)=x23;     yn(nm,3)=y23;     zn(nm,3)=z23;     efln(nm,3)=efl23;
  xn(nm,4)=x24;     yn(nm,4)=y24;     zn(nm,4)=z24;     efln(nm,4)=efl24;

  nm = nm+1;  % third sub-element

  xn(nm,1)=x14;     yn(nm,1)=y14;     zn(nm,1)=z14;     efln(nm,1)=efl14;
  xn(nm,2)=x24;     yn(nm,2)=y24;     zn(nm,2)=z24;     efln(nm,2)=efl24;
  xn(nm,3)=x34;     yn(nm,3)=y34;     zn(nm,3)=z34;     efln(nm,3)=efl34;
  xn(nm,4)=x(j,4);  yn(nm,4)=y(j,4);  zn(nm,4)=z(j,4);  efln(nm,4)=efl(j,4);

  nm = nm+1;  % fourth sub-element

  xn(nm,1)=x(j,1);  yn(nm,1)=y(j,1);  zn(nm,1)=z(j,1);  efln(nm,1)=efl(j,1);
  xn(nm,2)=x12;     yn(nm,2)=y12;     zn(nm,2)=z12;     efln(nm,2)=efl12;
  xn(nm,3)=x13;     yn(nm,3)=y13;     zn(nm,3)=z13;     efln(nm,3)=efl13;
  xn(nm,4)=x14;     yn(nm,4)=y14;     zn(nm,4)=z14;     efln(nm,4)=efl14;

  nm = nm+1;  % fifth sub-element

  xn(nm,1)=x34; yn(nm,1)=y34; zn(nm,1)=z34; efln(nm,1)=efl34;
  xn(nm,2)=x24; yn(nm,2)=y24; zn(nm,2)=z24; efln(nm,2)=efl24;
  xn(nm,3)=x14; yn(nm,3)=y14; zn(nm,3)=z14; efln(nm,3)=efl14;
  xn(nm,4)=x13; yn(nm,4)=y13; zn(nm,4)=z13; efln(nm,4)=efl13;

  nm = nm+1;  % sixth sub-element

  xn(nm,1)=x14; yn(nm,1)=y14; zn(nm,1)=z14; efln(nm,1)=efl14;
  xn(nm,2)=x12; yn(nm,2)=y12; zn(nm,2)=z12; efln(nm,2)=efl12;
  xn(nm,3)=x13; yn(nm,3)=y13; zn(nm,3)=z13; efln(nm,3)=efl13;
  xn(nm,4)=x24; yn(nm,4)=y24; zn(nm,4)=z24; efln(nm,4)=efl24;

  nm = nm+1;  % seventh sub-element

  xn(nm,1)=x12; yn(nm,1)=y12; zn(nm,1)=z12; efln(nm,1)=efl12;
  xn(nm,2)=x23; yn(nm,2)=y23; zn(nm,2)=z23; efln(nm,2)=efl23;
  xn(nm,3)=x13; yn(nm,3)=y13; zn(nm,3)=z13; efln(nm,3)=efl13;
  xn(nm,4)=x24; yn(nm,4)=y24; zn(nm,4)=z24; efln(nm,4)=efl24;
```

TABLE 8.3.1 Function *tetra4_sphere8* ($\rightarrow$ Continuing $\rightarrow$)

```
    nm = nm+1;   % eighth sub-element

    xn(nm,1)=x13; yn(nm,1)=y13; zn(nm,1)=z13; efln(nm,1)=efl13;
    xn(nm,2)=x24; yn(nm,2)=y24; zn(nm,2)=z24; efln(nm,2)=efl24;
    xn(nm,3)=x23; yn(nm,3)=y23; zn(nm,3)=z23; efln(nm,3)=efl23;
    xn(nm,4)=x34; yn(nm,4)=y34; zn(nm,4)=z34; efln(nm,4)=efl34;

  end % end of loop over current elements

%---

  ne = 8*ne;   % number of elements has increased
               % by a factor of eight

    for k=1:ne     % relabel the new points
                   % and put them in the master list
                   % project boundary nodes onto the unit sphere
       for l=1:4
          x(k,l) =   xn(k,l);
          y(k,l) =   yn(k,l);
          z(k,l) =   zn(k,l);
        efl(k,l) = efln(k,l);
         if(efl(k,l) == 1)
            rad = sqrt(x(k,l)^2+y(k,l)^2+z(k,l)^2);
            x(k,l)=x(k,l)/rad;
            y(k,l)=y(k,l)/rad;
            z(k,l)=z(k,l)/rad;
         end
       end
    end

end % end for of refinement loop
end % end if of refinement loop

%-----------------------------------------------------
% define the global nodes and the connectivity table
%-----------------------------------------------------

% the four nodes of the first element are entered manually

p(1,1) = x(1,1); p(1,2) = y(1,1); p(1,3) = z(1,1); gfl(1) = efl(1,1);
p(2,1) = x(1,2); p(2,2) = y(1,2); p(2,3) = z(1,2); gfl(2) = efl(1,2);
p(3,1) = x(1,3); p(3,2) = y(1,3); p(3,3) = z(1,3); gfl(3) = efl(1,3);
p(4,1) = x(1,4); p(4,2) = y(1,4); p(4,3) = z(1,4); gfl(4) = efl(1,4);

c(1,1) = 1;   % first  node of first element is global node 1
c(1,2) = 2;   % second node of first element is global node 2
c(1,3) = 3;   % third  node of first element is global node 3
c(1,4) = 4;   % fourth node of first element is global node 4
```

TABLE 8.3.1 Function *tetra4_sphere8* ($\rightarrow$ Continuing $\rightarrow$)

```
ng = 4;

%---
% loop over further elements
% Iflag=0 will signal a new global node
%---

eps = 0.00001;  % tolerance
for i=2:ne          % loop over the rest of the elements
 for j=1:4            % loop over element nodes

 Iflag=0;

 for k=1:ng

  if(abs(x(i,j)-p(k,1)) < eps)
   if(abs(y(i,j)-p(k,2)) < eps)
    if(abs(z(i,j)-p(k,3)) < eps)

     Iflag = 1;    % the node has been recorded previously
     c(i,j) = k;   % the jth local node of the ith element
                   %  is the kth global node
    end
   end
  end
 end

 if(Iflag==0)  % record the node
   ng = ng+1;
   p(ng,1) =   x(i,j);
   p(ng,2) =   y(i,j);
   p(ng,3) =   z(i,j);
   gfl(ng) = efl(i,j);
   c(i,j)  = ng;  % the jth local element node is a new global node
 end

 end
end  % end of loop over elements

%-----
% done
%-----

return;
```

TABLE 8.3.1 ($\rightarrow$ Continued) Function *tetra4_sphere8* divides a sphere into four-node tetrahedral elements by successively subdividing eight ancestral hard-coded elements.

The MATLAB function *delaunay3* performs the Delaunay tessellation prompted by the function call

$$c = \texttt{delaunay3(x,y,z)}$$

where x, y, and z are three input vectors holding the coordinates of the specified global nodes.[2] The familiar $N_E \times 4$ global connectivity matrix, c, holding the global labels of the N_E four-node tetrahedra arising from the Delaunay tessellation, is returned in the output. Care should be taken to ensure that the element node labels generated by this function yield positive element volumes. If this is not the case, the labels of an arbitrary pair of element nodes should be switched.

FSELIB function *tetra4_delaunay_cube*, listed in Table 8.3.2, calls the MAT-LAB function *delaunay3* to tessellate a cube into four-node tetrahedra based on a cubic lattice of grid points with N_x, N_y, and N_z divisions. The elements are visualized using the FSELIB function *tetra4_vis*, listed in Table 8.2.4. Alternatively, visualization can be performed using the MATLAB *tetramesh* function, prompted by the function call

$$c = \texttt{tetramesh(c,[x y z])}$$

where c is the $N_E \times 4$ global connectivity matrix. The element assembly generated for the parameters specified in the code is shown in Figure 8.3.1(c).

PROBLEM

8.3.1 *Delaunay tessellation*

Perform the Delaunay tessellation of set of nodes of your choice. Discuss the geometrical structure of the finite element assembly.

8.4   Finite element codes with four-node tetrahedra

Having discussed the geometrical structure and properties of four-node tetrahedral elements, and having developed pertinent domain discretization algorithms for spherical and square domains, we proceed to develop finite element codes in four stages: domain tessellation, assembly of the necessary global matrices from element matrices, solution of the final linear system, and visualization.

8.4.1   Laplace's equation

FSELIB code *laplt4_d*, listed in Table 8.4.1, solves Laplace's equation

$$\nabla^2 f = 0, \qquad\qquad (8.4.1)$$

---

[2]In recent versions of MATLAB, the function *delaunay3* has been superseded by the functions *delaunay* and *delaunaytri*. Information on the usage of these functions can be found at the MATLAB Web site.

```
function [ne,ng,p,c,efl,gfl] = tetra4_delaunay_cube

%===================================================
% Delaunay tessellation into four-node tetrahedra
%===================================================

%---------------------
% tessellation of a cube
% window and grid size
%---------------------

X1 = -1.0; X2 = 1.0;
Y1 = -1.0; Y2 = 1.0;
Z1 = -1.0; Z2 = 1.0;

Nx = 2; Ny = 2; Nz = 3;  % number of subdivisions

%--------
% prepare
%--------

Dx = (X2-X1)/Nx;
Dy = (Y2-Y1)/Ny;
Dz = (Z2-Z1)/Nz;

%-------------------------------
% arrange points on a mesh grid
% and set the boundary flag
%-------------------------------

ng = 0;

for k=1:Nz+1
 for j=1:Ny+1
  for i=1:Nx+1
   ng = ng+1;
   p(ng,1) = X1+(i-1.0)*Dx;
   p(ng,2) = Y1+(j-1.0)*Dy;
   p(ng,3) = Z1+(k-1.0)*Dz;
   gfl(ng) = 0;
   if(i==1 | i==Nx+1 | j==1 | j==Ny+1 | k==1 | k==Nz+1) gfl(ng)=1; end;
  end
 end
end

%----------------------
% Delaunay triangulation
%----------------------

for i=1:ng
 xdel(i) = p(i,1);
 ydel(i) = p(i,2);
 zdel(i) = p(i,3);
end
```

TABLE 8.3.2 Function *tetra4_delaunay_cube* (Continuing →)

```
c = delaunay3 (xdel,ydel,zdel);

%--------------------------------
% extract the number of elements
%--------------------------------

sc = size(c); ne = sc(1,1);

%--------------------------------
% set the element-node boundary flags
%--------------------------------

for i=1:ne
 efl(i,1) = gfl(c(i,1));
 efl(i,2) = gfl(c(i,2));
 efl(i,3) = gfl(c(i,3));
 efl(i,4) = gfl(c(i,4));
end

%--------------------------------
% ensure that volumes are positive
%--------------------------------

for i=1:ne

   for j=1:4
     xtmp(j) = p(c(i,j),1);
     ytmp(j) = p(c(i,j),2);
     ztmp(j) = p(c(i,j),3);
   end

   matrix=[1 1 1 1;
          xtmp(1) xtmp(2) xtmp(3) xtmp(4); ...
          ytmp(1) ytmp(2) ytmp(3) ytmp(4); ...
          ztmp(1) ztmp(2) ztmp(3) ztmp(4)];

   vlm = det(matrix)/6.0;

   if(vlm<0)
     save = c(i,1);
     c(i,1) = c(i,2);
     c(i,2) = save;
   end

end

%--------------------------------
% display the triangulation
%--------------------------------

figure(1)
hold on
axis equal
```

TABLE 8.3.2 Function *tetra4_delaunay_cube* ($\rightarrow$ Continuing $\rightarrow$)

```
option = 2;   % wire-frame
option = 5;   % painted solid white
option = 3;   % graded
option = 1;   % painted in color

volume = 0.0;

for i=1:ne

  vol(i) = 0.0;
  xmean = 0.0;
  ymean = 0.0;
  zmean = 0.0;

  for j=1:4
   xvis(j) = p(c(i,j),1); yvis(j) = p(c(i,j),2);
   zvis(j) = p(c(i,j),3);
   xmean = xmean+xvis(j); ymean = ymean+yvis(j);
   zmean = zmean+zvis(j);
  end

  if(xmean>0)
   vol(i) = tetra4_vis (xvis,yvis,zvis,zvis,option);
   end
  volume = volume+vol(i);

end

%----
% done
%-----

return
```

TABLE 8.3.2 ($\rightarrow$ Continued) Function *tetra4_delaunay_cube* performs the tessellation of a cube into four-node tetrahedra.

subject to the Dirichlet boundary condition specifying the surface distribution of the unknown function, f, over the boundary, as implemented in the code. The layout of the code is similar to those of codes *lapl3_d* and *lapl6_d* for Laplace equation in the plane with three- or six-node triangles, as discussed in Sections 4.4 and 5.3.1.

In the code listed in Table 8.4.1, the finite element grid is generated by one of the functions *tetra4_sphere12*, *tetra4_sphere8*, *tetra4_cube12*, or *tetra4_cube8* discussed in Section 8.3. The element diffusion matrices are computed by the function *edm_t4*, listed in Table 8.2.6. The solution is visualized by the function *plot_t4*, listed in Table 8.4.2.

The finite element solution for the boundary conditions implemented in the code involving a sinusoidal function is displayed in Figure 8.4.1.

```
%======================================
% Code lapl4_d
%
% Solution of Laplace's equation with
% the Dirichlet boundary condition
% using four-node tetrahedral elements
%======================================

%--------
% prepare
%-------

 figure(1)
 hold on
 axis equal

%-----------
% input data
%-----------

ndiv = 2;  % discretization level

%-----------
% discretize
%-----------

[ne,ng,p,c,efl,gfl] = tetra4_sphere12(ndiv);

%-------
% deform
%-------

defx = 0.50; defy = 0.50; defz = 0.50;

for i=1:ng
 p(i,1) = p(i,1)*(1.0+defx*p(i,1)^2);
 p(i,2) = p(i,2)*(1.0+defy*p(i,2)^2);
 p(i,3) = p(i,3)*(1.0+defz*p(i,3)^2);
end

%----------------------------------------
% specify the Dirichlet boundary condition
%----------------------------------------

for i=1:ng
 if(gfl(i)==1)
   bcd(i) = sin(pi*(p(i,1)+p(i,2)));    % example
 end
end

%--------------------------------------
% assemble the global diffusion matrix
%--------------------------------------

gdm = zeros(ng,ng); % initialize
```

TABLE 8.4.1   Code *laplt4_d* (Continuing →)

```
for l=1:ne              % loop over the elements

% compute the element diffusion matrix

j = c(l,1); x(1) = p(j,1); y(1) = p(j,2); z(1) = p(j,3);
j = c(l,2); x(2) = p(j,1); y(2) = p(j,2); z(2) = p(j,3);
j = c(l,3); x(3) = p(j,1); y(3) = p(j,2); z(3) = p(j,3);
j = c(l,4); x(4) = p(j,1); y(4) = p(j,2); z(4) = p(j,3);

[edm_elm] = edm_t4 (x,y,z);

    for i=1:4
      i1 = c(l,i);
      for j=1:4
        j1 = c(l,j);
        gdm(i1,j1) = gdm(i1,j1) + edm_elm(i,j);
      end
    end
end

%-------------------------------------------------
% set the right-hand side of the linear system
% and implement the Dirichlet boundary condition
%-------------------------------------------------

for i=1:ng
 b(i) = 0.0;
end

for j=1:ng
 if(gfl(j)==1)
    for i=1:ng
     b(i) = b(i) - gdm(i,j)*bcd(j);
     gdm(i,j) = 0; gdm(j,i) = 0;
    end
    gdm(j,j) = 1.0;
    b(j) = bcd(j);
 end
end

%-----------------------
% solve the linear system
%-----------------------

f = b/gdm';

%-----
% plot
%-----

plot_t4 (ne,ng,p,c,f)

%-----
% done
%-----
```

TABLE 8.4.1 ($\rightarrow$ Continued) Code *laplt4_d* solves Laplace's equation subject to the Dirichlet boundary condition using four-node tetrahedral elements.

```
function plot_t4 (ne,ng,p,c,f)

%==================================================
% Color mapped visualization of a function f in a
% domain discretized into four-node tetrahedra
%==================================================

%-----------------
% plotting option
%--------------

option = 4;
option = 1;
option = 2; % wireframe

%----------------------------------------------------
% compute the maximum and minimum of the function f
%----------------------------------------------------

fmin =   100.0;
fmax = -100.0;

for i=1:ng
 if(f(i) > fmax) fmax = f(i); end
 if(f(i) < fmin) fmin = f(i); end
end

range = 1.2*(fmax-fmin);
shift = fmin;

%---------------------------------------------
% shift the color index in the range (0, 1)
% and plot four-node elements
%---------------------------------------------

for l=1:ne

 j = c(1,1); xp(1)=p(j,1); yp(1)=p(j,2); zp(1)=p(j,3);
          cp(1) = (f(j)-shift)/range;
 j = c(1,2); xp(2)=p(j,1); yp(2)=p(j,2); zp(2)=p(j,3);
          cp(2) = (f(j)-shift)/range;
 j = c(1,3); xp(3)=p(j,1); yp(3)=p(j,2); zp(3)=p(j,3);
          cp(3) = (f(j)-shift)/range;
 j = c(1,4); xp(4)=p(j,1); yp(4)=p(j,2); zp(4)=p(j,3);
          cp(4) = (f(j)-shift)/range;

 volume = tetra4_vis (xp,yp,zp,cp,option);

end

%-----
% done
%-----

return;
```

TABLE 8.4.2 Function *plot_t4* visualizes the finite element solution for four-node tetrahedral elements.

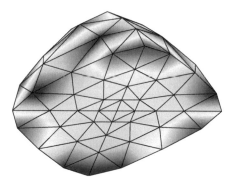

FIGURE 8.4.1   Finite element solution of Laplace's equation in three dimensions with
the Dirichlet boundary condition computed with four-node tetrahedral elements.

8.4.2   *Eigenvalues of the Laplacian operator*

An eigenfunction of the Laplacian operator, $f(x, y, z)$, satisfies the equation

$$\nabla^2 f + \lambda f = 0, \tag{8.4.2}$$

where λ is the corresponding eigenvalue, subject to the homogeneous Dirichlet
boundary condition, $f = 0$, around the entire boundary of the solution domain,
as discussed in Sections 4.5 and 5.3.2.

FSELIB code *laplt4_eig*, listed in Table 8.4.3, solves the eigenvalue problem in
three dimensions using the finite element method with four-node tetrahedra. The finite element grid is generated by one of the functions *tetra4_sphere12*, *tetra4_sphere8*,
tetra4_cube12, or *tetra4_cube8*, as discussed in Section 8.3. The element diffusion
matrices are computed by the function *edm_t4*, listed in Table 8.2.6, and the element mass matrices are computed by the function *emm_t4*, listed in Table 8.2.7.
The structure of the finite element code is similar to that for computing the eigenvalues of the Laplacian operator in two dimensions, discussed in Sections 4.5 and
5.3.2.

Numerical computations for a sphere of unit radius with $8, 64, 512$, and 4096
elements, differing by a factor of 8, predict that the smallest eigenvalue

$$30.0000, \qquad 14.8491, \qquad 11.2866, \qquad 10.3337, \tag{8.4.3}$$

respectively. As the number of elements increases, this sequence converges to the
exact value, which is known to be $\pi^2 = 9.8696$. The associated error is

$$20.10, \qquad 5.01, \qquad 1.42, \qquad 0.46, \tag{8.4.4}$$

respectively. As the number of elements increases by a factor of 8, the error decreases
approximately by a factor of 4. Consequently, the error scales with the square of
the element side length.

```
%=========================================
% Code laplt4_eig
%
% Eigenfunctions of Laplace's equation
% in a spherical-like domain
%=========================================

% input data
%-----------

ndiv = 2;  % discretization level

%-----------
% discretize
%-----------

  [ne,ng,p,c,efl,gfl] = tetra4_sphere8 (ndiv);

%---------------------------------------------------
% assemble the global diffusion and mass matrices
%---------------------------------------------------

gdm = zeros(ng,ng); % initialize
gmm = zeros(ng,ng); % initialize

for l=1:ne            % loop over the elements

% compute the element diffusion and mass matrices

j = c(l,1); x1 = p(j,1); y1 = p(j,2); z1 = p(j,3);
j = c(l,2); x2 = p(j,1); y2 = p(j,2); z2 = p(j,3);
j = c(l,3); x3 = p(j,1); y3 = p(j,2); z3 = p(j,3);
j = c(l,4); x4 = p(j,1); y4 = p(j,2); z4 = p(j,3);

[edm_elm, vlm_elm] = edm_t4 (x1,y1,z1,x2,y2,z2 ...
                            ,x3,y3,z3,x4,y4,z4);

[emm_elm] = emm_t4 (x1,y1,z1,x2,y2,z2 ...
                   ,x3,y3,z3,x4,y4,z4);
   for i=1:4
     i1 = c(l,i);
     for j=1:4
       j1 = c(l,j);
       gdm(i1,j1) = gdm(i1,j1) + edm_elm(i,j);
       gmm(i1,j1) = gmm(i1,j1) + emm_elm(i,j);
     end
   end

end

%--------------------
% reduce the matrices by removing equations
% corresponding to boundary nodes
%--------------------
```

TABLE 8.4.3   Code *laplt4_eig* (Continuing →)

```
Ic = 0;

for i=1:ng

 if(gfl(i)==0)

   Ic = Ic+1;
   map(Ic) = i;

   Jc = 0;

   for j=1:ng
     if(gfl(j)==0)
     Jc = Jc+1;
     A(Ic,Jc) = gdm(i,j);  B(Ic,Jc) = gmm(i,j);
     end
   end

 end

end

ngred = Ic;
%-----------------------
% compute the eigenvalues
%-----------------------

eigenvalues = eig(A,B);

%---
% done
%---
```

TABLE 8.4.3 ($\rightarrow$ Continued) Code *laplt4_eig* computes the eigenvalues of the Laplacian operator in three dimensions using four-node tetrahedral elements.

PROBLEMS

8.4.1 *Linear field*

Confirm that the code *laplt3_d* generates the exact solution for a linear field.

8.4.2 *Eigenvalues of the Laplacian operator inside a sphere*

Prove that the exact value for the smallest eigenvalue of the Laplacian operator inside the unit sphere is π^2 and derive the corresponding eigenfunction.

8.4.3 *Eigenvalues of the Laplacian operator inside a cube*

Discuss the convergence of the eigenvalues of the Laplacian inside a cube.

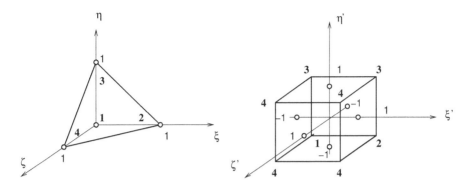

FIGURE 8.5.1 Mapping of the standard tetrahedron in the $\xi\eta\zeta$ space to the standard square in the $\xi'\eta'\zeta'$ space. The node labels are printed in bold.

8.5 Orthogonal polynomials over a tetrahedron

The standard tetrahedron in the $\xi\eta\zeta$ space can be mapped to a cube in the $\xi'\eta'\zeta'$ space, as illustrated in Figure 8.5.1. The mapping is mediated by the functions

$$\xi = \frac{1+\xi'}{2}\frac{1-\eta'}{2}\frac{1-\zeta'}{2}, \qquad \eta = \frac{1+\eta'}{2}\frac{1-\zeta'}{2}, \qquad \zeta = \frac{1+\zeta'}{2}, \qquad (8.5.1)$$

where

$$-1 \le \xi',\eta',\zeta' \le 1. \qquad (8.5.2)$$

The third barycentric coordinate is mapped according to the transformation rule

$$\omega \equiv 1-\xi-\eta-\zeta = \frac{1-\xi'}{2}\frac{1-\eta'}{2}\frac{1-\zeta'}{2}. \qquad (8.5.3)$$

When $\zeta' = -1$, corresponding to $\zeta = 0$, we recover the two-dimensional triangle transformations (4.2.39) and (4.2.40). For future reference, we note the relations

$$1-\zeta = \frac{1-\zeta'}{2}, \qquad 1-\eta-\zeta = \frac{1-\eta'}{2}\frac{1-\zeta'}{2}. \qquad (8.5.4)$$

The inverse mapping functions are given by

$$\xi' = \frac{2\xi}{1-\eta-\zeta} - 1, \qquad \eta' = 2\frac{\eta}{1-\zeta} - 1, \qquad \zeta' = 2\zeta - 1. \qquad (8.5.5)$$

When $\zeta = 0$ corresponding to $\zeta' = -1$, we recover the inverse two-dimensional triangle transformations (4.2.42).

The integral of a function $f(\xi,\eta,\zeta)$ over the volume of the tetrahedron in the $\xi\eta\zeta$ space can be expressed as an integral over the standard cube,

$$\iiint f(\xi,\eta,\zeta)\,\mathrm{d}\xi\,\mathrm{d}\eta\,\mathrm{d}\zeta = \int_{-1}^{1}\int_{-1}^{1}\int_{-1}^{1} f(\xi,\eta,\zeta)\,h_V'\,\mathrm{d}\xi'\,\mathrm{d}\eta'\,\mathrm{d}\zeta'. \qquad (8.5.6)$$

The metric coefficient of the transformation, h'_V, is equal to the determinant of the Jacobian matrix of the transformation,

$$h'_V = \det\left(\begin{bmatrix} \dfrac{\partial \xi}{\partial \xi'} & \dfrac{\partial \eta}{\partial \xi'} & \dfrac{\partial \zeta}{\partial \xi'} \\[2mm] \dfrac{\partial \xi}{\partial \eta'} & \dfrac{\partial \eta}{\partial \eta'} & \dfrac{\partial \zeta}{\partial \eta'} \\[2mm] \dfrac{\partial \xi}{\partial \zeta'} & \dfrac{\partial \eta}{\partial \zeta'} & \dfrac{\partial \zeta}{\partial \zeta'} \end{bmatrix} \right), \tag{8.5.7}$$

yielding

$$h'_V = \det\left(\begin{bmatrix} \frac{1}{8}(1-\eta')(1-\zeta') & 0 & 0 \\[2mm] -\frac{1}{8}(1+\xi')(1-\zeta') & \frac{1}{4}(1-\zeta') & 0 \\[2mm] -\frac{1}{8}(1+\xi')(1-\eta') & -\frac{1}{4}(1+\eta') & \frac{1}{2} \end{bmatrix} \right). \tag{8.5.8}$$

Computing the determinant and simplifying, we find that

$$h'_V = \frac{1}{64}(1-\eta')(1-\zeta')^2. \tag{8.5.9}$$

The integral over the cube on the right-hand side of (8.5.6) can be computed by applying an integration rule or a Gaussian quadrature for the individual one-dimensional integrals. This, however, is not necessarily the most efficient approach.

8.5.1 Karniadakis and Sherwin polynomials

The Proriol polynomials over the triangle discussed in Section 5.7 can be generalized into a broader family of polynomials that are orthogonal over the volume of a tetrahedron with a flat weighting function, as discussed by Karniadakis and Sherwin [32, 59, 60, 61]. The tetrahedral orthogonal polynomials are given by

$$\mathcal{P}_{klp} = L_k(\xi') \left(\frac{1-\eta'}{2}\right)^k \left(\frac{1-\zeta'}{2}\right)^k \mathcal{J}_l^{(2k+1,0)}(\eta') \left(\frac{1-\zeta'}{2}\right)^l \mathcal{J}_p^{(2k+2l+2,0)}(\zeta') \tag{8.5.10}$$

or

$$\mathcal{P}_{klp} = \left(\frac{1-\eta'}{2}\right)^k \left(\frac{1-\zeta'}{2}\right)^{k+l} L_k(\xi')\, \mathcal{J}_l^{(2k+1,0)}(\eta')\, \mathcal{J}_p^{(2k+2l+2,0)}(\zeta'), \tag{8.5.11}$$

where L_q is a Legendre polynomial and $\mathcal{J}_q^{(r,s)}$ is a Jacobi polynomial, as discussed in Appendix B.

Recalling the transformation rules (8.5.4) and noting the presence of the denominators on the right-hand sides of the transformation rule for ξ' and η', as shown in (8.5.5), we find that $\mathcal{P}_{klp}$ is a kth-degree polynomial in ξ, a $(k+l)$-degree polynomial

$$\mathcal{P}_{000} = 1$$
$$\mathcal{P}_{100} = 2\,\xi + \eta + \zeta - 1$$
$$\mathcal{P}_{010} = 3\,\eta + \zeta - 1$$
$$\mathcal{P}_{001} = 4\,\zeta - 1$$
$$\mathcal{P}_{200} = 6\,\xi^2 + \eta^2 + \zeta^2 + 6\,\xi\eta + 6\,\xi\zeta + 2\,\eta\zeta - 6\,\xi - 2\,\eta - 2\,\zeta + 1$$
$$\mathcal{P}_{110} = 5\,\eta^2 + \zeta^2 + 10\,\xi\eta + 2\,\xi\zeta + 6\,\eta\zeta - 2\,\xi - 6\,\eta - 2\,\zeta + 1$$
$$\mathcal{P}_{101} = 6\,\zeta^2 + 12\,\xi\zeta + 6\,\eta\zeta - 2\,\xi - \eta - 7\,\zeta + 1$$
$$\mathcal{P}_{020} = 10\,\eta^2 + \zeta^2 + 8\,\eta\zeta - 8\,\eta - 2\,\zeta + 1$$
$$\mathcal{P}_{011} = 6\,\zeta^2 + 18\,\eta\zeta - 3\,\eta - 7\,\zeta + 1$$
$$\mathcal{P}_{002} = 15\,\zeta^2 - 10\,\zeta + 1$$

TABLE 8.5.1 Explicit expressions for orthogonal polynomials over a tetrahedron.

in η, and a $(k + l + p)$-degree polynomial in ζ. Explicitly, the first few tetrahedral orthogonal polynomials are given in Table 8.5.1.

To prove orthogonality, we define the integral of the product of the klp and qrs polynomials over the volume of the tetrahedron,

$$\chi_{klp,qrs} \equiv \iiint \mathcal{P}_{klp}\, \mathcal{P}_{qrs}\, \mathrm{d}\xi\, \mathrm{d}\eta\, \mathrm{d}\zeta. \tag{8.5.12}$$

Using the integration formula (8.5.6), we obtain

$$\chi_{klp,qrs} = \int_{-1}^{1} \int_{-1}^{1} \int_{-1}^{1} \mathcal{P}_{klp}\, \mathcal{P}_{qrs}\, \frac{(1 - \eta')\,(1 - \zeta')^2}{64}\, \mathrm{d}\xi'\, \mathrm{d}\eta'\, \mathrm{d}\zeta', \tag{8.5.13}$$

yielding

$$\chi_{klp,qrs} = \frac{1}{8} \int_{-1}^{1} \mathrm{L}_k(\xi')\, \mathrm{L}_q(\xi')\, \mathrm{d}\xi' \cdot \int_{-1}^{1} \mathcal{J}_l^{2k+1,0}(\eta')\, \mathcal{J}_r^{2q+1,0}(\eta') \left(\frac{1 - \eta'}{2}\right)^{k+q+1} \mathrm{d}\eta'$$
$$\cdot \int_{-1}^{1} \mathcal{J}_p^{2k+2l+2,0}(\zeta')\, \mathcal{J}_s^{2q+2r+2,0}(\zeta') \left(\frac{1 - \zeta'}{2}\right)^{k+l+q+r+2} \mathrm{d}\eta'. \tag{8.5.14}$$

The integral with respect to ξ' is non-zero only if $k = q$. When $k = q$, the integral with respect to η' is non-zero only if $l = r$. When $k = q$ and $l = r$, the integral with respect to ζ' is non-zero only if $p = s$. Overall, the integral is non-zero only when $k = q$, $l = r$, and $p = s$, which shows that the polynomials are orthogonal. The self-projection integral is given by

$$\mathcal{G}_{klp} \equiv \iiint \mathcal{P}_{klp}^2(\xi, \eta, \zeta)\, \mathrm{d}\xi\, \mathrm{d}\eta\, \mathrm{d}\zeta = \frac{1}{(2k+1)\,(2k+2l+2)\,(2k+2l+2p+3)}. \tag{8.5.15}$$

$$1 = \mathcal{P}_{000}$$

$$\xi = \frac{1}{12}\left(3\mathcal{P}_{000} + 6\mathcal{P}_{100} - 2\mathcal{P}_{010} - \mathcal{P}_{001}\right)$$

$$\eta = \frac{1}{12}\left(3\mathcal{P}_{000} + 4\mathcal{P}_{010} - \mathcal{P}_{001}\right)$$

$$\zeta = \frac{1}{4}\left(\mathcal{P}_{000} + \mathcal{P}_{001}\right)$$

$$\xi^2 = \frac{1}{90}\left(9\mathcal{P}_{000} + 30\mathcal{P}_{100} - 10\mathcal{P}_{010} - 5\mathcal{P}_{001}\right.$$
$$\left. +15\mathcal{P}_{200} - 9\mathcal{P}_{110} - 6\mathcal{P}_{101} + 3\mathcal{P}_{020} + 2\mathcal{P}_{011} + \mathcal{P}_{002}\right)$$

$$\eta^2 = \frac{1}{90}\left(9\mathcal{P}_{000} + 20\mathcal{P}_{010} - 5\mathcal{P}_{001} + 9\mathcal{P}_{020} - 4\mathcal{P}_{011} + \mathcal{P}_{002}\right)$$

$$\zeta^2 = \frac{1}{30}\left(3\mathcal{P}_{000} + 5\mathcal{P}_{001} + 2\mathcal{P}_{002}\right)$$

$$\xi\eta = \frac{1}{180}\left(9\mathcal{P}_{000} + 15\mathcal{P}_{100} + 5\mathcal{P}_{010} - 5\mathcal{P}_{001}\right.$$
$$\left. +18\mathcal{P}_{110} - 3\mathcal{P}_{101} - 9\mathcal{P}_{020} - \mathcal{P}_{011} + \mathcal{P}_{002}\right)$$

$$\xi\zeta = \frac{1}{180}\left(9\mathcal{P}_{000} + 15\mathcal{P}_{100} - 5\mathcal{P}_{010} + 5\mathcal{P}_{001} + 15\mathcal{P}_{101} - 5\mathcal{P}_{011} - 4\mathcal{P}_{002}\right)$$

$$\eta\zeta = \frac{1}{180}\left(9\mathcal{P}_{000} + 10\mathcal{P}_{010} + 5\mathcal{P}_{001} + 10\mathcal{P}_{011} - 4\mathcal{P}_{002}\right)$$

TABLE 8.5.2 Expressions for power products in terms of tetrahedral orthogonal polynomials.

For example, $\mathcal{G}_{000} = 1/6$, which is equal to the volume of the standard tetrahedron in the $\xi\eta\zeta$ space.

A polynomial in ξ, η, and ζ can be recast in terms of the orthogonal tetrahedral polynomials using the relations shown in Table 8.5.2.

8.5.2 Orthogonal expansion

The tetrahedral orthogonal polynomials provide us with a complete orthogonal basis. Any non-singular function defined over the standard tetrahedron in the $\xi\eta\zeta$ space, $f(\xi, \eta, \zeta)$, can be approximated with a complete mth-degree polynomial in ξ, η, and ζ, expressed in the form

$$f(\xi, \eta, \zeta) \simeq \sum_{k=0}^{m}\left[\sum_{l=0}^{m-k}\left(\sum_{p=0}^{m-k-l} a_{klp}\, \mathcal{P}_{klp}(\xi, \eta, \zeta)\right)\right], \tag{8.5.16}$$

where the triple sum is designed so that $k + l + p \le m$, and a_{klp} are appropriate coefficients.

The triple sum on the right-hand side of (8.5.16) can be arranged into an infinite family of Pascal triangles with increasing dimensions. The first contrived triangle represents the constant term,

$$\mathcal{P}_{000}, \tag{8.5.17}$$

the second triangle encapsulates linear functions

$$\mathcal{P}_{100}$$
$$\mathcal{P}_{010} \quad \mathcal{P}_{001}, \tag{8.5.18}$$

the third triangle encapsulates quadratic functions

$$\mathcal{P}_{200}$$
$$\mathcal{P}_{110} \quad \mathcal{P}_{101} \tag{8.5.19}$$
$$\mathcal{P}_{020} \quad \mathcal{P}_{011} \quad \mathcal{P}_{002},$$

and the $m + 1$ triangle encapsulates mth-order functions. The triangles can be stacked downward to yield Pascal's pyramid, where the constant function $\mathcal{P}_{000}$ is located at the apex.

The total number of coefficients in the mth-order expansion is

$$N = 1 + 3 + 6 + \cdots + \frac{1}{2}(m+1)(m+2) = \frac{1}{2}\sum_{i=0}^{m}(i+1)(i+2) \tag{8.5.20}$$

or

$$N = \frac{1}{2}\left(\sum_{i=0}^{m} i^2 + 3\sum_{i=0}^{m} i + 2(m+1)\right). \tag{8.5.21}$$

Computing the sums, we obtain

$$N = \frac{1}{2}\left(\frac{m(m+1)(2m+1)}{6} + 3\frac{m(m+1)}{2} + 2(m+1)\right) \tag{8.5.22}$$

and finally

$$N = \frac{1}{6}(m+1)(m^2 + 5m + 6) = \frac{1}{6}(m+1)(m+2)(m+3) \tag{8.5.23}$$

or

$$N = \binom{m+3}{3}, \tag{8.5.24}$$

where the tall parentheses denote the combinatorial.

Multiplying (8.5.16) by $\mathcal{P}_{qrs}$, integrating the product over the volume of the standard tetrahedron and using the orthogonality property, we find that the expansion coefficients are given by

$$a_{klp} = \frac{1}{\mathcal{G}_{klp}} \iiint f(\xi, \eta, \zeta)\, \mathcal{P}_{klp}(\xi, \eta, \zeta)\, \mathrm{d}\xi\, \mathrm{d}\eta\, \mathrm{d}\zeta, \qquad (8.5.25)$$

where $\mathcal{G}_{klp}$ is defined in (8.5.15).

In practice, the polynomial coefficients introduced in (8.5.16) are found by enforcing N interpolation conditions to generate a Vandermonde system of linear equations, as discussed in Section 5.5. The orthogonality of the basis functions guarantees the well-conditioning of the coefficient matrix.

PROBLEMS

8.5.1 *Volume of a tetrahedron*

Apply the integration formula (8.5.6) for $f = 1$ to show that the volume of the standard tetrahedron is equal to $1/6$.

8.5.2 *Mapping of a prismatic element*

Discuss the mapping of the six-node prismatic element shown in Figure 8.1.2 to the standard cube.

8.6   High-order and spectral tetrahedral elements

In Section 8.2, we discussed four-node tetrahedral elements and derived node interpolation functions that are linear in the physical coordinates, x, y, and z, as well as in the parametric coordinates, ξ, η, and ζ. Quadratic and higher-order expansions arise as generalizations.

A complete mth-order polynomial expansion of any suitable function, $f(\xi, \eta, \zeta)$, over the volume of the tetrahedron takes the form

$$\begin{aligned}
f(\xi, \eta, \zeta) = \quad & a_{000} \\
& + a_{100}\,\xi + a_{010}\,\eta + a_{001}\,\zeta \\
& + a_{200}\,\xi^2 + a_{110}\,\xi\eta + a_{101}\,\xi\zeta + a_{020}\,\eta^2 + a_{011}\,\eta\zeta + a_{002}\,\zeta^2 \\
& \cdots \quad \cdots \quad \cdots \quad \cdots \quad \cdots \quad \cdots \quad \cdots \quad \cdots \quad \cdots \\
& + a_{m,0,0}\,\xi^m + a_{m-1,1,0}\,\xi^{m-1}\eta + \cdots + a_{0,1,m-1}\,\eta\,\zeta^m + a_{0,0,m}\,\zeta^m, \qquad (8.6.1)
\end{aligned}$$

where a_{ijk} are suitable coefficients. Note that the sum of the indices, $i + j + k$, is constant across each row. The first row contains only one coefficient, the second row contains three coefficients, and the ith row contains $\frac{1}{2}(i+1)(i+2)$ coefficients. According to (8.5.24), the total number of coefficients up to the mth row is

$$N = \frac{1}{6}(m+1)(m+2)(m+3). \qquad (8.6.2)$$

For example, when $m = 2$, corresponding to the complete quadratic polynomial, the expansion contains 10 terms.

Pascal pyramid

The terms in each row of (8.6.1) can be arranged in a Pascal triangle with increasing dimensions. The first triangle is a point representing the constant term,

$$1$$

the second triangle encapsulates linear functions,

$$\xi$$
$$\eta \quad \zeta,$$

the third triangle encapsulates quadratic functions,

$$\xi^2$$
$$\xi\eta \quad \zeta\xi$$
$$\eta^2 \quad \eta\zeta \quad \zeta^2,$$

and the $m + 1$ triangle encapsulates mth-order functions. The triangles can be stacked to yield Pascal's pyramid, where the constant term is located at the apex. To compute the expansion coefficients, we require an equal number of conditions or constraints associated with interpolation nodes or expansion modes.

8.6.1   Uniform node distributions

One way to ensure that the number of interpolation nodes is equal to the number of coefficients in the complete mth-order expansion is to arrange the nodes as shown in Figure 8.6.1(a). In this arrangement, nodal points along the ξ, η, and ζ axes are distributed on a one-dimensional uniform grid defined by a set of $m + 1$ points,

$$v_i = \frac{i - 1}{m} \tag{8.6.3}$$

for $i = 1, \ldots, m + 1$. The face and interior nodes are identified by the intersection of planes that are parallel to the four faces of the tetrahedron at evenly spaced intervals, yielding Pascal's pyramid.

The nodes can be labeled by a trio of indices, (i, j, k), corresponding to the ξ, η, and ζ axes, as shown in Figure 8.6.1(a). For each value of the index i in the range $i = 1, \ldots, m + 1$, the index j takes values in the range $j = 1, \ldots, m + 2 - i$. For each doublet (i, j), the index k takes values in the range $k = 1, \ldots, m + 3 - i - j$. The labeling scheme can be permuted without prejudice.

A plane that is parallel to the slanted face of the tetrahedron in the $\xi\eta\zeta$ space corresponds to a constant value of the fourth barycentric coordinate,

$$\omega = 1 - \xi - \eta - \zeta, \tag{8.6.4}$$

(*a*)   (*b*)

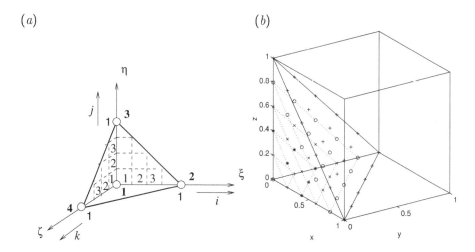

FIGURE 8.6.1   (*a*) Uniform node distribution over the standard tetrahedron in parametric space corresponding to a complete polynomial expansion. The nodes are
parametrized by three indices, i, j, and k. (*b*) Uniform grid for polynomial order $m = 6$ involving 84 nodes, generated by the FSELIB script *nodes_tetra_uni*.
Nodes plotted with the same symbol lie in slanted planes corresponding to a fixed
value of the barycentric coordinate, w and index l.

ranging from $w = 0$ at the slanted face to 1 at the origin, $\xi = 0$, $\eta = 0$, $\zeta = 0$. A
slanted plane hosting nodes is identified by the index

$$l = m + 4 - i - j - k, \tag{8.6.5}$$

which decreases from the value $l = 1$ along the slanted face to the value $m + 1$ at
the origin.

FSELIB script *nodes_tetra_uni*, listed in abbreviated form in Table 8.6.1, generates and displays the nodes for a specified polynomial order, m. An example for
$m = 6$ involving 84 nodes, $N = 84$, is shown in Figure 8.6.1(*b*). Nodes marked with
the same symbol lie in slanted planes corresponding to constant values of w and
associated label, l.

Node interpolation functions

The node interpolation functions can be deduced by generalizing the rule discussed
in Section 5.8.2 for triangular elements. Consider the ith node of a tetrahedron,
and assume that m planes can be found passing through all nodes, except for the
ith node. If these planes are described by linear functions

$$\mathcal{F}_j(\xi, \eta, \zeta) = A_j \xi + B_j \eta + C_j \zeta + D_j = 0 \tag{8.6.6}$$

```
%=====================================================
% Code nodes_tetra
%
% Uniform node distribution in the tetrahedron
% corresponding to a complete mth-degree polynomial
%=====================================================

m = 5     % example

%---------------------
% uniform master grid
%---------------------

for i=1:m+1
  v(i) = (i-1.0)/m;
end

%----------------------------
% deploy and count the nodes
%----------------------------

count = 0;

for i=1:m+1;
 xp = v(i);

 for j=1:m+2-i
  yp = v(j);

   for k=1:m+3-i-j
    zp = v(j);

    count = count+1;
    l = m+4-i-j-k;
    if(l==1) plot3(xp,yp,zp,'+'); end
    if(l==2) plot3(xp,yp,zp,'o'); end
    if(l==3) plot3(xp,yp,zp,'x'); end
    if(l==4) plot3(xp,yp,zp,'*'); end
    if(l==5) plot3(xp,yp,zp,'+'); end
    if(l==6) plot3(xp,yp,zp,'o'); end
    if(l==7) plot3(xp,yp,zp,'x'); end
    if(l==8) plot3(xp,yp,zp,'*'); end
    if(l==9) plot3(xp,yp,zp,'+'); end

  end
 end
end

......
```

TABLE 8.6.1 Abbreviation of the script *nodes_tetra* generating and displaying a uniform node distribution over the tetrahedron for a specified polynomial degree, m. The six dots at the end denote unprinted code that draws the edges.

for $j = 1, \ldots, m$, where A_j, B_j, C_j, and D_j are constant coefficients, then the ith node interpolation function is given by

$$\psi_i(\xi, \eta, \zeta) = \prod_{j=1}^{m} \frac{\mathcal{F}_j(\xi, \eta, \zeta)}{\mathcal{F}_j(\xi_i, \eta_i, \zeta_i)}. \tag{8.6.7}$$

The cardinal node interpolation functions for the uniform grid can be constructed using this formula based on the observation that, for each node, all other nodes lie in preceding horizontal and vertical planes and in subsequent slanted planes.

Using this rule, we find that the interpolation function corresponding to the (i, j, k) node can be expressed as the product of four functions,

$$\psi_{ijk}(\xi, \eta, \zeta) = \Xi_i^{(i-1)}(\xi) \cdot H_j^{(j-1)}(\eta) \cdot Z_k^{(k-1)}(\zeta) \cdot Y_l^{(l-1)}(\omega), \tag{8.6.8}$$

where $\Xi_i^{(i-1)}(\xi)$ is an $(i-1)$-degree polynomial, defined such that

$$\Xi_1^{(0)}(\xi) = 1, \qquad \Xi_i^{(i-1)}(\xi) = \frac{(\xi - v_1)(\xi - v_2) \cdots (\xi - v_{i-2})(\xi - v_{i-1})}{(v_i - v_1)(v_i - v_2) \cdots (v_i - v_{i-2})(v_i - v_{i-1})} \tag{8.6.9}$$

for $i = 2, \ldots, m+1$, $H_j^{(j-1)}(\eta)$ is a $(j-1)$-degree polynomial, defined such that

$$H_1^{(0)}(\eta) = 1, \qquad H_j^{(j-1)}(\eta) = \frac{(\eta - v_1)(\eta - v_2) \cdots (\eta - v_{j-2})(\eta - v_{j-1})}{(v_j - v_1)(v_j - v_2) \cdots (v_j - v_{j-2})(v_j - v_{j-1})} \tag{8.6.10}$$

for $j = 2, \ldots, m+1$, $Z_k^{(k-1)}(\eta)$ is a $(k-1)$-degree polynomial, defined such that

$$Z_1^{(0)}(\eta) = 1, \qquad Z_k^{(k-1)}(\eta) = \frac{(\zeta - v_1)(\zeta - v_2) \cdots (\zeta - v_{k-2})(\zeta - v_{k-1})}{(v_k - v_1)(v_k - v_2) \cdots (v_k - v_{k-2})(v_k - v_{k-1})} \tag{8.6.11}$$

for $k = 2, \ldots, m+1$, $Y_l^{(l-1)}(\omega)$ is an $(l-1)$-degree polynomial, defined such that

$$Y_1^{(0)}(\omega) = 1, \qquad Y_l^{(l-1)}(\omega) = \frac{(\omega - v_1)(\omega - v_2) \cdots (\omega - v_{l-2})(\omega - v_{l-1})}{(v_l - v_1)(v_l - v_2) \cdots (v_l - v_{l-2})(v_l - v_{l-1})} \tag{8.6.12}$$

for $l = 2, \ldots, m+1$.

Now we may confirm that $\psi_{ijk}(\xi, \eta, \zeta)$ is a polynomial of degree

$$(i-1) + (j-1) + (k-1) + (l-1) = i + j + k + l - 4 = m \tag{8.6.13}$$

with respect to ξ, η, and ζ, as required, and that all cardinal interpolation conditions are met.

8.6.2   Arbitrary node distributions

To compute an mth-degree node interpolation function for an arbitrary node distribution over the tetrahedron, we may express it as a sum of monomial products of tetrahedral orthogonal polynomials and compute the N expansion coefficients by solving a system of linear equations that arise by enforcing the cardinal interpolation conditions. Compiling these systems, we derive an expression for the vector of nodal interpolation functions in terms of the Vandermonde matrix.

Applying Cramer's rule, we find that each node interpolation function can be expressed as the ratio of the determinants of two generalized Vandermonde matrices, as shown in (5.5.34). The node distribution that maximizes the magnitude of the determinant of the Vandermonde matrix in the denominator, subject to the restriction that all nodes lie inside over the faces, or at the vertices of the tetrahedron, comprises the highly desirable Fekete set.

8.6.3   Spectral node distributions

To guarantee that the interpolating function is a complete mth-degree polynomial of two barycentric coordinates over each face, and thus facilitate enforcing the C^0 continuity condition of the finite element expansion, we may assign a group of interpolation nodes to the faces of the tetrahedron by one of the methods discussed in Section 5.5 for the triangle. The remaining nodes are arranged inside the tetrahedral element according to a new set of criteria.

The nodal set includes $N_v = 4$ vertex nodes, $N_e = 6 \times (m - 1)$ non-vertex edge nodes, and $N_f = 4 \times \frac{1}{2}(m - 1)(m - 2)$ interior face nodes, adding up to

$$N_s = 2(m^2 + 1) \tag{8.6.14}$$

surface nodes. The number of interior nodes is

$$N_i = N - N_s = \frac{1}{6}(m - 3)(m - 2)(m - 1) = \binom{m - 1}{3}, \tag{8.6.15}$$

which is precisely equal to the number of nodes corresponding to the $(m - 4)$-order complete polynomial expansion. Note that interior nodes are present only for polynomial orders $m \geq 4$.

In the case of the linear expansion, $m = 1$, we assign four nodes at the four element vertices, as discussed in Section 8.2. In the case of the quadratic expansion, $m = 2$, described by 10 nodes, we adopt the uniform node distribution discussed in Section 8.6.1. In the case of the cubic expansion, $m = 3$, we distribute the 20 interpolation nodes at the element faces using one of the methods discussed in Section 5.5 for the triangle.

In the distribution illustrated in Figure 8.6.2(a) for $m = 3$, the edge nodes are distributed at positions corresponding to the zeros of the fourth-degree completed

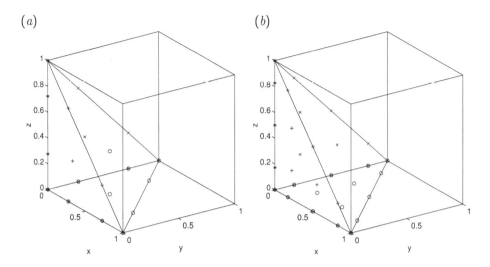

FIGURE 8.6.2   Lobatto node distributions for (a) a 20-node tetrahedron supporting a complete cubic expansion, $m = 3$, and (b) a 35-node tetrahedron supporting a complete quartic expansion, $m = 4$. For clarity, only the face nodes in the three orthogonal planes are shown in (b).

Lobatto polynomial,

$$\widehat{\mathrm{Lo}}_4(t) \equiv (1 - t^2)\,\mathrm{Lo}_2(t), \tag{8.6.16}$$

and the face nodes are positioned at the face centroids, where $\mathrm{Lo}_2(t)$ is the second-degree Lobatto polynomial. The nodal set illustrated in Figure 8.6.2(b) for $m = 4$ consists of 35 nodes, including 34 surface nodes distributed at positions corresponding to the Lobatto triangle grid. Common sense and symmetry considerations suggest placing the remaining solitary interior node at the element centroid located at $\xi = \eta = \zeta = 1/4$.

Chen and Babuška [9] computed node distributions by maximizing the magnitude of the determinant of the generalized Vandermonde matrix in the spirit of the Fekete set, or by minimizing the magnitude of the modified Lebesgue function

$$\left(\iiint \sum_{i=1}^{N} |\psi_i(\xi, \eta, \zeta)|^2 \, d\xi \, d\eta \, d\zeta \right)^{1/2}, \tag{8.6.17}$$

where the integral is computed over the volume of the standard tetrahedron, subject to the condition that the nodal sets observe the geometrical symmetries of the tetrahedron and the face nodes are distributed as in the case of two-dimensional interpolation discussed in Section 5.5. Hesthaven and Teng [29] performed a similar computation by minimizing instead an electrostatic potential.

Lobatto tetrahedral set

Luo and Pozrikidis [38] proposed a tetrahedral Lobatto grid as an extension of the Lobatto triangle grid discussed in Section 5.8.3, as follows:

- On the $\xi\eta$ face, nodes are distributed at

$$\xi = \frac{1}{3}\left(1 + 2\,v_i - v_j - v_l\right), \qquad \eta = \frac{1}{3}\left(1 - v_i + 2v_j - v_l\right), \qquad \zeta = 0$$

(8.6.18)

  for $i = 1, \ldots, m+1$ and $j = 1, 2, \ldots, m+2-i$, where $l = m+3-i-j$.

- On the $\eta\zeta$ face, nodes are distributed at

$$\xi = 0, \qquad \eta = \frac{1}{3}\left(1 + 2v_j - v_k - v_l\right), \qquad \zeta = \frac{1}{3}\left(1 - v_j + 2v_k - v_l\right)$$

(8.6.19)

  for $j = 1, \ldots, m$ and $k = 2, \ldots, m+2-j$, where $l = m+3-j-k$.

- On the $\zeta\xi$ face, nodes are distributed at

$$\xi = \frac{1}{3}\left(1 + 2v_i - v_k - v_l\right), \qquad \eta = 0, \qquad \zeta = \frac{1}{3}\left(1 - v_i + 2v_k - v_l\right)$$

(8.6.20)

  for $i = 2, \ldots, m$ and $k = 2, \ldots, m+2-i$, where $l = m+3-i-k$.

- On the slanted face, nodes are distributed at

$$\xi = \frac{1}{3}\left(1 + 2v_i - v_j - v_l\right), \quad \eta = \frac{1}{3}\left(1 - v_i + 2v_j - v_l\right), \quad \zeta = 1 - \xi - \eta$$

(8.6.21)

  for $i = 2, \ldots, m$ and $j = 2, \ldots, m+1-i$, where $l = m+3-i-j$.

- Interior nodes are distributed at positions

$$\xi = \frac{1}{4}\left(1 + 3\,v_i - v_j - v_k - v_l\right), \qquad \eta = \frac{1}{4}\left(1 - v_i + 3\,v_j - v_k - v_l\right),$$

$$\zeta = \frac{1}{4}\left(1 - v_i - v_j + 3\,v_k - v_l\right)$$

(8.6.22)

  for $i = 2, \ldots, m$, $j = 2, \ldots, m+1-i$, and $k = 2, \ldots, m+2-i-j$, where $l = m+4-i-j-k$. Note that the range of subscripts restricts the nodes inside the tetrahedron.

Node distributions over the canonical orthogonal and regular tetrahedra are shown in Figure 8.6.3 for $m = 5$. The properties of this heuristic set compare favorably with those of more rigorous sets arising from function minimization.

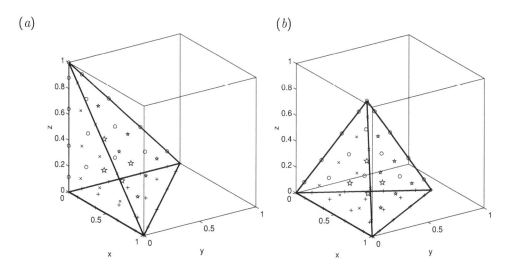

FIGURE 8.6.3   Lobatto tetrahedral node distribution for $m = 5$ over (a) the orthogonal and (b) corresponding regular tetrahedron.

8.6.4   Gradient of the element node interpolation functions

The gradients of the element node interpolation functions, $\nabla \psi_i$, are computed using the relations in (8.2.20), repeated below for convenience,

$$\frac{\partial \mathbf{x}}{\partial \xi} \cdot \nabla \psi_i = \frac{\partial \psi_i}{\partial \xi}, \qquad \frac{\partial \mathbf{x}}{\partial \eta} \cdot \nabla \psi_i = \frac{\partial \psi_i}{\partial \eta}, \qquad \frac{\partial \mathbf{x}}{\partial \zeta} \cdot \nabla \psi_i = \frac{\partial \psi_i}{\partial \zeta}. \qquad (8.6.23)$$

In the case of isoparametric interpolation, the derivatives on the left-hand sides are computed based on the relation

$$\mathbf{x} = \sum_{j=1}^{N} \mathbf{x}_j \, \psi_j, \qquad (8.6.24)$$

where $\mathbf{x}_j$ is the position of the jth element node. A specific application will be discussed in Section 8.7 for 10-node tetrahedra.

If the position vector, $\mathbf{x}$, is mapped using the linear functions discussed in Section 8.2 based on the four vertices of the tetrahedron alone, whereas a function of interest is interpolated using the quadratic functions shown in Table 8.5.2, then $\nabla \psi_i$ is found by solving system (8.2.23). In that case, the interpolation is super-parametric.

8.6.5   Numerical integration

The integral of a function, $f(x, y, z)$, over a tetrahedron is given by

$$\iiint f(x, y, z) \, dx \, dy \, dz = \iiint f(\xi, \eta, \zeta) \, h_V \, d\xi \, d\eta \, d\zeta, \qquad (8.6.25)$$

where h_V is the volume metric coefficient and the integral on the right-hand side is computed over the canonical tetrahedron in the $\xi\eta\zeta$ domain. The integration can be performed using a Gaussian integration quadrature, yielding

$$\iiint f(\xi, \eta, \zeta) \, h_V \, \mathrm{d}\xi \, \mathrm{d}\eta \, \mathrm{d}\zeta \simeq \frac{1}{6} \sum_{l=1}^{N_Q} f(\xi_l, \eta_l, \zeta_l) \, h_V \, w_l, \qquad (8.6.26)$$

where N_Q is a chosen number of base points, ξ_l, η_l, ζ_l are barycentric coordinates of the base points, and w_l are corresponding integration weights. The sum of the integration weights is equal to unity for any number of base points, N_Q, so that the exact result is recovered when f and h_V are constant.

FSELIB function *gauss_tetra24*, listed in Table 8.6.2, evaluates the quadrature base points and integration weights for $N_Q = 24$ (Keast [33]).[3] The application of this quadrature will be demonstrated in Section 8.7 for 10-node quadratic tetrahedra. Other quadrature rules with a different number of base points can be employed.

PROBLEMS

8.6.1 *Pascal pyramid*

Display the structure of the fourth and fifth triangles in the Pascal pyramid involving cubic and quartic monomial products.

8.6.2 *Quadrature for a tetrahedron*

List the coordinates of the base points and associated weights for $N_Q = 1$ and 4.

8.7   10-node quadratic tetrahedra

A complete quadratic expansion, corresponding to expansion order $m = 2$, is supported by 10 interpolation nodes, as shown in Figure 8.7.1. For convenience, the nodes can be numbered by a single index varying in the range 1–10, as shown in Figure 8.7.1. The first four nodes lie at the element vertices, and the last six nodes lie along the element edges.

The general shape of the tetrahedron in physical space is shown in Figure 8.7.1(a), and the standard shape in parametric space is shown in Figure 8.7.1(b). Note that the edges of the tetrahedron in physical space are not necessarily straight. The interpolated geometry of the element edges is determined implicitly by the quadratic mapping function associated with the 10 nodes.

For convenience, we map the edge nodes of the tetrahedron in physical space to the mid-points of the edges in parameter space, as shown in Figure 8.7.1, keeping in mind that distortions may occur when an edge node is located near a vertex node.

---

[3]http://people.sc.fsu.edu/~jburkardt/datasets/quadrature_rules_tet/quadrature_rules_tet.html

```
function [xiq,etq,ztq,weq] = gauss_tetra24

gauss = [
  0.35619138622254   0.21460287125915   0.21460287125915 0.03992275025816
  0.21460287125915   0.21460287125915   0.21460287125915 0.03992275025816
  0.21460287125915   0.21460287125915   0.35619138622254 0.03992275025816
  0.21460287125915   0.35619138622254   0.21460287125915 0.03992275025816
  0.87797812439616   0.04067395853461   0.04067395853461 0.01007721105532
  0.04067395853461   0.04067395853461   0.04067395853461 0.01007721105532
  0.04067395853461   0.04067395853461   0.87797812439616 0.01007721105532
  0.04067395853461   0.87797812439616   0.04067395853461 0.01007721105532
  0.03298632957317   0.32233789014227   0.32233789014227 0.05535718154365
  0.32233789014227   0.32233789014227   0.32233789014227 0.05535718154365
  0.32233789014227   0.32233789014227   0.03298632957317 0.05535718154365
  0.32233789014227   0.03298632957317   0.32233789014227 0.05535718154365
  0.26967233145831   0.06366100187501   0.06366100187501 0.04821428571428
  0.06366100187501   0.26967233145831   0.06366100187501 0.04821428571428
  0.06366100187501   0.06366100187501   0.26967233145831 0.04821428571428
  0.60300566479164   0.06366100187501   0.06366100187501 0.04821428571428
  0.06366100187501   0.60300566479164   0.06366100187501 0.04821428571428
  0.06366100187501   0.06366100187501   0.60300566479164 0.04821428571428
  0.06366100187501   0.26967233145831   0.60300566479164 0.04821428571428
  0.26967233145831   0.60300566479164   0.06366100187501 0.04821428571428
  0.60300566479164   0.06366100187501   0.26967233145831 0.04821428571428
  0.06366100187501   0.60300566479164   0.26967233145831 0.04821428571428
  0.26967233145831   0.06366100187501   0.60300566479164 0.04821428571428
  0.60300566479164   0.26967233145831   0.06366100187501 0.04821428571428
];

xiq = gauss(:,1);
etq = gauss(:,2);
ztq = gauss(:,3);
weq = gauss(:,4);

%-----
% done
%-----

return
```

TABLE 8.6.2 Function *gauss_tetra24* evaluates the quadrature base points and integration weights for integrating over the volume of a tetrahedron. The first, second, and third columns are the ξ, η, and ζ coordinates. The fourth column contains the integration weights.

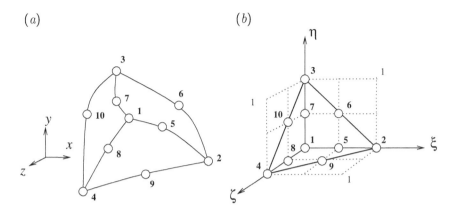

FIGURE 8.7.1   A 10-node tetrahedron supports a complete quadratic expansion, $m =$ 2. The node labels are printed in bold. (a) General shape of the tetrahedron in physical space and (b) standard shape in parametric space.

8.7.1   Node interpolation functions

Applying the expressions derived in Section 8.6.1 got uniform node distributions for $v_1 = 0$, $v_2 = 0.5$, and $v_3 = 1.0$, and setting $l = 6 - i - j - k$, we find that the interpolation function of node $(i, j, k) = (1, 1, 1)$, corresponding to $l = 3$, labeled node 1 in Figure 8.7.1, is given by

$$\psi_1 \equiv \psi_{1,1,1} = \Xi_1^{(0)}(\xi) \cdot H_1^{(0)}(\eta) \cdot Z_1^{(0)}(\zeta) \cdot Y_3^{(2)}(\omega). \tag{8.7.1}$$

Making substitutions, we obtain

$$\psi_1 = 1 \cdot 1 \cdot 1 \cdot \frac{(\omega - 0.0)(\omega - 0.5)}{(1 - 0.0)(1 - 0.5)} = \omega\,(2\,\omega - 1), \tag{8.7.2}$$

where $\omega = 1 - \xi - \eta - \zeta$. Working in a similar fashion for the second node, we find that

$$\psi_2 \equiv \psi_{3,1,1} = \Xi_3^{(2)}(\xi) \cdot H_1^{(0)}(\eta) \cdot Z_1^{(0)}(\zeta) \cdot Y_1^{(0)}(\omega) \tag{8.7.3}$$

or

$$\psi_2 = \frac{(\xi - 0)(\xi - 0.5)}{(1 - 0)(1 - 0.5)} \cdot 1 \cdot 1 \cdot 1 = \xi\,(2\,\xi - 1). \tag{8.7.4}$$

The rest of the node interpolation functions can be computed working in a similar fashion, and the results are shown in Table 8.7.1.

FSELIB function *elmt10_interp*, listed in Table 8.7.2, computes the node interpolation functions and their gradients, and evaluates the volume metric coefficient, h_V, at a position corresponding to a specified trio of barycentric coordinates, (ξ, η, ζ). The input to this function includes the coordinates of the 10 element nodes and the barycentric coordinates of the evaluation point.

Node label	ξ	η	ζ	ω	ψ
1	0	0	0	1	$\psi_1 \equiv \psi_{1,1,1} = \omega\,(2\omega - 1)$
2	1	0	0	0	$\psi_2 \equiv \psi_{3,1,1} = \xi\,(2\xi - 1)$
3	0	1	0	0	$\psi_3 \equiv \psi_{1,3,1} = \eta\,(2\eta - 1)$
4	0	0	1	0	$\psi_4 \equiv \psi_{1,1,3} = \zeta\,(2\zeta - 1)$
5	0.5	0	0	0.5	$\psi_5 \equiv \psi_{2,1,1} = 4\,\xi\omega$
6	0.5	0.5	0	0	$\psi_6 \equiv \psi_{2,2,1} = 4\,\xi\eta$
7	0	0.5	0	0.5	$\psi_7 \equiv \psi_{1,2,1} = 4\,\eta\omega$
8	0	0	0.5	0.5	$\psi_8 \equiv \psi_{1,1,2} = 4\,\zeta\omega$
9	0.5	0	0.5	0.0	$\psi_9 \equiv \psi_{2,1,2} = 4\,\xi\zeta$
10	0	0.5	0.5	0.0	$\psi_{10} \equiv \psi_{1,2,2} = 4\,\eta\zeta$

TABLE 8.7.1 Element interpolation functions for a 10-node tetrahedron with mid-point edge nodes, where $\omega \equiv 1 - \xi - \eta - \zeta$. The sum of all functions is equal to unity.

8.7.2 Element diffusion and mass matrices

FSELIB function *edmm_t10*, listed in Table 8.7.3, computes the element diffusion and mass matrices. The input to this function includes the coordinates of the 10 element nodes. The gradients of the node interpolation functions are computed according to the procedure discussed in Section 8.6.4. The integration is performed using a 24-point integration quadrature, as discussed in Section 8.6.5. For the reasons discussed in Section 1.2 and elsewhere, the element diffusion matrix is singular.

8.7.3 Domain discretization

FSELIB function *tetra10_sphere8*, listed in Table 8.7.4, discretizes the unit sphere into 10-node tetrahedra. The algorithm subdivides successively each one of the eight hard-coded ancestral elements shown in the first frame of Figure 8.7.2(a) into eight descendant tetrahedra. In the process of discretization, the boundary nodes are projected in the radial direction onto the unit sphere.

Each time a subdivision is carried out, the number of elements increases by a factor of eight. The element layout for three discretization levels is shown in Figure 8.7.2(a). For clarity, only elements in the left hemisphere, $x > 0$, are displayed.

FSELIB function *tetra10_cube8*, not listed in the text, discretizes the unit cube using a similar method. The algorithm successively subdivides each one of the 12 ancestral tetrahedra shown in the first frame of Figure 8.7.2(b) into eight descendant tetrahedra. The element layout for four discretization levels is shown in Figure 8.7.2(b).

```
function [psi, gpsi, hv] = elmt10_interp ...
  ...
      (x1,y1, x2,y2, x3,y3, x4,y4, x5,y5, x6,y6 ...
      ,x7,y7, x8,y8, x9,y9, x10,y10, xi,et,zt )

%================================================
% Evaluation of the volume metric coefficient, hv,
% and computation of the basis functions and
% their gradients, over a 10-node tetrahedron
%================================================

%--------
% prepare
%--------

om = 1.0-xi-et-zt;

%---------------------------
% compute the basis functions
%---------------------------

psi(1) = om*(2*om-1.0);
psi(2) = xi*(2*xi-1.0);
psi(3) = et*(2*et-1.0);
psi(4) = zt*(2*zt-1.0);
psi(5) = 4.0*xi*om;
psi(6) = 4.0*xi*et;
psi(7) = 4.0*et*om;
psi(8) = 4.0*zt*om;
psi(9) = 4.0*xi*zt;
psi(10)= 4.0*et*zt;

%-----------------------------------------------------
% compute the xi derivatives of the basis functions
%-----------------------------------------------------

dps(1) =-4*om+1.0;
dps(2) = 4*xi-1.0
dps(3) = 0.0
dps(4) = 0.0
dps(5) = 4*om-4*xi;
dps(6) = 4*et;
dps(7) =-4*et;
dps(8) =-4*zt;
dps(9) = 4*zt;
dps(10)= 0.0;

%-----------------------------------------------------
% compute the eta derivatives of the basis functions
%-----------------------------------------------------
```

TABLE 8.7.2 Function *elm10_interp* (Continuing →)

```
pps(1)  = -4*om+1.0;
pps(2)  = 0.0;
pps(3)  = 4*et-1.0;
pps(4)  = 0.0;
pps(5)  =-4*xi;
pps(6)  = 4*xi;
pps(7)  = 4*om-4*et;
pps(8)  =-4*zt;
pps(9)  = 0.0;
pps(10)= 4*zt;

%------------------------------------------------------
% compute the zeta derivatives of the basis functions
%------------------------------------------------------

qps(1)  =-4*om+1.0;
qps(2)  = 0.0;
qps(3)  = 0.0;
qps(4)  = 4*zt-1.0;
qps(5)  = -4*xi;
qps(6)  = 0.0;
qps(7)  =-4*et;
qps(8)  = 4*om-4*zt;
qps(9)  = 4*xi;
qps(10)= 4*et;

%------------------------------------------------
% compute the xi, eta, and zeta derivatives of x
%------------------------------------------------

DxDxi = x1*dps(1) + x2*dps(2) + x3*dps(3) ...
      + x4*dps(4) + x5*dps(5) + x6*dps(6) ...
      + x7*dps(7) + x8*dps(8) + x9*dps(9) + x10*dps(10);

DyDxi = y1*dps(1) + y2*dps(2) + y3*dps(3) ...
      + y4*dps(4) + y5*dps(5) + y6*dps(6) ...
      + y7*dps(7) + y8*dps(8) + y9*dps(9) + y10*dps(10);

DzDxi = z1*dps(1) + z2*dps(2) + z3*dps(3) ...
      + z4*dps(4) + z5*dps(5) + z6*dps(6) ...
      + z7*dps(7) + z8*dps(8) + z9*dps(9) + z10*dps(10);

DxDet = x1*pps(1) + x2*pps(2) + x3*pps(3) ...
      + x4*pps(4) + x5*pps(5) + x6*pps(6) ...
      + x7*pps(7) + x8*pps(8) + x9*pps(9) + x10*pps(10);

DyDet = y1*pps(1) + y2*pps(2) + y3*pps(3) ...
      + y4*pps(4) + y5*pps(5) + y6*pps(6) ...
      + y7*pps(7) + y8*pps(8) + y9*pps(9) + y10*pps(10);

DzDet = z1*pps(1) + z2*pps(2) + z3*pps(3) ...
      + z4*pps(4) + z5*pps(5) + z6*pps(6) ...
      + z7*pps(7) + z8*pps(8) + z9*pps(9) + z10*pps(10);
```

TABLE 8.7.2 Function *elm10_interp* ($\rightarrow$ Continuing $\rightarrow$)

```
DxDzt = x1*qps(1) + x2*qps(2) + x3*qps(3) ...
        + x4*qps(4) + x5*qps(5) + x6*qps(6) ...
        + x7*qps(7) + x8*qps(8) + x9*qps(9) + x10*qps(10);

DyDzt = y1*qps(1) + y2*qps(2) + y3*qps(3) ...
        + y4*qps(4) + y5*qps(5) + y6*qps(6) ...
        + y7*qps(7) + y8*qps(8) + y9*qps(9) + y10*qps(10);

DzDzt = z1*qps(1) + z2*qps(2) + z3*qps(3) ...
        + z4*qps(4) + z5*qps(5) + z6*qps(6) ...
        + z7*qps(7) + z8*qps(8) + z9*qps(9) + z10*qps(10);

%----------------------------
% compute the volume metric hv
%----------------------------

Jac = [DxDxi DxDet DxDzt;
       DyDxi DyDet DyDzt;
       DzDxi DzDet DzDzt];

hv = det(Jac);

%----------------------------------
% compute the gradient of the basis functions
% by solving two linear equations:
%
% dx/dxi . grad = d psi/dxi
% dx/det . grad = d psi/det
% dx/dzt . grad = d psi/dzt
%----------------------------------

 for i=1:10
  rhs = [dps(i),pps(i),qps(i)];
  sln = rhs/Jac;
  gpsi(i,1) = sln(1);
  gpsi(i,2) = sln(2);
  gpsi(i,3) = sln(3);
 end

%-----
% done
%-----

return;
```

TABLE 8.7.2 ($\rightarrow$ Continued) Function *elm10_interp* evaluates the volume metric co-efficient, h_V, and computes the node interpolation functions and their gradients over a 10-node tetrahedron.

```
function [edm, emm, volel] = edmm_t10 ...
 ...
    (x1,y1,z1, x2,y2,z2, x3,y3,z3 ...
   , x4,y4,z4, x5,y5,z5, x6,y6,z6 ...
    ,x7,y7,z7, x8,y8,z8, x9,y9,z9, x10,y10,z10)

%========================================================
% Evaluation of the element diffusion and mass matrices
% for a 10-node tetrahedron,
% using a Gauss-tetrahedron integration quadrature
%
% volel: element volume
%========================================================

%-----------------------------
% read the triangle quadrature
%-----------------------------

NQ = 24;

[xi, eta, zeta, w] = gauss_tetra24;

%-----------------------------------------------------
% initialize the element diffusion and mass matrices
%-----------------------------------------------------

 for k=1:10
  for l=1:10
    edm(k,l) = 0.0;
    emm(k,l) = 0.0;
  end
 end

volel = 0.0;  % element volume (optional)

%-----------------------
% perform the quadrature
%-----------------------

for i=1:NQ

[psi, gpsi, hv] = elmt10_interp ...
 ...
    (x1,y1,z1, x2,y2,z2, x3,y3,z3 ...
    ,x4,y4,z4, x5,y5,z5, x6,y6,z6 ...
    ,x7,y7,z7, x8,y8,z8, x9,y9,z9 ...
    ,x10,y10,z10 ...
    ,xi(i), eta(i), zeta(i));

 cf = hv*w(i)/6.0;
```

TABLE 8.7.3 Function *edmm_t10* (Continuing →)

```
for k=1:10
  for l=1:10
   edm(k,l) = edm(k,l) + (gpsi(k,1)*gpsi(l,1) ...
                       +  gpsi(k,2)*gpsi(l,2) ...
                       +  gpsi(k,3)*gpsi(l,3))*cf;
   emm(k,l) = emm(k,l) + psi(k)*psi(l)*cf;
  end
 end

 volel = volel + cf;

end

%-----
% done
%-----

return;
```

TABLE 8.7.3 ($\rightarrow$ Continued) Function *edmm_t10* evaluates the element diffusion and mass matrices over a 10-node tetrahedron.

```
function [ne,ng,p,c,efl,gfl] = tetra10_sphere8 (ndiv)

%=========================================================
% Discretize a sphere into 10-node tetrahedral elements
% by successive subdivisions of an 8-element structure
% into 8 descendant elements
%
% LEGEND:
%
% efl: element flag
% gfl: global node flag
%=========================================================

act = 1.0;

%----------------------------------------
% parental structure with eight elements
%----------------------------------------

ne = 8;

x(1,1)= 0.0; y(1,1)= 0.0; z(1,1)= 0.0; efl(1,1)=0;  % first element
x(1,2)= act; y(1,2)= 0.0; z(1,2)= 0.0; efl(1,2)=1;
x(1,3)= 0.0; y(1,3)= act; z(1,3)= 0.0; efl(1,3)=1;
x(1,4)= 0.0; y(1,4)= 0.0; z(1,4)= act; efl(1,4)=1;

x(2,1)= 0.0; y(2,1)= 0.0; z(2,1)= 0.0; efl(2,1)=0;  % second element
x(2,2)= 0.0; y(2,2)=-act; z(2,2)= 0.0; efl(2,2)=1;
x(2,3)= act; y(2,3)= 0.0; z(2,3)= 0.0; efl(2,3)=1;
x(2,4)= 0.0; y(2,4)= 0.0; z(2,4)= act; efl(2,4)=1;

x(3,1)= 0.0; y(3,1)= 0.0; z(3,1)= 0.0; efl(3,1)=0;  % third element
x(3,2)= 0.0; y(3,2)= act; z(3,2)= 0.0; efl(3,2)=1;
x(3,3)=-act; y(3,3)= 0.0; z(3,3)= 0.0; efl(3,3)=1;
x(3,4)= 0.0; y(3,4)= 0.0; z(3,4)= act; efl(3,4)=1;

x(4,1)= 0.0; y(4,1)= 0.0; z(4,1)= 0.0; efl(4,1)=0;  % fourth element
x(4,2)=-act; y(4,2)= 0.0; z(4,2)= 0.0; efl(4,2)=1;
x(4,3)= 0.0; y(4,3)=-act; z(4,3)= 0.0; efl(4,3)=1;
x(4,4)= 0.0; y(4,4)= 0.0; z(4,4)= act; efl(4,4)=1;

x(5,1)= 0.0; y(5,1)= 0.0; z(5,1)= 0.0; efl(5,1)=0;  % fifth element
x(5,2)= 0.0; y(5,2)= act; z(5,2)= 0.0; efl(5,3)=1;
x(5,3)= act; y(5,3)= 0.0; z(5,3)= 0.0; efl(5,2)=1;
x(5,4)= 0.0; y(5,4)= 0.0; z(5,4)=-act; efl(5,4)=1;

x(6,1)= 0.0; y(6,1)= 0.0; z(6,1)= 0.0; efl(6,1)=0;  % sixth element
x(6,2)= act; y(6,3)= 0.0; z(6,2)= 0.0; efl(6,2)=1;
x(6,3)= 0.0; y(6,3)=-act; z(6,3)= 0.0; efl(6,3)=1;
x(6,4)= 0.0; y(6,4)= 0.0; z(6,4)=-act; efl(6,4)=1;
```

TABLE 8.7.4 Function *tetra10_sphere8* (Continuing →)

```
x(7,1)= 0.0;  y(7,1)= 0.0;  z(7,1)= 0.0;  efl(7,1)=0;  % seventh element
x(7,2)=-act;  y(7,2)= 0.0;  z(7,2)= 0.0;  efl(7,2)=1;
x(7,3)= 0.0;  y(7,3)= act;  z(7,3)= 0.0;  efl(7,3)=1;
x(7,4)= 0.0;  y(7,4)= 0.0;  z(7,4)=-act;  efl(7,4)=1;

x(8,1)= 0.0;  y(8,1)= 0.0;  z(8,1)= 0.0;  efl(8,1)=0;  % eighth element
x(8,2)= 0.0;  y(8,2)=-act;  z(8,2)= 0.0;  efl(8,2)=1;
x(8,3)=-act;  y(8,3)= 0.0;  z(8,3)= 0.0;  efl(8,3)=1;
x(8,4)= 0.0;  y(8,4)= 0.0;  z(8,4)=-act;  efl(8,4)=1;

%-------------------
% element mid-points
%-------------------

for i=1:ne

  x(i,5)   = 0.5*(x(i,1)+x(i,2));
  y(i,5)   = 0.5*(y(i,1)+y(i,2));
  z(i,5)   = 0.5*(z(i,1)+z(i,2));
  efl(i,5) = 0;

  x(i,6)   = 0.5*(x(i,2)+x(i,3));
  y(i,6)   = 0.5*(y(i,2)+y(i,3));
  z(i,6)   = 0.5*(z(i,2)+z(i,3));
  efl(i,6) = 1;

  x(i,7)   = 0.5*(x(i,1)+x(i,3));
  y(i,7)   = 0.5*(y(i,1)+y(i,3));
  z(i,7)   = 0.5*(z(i,1)+z(i,3));
  efl(i,7) = 0;

  x(i,8)   = 0.5*(x(i,1)+x(i,4));
  y(i,8)   = 0.5*(y(i,1)+y(i,4));
  z(i,8)   = 0.5*(z(i,1)+z(i,4));
  efl(i,8) = 0;

  x(i,9)   = 0.5*(x(i,2)+x(i,4));
  y(i,9)   = 0.5*(y(i,2)+y(i,4));
  z(i,9)   = 0.5*(z(i,2)+z(i,4));
  efl(i,9) = 1;

  x(i,10)  = 0.5*(x(i,3)+x(i,4));
  y(i,10)  = 0.5*(y(i,3)+y(i,4));
  z(i,10)  = 0.5*(z(i,3)+z(i,4));
  efl(i,10) = 1;

end

%----------------
% refinement loop
%----------------
```

TABLE 8.7.4 Function *tetra10_sphere8* ($\rightarrow$ Continuing $\rightarrow$)

```
if(ndiv > 0)

for i=1:ndiv

  nm = 0; % count the new elements arising by each refinement loop
          % eight elements will be generated in each pass

  for j=1:ne    % loop over current elements

    % assign vertex nodes to sub-elements
    % these will become the "new" elements

    nm = nm+1;                            % first sub-element

    xn(nm,1)=x(j,7);  yn(nm,1)=y(j,7);   zn(nm,1)=z(j,7);
                                          efln(nm,1)=efl(j,7);
    xn(nm,2)=x(j,6);  yn(nm,2)=y(j,6);   zn(nm,2)=z(j,6);
                                          efln(nm,2)=efl(j,6);
    xn(nm,3)=x(j,3);  yn(nm,3)=y(j,3);   zn(nm,3)=z(j,3);
                                          efln(nm,3)=efl(j,3);
    xn(nm,4)=x(j,10); yn(nm,4)=y(j,10);  zn(nm,4)=z(j,10);
                                          efln(nm,4)=efl(j,10);

    xn(nm,5) = 0.5*(xn(nm,1)+xn(nm,2));
    yn(nm,5) = 0.5*(yn(nm,1)+yn(nm,2));
    zn(nm,5) = 0.5*(zn(nm,1)+zn(nm,2)); efln(nm,5) = 0;

    xn(nm,6) = 0.5*(xn(nm,2)+xn(nm,3));
    yn(nm,6) = 0.5*(yn(nm,2)+yn(nm,3));
    zn(nm,6) = 0.5*(zn(nm,2)+zn(nm,3)); enfln(nm,6) = 0;

    xn(nm,7) = 0.5*(xn(nm,1)+xn(nm,3));
    yn(nm,7) = 0.5*(yn(nm,1)+yn(nm,3));
    zn(nm,7) = 0.5*(zn(nm,1)+zn(nm,3)); efln(nm,7) = 0;

    xn(nm,8) = 0.5*(xn(nm,1)+xn(nm,4));
    yn(nm,8) = 0.5*(yn(nm,1)+yn(nm,4));
    zn(nm,8) = 0.5*(zn(nm,1)+zn(nm,4)); efln(nm,8) = 0;

    xn(nm,9) = 0.5*(xn(nm,2)+xn(nm,4));
    yn(nm,9) = 0.5*(yn(nm,2)+yn(nm,4));
    zn(nm,9) = 0.5*(zn(nm,2)+zn(nm,4)); efln(nm,9) = 0;

    xn(nm,10) = 0.5*(xn(nm,3)+xn(nm,4));
    yn(nm,10) = 0.5*(yn(nm,3)+yn(nm,4));
    zn(nm,10) = 0.5*(zn(nm,3)+zn(nm,4)); efln(nm,10) = 0;

    if(efln(nm,1)==1 & efln(nm,2)==1) efln(nm,5)  = 1; end
    if(efln(nm,2)==1 & efln(nm,3)==1) efln(nm,6)  = 1; end
    if(efln(nm,1)==1 & efln(nm,3)==1) efln(nm,7)  = 1; end
    if(efln(nm,1)==1 & efln(nm,4)==1) efln(nm,8)  = 1; end
```

TABLE 8.7.4 Function *tetra10_sphere8* ($\rightarrow$ Continuing $\rightarrow$)

```
if(efln(nm,2)==1 & efln(nm,4)==1) efln(nm,9)  = 1; end
if(efln(nm,3)==1 & efln(nm,4)==1) efln(nm,10) = 1; end

nm = nm+1;                 % second sub-element

xn(nm,1)=x(j,5); yn(nm,1)=y(j,5); zn(nm,1)=z(j,5);
                                 efln(nm,1)=efl(j,5);
xn(nm,2)=x(j,2); yn(nm,2)=y(j,2); zn(nm,2)=z(j,2);
                                 efln(nm,2)=efl(j,2);
xn(nm,3)=x(j,6); yn(nm,3)=y(j,6); zn(nm,3)=z(j,6);
                                 efln(nm,3)=efl(j,6);
xn(nm,4)=x(j,9); yn(nm,4)=y(j,9); zn(nm,4)=z(j,9);
                                 efln(nm,4)=efl(j,9);

xn(nm,5) = 0.5*(xn(nm,1)+xn(nm,2));
yn(nm,5) = 0.5*(yn(nm,1)+yn(nm,2));
zn(nm,5) = 0.5*(zn(nm,1)+zn(nm,2)); efln(nm,5) = 0;

xn(nm,6) = 0.5*(xn(nm,2)+xn(nm,3));
yn(nm,6) = 0.5*(yn(nm,2)+yn(nm,3));
zn(nm,6) = 0.5*(zn(nm,2)+zn(nm,3)); enfln(nm,6) = 0;

xn(nm,7)  = 0.5*(xn(nm,1)+xn(nm,3));
yn(nm,7)  = 0.5*(yn(nm,1)+yn(nm,3));
zn(nm,7)  = 0.5*(zn(nm,1)+zn(nm,3)); efln(nm,7) = 0;

xn(nm,8)  = 0.5*(xn(nm,1)+xn(nm,4));
yn(nm,8)  = 0.5*(yn(nm,1)+yn(nm,4));
zn(nm,8)  = 0.5*(zn(nm,1)+zn(nm,4)); efln(nm,8) = 0;

xn(nm,9)  = 0.5*(xn(nm,2)+xn(nm,4));
yn(nm,9)  = 0.5*(yn(nm,2)+yn(nm,4));
zn(nm,9)  = 0.5*(zn(nm,2)+zn(nm,4)); efln(nm,9) = 0;

xn(nm,10) = 0.5*(xn(nm,3)+xn(nm,4));
yn(nm,10) = 0.5*(yn(nm,3)+yn(nm,4));
zn(nm,10) = 0.5*(zn(nm,3)+zn(nm,4)); efln(nm,10) = 0;

if(efln(nm,1)==1 & efln(nm,2)==1) efln(nm,5)  = 1; end
if(efln(nm,2)==1 & efln(nm,3)==1) efln(nm,6)  = 1; end
if(efln(nm,1)==1 & efln(nm,3)==1) efln(nm,7)  = 1; end
if(efln(nm,1)==1 & efln(nm,4)==1) efln(nm,8)  = 1; end
if(efln(nm,2)==1 & efln(nm,4)==1) efln(nm,9)  = 1; end
if(efln(nm,3)==1 & efln(nm,4)==1) efln(nm,10) = 1; end

nm = nm+1;  % third sub-element

xn(nm,1)=x(j,8);  yn(nm,1)=y(j,8);  zn(nm,1)=z(j,8);
                                 efln(nm,1)=efl(j,8);
xn(nm,2)=x(j,9);  yn(nm,2)=y(j,9);  zn(nm,2)=z(j,9);
                                 efln(nm,2)=efl(j,9);
```

TABLE 8.7.4 Function *tetra10_sphere8* ($\rightarrow$ Continuing $\rightarrow$)

```
xn(nm,3)=x(j,10); yn(nm,3)=y(j,10); zn(nm,3)=z(j,10);
                                    efln(nm,3)=efl(j,10);
xn(nm,4)=x(j,4);   yn(nm,4)=y(j,4);   zn(nm,4)=z(j,4);
                                    efln(nm,4)=efl(j,4);

xn(nm,5)  = 0.5*(xn(nm,1)+xn(nm,2));
yn(nm,5)  = 0.5*(yn(nm,1)+yn(nm,2));
zn(nm,5)  = 0.5*(zn(nm,1)+zn(nm,2)); efln(nm,5) = 0;

xn(nm,6)  = 0.5*(xn(nm,2)+xn(nm,3));
yn(nm,6)  = 0.5*(yn(nm,2)+yn(nm,3));
zn(nm,6)  = 0.5*(zn(nm,2)+zn(nm,3)); enfln(nm,6) = 0;

xn(nm,7)  = 0.5*(xn(nm,1)+xn(nm,3));
yn(nm,7)  = 0.5*(yn(nm,1)+yn(nm,3));
zn(nm,7)  = 0.5*(zn(nm,1)+zn(nm,3)); efln(nm,7) = 0;

xn(nm,8)  = 0.5*(xn(nm,1)+xn(nm,4));
yn(nm,8)  = 0.5*(yn(nm,1)+yn(nm,4));
zn(nm,8)  = 0.5*(zn(nm,1)+zn(nm,4)); efln(nm,8) = 0;

xn(nm,9)  = 0.5*(xn(nm,2)+xn(nm,4));
yn(nm,9)  = 0.5*(yn(nm,2)+yn(nm,4));
zn(nm,9)  = 0.5*(zn(nm,2)+zn(nm,4)); efln(nm,9) = 0;

xn(nm,10) = 0.5*(xn(nm,3)+xn(nm,4));
yn(nm,10) = 0.5*(yn(nm,3)+yn(nm,4));
zn(nm,10) = 0.5*(zn(nm,3)+zn(nm,4)); efln(nm,10) = 0;

if(efln(nm,1)==1 & efln(nm,2)==1) efln(nm,5)  = 1; end
if(efln(nm,2)==1 & efln(nm,3)==1) efln(nm,6)  = 1; end
if(efln(nm,1)==1 & efln(nm,3)==1) efln(nm,7)  = 1; end
if(efln(nm,1)==1 & efln(nm,4)==1) efln(nm,8)  = 1; end
if(efln(nm,2)==1 & efln(nm,4)==1) efln(nm,9)  = 1; end
if(efln(nm,3)==1 & efln(nm,4)==1) efln(nm,10) = 1; end

nm = nm+1;  % fourth sub-element

xn(nm,1)=x(j,1); yn(nm,1)=y(j,1); zn(nm,1)=z(j,1);
                                    efln(nm,1)=efl(j,1);
xn(nm,2)=x(j,5); yn(nm,2)=y(j,5); zn(nm,2)=z(j,5);
                                    efln(nm,2)=efl(j,5);
xn(nm,3)=x(j,7); yn(nm,3)=y(j,7); zn(nm,3)=z(j,7);
                                    efln(nm,3)=efl(j,7);
xn(nm,4)=x(j,8); yn(nm,4)=y(j,8); zn(nm,4)=z(j,8);
                                    efln(nm,4)=efl(j,8);

xn(nm,5)  = 0.5*(xn(nm,1)+xn(nm,2));
yn(nm,5)  = 0.5*(yn(nm,1)+yn(nm,2));
zn(nm,5)  = 0.5*(zn(nm,1)+zn(nm,2)); efln(nm,5) = 0;
```

TABLE 8.7.4 Function *tetra10_sphere8* ($\rightarrow$ Continuing $\rightarrow$)

```
xn(nm,6)  = 0.5*(xn(nm,2)+xn(nm,3));
yn(nm,6)  = 0.5*(yn(nm,2)+yn(nm,3));
zn(nm,6)  = 0.5*(zn(nm,2)+zn(nm,3)); enfln(nm,6) = 0;

xn(nm,7)  = 0.5*(xn(nm,1)+xn(nm,3));
yn(nm,7)  = 0.5*(yn(nm,1)+yn(nm,3));
zn(nm,7)  = 0.5*(zn(nm,1)+zn(nm,3)); efln(nm,7) = 0;

xn(nm,8)  = 0.5*(xn(nm,1)+xn(nm,4));
yn(nm,8)  = 0.5*(yn(nm,1)+yn(nm,4));
zn(nm,8)  = 0.5*(zn(nm,1)+zn(nm,4)); efln(nm,8) = 0;

xn(nm,9)  = 0.5*(xn(nm,2)+xn(nm,4));
yn(nm,9)  = 0.5*(yn(nm,2)+yn(nm,4));
zn(nm,9)  = 0.5*(zn(nm,2)+zn(nm,4)); efln(nm,9) = 0;

xn(nm,10) = 0.5*(xn(nm,3)+xn(nm,4));
yn(nm,10) = 0.5*(yn(nm,3)+yn(nm,4));
zn(nm,10) = 0.5*(zn(nm,3)+zn(nm,4)); efln(nm,10) = 0;

if(efln(nm,1)==1 & efln(nm,2)==1) efln(nm,5)  = 1; end
if(efln(nm,2)==1 & efln(nm,3)==1) efln(nm,6)  = 1; end
if(efln(nm,1)==1 & efln(nm,3)==1) efln(nm,7)  = 1; end
if(efln(nm,1)==1 & efln(nm,4)==1) efln(nm,8)  = 1; end
if(efln(nm,2)==1 & efln(nm,4)==1) efln(nm,9)  = 1; end
if(efln(nm,3)==1 & efln(nm,4)==1) efln(nm,10) = 1; end

nm = nm+1;  % fifth sub-element

xn(nm,1)=x(j,10); yn(nm,1)=y(j,10); zn(nm,1)=z(j,10);
                                    efln(nm,1)=efl(j,10);
xn(nm,2)=x(j,9);  yn(nm,2)=y(j,9);  zn(nm,2)=z(j,9);
                                    efln(nm,2)=efl(j,9);
xn(nm,3)=x(j,8);  yn(nm,3)=y(j,8);  zn(nm,3)=z(j,8);
                                    efln(nm,3)=efl(j,8);
xn(nm,4)=x(j,7);  yn(nm,4)=y(j,7);  zn(nm,4)=z(j,7);
                                    efln(nm,4)=efl(j,7);

xn(nm,5)  = 0.5*(xn(nm,1)+xn(nm,2));
yn(nm,5)  = 0.5*(yn(nm,1)+yn(nm,2));
zn(nm,5)  = 0.5*(zn(nm,1)+zn(nm,2)); efln(nm,5) = 0;

xn(nm,6)  = 0.5*(xn(nm,2)+xn(nm,3));
yn(nm,6)  = 0.5*(yn(nm,2)+yn(nm,3));
zn(nm,6)  = 0.5*(zn(nm,2)+zn(nm,3)); enfln(nm,6) = 0;

xn(nm,7)  = 0.5*(xn(nm,1)+xn(nm,3));
yn(nm,7)  = 0.5*(yn(nm,1)+yn(nm,3));
zn(nm,7)  = 0.5*(zn(nm,1)+zn(nm,3)); efln(nm,7) = 0;
```

TABLE 8.7.4 Function *tetra10_sphere8* ($\rightarrow$ Continuing $\rightarrow$)

```
xn(nm,8)  = 0.5*(xn(nm,1)+xn(nm,4));
yn(nm,8)  = 0.5*(yn(nm,1)+yn(nm,4));
zn(nm,8)  = 0.5*(zn(nm,1)+zn(nm,4));  efln(nm,8) = 0;

xn(nm,9)  = 0.5*(xn(nm,2)+xn(nm,4));
yn(nm,9)  = 0.5*(yn(nm,2)+yn(nm,4));
zn(nm,9)  = 0.5*(zn(nm,2)+zn(nm,4));  efln(nm,9) = 0;

xn(nm,10) = 0.5*(xn(nm,3)+xn(nm,4));
yn(nm,10) = 0.5*(yn(nm,3)+yn(nm,4));
zn(nm,10) = 0.5*(zn(nm,3)+zn(nm,4));  efln(nm,10) = 0;

if(efln(nm,1)==1 & efln(nm,2)==1) efln(nm,5)  = 1; end
if(efln(nm,2)==1 & efln(nm,3)==1) efln(nm,6)  = 1; end
if(efln(nm,1)==1 & efln(nm,3)==1) efln(nm,7)  = 1; end
if(efln(nm,1)==1 & efln(nm,4)==1) efln(nm,8)  = 1; end
if(efln(nm,2)==1 & efln(nm,4)==1) efln(nm,9)  = 1; end
if(efln(nm,3)==1 & efln(nm,4)==1) efln(nm,10) = 1; end

nm = nm+1;  % sixth sub-element

xn(nm,1)=x(j,8); yn(nm,1)=y(j,8); zn(nm,1)=z(j,8);
                              efln(nm,1)=efl(j,8);
xn(nm,2)=x(j,5); yn(nm,2)=y(j,5); zn(nm,2)=z(j,5);
                              efln(nm,2)=efl(j,5);
xn(nm,3)=x(j,7); yn(nm,3)=y(j,7); zn(nm,3)=z(j,7);
                              efln(nm,3)=efl(j,7);
xn(nm,4)=x(j,9); yn(nm,4)=y(j,9); zn(nm,4)=z(j,9);
                              efln(nm,4)=efl(j,9);

xn(nm,5)  = 0.5*(xn(nm,1)+xn(nm,2));
yn(nm,5)  = 0.5*(yn(nm,1)+yn(nm,2));
zn(nm,5)  = 0.5*(zn(nm,1)+zn(nm,2));  efln(nm,5) = 0;

xn(nm,6)  = 0.5*(xn(nm,2)+xn(nm,3));
yn(nm,6)  = 0.5*(yn(nm,2)+yn(nm,3));
zn(nm,6)  = 0.5*(zn(nm,2)+zn(nm,3));  enfln(nm,6) = 0;

xn(nm,7)  = 0.5*(xn(nm,1)+xn(nm,3));
yn(nm,7)  = 0.5*(yn(nm,1)+yn(nm,3));
zn(nm,7)  = 0.5*(zn(nm,1)+zn(nm,3));  efln(nm,7) = 0;

xn(nm,8)  = 0.5*(xn(nm,1)+xn(nm,4));
yn(nm,8)  = 0.5*(yn(nm,1)+yn(nm,4));
zn(nm,8)  = 0.5*(zn(nm,1)+zn(nm,4));  efln(nm,8) = 0;
```

TABLE 8.7.4 Function *tetra10_sphere8* ($\rightarrow$ Continuing $\rightarrow$)

```
xn(nm,9)  = 0.5*(xn(nm,2)+xn(nm,4));
yn(nm,9)  = 0.5*(yn(nm,2)+yn(nm,4));
zn(nm,9)  = 0.5*(zn(nm,2)+zn(nm,4)); efln(nm,9)  = 0;

xn(nm,10) = 0.5*(xn(nm,3)+xn(nm,4));
yn(nm,10) = 0.5*(yn(nm,3)+yn(nm,4));
zn(nm,10) = 0.5*(zn(nm,3)+zn(nm,4)); efln(nm,10) = 0;

if(efln(nm,1)==1 & efln(nm,2)==1) efln(nm,5)  = 1; end
if(efln(nm,2)==1 & efln(nm,3)==1) efln(nm,6)  = 1; end
if(efln(nm,1)==1 & efln(nm,3)==1) efln(nm,7)  = 1; end
if(efln(nm,1)==1 & efln(nm,4)==1) efln(nm,8)  = 1; end
if(efln(nm,2)==1 & efln(nm,4)==1) efln(nm,9)  = 1; end
if(efln(nm,3)==1 & efln(nm,4)==1) efln(nm,10) = 1; end

nm = nm+1;  % seventh sub-element

xn(nm,1)=x(j,5); yn(nm,1)=y(j,5); zn(nm,1)=z(j,5);
                              efln(nm,1)=efl(j,5);
xn(nm,2)=x(j,6); yn(nm,2)=y(j,6); zn(nm,2)=z(j,6);
                              efln(nm,2)=efl(j,6);
xn(nm,3)=x(j,7); yn(nm,3)=y(j,7); zn(nm,3)=z(j,7);
                              efln(nm,3)=efl(j,7);
xn(nm,4)=x(j,9); yn(nm,4)=y(j,9); zn(nm,4)=z(j,9);
                              efln(nm,4)=efl(j,9);

xn(nm,5)  = 0.5*(xn(nm,1)+xn(nm,2));
yn(nm,5)  = 0.5*(yn(nm,1)+yn(nm,2));
zn(nm,5)  = 0.5*(zn(nm,1)+zn(nm,2)); efln(nm,5)  = 0;

xn(nm,6)  = 0.5*(xn(nm,2)+xn(nm,3));
yn(nm,6)  = 0.5*(yn(nm,2)+yn(nm,3));
zn(nm,6)  = 0.5*(zn(nm,2)+zn(nm,3)); enfln(nm,6) = 0;

xn(nm,7)  = 0.5*(xn(nm,1)+xn(nm,3));
yn(nm,7)  = 0.5*(yn(nm,1)+yn(nm,3));
zn(nm,7)  = 0.5*(zn(nm,1)+zn(nm,3)); efln(nm,7)  = 0;

xn(nm,8)  = 0.5*(xn(nm,1)+xn(nm,4));
yn(nm,8)  = 0.5*(yn(nm,1)+yn(nm,4));
zn(nm,8)  = 0.5*(zn(nm,1)+zn(nm,4)); efln(nm,8)  = 0;

xn(nm,9)  = 0.5*(xn(nm,2)+xn(nm,4));
yn(nm,9)  = 0.5*(yn(nm,2)+yn(nm,4));
zn(nm,9)  = 0.5*(zn(nm,2)+zn(nm,4)); efln(nm,9)  = 0;

xn(nm,10) = 0.5*(xn(nm,3)+xn(nm,4));
yn(nm,10) = 0.5*(yn(nm,3)+yn(nm,4));
zn(nm,10) = 0.5*(zn(nm,3)+zn(nm,4)); efln(nm,10) = 0;
```

TABLE 8.7.4 Function *tetra10_sphere8* ($\rightarrow$ Continuing $\rightarrow$)

```
if(efln(nm,1)==1 & efln(nm,2)==1) efln(nm,5)  = 1; end
if(efln(nm,2)==1 & efln(nm,3)==1) efln(nm,6)  = 1; end
if(efln(nm,1)==1 & efln(nm,3)==1) efln(nm,7)  = 1; end
if(efln(nm,1)==1 & efln(nm,4)==1) efln(nm,8)  = 1; end
if(efln(nm,2)==1 & efln(nm,4)==1) efln(nm,9)  = 1; end
if(efln(nm,3)==1 & efln(nm,4)==1) efln(nm,10) = 1; end

nm = nm+1;  % eighth sub-element

xn(nm,1)=x(j,7);   yn(nm,1)=y(j,7);   zn(nm,1)=z(j,7);
                                  efln(nm,1)=efl(j,7);
xn(nm,2)=x(j,9);   yn(nm,2)=y(j,9);   zn(nm,2)=z(j,9);
                                  efln(nm,2)=efl(j,9);
xn(nm,3)=x(j,6);   yn(nm,3)=y(j,6);   zn(nm,3)=z(j,6);
                                  efln(nm,3)=efl(j,6);
xn(nm,4)=x(j,10);  yn(nm,4)=y(j,10);  zn(nm,4)=z(j,10);
                                  efln(nm,4)=efl(j,10);

xn(nm,5)  = 0.5*(xn(nm,1)+xn(nm,2));
yn(nm,5)  = 0.5*(yn(nm,1)+yn(nm,2));
zn(nm,5)  = 0.5*(zn(nm,1)+zn(nm,2)); efln(nm,5) = 0;

xn(nm,6)  = 0.5*(xn(nm,2)+xn(nm,3));
yn(nm,6)  = 0.5*(yn(nm,2)+yn(nm,3));
zn(nm,6)  = 0.5*(zn(nm,2)+zn(nm,3)); enfln(nm,6) = 0;

xn(nm,7)  = 0.5*(xn(nm,1)+xn(nm,3));
yn(nm,7)  = 0.5*(yn(nm,1)+yn(nm,3));
zn(nm,7)  = 0.5*(zn(nm,1)+zn(nm,3)); efln(nm,7) = 0;

xn(nm,8)  = 0.5*(xn(nm,1)+xn(nm,4));
yn(nm,8)  = 0.5*(yn(nm,1)+yn(nm,4));
zn(nm,8)  = 0.5*(zn(nm,1)+zn(nm,4)); efln(nm,8) = 0;

xn(nm,9)  = 0.5*(xn(nm,2)+xn(nm,4));
yn(nm,9)  = 0.5*(yn(nm,2)+yn(nm,4));
zn(nm,9)  = 0.5*(zn(nm,2)+zn(nm,4)); efln(nm,9) = 0;

xn(nm,10) = 0.5*(xn(nm,3)+xn(nm,4));
yn(nm,10) = 0.5*(yn(nm,3)+yn(nm,4));
zn(nm,10) = 0.5*(zn(nm,3)+zn(nm,4)); efln(nm,10) = 0;

if(efln(nm,1)==1 & efln(nm,2)==1) efln(nm,5)  = 1; end
if(efln(nm,2)==1 & efln(nm,3)==1) efln(nm,6)  = 1; end
if(efln(nm,1)==1 & efln(nm,3)==1) efln(nm,7)  = 1; end
if(efln(nm,1)==1 & efln(nm,4)==1) efln(nm,8)  = 1; end
if(efln(nm,2)==1 & efln(nm,4)==1) efln(nm,9)  = 1; end
if(efln(nm,3)==1 & efln(nm,4)==1) efln(nm,10) = 1; end
```

TABLE 8.7.4 Function *tetra10_sphere8* ($\rightarrow$ Continuing $\rightarrow$)

```
end % end of loop over current elements
%---

ne = 8*ne;   % number of elements has increased
             % by a factor of eight

for k=1:ne      % relabel the new points and put them in a master list
                % project boundary nodes onto a sphere of radius "act"
    for l=1:10
       x(k,l) = xn(k,l); y(k,l) = yn(k,l); z(k,l) = zn(k,l);
      efl(k,l) = efln(k,l);
       if( efl(k,l)==1 )
         rad = sqrt(x(k,l)^2+y(k,l)^2+z(k,l)^2);
         fc = act/rad;
         x(k,l) = fc*x(k,l); y(k,l) = fc*y(k,l); z(k,l) = fc*z(k,l);
       end
    end

%---
    if( efl(k,5)==0 )
     x(k,5) = 0.5*(x(k,1)+x(k,2)); y(k,5) = 0.5*(y(k,1)+y(k,2));
     z(k,5) = 0.5*(z(k,1)+z(k,2));
    end
    if( efl(k,6)==0 )
     x(k,6) = 0.5*(x(k,2)+x(k,3)); y(k,6) = 0.5*(y(k,2)+y(k,3));
     z(k,6) = 0.5*(z(k,2)+z(k,3));
    end
    if( efl(k,7)==0 )
     x(k,7) = 0.5*(x(k,1)+x(k,3)); y(k,7) = 0.5*(y(k,1)+y(k,3));
     z(k,7) = 0.5*(z(k,1)+z(k,3));
    end
%---
    if( efl(k,8)==0 )
     x(k,8) = 0.5*(x(k,1)+x(k,4)); y(k,8) = 0.5*(y(k,1)+y(k,4));
     z(k,8) = 0.5*(z(k,1)+z(k,4));
    end
    if( efl(k,9)==0 )
     x(k,9) = 0.5*(x(k,2)+x(k,4)); y(k,9) = 0.5*(y(k,2)+y(k,4));
     z(k,9) = 0.5*(z(k,2)+z(k,4));
    end
    if( efl(k,10)==0 )
     x(k,10) = 0.5*(x(k,3)+x(k,4)); y(k,10) = 0.5*(y(k,3)+y(k,4));
     z(k,10) = 0.5*(z(k,3)+z(k,4));
    end

end % of look over k
```

TABLE 8.7.4 Function *tetra10_sphere8* ($\rightarrow$ Continuing $\rightarrow$)

```
end % end for of refinement loop
end % end if of refinement loop

%------------------------------------------------------
% define the global nodes and the connectivity table
%------------------------------------------------------

% 10 nodes of the first element are entered manually

p(1,1) =x(1,1);   p(1,2) =y(1,1);   p(1,3) =z(1,1);   gfl(1)=efl(1,1);
p(2,1) =x(1,2);   p(2,2) =y(1,2);   p(2,3) =z(1,2);   gfl(2)=efl(1,2);
p(3,1) =x(1,3);   p(3,2) =y(1,3);   p(3,3) =z(1,3);   gfl(3)=efl(1,3);
p(4,1) =x(1,4);   p(4,2) =y(1,4);   p(4,3) =z(1,4);   gfl(4)=efl(1,4);
p(5,1) =x(1,5);   p(5,2) =y(1,5);   p(5,3) =z(1,5);   gfl(5)=efl(1,5);
p(6,1) =x(1,6);   p(6,2) =y(1,6);   p(6,3) =z(1,6);   gfl(6)=efl(1,6);
p(7,1) =x(1,7);   p(7,2) =y(1,7);   p(7,3) =z(1,7);   gfl(7)=efl(1,7);
p(8,1) =x(1,8);   p(8,2) =y(1,8);   p(8,3) =z(1,8);   gfl(8)=efl(1,8);
p(9,1) =x(1,9);   p(9,2) =y(1,9);   p(9,3) =z(1,9);   gfl(9)=efl(1,9);
p(10,1)=x(1,10);  p(10,2)=y(1,10);  p(10,3)=z(1,10);  gfl(10)=efl(1,10);

c(1,1)  = 1;  % first  node of first element is global node 1
c(1,2)  = 2;  % second node of first element is global node 2
c(1,3)  = 3;  % third  node of first element is global node 3
c(1,4)  = 4;  % fourth node of first element is global node 4
c(1,5)  = 5;
c(1,6)  = 6;
c(1,7)  = 7;
c(1,8)  = 8;
c(1,9)  = 9;
c(1,10) = 10;

ng = 10;

%---
% loop over further elements
% Iflag=0 will signal a new global node
%---

eps = 0.00001;  % tolerance
```

TABLE 8.7.4 Function *tetra10_sphere8* ($\rightarrow$ Continuing $\rightarrow$)

```
for i=2:ne        % loop over the rest of the elements
 for j=1:10        % loop over element nodes

 Iflag=0;
 for k=1:ng

  if(abs(x(i,j)-p(k,1)) < eps)
   if(abs(y(i,j)-p(k,2)) < eps)
    if(abs(z(i,j)-p(k,3)) < eps)

    Iflag = 1;    % the node has been recorded previously
    c(i,j) = k;    % the jth local node of element i is the kth global node

    end
   end
  end

 end
 if(Iflag==0)  % record the node
  ng = ng+1;
   p(ng,1) =   x(i,j);
   p(ng,2) =   y(i,j);
   p(ng,3) =   z(i,j);
  gfl(ng)    = efl(i,j);
  c(i,j) = ng;   % the jth local node of element is the new global node
 end

 end
end   % end of loop over elements

%-----
% done
%-----

return;
```

Table 8.7.4 ($\rightarrow$ Continued) Function *tetra10_sphere8* discretizes a sphere into 10-node tetrahedral elements by successively subdividing an assembly of eight ancestral hard-coded elements (root set).

(*a*)

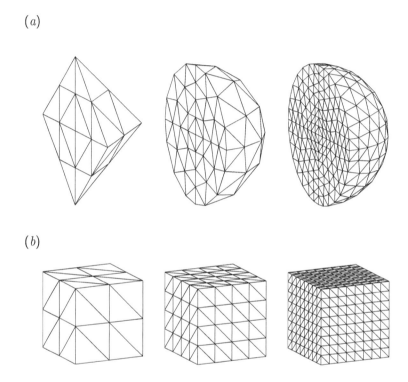

(*b*)

FIGURE 8.7.2　(*a*) Discretization of a sphere into 10-node tetrahedra generated by the
　　FSELIB function *tetra10_sphere8* for $ndiv = 0$ (root set), 1, and 2 (right). For
　　clarity, only half of the sphere is shown. (*b*) Discretization of a cube into 10-node
　　tetrahedra generated by the FSELIB function *tetra10_cube8* for $ndiv = 0$ (root
　　set), 1, and 2 (right).

8.7.4   Laplace's equation

FSELIB code *laplt10_d*, listed in Table 8.7.5, solves Laplace's equation

$$\nabla^2 f = 0. \tag{8.7.5}$$

The Dirichlet boundary condition specifying the surface distribution of the unknown function, f, is prescribed, as shown in the code.

The finite element grid is generated by the FSELIB function *tetra10_sphere8* or the function *tetra10_cube8*, as discussed in Section 8.7.3. The element diffusion matrices are computed by the function *edmm_t10*, listed in Table 8.7.3. The finite element solution is visualized by the function *plot_t10*, listed in Table 8.7.6. Function *plot_t10* calls the function *tetra10_vis*, listed in Table 8.7.7, to visualize the individual elements.

The finite element solution for the boundary condition implemented in the code is displayed in Figure 8.7.3.

8.7.5   Eigenvalues of the Laplacian operator

FSELIB code *laplt10_eig*, listed in Table 8.7.8, computes the eigenvalues, λ, and corresponding eigenfunctions of the Laplacian operator in three-dimensional space, $f(x, y, z)$, satisfying the equation

$$\nabla^2 f + \lambda f = 0. \tag{8.7.6}$$

The finite element grid for a sphere is generated by the function *tetra10_sphere8* discussed in Section 8.7.3. The element diffusion and mass matrices are computed by the function *edmm_t10*, listed in Table 8.7.3. The general structure of the finite element code is similar to that listed in Table 8.4.2 for four-node tetrahedra.

Numerical computations for a sphere with 8 and 64 elements, differing by a factor of 8, corresponding to discretization levels $ndiv = 0$ or 1, predict that the smallest eigenvalue is, respectively, 30 and 10.1009. As the number of elements increases, the results converge to the exact value, $\pi^2 = 9.8696$.

PROBLEMS

8.7.1 *Linear field*

Confirm that the code *laplt10_d* generates the exact solution for a linear field along the x, y, or z axis.

8.7.2 *Eigenvalues*

Study the convergence of the eigenvalues of the Laplacian operator for a sphere.

```
%==========================================
% Code lapl10_d
%
% Solution of Laplace's equation with
% the Dirichlet boundary condition
% using 10-node tetrahedral elements
%==========================================

%-----------
% input data
%-----------

ndiv = 2;  % discretization level

%-----------
% discretize
%-----------

[ne,ng,p,c,efl,gfl] = tetra10_sphere8 (ndiv);

%--------------------------------------------------
% specify the Dirichlet boundary condition (bcd)
%--------------------------------------------------

for i=1:ng
 bcd(i) = 0.0;   % default
 if(gfl(i)==1)
   bcd(i) = sin(pi*(p(i,1)+p(i,2)));    % example
 end
end

%--------------------------------------
% assemble the global diffusion matrix
%--------------------------------------

gdm = zeros(ng,ng); % initialize

volume = 0.0;

for l=1:ne            % loop over the elements

% compute the element diffusion matrix

j = c(l,1);  x1 = p(j,1); y1 = p(j,2); z1 = p(j,3);
j = c(l,2);  x2 = p(j,1); y2 = p(j,2); z2 = p(j,3);
j = c(l,3);  x3 = p(j,1); y3 = p(j,2); z3 = p(j,3);
j = c(l,4);  x4 = p(j,1); y4 = p(j,2); z4 = p(j,3);
j = c(l,5);  x5 = p(j,1); y5 = p(j,2); z5 = p(j,3);
j = c(l,6);  x6 = p(j,1); y6 = p(j,2); z6 = p(j,3);
j = c(l,7);  x7 = p(j,1); y7 = p(j,2); z7 = p(j,3);
j = c(l,8);  x8 = p(j,1); y8 = p(j,2); z8 = p(j,3);
j = c(l,9);  x9 = p(j,1); y9 = p(j,2); z9 = p(j,3);
j = c(l,10); x10= p(j,1); y10= p(j,2); z10= p(j,3);
```

TABLE 8.7.5   Code *laplt10_d* (Continuing →)

```
[edm_elm, emm_elm, vlm_elm] = edmm_t10 ...
...
   (x1,y1,z1, x2,y2,z2, x3,y3,z3, x4,y4,z4, x5,y5,z5, x6,y6,z6 ...
   ,x7,y7,z7, x8,y8,z8, x9,y9,z9, x10,y10,z10);

 volume = volume + vlm_elm;

   for i=1:10
     i1 = c(1,i);
     for j=1:10
       j1 = c(1,j);
       gdm(i1,j1) = gdm(i1,j1) + edm_elm(i,j);
     end
   end

end

%-----------------------------------------------
% set the right-hand side of the linear system
% and implement the Dirichlet boundary condition
%-----------------------------------------------

for i=1:ng
 rhs(i) = 0.0;
end

for j=1:ng
 if(gfl(j)==1)
   for i=1:ng
     rhs(i) = rhs(i) - gdm(i,j)*bcd(j);
     gdm(i,j) = 0.0; gdm(j,i) = 0.0;
   end
   gdm(j,j) = 1.0;
   rhs(j) = bcd(j);
 end
end

%-----------------------
% solve the linear system
%-----------------------

f = rhs/gdm';

%-----
% plot
%-----

plot_t10 (ne,ng,p,c,f)

%-----
% done
%-----
```

TABLE 8.7.5   ($\rightarrow$ Continued) Code *laplt10_d* solves Laplace's equation subject to the
Dirichlet boundary condition using 10-node tetrahedral elements.

```
function plot_t10 (ne,ng,p,c,f)

%==================================================
% Color mapped visualization of a function, f,
% in a domain discretized into 10-node tetrahedra
%==================================================

%-----------------
% plotting option
%--------------

option = 2; % wireframe
option = 1; % colors
option = 4; % graded

%------------------------------------------------------
% compute the maximum and minimum of the function f
%------------------------------------------------------

fmin =   100.0;
fmax = -100.0;

for i=1:ng
 if(f(i) > fmax) fmax = f(i); end
 if(f(i) < fmin) fmin = f(i); end
end

range = 1.2*(fmax-fmin);
shift = fmin;

%-------------------------------------------------
% shift the color index in the range (0, 1)
% and plot four-node elements
%-------------------------------------------------

for l=1:ne

 j = c(l,1); xp(1)=p(j,1); yp(1)=p(j,2); zp(1)=p(j,3);
             cp(1)=(f(j)-shift)/range;
 j = c(l,2); xp(2)=p(j,1); yp(2)=p(j,2); zp(2)=p(j,3);
             cp(2)=(f(j)-shift)/range;
 j = c(l,3); xp(3)=p(j,1); yp(3)=p(j,2); zp(3)=p(j,3);
             cp(3)=(f(j)-shift)/range;
 j = c(l,4); xp(4)=p(j,1); yp(4)=p(j,2); zp(4)=p(j,3);
             cp(4)=(f(j)-shift)/range;
 j = c(l,5); xp(5)=p(j,1); yp(5)=p(j,2); zp(5)=p(j,3);
             cp(5)=(f(j)-shift)/range;
 j = c(l,6); xp(6)=p(j,1); yp(6)=p(j,2); zp(6)=p(j,3);
             cp(6)=(f(j)-shift)/range;
 j = c(l,7); xp(7)=p(j,1); yp(7)=p(j,2); zp(7)=p(j,3);
             cp(7)=(f(j)-shift)/range;
```

TABLE 8.7.6   Function *plot_t10* (Continuing →)

```
j = c(1,8);  xp(8)=p(j,1);  yp(8)=p(j,2);  zp(8)=p(j,3);
             cp(8)=(f(j)-shift)/range;
j = c(1,9);  xp(9)=p(j,1);  yp(9)=p(j,2);  zp(9)=p(j,3);
             cp(9)=(f(j)-shift)/range;
j = c(1,10);xp(10)=p(j,1);  yp(10)=p(j,2);  zp(10)=p(j,3);
             cp(10)=(f(j)-shift)/range;

volume = tetra10_vis (xp,yp,zp,cp,option);

end

%-----
% done
%-----

return;
```

TABLE 8.7.6   ($\rightarrow$ Continued) Function *plot_t10* visualizes the finite element solution for 10-node tetrahedral elements.

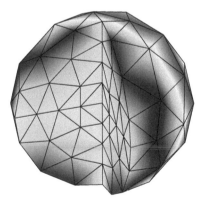

FIGURE 8.7.3   Finite element solution of Laplace's equation with the Dirichlet boundary condition computed with 10-node tetrahedral elements.

```
function volume = tetra10_vis (x,y,z,f,option)

%=================================
% visualize a 10-node tetrahedron
%=================================

%---
% paint the faces:
%---

for i=1:16

  if(i==1)
    xp(1)=x(1); yp(1)=y(1); zp(1)=z(1); fp(1)=f(1);
    xp(2)=x(5); yp(2)=y(5); zp(2)=z(5); fp(2)=f(5);
    xp(3)=x(7); yp(3)=y(7); zp(3)=z(7); fp(3)=f(7);
    xp(4)=x(1); yp(4)=y(1); zp(4)=z(1); fp(4)=f(1);
  elseif(i==2)
    xp(1)=x(2); yp(1)=y(2); zp(1)=z(2); fp(1)=f(2);
    xp(2)=x(6); yp(2)=y(6); zp(2)=z(6); fp(2)=f(6);
    xp(3)=x(5); yp(3)=y(5); zp(3)=z(5); fp(3)=f(5);
    xp(4)=x(2); yp(4)=y(2); zp(4)=z(2); fp(4)=f(2);
  elseif(i==3)
    xp(1)=x(3); yp(1)=y(3); zp(1)=z(3); fp(1)=f(3);
    xp(2)=x(7); yp(2)=y(7); zp(2)=z(7); fp(2)=f(7);
    xp(3)=x(6); yp(3)=y(6); zp(3)=z(6); fp(3)=f(6);
    xp(4)=x(3); yp(4)=y(3); zp(4)=z(3); fp(4)=f(3);
  elseif(i==4)
    xp(1)=x(5); yp(1)=y(5); zp(1)=z(5); fp(1)=f(5);
    xp(2)=x(6); yp(2)=y(6); zp(2)=z(6); fp(2)=f(6);
    xp(3)=x(7); yp(3)=y(7); zp(3)=z(7); fp(3)=f(7);
    xp(4)=x(5); yp(4)=y(5); zp(4)=z(5); fp(4)=f(5);
  elseif(i==5)
    xp(1)=x(1); yp(1)=y(1); zp(1)=z(1); fp(1)=f(1);
    xp(2)=x(5); yp(2)=y(5); zp(2)=z(5); fp(2)=f(5);
    xp(3)=x(8); yp(3)=y(8); zp(3)=z(8); fp(3)=f(8);
    xp(4)=x(1); yp(4)=y(1); zp(4)=z(1); fp(4)=f(1);
  elseif(i==6)
    xp(1)=x(2); yp(1)=y(2); zp(1)=z(2); fp(1)=f(2);
    xp(2)=x(9); yp(2)=y(9); zp(2)=z(9); fp(2)=f(9);
    xp(3)=x(5); yp(3)=y(5); zp(3)=z(5); fp(3)=f(5);
    xp(4)=x(2); yp(4)=y(2); zp(4)=z(2); fp(4)=f(2);
  elseif(i==7)
    xp(1)=x(4); yp(1)=y(4); zp(1)=z(4); fp(1)=f(4);
    xp(2)=x(8); yp(2)=y(8); zp(2)=z(8); fp(2)=f(8);
    xp(3)=x(9); yp(3)=y(9); zp(3)=z(9); fp(3)=f(9);
    xp(4)=x(4); yp(4)=y(4); zp(4)=z(4); fp(4)=f(4);
  elseif(i==8)
    xp(1)=x(5); yp(1)=y(5); zp(1)=z(5); fp(1)=f(5);
    xp(2)=x(9); yp(2)=y(9); zp(2)=z(9); fp(2)=f(9);
    xp(3)=x(8); yp(3)=y(8); zp(3)=z(8); fp(3)=f(8);
    xp(4)=x(5); yp(4)=y(5); zp(4)=z(5); fp(4)=f(5);
```

TABLE 8.7.7   Function *tetra10_vis* (Continuing →)

```
elseif(i==9)
  xp(1)=x(1);  yp(1)=y(1);  zp(1)=z(1);  fp(1)=f(1);
  xp(2)=x(8);  yp(2)=y(8);  zp(2)=z(8);  fp(2)=f(8);
  xp(3)=x(7);  yp(3)=y(7);  zp(3)=z(7);  fp(3)=f(7);
  xp(4)=x(1);  yp(4)=y(1);  zp(4)=z(1);  fp(4)=f(1);
elseif(i==10)
  xp(1)=x(8);   yp(1)=y(8);   zp(1)=z(8);   fp(1)=f(8);
  xp(2)=x(4);   yp(2)=y(4);   zp(2)=z(4);   fp(2)=f(4);
  xp(3)=x(10);  yp(3)=y(10);  zp(3)=z(10);  fp(3)=f(10);
  xp(4)=x(8);   yp(4)=y(8);   zp(4)=z(8);   fp(4)=f(8);
elseif(i==11)
  xp(1)=x(3);   yp(1)=y(3);   zp(1)=z(3);   fp(1)=f(3);
  xp(2)=x(7);   yp(2)=y(7);   zp(2)=z(7);   fp(2)=f(7);
  xp(3)=x(10);  yp(3)=y(10);  zp(3)=z(10);  fp(3)=f(10);
  xp(4)=x(3);   yp(4)=y(3);   zp(4)=z(3);   fp(4)=f(3);
elseif(i==12)
  xp(1)=x(7);   yp(1)=y(7);   zp(1)=z(7);   fp(1)=f(7);
  xp(2)=x(8);   yp(2)=y(8);   zp(2)=z(8);   fp(2)=f(8);
  xp(3)=x(10);  yp(3)=y(10);  zp(3)=z(10);  fp(3)=f(10);
  xp(4)=x(7);   yp(4)=y(7);   zp(4)=z(7);   fp(4)=f(7);
elseif(i==13)
  xp(1)=x(2);  yp(1)=y(2);  zp(1)=z(2);  fp(1)=f(2);
  xp(2)=x(6);  yp(2)=y(6);  zp(2)=z(6);  fp(2)=f(6);
  xp(3)=x(9);  yp(3)=y(9);  zp(3)=z(9);  fp(3)=f(9);
  xp(4)=x(2);  yp(4)=y(2);  zp(4)=z(2);  fp(4)=f(2);
elseif(i==14)
  xp(1)=x(6);   yp(1)=y(6);   zp(1)=z(6);   fp(1)=f(6);
  xp(2)=x(3);   yp(2)=y(3);   zp(2)=z(3);   fp(2)=f(3);
  xp(3)=x(10);  yp(3)=y(10);  zp(3)=z(10);  fp(3)=f(10);
  xp(4)=x(6);   yp(4)=y(6);   zp(4)=z(6);   fp(4)=f(6);
elseif(i==15)
  xp(1)=x(4);   yp(1)=y(4);   zp(1)=z(4);   fp(1)=f(4);
  xp(2)=x(9);   yp(2)=y(9);   zp(2)=z(9);   fp(2)=f(9);
  xp(3)=x(10);  yp(3)=y(10);  zp(3)=z(10);  fp(3)=f(10);
  xp(4)=x(4);   yp(4)=y(4);   zp(4)=z(4);   fp(4)=f(4);
elseif(i==16)
  xp(1)=x(9);   yp(1)=y(9);   zp(1)=z(9);   fp(1)=f(9);
  xp(2)=x(6);   yp(2)=y(6);   zp(2)=z(6);   fp(2)=f(6);
  xp(3)=x(10);  yp(3)=y(10);  zp(3)=z(10);  fp(3)=f(10);
  xp(4)=x(9);   yp(4)=y(9);   zp(4)=z(9);   fp(4)=f(9);
end

if(option==1)    % solid

  if(i==1)  patch(xp,yp,zp,'r'); end
  if(i==2)  patch(xp,yp,zp,'b'); end
  if(i==3)  patch(xp,yp,zp,'y'); end
  if(i==4)  patch(xp,yp,zp,'g'); end
  if(i==5)  patch(xp,yp,zp,'r'); end
  if(i==6)  patch(xp,yp,zp,'b'); end
  if(i==7)  patch(xp,yp,zp,'y'); end
  if(i==8)  patch(xp,yp,zp,'g'); end
```

TABLE 8.7.7   Function *tetra10_vis* ($\rightarrow$ Continuing $\rightarrow$)

```
      if(i==9)  patch(xp,yp,zp,'r'); end
      if(i==10) patch(xp,yp,zp,'b'); end
      if(i==11) patch(xp,yp,zp,'y'); end
      if(i==12) patch(xp,yp,zp,'g'); end
      if(i==13) patch(xp,yp,zp,'r'); end
      if(i==14) patch(xp,yp,zp,'b'); end
      if(i==15) patch(xp,yp,zp,'y'); end
      if(i==16) patch(xp,yp,zp,'g'); end

  elseif(option==2)     % wireframe

    plot3(xp,yp,zp,'k');

  elseif(option==3)   % graded

    cp(1)=0.5+0.3*xp(1);
    cp(2)=0.5+0.3*xp(2);
    cp(3)=0.5+0.3*xp(3);
    cp(4)=0.5+0.3*xp(4);

    cp(1)=0.5+0.3*zp(1);
    cp(2)=0.5+0.3*zp(2);
    cp(3)=0.5+0.3*zp(3);
    cp(4)=0.5+0.3*zp(4);

    patch(xp,yp,zp,cp);

   elseif(option==4)  % graded

    patch(xp,yp,zp,fp);

   elseif(option==5)    % solid white

    patch(xp,yp,zp,'w');

   end
end

%-------------------
% compute the volume
%-------------------

matrix=[1 1 1 1;
        x(1) x(2) x(3) x(4); ...
        y(1) y(2) y(3) y(4); ...
        z(1) z(2) z(3) z(4)];

volume = det(matrix)/6.0;

%-----
% done
%-----
return
```

TABLE 8.7.7   ($\rightarrow$ Continued) Function *tetra10_vis* visualizes a function over a 10-node tetrahedron.

```
%==========================================
% Code laplt10_eig
%
% Eigenfunctions of Laplace's equation
% in a spherical-like domain
% using 10-node tetrahedra
%==========================================

%-----------
% input data
%-----------

ndiv = 1;   % discretization level

%-----------
% discretize
%-----------

[ne,ng,p,c,efl,gfl] = tetra10_sphere8 (ndiv);

%------------------------------------
% assemble the global diffusion and mass matrices
%------------------------------------

gdm = zeros(ng,ng); % initialize
gmm = zeros(ng,ng); % initialize

volume = 0;

for l=1:ne          % loop over the elements

% compute the element diffusion and mass matrices

j = c(l,1);  x1 = p(j,1);  y1 = p(j,2);  z1 = p(j,3);
j = c(l,2);  x2 = p(j,1);  y2 = p(j,2);  z2 = p(j,3);
j = c(l,3);  x3 = p(j,1);  y3 = p(j,2);  z3 = p(j,3);
j = c(l,4);  x4 = p(j,1);  y4 = p(j,2);  z4 = p(j,3);
j = c(l,5);  x5 = p(j,1);  y5 = p(j,2);  z5 = p(j,3);
j = c(l,6);  x6 = p(j,1);  y6 = p(j,2);  z6 = p(j,3);
j = c(l,7);  x7 = p(j,1);  y7 = p(j,2);  z7 = p(j,3);
j = c(l,8);  x8 = p(j,1);  y8 = p(j,2);  z8 = p(j,3);
j = c(l,9);  x9 = p(j,1);  y9 = p(j,2);  z9 = p(j,3);
j = c(l,10); x10= p(j,1);  y10= p(j,2);  z10= p(j,3);

[edm_elm, emm_elm, volel] = edmm_t10 ...
...
   (x1,y1,z1, x2,y2,z2, x3,y3,z3 ...
   ,x4,y4,z4, x5,y5,z5, x6,y6,z6 ...
   ,x7,y7,z7, x8,y8,z8, x9,y9,z9, x10,y10,z10);
```

TABLE 8.7.8   Code *laplt10_eig* (Continuing →)

```
    for i=1:10
      i1 = c(1,i);
      for j=1:10
        j1 = c(1,j);
        gdm(i1,j1) = gdm(i1,j1) + edm_elm(i,j);
        gmm(i1,j1) = gmm(i1,j1) + emm_elm(i,j);
      end
    end
    volume = volume + volel;
end

%--------------------
% clip the matrices by removing those equations
% corresponding to boundary nodes
%--------------------

Ic=0;

for i=1:ng

 if(gfl(i)==0)

  Ic = Ic+1;
  map(Ic) = i;

  Jc=0;
  for j=1:ng
   if(gfl(j)==0)
    Jc = Jc+1;
    A(Ic,Jc) = gdm(i,j);
    B(Ic,Jc) = gmm(i,j);
   end
  end

 end

end

ngred = Ic;

%------------------------
% compute the eigenvalues
%------------------------

eigenvalues = eig(A,B);

%-----
% done
%-----
```

Table 8.7.8   ($\rightarrow$ Continued) Code *laplt10_eig* computes the eigenvalues of the Laplacian operator using 10-node tetrahedral elements.

8.8 Modal expansions in a tetrahedron

Modal expansions can be implemented as discussed in Section 5.9 for the triangle. Sherwin and Karniadakis [32, 59, 60, 61] proposed approximating a function of interest, $f(\xi, \eta, \zeta)$, defined over the volume of the standard tetrahedron in the $\xi\eta\zeta$ space with an mth-degree polynomial,

$$f(\xi, \eta, \zeta) \simeq F_v + F_e + F_f + F_i, \tag{8.8.1}$$

where

$$F_v(\xi, \eta, \zeta) = f_1\, \zeta_1^v(\xi, \eta, \zeta) + f_2\, \zeta_2^v(\xi, \eta, \zeta) + f_3\, \zeta_3^v(\xi, \eta, \zeta) + f_4\, \zeta_4^v(\xi, \eta, \zeta) \tag{8.8.2}$$

is the *vertex part*,

$$
F_e(\xi, \eta, \zeta) = \sum_{i=1}^{m-1} c_i^{12} \zeta_i^{12}(\xi, \eta, \zeta) + \sum_{i=1}^{m-1} c_i^{13} \zeta_i^{13}(\xi, \eta, \zeta) + \sum_{i=1}^{m-1} c_i^{23} \zeta_i^{23}(\xi, \eta, \zeta)
$$
$$
+ \sum_{i=1}^{m-1} c_i^{14} \zeta_i^{14}(\xi, \eta, \zeta) + \sum_{i=1}^{m-1} c_i^{24} \zeta_i^{24}(\xi, \eta, \zeta) + \sum_{i=1}^{m-1} c_i^{34} \zeta_i^{34}(\xi, \eta, \zeta) \tag{8.8.3}
$$

is the *edge part*, present when $m \geq 2$,

$$
F_f(\xi, \eta, \zeta) = \sum_{i=1}^{m-2} \left(\sum_{j=1}^{m-i-1} c_{ij}^{123} \zeta_{ij}^{123}(\xi, \eta, \zeta) \right) + \sum_{i=1}^{m-2} \left(\sum_{j=1}^{m-i-1} c_{ij}^{134} \zeta_{ij}^{134}(\xi, \eta, \zeta) \right)
$$
$$
+ \sum_{i=1}^{m-2} \left(\sum_{j=1}^{m-i-1} c_{ij}^{142} \zeta_{ij}^{142}(\xi, \eta, \zeta) \right) + \sum_{i=1}^{m-2} \left(\sum_{j=1}^{m-i-1} c_{ij}^{234} \zeta_{ij}^{234}(\xi, \eta, \zeta) \right) \tag{8.8.4}
$$

is the *face part*, present when $m \geq 3$, and

$$
F_i(\xi, \eta) = \sum_{i=1}^{m-3} \left[\sum_{j=1}^{m-i-2} \left(\sum_{k=1}^{m-i-j-1} c_{ijk}\, \zeta_{ijk}(\xi, \eta, \zeta) \right) \right] \tag{8.8.5}
$$

is the *interior part*, present when $m \geq 4$. Each part is comprised of corresponding modes represented by interpolation functions multiplied by generally time-depended coefficients, sometimes also called degrees of freedom.

The modal expansion includes $N_v = 4$ vertex modes, $N_e = 6 \times (m-1)$ edge modes, $N_f = 4 \times \frac{1}{2}(m-1)(m-2)$ face modes, and $N_i = \frac{1}{6}(m-3)(m-2)(m-1)$ interior modes. The total number of modes is

$$N = N_v + N_e + N_f + N_i. \tag{8.8.6}$$

Making substitutions, we obtain

$$N = 4 + 6\,(m-1) + 2\,(m-1)(m-2) + \frac{1}{6}(m-3)(m-2)(m-1), \tag{8.8.7}$$

yielding

$$N = \frac{1}{6}(m+1)(m+2)(m+3), \qquad (8.8.8)$$

which is precisely equal to the number of terms in the complete mth-order polynomial expansion in ξ, η, and ζ. In fact, the modes are designed so that the modal expansion expressed by (5.9.1) is a complete mth-degree polynomial in ξ, η, and ζ. The four-component modal decomposition is motivated by ease in enforcing continuity of the finite element expansion at nodes, edges, and faces shared by adjacent elements, as discussed in Section 5.9 for the triangle.

Vertex modes

The coefficients f_1, f_2, f_3, and f_4 in (8.8.2) are the values of the function $f(\xi, \eta, \zeta, t)$ at the four vertex nodes labeled 1, 2, 3, and 4. To ensure C^0 continuity of the finite element expansion, these values must be shared by neighboring elements at the common nodes. The corresponding cardinal interpolation functions, ζ_1^v, ζ_2^v, ζ_3^v, and ζ_4^v, are identical to the linear nodal interpolation functions given in (8.2.3),

$$\zeta_1^v = 1 - \xi - \eta - \zeta, \qquad \zeta_2^v = \xi, \qquad \zeta_3^v = \eta, \qquad \zeta_4^v = \zeta. \qquad (8.8.9)$$

Edge modes

The coefficients c_i^{12}, c_i^{13}, c_i^{23} c_i^{14}, c_i^{24}, and c_i^{34} in (8.8.3) are associated with six families of the edge interpolation functions, ζ_i^{12}, ζ_i^{13}, ζ_i^{23}, ζ_i^{14}, ζ_i^{24}, and ζ_i^{34}, where $i = 1, \ldots, m - 1$. To ensure C^0 continuity of the finite element expansion, these coefficients must be shared by neighboring elements at the common edges. The edge modes take the value of zero along all edges, except for the superscripted edge. Working as in Section 5.9 for the triangle, we derive the specific expressions

$$\zeta_i^{12} = \frac{1-\xi'}{2}\frac{1+\xi'}{2}\left(\frac{1-\eta'}{2}\right)^{i+1}\left(\frac{1-\zeta'}{2}\right)^{i+1}\mathrm{Lo}_{i-1}(\xi'),$$

$$\zeta_i^{13} = \frac{1+\eta'}{2}\frac{1-\xi'}{2}\frac{1-\eta'}{2}\left(\frac{1-\zeta'}{2}\right)^{i+1}\mathrm{Lo}_{i-1}(\eta'),$$

$$\zeta_i^{23} = \frac{1+\xi'}{2}\frac{1+\eta'}{2}\frac{1-\eta'}{2}\left(\frac{1-\zeta'}{2}\right)^{i+1}\mathrm{Lo}_{i-1}(\eta'),$$

$$\zeta_i^{14} = \frac{1-\xi'}{2}\frac{1-\eta'}{2}\frac{1-\zeta'}{2}\frac{1+\zeta'}{2}\mathrm{Lo}_{i-1}(\zeta'), \qquad (8.8.10)$$

$$\zeta_i^{24} = \frac{1+\xi'}{2}\frac{1-\eta'}{2}\frac{1-\zeta'}{2}\frac{1+\zeta'}{2}\mathrm{Lo}_{i-1}(\zeta'),$$

$$\zeta_i^{34} = \frac{1+\eta'}{2}\frac{1-\zeta'}{2}\frac{1+\zeta'}{2}\mathrm{Lo}_{i-1}(\zeta')$$

for $i = 1, \ldots, m - 1$, where Lo_k is a Lobatto polynomial.

Face modes

The coefficients c_{ij}^{pqr} in (8.8.4) are associated with four families of face modes represented by the functions ζ_{ij}^{pqr}, where

$$i = 1, \ldots, m-1, \qquad j = 1, \ldots, m-i-1. \qquad (8.8.11)$$

To ensure C^0 continuity of the finite element expansion, these coefficients must be shared by neighboring elements at the common faces. The face modes take the value of zero over all faces, except for the superscripted face. Karniadakis and Sherwin [32, 59, 60, 61] define the face modes as

$$\zeta_{ij}^{123} = \frac{1-\xi'}{2} \frac{1+\xi'}{2} \mathrm{Lo}_{i-1}(\xi') \left(\frac{1-\eta'}{2}\right)^{i+1} \frac{1+\eta'}{2} \mathcal{J}_{j-1}^{(2i+1,1)}(\eta') \left(\frac{1-\zeta'}{2}\right)^{i+j+1},$$

$$\zeta_{ij}^{124} = \frac{1-\xi'}{2} \frac{1+\xi'}{2} \mathrm{Lo}_{i-1}(\xi') \left(\frac{1-\eta'}{2}\right)^{i+1} \mathcal{J}_{j-1}^{(2i+1,1)}(\eta') \left(\frac{1-\zeta'}{2}\right)^{i1} \frac{1+\zeta'}{2},$$

$$\zeta_{ij}^{134} = \frac{1-\xi'}{2} \frac{1-\eta'}{2} \frac{1+\eta'}{2} \mathrm{Lo}_{i-1}(\eta') \left(\frac{1-\zeta'}{2}\right)^{i+1} \frac{1+\zeta'}{2} \mathcal{J}_{j-1}^{(2i+1,1)}(\eta'),$$

$$\zeta_{ij}^{234} = \frac{1+\xi'}{2} \frac{1-\eta'}{2} \frac{1+\eta'}{2} \mathrm{Lo}_{i-1}(\eta') \left(\frac{1-\zeta'}{2}\right)^{i+1} \frac{1+\zeta'}{2} \mathcal{J}_{j-1}^{(2i+1,1)}(\eta'), \quad (8.8.12)$$

where $\mathcal{J}_{\alpha}^{(\beta,\gamma)}$ are the Jacobi polynomials discussed in Section B.8, Appendix B.

Interior modes

The interior modes, also called bubble modes, are zero over all faces of the tetrahedron and non-zero only inside the element. Requiring partial orthogonality, we find that

$$\zeta_{ijk} = \frac{1+\xi'}{2} \frac{1-\xi'}{2} \mathrm{Lo}_{i-1}(\xi') \frac{1+\eta'}{2} \left(\frac{1-\eta'}{2}\right)^{i+1}$$

$$\times \mathcal{J}_{j-1}^{(2i+1,1)}(\eta') \left(\frac{1-\zeta'}{2}\right)^{i+j+1} \frac{1+\zeta'}{2} \mathcal{J}_{k-1}^{(2i+2j+1,1)}(\zeta'). \qquad (8.8.13)$$

For example, when $m = 4$, we obtain a single bubble mode corresponding to $i = 1, j = 1$, and $k = 1$, given by

$$\zeta_{111} = \xi\eta\zeta\omega. \qquad (8.8.14)$$

Implementation of the modal expansion

The Galerkin finite element method for the modal expansion is implemented as discussed earlier in this chapter for the nodal expansion. The main difference is that the element node interpolation functions are replaced by the non-cardinal, modal interpolation functions, and the interior nodal values are replaced by the

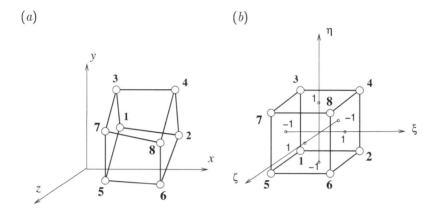

FIGURE 8.9.1   Illustration of an eight-node hexahedral element with 12 straight edges in the (a) physical and (b) parametric space. The element node labels are printed in bold.

corresponding modal expansion coefficients. The numerical computation of the mass and diffusion matrices, consisting of integrals of pairwise products of the modal functions and their gradients, is discussed in detail by Sherwin and Karniadakis [32, 60, 61].

PROBLEM

8.8.1 *Bubble modes*

Explain how the bubble mode shown in (8.8.14) follows from (8.8.13).

8.9   Hexahedral elements

Hexahedral brick-like elements have six planar or curved faces intersecting at 12 straight or curved edges.[4] These elements are the counterparts of the quadrilateral elements in two dimensions discussed in Sections 4.10 and 5.11.

The simplest hexahedral element has 12 straight edges defined by 8 geometrical vertex nodes,

$$\mathbf{x}_i^{\mathrm{E}} = (\, x_i^{\mathrm{E}}, y_i^{\mathrm{E}}, z_i^{\mathrm{E}} \,) \qquad\qquad (8.9.1)$$

for $i = 1, \ldots, 8$, as shown in Figure 8.9.1(a). The superscript E emphasizes that these are local or element nodes, which can be mapped to the unique global nodes through a connectivity table.

---

[4]The word *hexahedron* derives from the Greek work $\epsilon\xi\alpha\epsilon\delta\rho o\nu$, which consists of the word $\epsilon\xi\iota$, meaning *six*, and the word "$\epsilon\delta\rho\alpha$," meaning *base* or *side*.

```
% vertex coordinates

x(1) = 0.0; y(1) = 0.0; z(1) = 0.00;
x(2) = 0.9; y(2) = 0.2; z(2) = 0.02;
x(3) = 0.1; y(3) = 1.1; z(3) =-0.05;
x(4) = 0.8; y(4) = 1.0; z(4) = 0.02;
x(5) = 0.1; y(5) = 0.1; z(5) = 1.01;
x(6) = 0.8; y(6) = 0.6; z(6) = 1.10;
x(7) = 0.2; y(7) = 0.9; z(7) = 0.90;
x(8) = 0.7; y(8) = 0.8; z(8) = 0.80;

% face node labels

s(1,1) = 1; s(1,2) = 2; s(1,3) = 4; s(1,4) = 3;
s(2,1) = 1; s(2,2) = 3; s(2,3) = 7; s(2,4) = 5;
s(3,1) = 6; s(3,2) = 5; s(3,3) = 7; s(3,4) = 8;
s(4,1) = 2; s(4,2) = 6; s(4,3) = 8; s(4,4) = 4;
s(5,1) = 3; s(5,2) = 4; s(5,3) = 8; s(5,4) = 7;
s(6,1) = 2; s(6,2) = 6; s(6,3) = 5; s(6,4) = 1;

% paint the faces

for i=1:6

  for j=1:4
   xp(j) = x(s(i,j)); yp(j) = y(s(i,j));
   zp(j) = z(s(i,j));
  end

  if(i==1) patch(xp,yp,zp,'r'); end
  if(i==2) patch(xp,yp,zp,'b'); end
  if(i==3) patch(xp,yp,zp,'y'); end
  if(i==4) patch(xp,yp,zp,'g'); end
  if(i==5) patch(xp,yp,zp,'r'); end
  if(i==6) patch(xp,yp,zp,'b'); end

end
```

TABLE 8.9.1   FSELIB script *hexa8* generates and displays a hexahedral element defined by eight nodes.

FSELIB script *hexa8*, listed in Table 8.9.1, visualizes the element by painting
the faces as shown in Figure 8.9.2.

8.9.1   Parametric representation

To describe a eight-node hexahedral element in parametric form, we map it from the
physical xyz space to the standard cube whose vertices are located at the points
$(\pm 1, \pm 1, \pm 1)$ in the $\xi \eta \zeta$ space, as shown in Figure 8.9.1(b). The volume of the
parametric cube is equal to eight dimensionless units.

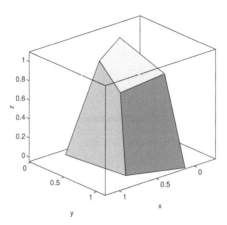

FIGURE 8.9.2   Visualization of a hexahedral element generated by the FSELIB script
         hexa8.

The mapping from the physical to the parametric space is mediated by the
function

$$\mathbf{x} = \sum_{i=1}^{8} \mathbf{x}_i^E \, \psi_i(\xi, \eta, \zeta),$$ (8.9.2)

where $\psi_i(\xi, \eta, \zeta)$ are element interpolation functions satisfying familiar cardinal
properties: $\psi_i = 1$ at the ith element node, and $\psi_i = 0$ at the other seven element nodes for $i = 1$–8.

Working as in Section 4.10 for the four-node quadrilateral element, we derive
the element interpolation functions

$$\psi_1 = \frac{1-\xi}{2}\frac{1-\eta}{2}\frac{1-\zeta}{2}, \qquad \psi_2 = \frac{1+\xi}{2}\frac{1-\eta}{2}\frac{1-\zeta}{2},$$

$$\psi_3 = \frac{1-\xi}{2}\frac{1+\eta}{2}\frac{1-\zeta}{2}, \qquad \psi_4 = \frac{1+\xi}{2}\frac{1+\eta}{2}\frac{1-\zeta}{2},$$

$$\psi_5 = \frac{1-\xi}{2}\frac{1-\eta}{2}\frac{1+\zeta}{2}, \qquad \psi_6 = \frac{1+\xi}{2}\frac{1-\eta}{2}\frac{1+\zeta}{2}, \qquad (8.9.3)$$

$$\psi_7 = \frac{1-\xi}{2}\frac{1+\eta}{2}\frac{1+\zeta}{2}, \qquad \psi_8 = \frac{1+\xi}{2}\frac{1+\eta}{2}\frac{1+\zeta}{2},$$

corresponding to the node numbering scheme defined in Figure 8.9.1.

8.9.2   *Integral over the volume of the hexahedron*

The Jacobian matrix of the mapping from the physical xyz space to the parametric
$\xi\eta\zeta$ space is defined in (8.2.9). The determinant of the Jacobian matrix is the
volume metric coefficient, h_V, defined in (8.2.10).

Substituting the quadratic mapping shown in (8.9.3), we derive a nonlinear expression for h_V in ξ, η, and ζ. The metric coefficient is constant only if the hexahedral element in physical space is orthogonal. In contrast, we recall that the corresponding metric coefficient of the four-node tetrahedron is constant, independent of the element shape.

Using elementary calculus, we find that the integral of a function, $f(x, y, z)$, over the physical hexahedron in the xyz space can be expressed as an integral over the cube in the parametric $\xi\eta\zeta$ space, in the form

$$\iiint f(x, y, z)\, \mathrm{d}x\, \mathrm{d}y\, \mathrm{d}z = \iiint f(\xi, \eta, \zeta)\, h_V\, \mathrm{d}\xi\, \mathrm{d}\eta\, \mathrm{d}\zeta. \tag{8.9.4}$$

The volume integral over the cube can be computed by the triple application of a one-dimensional integration quadrature.

For example, adopting the Lobatto quadrature shown in (3.4.14), we obtain the approximation

$$\begin{aligned}
&\iiint f(\xi, \eta, \zeta)\, h_V\, \mathrm{d}\xi\, \mathrm{d}\eta\, \mathrm{d}\zeta \\
&\simeq \sum_{p_1=1}^{k_1+1} \sum_{p_2=1}^{k_2+1} \sum_{p_3=1}^{k_3+1} f(t_{p_1}, t_{p_2}, t_{p_3})\, h_V(t_{p_1}, t_{p_2}, t_{p_3})\, w_{p_1}\, w_{p_2}\, w_{p_3},
\end{aligned} \tag{8.9.5}$$

where k_1, k_2, and k_3 are three specified integer quadrature orders. The integration is exact for particular classes of functions that depend on k_1, k_2, and k_3.

8.9.3   High-order and spectral hexahedral elements

Hexahedral elements with a higher number of edge, face, and interior nodes can be developed, as discussed in Section 5.11 for quadrilateral elements.

Grid nodes via tensor-product expansions

In one class of hexahedral elements, the interpolation nodes are distributed on a $(m_1 + 1) \times (m_2 + 1) \times (m_3 + 1)$ Cartesian grid covering the standard cube in the $\xi\eta\zeta$ space. The nodes are identified by the coordinate triplet (ξ_i, η_j, ζ_k), where $i = 1, \ldots, m_1 + 1$, $j = 1, \ldots, m_2 + 1$, and $k = 1, \ldots, m_3 + 1$.

To ensure that common nodes are introduced at the element vertices and along the edges, and thereby ensure the C^0 continuity condition of the finite element expansion, we require that

$$\begin{aligned}
\xi_1 &= -1, & \xi_{m_1+1} &= 1, \\
\eta_1 &= -1, & \eta_{m_2+1} &= 1, \\
\zeta_1 &= -1, & \zeta_{m_3+1} &= 1.
\end{aligned} \tag{8.9.6}$$

The cardinal interpolation function of the (i, j, k) node is given by the tensor product

$$\psi_{ijk}(\xi, \eta, \zeta) \simeq \mathcal{L}_i(\xi) \cdot \mathcal{M}_j(\eta) \cdot \mathcal{N}_k(\zeta), \tag{8.9.7}$$

where

$$\mathcal{L}_i(\xi) = \frac{(\xi - \xi_1)(\xi - \xi_2) \cdots (\xi - \xi_{i-1})(\xi - \xi_{i+1}) \cdots (\xi - \xi_{m_1+1})}{(\xi_i - \xi_1)(\xi_i - \xi_2) \cdots (\xi_i - \xi_{i-1})(\xi_i - \xi_{i+1}) \cdots (\xi_i - \xi_{m_1+1})} \tag{8.9.8}$$

is an m_1-degree Lagrange interpolating polynomial defined with respect to the ξ grid lines,

$$\mathcal{M}_j(\eta) = \frac{(\eta - \eta_1)(\eta - \eta_2) \cdots (\eta - \eta_{j-1})(\eta - \eta_{j+1}) \cdots (\eta - \eta_{m_2+1})}{(\eta_j - \eta_1)(\eta_j - \eta_2) \cdots (\eta_j - \eta_{j-1})(\eta_j - \eta_{j+1}) \cdots (\eta_j - \eta_{m_2+1})} \tag{8.9.9}$$

is an m_2-degree Lagrange interpolating polynomial defined with respect to the η grid lines, and

$$\mathcal{N}_k(\zeta) = \frac{(\zeta - \zeta_1)(\zeta - \zeta_2) \cdots (\zeta - \zeta_{k-1})(\zeta - \zeta_{k+1}) \cdots (\zeta - \zeta_{m_3+1})}{(\zeta_k - \zeta_1)(\zeta_k - \zeta_2) \cdots (\zeta_k - \zeta_{k-1})(\zeta_k - \zeta_{z+1}) \cdots (\zeta_z - \zeta_{m_3+1})} \tag{8.9.10}$$

is an m_3-degree Lagrange interpolating polynomial defined with respect to the ζ grid lines. Substituting these expressions into (8.9.7), we obtain a polynomial expansion in (ξ, η, ζ) involving $(m_1 + 1) \times (m_2 + 1) \times (m_3 + 1)$ parameters.

Spectral node distributions

For best interpolation accuracy and enhanced numerical stability, the interior grid lines, ξ_i for $i = 2, \ldots, m_1$, should be placed at the zeros of the $(m_1 - 1)$-degree Lobatto polynomial, the interior grid lines η_j for $j = 2, \ldots, m_2$, should be placed at the zeros of the $(m_2 - 1)$-degree Lobatto polynomial, and the interior grid lines ζ_k for $k = 2, \ldots, m_3$, should be placed at the zeros of the $(m_3 - 1)$-degree Lobatto polynomial, yielding a spectral node distribution. When this is done, the interpolation error decreases spectrally, that is, faster than algebraically with respect to the polynomial order m_1, m_2, or m_3. The spectral interpolation properties in three dimensions derive from those of its one-dimensional constituents, as discussed in Chapter 5.

8.9.4 Modal expansion

In the modal expansion, a function of interest defined over the standard cube, $f(\xi, \eta, \zeta, t)$, is approximated with an mth-degree polynomial as

$$f(\xi, \eta, \zeta) \simeq F_v + F_e + F_f + F_i, \tag{8.9.11}$$

where

$$\begin{aligned} F_v(\xi, \eta, \zeta) = f_1 \, \zeta_1^v(\xi, \eta, \zeta) + f_2 \, \zeta_2^v(\xi, \eta, \zeta) \\ + f_3 \, \zeta_3^v(\xi, \eta, \zeta) + f_4 \, \zeta_4^v(\xi, \eta, \zeta) \\ + f_5 \, \zeta_5^v(\xi, \eta, \zeta) + f_6 \, \zeta_6^v(\xi, \eta, \zeta) \\ + f_7 \, \zeta_7^v(\xi, \eta, \zeta) + f_8 \, \zeta_8^v(\xi, \eta, \zeta) \end{aligned} \tag{8.9.12}$$

is the *vertex part*,

$$
\begin{aligned}
F_e(\xi, \eta, \zeta) = &\sum_{i=1}^{m_1-1} c_i^{12} \zeta_i^{12}(\xi, \eta, \zeta) + \sum_{i=1}^{m_1-1} c_i^{34} \zeta_i^{34}(\xi, \eta, \zeta) \\
&+ \sum_{i=1}^{m_1-1} c_i^{56} \zeta_i^{56}(\xi, \eta, \zeta) + \sum_{i=1}^{m_1-1} c_i^{78} \zeta_i^{78}(\xi, \eta, \zeta) \\
&+ \sum_{i=1}^{m_2-1} c_i^{13} \zeta_i^{13}(\xi, \eta, \zeta) + \sum_{i=1}^{m_2-1} c_i^{24} \zeta_i^{24}(\xi, \eta, \zeta) \\
&+ \sum_{i=1}^{m_2-1} c_i^{57} \zeta_i^{57}(\xi, \eta, \zeta) + \sum_{i=1}^{m_2-1} c_i^{68} \zeta_i^{68}(\xi, \eta, \zeta) \\
&+ \sum_{i=1}^{m_3-1} c_i^{15} \zeta_i^{15}(\xi, \eta, \zeta) + \sum_{i=1}^{m_3-1} c_i^{26} \zeta_i^{26}(\xi, \eta, \zeta) \\
&+ \sum_{i=1}^{m_3-1} c_i^{37} \zeta_i^{37}(\xi, \eta, \zeta) + \sum_{i=1}^{m_3-1} c_i^{48} \zeta_i^{48}(\xi, \eta, \zeta)
\end{aligned} \tag{8.9.13}
$$

is the *edge part*, present when $m_1 \geq 2$, or $m_2 \geq 2$, or $m_3 \geq 2$,

$$
\begin{aligned}
F_f(\xi, \eta, \zeta) = &\sum_{i=1}^{m_1-1} \left(\sum_{j=1}^{m_2-1} c_{ij}^{1234} \, \zeta_{ij}^{1234}(\xi, \eta, \zeta) \right) + \sum_{i=1}^{m_1-1} \left(\sum_{j=1}^{m_2-1} c_{ij}^{5678} \, \zeta_{ij}^{134}(\xi, \eta, \zeta) \right) \\
&+ \sum_{i=1}^{m_1-1} \left(\sum_{j=1}^{m_3-1} c_{ij}^{1256} \, \zeta_{ij}^{1256}(\xi, \eta, \zeta) \right) + \sum_{i=1}^{m_1-1} \left(\sum_{j=1}^{m_3-1} c_{ij}^{3478} \, \zeta_{ij}^{3478}(\xi, \eta, \zeta) \right) \\
&+ \sum_{i=1}^{m_2-1} \left(\sum_{j=1}^{m_3-1} c_{ij}^{1357} \, \zeta_{ij}^{1357}(\xi, \eta, \zeta) \right) + \sum_{i=1}^{m_2-1} \left(\sum_{j=1}^{m_3-1} c_{ij}^{2468} \, \zeta_{ij}^{2468}(\xi, \eta, \zeta) \right)
\end{aligned} \tag{8.9.14}
$$

is the *face part*, present when $m_1 \geq 2$, or $m_2 \geq 2$, or $m_3 \geq 2$, and

$$
F_i(\xi, \eta) = \sum_{i=1}^{m_1-1} \left[\sum_{j=1}^{m_2-1} \left(\sum_{k=1}^{m_3-1} c_{ijk} \, \zeta_{ijk}(\xi, \eta, \zeta) \right) \right] \tag{8.9.15}
$$

is the *interior part* present when $m_1 \geq 2$, or $m_2 \geq 2$, or $m_3 \geq 2$.

Adding the $N_v = 8$ vertex modes, the

$$
N_e = 4 \times (m_1 - 1) + 4 \times (m_2 - 1) + 4 \times (m_3 - 1) \tag{8.9.16}
$$

edge modes, the

$$
N_f = 2 \times (m_1 - 1)(m_2 - 1) + 2 \times (m_1 - 1)(m_2 - 1) + 2 \times (m_2 - 1)(m_3 - 1) \tag{8.9.17}
$$

face modes, and the

$$N_i = (m_1 - 1)(m_2 - 1)(m_3 - 1) \tag{8.9.18}$$

interior nodes, we obtain the total number of modes,

$$N = N_v + N_e + N_f + N_i = (m_1 + 1)(m_2 + 1)(m_3 + 1), \tag{8.9.19}$$

which is precisely equal to that involved in the tensor-product expansion discussed in Section 6.4.3.

Vertex modes

The coefficients f_1–f_8 in (8.9.12) are the values of the function $f(\xi, \eta, \zeta)$ at the eight vertex nodes shown in Figure 8.9.1. To ensure C^0 continuity of the finite element expansion, these values must be shared by neighboring elements with common vertices. The associated modal functions, ζ_i^v, where $i = 1, \ldots, 8$, are identical to the linear nodal interpolation functions given in (8.9.3).

Edge modes

The coefficients c_i^{pq} in (8.9.13) are associated with 12 families of edge interpolation functions ζ_i^{pq}, where $i = 1, \ldots, m_1 - 1$, or $i = 1, \ldots, m_2 - 1$, or $i = 1, \ldots, m_3 - 1$. To ensure C^0 continuity of the finite element expansion, these coefficients must be shared by neighboring elements at the common edges. The edge modes take the value of zero along all edges, except along the superscripted edge. Working as in Section 5.11.4, we obtain

$$\zeta_i^{12} = \frac{1-\xi}{2} \frac{1+\xi}{2} \frac{1-\eta}{2} \mathrm{Lo}_{i-1}(\xi) \frac{1-\zeta}{2},$$

$$\zeta_i^{34} = \frac{1-\xi}{2} \frac{1+\xi}{2} \frac{1+\eta}{2} \mathrm{Lo}_{i-1}(\xi) \frac{1-\zeta}{2},$$

$$\zeta_i^{56} = \frac{1-\xi}{2} \frac{1+\xi}{2} \frac{1-\eta}{2} \mathrm{Lo}_{i-1}(\xi) \frac{1+\zeta}{2}, \tag{8.9.20}$$

$$\zeta_i^{78} = \frac{1-\xi}{2} \frac{1+\xi}{2} \frac{1+\eta}{2} \mathrm{Lo}_{i-1}(\xi) \frac{1+\zeta}{2}$$

for $i = 1, \ldots, m_1 - 1$,

$$\zeta_i^{13} = \frac{1-\xi}{2} \frac{1-\eta}{2} \frac{1+\eta}{2} \mathrm{Lo}_{i-1}(\eta) \frac{1-\zeta}{2},$$

$$\zeta_i^{24} = \frac{1+\xi}{2} \frac{1-\eta}{2} \frac{1+\eta}{2} \mathrm{Lo}_{i-1}(\eta) \frac{1-\zeta}{2},$$

$$\zeta_i^{57} = \frac{1-\xi}{2} \frac{1-\eta}{2} \frac{1+\eta}{2} \mathrm{Lo}_{i-1}(\eta) \frac{1+\zeta}{2}, \tag{8.9.21}$$

$$\zeta_i^{68} = \frac{1+\xi}{2} \frac{1-\eta}{2} \frac{1+\eta}{2} \mathrm{Lo}_{i-1}(\eta) \frac{1+\zeta}{2}$$

for $i = 1, \ldots, m_2 - 1$, and

$$\zeta_i^{15} = \frac{1 - \xi}{2} \frac{1 - \eta}{2} \frac{1 - \zeta}{2} \frac{1 + \zeta}{2} \mathrm{Lo}_{i-1}(\zeta),$$

$$\zeta_i^{26} = \frac{1 + \xi}{2} \frac{1 - \eta}{2} \frac{1 - \zeta}{2} \frac{1 + \zeta}{2} \mathrm{Lo}_{i-1}(\zeta),$$

$$\zeta_i^{37} = \frac{1 - \xi}{2} \frac{1 + \eta}{2} \frac{1 - \zeta}{2} \frac{1 + \zeta}{2} \mathrm{Lo}_{i-1}(\zeta),$$

(8.9.22)

$$\zeta_i^{48} = \frac{1 + \xi}{2} \frac{1 + \eta}{2} \frac{1 - \zeta}{2} \frac{1 + \zeta}{2} \mathrm{Lo}_{i-1}(\zeta)$$

for $i = 1, \ldots, m_3 - 1$.

Face modes

The coefficients c_{ij}^{pqrs} in (8.9.14) are associated with six families of face modes represented by the functions ζ_{ij}^{pqrs}, where i and j range over the limits stated in the sums. To ensure C^0 continuity of the finite element expansion, these coefficients must be shared by neighboring elements at the common faces. The face modes take the value of zero over all faces, except for the superscripted face. Requiring partial orthogonality, we find that

$$\zeta_{ij}^{1234} = \frac{1 - \xi}{2} \frac{1 + \xi}{2} \mathrm{Lo}_{i-1}(\xi) \frac{1 - \eta}{2} \frac{1 + \eta}{2} \mathrm{Lo}_{j-1}(\eta) \frac{1 - \zeta}{2},$$

(8.9.23)

$$\zeta_{ij}^{5678} = \frac{1 - \xi}{2} \frac{1 + \xi}{2} \mathrm{Lo}_{i-1}(\xi) \frac{1 - \eta}{2} \frac{1 + \eta}{2} \mathrm{Lo}_{j-1}(\eta) \frac{1 + \zeta}{2}$$

for $i = 1, \ldots, m_1 - 1$ and $j = 1, \ldots, m_2 - 1$,

$$\zeta_{ij}^{1256} = \frac{1 - \xi}{2} \frac{1 + \xi}{2} \mathrm{Lo}_{i-1}(\xi), \frac{1 - \eta}{2} \frac{1 - \zeta}{2} \frac{1 + z\eta}{2} \mathrm{Lo}_{j-1}(\zeta),$$

(8.9.24)

$$\zeta_{ij}^{3478} = \frac{1 - \xi}{2} \frac{1 + \xi}{2} \mathrm{Lo}_{i-1}(\xi) \frac{1 + \eta}{2} \frac{1 - \zeta}{2} \frac{1 + \zeta}{2} \mathrm{Lo}_{j-1}(\zeta)$$

for $i = 1, \ldots, m_1 - 1$ and $j = 1, \ldots, m_3 - 1$, and

$$\zeta_{ij}^{1357} = \frac{1 - \xi}{2} \frac{1 - \eta}{2} \frac{1 + \eta}{2} \mathrm{Lo}_{i-1}(\eta) \frac{1 - \zeta}{2} \frac{1 + \zeta}{2} \mathrm{Lo}_{j-1}(\zeta),$$

(8.9.25)

$$\zeta_{ij}^{2468} = \frac{1 + \xi}{2} \frac{1 - \eta}{2} \frac{1 + \eta}{2} \mathrm{Lo}_{i-1}(\eta) \frac{1 - \zeta}{2} \frac{1 + \zeta}{2} \mathrm{Lo}_{j-1}(\zeta)$$

for $i = 1, \ldots, m_2 - 1$ and $j = 1, \ldots, m_3 - 1$.

Interior modes

The interior or bubble modes are zero over all faces and non-zero only inside the element. Enforcing the orthogonality condition, we find that

$$\zeta_{ijk} = \frac{1-\xi}{2}\frac{1+\xi}{2}\operatorname{Lo}_{i-1}(\xi)\,\frac{1-\eta}{2}\frac{1+\eta}{2}\operatorname{Lo}_{j-1}(\eta)\,\frac{1-\zeta}{2}\frac{1+\eta}{2}\operatorname{Lo}_{k-1}(\zeta)$$

$$(8.9.26)$$

or

$$\zeta_{ijk} = \operatorname{Lo}_{i-1}(\xi)\,\operatorname{Lo}_{j-1}(\eta)\,\operatorname{Lo}_{k-1}(\zeta)\,\frac{1-\xi}{2}\frac{1+\xi}{2}\frac{1-\eta}{2}\left(\frac{1+\eta}{2}\right)^2\frac{1-\zeta}{2}$$

$$(8.9.27)$$

for $i = 1, \ldots, m_1 - 1$, $j = 1, \ldots, m_2 - 1$, and $k = 1, \ldots, m_3 - 1$.

Implementation of the modal expansion

The Galerkin finite element method for the modal expansion is implemented as discussed in Section 8.3 for tetrahedral elements. The element mass and diffusion matrices, consisting of integrals of pairwise products of the modal functions and their derivatives, are computed numerically using appropriate Gaussian quadratures or other integration rules.

PROBLEMS

8.9.1 *Computation of an integral over a hexahedral element*

Write a MATLAB function that uses the triple Lobatto quadrature to evaluate the integral of a given function, $f(\xi, \eta, \zeta)$, over a hexahedral element. The input to the function should include the position of the eight vertices, $\mathbf{x}_i$ for $i = 1$–8. Call the function with $f(\xi, \eta, \zeta) = 1$ and confirm that the answer is equal to the volume of the hexahedron.

8.9.2 *Modal expansion*

Write the specific expressions of the 12 modes corresponding to expansion orders $m_1 = 1$, $m_2 = 1$, and $m_3 = 2$.

Appendices

Mathematical supplement

A

In this appendix, we present a summary of fundamental mathematical concepts, notations, and definitions pertinent to finite and spectral element methods.

A.1 Index notation

In index notation, a vector in the Nth-dimensional space, $\mathbf{u}$, is denoted as u_i, where i is a free index taking values in the range $i = 1, \ldots, N$. A two-dimensional matrix, $\mathbf{A}$, is denoted as A_{ij}, where the free indices, i and j, run over appropriate ranges determined by the matrix dimensions. A three-dimensional matrix, $\mathbf{T}$, is denoted as T_{ijk}. Similar notation applies to higher-dimensional matrices.

Repeated index summation convention

Einstein's repeated-index summation convention states that, if a subscript appears twice in an expression involving products, then summation over that subscript is implied in its range. Under this convention,

$$u_i \, v_i \equiv u_1 \, v_1 + u_2 \, v_2 + \cdots + u_N \, v_N, \tag{A.1.1}$$

and

$$A_{ii} \equiv A_{11} + A_{22} + \cdots + A_{NN}, \tag{A.1.2}$$

where N is the maximum value of the index i. Expression (A.1.1) defines the inner product of a pair of vectors, $\mathbf{u}$ and $\mathbf{v}$. Expression (A.1.2) defines the trace of a matrix, $\mathbf{A}$.

A.2 Kronecker's delta

Kronecker's delta, δ_{ij}, represents the identity or unit matrix: $\delta_{ij} = 1$ if $i = j$, and $\delta_{ij} = 0$ if $i \neq j$. Using this definition, we find that

$$u_i \delta_{ij} = u_j, \quad A_{ij}\delta_{jk} = A_{ik}, \quad \delta_{ij}A_{jk}\delta_{kl} = A_{il}, \quad \delta_{ij}\delta_{jk}\delta_{kl} = \delta_{il}, \tag{A.2.1}$$

and

$$\delta_{ii} = N, \tag{A.2.2}$$

subject to the Einstein summation convention, where the index i runs from 1 to N.

If x_i is a set of N independent variables, then

$$\frac{\partial x_i}{\partial x_j} = \delta_{ij}, \qquad \frac{\partial x_i}{\partial x_i} = N, \qquad \text{(A.2.3)}$$

subject to the Einstein summation convention, which is tacitly assumed in the remainder of this appendix.

A.3   Alternating tensor

The alternating tensor, ϵ_{ijk} for $i, j, k = 1, 2, 3$, is defined such that $\epsilon_{ijk} = 0$ if any two indices have the same value, $\epsilon_{ijk} = 1$ if the indices are arranged in one of the three cyclic orders 123, 231, or 312, and $\epsilon_{ijk} = -1$ otherwise. Thus,

$$\epsilon_{132} = -1, \qquad \epsilon_{122} = 0, \qquad \epsilon_{ijj} = 0, \qquad \epsilon_{ijk} = \epsilon_{jki} = \epsilon_{kij}. \qquad \text{(A.3.1)}$$

Two useful properties of the alternating tensor are

$$\epsilon_{ijk}\,\epsilon_{ijl} = 2\,\delta_{kl}, \qquad \epsilon_{ijk}\,\epsilon_{ilm} = \delta_{jl}\,\delta_{km} - \delta_{jm}\,\delta_{kl}, \qquad \text{(A.3.2)}$$

subject to the repeated index summation convention.

A.4   Two- and three-dimensional vectors

The inner or dot product of a pair of two- or three-dimensional vectors, $\mathbf{a}$ and $\mathbf{b}$, is a scalar defined as

$$s \equiv \mathbf{a} \cdot \mathbf{b} = a_i\,b_i, \qquad \text{(A.4.1)}$$

subject to the Einstein summation convention. From a geometrical point of view, the inner product is the product of the lengths of the two vectors and the cosine of the angle θ subtended between the two vectors in their plane,

$$\mathbf{a} \cdot \mathbf{b} = |\mathbf{a}||\mathbf{b}|\cos\theta. \qquad \text{(A.4.2)}$$

If the inner product is zero, then $\theta = \pi/2$ and the two vectors $\mathbf{a}$ and $\mathbf{b}$ are orthogonal.

The outer product of an *ordered* pair of two- or three-dimensional vectors, $\mathbf{a}$ and $\mathbf{b}$, is a new vector, denoted as $\mathbf{a} \times \mathbf{b}$, given by the determinant of a matrix,

$$\mathbf{a} \times \mathbf{b} \equiv \det\left(\begin{bmatrix} \mathbf{e}_1 & \mathbf{e}_2 & \mathbf{e}_3 \\ a_1 & a_2 & a_3 \\ b_1 & b_2 & b_3 \end{bmatrix}\right), \qquad \text{(A.4.3)}$$

where $\mathbf{e}_i$ is the unit vector in the ith Cartesian direction and a_i, b_i are the corresponding vector components. The magnitude of the outer product is equal to the product of the lengths of the vectors, $\mathbf{a}$ and $\mathbf{b}$, and the sine of the angle subtended between these vectors in their plane.

$$|\mathbf{a} \times \mathbf{b}| = |\mathbf{a}| \, |\mathbf{b}| \, |\sin \theta|. \tag{A.4.4}$$

The outer product is oriented normal to the plane of $\mathbf{a}$ and $\mathbf{b}$, Its direction is determined by the right-handed rule applied to the ordered triplet $\mathbf{a}, \mathbf{b}$, and $\mathbf{c}$. If the outer product is the null vector, the two vectors $\mathbf{a}$ and $\mathbf{b}$ are parallel.

In index notation, the ith component of the outer product is given by

$$(\mathbf{a} \times \mathbf{b})_i = \epsilon_{ijk} \, a_j b_k, \tag{A.4.5}$$

where ϵ_{ijk} is the alternating tensor.

The triple scalar product of an ordered triplet of vectors, $\mathbf{a}$, $\mathbf{b}$, and $\mathbf{c}$, is the scalar

$$s \equiv (\mathbf{a} \times \mathbf{b}) \cdot \mathbf{c} = (\mathbf{b} \times \mathbf{c}) \cdot \mathbf{a} = (\mathbf{c} \times \mathbf{a}) \cdot \mathbf{b}, \tag{A.4.6}$$

which is equal to the determinant of a matrix,

$$s = \det\left(\begin{bmatrix} a_1 & a_2 & a_3 \\ b_1 & b_2 & b_3 \\ c_1 & c_2 & c_3 \end{bmatrix} \right), \tag{A.4.7}$$

where a_i, b_i, and c_i are the Cartesian vector components. If any pair of vectors are parallel, the triple mixed product is zero.

A.5 Del or nabla operator

The Cartesian components of the *del* or *nabla* operator, ∇, are the partial derivatives with respect to the corresponding coordinates,

$$\nabla \equiv \left(\frac{\partial}{\partial x}, \ \frac{\partial}{\partial y}, \ \frac{\partial}{\partial z} \right). \tag{A.5.1}$$

In two dimensions, only the x and y derivatives appear.

A.6 Gradient and divergence

If f is a scalar function of position, $\mathbf{x}$, then its gradient, ∇f, is a vector defined as

$$\nabla f = \mathbf{e}_x \frac{\partial f}{\partial x} + \mathbf{e}_y \frac{\partial f}{\partial y} + \mathbf{e}_z \frac{\partial f}{\partial z}, \tag{A.6.1}$$

where $\mathbf{e}_i$ is the unit vector in ith direction.

The divergence of a vector function of position, $\mathbf{F} = (F_x, F_y, F_z)$, is a scalar defined as

$$\nabla \cdot \mathbf{F} = \frac{\partial F_i}{\partial x_i} = \frac{\partial F_x}{\partial x} + \frac{\partial F_y}{\partial y} + \frac{\partial F_z}{\partial z}. \tag{A.6.2}$$

In two dimensions, only the x and y derivatives appear. If $\nabla \cdot \mathbf{F}$ vanishes at every point, then the vector field $\mathbf{F}$ is called *solenoidal*.

The Laplacian of a scalar function of position, f, is a scalar defined as

$$\nabla \cdot (\nabla f) \equiv \nabla^2 f = \frac{\partial^2 f}{\partial x^2} + \frac{\partial^2 f}{\partial y^2} + \frac{\partial^2 f}{\partial z^2}. \tag{A.6.3}$$

Note that the Laplacian is equal to the divergence of the gradient. In two dimensions, only the x and y derivatives appear.

The gradient of a vector field, $\mathbf{F}$, denoted as $\mathbf{U} \equiv \nabla \mathbf{F}$, is a two-dimensional matrix with elements

$$U_{ij} = \frac{\partial F_j}{\partial x_i}. \tag{A.6.4}$$

The divergence of $\mathbf{F}$ is equal to the trace of $\mathbf{U}$, which is defined as the sum of the diagonal elements of $\mathbf{U}$.

The curl of a vector field, $\mathbf{F}$, denoted as $\nabla \times \mathbf{F}$, is another vector field computed according to the usual rules of the outer vector product, treating the del operator as a regular vector, yielding

$$\nabla \times \mathbf{F} = \mathbf{e}_x \left(\frac{\partial F_z}{\partial y} - \frac{\partial F_y}{\partial z} \right) + \mathbf{e}_y \left(\frac{\partial F_x}{\partial z} - \frac{\partial F_z}{\partial x} \right) + \mathbf{e}_z \left(\frac{\partial F_y}{\partial x} - \frac{\partial F_x}{\partial y} \right). \tag{A.6.5}$$

In index notation, the ith component of the curl is

$$(\nabla \times \mathbf{F})_i = \epsilon_{ijk} \frac{\partial F_k}{\partial x_j}, \tag{A.6.6}$$

where ϵ_{ijk} is the alternating tensor.

A.7   Vector identities

Let f be a scalar function, and $\mathbf{F}$ and $\mathbf{G}$ be two vector functions of position. It can be shown that

$$\nabla \cdot (f\,\mathbf{F}) = f\,\nabla \cdot \mathbf{F} + \mathbf{F} \cdot \nabla f, \tag{A.7.1}$$

$$\nabla\,(\mathbf{F} \cdot \mathbf{G}) = \mathbf{F} \cdot \nabla \mathbf{G} + \mathbf{G} \cdot \nabla \mathbf{F} + \mathbf{F} \times (\nabla \times \mathbf{G}) + \mathbf{G} \times (\nabla \times \mathbf{F}), \tag{A.7.2}$$

$$\nabla \cdot (\mathbf{F} \times \mathbf{G}) = \mathbf{G} \cdot \nabla \times \mathbf{F} - \mathbf{F} \cdot \nabla \times \mathbf{G}, \tag{A.7.3}$$

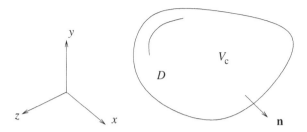

FIGURE A.8.1   Illustration of an arbitrary control volume in space, V_c, bounded by a surface, D, used to derive the divergence theorem. The unit normal vector, $\mathbf{n}$, points outward.

$$\nabla \times (\mathbf{F} \times \mathbf{G}) = \mathbf{F} \nabla \cdot \mathbf{G} - \mathbf{G} \nabla \cdot \mathbf{F} + \mathbf{G} \cdot \nabla \mathbf{F} - \mathbf{F} \cdot \nabla \mathbf{G}, \quad (\text{A.7.4})$$

$$\nabla \times (\nabla f) = \mathbf{0}, \quad (\text{A.7.5})$$

$$\nabla \cdot (\nabla \times \mathbf{F}) = 0, \quad (\text{A.7.6})$$

$$\nabla \times (\nabla \times \mathbf{F}) = \nabla (\nabla \cdot \mathbf{F}) - \nabla^2 \mathbf{F}. \quad (\text{A.7.7})$$

The Laplacian of the vector field in the last equation is a vector whose Cartesian components are the Laplacians of the individual vector components. These identities can be proved working in index notation.

A.8   Gauss divergence theorem

Let V_c be an arbitrary control volume bounded by a closed surface, D, and $\mathbf{n}$ be the unit vector that is normal to D pointing *outward*, as shown in Figure A.8.1. The Gauss divergence theorem states that the volume integral of the divergence of a differentiable vector function of position, $\mathbf{F} = (F_x, F_y, F_z)$, over V_c, is equal to the flow rate of $\mathbf{F}$ across D,

$$\iiint_{V_c} \nabla \cdot \mathbf{F} \, dV = \iint_D \mathbf{F} \cdot \mathbf{n} \, dS. \quad (\text{A.8.1})$$

The inner product, $\mathbf{F} \cdot \mathbf{n}$, is identified as the *surface flux*. We emphasize that the unit normal vector $\mathbf{n}$ points *outward* from D.

Making the three sequential selections $\mathbf{F} = (f,\, 0,\, 0)$, $\mathbf{F} = (0,\, f,\, 0)$, and $\mathbf{F} = (0,\, 0,\, f)$, we obtain the vector form of the divergence theorem

$$\iiint_{V_c} \nabla f \, dV = \iint_D f \mathbf{n} \, dS, \quad (\text{A.8.2})$$

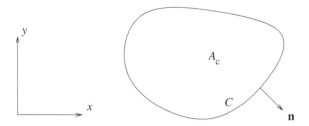

FIGURE A.9.1   Illustration of an arbitrary control area in the xy plane, A_c, bounded
  by a contour, C, used to derive the divergence theorem in the plane. The unit
  normal vector, $\mathbf{n}$, points outward.

where f is a differentiable scalar function. The particular choice $f = x$, $f = y$,
or $f = z$ yield the volume of V_c in terms of a surface integral of the x, y, or z
component of the normal vector.

Setting $\mathbf{F} = \mathbf{a} \times \mathbf{G}$, where $\mathbf{a}$ is a constant vector and $\mathbf{G}$ is a differentiable
function, and then discarding the arbitrary constant $\mathbf{a}$, we obtain the identity

$$\iiint_{V_c} \nabla \times \mathbf{G} \, dV = \iint_D \mathbf{n} \times \mathbf{G} \, dS. \tag{A.8.3}$$

Note that the vector $\mathbf{n} \times \mathbf{G}$ is tangential to D.

A.9 Gauss divergence theorem in the plane

Let A_c be an arbitrary control area in the xy plane bounded by a closed contour,
C, and $\mathbf{n}$ be the unit vector normal to C pointing *outward*, as shown in Figure
A.9.1. The Gauss divergence theorem in the plane states that the areal integral
of the divergence of any two-dimensional differentiable vector function of position,
$\mathbf{F} = (F_x, F_y)$, over A_c, is equal to the flow rate of $\mathbf{F}$ across C,

$$\iint_{A_c} \nabla \cdot \mathbf{F} \, dA = \oint_C \mathbf{F} \cdot \mathbf{n} \, dl, \tag{A.9.1}$$

where l is the arc length along C. We emphasize that the unit normal vector $\mathbf{n}$
points *outward*.

Making the sequential choices $\mathbf{F} = (f, 0)$ and $\mathbf{F} = (0, f)$, we obtain the vector
form of the divergence theorem

$$\iint_{A_c} \nabla f \, dA = \oint_C f \mathbf{n} \, dl, \tag{A.9.2}$$

where f is a differentiable scalar function. The particular choice $f = x$ or $f = y$
yields the area of A_c in terms of a line integral of the x or y component of the
normal vector.

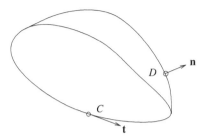

FIGURE A.10.1   Illustration of an arbitrary open surface, D, bounded by a closed loop, C, used to derive Stokes's theorem.

A.10   Stokes's theorem

Let C be an arbitrary closed loop with unit tangent vector $\mathbf{t}$, and D be an arbitrary surface bounded by C with unit normal vector $\mathbf{n}$, as shown in Figure A.10.1. The unit normal vector, $\mathbf{n}$, is oriented according to the right-handed rule with respect to $\mathbf{t}$ and with reference to a designated side of D. As we look at the designated side of D, the normal vector points against us, while the tangent vector describes a counterclockwise path.

Stokes's theorem states that the circulation of a differentiable vector function of position, $\mathbf{F}$, along C, is equal to the flow rate of the curl of $\mathbf{F}$ across D,

$$\oint_C \mathbf{F} \cdot \mathbf{t}\, \mathrm{d}l = \iint_D (\nabla \times \mathbf{F}) \cdot \mathbf{n}\, \mathrm{d}S, \qquad (\text{A.10.1})$$

where l is the arc length along C.

Setting $\mathbf{F} = \mathbf{a} \times \mathbf{G}$, where $\mathbf{a}$ is a constant vector and $\mathbf{G}$ is a differentiable function, expanding curl on the right-hand side of (A.10.1), and then discarding the arbitrary constant $\mathbf{a}$, we obtain a new identity,

$$\oint_C \mathbf{G} \times \mathbf{t}\, \mathrm{d}l = \iint_D \left[\mathbf{n}\nabla \cdot \mathbf{G} - (\nabla \mathbf{G}) \cdot \mathbf{n} \right] \mathrm{d}S. \qquad (\text{A.10.2})$$

Orthogonal polynomials

B

Orthogonal polynomials play an important role in the theory and practice of function interpolation and approximation, numerical integration, and numerical solution of differential equations by spectral expansions and orthogonal collocation. In this appendix, we summarize the basic theory of orthogonal polynomials in one dimension and review their salient properties pertinent to finite and spectral element methods. Extensive discussions can be found in mathematical handbooks and monographs by Sansone [56], Szegö [66], Krylov [35], Stroud and Secrest [63], Freud [22], Dahlquist and Björck [14], and Chihara [10] (e.g., Pozrikidis [47]).

B.1 Definitions and basic properties

Consider an infinite triangular family of polynomials, $p_n(t)$ for $n = 0, 1, \ldots$, where $p_n(t)$ is an nth-degree polynomial of an independent variable, t. The leading-order coefficient of $p_n(t)$ is denoted by A_n. Explicitly, the family takes the form

$$
\begin{aligned}
&p_0(t) = A_0, \\
&p_1(t) = A_1\, t + A_{1,0}, \\
&p_2(t) = A_2\, t^2 + A_{2,1}\, t + A_{2,0}, \\
&\cdots, \\
&p_n(t) = A_n\, t^n + A_{n,n-1}\, t^{n-1} + A_{n,n-2}\, t^{n-2} + \cdots + A_{n,0}, \\
&\cdots,
\end{aligned}
\tag{B.1.1}
$$

where $A_{i,j}$ are polynomial coefficients. For convenience, we have denoted the leading-power coefficient as $A_n \equiv A_{n,n}$.

If the mutual weighted projection of any pair of these polynomials over a specified interval, $[c, d]$, satisfies the orthogonality condition

$$
(p_i, p_j) \equiv \int_c^d p_i(t)\, p_j(t)\, w(t)\, \mathrm{d}t = \begin{cases} \mathcal{G}_i & \text{if } i = j, \\ 0 & \text{if } i \neq j, \end{cases}
\tag{B.1.2}
$$

then the triangular family is orthogonal, where $w(t)$ is a positive weighting function and $\mathcal{G}_i$ are constants. If, in addition, $\mathcal{G}_i = 1$ for any i, the family is *orthonormal*.

683

For every weighting function, $w(t)$, that takes non-negative values inside the specified interval, $[c, d]$, there is a unique corresponding family of orthonormal polynomials defined in that interval. In practice, the members of the family can be computed using the Gram–Schmidt orthogonalization method discussed in Section B.1.8.

The distinguishing properties of several common families of orthogonal polynomials are summarized in Table B.1.1. Additional properties of the Legendre, Lobatto, Chebyshev, and Jacobi polynomials are discussed in Sections B.5–B.8.

B.1.1   Orthogonality against lower-degree polynomials

The equations in (B.1.1) allow us to express any monomial power, t^n for $n = 0, 1, \ldots,$ in terms of the first $n + 1$ orthogonal polynomials of a chosen family as

$$1 = \frac{p_0}{A_0},$$

$$t = \frac{p_1 - A_{1,0}}{A_1} = \frac{1}{A_1}\left(p_1 - \frac{A_{1,0}}{A_0}\,p_0\right), \qquad (B.1.3)$$

$$\ldots \ .$$

It is evident that we can write

$$1 = B_0\,p_0(t),$$

$$t = B_1\,p_1(t) + B_{1,0}\,p_0(t), \qquad (B.1.4)$$

$$\ldots \,,$$

$$t^n = B_n\,p_n(t) + B_{n,n-1}\,p_{n-1}(t) + B_{n,n-2}\,p_{n-2}(2) + \cdots + B_{n,0}\,p_0(t),$$

where $B_{i,j}$ are expansion coefficients related to the polynomial coefficients, $A_{i,j}$. For simplicity, we have denoted $B_n \equiv B_{n,n}$.

The expressions in (B.1.4) can be used to represent uniquely any mth-degree polynomial, $Q_m(t)$, as a linear combination of the first $m+1$ orthogonal polynomials of any chosen family,

$$p_0(t), \quad p_1(t), \quad \ldots, \quad p_m(t), \qquad (B.1.5)$$

as

$$Q_m(t) = c_0\,p_0(t) + c_1\,p_1(t) + \cdots + c_m\,p_m(t), \qquad (B.1.6)$$

where c_i are appropriate expansion coefficients. Using the orthogonality property (B.1.2), we find that

$$c_i = \frac{1}{G_i}\int_c^d Q_m(t)\,p_i(t)\,w(t)\,\mathrm{d}t \equiv \frac{1}{G_i}\,(p_i, Q_m) \qquad (B.1.7)$$

Family	*Symbol*	$[c, d]$	$w(t)$	*Normalization*	$\mathcal{G}_i$
Chebyshev	$\mathrm{T}_i(t)$	$[-1, 1]$	$\dfrac{1}{\sqrt{1 - t^2}}$	$\mathrm{T}_i(1) = 1$	†
Chebyshev second kind	$\mathcal{T}_i(t)$	$[-1, 1]$	$\sqrt{1 - t^2}$	$\mathcal{T}_i(1) = i + 1$	$\frac{1}{2}\pi$
Hermite	$\mathrm{H}_i(t)$	$(-\infty, \infty)$	$\exp(-t^2)$	$A_i = 2^i$	$\sqrt{\pi}\, 2^i\, i!$
Jacobi ††	$\mathrm{J}_i^{(\alpha,\beta)}(t)$	$[-1, 1]$	$(1 - t)^\alpha (1 + t)^\beta$	†††	††††
Laguerre	$\mathcal{L}_i(t)$	$[0, \infty)$	$\exp(-t)$	$A_i = (-1)^i$	$(i!)^2$
Legendre	$\mathrm{L}_i(t)$	$[-1, 1]$	1	$\mathrm{L}_i(1) = 1$	$\frac{2}{2i+1}$
Lobatto	$\mathrm{Lo}_i(t)$	$[-1, 1]$	$1 - t^2$	$\mathrm{Lo}_i(t) = \mathrm{L}'_{i+1}$	$\frac{2\,(i+1)(i+2)}{2i+3}$
Radau	$\mathrm{R}_i(t)$	$[-1, 1]$	$1 + t$	$\mathrm{R}_i(1) = 1$	$\frac{2}{i+1}$

† $\mathcal{G}_0 = \pi$, and $\mathcal{G}_i = \frac{1}{2}\pi$ for $i \geq 1$.

†† It must be $\alpha, \beta > -1$. The Legendre polynomials arise from the Jacobi polynomials for $\alpha = 0$ and $\beta = 0$, that is, $\mathrm{L}_i(t) = \mathrm{J}_i^{(0,0)}(t)$. The Lobatto polynomials arise from the Jacobi polynomials for $\alpha = 1$ and $\beta = 1$, as shown in (B.6.5). The Chebyshev polynomials arise from the Jacobi polynomials for $\alpha = -1/2$ and $\beta = -1/2$, as shown in (B.7.16).

$$\text{†††} \quad \mathrm{J}_i^{\alpha,\beta}(1) = \frac{\Gamma(i+1+\alpha)}{\Gamma(i+1)\Gamma(\alpha+1)} \qquad\qquad \text{††††} \quad \mathcal{G}_i = \frac{2^{\alpha+\beta+1}}{i!}\,\frac{\Gamma(i+1+\alpha)\,\Gamma(i+1+\beta)}{(2i+1+\alpha+\beta)\,\Gamma(i+1+\alpha+\beta)},$$

TABLE B.1.1   Distinguishing properties of common families of orthogonal polynomials. In the expressions for the Jacobi polynomials, Γ is the Gamma function; if m is a positive integer, $\Gamma(m + 1) = m!$ (e.g., Abramowitz & Stegun [1]).

for $i = 1, \ldots, m$, where the parentheses denote the projection defined in (B.1.2),

$$(p_i, Q_m) \equiv \int_c^d p_i(t) \, Q_m(t) \, w(t) \, \mathrm{d}t. \tag{B.1.8}$$

Using expansion (B.1.6), we find that the polynomial $p_i(t)$ is orthogonal against any lower-degree polynomial, $Q_m(t)$, with $m < i$, that is,

$$(p_i, Q_m) = 0 \tag{B.1.9}$$

for $m < i$.

The orthogonality property (B.1.9) provides us with a practical method of computing the coefficients of the monomial powers in the orthogonal polynomial $p_{m+1}(t)$. First, we assign an arbitrary value to the leading-order coefficient, A_{m+1}. Second, we apply (B.1.9) for a set of $m + 1$ linearly independent test polynomials, $Q_m(t)$. Finally, we solve the linear system of equations thus derived for the rest of the monomial coefficients.

As an example, we choose $m = 0$, express $p_1(t)$ as shown in (B.1.1) with $A_1 = 1$, stipulate that $w(t) = 1$, $c = -1$, and $d = 1$, and select the test polynomial $Q_0(t) = 1$. Applying (B.1.9), we obtain

$$(p_1, Q_0) \equiv \int_{-1}^1 (t + A_{1,0}) \, \mathrm{d}t = 0, \tag{B.1.10}$$

which requires that $A_{1,0} = 0$, yielding the first-degree Legendre polynomial, $p_1(t) = A_1 t = t$.

B.1.2   Roots of orthogonal polynomials

Any nth-degree orthogonal polynomial, $p_n(t)$, has n distinct real roots that lie in their domain of definition, $[c, d]$. Thus, the graph of the polynomial, $p_n(t)$, does not cross the t axis outside the interval $[c, d]$.

The roots of any nth-degree orthogonal polynomial, $p_n(t)$, interleave those of its sibling $(n - 1)$-degree polynomial, $p_{n-1}(t)$. That is, there is exactly one root of the latter polynomial between any two consecutive roots of the former polynomial.

B.1.3   Discrete orthogonality

For every orthogonal polynomial, $p_l(t)$, a set of l points can be found, t_i for $i = 1, \ldots, l$, so that the following discrete orthogonality holds,

$$\sum_{i=1}^l p_n(t_i) \, p_m(t_i) = 0 \tag{B.1.11}$$

for $n \neq m$, when $n, m < l$.

B.1.4 Gram polynomials

We can allow the weighing function, $w(t)$, to be a sum of weighed Dirac's one-dimensional delta functions, δ, centered at $k+1$ points, t_l for $l = 1, \ldots, k+1$, distributed in an arbitrary fashion in the interval $[c, d]$, so that

$$w(t) = \sum_{l=1}^{k+1} w_l \, \delta(t - t_l), \tag{B.1.12}$$

where w_l are specified weighting coefficients. The orthogonality condition (B.1.2) then becomes

$$(p_i, p_j) \equiv \sum_{l=1}^{k+1} p_i(t_l) \, p_j(t_l) \, w_l = \begin{cases} \mathcal{G}_i & \text{if } i = j, \\ 0 & \text{if } i \neq j. \end{cases} \tag{B.1.13}$$

It can be shown that the corresponding family contains only a finite set of $l + 1$ polynomials,

$$p_0(t), \qquad p_1(t), \qquad \ldots, \qquad p_N(t), \tag{B.1.14}$$

while higher-order polynomials are identically zero. An example is the family of Gram polynomials, corresponding to evenly spaced nodes t_l, defined in the interval $[-1, 1]$ with equal weighting coefficients, $w_l = 1$.

B.1.5 Recursion relation

Any family of orthogonal polynomials satisfies a recursion relation that allows us to evaluate an orthogonal polynomial in terms of its lower-degree siblings,

$$\begin{aligned} p_0(t) &= A_0, \\ p_1(t) &= \alpha_0 \, (t - \beta_0) \, p_0(t), \\ &\ldots, \\ p_{i+1}(t) &= \alpha_i \, (t - \beta_i) \, p_i(t) - \gamma_i \, p_{i-1}(t), \\ &\ldots \end{aligned} \tag{B.1.15}$$

for $i = 1, \ldots$, where:

- The coefficients α_i are given by

$$\alpha_i = \frac{A_{i+1}}{A_i} \tag{B.1.16}$$

for $i \geq 0$. The leading-term coefficients, A_i, defined in (B.1.1), can be arbitrary.

- The coefficients β_i are given by

$$\beta_i = \frac{1}{\mathcal{G}_i} \left(p_i, \, t \, p_i \right) \tag{B.1.17}$$

for $i \geq 0$.

- The coefficients γ_i are given by

$$\gamma_i = \frac{\alpha_i}{\mathcal{G}_{i-1}} \left(p_i, t\, p_{i-1} \right) = \frac{\alpha_i\, \mathcal{G}_i}{\alpha_{i-1}\, \mathcal{G}_{i-1}} \tag{B.1.18}$$

for $i \geq 1$.

The constants $\mathcal{G}_i$ and the projection operator represented by the parentheses are defined in equation (B.1.2).

If the graph of the weighting function $w(t)$ is symmetric about the point $t = \beta$, then $\beta_i = \beta$ for any i.

As an example, we choose the flat weighting function, $w(t) = 1$, set $c = -1$, and $d = 1$, specify that $p_0(t) = A_0 = 1$, find that $\mathcal{G}_0 = 2$, and stipulate that $A_1 = 1$. Applying (B.1.16), we obtain $\alpha_0 = A_1/A_0 = 1$. Applying (B.1.17), we obtain

$$\beta_0 = \frac{1}{2}\, (1, t) = \frac{1}{2} \int_{-1}^{1} t\, \mathrm{d}t = 0, \tag{B.1.19}$$

yielding $p_1(t) = t$, which is the first-degree Legendre polynomial discussed in Section B.5.

Derivation of the recursion relation

To derive the recursion relations, we note that the right-hand side of the recursion relation

$$p_{i+1}(t) = \alpha_i \left(t - \beta_i \right) p_i(t) - \gamma_i\, p_{i-1}(t) \tag{B.1.20}$$

is an $(i + 1)$-degree polynomial in t, introduce the expansion

$$\alpha_i \left(t - \beta_i \right) p_i(t) - \gamma_i\, p_{i-1}(t)$$
$$= c_0\, p_0(t) + c_1\, p_1(t) + \cdots + c_{i+1}\, p_{i+1}(t) = p_{i+1}(t), \tag{B.1.21}$$

and require that

$$c_0 = c_1 = \cdots = c_{i-1} = c_i = 0, \qquad c_{i+1} = 1. \tag{B.1.22}$$

Using (B.1.7) and exploiting the orthogonality property, we find that

$$c_{i+1} = \frac{1}{\mathcal{G}_{i+1}} \int_c^d \left[\alpha_i \left(t - \beta_i \right) p_i(t) - \gamma_i\, p_{i-1}(t) \right] p_{i+1}(t)\, w(x)\, \mathrm{d}t. \tag{B.1.23}$$

Exploiting the orthogonality property, we obtain

$$c_{i+1} = \frac{\alpha_i}{\mathcal{G}_{i+1}} \int_c^d t\, p_i(t)\, p_{i+1}(t)\, w(x)\, \mathrm{d}t \equiv \frac{\alpha_i}{\mathcal{G}_{i+1}} \left(t\, p_i, p_{i+1} \right). \tag{B.1.24}$$

Setting $c_{i+1} = 1$, we obtain the interesting property

$$\mathcal{G}_{i+1} = \alpha_i \left(t\, p_i, \, p_{i+1} \right). \tag{B.1.25}$$

Working in a similar fashion, we find that

$$c_i = \frac{1}{\mathcal{G}_i} \int_c^d \left[\alpha_i \left(t - \beta_i \right) p_i(t) - \gamma_i\, p_{i-1}(t) \right] p_i(t)\, w(x)\, dt \tag{B.1.26}$$

and then

$$c_i = \frac{\alpha_i}{\mathcal{G}_i} \left(\int_c^d t\, p_i^2(t)\, dt - \beta_i \mathcal{G}_i \right) \equiv \frac{\alpha_i}{\mathcal{G}_i} \left(\left(p_i,\, t p_i \right) - \beta_i\, \mathcal{G}_i \right). \tag{B.1.27}$$

Setting $c_i = 0$, we recover (B.1.17).

The next coefficient is given by

$$c_{i-1} = \frac{1}{\mathcal{G}_{i-1}} \int_c^d \left[\alpha_i \left(t - \beta_i \right) p_i(t) - \gamma_i\, p_{i-1}(t) \right] p_{i-1}(t)\, w(x)\, dt, \tag{B.1.28}$$

yielding

$$c_{i-1} = \frac{1}{\mathcal{G}_{i-1}} \left(\alpha_i \left(p_i, t\, p_{i-1} \right) - \gamma_i\, \mathcal{G}_{i-1} \right). \tag{B.1.29}$$

Setting $c_{i-1} = 0$ and using (B.1.25), we recover (B.1.18).

Continuing in this fashion, we find that the remaining coefficients on the right-hand side of (B.1.21), $c_{i-2}, \ldots, c_0$, are zero, as required.

B.1.6   Evaluation as the determinant of a tridiagonal matrix

The recursion relations in (B.1.15) suggest that the polynomial $p_{n+1}(t)$ can be evaluated in terms of the determinant of an $(n+1) \times (n+1)$ tridiagonal matrix, $\mathbf{T}$, as

$$p_{n+1}(t) = A_0 \det[\mathbf{T}(t)] \tag{B.1.30}$$

for $n \geq 0$. The diagonal components of this matrix are given by

$$T_{i,i} = \alpha_{i-1} \left(t - \beta_{i-1} \right) \tag{B.1.31}$$

for $i = 1, \ldots, n+1$. The sub- and super-diagonal components satisfy the relation

$$T_{i,i-1}\, T_{i-1,i} = \gamma_{i-1} \tag{B.1.32}$$

for $i = 2, \ldots, n + 1$. Setting for convenience the super-diagonal elements equal to unity, $T_{i-1,i} = 1$, we obtain the matrix

$$
\mathbf{T} \equiv
\begin{bmatrix}
\alpha_0 (t - \beta_0) & 1 & 0 & 0 & \cdots \\
\gamma_1 & \alpha_1 (t - \beta_1) & 1 & 0 & \cdots \\
0 & \gamma_2 & \alpha_2 (t - \beta_2) & 1 & \cdots \\
\vdots & \vdots & \vdots & \ddots & \vdots \\
0 & 0 & \cdots & \cdots & \cdots \\
0 & 0 & \cdots & \cdots & \cdots \\
0 & 0 & \cdots & \cdots & \cdots
\end{bmatrix}
\longrightarrow
\tag{B.1.33}
$$

$$
\longrightarrow
\begin{bmatrix}
\cdots & 0 & 0 & 0 & 0 & 0 \\
\cdots & 0 & 0 & 0 & 0 & 0 \\
\cdots & 0 & 0 & 0 & 0 & 0 \\
\cdots & \vdots & \ddots & \vdots & \vdots & \vdots \\
\cdots & 0 & \gamma_{n-2} & \alpha_{n-2} (t - \beta_{n-2}) & 1 & 0 \\
\cdots & 0 & 0 & \gamma_{n-1} & \alpha_{n-1} (t - \beta_{n-1}) & 1 \\
\cdots & 0 & 0 & 0 & \gamma_n & \alpha_n (t - \beta_n)
\end{bmatrix}.
$$

Alternatively, the subdiagonal elements can be split into two equal parts, $\sqrt{\gamma_{i-1}}$, placed along the sub- and super-diagonal lines, yielding a symmetric tridiagonal matrix with guaranteed real eigenvalues.

B.1.7 Clenshaw's algorithm

The fastest way of evaluating an arbitrary expansion in orthogonal polynomials,

$$
Q_N(t) = \sum_{i=0}^{N} c_i \, p_i(t),
\tag{B.1.34}
$$

is by Clenshaw's algorithm listed in Table B.1.2, where c_i are given coefficients and the coefficients α_i, β_i, and γ_i are defined in the equations given in (B.1.15).

B.1.8 Gram–Schmidt orthogonalization

The coefficients of a chosen family of orthogonal polynomials can be computed by the Gram–Schmidt orthogonalization process, best known for generating a family of N mutually orthogonal M-dimensional vectors, $\mathbf{u}_i$, from an arbitrary set of N linearly independent M-dimensional vectors, $\mathbf{v}_i$ for $i = 1, \ldots, N$, where $M \leq N$. The orthogonal set of vectors is constructed working as follows:

1. We begin by setting $\mathbf{u}_1 = \mathbf{v}_1$.

2. We require that $\mathbf{u}_2$ lies in the plane of $\mathbf{u}_1$ and $\mathbf{v}_2$, and is orthogonal to $\mathbf{u}_1$.

3. We require that $\mathbf{u}_3$ lies in the space of $\mathbf{u}_1$, $\mathbf{u}_2$, $\mathbf{v}_3$, and is orthogonal to $\mathbf{u}_1$ and $\mathbf{u}_2$.

$d_N = c_N$

$d_{N-1} = \alpha_{N-1} (t - \beta_{N-1}) d_N + c_{N-1}$

Do $k = N - 2, N - 1, \ldots, 0$

$\qquad d_k = \alpha_k (t - \beta_k) d_{k+1} - \gamma_{k+1} d_{k+2} + c_{k-1}$

End Do

$Q_N(t) = A_0 d_0$

TABLE B.1.2   Clenshaw's algorithm for evaluating an expansion of orthogonal polynomials.

4. We continue in this fashion until the desired set is complete.

The specifics of the algorithm are shown in Table B.1.3.

To generate orthogonal polynomials, we introduce a family of monomials playing the role of the linearly independent vectors, $\mathbf{v}_i$,

$$q_0 = 1, \qquad q_1 = x, \qquad q_2 = x^2, \qquad q_3 = x^3, \qquad \cdots \quad , \tag{B.1.35}$$

and then construct the orthogonal polynomials as follows:

$$p_0 = q_0,$$
$$p_1 = q_1 - \alpha_{1,0} p_0,$$
$$p_2 = q_2 - \alpha_{2,1} p_1 - \alpha_{2,0} p_0,$$
$$\cdots, \tag{B.1.36}$$

where the coefficients, $\alpha_{n,m}$, are such that the orthogonality condition (B.1.2) is fulfilled. In practice, we apply the algorithm shown in Table B.1.3 with the inner vector product designated by the centered dot replaced by the projection integral defined in (B.1.2).

The polynomial coefficients, $A_{i,j}$, defined in (B.1.1), arise by setting $A_i = 1$ and computing

$$A_{i,j} = \sum_{m=0}^{i-1} \alpha_{i,m} A_{m,j} \tag{B.1.37}$$

for $i > 0$ and $j = 0, 1, \ldots, i - 1$. After the coefficients have been computed for each i, scaling is performed to agree with a desired normalization.

$\mathbf{u}_1 = \mathbf{v}_1$

Do $j = 2, \ldots, N$

    Do $i = 1, \ldots, j - 1$

$$\alpha_{i,j} = \frac{\mathbf{u}_i \cdot \mathbf{v}_j}{\mathbf{u}_i \cdot \mathbf{u}_i}$$

    End Do

$$\mathbf{u}_j = \mathbf{v}_j - \sum_{i=1}^{j-1} \alpha_{i,j}\, \mathbf{u}_i$$

End Do

TABLE B.1.3 Gram–Schmidt orthogonalization of a set of linearly independent vectors, $\mathbf{v}_i$, generating a set of mutually orthogonal vectors, $\mathbf{u}_i$.

B.1.9 Orthonormal polynomials

The members of a family of orthogonal polynomials can be normalized to yield a corresponding family of orthonormal polynomials, designated by a hat,

$$\widehat{p}_i(t) \equiv \frac{1}{\sqrt{\mathcal{G}_i}}\, p_i(t), \tag{B.1.38}$$

satisfying the orthonormality condition

$$(\widehat{p}_i, \widehat{p}_j) \equiv \int_c^d \widehat{p}_i(t)\, \widehat{p}_j(t)\, w(t)\, \mathrm{d}t = \begin{cases} 1 & \text{if } i = j, \\ 0 & \text{if } i \neq j. \end{cases} \tag{B.1.39}$$

By analogy with (B.1.1), the nth-degree orthonormal polynomial can be expressed in the form

$$\widehat{p}_n(t) = \hat{A}_n\, t^n + \hat{A}_{n,n-1}\, t^{n-1} + \hat{A}_{n,n-2}\, t^{n-2} + \cdots + \hat{A}_{n,0}, \tag{B.1.40}$$

where $\hat{A}_{i,j}$ are polynomial coefficients; to simplify the notation, we have denoted $\hat{A}_n \equiv \hat{A}_{n,n}$. By definition then,

$$\hat{A}_n = \frac{A_n}{\sqrt{\mathcal{G}_n}}. \tag{B.1.41}$$

B.1.10 Christoffel–Darboux formula

Let us introduce a new independent variable, τ, and multiply the recursion (B.1.20) by $p_i(\tau)$ to obtain

$$p_{i+1}(t)\, p_i(\tau) = \alpha_i\, (t - \beta_i)\, p_i(t)\, p_i(\tau) - \gamma_i\, p_{i-1}(t)\, p_i(\tau). \tag{B.1.42}$$

Interchanging the roles of t and τ, we obtain

$$p_{i+1}(\tau)\,p_i(t) = \alpha_i\,(\tau - \beta_i)\,p_i(\tau)\,p_i(t) - \gamma_i\,p_{i-1}(\tau)\,p_i(t). \tag{B.1.43}$$

Formulating the difference between the last two equations and rearranging, we obtain

$$
\begin{aligned}
p_{i+1}(t)\,p_i(\tau) - p_{i+1}(\tau)\,p_i(t) &= \alpha_i\,(t - \tau)\,p_i(t)\,p_i(\tau) \\
&\quad + \gamma_i\,\big[\,p_i(t)\,p_{i-1}(\tau) - p_i(\tau)\,p_{i-1}(t)\,\big],
\end{aligned} \tag{B.1.44}
$$

which can be recast into the compact form

$$\alpha_i\,(t - \tau)\,p_i(t)\,p_i(\tau) = \Pi_i(t, \tau) - \gamma_i\,\Pi_{i-1}(t, \tau), \tag{B.1.45}$$

where

$$\Pi_i(t, \tau) \equiv p_{i+1}(t)\,p_i(\tau) - p_{i+1}(\tau)\,p_i(t). \tag{B.1.46}$$

Now substituting the second expression in (B.1.18) for γ_i

$$\gamma_i = \frac{\alpha_i\,\mathcal{G}_i}{\alpha_{i-1}\,\mathcal{G}_{i-1}}, \tag{B.1.47}$$

and rearranging, we obtain

$$\frac{1}{\mathcal{G}_i}\,(t - \tau)\,p_i(t)\,p_i(\tau) = \frac{\Pi_i(t, \tau)}{\alpha_i\,\mathcal{G}_i} - \frac{\Pi_{i-1}(t, \tau)}{\alpha_{i-1}\,\mathcal{G}_{i-1}}. \tag{B.1.48}$$

Applying this relation for $i = 0, 1, \ldots, n$, with the understanding that $\Pi_{-1} = 0$, adding the equations thus obtained, simplifying, and recalling that $\alpha_n \equiv A_{n+1}/A_n$, we derive the Christoffel–Darboux formula

$$(t - \tau)\sum_{i=0}^{n} \frac{1}{\mathcal{G}_i}\,p_i(t)\,p_i(\tau) = \frac{A_n}{A_{n+1}\,\mathcal{G}_n}\,\Pi_n(t, \tau), \tag{B.1.49}$$

where $n \geq 0$.

Replacing the orthogonal polynomials and their leading-order coefficients with their orthonormal counterparts defined in (B.1.38) and (B.1.41), denoted by a hat, we derive the simpler expression

$$(t - \tau)\sum_{i=0}^{n} \widehat{p}_i(t)\,\widehat{p}_i(\tau) = \frac{\hat{A}_n}{\hat{A}_{n+1}}\,\widehat{\Pi}_n(t, \tau), \tag{B.1.50}$$

where

$$\widehat{\Pi}_i(t, \tau) \equiv \widehat{p}_{i+1}(t)\,\widehat{p}_i(\tau) - \widehat{p}_{i+1}(\tau)\,\widehat{p}_i(t). \tag{B.1.51}$$

The Christoffel–Darboux formula is useful in deriving Gaussian integration quadratures, as discussed in Section B.2.

B.2   Gaussian integration quadratures

Consider the weighted definite integral of a $(2N+1)$-degree polynomial, $Q_{2N+1}(t)$, between two specified integration limits, c and d,

$$\mathcal{I} \equiv \int_c^d Q_{2N+1}(t)\, w(t)\, \mathrm{d}t, \tag{B.2.1}$$

where $w(t)$ is a specified weighting function. Without loss of generality, we can select a group of $N+1$ interpolation nodes, $t_1, \ldots, t_{N+1}$, and express $Q_{2N+1}(t)$ in the form

$$Q_{2N+1}(t) = P_N(t) + \Xi_N(t)\,(t-t_1)(t-t_2)\cdots(t-t_{N+1}) \tag{B.2.2}$$

without any approximation, where $P_N(t)$ is the Nth-degree interpolating polynomial, as discussed in Appendix D, and $\Xi_N(t)$ is an Nth-degree polynomial. Note that the second term on the right-hand side of (B.2.2) is a $(2N+1)$-degree polynomial.

The integration error incurred by replacing Q_{2N+1} with the lower-degree interpolating polynomial, P_N, is

$$E = \int_c^d \Xi_N(t)\,(t-t_1)(t-t_2)\cdots(t-t_{N+1})\, w(t)\, \mathrm{d}t. \tag{B.2.3}$$

This error will be zero, provided that the $N+1$ nodes, t_i, are identified with the zeros of the $(N+1)$-degree polynomial, $p_{N+1}(t)$, that belongs to the family of orthogonal polynomials defined over the interval $[c, d]$ with weighting function $w(t)$.

To prove this assertion, we observe that the product of the $N+1$ monomials, $(t-t_i)$, is proportional to $p_{N+1}(t)$, and invoke the orthogonality of $p_{N+1}(t)$ against lower-degree polynomials, as discussed in Section B.1. Consequently, the definite integral of the $(2N+1)$-degree polynomial, $Q_{2N+1}(t)$, is equal to the definite integral of the Nth-degree interpolating polynomial, $P_N(t)$.

To formalize the numerical method, we consider the weighted definite integral of a non-singular function, $f(t)$,

$$\mathcal{J} \equiv \int_c^d f(t)\, w(t)\, \mathrm{d}t, \tag{B.2.4}$$

and express the Nth-degree interpolating polynomial of $f(t)$ in terms of the Lagrange interpolating polynomials,

$$P_N(t) = \sum_{i=1}^{N+1} f(t_i)\, l_{N,i}(t), \tag{B.2.5}$$

where

$$l_{N,i}(t) = \frac{(t - t_1)(t - t_2) \cdots (t - t_{i-1})(t - t_{i+1}) \cdots (t - t_{N+1})}{(t_i - t_1)(t_i - t_2) \cdots (t_i - t_{i-1})(t_i - t_{i+1}) \cdots (t_i - t_{N+1})}. \tag{B.2.6}$$

Performing the integration, we derive the Gaussian quadrature formula

$$\int_c^d f(t)\, w(t)\, dt \simeq \int_c^d P_N(t)\, w(t)\, dt = \sum_{i=1}^{N+1} f(t_i)\, w_i, \tag{B.2.7}$$

where

$$w_i = \int_c^d l_{N,i}(t)\, w(t)\, dt \tag{B.2.8}$$

are the integration weights. If $f(t)$ is a $(2N+1)$-degree polynomial, the quadrature is exact.

B.2.1   Evaluation of the integration weights

Using the representation of the Lagrange interpolation functions in terms of the generating function, as shown in equation (D.2.10), Appendix D, we write

$$l_{N,i}(t) = \frac{p_{N+1}(t)}{(t - t_i)\, p'_{N+1}(t_i)}, \tag{B.2.9}$$

which shows that the integration weights are given by

$$w_i = \frac{1}{p'_{N+1}(t_i)} \int_c^d \frac{p_{N+1}(t)}{t - t_i}\, w(t)\, dt \tag{B.2.10}$$

for $i = 1, \ldots, N + 1$.

To compute the integral on the right-hand side of (B.2.10), we apply the Christoffel–Darboux formula (B.1.49) for $n = N + 1$ and $\tau = t_i$, and find that

$$(t - t_i) \sum_{j=0}^{N} \frac{1}{\mathcal{G}_j}\, p_j(t)\, p_j(t_i) = \frac{A_{N+1}}{A_{N+2}\, \mathcal{G}_{N+1}}\, \Pi_{N+1}(t, t_i). \tag{B.2.11}$$

Observing that $p_{N+1}(t_i) = 0$, and thus

$$\Pi_{N+1}(t, t_i) \equiv p_{N+2}(t)\, p_{N+1}(t_i) - p_{N+2}(t_i)\, p_{N+1}(t) = -p_{N+2}(t_i)\, p_{N+1}(t), \tag{B.2.12}$$

we find that

$$\sum_{j=0}^{N} \frac{1}{\mathcal{G}_j}\, p_j(t)\, p_j(t_i) = -\frac{A_{N+1}\, p_{N+2}(t_i)}{A_{N+2}\, \mathcal{G}_{N+1}}\, \frac{p_{N+1}(t)}{t - t_i}. \tag{B.2.13}$$

Next, we multiply both sides of this equation by the weighting function, $w(t)$, and integrate the product over the domain of definition of the orthogonal polynomials to obtain

$$\int_c^d \frac{p_{N+1}(t)}{t - t_i}\, dt = -\frac{A_{N+2}\, \mathcal{G}_{N+1}}{A_{N+1}\, p_{N+2}(t_i)} \sum_{j=0}^N \frac{1}{\mathcal{G}_j}\, p_j(t_i) \int_c^d p_j(t)\, w(t)\, dt. \qquad (\text{B.2.14})$$

Because of orthogonality, only the first term in the sum corresponding $j = 0$ makes a non-zero contribution, and this contribution is equal to unity. Thus,

$$\int_c^d \frac{p_{N+1}(t)}{t - t_i}\, dt = -\frac{A_{N+2}\, \mathcal{G}_{N+1}}{A_{N+1}\, p_{N+2}(t_i)} \qquad (\text{B.2.15})$$

and

$$w_i = -\frac{A_{N+2}\, \mathcal{G}_{N+1}}{A_{N+1}} \frac{1}{p'_{N+1}(t_i)\, p_{N+2}(t_i)} \qquad (\text{B.2.16})$$

for $i = 1, \ldots, N+1$.

In terms of the associated orthonormal polynomials, indicated by a hat, we have

$$w_i = -\frac{\widehat{A}_{N+2}}{\widehat{A}_{N+1}} \frac{1}{\widehat{p}'_{N+1}(t_i)\, \widehat{p}_{N+2}(t_i)}. \qquad (\text{B.2.17})$$

B.2.2   Standard Gaussian quadratures

Standard integration quadratures arise from different families of orthogonal polynomials corresponding to different weighting functions and integration intervals.

Gauss–Legendre quadrature

The Gauss–Legendre quadrature applies to the integration interval $[-1, 1]$, with a flat weighting function $w(t) = 1$, yielding

$$\int_{-1}^1 f(t)\, dt \simeq \sum_{i=1}^{N+1} f(t_i)\, w_i. \qquad (\text{B.2.18})$$

The zeros of the $(N+1)$-degree Legendre polynomial, t_i, and corresponding weights, w_i, are listed in mathematical handbooks (e.g., Abramowitz & Stegun [1], Pozrikidis [47], see also Section B.5).

Gauss–Chebyshev quadrature

The Gauss–Chebyshev quadrature applies to the integration interval $[-1, 1]$ with a singular but integrable weighting function, $w(t) = 1/\sqrt{1 - t^2}$, yielding

$$\int_{-1}^1 f(t)\, \frac{1}{\sqrt{1 - t^2}}\, dt = \sum_{i=1}^{N+1} f(t_i)\, w_i. \qquad (\text{B.2.19})$$

The zeros of the $(N+1)$-degree Chebyshev polynomial, t_i, are given by

$$t_i = \cos\left[\left(i - \frac{1}{2}\right)\frac{\pi}{N+1}\right] \tag{B.2.20}$$

for $i = 1, \ldots, N+1$, and the corresponding weights are given by

$$w_i = \frac{\pi}{N+1} \tag{B.2.21}$$

(see also Section B.2.7).

B.3   Lobatto integration quadrature

Consider the following definite integral of a $(2N-1)$-degree polynomial, $Q_{2N-1}(t)$,

$$\mathcal{I} \equiv \int_{-1}^{1} Q_{2N-1}(t)\,\mathrm{d}t. \tag{B.3.1}$$

To evaluate this integral, we select a set of $N+1$ interpolation nodes, $t_1, \ldots, t_{N+1}$, stipulate that $t_1 = -1$ and $t_{N+1} = 1$, and express $Q_{2N-2}(t)$ in the form

$$Q_{2N-1}(t) = P_N(t) + \Xi_{N-2}(t)\left(1 - t^2\right)(t - t_2) \cdots (t - t_N) \tag{B.3.2}$$

without any approximation, where $P_N(t)$ is the Nth-degree interpolating polynomial discussed in Appendix D, and $\Xi_{N-2}(t)$ is a $(N-2)$-degree polynomial. Note that the product of the $(N-2)$-degree polynomial and the $(N+1)$-degree polynomial in the second term on the right-hand side of (B.3.2) generates a $(2N-1)$-degree polynomial.

The stipulations that $t_1 = -1$ and $t_{N+1} = 1$ distinguish the Lobatto integration quadrature presently considered from the Gauss–Lobatto integration quadrature falling under the auspices of the standard Gaussian quadrature, as discussed in Section B.2.

The integration error incurred by replacing $Q_{2N-1}(t)$ with the interpolating polynomial, $P_N(t)$, is

$$E = \int_{-1}^{1} \Xi_{N-2}(t)\left(1 - t^2\right)(t - t_2) \cdots (t - t_N)\,\mathrm{d}t. \tag{B.3.3}$$

Following the discussion of Section B.2, we conclude that this error will be zero, provided that the $N-1$ interior nodes, t_i for $i = 2, \ldots, N$, coincide with the zeros of the $(N-1)$-degree Lobatto polynomial, $Lo_{N-1}(t)$, corresponding to the weighting function $w(t) = 1 - t^2$.

To formalize a numerical method, we consider the definite integral of a non-singular function, $f(t)$,

$$\mathcal{J} \equiv \int_{-1}^{1} f(t)\,\mathrm{d}t, \tag{B.3.4}$$

and express the Nth-degree interpolating polynomial of $f(t)$ as shown in (B.2.5), where the Lagrange interpolating polynomials are given in (B.2.6) and the abscissas, t_i, are the zeros of the $(N+1)$-degree completed Lobatto polynomial indicated by a caret,

$$\widehat{\mathrm{Lo}}_{N+1}(t) \equiv (1-t^2)\,\mathrm{Lo}_{N-1}(t). \tag{B.3.5}$$

Performing the integration, we derive the Lobatto quadrature formula

$$\int_{-1}^{1} f(t)\,\mathrm{d}t \simeq \int_{-1}^{1} P_N(t)\,\mathrm{d}t = \sum_{i=1}^{N+1} f(t_i)\,w_i, \tag{B.3.6}$$

where

$$w_i = \int_{-1}^{1} l_{N,i}(t)\,\mathrm{d}t \tag{B.3.7}$$

are the integration weights. If $f(t)$ is a $(2N-1)$-degree polynomial, the quadrature is exact.

Evaluation of the integration weights

To compute the integration weights, we express the Lagrange interpolating polynomials as shown in (B.2.9), and use the property

$$\left[(t^2-1)\,\mathrm{Lo}_{N-1}(t)\right]' = N(N+1)\,\mathrm{L}_N(t) \tag{B.3.8}$$

to find that

$$l_{N,i}(t) = \frac{1}{N(N+1)\,\mathrm{L}_N(t_i)} \frac{(t^2-1)\,\mathrm{Lo}_{N-1}(t)}{t-t_i}, \tag{B.3.9}$$

where $\mathrm{L}_N(t)$ is a Legendre polynomial.

The first integration weight can be computed as

$$w_1 = \int_{-1}^{1} l_{N,1}(t)\,\mathrm{d}t = \frac{1}{N\,(N+1)\,\mathrm{L}_N(-1)} \int_{-1}^{1} (t-1)\,\mathrm{Lo}_{N-1}(t)\,\mathrm{d}t, \tag{B.3.10}$$

yielding

$$w_1 = \frac{1}{N\,(N+1)\,\mathrm{L}_N(-1)} \int_{-1}^{1} (t-1)\,\mathrm{L}'_N(t)\,\mathrm{d}t, \tag{B.3.11}$$

then

$$w_1 = \frac{1}{N\,(N+1)\,\mathrm{L}_N(-1)} \left\{ \left[(t-1)\,\mathrm{L}_N(t)\right]_{-1}^{1} - \int_{-1}^{1} \mathrm{L}_N(t)\,\mathrm{d}t \right\}, \tag{B.3.12}$$

and finally

$$w_1 = \frac{2}{N(N+1)}. \tag{B.3.13}$$

The last integration weight, w_{N+1}, can be computed in a similar fashion, yielding precisely the same result. The intermediate integration weights can be computed using the Christoffel–Darboux formula discussed in Section B.1.10.

In summary, the integration weights are given by

$$w_1 = w_{N+1} = \frac{2}{N(N+1)}, \qquad w_i = \frac{2}{N(N+1)} \frac{1}{L_N^2(t_i)} \tag{B.3.14}$$

for $i = 2, \ldots, N$, where L_N is a Legendre polynomial. Numerical values for the base points, t_i, and weights are given in Table 4.2.2.

B.4   Chebyshev integration quadrature

Consider the following *weighted* definite integral of an $(2N - 1)$-degree polynomial, $Q_{2N-1}(t)$,

$$\mathcal{I} \equiv \int_{-1}^{1} Q_{2N-1}(t) \frac{1}{\sqrt{1 - t^2}} \, dt. \tag{B.4.1}$$

To evaluate this integral, we select a set of $N + 1$ interpolation nodes, $t_1, \ldots, t_{N+1}$, stipulate that $t_1 = -1$ and $t_{N+1} = 1$, and express $Q_{2N-1}(t)$ in the form

$$Q_{2N-1}(t) = P_N(t) + \Xi_{N-2}(t) (1 - t^2) (t - t_2) \cdots (t - t_N) \tag{B.4.2}$$

without any approximation, where $P_N(t)$ is the Nth-degree interpolating polynomial, as discussed in Appendix D, and $\Xi_{N-2}(t)$ is an $(N - 2)$-degree polynomial. The stipulations $t_1 = -1$ and $t_{N+1} = 1$ distinguish the Chebyshev integration quadrature presently considered from the Gauss–Chebyshev integration quadrature discussed in Section B.2.2.

By definition, the integration error incurred by replacing the polynomial $Q_{2N-1}(t)$ with the interpolating polynomial $P_N(t)$ is

$$E = \int_{-1}^{1} \Xi_{N-2}(t) (1 - t^2)(t - t_2) \cdots (t - t_N) \frac{1}{\sqrt{1 - t^2}} \, dt. \tag{B.4.3}$$

We will prove that, if the $N - 1$ unspecified nodes, $t_2, \ldots, t_N$, are identified with the zeros of the $(N - 1)$-degree Chebyshev polynomial of the second kind, $\mathcal{T}_{N-1}(t)$, the integration error is precisely zero. To demonstrate this assertion, we restate the error in the form

$$E = \frac{1}{2^N} \int_{-1}^{1} \Xi_{N-2}(t) (1 - t^2) \, \mathcal{T}_{N-1}(t) \frac{1}{\sqrt{1 - t^2}} \, dt$$

$$= \frac{1}{2^N} \int_{-1}^{1} \Xi_{N-2}(t) \, \mathcal{T}_{N-1}(t) \sqrt{1 - t^2} \, dt, \tag{B.4.4}$$

which is zero in light of the orthogonality of the $(N-1)$-degree Chebyshev polynomial of the second kind against any lower-degree polynomial with weighting function $w(t) = \sqrt{1-t^2}$.

The emerging Chebyshev integration quadrature takes the form

$$\int_{-1}^{1} f(t) \, \frac{1}{\sqrt{1-t^2}} \, dt \simeq \sum_{i=1}^{N+1} f(t_i) \, w_i, \tag{B.4.5}$$

where $f(t)$ is a non-singular function and

$$t_i = \cos\left(\frac{i-1}{N}\,\pi\right) \tag{B.4.6}$$

for $i = 1, \ldots, N+1$, are the zeros of the completed Chebyshev polynomial of the second kind,

$$\widehat{\mathcal{T}}_{N+1}(t) \equiv (1-t^2)\,\mathcal{T}_{N-1}(t). \tag{B.4.7}$$

The integration weights are found to be

$$w_1 = w_{N+1} = \frac{\pi}{2N}, \qquad w_i = \frac{\pi}{N} \tag{B.4.8}$$

for $i = 2, \ldots, N$. If $f(t)$ is a $(2N-1)$-degree polynomial, the quadrature is exact.

B.5 Legendre polynomials

In this section, we summarize the salient properties of the Legendre polynomials.

- *Domain of definition:* $[-1, 1]$.

- *Members:* The first few members are listed in Table 3.3.6.

- *Explicit formula (Rodrigues):*

$$L_i(t) = \frac{1}{2^i i!} \frac{d^i (t^2-1)^i}{dt^i} = \frac{(2i)!}{2^i (i!)^2} t^i + \cdots$$

$$= \frac{1}{2^i} \sum_{m=0}^{[i/2]} (-1)^m \begin{pmatrix} i \\ m \end{pmatrix} \begin{pmatrix} 2(i-m) \\ i \end{pmatrix} t^{i-2m}, \tag{B.5.1}$$

where $[i/2]$ denotes the integral part and the tall parentheses denote the combinatorial,

$$\begin{pmatrix} i \\ m \end{pmatrix} = \frac{i!}{m! \, (i-m)!}. \tag{B.5.2}$$

- *Leading-power coefficient:*

$$A_i = \frac{(2i)!}{2^i (i!)^2}.$$ (B.5.3)

- *Generating function:*

$$\frac{1}{\sqrt{1 - 2t\eta + \eta^2}} = \sum_{i=0}^{\infty} L_i(t)\, \eta^i,$$ (B.5.4)

for $-1 < t < 1$ and $|\eta| < 1$.

- *Standard normalization:*

$$L_i(1) = 1.$$ (B.5.5)

- *Orthogonality:*

$$\int_{-1}^{1} L_i(t)\, L_j(t)\, \mathrm{d}t = \frac{2}{2i+1}\, \delta_{ij},$$ (B.5.6)

where δ_{ij} is Kronecker's delta.

- *Recursion relation:*

$$L_{i+1}(t) = \frac{2i+1}{i+1}\, t\, L_i(t) - \frac{i}{i+1}\, L_{i-1}(t).$$ (B.5.7)

With reference to (B.1.15),

$$A_0 = 1, \qquad \alpha_i = \frac{2i+1}{i+1} \quad \text{for } i = 0, 1, \dots,$$

$$\beta_i = 0 \quad \text{for } i = 0, 1, \dots, \qquad \gamma_i = \frac{i}{i+1}, \quad \text{for } i = 1, \dots.$$ (B.5.8)

- *Recursion relations for the first derivative:*

$$(1 - t^2)\, L'_{i+1} = (i+1)\,(-t\, L_{i+1} + L_i) = (i+2)\,(-L_{i+2} + t\, L_{i+1})$$
$$= \frac{(i+1)(i+2)}{2i+3}\,(L_i - L_{i+2}),$$ (B.5.9)

and

$$L'_{i+1} - L'_{i-1} = (2i+1)\, L_i.$$ (B.5.10)

- *Differential equations:*

$$\left[(t^2 - 1)\, L'_i(t)\right]' = i\,(i+1)\, L_i(t)$$ (B.5.11)

and

$$(1 - t^2)\, L''_i(t) - 2t\, L'_i(t) + i(i+1)\, L_i(t) = 0.$$ (B.5.12)

- *Zero-mean property:*

  Applying the orthogonality property with $j = 0$, we find that

  $$\int_{-1}^{1} L_i(t)\, dt = 0, \quad \text{for} \quad i = 1, \ldots \quad . \tag{B.5.13}$$

- *Range of variation:*

  $$|L_i(t)| \le 1, \quad -1 \le t \le 1. \tag{B.5.14}$$

- *Relation to the Jacobi polynomials:*

  The Legendre polynomials are related to the Jacobi polynomials by

  $$L_i(t) = J_i^{(0,0)}(t). \tag{B.5.15}$$

- *Gauss–Legendre integration weights:*

  $$w_i = \frac{1}{L_m'(t_i)} \int_{-1}^{1} \frac{L_m(t)}{t - t_i}\, dt = \frac{2}{1 - t_i^2} \frac{1}{L_m'^2(t_i)} = -\frac{2}{m+1} \frac{1}{L_m'(t_i)\, L_{m+1}(t_i)}$$

  $$= \frac{2}{m} \frac{1}{L_m'(t_i)\, L_{m-1}(t_i)} = \frac{2}{(m+1)^2} \frac{1 - t_i^2}{L_{m+1}^2(t_i)}, \tag{B.5.16}$$

  where t_i are the zeros of $L_m(t)$ for $i = 1, \ldots, m$.

B.6 Lobatto polynomials

Since the Lobatto polynomials, Lo_i, are the derivatives of the Legendre polynomials, L_i, according to the relation

$$Lo_i(t) \equiv L_{i+1}'(t), \tag{B.6.1}$$

many of their properties derive from those listed in Section B.6.5 for the Legendre polynomials. A summary and further properties include the following:

- *Domain of definition:* $[-1, 1]$.

- *Members:* The first few members are listed in Table 3.3.1.

- *Orthogonality:*

  $$\int_{-1}^{1} Lo_i(t)\, Lo_j(t)\, (1 - t^2)\, dt = \frac{2\,(i+1)(i+2)}{2i+3}\, \delta_{ij}, \tag{B.6.2}$$

  where δ_{ij} is Kronecker's delta.

- *Zero-mean property:*

  Applying the orthogonality property with $j = 0$, we find that

  $$\int_{-1}^{1} \mathrm{Lo}_i(t)\,(1 - t^2)\,\mathrm{d}t = 0, \quad \text{for} \quad i = 1, \ldots. \tag{B.6.3}$$

- *Lobatto integration weights:*

  $$w_i = \frac{2}{m(m+1)}\,\frac{1}{\mathrm{L}_m^2(t_i)}, \tag{B.6.4}$$

  where t_i are the zeros of $\mathrm{Lo}_{m-1}(t)$ for $i = 1, \ldots, m - 1$.

- *Relation to the Jacobi polynomials:*

  The Lobatto polynomials are related to the Jacobi polynomials by

  $$\mathrm{Lo}_i(t) = \frac{i + 2}{2}\,\mathrm{J}_i^{(1,1)}(t). \tag{B.6.5}$$

As an exercise, we demonstrate the orthogonality property (B.6.2) based on the properties of the Legendre polynomials. From the definition of the Lobatto polynomials, we write

$$(\mathrm{Lo}_i, \mathrm{Lo}_j) \equiv \int_{-1}^{1} \mathrm{Lo}_i(t)\,\mathrm{Lo}_j(t)\,(1 - t^2)\,\mathrm{d}t = \int_{-1}^{1} \mathrm{L}_{i+1}'(t)\,\mathrm{L}_{j+1}'(t)\,(1 - t^2)\,\mathrm{d}t. \tag{B.6.6}$$

Integrating by parts, we obtain

$$(\mathrm{Lo}_i, \mathrm{Lo}_j) = \left[\mathrm{L}_{i+1}(t)\,\mathrm{L}_{j+1}'(t)\,(1 - t^2)\right]_{-1}^{1} - \int_{-1}^{1} \mathrm{L}_{i+1}(t)\left(\mathrm{L}_{j+1}'(t)\,(1 - t^2)\right)'\,\mathrm{d}t. \tag{B.6.7}$$

The first term on the right-hand side is zero. Using (B.5.11), we find that the second term can be simplified to give

$$(\mathrm{Lo}_i, \mathrm{Lo}_j) = (i + 1)(i + 2)\int_{-1}^{1} \mathrm{L}_{i+1}(t)\,\mathrm{L}_{j+1}(t)\,\mathrm{d}t. \tag{B.6.8}$$

Finally, we use the orthogonality property (B.5.6) to obtain (B.6.2).

B.7   Chebyshev polynomials

In this section, we summarize the salient properties of the Chebyshev polynomials of the first kind, $T_i(t)$, simply called the Chebyshev polynomials.

- *Domain of definition:* $[-1, 1]$.

- *Members:*

The first few members are:

$$
\begin{aligned}
T_0 &= 1, \\
T_1 &= t, \\
T_2 &= 2\,t^2 - 1, \\
T_3 &= 4\,t^3 - 3\,t, \\
T_4 &= 8\,t^4 - 8\,t^2 + 1, \\
T_5 &= 16\,t^5 - 20\,t^3 + 5\,t,
\end{aligned}
$$

$$
\cdots
$$

$$
T_i = 2^{i-1}\,t^i - i\,2^{i-3}\,t^{i-2} + \cdots \quad . \tag{B.7.1}
$$

- *Explicit formula (Rodrigues):*

$$
T_i(t) = \cos(i \arccos t) = \frac{\sqrt{\pi}}{2^{i+1}\,\Gamma(i + \frac{1}{2})}\,\sqrt{1 - t^2}\,\frac{d^i (t^2 - 1)^{i - \frac{1}{2}}}{dt^i}
$$

$$
= \frac{1}{2}\,i \sum_{m=0}^{[i/2]} (-1)^m \frac{(i - m - 1)!}{m!\,(i - 2m)!}\,(2\,t)^{i - 2m}, \tag{B.7.2}
$$

where $[i/2]$ denotes the integral part.

- *Generating function:*

$$
\frac{1 - t\eta}{1 - 2\,t\,\eta + \eta^2} = \sum_{i=0}^{\infty} T_i(t)\,\eta^i, \tag{B.7.3}
$$

for $-1 < t < 1$ and $|\eta| < 1$.

- *Second explicit formula:*

$$
T_i(t) = \frac{1}{2}\,\left(z^i + \frac{1}{z^i}\right), \tag{B.7.4}
$$

where z satisfies

$$
z^2 - 2\,tz + 1 = 0. \tag{B.7.5}
$$

- *Standard normalization:*

$$
T_i(1) = 1. \tag{B.7.6}
$$

- *Orthogonality:*

$$\int_{-1}^{1} T_i(t)\, T_j(t)\, \frac{1}{\sqrt{1-t^2}}\, dt = \begin{cases} \pi & \text{if } i = j = 0, \\ \frac{1}{2}\pi & \text{if } i = j \neq 0, \\ 0 & \text{if } i \neq j. \end{cases} \tag{B.7.7}$$

- *Recursion relation:*

$$T_{i+1}(t) = 2t\, T_i(t) - T_{i-1}(t). \tag{B.7.8}$$

  With reference to (B.1.15),

$$\begin{array}{llll} A_0 = 1, & \alpha_0 = 1, & \alpha_i = 2 & \text{for } i = 1, \ldots, \\ \beta_i = 0 & \text{for } i = 0, \ldots, & \gamma_i = 1 & \text{for } i = 1, \ldots \end{array} \tag{B.7.9}$$

- *Recursion relations for the first derivative:*

$$(1 - t^2)\, T'_{i+1} = (i+1)\,(-t\, T_{i+1} + T_i), \tag{B.7.10}$$

  and

$$\frac{T'_{i+1}}{i+1} - \frac{T'_{i-1}}{i-1} = 2\, T_i. \tag{B.7.11}$$

- *Zeros:*

  $T_i(t)$ has i zeros in the interval $[-1, 1]$, given by the Chebyshev abscissas,

$$t_k = \cos\left[\frac{\pi}{i}\left(k - \frac{1}{2}\right)\right] \tag{B.7.12}$$

  for $k = 1, \ldots, i$.

- *Discrete orthogonality:*

  If x_k for $k = 1, \ldots, m+1$ are the zeros of $T_{m+1}(t)$, then

$$\sum_{k=1}^{m+1} T_i(x_k)\, T_j(x_k) = \begin{cases} m+1 & \text{if } i = j = 0, \\ \frac{1}{2}(m+1) & \text{if } i = j \neq 0, \\ 0 & \text{if } i \neq j \end{cases} \tag{B.7.13}$$

  for $0 \le i, j \le m$.

- *Differential equation:*

$$(1 - t^2)\, T''_i(t) - t\, T'_i(t) + i^2\, T_i(t) = 0. \tag{B.7.14}$$

- *Range of variation:*

$$|T_i(t)| \le 1, \quad -1 \le t \le 1. \tag{B.7.15}$$

- *Minimax property:*

  Of all nth-degree polynomials with leading-power coefficient equal to 1, the scaled Chebyshev polynomial $2^{1-n}T_n(t)$ has the smallest maximum norm in the interval $[-1, 1]$.

- *Relation to the Jacobi polynomials:*

  The Chebyshev polynomials are related to the Jacobi polynomials by

  $$T_i(t) = \frac{n!\sqrt{\pi}}{\Gamma(i + \frac{1}{2})}\, \mathrm{J}_i^{(-1/2,\,-1/2)}(t). \tag{B.7.16}$$

B.8   Jacobi polynomials

- *Standard notation:* $\mathrm{J}_i^{(\alpha,\beta)}$, where $\alpha, \beta > -1$.

- *Domain of definition:* $[-1, 1]$.

- *Members:*

  The first few members are:

  $$\mathrm{J}_0^{(\alpha,\beta)}(t) = 1,$$
  $$\mathrm{J}_1^{(\alpha,\beta)}(t) = \frac{1}{2}\,(\alpha + \beta + 2)\,t + \frac{1}{2}\,(\alpha - \beta),$$

  $$\mathrm{J}_2^{(\alpha,\beta)}(t) = \frac{1}{8}\,(\alpha + \beta + 3)\,(\alpha + \beta + 4)\,t^2 + \frac{1}{4}\,(\alpha - \beta)\,(\alpha + \beta + 3)\,t$$
  $$+ \frac{1}{8}\,[\,(\alpha - \beta)^2 - (\alpha + \beta + 4)\,]. \tag{B.8.1}$$

- *Explicit formula (Rodrigues):*

  $$\mathrm{J}_i^{(\alpha,\beta)}(t) = \frac{(-1)^i}{2^i i!}(1 - t)^{-\alpha}\,(1 + t)^{-\beta}\,\frac{\mathrm{d}^i\,[\,(1 - t)^{i+\alpha}(1 + t)^{i+\beta}\,]}{\mathrm{d}t^i}. \tag{B.8.2}$$

- *Reciprocal relation:*

  $$\mathrm{J}_i^{(\alpha,\beta)}(t) = (-1)^i\,\mathrm{J}_i^{(\beta,\alpha)}(-t). \tag{B.8.3}$$

- *Standard normalization:*

  $$\mathrm{J}_i^{(\alpha,\beta)}(1) = \frac{\Gamma(i + \alpha + 1)}{\Gamma(i + 1)\,\Gamma(\alpha + 1)} = \left(\begin{array}{c} i + \alpha \\ i \end{array}\right), \tag{B.8.4}$$

  $$\mathrm{J}_i^{(\alpha,\beta)}(-1) = (-1)^i\,\frac{\Gamma(i + \beta + 1)}{\Gamma(i + 1)\,\Gamma(\beta + 1)} = (-1)^i\left(\begin{array}{c} i + \beta \\ i \end{array}\right). \tag{B.8.5}$$

- *Orthogonality:*

$$\int_{-1}^{1} \mathrm{J}_i^{(\alpha,\beta)}(t) \, \mathrm{J}_j^{(\alpha,\beta)}(t) \, (1-t)^{\alpha} \, (1+t)^{\beta} \, dt$$

$$= \frac{2^{\alpha+\beta+1}}{2i+\alpha+\beta+1} \, \frac{\Gamma(i+\alpha+1)\,\Gamma(i+\beta+1)}{i!\,\Gamma(i+\alpha+\beta+1)} \, \delta_{ij}, \qquad \text{(B.8.6)}$$

where δ_{ij} is Kronecker's delta.

- *Recursion relation:*

With reference to (B.1.15),

$$A_0 = 1,$$

$$\alpha_i = \frac{(2i+\alpha+\beta+1)(2i+\alpha+\beta+2)}{2(i+1)(i+\alpha+\beta+1)} \qquad \text{for } i = 0, 1, \ldots,$$

$$\beta_i = \frac{\beta^2 - \alpha^2}{(2i+\alpha+\beta)(2i+\alpha+\beta+2)} \qquad \text{for } i = 0, 1, \ldots, \qquad \text{(B.8.7)}$$

$$\gamma_i = \frac{(i+\alpha)(i+\beta)(2i+\alpha+\beta+2)}{(i+1)(i+\alpha+\beta+1)(2i+\alpha+\beta)} \qquad \text{for } i = 1, \ldots \quad .$$

- *Differential equations:*

$$\left[(1-t)^{1+\alpha}(1+t)^{1+\beta} \, y'(t) \right]' = -i\,(i+\alpha+\beta+1)\,y(t) \qquad \text{(B.8.8)}$$

and

$$(1-t^2)\,y''(t) - [\alpha - \beta + (\alpha+\beta+2)\,t]\,y'(t) + i\,(i+\alpha+\beta+1)\,y(t) = 0, \qquad \text{(B.8.9)}$$

where $y = \mathrm{J}_i^{(\alpha,\beta)}(t)$.

- *Generating function:*

$$\frac{2^{\alpha+\beta}}{\mathcal{R}\,(1-\eta+\mathcal{R})^{\alpha}\,(1+\eta+\mathcal{R})^{\beta}} = \sum_{i=0}^{\infty} \mathrm{J}_i^{(\alpha,\beta)}(t)\,\eta^i \qquad \text{(B.8.10)}$$

for $-1 < t < 1$ and $|\eta| < 1$, where

$$\mathcal{R} = \sqrt{1 - 2t\,\eta + \eta^2}. \qquad \text{(B.8.11)}$$

C

Linear solvers

In the finite element codes discussed in this book, the systems of linear algebraic equations arising from the implementation of the Galerkin finite element method were solved either by the Thomas algorithm for tridiagonal and pentadiagonal systems or by invoking a MATLAB function implemented by a vector-by-matrix division for more general systems. In practice, linear systems arising in finite element applications can have large or exorbitant dimensions, on the order of hundreds of thousands or even higher. To solve such large systems, we employ general-purpose or custom-made algorithms that exploit the sparsity and possible symmetry of the coefficient matrix to reduce memory storage requirements and the number of floating point operations per second (FLOPS).

In scientific computing, hardware performance is measured in units of megaflops, (MFLOPS) which is equal to 10^6 flops, gigaflops (GFLOPS), which is equal to 10^9 FLOPS, teraflops (TFLOPS), which is equal to 10^{12} FLOPS, and petaflops (PFLOPS), which is equal to 10^{15} FLOPS. An efficient desktop computer can perform 4 floating point operations per clock cycle, which amounts to a few GFLOPS. At the time of this writing, the fastest supercomputer can perform 33.86 PFLOPS.

In Sections C.1–C.3, we present a general overview of selected methods for solving systems of linear equations, and demonstrate their MATLAB implementation. In Section C.4, we present a summary of topics concerning matrix storage and manipulation in finite element applications.

C.1   Gauss elimination

Gauss elimination is the most popular *direct* method for solving systems of linear equations with small or moderate size. Consider a system of N linear algebraic equations for N unknowns, $x_1, x_2, \ldots, x_N$,

$$
\begin{aligned}
A_{1,1}\, x_1 + A_{1,2}\, x_2 + \cdots + A_{1,N-1}\, x_{N-1} + A_{1,N}\, x_N &= b_1, \\
A_{2,1}\, x_1 + A_{2,2}\, x_2 + \cdots + A_{2,N-1}\, x_{N-1} + A_{2,N}\, x_N &= b_2, \\
\ldots & \\
A_{N,1}\, x_1 + A_{N,2}\, x_2 + \cdots + A_{N,N-1}\, x_{N-1} + A_{N,N}\, x_N &= b_N,
\end{aligned}
\tag{C.1.1}
$$

where $A_{i,j}$ are given coefficients and b_i are given constants, for $i, j = 1, \ldots, N$. In matrix notation, the system (C.1.1) takes the compact form

$$\mathbf{A} \cdot \mathbf{x} = \mathbf{b}, \tag{C.1.2}$$

where $\mathbf{A}$ is an $N \times N$ coefficient matrix,

$$\mathbf{A} = \begin{bmatrix} A_{1,1} & A_{1,2} & \cdots & A_{1,N-1} & A_{1,N} \\ A_{2,1} & A_{2,2} & \cdots & A_{2,N-1} & A_{2,N} \\ \vdots & \vdots & \ddots & \vdots & \vdots \\ A_{N-1,1} & A_{N-1,2} & \cdots & A_{N-1,N-1} & A_{N-1,N} \\ A_{N,1} & A_{N,2} & \cdots & A_{N,N-1} & A_{N,N} \end{bmatrix}, \tag{C.1.3}$$

and $\mathbf{b}$ is an N-dimensional vector,

$$\mathbf{b} = \begin{bmatrix} b_1 \\ b_2 \\ \vdots \\ b_{N-1} \\ b_N \end{bmatrix}. \tag{C.1.4}$$

In the method of Gauss elimination, the original system (C.1.2) is reduced to an upper triangular system,

$$\mathbf{U} \cdot \mathbf{x} = \mathbf{y}, \tag{C.1.5}$$

where $\mathbf{U}$ is an upper triangular matrix,

$$\mathbf{U} = \begin{bmatrix} U_{1,1} & U_{1,2} & \cdots & U_{1,N-1} & U_{1,N} \\ 0 & U_{2,2} & \cdots & U_{2,N-1} & U_{2,N} \\ \vdots & \vdots & \ddots & \vdots & \vdots \\ 0 & 0 & \cdots & U_{N-1,N-1} & U_{N-1,N} \\ 0 & 0 & \cdots & 0 & U_{N,N} \end{bmatrix}, \tag{C.1.6}$$

and $\mathbf{y}$ is a properly constructed right-hand side. The elimination is followed by back-substitution, which involves solving the last equation of (C.1.5) for x_N, and substituting its value into all previous equations to eliminate this unknown. Next, we solve the penultimate equation for x_{N-1}, and substitute its value into all previous equations to eliminate this unknown. Moving backward in this fashion, we compute all unknowns, up to the first unknown, x_1.

As a bonus, Gauss elimination simultaneously generates a lower diagonal matrix, $\mathbf{L}$, with *ones* along the diagonal,

$$\mathbf{L} = \begin{bmatrix} 1 & 0 & \cdots & 0 & 0 \\ L_{2,1} & 1 & \cdots & 0 & 0 \\ \vdots & \vdots & \ddots & \vdots & \vdots \\ L_{N-1,1} & L_{N-1,2} & \cdots & 1 & 0 \\ L_{N,1} & L_{N,2} & \cdots & 0 & 1 \end{bmatrix}, \tag{C.1.7}$$

so that

$$\mathbf{A} = \mathbf{L} \cdot \mathbf{U}. \tag{C.1.8}$$

Thus, Gauss elimination can be used to perform the Doolittle LU decomposition of the coefficient matrix, $\mathbf{A}$, and can be used exclusively for that purpose and without reference to a system of linear equations. When the option of row pivoting is enabled, as discussed in Section C.1.1, the method can be used to perform the Doolittle LU decomposition of a matrix that arises by interchanging rows of $\mathbf{A}$.

The basic idea behind Gauss elimination is to solve the first equation in (C.1.1) for the first unknown, x_1, and use the expression for x_1 thus obtained to eliminate this unknown from all subsequent equations. We then retain the first equation as is, and replace all subsequent equations with their descendants that do not contain x_1. In the second stage, we solve the second equation for the second unknown, x_2, and use the expression for x_2 thus obtained to eliminate this unknown from all subsequent equations. We then retain the first and second equations, and replace all subsequent equations with their descendants that do not contain x_1 or x_2. Continuing in this manner, we arrive at the last equation, which contains only the last unknown, x_N, and this concludes the process of elimination.

C.1.1 Pivoting

Immediately before the mth equation has been solved for the mth unknown, the linear system appears as

$$
\begin{bmatrix}
A_{1,1}^{(m)} & A_{1,2}^{(m)} & \cdots & & \cdots & A_{1,N}^{(m)} \\
0 & A_{2,2}^{(m)} & \cdots & & \cdots & A_{2,N}^{(m)} \\
0 & 0 & \ddots & \cdots & & \\
0 & 0 & A_{m-1,m-1}^{(m)} & A_{m-1,m}^{(m)} & \cdots & A_{m-1,N}^{(m)} \\
0 & \cdots & 0 & A_{m,m}^{(m)} & \vdots & A_{m,N}^{(m)} \\
0 & \cdots & 0 & \cdots & & \\
0 & \cdots & 0 & A_{N,m}^{(m)} & \cdots & A_{N,N}^{(m)}
\end{bmatrix}
\cdot
\begin{bmatrix}
x_1 \\ x_2 \\ x_3 \\ \vdots \\ x_{N-1} \\ x_N
\end{bmatrix}
=
\begin{bmatrix}
b_1^{(m)} \\ b_2^{(m)} \\ b_3^{(m)} \\ \vdots \\ b_{N-1}^{(m)} \\ b_N^{(m)}
\end{bmatrix},
$$

$$\tag{C.1.9}$$

where $m = 1, \ldots, N - 1$, $A_{i,j}^{(m)}$ are intermediate coefficients, and $b_i^{(m)}$ are intermediate right-hand sides. The first equation in (C.1.9) is identical to the first equation of the original system (C.1.1) at any stage, m. Subsequent equations are different, except at the first step corresponding to $m = 1$.

An apparent failure occurs when the diagonal element $A_{m,m}^{(m)}$ is nearly or precisely equal to zero, as we may no longer solve the mth equation in (C.1.9) for x_m, as required. However, the breakdown of the algorithm does *not* necessarily mean that the system of equations does not have a solution. To circumvent this difficulty, we simply rearrange the equations or relabel the unknowns to bring the mth unknown to the mth equation using the method of *pivoting*. If there is no way we can make this happen, the matrix $\mathbf{A}$ is singular and the linear system has either no solution or an infinite number of solutions.

In the method of *row pivoting*, potential difficulties are bypassed by switching the mth equation in the system (C.1.9) with the subsequent kth equation, where $k > m$, and the value of k is chosen so that $|A_{k,m}^{(m)}|$ is the maximum value of the elements in the mth column below the diagonal,

$$A_{i,m}^{(m)} \qquad \text{for} \quad i \geq m. \tag{C.1.10}$$

If $A_{i,m}^{(m)} = 0$ for any $i \geq m$, then the system under consideration does not have a unique solution, which means that the matrix $\mathbf{A}$ is singular.

Pivoting is not necessary for systems with diagonally dominant coefficient matrices.[1] Pivoting is not necessary for systems with symmetric and positive-definite coefficient matrices. Pivoting prohibits the processor parallelization.

It appears that pivoting should be enabled only when the magnitude of the intermediate element, $A_{m,m}^{(m)}$, is smaller than a specified threshold. However, in practice, we want to make $|A_{m,m}^{(m)}|$ as large as possible in order to reduce the round-off error associated with the floating point representation, and thus enable pivoting even when $|A_{m,m}^{(m)}|$ is not necessarily small.

C.1.2 Implementation

To implement the method of Gauss elimination with row pivoting, we proceed according to the following steps:

Setup

Formulate the $N \times (N + 1)$ partitioned augmented matrix

$$\mathbf{C}^{(1)} \equiv \left[\, \mathbf{A} \,\middle|\, \mathbf{b} \,\right], \tag{C.1.11}$$

and introduce an $N \times N$ matrix, $\mathbf{L}$, whose elements are initialized to zero.

First pass

Eliminate x_1 from the second to the last equation working as follows:

1. Find the location of the element with the maximum norm in the first column of $\mathbf{C}^{(1)}$, that is, search for the maximum norm of the elements $|C_{i,1}^{(1)}|$, for

---

[1] By definition, the magnitude of each diagonal element of a diagonally dominant matrix is larger than the sum of the magnitudes of the remaining elements in the corresponding row.

$i = 1, \ldots, N$. Assume that this is equal to $|C_{k,1}^{(1)}|$, corresponding to the kth row.

2. If $k = 1$, skip this step. Otherwise, switch the first row with the kth row of $\mathbf{C}^{(1)}$ and the first row with the kth row of $\mathbf{L}$.

3. Compute the first column of $\mathbf{L}$ below the diagonal by setting $L_{i,1} = C_{i,1}^{(1)}/C_{1,1}^{(1)}$ for $i = 2, \ldots, N$.

4. Subtract from the ith row of $\mathbf{C}^{(1)}$ the first row multiplied by $L_{i,1}$ for $i = 2, \ldots, N$ to obtain a new augmented matrix,

$$\mathbf{C}^{(2)} \equiv [\mathbf{A}^{(2)}|\mathbf{b}^{(2)}]. \qquad (C.1.12)$$

Second pass

Eliminate x_2 from the third to the last equation working as follows:

1. Find the location of the element with the maximum norm in the second column of $\mathbf{C}^{(2)}$ below the diagonal, that is, search for the maximum norm of the elements $|C_{i,2}^{(2)}|$ for $i = 2, \ldots, N$. Assume that this is equal to $|C_{k,2}^{(2)}|$, corresponding to the kth row.

2. If $k = 2$, skip this step. Otherwise, switch the second row with the kth row of $\mathbf{C}^{(2)}$ and the second row with the kth row of $\mathbf{L}$.

3. Compute the second column of $\mathbf{L}$ below the diagonal by setting $L_{i,2} = C_{i,2}^{(2)}/C_{2,2}^{(2)}$ for $i = 3, \ldots, N$.

4. Subtract from the ith row of $\mathbf{C}^{(2)}$ the second row multiplied by $L_{i,2}$ for $i = 3, \ldots, N$ to obtain a new augmented matrix,

$$\mathbf{C}^{(3)} \equiv [\mathbf{A}^{(3)}|\mathbf{b}^{(3)}]. \qquad (C.1.13)$$

mth pass

Eliminate x_m from the $m+1$ to the last equation working as follows:

1. Find the location of the element with the maximum norm in the mth column of $\mathbf{C}^{(m)}$ below the diagonal, that is, search for the maximum norm of the elements $|C_{i,m}^{(m)}|$ for $i = m, \ldots, N$. Assume that this is equal to $|C_{k,m}^{(m)}|$, corresponding to the kth row.

2. If $k = m$, skip this step. Otherwise, switch the mth row with the kth row of $\mathbf{C}^{(m)}$ and the mth row with the kth row of $\mathbf{L}$.

3. Compute the mth column of $\mathbf{L}$ below the diagonal by setting $L_{i,m} = C_{i,m}^{(m)}/C_{m,m}^{(m)}$ for $i = m+1, \ldots, N$.

4. Subtract from the ith row of $\mathbf{C}^{(m)}$ the mth row multiplied by $L_{i,m}$ for $i = m + 1, \ldots, N$ to obtain a new augmented matrix,

$$\mathbf{C}^{(m+1)} \equiv [\mathbf{A}^{(m+1)} | \mathbf{b}^{(m+1)}]. \tag{C.1.14}$$

$N - 1$ *pass*

At the end of the $N - 1$ pass, corresponding to $m = N - 1$, the augmented matrix has the form

$$\mathbf{C}^{(N)} = \left[\mathbf{A}^{(N)} \,\middle|\, \mathbf{b}^{(N)} \right], \tag{C.1.15}$$

where $\mathbf{A}^{(N)} \equiv \mathbf{U}$ is the upper triangular matrix shown in (C.1.5).

Backward substitution

Finally, we solve the upper triangular system

$$\mathbf{U} \cdot \mathbf{x} = \mathbf{b}^{(N)} \tag{C.1.16}$$

by backward substitution to extract the solution of the original system of equations $\mathbf{A} \cdot \mathbf{x} = \mathbf{b}$. This is done by solving the last equation in (C.1.16) for the last unknown, x_N. Once this is available, we solve the penultimate equation for x_{N-1}. Continuing backward in this manner, we finally compute x_1.

Complete the matrix L *(optional)*

Set the diagonal elements of the matrix $\mathbf{L}$ equal to 1. It can be shown by straightforward algebraic manipulations that the matrices $\mathbf{L}$ and $\mathbf{U}$ provide us with the *LU* decomposition of the matrix $\mathbf{A}$,

$$\mathbf{L} \cdot \mathbf{U} = \mathbf{A}^{\mathrm{mod}}, \tag{C.1.17}$$

where the matrix $\mathbf{A}^{\mathrm{mod}}$ is identical to $\mathbf{A}$, except that the rows may have been interchanged due to pivoting. If pivoting has been disabled, $\mathbf{A}^{\mathrm{mod}} = \mathbf{A}$.

Compute the determinant (optional)

The determinant of the matrix $\mathbf{A}$ follows from the expression

$$\det(\mathbf{A}) = \pm\det(\mathbf{A}^{\mathrm{mod}}) = \pm\det(\mathbf{L}) \cdot \det(\mathbf{U}) = \pm U_{1,1} \, U_{2,2} \cdots U_{N,N}. \tag{C.1.18}$$

The plus sign applies in the absence of pivoting or when an even number of row interchanges have been done due to pivoting, and the minus sign otherwise.

C.1.3 Symmetric matrices

If the original matrix of the linear system, $\mathbf{A}$, is symmetric and pivoting is not enabled at the risk of having to divide by zero or foster the growth of round-off error, the lower square diagonal $(N - m + 1) \times (N - m + 1)$ block of the matrix displayed in equation (C.1.9) will remain symmetric for any value of m. This observation can be exploited for storing, and working only with the upper or lower triangular parts of the evolving coefficient matrix, thereby economizing the computations.

Specifically, for any value of m, we may work only with the elements of the augmented matrix on or above the diagonal, and set the values of the elements below the diagonal equal to their symmetric counterparts, as required. This modification reduces the number of operations nearly by a factor of 2. When $\mathbf{A}$ is positive definite, pivoting is not required and economization can be implemented without a risk.

C.1.4 Computational cost

The cost of Gauss elimination scales with $\frac{1}{3}N^3$, where N is the system size. This means that, if the system size is increased by a factor of 2, the computational cost is raised by a factor of 8. For systems of large size, Gauss elimination requires a prohibitive computational time. For example, when $N = 100,000$, a desktop computer that performs 1 GFLOPS requires a CPU time on the order of $\frac{1}{3} 10^{15} \times 10^{-9}\mathrm{s} \sim 92\,\mathrm{h}$, that is approximately four days and nights.

C.1.5 Gauss elimination code

FSELIB function *gel*, listed in Table C.1.1, implements the Gauss elimination algorithm with optional row pivoting, and also allows for the possibility for expedited solution in the case of a symmetric coefficient matrix.

The driver script *gel_dr*, listed in Table C.1.2, reads the coefficient matrix and right-hand side of the linear system from file *mat_vec.dat* and calls *gel* to compute the solution. To verify the solution, we type at the MATLAB prompt the statement:

```
>> A*x'-rhs'
```

and confirm that the answer is a very small residual vector whose magnitude is comparable to the round-off error.

C.1.6 Multiple right-hand sides

A simple modification of the basic algorithm allows us to solve at once multiple systems of equations with the same coefficient matrix but different right-hand sides,

$$\mathbf{A} \cdot \mathbf{x}^{(j)} = \mathbf{b}^{(j)},$$

(C.1.19)

```
function [x, ...
          l,u,det, ...
          Istop] = gel (n,a,rhs,Ipvt,Isym)

%=========================================
% Solution of the nxn linear system:
%    a x = rhs
% by Gauss elimination with row pivoting
%
% If Ipvt = 1, row pivoting is enabled
% If Isym = 1, matrix a is symmetric
%
% c: extended coefficient matrix
% l: lower triangular matrix
% u: upper triangular matrix
%
% det: determinant of a
%=========================================

%-----------
% initialize
%-----------

Istop  = 0;           % error flag
Icount = 0;           % counts row interchanges in pivoting
eps = 0.00000000001;  % tolerance

%-----------
% pivoting is not done for a symmetric system
%-----------

if(Isym==1)
 disp(' gel: system is symmetric; pivoting is disabled')
 Ipvt = 0;
end

%--------
% prepare
%--------

na = n-1;
n1 = n+1;

%------------------
% initialize l and c
%------------------

for i=1:n
  for j=1:n
    l(i,j) = 0.0;
    c(i,j) = a(i,j);
  end
end
```

TABLE C.1.1   Function *gel* (Continuing →)

```
  c(i,n1) = rhs(i);
end

%---------------------
% begin row reductions
%---------------------

for m=1:na   % outer loop for working row

   ma = m-1; m1 = m+1;

%------------------------
% pivoting module:
%
% search the ith column
% for the largest element
%------------------------

if(Ipvt==1)   % pivoting will be done if Ipvt=1

  Ipv = m; pivot = abs(c(m,m));

    for j=m1:n
      if(abs(c(j,m)) > pivot)
       Ipv = j; pivot = abs(c(j,m));
      end
    end

    if(pivot < eps)
       disp ('gel: trouble in station 1')
       Istop = 1;
       return
    end

%------------------------------------
% switch the working row with the row
% containing the pivot element;
% also switch rows in l
%------------------------------------

    if(Ipv ~= m)

     for j=m:n1
       save = c(m,j);
       c(m,j) = c(Ipv,j); c(Ipv,j) = save;
     end

     for j=1:ma
       save = l(m,j);
       l(m,j) = l(Ipv,j); l(Ipv,j) = save;
     end
     Icount = Icount+1;    % increase the pivoting counter

    end

end   % end of pivoting module
```

TABLE C.1.1 Function *gel* ($\rightarrow$ Continuing $\rightarrow$)

```
%----------------------------------------
% reduce column i beneath element c(m,m)
%----------------------------------------

 for i=m1:n

  l(i,m) = c(i,m)/c(m,m);

  if(Isym==1)
   l(i,m) = c(m,i)/c(m,m); ilow = i;
  else
   l(i,m) = c(i,m)/c(m,m); ilow = m1;
  end

  c(i,m) = 0.0;

  for j=ilow:n1
    c(i,j) = c(i,j)-l(i,m)*c(m,j);
  end

 end

%---
% end of outer loop for working row:
%---

end

%----------------------------------
% check the last diagonal element
% for a singularity
%----------------------------------

if(abs(c(n,n)) < eps)

   disp('gel: trouble in station 2')
   Istop = 1;
   return;

end

%----------------------
% complete the matrix l
%----------------------

for i=1:n
  l(i,i) = 1.0;
end

%--------------------
% define the matrix u
%--------------------
```

TABLE C.1.1 Function *gel* ($\rightarrow$ Continuing $\rightarrow$)

```
for i=1:n
  for j=1:n
    u(i,j) = c(i,j);
  end
end

%------------------------------------
% perform back-substitution to solve
% the reduced system
% using the upper triangular matrix c
%------------------------------------

x(n) = c(n,n1)/c(n,n);

for i=na:-1:1

  sum = c(i,n1);
  for j=i+1:n
    sum = sum-c(i,j)*x(j);
  end
  x(i) = sum/c(i,i);

end

%---------------------------
% compute the determinant as:
%
% det(a) = (+-) det(l)*det(u)
%---------------------------

det = 1.0;

for i=1:n
   det = det*c(i,i);
end

if(Ipvt == 1)
  for i=1:Icount
    det = -det;
  end
end

%-----
% done
%-----

return
```

TABLE C.1.1 ($\rightarrow$ Continued) Function *gel* solves a linear system of equations by the method of Gauss elimination with optional row pivoting, and also performs the LU decomposition and computes the determinant of the coefficient matrix. An option for an expeditious calculation in the case of a symmetric coefficient matrix can be selected.

```
%================================
% Driver for Gauss Elimination
%================================

file1 = fopen('mat_vec.dat');
 N   = fscanf(file1,'%f',[1,1]);
 A   = fscanf(file1,'%f',[N,N]);
 rhs = fscanf(file1,'%f',[1,N]);
fclose(file1);

A = A';  % because A was read column-wise,
         % replace with the transpose

Ipvt=1; % pivoting enabled (0 to disable)
Isym=0; % system is not symmetric

[x,l,u,det,Istop] = gel (N,A,rhs,Ipvt,Isym);

disp ('Solution:'); x
```

TABLE C.1.2   Driver script for Gauss elimination with possible row pivoting and an option for symmetric systems.   The data are read from an input file named *mat_vec.dat*.

where $j = 1, \ldots, p$ and p is the number of systems to be solved. In the standard method of Gauss elimination, the primary system $\mathbf{A} \cdot \mathbf{x}^{(j)} = \mathbf{b}^{(j)}$ is reduced to the modified system $\mathbf{U} \cdot \mathbf{x}^{(j)} = \mathbf{c}^{(j)}$, which is then solved by backward substitution. A key observation is that the computation of the vectors $\mathbf{c}^{(j)}$ can be done simultaneously, working with an $N \times (N + p)$ partitioned augmented matrix,

$$\mathbf{C}^{(1)} \equiv \left[\mathbf{A} \,|\, \mathbf{b}^{(1)} \,|\, \mathbf{b}^{(2)} \,|\, \cdots \,|\, \mathbf{b}^{(p)} \right]. \tag{C.1.20}$$

At the end of the $N - 1$ pass, the augmented matrix $\mathbf{C}^{(N)}$ will have a partially block-triangular form,

$$\mathbf{C}^{(N)} = \left[\mathbf{A}^{(N)} \,|\, \mathbf{D}^{(N)} \right], \tag{C.1.21}$$

where $\mathbf{A}^{(N)} \equiv \mathbf{U}$ is an upper triangular matrix. The p columns of the matrix $\mathbf{D}^{(N)}$ contain the evolved right-hand sides, $\mathbf{c}^{(j)}$, where

$$\mathbf{U} \cdot \mathbf{x}^{(j)} = \mathbf{c}^{(j)}. \tag{C.1.22}$$

The algorithm concludes with p back substitutions.

C.1.7   Computation of the inverse

The jth column of the inverse of a square matrix, $\mathbf{A}$, denoted by $\mathbf{x}^{(j)}$, satisfies the linear system

$$\mathbf{A} \cdot \mathbf{x}^{(j)} = \mathbf{e}^{(j)} \qquad\qquad (\text{C.1.23})$$

for $j = 1, \ldots N$, where all components of the vector $\mathbf{e}^{(j)}$ are zero, except for the jth component that is equal to unity. The computation of the vectors $\mathbf{x}^{(j)}$ can be done in a compact manner, working with a $N \times 2N$ augmented matrix,

$$\mathbf{C}^{(1)} \equiv \left[\, \mathbf{A} \,|\, \mathbf{I} \,\right], \qquad\qquad (\text{C.1.24})$$

where $\mathbf{I}$ is the $N \times N$ identity matrix. At the end of the $N - 1$ pass, the augmented matrix, $\mathbf{C}^{(N)}$, has the form shown in (C.1.21). The algorithm concludes with p back substitutions.

C.1.8   Gauss–Jordan reduction

The method Gauss–Jordan reduction is a first cousin once removed of the method of Gauss elimination, involving the following steps:

1. Divide the first equation by $A^{(1)}_{1,1}$, solve it for the first unknown, x_1, and use the expression thus obtained to eliminate x_1 from *all* subsequent equations.

2. Divide the second equation by $A^{(2)}_{2,2}$, solve it for the second unknown, x_2, and use the expression thus obtained to eliminate x_2 from the *first and all subsequent equations*.

3. Continue in this manner, until the last equation contains only the last unknown. At that point, the evolved coefficient matrix will be diagonal, with all diagonal elements equal to unity. Consequently, the solution will be displayed on the right-hand side.

To implement the method, we formulate an augmented matrix,

$$\mathbf{C}^{(1)} \equiv \left[\, \mathbf{A} \,|\, \mathbf{b} \,\right], \qquad\qquad (\text{C.1.25})$$

and proceed as in the method of Gauss elimination with straightforward modifications. At the end of the $N - 1$ pass, corresponding to $m = N - 1$, the evolved matrix $C^{(N)}$ has the forms

$$\mathbf{C}^{(N)} = \left[\, \mathbf{I} \,|\, \mathbf{b}^{(N)} \,\right], \qquad\qquad (\text{C.1.26})$$

where $\mathbf{b}^{(N)}$ is the required solution. Row pivoting is implemented as in Gauss elimination.

It may appear that the method of Gauss–Jordan elimination is competitive with, if not preferable over, the method of Gauss elimination. However, counting the number of operations shows that the method is slower than the standard Gauss elimination, roughly by a factor of 3. Nevertheless, the robustness of Gauss–Jordan reduction makes it attractive as a benchmark for solving systems with small or moderate size.

C.2   Iterative methods based on matrix splitting

The need to solve systems of large size has motivated the development of a host of powerful methods for general-purpose and specialized applications. Iterative solution procedures are typically used for sparse systems arising in finite difference and finite element implementations.

In one class of iterative methods, the coefficient matrix, $\mathbf{A}$, is split into two matrices,

$$\mathbf{A} = \mathbf{B} - \mathbf{C}, \tag{C.2.1}$$

and the system $\mathbf{A} \cdot \mathbf{x} = \mathbf{b}$ is recast into the form

$$\mathbf{B} \cdot \mathbf{x} = \mathbf{C} \cdot \mathbf{x} + \mathbf{b}. \tag{C.2.2}$$

The algorithm involves guessing the solution, $\mathbf{x}$, computing the right-hand side of (C.2.2), and solving for the vector $\mathbf{x}$ on the left-hand side. The computation is repeated until the vector $\mathbf{x}$ used to compute the right-hand side of (C.2.2) is virtually identical to that arising from solving the linear system, within a preset tolerance.

The calculations are carried out based on successive substitutions according to the formula

$$\mathbf{B} \cdot \mathbf{x}^{(k+1)} = \mathbf{C} \cdot \mathbf{x}^{(k)} + \mathbf{b}, \tag{C.2.3}$$

where the superscript (k) denotes the kth iteration, for $k = 0, 1, \ldots$, and $\mathbf{x}^{(0)}$ is the initial guess. A more explicit interpretation of the iterative method emerges by recasting (C.2.3) into the form

$$\mathbf{B} \cdot \boldsymbol{\epsilon}^{(k)} = -\mathbf{A} \cdot \mathbf{x}^{(k)} + \mathbf{b}, \tag{C.2.4}$$

where

$$\boldsymbol{\epsilon}^{(k)} \equiv \mathbf{x}^{(k+1)} - \mathbf{x}^{(k)} \tag{C.2.5}$$

is the correction. If $\mathbf{x}^{(k)}$ is the desired solution, the right-hand side is zero, and so is the correction.

The main advantage of the iterative approach is that, if the splitting shown in (C.2.1) is done judiciously, solving (C.2.3) for $\mathbf{x}^{(k+1)}$ is much easier than solving (C.1.2) for $\mathbf{x}$ on the left-hand side. Thus, even though an indefinite number of iteration is carried out, the recursive method can be significantly more efficient than the direct method.

Convergence

To study the convergence of the iterations, we recast (C.2.3) into the form

$$\mathbf{x}^{(k+1)} = \mathbf{P} \cdot \mathbf{x}^{(k)} + \mathbf{B}^{-1} \cdot \mathbf{b}, \tag{C.2.6}$$

where

$$\mathbf{P} \equiv \mathbf{B}^{-1} \cdot \mathbf{C} \tag{C.2.7}$$

is a *projection matrix*. Theoretical analysis shows that the iterations will converge for any initial guess, provided that the spectral radius of the projection matrix is less than unity, that is, the magnitude of each real or complex eigenvalue of $\mathbf{P}$ is less than unity (e.g., Pozrikidis [47]).

If the dominant eigenvalue of the projection matrix is available, the rate of convergence of the iterations can be improved by applying Wielandt's method of spectrum deflation (e.g., Pozrikidis [47]).

C.2.1 Jacobi's method

In Jacobi's method, the matrix $\mathbf{A}$ is split into the diagonal part, $\mathbf{D}$, where $D_{ii} = A_{ii}$ and $D_{ij} = 0$ for $i \neq j$, and a remainder with zero diagonals,

$$\mathbf{B} = \mathbf{D}, \qquad \mathbf{C} = \mathbf{D} - \mathbf{A}. \tag{C.2.8}$$

The iterations are based on the formula

$$x_i^{(k+1)} = \frac{1}{A_{ii}} \left(b_i - \sum_{j=1}^{N}{}' A_{ij}\, x_j^{(k)} \right), \tag{C.2.9}$$

where the prime indicates that the term $j = i$ is excluded from the sum. The underlying projection matrix is

$$\mathbf{P} = \mathbf{I} - \mathbf{D}^{-1} \cdot \mathbf{A}, \tag{C.2.10}$$

where the inverse diagonal matrix, $\mathbf{D}^{-1}$, contains the inverses of the diagonals.

C.2.2 Gauss–Seidel method

The Gauss–Seidel method differs from Jacobi's method in that the newly updated values of the solution replace the old values as soon as they are available. Specifically, the iterations are based on the formula

$$x_i^{(k+1)} = \frac{1}{A_{ii}} \left(b_i - \sum_{j=1}^{i-1} A_{ij}\, x_j^{(k+1)} - \sum_{j=i+1}^{N} A_{ij}\, x_j^{(k)} \right). \tag{C.2.11}$$

Cursory inspection reveals that the coefficient matrix is split into the following components:

$$\mathbf{B} = \mathbf{L} + \mathbf{D} \qquad \mathbf{C} = \mathbf{L} + \mathbf{D} - \mathbf{A}, \qquad\qquad (\text{C.2.12})$$

where:

- $\mathbf{D}$ is the diagonal part of $\mathbf{A}$.

- $\mathbf{L}$ is the strictly lower triangular part of $\mathbf{A}$, with zeros along the diagonal.

- $\mathbf{C}$ is the negative of the strictly upper triangular part of $\mathbf{A}$, with zeros along the diagonal, $\mathbf{A} = \mathbf{L} + \mathbf{D} - \mathbf{C}$.

The underlying projection matrix is

$$\mathbf{P} = -(\mathbf{L} + \mathbf{D})^{-1} \cdot \mathbf{C} = \mathbf{I} - (\mathbf{L} + \mathbf{D})^{-1} \cdot \mathbf{A}, \qquad\qquad (\text{C.2.13})$$

where the inverse of the lower triangular matrix $(\mathbf{L} + \mathbf{D})^{-1}$ is also lower triangular.

C.2.3 Successive over-relaxation (SOR)

The successive over-relaxation method is based on the following modification of the Gauss–Seidel splitting matrices shown in (C.2.12):

$$\mathbf{B} = \omega\,\mathbf{L} + \mathbf{D}, \qquad\qquad \mathbf{C} = \omega\,\mathbf{L} + \mathbf{D} - \mathbf{A} = (\omega - 1)\,\mathbf{L} - \mathbf{U}, \qquad (\text{C.2.14})$$

where ω is an adjustable parameter. When $\omega = 1$, we recover the Gauss–Seidel method. The iterations are based on the formula

$$x_i^{(k+1)} = (1 - \omega)\,x_i^{(k)} + \frac{\omega}{A_{ii}}\left(b_i - \sum_{j=1}^{i-1} A_{ij}\,x_j^{(k+1)} - \sum_{j=i+1}^{N} A_{ij}\,x_j^{(k)}\right). \quad (\text{C.2.15})$$

It can be shown that a necessary condition for the iterations to converge is $0 \le \omega \le 2$ (e.g., Pozrikidis [47]).

C.2.4 Operator- and grid-based splitting

In practical applications, the decomposition shown in (C.2.1) can be dictated by the splitting of differential or integral operators involved, or else motivated by the specifics of the numerical implementation. For example, Jacobi's method can be generalized into a block-diagonal splitting method where the diagonal blocks correspond to different parts of the solution domain. In other applications, the coefficient matrix can be split into a tridiagonal part and the remainder, and the iterations can be carried out using Thomas's algorithm. Creativity and imagination go a long way for the efficient solution of linear systems in physical applications.

C.3   Iterative methods based on path search

A powerful class of iterative methods search for the solution vector, $\mathbf{x}$, by making steps in the N-dimensional space toward carefully selected or optimal directions. The general strategy is to select or dynamically compose a set of search directions expressed by a finite or infinite collection of vectors,

$$\mathbf{p}^{(1)}, \qquad \mathbf{p}^{(2)}, \qquad \mathbf{p}^{(3)}, \qquad \ldots, \qquad (\text{C.3.1})$$

and then advance a guessed solution stepwise, according to the equation

$$\mathbf{x}^{(k)} = \mathbf{x}^{(k-1)} + \alpha_k \, \mathbf{p}^{(k)} \qquad (\text{C.3.2})$$

for $k \geq 1$, where α_k are appropriate coefficients expressing the length of each step. The evolved solution at the end of the kth step is

$$\mathbf{x}^{(k)} = \mathbf{x}^{(0)} + \sum_{i=1}^{k} \alpha_i \, \mathbf{p}^{(i)} \qquad (\text{C.3.3})$$

for $k \geq 1$, where $\mathbf{x}^{(0)}$ is the initial guess.

C.3.1   Symmetric and positive-definite matrices

Consider a square $N \times N$ matrix $\mathbf{A}$, and choose an N-dimensional vector, $\mathbf{x}$. If the scalar number $\mathbf{x} \cdot \mathbf{A} \cdot \mathbf{x}$ is positive for any non-null $\mathbf{x}$, then the matrix $\mathbf{A}$ is called positive definite.

Now consider a linear system, $\mathbf{A} \cdot \mathbf{x} = \mathbf{b}$. We will show that, if the coefficient matrix $\mathbf{A}$ is symmetric and positive definite, computing the solution of the linear system is equivalent to finding a vector, $\mathbf{X}$, that minimizes the scalar quadratic form

$$\mathcal{F}(\mathbf{x}) = \frac{1}{2} \mathbf{x} \cdot \mathbf{A} \cdot \mathbf{x} - \mathbf{b} \cdot \mathbf{x}. \qquad (\text{C.3.4})$$

The equivalence is evident for a single equation, $a\,x = b$, where the quadratic form reduces to

$$\mathcal{F}(x) = x \left(\frac{1}{2} \, a\,x - b \right). \qquad (\text{C.3.5})$$

The minimum value of $\mathcal{F}(x)$ clearly occurs at the point $x = X$ where $\partial \mathcal{F} / \partial x = aX - b = 0$.

To demonstrate the equivalence for a higher number of equations, we compute the gradient of $\mathcal{F}(\mathbf{x})$, defined as the vector of partial derivatives with respect to x_i, where $i = 1, \ldots, N$. Taking advantage of the symmetry of $\mathbf{A}$, we write

$$\frac{\partial \mathcal{F}}{\partial x_i} - \frac{1}{2} A_{ki}\,x_k + \frac{1}{2} A_{ik}\,x_k - b_i = A_{ik}\,x_k - b_i, \qquad (\text{C.3.6})$$

and note that all derivatives are zero when $\mathbf{x} = \mathbf{X}$, where $\mathbf{A} \cdot \mathbf{X} = \mathbf{b}$. Thus, $\mathbf{X}$ is a minimum, a maximum, or a saddle point of the quadratic form, $\mathcal{F}(\mathbf{x})$. To assess which one it is, we expand $\mathcal{F}(\mathbf{x})$ in a Taylor series about the critical point $\mathbf{X}$, observe that all but the constant and quadratic terms vanish, and obtain the exact representation

$$\mathcal{F}(\mathbf{x}) = \frac{1}{2} \left(\hat{\mathbf{x}} \cdot \mathbf{A} \cdot \hat{\mathbf{x}} - \mathbf{b} \cdot \mathbf{X} \right), \tag{C.3.7}$$

where $\hat{\mathbf{x}} \equiv \mathbf{x} - \mathbf{X}$ is the distance from the critical point. Because the matrix $\mathbf{A}$ has been assumed positive definite, the first term on the right-hand side is positive for any $\hat{\mathbf{x}}$, and this guarantees that $\mathcal{F}(\mathbf{x})$ attains the minimum value when $\mathbf{x} = \mathbf{X}$.

In summary, we have reduced the problem of computing the solution of the equation $\mathbf{A} \cdot \mathbf{x} = \mathbf{b}$ to the problem of finding the minimum of a quadratic form. For a certain class of matrices, $\mathbf{A}$, the solution of the minimization problem can be found with much less effort than that required by direct or iterative methods discussed in previous sections.

Steepest-descent search

In the steepest-descent search, we make an initial guess, $\mathbf{x}^{(0)}$, and then improve it by stepping in the direction where the quadratic form $\mathcal{F}(\mathbf{x})$ changes most rapidly. The steepest-descent direction is aligned with the residual vector

$$\mathbf{r} = -\nabla \mathcal{F} = -\mathbf{A} \cdot \mathbf{x} + \mathbf{b} \tag{C.3.8}$$

evaluated at $\mathbf{x}^{(0)}$, denoted by

$$\mathbf{r}^{(0)} \equiv -\mathbf{A} \cdot \mathbf{x}^{(0)} + \mathbf{b}. \tag{C.3.9}$$

Thus, the first search direction is $\mathbf{p}^{(1)} = \mathbf{r}^{(0)}$.

The question of how long a distance we should travel is pending. To answer this question, we note that, as we travel in the direction of $\mathbf{r}^{(0)}$, the vector $\mathbf{x}$ is described by

$$\mathbf{x} = \mathbf{x}^{(0)} + \alpha \, \mathbf{r}^{(0)}, \tag{C.3.10}$$

where α is an unspecified scalar parameter. The value of the quadratic form along this path is

$$\mathcal{F}(\mathbf{x}) = \frac{1}{2} \left(\mathbf{x}^{(0)} + \alpha \, \mathbf{r}^{(0)} \right) \cdot \mathbf{A} \cdot \left(\mathbf{x}^{(0)} + \alpha \, \mathbf{r}^{(0)} \right) - \mathbf{b} \cdot \left(\mathbf{x}^{(0)} + \alpha \, \mathbf{r}^{(0)} \right), \tag{C.3.11}$$

which is a quadratic function of α. We want to stop traveling when $\mathcal{F}$ has reached a minimum, that is, at the point where $\partial \mathcal{F} / \partial \alpha = 0$. Taking the partial derivative of the right-hand side of (C.3.11) with respect to α and setting the resulting expression equal to zero, we find the optimal value

$$\alpha_1 = \frac{\mathbf{r}^{(0)} \cdot \mathbf{r}^{(0)}}{\mathbf{r}^{(0)} \cdot \mathbf{A} \cdot \mathbf{r}^{(0)}}, \tag{C.3.12}$$

which yields the improved position

$$\mathbf{x}^{(1)} = \mathbf{x}^{(0)} + \alpha_1 \, \mathbf{r}^{(0)}. \tag{C.3.13}$$

The minimization process is subsequently repeated in a search direction determined by the current residual, $\mathbf{p}^{(k)} = \mathbf{r}^{(k-1)}$, where

$$\alpha_k = \frac{\mathbf{p}^{(k)} \cdot \mathbf{p}^{(k)}}{\mathbf{p}^{(k)} \cdot \mathbf{A} \cdot \mathbf{p}^{(k)}}, \tag{C.3.14}$$

until the minimum has been reached within a specified tolerance.

It is instructive to test the performance of the method for a single equation, $ax = b$. Applying the preceding formulas, we find that $r^{(0)} = -a\,x^{(0)} + b$ and $\alpha_1 = 1/a$, which provides us with the exact solution $x^{(1)} = b/a$ in a single step. Unfortunately, this excellent performance does not hold for two or more equations.

In general, while the method is guaranteed to converge, the rate of convergence can be prohibitively slow. Physically, when the graph of the quadratic form has narrow valleys, successive approximations bounce off opposite sides, slowly approaching the trough.

Method of conjugate gradients

Without loss of generality, we may begin the search from the origin by setting $\mathbf{x}^{(0)} = \mathbf{0}$. Our goal is to compute the search directions so that the exact solution, $\mathbf{X}$, is found exactly after N steps, that is,

$$\mathbf{X} = \alpha_1 \, \mathbf{p}^{(1)} + \alpha_2 \, \mathbf{p}^{(2)} + \cdots + \alpha_N \, \mathbf{p}^{(N)}, \tag{C.3.15}$$

where N is the system size. We will see that this is an ambitious yet achievable goal.

To address the question of how long a distance we should travel at the kth step, we note that, as we travel in the direction of $\mathbf{p}^{(k)}$, the vector $\mathbf{x}$ is described by

$$\mathbf{x} = \mathbf{x}^{(k-1)} + \alpha_k \, \mathbf{p}^{(k)} \tag{C.3.16}$$

and the value of the quadratic form along this path is

$$\mathcal{F}(\mathbf{x}) = \frac{1}{2} \left(\mathbf{x}^{(k-1)} + \alpha_k \, \mathbf{p}^{(k)} \right) \cdot \mathbf{A} \cdot \left(\mathbf{x}^{(k-1)} + \alpha_k \, \mathbf{p}^{(k)} \right) - \mathbf{b} \cdot \left(\mathbf{x}^{(k-1)} + \alpha_k \, \mathbf{p}^{(k)} \right), \tag{C.3.17}$$

which is a quadratic function of α_k. Setting $\partial \mathcal{F}/\partial \alpha_k = 0$ to ensure an optimal stopping point, we find that

$$\alpha_k = \frac{\mathbf{p}^{(k)} \cdot \mathbf{r}^{(k-1)}}{\mathbf{p}^{(k)} \cdot \mathbf{A} \cdot \mathbf{p}^{(k)}}, \tag{C.3.18}$$

which is inclusive of (C.3.14).

The distinguishing feature of the method of conjugate gradients is that the N search directions are required to be A-conjugate to one another, which means that

$$\mathbf{p}^{(i)} \cdot \mathbf{A} \cdot \mathbf{p}^{(j)} = 0 \qquad (C.3.19)$$

for $i \neq j$. We will see that this constraint arises naturally in the minimization of the quadratic form with respect to the search directions at every stage. It is important to note that the conjugation constraint (C.3.19) does not uniquely determine the search directions, but rather imposes a set of restrictions among them.

Assuming that relation (C.3.19) is fulfilled, we take the inner product of both sides of (C.3.15) with the vector $\mathbf{p}^{(k)} \cdot \mathbf{A}$, and obtain

$$\alpha_k = \frac{\mathbf{p}^{(k)} \cdot \mathbf{A} \cdot \mathbf{X}}{\mathbf{p}^{(k)} \cdot \mathbf{A} \cdot \mathbf{p}^{(k)}} = \frac{\mathbf{p}^{(k)} \cdot \mathbf{b}}{\mathbf{p}^{(k)} \cdot \mathbf{A} \cdot \mathbf{p}^{(k)}}, \qquad (C.3.20)$$

which appears to differ from (C.3.14) and its generalization (C.3.18).

If fact, the two formulas are identical. To show this, we introduce the residual at the kth step,

$$\mathbf{r}^{(k)} \equiv -\mathbf{A} \cdot \mathbf{x}^{(k)} + \mathbf{b} = -\mathbf{A} \cdot \sum_{j=1}^{k} \alpha_j \, \mathbf{p}^{(j)} + \mathbf{b}. \qquad (C.3.21)$$

Taking the inner product of both sides with the vector $\mathbf{p}^{(i)}$, where $i \leq k$, and using the A-conjugation condition and (C.3.20), we find that

$$\mathbf{p}^{(i)} \cdot \mathbf{r}^{(k)} = 0 \qquad (C.3.22)$$

for $i \leq k$, which shows that any vector that can be expressed as a linear combination of $\mathbf{p}^{(i)}$ with $i \leq k$ is orthogonal to the current residual, $\mathbf{r}^{(k)}$. We proceed by expressing the kth residual as

$$\mathbf{r}^{(k)} = \mathbf{r}^{(k-1)} - \alpha_k \, \mathbf{A} \cdot \mathbf{p}^{(k)}, \qquad (C.3.23)$$

take the inner product of both sides with $\mathbf{p}^{(k)}$, note that $\mathbf{p}^{(k)} \cdot \mathbf{r}^{(k)} = 0$, and find that

$$\mathbf{p}^{(k)} \cdot \mathbf{r}^{(k-1)} = \alpha_k \, \mathbf{p}^{(k)} \cdot \mathbf{A} \cdot \mathbf{p}^{(k)}, \qquad (C.3.24)$$

which reproduces (C.3.18).

It remains to show that the exact solution will be found after N steps. To prove this, all we have to do is ensure that, of all vectors that can be written as linear combinations of $\mathbf{p}^{(i)}$ with $i \leq k$, the vector

$$\mathbf{x}^{(k)} = \alpha_1 \, \mathbf{p}^{(1)} + \cdots + \alpha_k \, \mathbf{p}^{(k)} \qquad (C.3.25)$$

with coefficients computed from (C.3.20) minimizes the quadratic form (C.3.4). With this goal in mind, we consider the perturbed iterant $\mathbf{x}^{(k)} + \mathbf{d}$, where the vector $\mathbf{d}$ can be expressed as a linear combination of the set $\mathbf{p}^{(i)}$ for $i \leq k$, and compute

$$\mathcal{F}(\mathbf{x}^{(k)} + \mathbf{d}) = \frac{1}{2}\left(\mathbf{x}^{(k)} + \mathbf{d}\right) \cdot \mathbf{A} \cdot \left(\mathbf{x}^{(k)} + \mathbf{d}\right) - \mathbf{b} \cdot \left(\mathbf{x}^{(k)} + \mathbf{d}\right),$$

$$= \frac{1}{2}\mathbf{x}^{(k)} \cdot \mathbf{A} \cdot \mathbf{x}^{(k)} - \mathbf{b} \cdot \mathbf{x}^{(k)} + \frac{1}{2}\mathbf{d} \cdot \mathbf{A} \cdot \mathbf{d} + \mathbf{d} \cdot (\mathbf{A} \cdot \mathbf{x}^{(k)} - \mathbf{b}). \quad \text{(C.3.26)}$$

The last term is zero because of property (C.3.22), and the penultimate term is non-negative because $\mathbf{A}$ is positive definite. Consequently, the functional $\mathcal{F}$ reaches its minimum value when $\mathbf{d} = \mathbf{0}$. The choice of search directions is still pending.

In the original method of conjugate gradients developed by Hestenes and Stiefel [27], the current direction, $\mathbf{p}^{(k)}$, is aligned as much as possible with the current residual, $\mathbf{r}^{(k-1)}$, subject to the A-conjugation constraint. The objective is to make the magnitude of the numerator in (C.3.18) as large as possible, and thereby move toward the minimum of the quadratic form at the fastest possible rate. In that case, it can be shown that the residual vectors $\mathbf{r}^{(k)}$ for a non-singular matrix $\mathbf{A}$ form an orthogonal basis of the Krylov space spanned by the vectors

$$\mathbf{r}^{(0)}, \qquad \mathbf{A} \cdot \mathbf{r}^{(0)}, \qquad \mathbf{A}^2 \cdot \mathbf{r}^{(0)}, \qquad \dots, \qquad \text{(C.3.27)}$$

that is,

$$\mathbf{r}^{(i)} \cdot \mathbf{r}^{(k)} = 0 \qquad \text{(C.3.28)}$$

for $i \neq k$. The method is implemented according to the algorithm listed in Table C.3.1, which is programmed in the FSELIB function *cg*, listed in Table C.3.2.

The following driver script *cg_dr* reads the coefficient matrix and right-hand side from the file *mat_s_vec.dat*, and calls **cg** to compute the solution:

```
file1 = fopen('mat_s_vec.dat');
N    = fscanf(file1,'%f',[1,1]);
A    = fscanf(file1,'%f',[N,N]);
rhs = fscanf(file1,'%f',[1,N]);
fclose(file1);
[sln] = cg (N, A, rhs);
disp ('Solution:'); sln
```

$\mathbf{x}^{(0)} = \mathbf{0}$

Do $k = 1, \ldots, N$

    If $k = 1$

        $\mathbf{p}^{(1)} = \mathbf{r}^{(0)} = \mathbf{b}$

    Else If $k > 1$

$$\beta_k = \frac{\mathbf{r}^{(k-1)} \cdot \mathbf{r}^{(k-1)}}{\mathbf{r}^{(k-2)} \cdot \mathbf{r}^{(k-2)}}$$

        $\mathbf{p}^{(k)} = \mathbf{r}^{(k-1)} + \beta_k \, \mathbf{p}^{(k-1)}$

    End if

$$\alpha_k = \frac{\mathbf{r}^{(k-1)} \cdot \mathbf{r}^{(k-1)}}{\mathbf{p}^{(k)} \cdot \mathbf{A} \cdot \mathbf{p}^{(k)}}$$

    $\mathbf{x}^{(k)} = \mathbf{x}^{(k-1)} + \alpha_k \, \mathbf{p}^{(k)}$

    $\mathbf{r}^{(k)} = \mathbf{r}^{(k-1)} - \alpha_k \, \mathbf{A} \cdot \mathbf{p}^{(k)}$

End Do

TABLE C.3.1   A conjugate-gradients algorithm for solving a linear system , $\mathbf{A} \cdot \mathbf{x} = \mathbf{b}$, with a real, symmetric, and positive definite coefficient matrix $\mathbf{A}$.

The method of conjugate gradients generates a sequence of vectors that approximate the solution of a linear system with increasing accuracy. However, the reduction in error may be uneven through the iterations, as some steps may improve the solution a little, and others a great deal. In practice, we want to make a number of steps that is substantially less than the system size, and yet obtain a good approximation to the solution. If large corrections are made at the beginning, we are fortunate; but if large corrections are made at the end, we are unfortunate. The first scenario occurs when the condition number of the matrix, identified as the magnitude of the ratio of the largest to the smallest eigenvalue, is sufficiently small.

To reduce the condition number, we precondition the linear system before applying the numerical method. This is done by multiplying both sides with the inverse of a preconditioning matrix, $\mathbf{C}^{-1}$, to obtain a preconditioned system,

$$\mathbf{A}^P \cdot \mathbf{y} = \mathbf{b}^P, \tag{C.3.29}$$

```
function [sln] = cg (n, a, rhs)

%=====================================================
%  Solution of a linear symmetric positive-definite
%  system by the method of conjugate gradients
%
%  The search vectors are chosen by the method of
%  Hestenes and Steifel (1952)
%
%  SYMBOLS:
%
%  a .... symmetric positive definite matrix
%  n .... size (rows/columns) of matrix a
%  rhs .. right hand side vector (e.g., b, as in Ax=b)
%  x .... evolving solution vector
%  sln .. final solution vector
%  p .... search directions
%  r .... residual vectors
%  alpha. scale parameter for solution update
%  beta.. scale parameter for search direction
%
%=====================================================

%-----------------------------------------------------------
%  set the initial values of the vectors x and r (step 0)
%  set the first-step value of p
%-----------------------------------------------------------

for i=1:n
  x0(i) = 0.0;
  r0(i) = rhs(i);
  p(1,i) = rhs(i);
end

%-------------------------------
%  compute the sums used in alpha
%-------------------------------

alpha_num = 0.0; alpha_den = 0.0;

for i=1:n

  alpha_num = alpha_num + r0(i)*r0(i);
  for j=1:n
    alpha_den = alpha_den + p(1,i)*a(i,j)*p(1,j);
  end

end

alpha = alpha_num/alpha_den;

%-----------------------------------------------------------
%  set first step values of alpha and vectors x and r
%-----------------------------------------------------------
```

TABLE C.3.2 Function *cg* (Continuing →)

```
for i=1:n
  x(1,i) = x0(i)+alpha*p(1,i);
  r(1,i) = r0(i);
  for j=1:n
    r(1,i) = r(1,i)-alpha*a(i,j)*p(1,j);
  end
end

%------------------------------------------
% loop through the remaining search vectors
% 2 to n, and compute alpha, beta,
% and the vectors p, x, and r
%------------------------------------------

for k=2:n        %  outer loop over search directions

%---
% sums used in beta
%---

beta_num = 0.0; beta_den = 0.0D0;

for i=1:n

  beta_num = beta_num + r(k-1,i)^2;
  if(k==2)
    beta_den = beta_den + r0(i)^2;
  else
    beta_den = beta_den + r(k-2,i)^2;
  end

end

beta = beta_num/beta_den;

for i=1:n
  p(k,i) = r(k-1,i)+beta*p(k-1,i);
end

%---
% compute the sums used in alpha
%---

alpha_num = beta_num;
alpha_den = 0.0;

for i=1:n
  for j=1:n
    alpha_den = alpha_den + p(k,i)*a(i,j)*p(k,j);
  end
end

alpha = alpha_num/alpha_den;
```

TABLE C.3.2 Function *cg* ($\to$ Continuing $\to$)

```
%---
% compute the kth iterations of vectors x and r
%---

for i=1:n
   x(k,i) = x(k-1,i)+alpha*p(k,i);
   r(k,i) = r(k-1,i);
   for j=1:n
      r(k,i)=r(k,i)-alpha*a(i,j)*p(k,j);
   end
end

end                    % end of outer loop

%--------------------
% extract the solution
%--------------------

for i=1:n
   sln(i) = x(n,i);
end

%-----
% done
%-----

return
```

TABLE C.3.2 ($\rightarrow$ Continued) Function *cg* solves a symmetric and positive definite system of linear equations by the method of conjugate gradients.

where

$$\mathbf{A}^P = \mathbf{C}^{-1} \cdot \mathbf{A} \cdot \mathbf{C}^{-1}, \qquad \mathbf{y} = \mathbf{C} \cdot \mathbf{x}, \qquad \mathbf{b}^P = \mathbf{C}^{-1} \cdot \mathbf{b}. \qquad (C.3.30)$$

The algorithm involves solving system (C.3.29) by the method of conjugate gradients for $\mathbf{y}$, and then recovering the solution, $\mathbf{x}$, by inverting the second equation in (C.3.30), which is done by a matrix-vector multiplication, $\mathbf{x} = \mathbf{C}^{-1} \cdot \mathbf{y}$. Note that the matrix $\mathbf{C}$ does not have to be available.

C.3.2 General methods

The method of conjugate gradients can be generalized into the method of biconjugate gradients, which is applicable to non-symmetric and non-positive-definite systems (e.g., Pozrikidis [47]). A popular alternative is the method of generalized minimal residuals (GMRES) (Saad & Schultz [54]). In the method of conjugate gradients, the residuals form an orthogonal set of the Krylov space. In the GMRES method, an orthonormal basis consisting of the vectors $\mathbf{v}^{(i)}$ is generated explicitly at every step using the Gram–Schmidt orthogonalization process (e.g., Pozrikidis

[47]). Specifically, the GMRES sequence is constructed according to the formula

$$\mathbf{x}^{(k)} = \mathbf{x}^{(0)} + \alpha_1 \mathbf{v}^{(1)} + \alpha_2 \mathbf{v}^{(2)} + \cdots + \alpha_k \mathbf{v}^{(k)}, \qquad (\text{C.3.31})$$

where the coefficients α_i are chosen at every step to minimize the norm of the residual $|\mathbf{A} \cdot \mathbf{x}^{(k)} - \mathbf{b}|$.

C.4   Finite element system solvers

Finite-element implementations usually culminate in linear systems with banded coefficient matrices with large or excessive dimensions. In some cases, the coefficient matrices are symmetric and positive-definite and the solution can be found reliably, efficiently, and economically using compact implementations of the method of Gauss elimination, LU decomposition without pivoting, and the iterative methods discussed previously in this section. For systems with large size, economizing storage and ensuring computational efficiency are of primary concern.

Significant gains in efficiency can be achieved with the implementation of bandwidth and skyline storage strategies and frontal solution algorithms (e.g., Harbani & Engelman [26], Hood [31]). In skyline storage, a dedicated skyline index matrix is introduced containing the indices of the highest and lowest elements in each column of the coefficient matrix. The matrix components are then stored in a partitioned one-dimensional skyline array. If the coefficient matrix is symmetric, the skyline index matrix contains the location of the diagonal elements in the skyline array. Methods for solving linear equations arising in finite element applications are reviewed by Schwarz ([58], Chapter 4).

Function interpolation

D

Given the values of a function, $f(x)$, at $N+1$ data points with abscissas located at x_i, where $i = 1, \ldots, N+1$, we want to interpolate at an intermediate point, that is, we want to compute the function at an arbitrary point, x. To carry out the interpolation, we replace the unknown function, $f(x)$, with a standard smooth function, $g(x)$, whose graph passes through the data points, as illustrated in Figure D.1. This means that the interpolating function, $g(x)$, is required to satisfy $N+1$ interpolation or matching conditions,

$$g(x_i) = f(x_i)$$

for $i = 1, \ldots, N+1$, with no additional stipulations or constraints. Once the interpolating function $g(x)$ is available, it can be differentiated or integrated to yield approximations to derivatives or to the definite integral of the interpolated function, $f(x)$, in terms of the data, $f(x_i)$. Differentiation provides us with finite difference approximations, and integration provides us with numerical integration rules and quadratures.

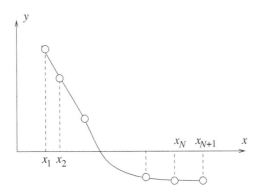

FIGURE D.1   Interpolation of a function through a set of $N+1$ data points. In polynomial interpolation, the function is approximated with an Nth-degree polynomial whose graph passes through the data points.

In this appendix, we summarize the basic theory of polynomial interpolation. The interested reader can find further discussion in dedicated monographs and texts on numerical methods (i.e., Davis [15], Pozrikidis [47]).

D.1 The interpolating polynomial

The most popular choice of an interpolating function, $g(x)$, is the Nth degree polynomial,

$$P_N(x) = a_1 x^N + a_2 x^{N-1} + a_3 x^{N-2} + \cdots + a_N x + a_{N+1}. \qquad (D.1.1)$$

The interpolation condition $g(x_i) = f(x_i)$ requires that the graph of the polynomial passes through the $N + 1$ data points,

$$P_N(x_i) = f(x_i) \qquad (D.1.2)$$

for $i = 1, \ldots, N + 1$. The problem is reduced to computing the $N + 1$ coefficients, $a_1, a_2, \ldots, a_{N+1}$, so that $P_N(x)$ satisfies the $N + 1$ interpolation conditions (D.1.2). Having completed this computation, we can evaluate $P_N(x)$ for any specified value of the independent variable, x, and thus obtain an approximation to the interpolated function, $f(x)$.

D.1.1 Vandermonde matrix

The most direct method of computing the polynomial coefficients is by enforcing the constraints (D.1.2) at the $N + 1$ data points to derive a system of $N + 1$ linear equations for the polynomial coefficients,

$$
\begin{aligned}
a_1 x_1^N + a_2 x_1^{N-1} + a_3 x_1^{N-2} + \cdots + a_N x_1 + a_{N+1} &= f(x_1), \\
a_1 x_2^N + a_2 x_2^{N-1} + a_3 x_2^{N-2} + \cdots + a_N x_2 + a_{N+1} &= f(x_2), \\
\cdots & \qquad \cdots \\
a_1 x_N^N + a_2 x_N^{N-1} + a_3 x_N^{N-2} + \cdots + a_N x_N + a_{N+1} &= f(x_N), \\
a_1 x_{N+1}^N + a_2 x_{N+1}^{N-1} + a_3 x_{N+1}^{N-2} + \cdots + a_N x_{N+1} + a_{N+1} &= f(x_{N+1}).
\end{aligned}
\qquad (D.1.3)
$$

In terms of the polynomial coefficient vector, $\mathbf{a}$, and data vector, $\mathbf{f}$, defined as

$$
\mathbf{a} \equiv
\begin{bmatrix}
a_{N+1} \\
a_N \\
\vdots \\
a_2 \\
a_1
\end{bmatrix},
\qquad
\mathbf{f} \equiv
\begin{bmatrix}
f(x_1) \\
f(x_2) \\
\vdots \\
f(x_N) \\
f(x_{N+1})
\end{bmatrix},
\qquad (D.1.4)
$$

the linear system takes the form

$$\mathbf{V}^T \cdot \mathbf{a} = \mathbf{f}, \qquad (D.1.5)$$

where the superscript T denotes the matrix transpose, and

$$
\mathbf{V}(x_1, x_2, \ldots, x_{N+1}) \equiv
\begin{bmatrix}
1 & 1 & 1 & \ldots & 1 & 1 \\
x_1 & x_2 & x_3 & \ldots & x_N & x_{N+1} \\
\ldots & \ldots & \ldots & \ldots & \ldots & \ldots \\
x_1^{N-1} & x_2^{N-1} & x_3^{N-1} & \ldots & x_N^{N-1} & x_{N+1}^{N-1} \\
x_1^{N} & x_2^{N} & x_3^{N} & \ldots & x_N^{N} & x_{N+1}^{N}
\end{bmatrix}
\tag{D.1.6}
$$

is the $(N + 1) \times (N + 1)$ *Vandermonde matrix*. Note that the abscissas, x_i, do not have to be arranged in any particular order, for example, in ascending or descending order, and can be arbitrarily juxtaposed.

Determinant of the Vandermonde matrix

The determinant of the Vandermonde matrix is equal to the product of all possible ordered differences between two distinct abscissas,

$$
\det(\mathbf{V}) = \prod_{i=1}^{N} \prod_{j=i+1}^{N+1} (x_j - x_i),
\tag{D.1.7}
$$

where Π denotes the product. Note that the right-hand side of (D.1.7) consists of

$$
N + \cdots + 2 + 1 = \frac{1}{2} N(N + 1)
\tag{D.1.8}
$$

multiplicative factors. When $N = 1$, we find that $\det(\mathbf{V}) = x_2 - x_1$. When $N = 2$, we find that $\det(\mathbf{V}) = (x_2 - x_1)(x_3 - x_1)(x_3 - x_2)$. Expression (D.1.7) demonstrates that, as long as the abscissas are distinct, $x_i \neq x_j$, the determinant of the Vandermonde matrix is non-zero the matrix is non-singular, the solution of the linear system (D.1.5) is unique, and the interpolation problem is well posed.

To prove (D.1.7), we observe that $\det[\mathbf{V}(x_1, x_2, \ldots, x_N, x_{N+1})]$ is an Nth-degree polynomial in x_{N+1}. Recalling that the determinant of a matrix is zero when two columns of a matrix are repeated, we find that the roots of this polynomial are equal to $x_1, x_2, \ldots, x_N$. This observation motivates the recursion relation

$$
\det[\mathbf{V}(x_1, \ldots, x_N, x_{N+1})]
$$
$$
= (x_{N+1} - x_1)(x_{N+1} - x_2) \cdots (x_{N+1} - x_N) \det[\mathbf{V}(x_1, \ldots, x_N)], \quad (D.1.9)
$$

which emerges by expanding the determinant of the Vandermonde matrix in terms of the minors of the last column. Repeated application of the recursion formula leads us to (D.1.7).

Without loss of generality, we may arrange the abscissas in ascending order, so that

$$
x_1 < x_2 < x_3 < \cdots < x_N < x_{N+1}.
\tag{D.1.10}
$$

Moreover, we can rescale the x axis so that $x_1 = 0$ and $x_{N+1} = 1$. Let us then hold the position of the first and last points fixed and start increasing the polynomial order, N, from the value of one, corresponding to two data points, to higher values. Since the distances between consecutive data points are being reduced, the factors on the left-hand side of (D.1.7) become decreasingly small. In practice, the determinant of the Vandermonde matrix becomes exceedingly small and the linear system (D.1.5) becomes nearly singular even at moderate values of N. This behavior places a serious limitation on the usefulness of the Vandermonde matrix approach.

The aforementioned difficulty concerning the possibly nearly singular nature of the Vandermonde matrix does not imply that the problem of polynomial interpolation is ill-posed, in the sense that the results are hopelessly sensitive to the numerical round-off error. We must simply find another way of computing the interpolating polynomial, which can be done using several viable alternatives. When a numerical method does not work, we must not immediately blame the problem.

Fekete points

An intriguing property of the Vandermonde matrix relates to the zeros of the Lobatto polynomials discussed in Section B.6, Appendix B. To demonstrate this connection, we hold the first and last interpolation points fixed at the standard values $x_1 = -1$ and $x_{N+1} = 1$, and search for the positions of the $N-1$ intermediate points that maximize the magnitude of the determinant of the Vandermonde matrix. By definition, these are the Fekete points for a finite interval of the x axis.

In Section D.5, we will show that the Fekete points coincide with the well-known zeros of the $(N-1)$-degree Lobatto polynomial $\mathrm{Lo}_{N-1}(x)$, discussed in Section B.6, Appendix B. Corresponding Fekete points can be defined in higher dimensions for specified geometrical domains, as discussed in Chapters 5 and 8.

D.1.2 Generalized Vandermonde matrix

The Nth-degree interpolating polynomial can be expressed as a sum of $N+1$ chosen basis functions,

$$P_N(x) = b_1\, \phi_{N+1}(x) + b_2\, \phi_N(x) + \cdots + b_N\, \phi_2(x) + b_{N+1}\, \phi_1(x), \quad \text{(D.1.11)}$$

where $\phi_i(x)$, $i = 1, \ldots, N+1$ is a set of Nth-degree polynomials that constitute a complete base for all Nth-degree polynomials. Completeness guarantees that any Nth-degree polynomial can be expressed as shown in (D.1.11) with appropriate coefficients, b_i. The standard representation (D.1.1) corresponds to the monomial base $\phi_i(x) = x^{i-1}$. Other possible bases are provided by the orthogonal polynomials discussed in Appendix B.

To compute the expansion coefficients, b_i, we require the interpolation conditions (D.1.2), and obtain a linear system,

$$\mathbf{V}_\phi^T \cdot \mathbf{b} = \mathbf{f}, \quad \text{(D.1.12)}$$

where

$$
\mathbf{V}_\phi \equiv
\begin{bmatrix}
\phi_1(x_1) & \phi_1(x_2) & \cdots & \phi_1(x_N) & \phi_1(x_{N+1}) \\
\phi_2(x_1) & \phi_2(x_2) & \cdots & \phi_2(x_N) & \phi_2(x_{N+1}) \\
\cdots & \cdots & \cdots & \cdots & \cdots \\
\phi_N(x_1) & \phi_N(x_2) & \cdots & \phi_N(x_N) & \phi_N(x_{N+1}) \\
\phi_{N+1}(x_1) & \phi_{N+1}(x_2) & \cdots & \phi_{N+1}(x_N) & \phi_{N+1}(x_{N+1})
\end{bmatrix}
\tag{D.1.13}
$$

is the $(N+1) \times (N+1)$ generalized Vandermonde matrix, and

$$
\mathbf{b} \equiv
\begin{bmatrix}
b_{N+1} \\
b_N \\
\vdots \\
b_2 \\
b_1
\end{bmatrix},
\qquad
\mathbf{f} \equiv
\begin{bmatrix}
f(x_1) \\
f(x_2) \\
\vdots \\
f(x_N) \\
f(x_{N+1})
\end{bmatrix}
\tag{D.1.14}
$$

are the coefficient vector, and data vector. In the case of the monomial base, $\phi_i(x) = x^{i-1}$, the generalized Vandermonde matrix reduces to the standard Vandermonde matrix displayed in (D.1.6). With a suitable choice of the polynomial base provided by the families of orthogonal polynomials discussed in Appendix B, the generalized Vandermonde system is well-conditioned even for high polynomial orders, N.

D.1.3   Newton interpolation

In Newton's interpolation method, the interpolating polynomial is expressed as a linear combination of a family of factorized triangular polynomials defined with reference to the first N interpolation nodes, x_i for $i = 1, \ldots, N$, as

$$
\begin{aligned}
P_N(x) = \ &c_0 \\
&+ c_1(x - x_1) \\
&+ c_2(x - x_1)(x - x_2) \\
&\cdots \\
&+ c_N(x - x_1)(x - x_2) \cdots (x - x_N),
\end{aligned}
\tag{D.1.15}
$$

where c_i, for $i = 0, 1, \ldots, N$ are $N + 1$ unknown coefficients.

Applying the interpolation condition (D.1.2) for $i = 1$, we obtain

$$
c_0 = f(x_1).
\tag{D.1.16}
$$

Applying next (D.1.2) for $i = 2$ and rearranging, we obtain

$$
c_1 = \frac{f(x_2) - c_0}{x_2 - x_1}.
\tag{D.1.17}
$$

The rest of the coefficients can be computed in a similar manner based on the recursive formula

$$
c_i = \frac{f(x_{i+1}) - c_0 - \sum_{j=1}^{i-1} c_j \left(\prod_{k=1}^{j} (x_{i+1} - x_k) \right)}{\prod_{k=1}^{i} (x_{i+1} - x_k)}.
\tag{D.1.18}
$$

In practice, the coefficients are computed using an algorithmic version of Newton's divided-difference table (e.g., Pozrikidis [47]).

D.2 Lagrange interpolation

Lagrange interpolation provides us with the interpolating polynomial in a manner that circumvents the explicit computation of the polynomial coefficients. To implement the method, we introduce a family of Nth-degree Lagrange polynomials associated with the given abscissas, x_i. The Lagrange polynomial associated with the ith abscissa, denoted as $l_{N,i}(x)$, is defined in terms of the abscissas of all data points, x_j, where $j = 1, \ldots, N + 1$. By construction, $l_{N,i}(x)$ is equal to zero at all data points, except at the ith data point where it is equal to unity, that is,

$$l_{N,i}(x_j) = \delta_{ij}, \tag{D.2.1}$$

where δ_{ij} is Kronecker's delta representing the identity matrix. One may readily verify that the Nth-degree polynomial

$$l_{N,i}(x) = \frac{(x - x_1)(x - x_2) \cdots (x - x_{i-1})(x - x_{i+1}) \cdots (x - x_{N+1})}{(x_i - x_1)(x_i - x_2) \cdots (x_i - x_{i-1})(x_i - x_{i+1}) \cdots (x_i - x_{N+1})} \tag{D.2.2}$$

satisfies this cardinal condition. As a mnemonic rule, we note that the multiplicative factor $(x - x_i)$ is missing from the numerator, and the corresponding troublesome null factor $(x_i - x_i)$ is missing from the denominator. Moreover, the denominator is a constant, whereas the numerator is an Nth-degree polynomial in x.

The interpolating polynomial can be expressed in terms of the Lagrange polynomials as

$$P_N(x) = \sum_{i=1}^{N+1} f(x_i) \, l_{N,i}(x). \tag{D.2.3}$$

Property (D.2.1) ensures that the right-hand side of (D.2.3) satisfies the matching conditions (D.1.2), and is thus the desired interpolating polynomial. To confirm this, we compute

$$P_N(x_j) = \sum_{i=1}^{N+1} f(x_i) \, l_{N,i}(x_j) = \sum_{i=1}^{N+1} f(x_i) \, \delta_{ij} = f(x_j), \tag{D.2.4}$$

as required.

D.2.1 *Cauchy relations*

If the interpolated function $f(x)$ is an Nth- or lower-degree polynomial, then the interpolating polynomial given in (D.2.3) is exact by construction.

Choosing $f(x) = P_N(x) = (x - a)^m$, where a is a constant and $m = 0, 1, \ldots, N$, we obtain the identity

$$(x - a)^m = \sum_{i=1}^{N+1} (x_i - a)^m \, l_{N,i}(x) \tag{D.2.5}$$

for $0 \le m \le N$. Setting $a = x$, we derive the Cauchy relations

$$\sum_{i=1}^{N+1} l_{N,i}(x) = 1 \tag{D.2.6}$$

and

$$\sum_{i=1}^{N+1} (x_i - x)^m \, l_{N,i}(x) = 0 \tag{D.2.7}$$

for $m = 1, \ldots, N$. Relation (D.2.6) ensures that, if $f(x_i)$ are constant, then $P_N(x)$ is also a constant.

D.2.2 *Representation in terms of a generating polynomial*

An alternative representation of the Nth-degree Lagrange interpolating polynomials emerges by introducing the $(N + 1)$-degree generating polynomial

$$\Phi_{N+1}(x) = (x - x_1)(x - x_2) \cdots (x - x_N)(x - x_{N+1}), \tag{D.2.8}$$

which is zero at *all* data points. Straightforward product differentiation shows that

$$\Phi'_{N+1}(x_i) = (x_i - x_1) \cdots (x_i - x_{i-1})(x_i - x_{i+1}) \cdots (x_i - x_{N+1}), \tag{D.2.9}$$

where a prime denotes a derivative with respect to x. Accordingly, the ith Lagrange interpolation polynomial can be expressed in the form

$$l_{N,i}(x) = \frac{1}{\Phi'_{N+1}(x_i)} \frac{\Phi_{N+1}(x)}{x - x_i}. \tag{D.2.10}$$

In terms of the *barycentric weights*,

$$c_i \equiv \frac{1}{\Phi'_{N+1}(x_i)}, \tag{D.2.11}$$

we obtain

$$l_{N,i}(x) = c_i \frac{\Phi_{N+1}(x)}{x - x_i}. \tag{D.2.12}$$

Substituting this expression into the Cauchy formula (D.2.6) and rearranging, we obtain

$$\frac{1}{\Phi_{N+1}(x)} = \sum_{i=1}^{N+1} \frac{c_i}{x - x_i}, \tag{D.2.13}$$

which shows that the barycentric weights arise by expanding the inverse of the generating function into a sum of partial fractions.

The interpolating polynomial can be computed using the computationally efficient formula

$$P_N(x) = \Phi_{N+1}(x) \sum_{i=1}^{N+1} f(x_i) \frac{c_i}{x - x_i}. \tag{D.2.14}$$

Using (D.2.13), we obtain the alternative form

$$P_N(x) = \frac{\sum_{i=1}^{N+1} f(x_i) \frac{c_i}{x - x_i}}{\sum_{i=1}^{N+1} \frac{c_i}{x - x_i}}. \tag{D.2.15}$$

D.2.3   First derivative and the node differentiation matrix

Differentiating expression (D.2.3), we obtain an expression for the first derivative of the interpolating polynomial in terms of the data,

$$\frac{\mathrm{d}P_N(x)}{\mathrm{d}x} = \sum_{i=1}^{N+1} f(x_i) \, \frac{\mathrm{d}l_{N,i}(x)}{\mathrm{d}x}. \tag{D.2.16}$$

Differentiating expression (D.2.2) and using the rules of product differentiation, we find that

$$\frac{\mathrm{d}l_{N,i}(x)}{\mathrm{d}x} = \frac{1}{\Phi'_{N+1}(x_i)} \frac{\Phi_{N+1}(x)}{x - x_i} \, S_i(x) = l_{N,i}(x) \, S_i(x), \tag{D.2.17}$$

where

$$S_i(x) \equiv \frac{1}{x - x_1} + \cdots + \frac{1}{x - x_{i-1}} + \frac{1}{x - x_{i+1}} + \cdots + \frac{1}{x - x_{N+1}}. \tag{D.2.18}$$

We conclude that the Lagrange polynomials satisfy the differential equation

$$\frac{\mathrm{d}\ln l_{N,i}}{\mathrm{d}x} = S_i(x). \tag{D.2.19}$$

Alternatively, differentiating expression (D.2.10) using the rules of quotient differentiation, we obtain the expression

$$\frac{\mathrm{d}l_{N,i}(x)}{\mathrm{d}x} = \frac{\Phi'_{N+1}(x)(x - x_i) - \Phi_{N+1}(x)}{(x - x_i)^2 \, \Phi'_{N+1}(x_i)}. \tag{D.2.20}$$

Evaluating the derivatives on the right-hand side at the interpolation nodes and observing that $\Phi_{N+1}(x_j) = 0$, we obtain the *node differentiation matrix*

$$d_{ij} \equiv \left(\frac{\mathrm{d}l_{N,i}}{\mathrm{d}x} \right)_{x=x_j} = \begin{cases} \dfrac{\Phi'_{N+1}(x_j)}{(x_j - x_i)\,\Phi'_{N+1}(x_i)} & \text{if } i \neq j, \\[3mm] \dfrac{\Phi''_{N+1}(x_i)}{2\,\Phi'_{N+1}(x_i)} & \text{if } i = j. \end{cases} \tag{D.2.21}$$

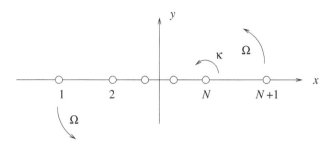

FIGURE D.2.1   A collection of $N+1$ point vortices with identical strengths located at the zeros of the $(N+1)$-degree Hermite polynomial rotates as a whole with angular velocity Ω.

The expression for $i = j$ arises by using the l'Hôpital rule to evaluate the right-hand side of (D.2.20).

Substituting the definition of the generating function, $\Phi_{N+1}(x)$, and carrying out straightforward differentiations, we obtain the explicit formulas

$$d_{ij} = \frac{1}{x_j - x_i} \frac{(x_j - x_1) \cdots (x_j - x_{j-1})(x_j - x_{j+1}) \cdots (x_j - x_{N+1})}{(x_i - x_1) \cdots (x_i - x_{i-1})(x_i - x_{i+1}) \cdots (x_i - x_{N+1})} \quad \text{(D.2.22)}$$

for $i \neq j$, and

$$d_{ii} \equiv \left(\frac{\mathrm{d}l_{N,i}}{\mathrm{d}x} \right)_{x=x_i} = \frac{\Phi''_{N+1}(x_i)}{2\,\Phi'_{N+1}(x_i)} = S_i(x_i), \quad \text{(D.2.23)}$$

where

$$S_i(x_i) = \frac{1}{x_i - x_1} + \cdots + \frac{1}{x_i - x_{i-1}} + \frac{1}{x_i - x_{i+1}} + \cdots + \frac{1}{x_i - x_{N+1}}. \quad \text{(D.2.24)}$$

Expression (D.2.23) arises also simply by evaluating (D.2.17) at $x = x_i$.

For example, when $N = 1$, we find that $d_{11} = d_{12} = 1/(x_1 - x_2)$, and $d_{21} = d_{22} = 1/(x_2 - x_1)$.

Point vortex motion

The sum representation of the diagonal component of the node differentiation matrix shown in (D.2.24) suggests a connection with a class of problems in mathematical physics involving multi-particle and multi-body interactions, including vortex motion in an ideal fluid (e.g., Aref et al. [3], Pozrikidis [46]).

Consider the mutually induced motion of a collection of $N+1$ collinear point vortices with identical strength, κ, situated along the x axis at a particular instant, as shown in Figure D.2.1. Using the equations of inviscid hydrodynamics, we find that the y velocity component of the ith point vortex is given by

$$v_i = \frac{\kappa}{2\pi} S_i(x_i) = \frac{\kappa}{2\pi} \frac{\Phi''_{N+1}(x_i)}{2\,\Phi'_{N+1}(x_i)} \tag{D.2.25}$$

(e.g., Pozrikidis [46]). If the point vortex array rotates as a whole around the origin with angular velocity Ω, then

$$v_i = \Omega\, x_i \tag{D.2.26}$$

and thus

$$\frac{\kappa}{2\pi} \frac{\Phi''_{N+1}(x_i)}{2\,\Phi'_{N+1}(x_i)} = \Omega\, x_i \tag{D.2.27}$$

for $i = 1, \ldots, N + 1$, which shows that x_i are the roots of the $(N + 1)$-degree polynomial

$$\Pi_{N+1}(x) \equiv \Phi''_{N+1}(x) - \frac{4\pi\Omega}{\kappa}\, x\, \Phi'_{N+1}(x). \tag{D.2.28}$$

Accordingly, we can write

$$\Pi_{N+1}(x) = c\, \Phi_{N+1}(x), \tag{D.2.29}$$

where c is a constant. To ensure that the coefficients of the highest power of x on the right-hand sides of the last two expressions are the same, we set $c = -4(N+1)\pi\Omega/\kappa$ and derive the differential equation

$$\Phi''_{N+1}(x) - \frac{4\pi\,\Omega}{\kappa}\, x\, \Phi'_{N+1}(x) + \frac{4\pi\Omega}{\kappa}\,(N + 1)\, \Phi_{N+1}(x) = 0. \tag{D.2.30}$$

In terms of the dimensionless position,

$$\widehat{x} \equiv x\, \sqrt{\frac{2\pi\Omega}{\kappa}}, \tag{D.2.31}$$

equation (D.2.30) reduces to the Hermite equation

$$\Phi''_{N+1}(\widehat{x}) - 2\widehat{x}\, \Phi'_{N+1}(\widehat{x}) + 2\,(N + 1)\, \Phi_{N+1}(\widehat{x}) = 0, \tag{D.2.32}$$

where a prime now designates a derivative with respect to $\widehat{x}$. The solution of this equation is the $(N + 1)$-degree Hermite polynomial, whose roots, corresponding to the positions of the point vortex assembly, are available in analytical or tabular form (e.g., Pozrikidis [47]).

D.2.4   Representation in terms of the Vandermonde matrix

Applying identity (D.2.5) for $a = 0$, we find that

$$x^m = \sum_{i=1}^{N+1} x_i^m \, l_{N,i}(x) \tag{D.2.33}$$

for $m = 0, 1, \ldots, N$, which can be rearranged into a linear system,

$$\mathbf{V} \cdot \begin{bmatrix} l_{N,1}(x) \\ l_{N,2}(x) \\ \vdots \\ l_{N,N}(x) \\ l_{N,N+1}(x) \end{bmatrix} = \begin{bmatrix} 1 \\ x \\ \vdots \\ x^{N-1} \\ x^N \end{bmatrix}, \tag{D.2.34}$$

where $\mathbf{V}$ is the Vandermonde matrix defined in (D.1.6). Computing the solution by Cramer's rule, we find that

$$l_{N,i}(x) = \frac{\det[\mathbf{V}(x_1, x_2, \ldots, x_{i-1}, x, x_{i+1}, \ldots, x_{N+1})]}{\det[\mathbf{V}(x_1, x_2, \ldots, x_{i-1}, x_i, x_{i+1}, \ldots, x_{N+1})]}. \tag{D.2.35}$$

Note that the argument x_i has been replaced by the independent variable x in the numerator on the right-hand side to give an Nth-degree polynomial. Thus, the ith Lagrange polynomial can be expressed as the ratio of the determinants of two $(N+1) \times (N+1)$ Vandermonde matrices.

To simplify the notation, we define

$$\mathcal{V}_i(x) \equiv \det[\mathbf{V}(x_1, x_2, \ldots, x_{i-1}, x, x_{i+1}, \ldots, x_{N+1})], \tag{D.2.36}$$

and write

$$l_{N,i}(x) = \frac{\mathcal{V}_i(x)}{\mathcal{V}}, \tag{D.2.37}$$

where

$$\mathcal{V} = \mathcal{V}_i(x_i) \tag{D.2.38}$$

is the determinant of the Vandermonde matrix, independent of i.

To confirm that (D.2.35) is an acceptable representation, we make three key observations:

1. The numerator in (D.2.35) is an Nth-degree polynomial in x. This can be demonstrated readily by writing the Laplace expansion of the determinant with respect to the row involving x.

2. The numerator in (D.2.35) is zero when $x = x_j$ for $j \neq i$, since two rows of the Vandermonde matrix become identical.

3. The right-hand side of (D.2.35) is equal to unity when $x = x_i$, since the numerator becomes equal to the denominator.

Differentiating (D.2.37) with respect to x and rearranging, we obtain

$$\frac{\mathrm{d}l_{N,i}(x)}{\mathrm{d}x} = \frac{1}{\mathcal{V}} \frac{\mathrm{d}\mathcal{V}_i(x)}{\mathrm{d}x}. \tag{D.2.39}$$

Evaluating this expression at $x = x_i$, we obtain

$$d_{ii} = \frac{1}{\mathcal{V}} \left(\frac{\mathrm{d}\mathcal{V}_i(x)}{\mathrm{d}x} \right)_{x=x_i} = \frac{1}{\mathcal{V}} \frac{\partial \mathcal{V}}{\partial x_i}, \tag{D.2.40}$$

where d_{ii} are the diagonal components of the node differentiation matrix defined in (D.2.21) and given in (D.2.23).

Fekete points

If the interpolation abscissas, x_i, are distributed such that the magnitude of the associated Lagrange polynomials is less than unity over a specified interpolation domain,

$$|l_{N,i}(x)| \leq 1, \tag{D.2.41}$$

then the numerator in (D.2.35) reaches the maximum of unity when $x = x_i$, whereupon it becomes equal to the denominator. Stated differently, the function $\mathcal{V}_i(x)$ reaches an extreme value when $x = x_i$, yielding

$$\left(\frac{\partial \mathcal{V}_i(x)}{\partial x} \right)_{x=x_i} = \frac{\partial \mathcal{V}}{\partial x_i} = 0. \tag{D.2.42}$$

Formula (D.2.40) then shows that the corresponding diagonal components of the node differentiation matrix are zero,

$$d_{ii} = 0. \tag{D.2.43}$$

The set of points that maximizes the magnitude of the determinant of the Vandermonde matrix are the Fekete points in one dimension. In Section D.5, we will show that, when $x_1 = -1$ and $x_{N+1} = 1$, the $N-1$ intermediate Fekete points x_i for $i = 2, \ldots, N$, are the well-known roots of the $(N-1)$-degree Lobatto polynomial, $\mathrm{Lo}_{N-1}(x)$, discussed in Section B.6, Appendix B.

Generalized Vandermonde matrix

Consider the representation of the interpolating polynomial in terms of a polynomial base, ϕ_i, as shown in (D.1.11). Working as previously in this section, we find that the Lagrange interpolating polynomials satisfy the counterpart of the linear system (D.2.34),

$$\mathbf{V}_\phi \cdot \begin{bmatrix} l_{N,1}(x) \\ l_{N,2}(x) \\ \vdots \\ l_{N,N}(x) \\ l_{N,N+1}(x) \end{bmatrix} = \begin{bmatrix} \phi_1(x) \\ \phi_2(x) \\ \vdots \\ \phi_N(x) \\ \phi_{N+1}(x) \end{bmatrix}, \tag{D.2.44}$$

where $\mathbf{V}_\phi$ is the generalized Vandermonde matrix defined in (D.1.13). Computing the solution by Cramer's rule, we find that

$$l_{N,i}(x) = \frac{\det[\mathbf{V}_\phi(x_1, x_2, \ldots, x_{i-1}, x, x_{i+1}, \ldots, x_{N+1})]}{\det[\mathbf{V}_\phi(x_1, x_2, \ldots, x_{i-1}, x_i, x_{i+1}, \ldots, x_{N+1})]}, \tag{D.2.45}$$

which is a generalization of (D.2.35).

Next, we denote

$$\mathcal{V}_\phi \equiv \det[\mathbf{V}_\phi(x_1, x_2, \ldots, x_{j-1}, x_j, x_{j+1}, \ldots, x_{N+1})] \tag{D.2.46}$$

and apply the chain rule of differentiation to obtain

$$\frac{\partial \mathcal{V}_\phi}{\partial x_j} = \sum_{i=1}^{N+1} \frac{\partial \mathcal{V}_\phi}{\partial \phi_i(x_j)} \frac{\partial \phi_i(x_j)}{\partial x_j}. \tag{D.2.47}$$

Using the Laplace expansion, we find that the derivative of the determinant of the generalized Vandermonde matrix with respect to the ij matrix element is given by

$$\frac{\partial \mathcal{V}_\varphi}{\partial \phi_i(x_j)} = (-1)^{i+j} M_{ij}, \tag{D.2.48}$$

where M_{ij} is the element minor. Accordingly,

$$\frac{\partial \mathcal{V}_\phi}{\partial x_j} = \sum_{i=1}^{N+1} (-1)^{i+j} M_{ij} \frac{\partial \phi_i(x_j)}{\partial x_j}. \tag{D.2.49}$$

If we identify the polynomial base functions, $\phi_i(x)$, with the Lagrange interpolation functions, $l_{N,i}(x)$, we will find that the generalized Vandermonde matrix reduces to the identity matrix, $\mathbf{V}_\phi = \mathbf{I}$. Consideration of (D.2.48) then shows that the matrix of partial derivatives of the generalized Vandermonde matrix with respect to any matrix element also reduces to the unit matrix and (D.2.47) simplifies to

$$\frac{\partial \mathcal{V}_\phi}{\partial x_j} = \frac{\partial l_{N,j}(x_j)}{\partial x_j} \tag{D.2.50}$$

(Taylor, Wingate, & Vincent [67]).

D.2.5   Lagrange polynomials corresponding to polynomial roots

Let us assume that an $(N + 1)$-degree polynomial,

$$Q_{N+1}(x) = a_1\, x^{N+1} + a_2\, x^N + \cdots + a_{N+1}\, x + a_{N+2}, \tag{D.2.51}$$

has $N + 1$ distinct roots, x_i for $i = 1, \ldots, N + 1$, where a_i are specified polynomial coefficients. For reasons that will become evident shortly, we express this polynomial in the form

$$Q_{N+1}(x) = a_1 x^{N+1} - P_N(x), \tag{D.2.52}$$

where

$$P_N(x) \equiv -a_2 x^N - \cdots - a_{N+1}\, x - a_{N+2} \tag{D.2.53}$$

is an Nth-degree polynomial.

Next, we formulate the Lagrange polynomials $l_{N,i}(x)$ associated with the roots, x_i, where $i = 1, \ldots, N + 1$, and introduce the *exact* representation

$$P_N(x) = \sum_{i=1}^{N+1} P_N(x_i)\, l_{N,i}(x). \tag{D.2.54}$$

By definition,

$$Q_{N+1}(x_i) = a_1 x_i^{N+1} - P_N(x_i) = 0 \tag{D.2.55}$$

or

$$P_N(x_i) = a_1 x_i^{N+1}, \tag{D.2.56}$$

and thus

$$P_N(x) = a_1 \sum_{i=1}^{N+1} x_i^{N+1} l_{N,i}(x). \tag{D.2.57}$$

Substituting this result into (D.2.52), we derive a representation in terms of the leading-power coefficient and the polynomial roots,

$$Q_{N+1}(x) = a_1 \left(x^{N+1} - \sum_{i=1}^{N+1} x_i^{N+1} l_{N,i}(x) \right), \tag{D.2.58}$$

which complements the more familiar representation

$$Q_{N+1}(x) = a_1 \prod_{i=1}^{N+1} (x - x_i). \tag{D.2.59}$$

Note that the right-hand side of (D.2.58) is *not* a polynomial function of the roots.

D.2.6   Lagrange polynomials for Hermite interpolation

The Lagrange interpolation method discussed previously in this appendix fails when the abscissa of a certain point, x_j, tends to the abscissa of another point, x_i. In this limit, the interpolating polynomial continues to exist only if $f(x_j)$ also tends to $f(x_i)$. This may occur if the graph of the function $f(x)$ tends to become horizontal at a point, x_i, which means that $f'(x_i) = 0$.

A generalization of these circumstances leads us to a special case of the Hermite interpolation problem, which can be stated as follows: find an Nth-degree polynomial, $P_N(x)$, so that

$$P_N(x_i) = f(x_i), \qquad (D.2.60)$$

and

$$\left(\frac{\mathrm{d}P_N}{\mathrm{d}x}\right)_{x_i} = 0, \quad \ldots, \quad \left(\frac{\mathrm{d}^{m_i-1}P_N}{\mathrm{d}x^{m_i-1}}\right)_{x_i} = 0 \qquad (D.2.61)$$

for $i = 1, \ldots, r$, where r is the number of distinct abscissas, m_i is the grade of the ith abscissa, under the stipulation that

$$m_1 + m_2 + \cdots + m_r = N + 1. \qquad (D.2.62)$$

When $r = N+1$ and $m_i = 1$ for all i, we recover the standard interpolation problem.

The requisite interpolating polynomial can be expressed in terms of the generalized Lagrange polynomials, $\ell_{N,i}$, as

$$P_N(x) = \sum_{i=1}^{r} f(x_i)\,\ell_{N,i}(x). \qquad (D.2.63)$$

The ith generalized Lagrange polynomial is required to satisfy the super-cardinal, Hermite-like properties

$$\ell_{N,i}(x_j) = \delta_{ij} \qquad (D.2.64)$$

and

$$\left(\frac{\mathrm{d}\ell_{N,i}}{\mathrm{d}x}\right)_{x_j} = 0, \quad \ldots, \quad \left(\frac{\mathrm{d}^{m_j-1}\ell_{N,i}}{\mathrm{d}x^{m_j-1}}\right)_{x_j} = 0 \qquad (D.2.65)$$

for $j = 1, \ldots, r$. If the interpolated function is constant, the interpolation is exact by construction for any r and m_i. This observation justifies the generalized Cauchy relation

$$\sum_{i=1}^{r} \ell_{N,i}(x) = 1. \qquad (D.2.66)$$

As in the case of the standard Lagrange polynomials, we may introduce the generating function

$$\Phi_{N+1}(x) \equiv (x - x_1)^{m_1}(x - x_2)^{m_2} \cdots (x - x_r)^{m_r}, \tag{D.2.67}$$

and express the generalized Lagrange polynomials in the form

$$\ell_{N,i}(x) = \chi_{m_i-1,i}(x) \frac{\Phi_{N+1}(x)}{(x - x_i)^{m_i}}, \tag{D.2.68}$$

where $\chi_{m_i-1,i}(x)$ are $(m_i - 1)$-degree polynomials arising from the partial fraction decomposition of the inverse of the generating function,

$$\frac{1}{\Phi_{N+1}(x)} = \frac{\chi_{m_1-1,1}(x)}{(x - x_1)^{m_1}} + \cdots + \frac{\chi_{m_r-1,r}(x)}{(x - x_r)^{m_r}}. \tag{D.2.69}$$

If $m_i = 1$, then

$$\chi_{0,i} \equiv c_i = \frac{1}{\Phi'_{N+1}(x_i)}, \tag{D.2.70}$$

where

$$c_i = \frac{1}{(x_i - x_1)^{m_1} \cdots (x_i - x_{i-1})^{m_{i-1}}(x_i - x_{i+1})^{m_{i+1}} \cdots (x_i - x_r)^{m_r}} \tag{D.2.71}$$

is a constant.

As an example, we consider interpolation with $r = 2$ distinct abscissas with grades $m_1 = 2$ and $m_2 = 1$, corresponding to $N+1 = 3$. The generating polynomial is given by

$$\Phi_3(x) = (x - x_1)^2 (x - x_2). \tag{D.2.72}$$

After straightforward calculations, we obtain the partial fraction decomposition

$$\frac{1}{\Phi_3(x)} = \frac{1}{(x_2 - x_1)^2} \left(\frac{-x + (2x_1 - x_2)}{(x - x_1)^2} + \frac{1}{x - x_2} \right), \tag{D.2.73}$$

which shows that

$$\chi_{1,1}(x) = \frac{-x + (2x_1 - x_2)}{(x_2 - x_1)^2}, \qquad \chi_{0,2} = \frac{1}{(x_2 - x_1)^2}. \tag{D.2.74}$$

The generalized Lagrange polynomials are given by

$$\ell_{2,1}(x) = \frac{-x + (2x_1 - x_2)}{(x_2 - x_1)^2} (x - x_2), \qquad \ell_{2,2}(x) = \frac{(x - x_1)^2}{(x_2 - x_1)^2}. \tag{D.2.75}$$

D.3   Error in polynomial interpolation

Unless the interpolated function, $f(x)$, happens to be a polynomial of degree equal to N or less, the interpolating polynomial $P_N(x)$ and $f(x)$ will not necessarily agree between the data points. The difference between the approximate and exact values is the interpolation error,

$$e(x) \equiv P_N(x) - f(x). \tag{D.3.1}$$

The mandatory satisfaction of the interpolation conditions (D.1.2) ensures that

$$e(x_i) = 0 \tag{D.3.2}$$

for $i = 1, \ldots N + 1$, that is, the function $e(x)$ has at least $N + 1$ zeros in the interpolation domain. However, in general, $e(x) \neq 0$ when $x \neq x_i$.

We will show that, when the function $f(x)$ is sufficiently smooth, the error incurred by the polynomial interpolation is given by

$$e(x) = -\frac{f^{(N+1)}(\xi)}{(N+1)!} \, \Phi_{N+1}(x), \tag{D.3.3}$$

where:

- $(N + 1)! \equiv 1 \cdot 2 \cdots N \cdot (N + 1)$ is the factorial.

- The $(N + 1)$-degree generating polynomial $\Phi_{N+1}(x)$ is defined in (D.2.8).

- $f^{(N+1)}(\xi)$ is the $(N + 1)$-derivative of the function f evaluated at a certain point ξ that lies somewhere inside the smallest interval that contains the abscissas of all data points.

Without loss of generality, we may assume that the abscissas are ordered so that $x_1 \leq \xi \leq x_{N+1}$. Since the precise location of ξ depends on the value of x, the value $f^{(N+1)}(\xi)$ is truly an implicit function of the independent variable, x.

To prove the error formula, we hold the value of x fixed, introduce a new function of a new independent variable, z, defined as

$$Q(z) \equiv \Phi_{N+1}(z) \, e(x) - \Phi_{N+1}(x) \, e(z), \tag{D.3.4}$$

and observe that

$$Q(x_i) = 0 \tag{D.3.5}$$

for $i = 1, \ldots, N + 1$, and also that

$$Q(x) = 0. \tag{D.3.6}$$

Because the function $Q(z)$ has at least $N+2$ zeros, it must attain at least $N+1$ local maxima and minima inside the interval $[x_1, x_{N+1}]$. Accordingly, the first derivative, $Q'(z)$, must have at least $N+1$ zeros, and the second derivative, $Q''(z)$, must have at least N zeros between x_1 and x_{N+1}. Continuing in this fashion, we find that the $N+1$ derivative, $Q^{(N+1)}(z)$, must have at least one zero between x_1 and x_{N+1}. One of these zeros is assumed to occur at $z = \xi$. By definition then,

$$Q^{(N+1)}(\xi) = \Phi_{N+1}^{(N+1)}(\xi)\, e(x) - \Phi_{N+1}^{(N+1)}(x)\, e^{(N+1)}(\xi) = 0. \tag{D.3.7}$$

Differentiating the right-hand side of (D.3.1) $N+1$ times with respect to x, noting that $P_N^{(N+1)}(x) = 0$, and evaluating the resulting expression at $x = \xi$, we find that

$$e^{(N+1)}(\xi) = -f^{(N+1)}(\xi). \tag{D.3.8}$$

The $N+1$ derivative of the Lagrange generating function is a constant,

$$\Phi_{N+1}^{(N+1)}(\xi) = (N+1)! \quad . \tag{D.3.9}$$

Substituting the last two expressions into (D.3.7) and solving for $e(x)$, we obtain the error formula (D.3.3).

D.3.1   Convergence and the Lebesgue constant

The max norm of a function $g(x)$, denoted as $||g(x)||$, is defined as the maximum of the absolute value of $g(x)$ over a specified interval, $a \leq x \leq b$. The preceding discussion suggests that the norm of the interpolation error, $||e(x)||$, depends on the location of the interpolation points. Of all Nth-degree polynomials approximating a function, $f(x)$, an optimal polynomial can be found, $P_N^{\mathrm{opt}}(x)$, that exhibits the minimum error $||e(x)||$, called the *minimax error* and denoted by $\rho_N[f(x)]$. This optimal polynomial is not necessarily an interpolating polynomial, that is, it does not necessarily agree with the interpolated function at $N+1$ data points.

Consider the set of all functions $f(t)$ with unit max norm, $||f(t)|| = 1$. The corresponding norm of the interpolation error is

$$||e(x)|| \equiv ||P_N(x) - f(x)|| = ||\big(P_N(x) - P_N^{\mathrm{opt}}(x)\big) + \big(P_N^{\mathrm{opt}}(x) - f(x)\big)||. \tag{D.3.10}$$

Consequently,

$$||e(x)|| \leq ||P_N(x) - P_N^{\mathrm{opt}}(x)|| + ||P_N^{\mathrm{opt}}(x) - f(x)|| \tag{D.3.11}$$

and

$$||e(x)|| \leq ||P_N(x) - P_N^{\mathrm{opt}}(x)|| + \rho_N[f(x)]. \tag{D.3.12}$$

To emphasize that the polynomial $P_N(x)$ approximates the function $f(x)$, we write it as $P_N(x, f)$. Next, we apply the Lebesgue lemma, and derive the inequality

$$||P_N(x, f) - P_N^{\text{opt}}(x)|| = ||P_N(x, f) - P_N(x, P_N^{\text{opt}})||$$
$$\leq ||P_N|| \cdot ||f(x) - P_N^{\text{opt}}(x)||, \qquad \text{(D.3.13)}$$

where

$$||P_N|| \equiv \max(||P_N(f(x))||), \qquad \text{(D.3.14)}$$

and the maximum is computed over all permissible functions, $f(x)$. Accordingly,

$$||e(x)|| \leq (1 + ||P_N||) \, \rho_N[f(x)]. \qquad \text{(D.3.15)}$$

It is important to note that, given N, $\rho_N[f(x)]$ is independent of the interpolation point distribution, x_i.

To derive a bound for the norm $||P_N||$, we express the interpolating polynomial in terms of the Lagrange polynomials using (D.2.63). Recalling the requirement $||f(x)|| = 1$, we write

$$||P_N|| \equiv \max\left(|\sum_{i=1}^{N+1} f(x_i) \, l_{N,i}(x)|\right) \leq \max\left(\sum_{i=1}^{N+1} |f(x_i)| \, |l_{N,i}(x)|\right) \leq \max\left(\mathcal{L}_N(x)\right),$$
$$\text{(D.3.16)}$$

where

$$\mathcal{L}_N(x) \equiv \sum_{i=1}^{N+1} |l_{N,i}(x)| \qquad \text{(D.3.17)}$$

is the Lebesgue function. The maximum value of the Lebesgue function is the Lebesgue constant,

$$\Lambda_N \equiv \max\left(\mathcal{L}_N(x)\right). \qquad \text{(D.3.18)}$$

Combining this definition with (D.3.15), we obtain

$$||e(x)|| \leq (1 + \Lambda_N) \, \rho_N[f(x)], \qquad \text{(D.3.19)}$$

which shows that the convergence properties of the interpolation depends on the functional dependence of the minimax error and Lebesgue constant on the polynomial order, N.

Behavior of the minimax error and Jackson's theorems

Jackson's first theorem places an upper bound on the minimax error, $\rho_N[f(x)]$. When interpolating a continuous function $f(x)$ inside the canonical interval $[-1, 1]$, the theorem states that

$$\rho_N[f(x)] \leq \left(1 + \frac{1}{2}\pi^2\right) w\left(\frac{1}{N}\right), \qquad \text{(D.3.20)}$$

where $w(\delta)$ is the *modulus of continuity* of $f(x)$, defined as

$$w(\delta) \equiv \max_{\left(|x_1 - x_2| \leq \delta\right)} |f(x_1) - f(x_2)| \tag{D.3.21}$$

(e.g., Davis [15], p. 338). For a function that satisfies the Lipschitz condition

$$|f(x_1) - f(x_2)| \leq A |x_1 - x_2|^\alpha \tag{D.3.22}$$

in the interval $-1 \leq x_1, x_2 \leq 1$, we find that $w(\delta) \leq A \delta^\alpha$ and thus

$$\rho_N[f(x)] \leq A \left(1 + \frac{1}{2}\pi^2\right) \frac{1}{N^\alpha}, \tag{D.3.23}$$

where $\alpha > 0$ is a positive exponent, A is a positive constant,

Jackson's second theorem states that, if the kth derivative of a function $f(x)$ exists in the interval $[-1, 1]$, that is, the kth derivative does not become infinite at a point in $[-1, 1]$, then

$$\rho_N[f(x)] \leq \left(1 + \frac{1}{2}\pi^2\right)^{k+1} \frac{e^k}{(1+k)} \frac{1}{N^k} w\left(\frac{1}{N-k}\right), \tag{D.3.24}$$

for $N > k$ (e.g., Rivlin [52], p. 23). The first theorem is a special case of the second theorem for $k = 0$.

Behavior of the Lebesgue constant

The Lebesgue constant is known to grow as N tends to infinity. Erdös's theorem places a lower bound on the slowest possible growth,

$$\Lambda_N > \frac{2}{\pi} \ln N + 1 - \beta, \tag{D.3.25}$$

where β is a positive constant. Thus, the Lebesgue constant grows at least as fast as $\ln N$.

In the case of evenly spaced points over a specified interval, the Lebesgue constant is known to grow rapidly with N, exhibiting the asymptotic behavior

$$\Lambda_N \sim \frac{2^N}{N \log N}. \tag{D.3.26}$$

In fact, numerical computations show that Λ_N increases from the value of 1 when $N = 1$, to 4.55 when $N = 6$, to 89.3 when $N = 12$, to 3171.1 when $N = 18$, to 137852.1 when $N = 24$ (e.g., Hesthaven [28]). This rapid growth is responsible for the deleterious Runge effect, as discussed in Section 3.1.2

In the next two sections, we will see that, when the interpolation nodes are distributed at the zeros of orthogonal polynomials, Λ_N increases much slower, at the nearly optimal logarithmic rate.

D.4   Chebyshev interpolation

Suppose that we have the luxury of distributing the $N + 1$ nodes in any way we desire over a certain closed interval of the x axis, $[a, b]$. We ask whether there is an optimal distribution that minimizes the interpolation error in some appropriate sense and, moreover, whether a systematic way of distributing the data points can be found so that, as N increases, the error decreases uniformly and at the fastest possible rate.

To address these questions, we introduce triangular families of orthogonal polynomials, as discussed in Appendix B. Briefly, each family consists of an infinite sequence of polynomials $p_i(t)$, $i = 0, 1, \ldots$, where p_0 is a constant, $p_1(t)$ is a first-degree polynomial, and $p_i(t)$ is an ith-degree polynomial of the independent variable t, defined over a certain interval $[c, d]$. By definition, the orthogonal polynomials of a certain family satisfy the orthogonality condition

$$(p_i, p_j) \equiv \int_c^d p_i(t)\, p_j(t)\, w(t)\, \mathrm{d}t = \begin{cases} \mathcal{G}_i & \text{if } i = j, \\ 0 & \text{if } i \neq j, \end{cases} \tag{D.4.1}$$

where $w(t)$ is a weighting function and $\mathcal{G}_i$ are constants. Any mth-degree polynomial can be expressed as a linear combination of the first $m + 1$ orthogonal polynomials of any chosen family, $p_0(t), p_1(t), \ldots, p_m(t)$.

Now we are in a position to tackle the issue of optimal positioning of the abscissas. As a preliminary, we introduce an independent variable, t, that is related to the independent variable, x, by the linear transformation

$$x = q(t) = \frac{b + a}{2} + \frac{b - a}{2} \frac{2t - c - d}{d - c}. \tag{D.4.2}$$

As t increases from c to d, x increases from a to b. Having introduced this transformation, we regard the function f as a function of t, writing

$$f(x) = f[q(t)] \equiv h(t). \tag{D.4.3}$$

Using formula (D.3.3), we find that the error incurred by the polynomial interpolation of $h(t)$ with $N + 1$ abscissas, t_i, corresponding to x_i, is given by

$$e(t) \equiv P_N(t) - h(t) = -\frac{h^{(N+1)}(\xi)}{(N + 1)!} (t - t_1)(t - t_2) \cdots (t - t_N)(t - t_{N+1}), \tag{D.4.4}$$

where the point ξ lies somewhere inside the interval $[c, d]$.

Next, we identify the interpolation nodes, $t_1, t_2, \ldots, t_{N+1}$, with the roots of an $(N + 1)$-degree polynomial defined in the interval $[c, d]$ whose coefficient of the leading power is equal to unity, denoted by

$$Q_{N+1}(t) = t^{N+1} + \cdots, \tag{D.4.5}$$

and obtain

$$e(t) = -\frac{h^{(N+1)}(t)}{(N+1)!} Q_{N+1}(t).$$ (D.4.6)

To standardize the problem, without loss of generality, we fix the end-points of the interpolation interval at $c = -1$ and $d = 1$, and search for the polynomial $Q_{N+1}(t)$ with the smallest maximum (minimax) norm. Our search is guided by the minimax property of the Chebyshev polynomials discussed in Section B.7, Appendix B. The requisite optimal polynomial is

$$Q_{N+1}(t) = \frac{1}{2^N} T_{N+1}(t),$$ (D.4.7)

where $T_{N+1}(t)$ is the $(N+1)$-degree Chebyshev polynomial,

$$T_{N+1}(t) = 2^N (t - t_1)(t - t_2) \cdots (t - t_{N+1})$$ (D.4.8)

and t_i are the zeros of the Chebyshev polynomial located at

$$t_i = \cos\left[\left(i - \frac{1}{2}\right)\frac{\pi}{N+1}\right]$$ (D.4.9)

for $i = 1, \ldots, N+1$. Substituting (D.4.7) into (D.4.6), we obtain

$$e(t) = -\frac{h^{(N+1)}(\xi)}{2^N(N+1)!} T_{N+1}(t).$$ (D.4.10)

Since the magnitude of the Chebyshev polynomials is less than unity for any value of t in the interval $[-1, 1]$, the magnitude of the interpolation error is at most on the order of the fraction on the right-hand side of (D.4.10). Unless the function $h(t)$ is ill-behaved, the error $e(t)$ is decreases rapidly as the polynomial order, N is raised.

More precisely, it can be shown that the error incurred by the Chebyshev interpolation is not much worse than that incurred by the minimax approximation,

$$\max[e(t)] \leq \left(\frac{2}{\pi} \log(N+1) + 2\right) \rho_N[h(t)],$$ (D.4.11)

where $\rho_N[h(t)]$ is the minimax error discussed in Section D.3 (e.g., Davis [15], Rivlin [53]). This inequality shows that the Lebesgue constant increases at the optimal logarithmic rate with respect to the polynomial order, N.

D.5 Lobatto interpolation

The linear transformation

$$x = \frac{x_{N+1} + x_1}{2} + \frac{x_{N+1} - x_1}{2} t$$ (D.5.1)

maps the interpolation interval with respect to the independent variable, x, to a canonical interpolation interval with respect to the independent variable, t, so that the first and last interpolation points, x_1 and x_{N+1}, are mapped to $t_1 = -1$ and $t_{N+1} = 1$. The interpolation can be carried out with respect to t in terms of the corresponding Nth-degree Lagrange polynomials, $l_{N,i}(t)$, where $i = 1, \ldots, N+1$.

For reasons that will become evident in hindsight, we require that, in addition to satisfying the $N + 1$ *mandatory* cardinal interpolation conditions

$$l_{N,i}(t_j) = \delta_{ij} \tag{D.5.2}$$

for $i, j = 1, \ldots, N+1$, the node distribution is such that the Lagrange polynomials also satisfy the $N - 1$ *optional* conditions

$$l'_{N,i}(t_i) = 0 \tag{D.5.3}$$

for $i = 2, \ldots, N$, where a prime denotes a derivative with respect to t. Property (D.5.3) ensures that the Lagrange polynomials reach the local maximum value of unity at the $N - 1$ interior nodes.

Using the node differentiation matrix (D.2.21) for $i = j$, we find that (D.5.3) implies that

$$\Phi''_{N+1}(t_i) = 0 \tag{D.5.4}$$

for $i = 2, \ldots, N$. Since $\Phi''_{N+1}(t)$ is an $(N - 1)$-degree polynomial, we can write

$$\Phi''_{N+1}(t) = c \, \frac{\Phi_{N+1}(t)}{(t-1)(t+1)} = -c \, \frac{\Phi_{N+1}(t)}{1 - t^2}, \tag{D.5.5}$$

where c is a constant. Because the coefficient of the highest power of t in $\Phi_{N+1}(t)$ is equal to unity, we must have

$$c = N(N + 1). \tag{D.5.6}$$

Consequently, the $(N + 1)$-degree polynomial $\Phi_{N+1}(t)$ satisfies the second-order differential equation

$$(1 - t^2) \, \Phi''_{N+1}(t) + N(N + 1) \, \Phi_{N+1}(t) = 0. \tag{D.5.7}$$

Next, we recall the differential equation (B.5.11), Appendix B, satisfied by the Legendre polynomials, and find that the solution of (D.5.7) is

$$\Phi_{N+1}(t) = \alpha_{N+1} \, (t^2 - 1) \, L'_N(t) = \alpha_{N+1} \, (t^2 - 1) \, \mathrm{Lo}_{N-1}(t), \tag{D.5.8}$$

where L_N is a Legendre polynomial, Lo_{N-1} is a Lobatto polynomial, and α_{N+1} is a constant designed so that the coefficient of the highest power of t in $\Phi_{N+1}(t)$ is equal to unity. Expression (D.5.8) shows that the interpolation nodes, t_i for $i = 2, \ldots, N$, are the zeros of the $(N - 1)$-degree Lobatto polynomial.

Spectral convergence

The significance of the preceding derivation becomes evident by considering the $2N$-degree polynomial

$$G_{2N}(t) \equiv l_{N,1}^2(t) + l_{N,2}^2(t) + \cdots + l_{N,N+1}^2(t) - 1. \tag{D.5.9}$$

Since the cardinal interpolation property (D.5.2) requires that

$$G_{2N}(t_i) = 0 \tag{D.5.10}$$

for $i = 1, \ldots, N+1$, we may write

$$G_{2N}(t) \equiv \Phi_{N+1}(t)\, \Psi_{N-1}(t), \tag{D.5.11}$$

where $\Psi_{N-1}(t)$ is an $(N-1)$-degree polynomial. Differentiating (D.5.9), we obtain

$$G_{2N}'(t) = 2 \sum_{i=1}^{N+1} l_{N,i}(t)\, l_{N,i}'(t), \tag{D.5.12}$$

which, in light of (D.5.3), shows that

$$G_{2N}'(t_i) = 0 \tag{D.5.13}$$

for $i = 2, \ldots, N$. Because the points t_i for $i = 2, \ldots, N$ are double roots of $G_{2N}(t)$, we may write

$$\Psi_{N-1}(t) = d\, \frac{\Phi_{N+1}(t)}{(t+1)(t-1)} = -d\, \frac{\Phi_{N+1}(t)}{1-t^2}, \tag{D.5.14}$$

where d is a constant. Thus,

$$G_{2N}(t) = -d\, \frac{\Phi_{N+1}^2(t)}{1-t^2} = -\beta\, (1-t^2)\, \mathrm{Lo}_{N-1}^2(t), \tag{D.5.15}$$

where β is a new constant.

To compute the constant, β, we evaluate (D.5.12) at $t = 1$ and use the cardinal interpolation property to find that

$$G_{2N}'(t = 1) = 2\, l_{N,N+1}'(t = 1) = \frac{\Phi_{N+1}''(t = 1)}{\Phi_{N+1}'(t = 1)}, \tag{D.5.16}$$

where the second expression arises from the node differentiation matrix shown in (D.2.21). Substituting the generating function from (D.5.8), we obtain

$$G_{2N}'(t = 1) = \left(\frac{[(t^2 - 1)\, \mathrm{Lo}_{N-1}(t)]''}{[(t^2 - 1)\, \mathrm{Lo}_{N-1}(t)]'} \right)_{t=1}. \tag{D.5.17}$$

Since the Lobatto polynomials satisfy the differential equation

$$\left[(t^2 - 1)\,\mathrm{Lo}_{N-1}(t)\right]' = N(N+1)\,\mathrm{L}_N(t), \tag{D.5.18}$$

we find that

$$G'_{2N}(t = 1) = \frac{\mathrm{L}'_N(t=1)}{\mathrm{L}_N(t=1)} = \frac{\mathrm{Lo}_{N-1}(t=1)}{\mathrm{L}_N(t=1)}, \tag{D.5.19}$$

where L_N is a Legendre polynomial. The Legendre polynomials are normalized such that $\mathrm{L}_N(1) = 1$. Evaluating (D.5.18) at $t = 1$, we find that

$$\mathrm{Lo}_{N-1}(t = 1) = \frac{1}{2}\,N(N+1) \tag{D.5.20}$$

and thus

$$G'_{2N}(t = 1) = \frac{1}{2}\,N(N+1). \tag{D.5.21}$$

In the next step, we differentiate the last expression in (D.5.15) and obtain

$$G'_{2N}(t = 1) = 2\beta\,\mathrm{Lo}^2_{N-1}(t = 1) = \beta\,\frac{1}{2}\,N^2(N+1)^2. \tag{D.5.22}$$

Combining the last two expressions, we obtain

$$\beta = \frac{1}{N(N+1)} \tag{D.5.23}$$

and derive the final expression

$$G_{2N}(t) = -\frac{1}{N(N+1)}\,(1 - t^2)\,\mathrm{Lo}^2_{N-1}(t), \tag{D.5.24}$$

which can be restated as

$$G_{2N}(t) = -\frac{1}{N(N+1)}\,(1 - t^2)\,\mathrm{L}'^2_N(t). \tag{D.5.25}$$

In the interval of interest, $-1 \le t \le 1$, the binomial $1 - t^2$ is non-negative and the right-hand side of (D.5.25) is non-positive. Accordingly,

$$G_{2N}(t) \le 0 \tag{D.5.26}$$

and thus

$$l^2_{N,1}(t) + l^2_{N,2}(t) + \cdots + l^2_{N,N+1}(t) \le 1, \tag{D.5.27}$$

which requires that

$$l^2_{N,i}(t) \le 1 \tag{D.5.28}$$

or

$$|l_{N,i}(t)| \leq 1 \qquad (D.5.29)$$

individually for $i = 1, \ldots, N + 1$ (Fejér [20, 21]). The maximum value of $|l_{N,i}(t)|$ is thus achieved when $t = t_i$.

Applying the Cauchy–Schwartz inequality, we find that

$$\left(\sum_{i=1}^{N+1} |l_{N,i}(t)| \right)^2 \leq (N + 1) \sum_{i=1}^{N+1} |l_{N,i}(t)|^2. \qquad (D.5.30)$$

Recalling the definition of the Lebesgue constant in (D.3.18), and using (D.5.28), we find that

$$\Lambda_N^2 \leq N + 1 \qquad \text{or} \qquad \Lambda_N \leq \sqrt{N + 1}. \qquad (D.5.31)$$

In fact, numerical investigation reveals the much stricter bound

$$\Lambda_N \leq \frac{2}{\pi} \log(N + 1) + 0.685, \qquad (D.5.32)$$

which allows for well-known proofs of uniform and exponential convergence of interpolation, numerical differentiation and integration (Bos [5], Hesthaven [28]).

Fekete points

By definition, the Fekete points for a finite interval maximize the magnitude of the determinant of the Vandermonde matrix defined in (D.1.6). Since any change in the polynomial base only multiplies the determinant by a constant factor that is independent of the point locations, the Fekete points also maximize the magnitude of the generalized Vandermonde matrix defined in (D.1.13) for any polynomial base.

Combining (D.2.45) with (D.5.29), we find that the following inequality holds true for the Lobatto point distribution,

$$\left| \frac{\det[\mathbf{V}_\phi(x_1, x_2, \ldots, x_{i-1}, x, x_{i+1}, \ldots, x_{N+1})]}{\det[\mathbf{V}_\phi(x_1, x_2, \ldots, x_{i-1}, x_i, x_{i+1}, \ldots, x_{N+1})]} \right| \leq 1 \qquad (D.5.33)$$

for $i = 2, \ldots, N$. Thus, the Lobatto points are identical to the Fekete points mapped to the interval $[-1, 1]$.

D.6 Interpolation in two and higher dimensions

Given the values of a function, $f(x, y)$, at N data points in the xy plane, (x_i, y_i) for $i = 1, \ldots, N$, we want to interpolate at an intermediate point, that is, we want to compute the value of the function at an arbitrary point, (x, y). To carry out the interpolation, we introduce a set of N linearly independent functions, $\phi_i(x, y)$ for $i =$

$1, \ldots, N$, and approximate the interpolated function, $f(x, y)$, with the interpolating function $g(x, y)$,

$$f(x, y) \simeq g(x, y) = c_1\, \phi_N(x, y) + c_2\, \phi_{N-1}(x, y)$$
$$+ \cdots + c_{N-1}\, \phi_2(x, y) + c_N\, \phi_1(x, y). \tag{D.6.1}$$

The coefficients, c_i, are computed to satisfy the interpolation condition

$$g(x_i, y_i) = f(x_i, y_i) \tag{D.6.2}$$

for $i = 1, \ldots, N$. Enforcing the interpolation conditions, we derive a system of linear equations,

$$\mathbf{V}_\phi^T \cdot \mathbf{c} = \mathbf{f}, \tag{D.6.3}$$

where

$$\mathbf{V}_\phi \equiv \begin{bmatrix} \phi_1(x_1, y_1) & \phi_1(x_2, y_2) & \cdots & \phi_1(x_N, y_N) \\ \phi_2(x_1, y_1) & \phi_2(x_2, y_2) & \cdots & \phi_2(x_N, y_N) \\ \vdots & \vdots & \vdots & \vdots \\ \phi_{N-1}(x_1, y_1) & \phi_{N-1}(x_2, y_2) & \cdots & \phi_{N-1}(x_N, y_N) \\ \phi_N(x_1, y_1) & \phi_N(x_2, y_2) & \cdots & \phi_N(x_N, y_N) \end{bmatrix} \tag{D.6.4}$$

is the $N \times N$ generalized Vandermonde matrix and

$$\mathbf{c} \equiv \begin{bmatrix} c_N \\ c_{N-1} \\ \vdots \\ c_2 \\ c_1 \end{bmatrix}, \qquad \mathbf{f} \equiv \begin{bmatrix} f(x_1, y_1) \\ f(x_2, y_2) \\ \vdots \\ f(x_{N-1}, y_{N-1}) \\ f(x_N, y_N) \end{bmatrix} \tag{D.6.5}$$

are the coefficient vector and data vector. The properties of the Vandermonde system in two dimensions, including existence and uniqueness of solution, are largely unknown. In an effort to maximize the determinant of the Vandermonde matrix, and thereby ensure a solution, we introduce the notion of the Fekete points in two dimensions for a specified interpolation domain.

If the basis functions, $\phi_i(x, y)$, are cardinal interpolation functions satisfying the property

$$\phi_i(x_j, y_j) = \delta_{ij} \tag{D.6.6}$$

for $i, j = 1, \ldots, N$, where δ_{ij} is Kronecker's delta, then the generalized Vandermonde matrix, $\mathbf{V}_\phi$, reduces to the identity matrix and the expansion coefficients represent the prescribed data, $c_i = f(x_i, y_i)$.

To approximate a function $f(x, y)$ with a complete mth degree polynomial in x and y, we may introduce N data points, where

$$N = \frac{1}{2}\,(m + 1)(m + 2), \tag{D.6.7}$$

and identify the basis functions, $\phi_i(x, y)$, with the set of monomial products, $x^k y^l$, to obtain

$$
\begin{aligned}
f(x, y) \simeq g(x, y) = a_{00} \\
+ a_{10}\, x + a_{01}\, y, \\
+ a_{20}\, x^2 + a_{11}\, x\, y + a_{02}\, y^2 \\
+ a_{30}\, \xi^3 + a_{21}\, x^2\, y + a_{12}\, x\, y^2 + a_{03}\, y^3
\end{aligned}
\tag{D.6.8}
$$

$$
\cdots \quad \cdots \quad \cdots \quad \cdots \quad \cdots \quad \cdots \quad \cdots \quad \cdots \quad \cdots
$$

$$
+ a_{m,0}\, x^m + a_{m-1,1}\, x^{m-1}\, y + \cdots + a_{1,m-1}\, x\, y^{m-1} + a_{0,m}\, y^m,
$$

where a_{ij} is a new set of coefficients that can be mapped to the previous coefficients, c_i. Other choices of basis functions include the Appell and Proriol polynomials discussed in Sections 5.6 and 5.7.

Interpolation in three and higher dimensions is carried out by similar methods using appropriate sets of basis functions, as discussed in Section 8.6.

Element grid generation $\boxed{E}$

Discretizing the solution domain into an element assembly by the process of grid generation or tessellation is an important aspect of a finite or spectral element implementation. In fact, grid generation for complicated geometries can be the most demanding, subtle, and expensive module of a finite or spectral element code.

Grids can be classified broadly as *structured*, meaning that the element vertices and edges can be indexed by simple algorithms and nearly all nodes have the same number of neighbors, or *unstructured*, meaning that their manipulation relies heavily on a non-obvious connectivity matrix for element and node identification. Structured grids are more restrictive in that transposing the labels of two elements results in an unstructured grid, but not *vice versa*.

One-dimensional grid generation

Element discretization in one dimension, discussed in Chapter 1 of this book, is straightforward compared to its two- and three-dimensional counterparts. In adaptive discretization, the algorithm is designed so that the more rapidly the solution varies in a certain interval, or the higher the magnitude of the local numerical error quantified by a sensible measure, the smaller the local element size and the denser the element distribution. For problems governed by one-dimensional equations defined over a curved, planar or three-dimensional line, the element size can be adjusted according to the local curvature using appropriate criteria to ensure adequate spatial resolution (e.g., Kwak & Pozrikidis [36]).

Two- and three-dimensional grid generation

Several algorithms are available for the programmable discretization of two- and three-dimensional rectangular or curved domains, including the method of successive subdivision and the method of Delaunay triangulation, discussed in Chapters 4 and 5 of this book, the advancing front method (AFM), and their hybrid implementations.

In the advancing front method, the boundary contour is discretized into one-dimensional elements defining the initial front (e.g., Kwak & Pozrikidis [36]). Triangles are then added inward so that each element has at least one edge at the initial front. When the initial front has been depleted, the process is repeated with a redefined new front until the whole of the discretization domain has been tiled. A similar method can be implemented in three dimensions starting with a surface front described by triangles and using tetrahedral elements to tessellate the solution domain.

Internet resources

Internet sites with numerous links to grid generation pages can be found at the following links:

- *Mesh generation and grid generation on the Web:*

`http://www-users.informatik.rwth-aachen.de/~roberts/meshgeneration.html`

- *Mesh generators:*

`http://www.engr.usask.ca/~macphed/finite/fe_resources/mesh.html`

- *Internet finite element resources:*

`http://www.engr.usask.ca/~macphed/finite/fe_resources/fe_resources.html`

- *Meshing research corner:*

`http://www.andrew.cmu.edu/user/sowen/mesh.html`

A number of high-quality grid generation codes listed in these sites are available in the public domain, subject to the terms of a GNU license agreement. Two examples are the *DistMesh* and the *triangle* codes.

DistMesh

DistMesh is a MATLAB code capable of generating unstructured triangular meshes with triangular and tetrahedral elements (Persson & Strang [43]). The source code is available from the Internet site

$$\texttt{http://www-math.mit.edu/~persson/mesh/}$$

The discretization code uses the Delaunay triangulation algorithm embedded in MATLAB for mesh generation and seeks to optimize the node locations by a force-based smoothing procedure, while the topology is regularly updated using the *delaunay* function. The boundary points are only allowed to move tangentially by projection using a distance function. The iterative procedure typically results in well proportioned meshes.

triangle

The award-winning discretization code *triangle*, written in C, generates exact Delaunay triangulations, constrained Delaunay triangulations, Voronoi diagrams, and quality conforming Delaunay triangulations (Shewchuk [62]). The source code is available from the Internet site

$$\texttt{http://www-2.cs.cmu.edu/~quake/triangle.html}$$

Glossary

F

This glossary is meant to complement the subject index by providing further definitions and references on specialized aspects of the finite element method.

Arbitrary Lagrangian-Eulerian formulation (ALE): A methodology for solving problems with time dependence using moving finite elements (see also *Lagrangian formulation* and *Eulerian formulation* in this Appendix). The nodal velocities are set arbitrarily yet judiciously depending on the physics of the problem under consideration. For example, nodes lying in a material interface may move with the normal velocity alone to track the motion of the interface, while avoiding tangential accumulation (e.g., Donea & Huerta [17]).

Babuška–Brezzi condition: A condition for the convergence of the finite element solution for the Stokes flow problem where a C^0 expansion is used for the velocity and a lower-order discontinuous expansion is used for the pressure (e.g., Zhang [72]).

Boundary-fitted mesh: The edges and faces of the finite elements are physical boundaries of the solution domain where boundary conditions are prescribed. Troublesome non-boundary-fitted meshes are sometimes called non-aligned.

Conformal mesh: The nodes, edges, and sides of neighboring elements are perfectly matched. Non-conformal meshes contain hanging nodes and overlapping zones.

Discontinuous Galerkin method (DGM): In this modification of the standard Galerkin method for convection problems governed by conservation laws, the finite element solution is allowed to be discontinuous across the element boundaries. This formulation is a hybrid of the standard Galerkin method and the finite volume method (FVM).

Eulerian formulation: The element nodes are stationary marker points embedded in a convected field, such as a fluid flow.

Free-surface problem: The solution domain consists of two distinct media whose interface is either stationary or evolves in time. A free-surface problem is an interfacial problem where one of the media is inactive, meaning that the requisite solution and its interfacial distribution have simple forms.

hp Spectral element method: Adaptive finite element method where the mesh size, h, and the order of the polynomial expansion over each element, p, are adjusted to generate a high-accuracy solution, while ensuring adequate spatial resolution (e.g., Karniadakis & Sherwin [32], Schwab [57]).

Infinite element: Not surprisingly, an infinite element extends to infinity. When the asymptotic behavior of the solution is known, the element interpolation functions are designed craftily to capture the far-field behavior in terms of *a priori* unknown coefficients.

Lagrangian formulation: The finite element nodes are material (Lagrangian) point particles moving with the medium velocity, and the finite element grid is convected with the instantaneous flow. In practice, this may result in severe grid distortion that can be prevented either by regridding or by employing the arbitrary Lagrangian–Eulerian formulation (ALE).

Least-squares finite element method (LSFEM): An alternative formulation where algebraic and ordinary differential equations for the nodal values arise by minimizing a least-squares functional instead of the Rayleigh functional associated with the Galerkin finite element method (GFEM).

Mortar finite element method: The solution domain is divided into different regimes, and the individual solutions are matched at generally non-conforming elements by means of mortar functions.

Moving-boundary problems: The boundary of the solution domain moves due to convection, accumulation, or dissolution. The boundary nodes are convected to describe the boundary motion.

Patch test: A patch consists of the union of all elements attached to a node, called the patch node. More generally, a patch is a well-defined sub-structure of a finite element grid that retains its identity as the grid is refined. The patch test is used to ensure that the solution of the problem under consideration over the participating elements, subject to a stipulated condition around the patch boundary, is consistent with expectation. In solid mechanics, the displacement, the traction, or both, are prescribed as boundary conditions.

Singular finite element method (SFEM): When the solution is expected to exhibit a singular behavior at a point or along a line, special singular elements are employed to capture the singularity. The coefficient of the singular term is included in the finite element expansion in lieu of a nodal value.

Streamwise Upwind Petrov–Galerkin method (SUPG): In this variation of the standard Galerkin method for solving the convection equation, the governing equation is projected onto functions that are biased in the direction of flow in lieu of upwind differencing.

MATLAB primer

G

MATLAB is a software product (application) developed by the *MathWorks* corporation, built to run on a variety of operating systems (OS), including UNIX and LINUX. Initially, MATLAB was developed as a virtual laboratory for matrix calculus and linear algebra. Today, MATLAB can be described both as a *programming language* and a *computing environment*.

MATLAB as a programming language

As a programming language, MATLAB is roughly equivalent, in some ways superior and in some ways inferior to traditional upper-level languages such as FORTRAN, Pascal, C, and C++. An attractive feature of MATLAB is the availability of a broad range of utility commands and intrinsic functions, most notably, graphics.

A simplifying feature of MATLAB is that the dimensions of vectors and matrices used in the calculations are assigned automatically and can be changed in the course of a computational session, thereby circumventing the need for memory declaration and allocation.

Although the graphics component makes MATLAB dependent on the graphics display library that accompanies the OS, the dependence is nearly transparent to the user and becomes evident only when a newer version of MATLAB fails to run or when an installed version of MATLAB stops working properly after the operating system has been upgraded.

MATLAB as a computing environment

As a computing environment, MATLAB is able to run indefinitely in its own workspace. A MATLAB session defined by the values of all initialized variables and graphical objects can be saved in a file and reinstated at a later time. In this sense, MATLAB is an operating system running inside the operating system empowering the computer.

To invoke MATLAB in a UNIX or LINUX operating system endowed with a window manager, we open a new command line terminal and issue the command: *matlab*. Assuming that the application is in the shell path, this launches MATLAB in some graphics or command line form. To invoke MATLAB in an operating system that does not offer a command line terminal, we launch the application (executable MATLAB binary file) by double-clicking on the MATLAB icon.

An alternative to MATLAB is the public domain application SCILAB. The SCILAB package includes a wealth of mathematical functions and allows for interfacing with various programming languages, including C and FORTRAN. SCILAB is able to run on most personal computers, UNIX and LINUX workstations.

G.1 Programming in MATLAB

Only elementary knowledge of computer programming is required to read and write MATLAB code. The code is written in one file or a collection of files, called the *source* or *program* files, using a standard file editor, such as the *vi* editor on UNIX and LINUX. The source code contains the main program, sometimes also called a script, and the necessary user-defined functions. The names of the program files must be endowed with the dot-m suffix (*.m*). Execution begins by typing the name of the file containing the main program in the MATLAB environment. Alternatively, the code can be typed one line at a time followed by the *Enter* keystroke in the MATLAB environment.

MATLAB is an interpreted language, which means that the instructions are translated into machine language and executed in real time, one at a time. In contrast, a source code written in FORTRAN or C must first be compiled to produce binary object files which are then linked together with the necessary system libraries to produce an executable binary image, sometimes called the binary file or application (app). In fact, the MATLAB distribution is an application residing in an appropriate directory. For efficiency, intrinsic MATLAB functions have been pre-compiled and do not have to be interpreted in the course of the execution.

G.1.1 Grammar and syntax

Following is a list of general rules regarding MATLAB grammar and syntax. When confronted with an error after issuing a command or during execution, this list should serve as a first check point:

- *Variables are (lower and upper) case sensitive.*
  The variable *echidna* is different than the variable *echiDna*. Similarly, the MATLAB command *return* is not equivalent to the erroneous command *Return*, which is not recognized by the MATLAB interpreter.

- *A variable must start with a letter.*
  A variable name is described by a string of up to 31 characters including letters, digits, and the underscore; punctuation marks are not allowed.

- *A string variable is enclosed by a single quote.*
  For example, we may define the string variable: *thinker_764 = 'Thucydides'*.

- *A command may begin at any position in a line.*
  A command may continue practically indefinitely in the same line. To continue a command to the next line, we put three dots at the end of the line.

- *Multiple commands can be typed in a single line.*
  Two or more commands may be placed in the same line, provided they are separated with a semi-colon (;).

- *Values, results, and output are displayed by default.*
  When a command is executed explicitly or by running a MATLAB code, MATLAB displays the numerical value assignment or the result of a calculation. To suppress the output, we put a semi-colon (;) at the end of the command.

- *Multiple blank spaces are ignored.*
  More than one empty space between words are ignored by the interpreter. However, numbers cannot be split in sections separated by blank spaces.

- *Indices must be positive.*
  Vectors and arrays must have positive and non-zero indices. Thus, the vector entry *v(-3)* is unacceptable. This annoying restriction can be circumvented in clever ways by shifting the range of the indices.

- *Comments are preceded by the % character.*
  A line beginning with the % character, or the tail end of a line after the % character, is a comment ignored by the MATLAB interpreter.

Mathematical symbols and special characters used in MATLAB interactive dialog and programming are listed in Table G.1.1. Basic logical control flow commands are listed in Table G.1.2. Basic Input/Output (I/O) commands, functions, and format are listed in Table G.1.3. Once the output format has been set, it remains in effect until changed.

G.1.2   Precision

MATLAB stores all numbers in the long format of the floating-point representation. This means that real numbers have a finite precision of roughly 16 significant digits and a range of definition varying approximately between 10^{-308} and 10^{+308} in absolute value. Numbers smaller than 10^{-308} or larger than 10^{+308} in absolute value cannot be accommodated.

MATLAB performs all computations in double precision. However, this should not be confused with the ability to view and print numbers with a specified number of significant figures using the commands listed at the bottom of Table G.1.3.

+	Plus
-	Minus
*	Matrix multiplication
.*	Array multiplication
^	Matrix power
.^	Array power
kron	Kronecker tensor product
\	Backslash or left division
/	Slash or right division
./	Array division
:	Colon
()	Parentheses
[]	Brackets
.	Decimal point
..	Parent directory
...	Line continuation
,	Comma
;	Semicolon, used to suppress the screen display
%	Indicates that the rest of the line is a comment
!	Exclamation mark
'	Matrix transpose
"	Quote
.'	Nonconjugated transpose
=	Set equal to
==	Equal
~=	Not equal
<	Less than
<=	Less than or equal to
>	Greater than
>=	Greater than or equal to
&	Logical *and*
|	Logical *or*
~	Logical *not*
xor	Logical *exclusive or*
i, j	Imaginary unit
pi	number $\pi = 3.14159265358\ldots$

TABLE G.1.1   MATLAB operators, symbols, special characters, and constants.

Control flow commands:

break	Terminate the execution
else	Use with the *if* statement
elseif	Use with the *if* statement
end	Terminate a *for* loop, a *while* loop, or an *if* block
error	Display a message and abort
for	Loop over commands a specific number of times
if	Conditionally execute commands
pause	Wait for user's response
return	Return to the MATLAB environment, invoking program or function
while	Repeat statements an indefinite number of times until a specified condition is met

TABLE G.1.2   MATLAB logical control flow commands and logical construct components.

G.1.3   MATLAB commands

Once invoked, MATLAB responds interactively to various commands, statements, and definitions issued by the user in its window. These are implemented by typing the corresponding name, single- or multi-line syntax, and then pressing the *Enter* key.

General utility and interactive-input MATLAB commands are listed in Table G.1.4. Issuing the command *demos* initiates various demonstrations and illustrative examples of MATLAB code, worthy of exploration.

To obtain a detailed explanation of a MATLAB command, statement, operator, or function, we may use the MATLAB *help* facility, which is the counterpart of the UNIX *man* facility.

For example, issuing the command *help break* in the MATLAB environment spawns the following description:

```
BREAK Terminate execution of WHILE or FOR loop.
    BREAK terminates the execution of FOR and WHILE loops.
    In nested loops, BREAK exits from the innermost loop only.
    If you use BREAK outside of a FOR or WHILE loop in a MATLAB
    script or function, it terminates the script or function at
    that point.  If BREAK is executed in an IF, SWITCH-CASE, or
```

I/O commands:

disp	Display numerical values or text
	Use as: *disp* *disp()* *disp('text')*
fclose	Close a file
fopen	Open a file
fread	Read binary data from a file
fwrite	Write binary data to a file
fgetl	Read a line from a file, discard newline character
fgets	Read a line from a file, keep newline character
fprintf	Write formatted data to a file using C language conventions
fscanf	Read formatted data from a file
feof	Test for end-of-file (EOF)
ferror	Inquire the I/O error status of a file
frewind	Rewind a file
fseek	Set file position indicator
ftell	Get file position indicator
sprintf	Write formatted data to string
sscanf	Read formatted string from file
csvread	Read from a file values separated by commas
csvwrite	Write into file values separated by commas
uigetfile	Retrieve the name of a file to open through dialog box
uiputfile	Retrieve the name of a file to write through dialog box

Interactive input:

input	Prompt for user input
keyboard	Invoke keyboard as though it were a script file
menu	Generate menu of choices for user input

Output format:

format short	fixed point with 4 decimal places (default)
format long	fixed point with 14 decimal places
format short e	scientific notation with 4 decimal places
format long e	scientific notation with 15 decimal places
format hex	hexadecimal format

TABLE G.1.3   MATLAB Input/Output (I/O) commands, functions, and format.

clear	Clear variables and functions from memory
demo	Run demos
exit	Terminate a MATLAB session
help	On-line documentation
load	Retrieve variables from a specified directory
save	Save workspace variables to a specified directory
saveas	Save figure or model using a specified format
size	Reveal the size of matrix
who	List current variables
quit	Quit a MATLAB session

TABLE G.1.4   General utility MATLAB commands used to control the MATLAB environment.

```
TRY-CATCH statement, it terminates the statement at that point.
```

The command *clear* is especially important, as it resets all variables to the uninitialized status, and thereby prevents the use of improper values defined or produced in a previous calculation. A detailed explanation of this command can be obtained by typing: *help clear.*

G.1.4   Elementary examples

In the following examples, the interactive usage of MATLAB is demonstrated by elementary examples. In these sessions, a line that begins with two *greater than* signs (>>) denotes the MATLAB command line where a definition is made, a function is invoked, or a command is issued. Unless stated otherwise, a line that does not begin with >> is MATLAB output. Recall that the command *clear* clears the memory content from previous definitions to prevent misappropriation.

• Numerical value assignment and addition:

```
>> a = 1
a =
      1
>> b = 2
b =
      2
>> c = a + b
c =
      3
```

- Numerical value assignment and subtraction:

```
>> clear
>> a=1; b=-3; c=a-b
c =
    4
```

- Number multiplication:

```
>> clear
>> a = 2.0; b =-3.5; c= a*b;
>> c
c =
    -7
```

Note that typing the name of the variable c displays its current value, in this case −7.

- Vector definition:

```
>> clear
>> v = [2 1]
v =
    2    1

>> v(1)
ans =
    2

>> v' % transpose
ans =
    2
    1
```

Note that typing v(1) displays the first component of the vector v as an answer. The comment transpose is ignored since it is preceded by the comment delimiter %. In fact, the result ans is a variable evaluated by MATLAB.

- Vector addition:

```
>> v = [1 2]; u = [-1, -2]; u+v
ans =
    0    0
```

- Matrix definition, addition, and multiplication:

```
>> a = [1 2; 3 4]
a =
    1    2
    3    4
```

```
>> b = [ [1 2]' [2 4]' ]
b =
     1     2
     2     4

>> a+b
ans =
     2     4
     5     8

>> c = a*b
c =
     5     10
    11     22
```

• Multiply a complex matrix by a complex vector:

```
>> a = [1+2i 2+3i; -1-i 1+i]
a =
    1.0000 + 2.0000i     2.0000 + 3.0000i
   -1.0000 - 1.0000i     1.0000 + 1.0000i

>> v = [1+i 1-i]
v =
    1.0000 + 1.0000i     1.0000 - 1.0000i

>> c = a*v'
c =
    2.0000 + 6.0000i
   -2.0000 + 2.0000i
```

Note that, by taking its transpose indicated by a prime, the row vector v becomes
a column vector that is conformable with the square matrix a.

• Print π:

```
>> format long
>> pi
ans =
    3.14159265358979
```

• for loop:

```
>> for j=-1:1
    j
   end
```

```
j =
      -1
j =
       0
j =
       1
```

In this example, the first three lines are entered by the user.

- if statement:

```
>> j=0;
>> i=1;
>> if i==j+1, disp 'case 1', end
case 1
```

- for loop:

```
>> n=3;
>> for i=n:-1:2
disp 'i='; disp (i), end
i=
      3
i=
      2
```

The loop is executed backward, starting at **n** with step of **-1**.

- if loop:

```
>> i=1; j=2;
>> if i==j+1;    disp 'case 1'
elseif i==j; disp 'case2'
else;        disp 'case3'
end
case3
```

In this example, all but the last line are entered by the user.

- while loop:

```
>> i=0;
>> while i<2, i=i+1; disp(i), end
     1

     2
```

The four statements in the while loop could be typed in separate lines; that is, the commas can be replaced by the *Enter* keystroke.

G.2   MATLAB functions

In scientific computing, a function is an evaluation procedure that receives single or multiple input and generates single or multiple output. MATLAB comes with an extensive library of intrinsic or embedded functions for numerical computation and data visualization.

General and specialized MATLAB mathematical functions are listed in Table G.2.1. MATLAB performs these functions based on algorithms, procedures, and approximations discussed in mathematical handbooks and texts of numerical methods (e.g., Pozrikidis [47]). To obtain specific information on the proper usage of a particular function, we may use the MATLAB help facility. If you are unsure about the proper syntax or reliability of a function, it is best to write your own code from first principles. It is both rewarding and instructive to duplicate a MATLAB function and create a personal library of user-defined functions based on control-flow commands.

User-defined functions

A user-defined MATLAB function is written by the programmer in a file whose name is the same as the name of the function. The file name must be suffixed with the standard MATLAB file identifier $(.m)$. Thus, a function named *koumbaros* must reside in a file named *koumbaros.m*, whose general structure is:

```
function [output1, output2, ...] = koumbaros(input1, input2,...)
   ......
return
```

The three dots indicate additional input and output arguments separated by commas; the six dots indicate additional lines of code. The output string, *output1, output2, ...*, consists of numbers, vectors, matrices, and string variables evaluated by the function by performing operations involving the input string, *input, input2,....* To execute this function in the MATLAB environment or invoke it from a program file, we issue the command:

```
[evaluate1, evaluate2, ...] = koumbaros(parameter1, parameter2,...)
```

After the function has been successfully executed, *evaluate1* takes the value of *output1*, *evaluate2* takes the value of *output2*, and the rest of the output variables take corresponding values.

If a function evaluates only one number, vector, matrix, character string, entity or object, then the function statement and corresponding function can be simplified to:

```
function evaluate =  koumbaros(input1, input2,...)
   ...
return
```

An example of a simple function residing in a function file named *bajanakis.m* is:

Standard:

abs	Absolute value
acos	Inverse cosine
acosh	Inverse hyperbolic cosine
acot	Inverse cotangent
acoth	Inverse hyperbolic cotangent
acsc	Inverse cosecant
acsch	Inverse hyperbolic cosecant
angle	Phase angle
asec	Inverse secant
asech	Inverse hyperbolic secant
asin	Inverse sine
asinh	Inverse hyperbolic sine
atan	Inverse tangent
atan2	Four quadrant inverse tangent
atanh	Inverse hyperbolic tangent
cart2pol	Cartesian to polar coordinate conversion
cart2sph	Cartesian to spherical coordinate conversion
conj	Complex conjugate
cos	Cosine
cosh	Hyperbolic cosine
cot	Cotangent
coth	Hyperbolic cotangent
csc	Cosecant
csch	Hyperbolic cosecant
exp	Exponential
expm	Matrix exponential
fix	Round toward zero
floor	Round toward minus infinity
gcd	Greatest common divisor
imag	Complex imaginary part
lcm	Least common multiple
log	Natural logarithm
log10	Common logarithm
pol2cart	Polar to Cartesian coordinate conversion

TABLE G.2.1 Standard and specialized MATLAB mathematical functions (Continuing →)

real	Complex real part
rem	Remainder after division
round	Round to the nearest integer
sec	Secant
sech	Hyperbolic secant
sign	Signum function
sin	Sine
sinh	Hyperbolic sine
sph2cart	Polar to Cartesian coordinate conversion
sqrt	Square root
tan	Tangent
tanh	Hyperbolic tangent

Specialized:

bessel	Bessel functions
besseli	Modified Bessel functions of the first kind
besselj	Bessel functions of the first kind
besselk	Modified Bessel functions of the second kind
bessely	Bessel functions of the second kind
beta	Beta function
betainc	Incomplete beta function
betaln	Logarithm of the beta function
ellipj	Jacobi elliptic functions
ellipke	Complete elliptic integral
erf	Error function
erfc	Complementary error function
erfcx	Scaled complementary error function
erfinv	Inverse error function
expint	Exponential integral
gamma	Gamma function
gammainc	Incomplete gamma function
gammaln	Logarithm of gamma function
legendre	Associated Legendre functions
log2	Dissect floating point numbers
pow2	Scale floating point numbers

TABLE G.2.1   Standard and specialized MATLAB mathematical functions ($\rightarrow$ Continuing $\rightarrow$)

| rat | Rational approximation |
| rats | Rational output |

Matrix and vector initialization:

eye	Identity matrix
ones	Matrix of ones
rand	Uniformly distributed random numbers and arrays
randn	Normally distributed random numbers and arrays
zeros	Matrix of zeros

TABLE G.2.1   ($\rightarrow$ Continued) Standard and specialized MATLAB mathematical functions.

```
function bname = bajanakis(isel)

  if(isel == 1)
   bname = 'phaedrus';
  elseif(isel == 2)
   bname = 'phaethon';
  else
   bname = 'alkiviadis';
  end

return
```

Numerous examples of user-defined functions can be found in the directories of FSELIB listed in the text.

G.3   Numerical methods

MATLAB encapsulates a general-purpose numerical methods library whose functions perform numerical linear algebra, solve algebraic equations, carry out function differentiation and integration, solve differential equations, and execute a variety of other tasks. Special-purpose libraries of interest to a particular discipline are accommodated in *toolboxes*. The theory underlying the numerical methods is discussed in texts on numerical methods and scientific computing (e.g., Pozrikidis [47]). Selected numerical methods functions are listed in Table G.3.1.

cat	Concatenate arrays
cond	Condition number of a matrix
det	Matrix determinant
eig	Matrix eigenvalues and eigenvectors
inv	Matrix inverse
lu	LU-decomposition of a matrix
ode23	Solution of ordinary differential equations by the second/third order Runge–Kutta method
ode45	Solution of ordinary differential equations by the fourth/fifth order Runge–Kutta–Fehlberg method
qr	QR-decomposition of a matrix, where Q is an orthogonal matrix, and R is an upper triangular (right) matrix
poly	Produces the characteristic polynomial of a matrix
quad	Function integration by Simpson's rule
root	Polynomial root finder
svd	Singular-value decomposition
trapz	Function integration by the trapezoidal rule
x = A\b	Solves the linear system $\mathbf{A} \cdot \mathbf{x} = \mathbf{b}$, where $\mathbf{A}$ is an $N \times N$ matrix, and $\mathbf{b}, \mathbf{x}$ are N-dimensional column vectors
x = b/A	Solves the linear system $\mathbf{x} \cdot \mathbf{A} = \mathbf{b}$, where $\mathbf{A}$ is an $N \times N$ matrix, and $\mathbf{b}, \mathbf{x}$ are N-dimensional row vectors

TABLE G.3.1   An assortment of general-purpose numerical methods MATLAB functions.

G.4   MATLAB graphics

A powerful feature of MATLAB is the ability to generate professional quality graphics, including animation. Graphics are displayed in dedicated windows that appear in response to graphics function calls. Graphics functions are listed in Tables G.4.1–G.4.3 in several categories. To generate a new graphics window, we use the *figure* function. To superimpose graphs, we use the *hold* function. To obtain a detailed description of a graphics function, use the *help* function.

Several graphics sessions followed by the graphics display generated by MATLAB are presented in the remainder of this section.

• Graph of the function: $f(x) = sin^3(\pi x)$

```
>> x=-1.0:0.01:1.0;    % define an array of abscissae
```

Two-dimensional (xy) graphs:

bar	Bar graph
comet	Animated comet plot
compass	Compass plot
errorbar	Error bar plot
feather	Feather plot
fplot	Plot a function
fill	Draw filled two-dimensional polygons
hist	Histogram plot
loglog	Log-log scale plot
plot	Linear plot
polar	Polar coordinate plot
rose	Angle histogram plot
semilogx	Semi-log scale plot, x-axis logarithmic
semilogy	Semi-log scale plot, y-axis logarithmic
stairs	Stair-step plot
stem	Stem plot for discrete sequence data

Graph annotation and operations:

grid	Grid lines
gtext	Mouse placement of text
legend	Add legend to plot
text	Text annotation
title	Graph title
xlabel	x-axis label
ylabel	y-axis label
zoom	Zoom in and out of a two-dimensional plot

TABLE G.4.1   Elementary and specialized xy graphs and commands for graph annotation and operations.

```
>> y = sin(pi*x).^3;     % note the .^ operator (Table G.1.1)
>> plot(x,y)
```

The graphics display is shown in Figure G.4.1(a).

• Graph of the Gaussian function: $f(x) = e^{-x^2}$

```
>> fplot('exp(-x^2)',[-5, 5])
```

Line and fill commands:

fill3	Draw filled three-dimensional polygons
plot3	Plot lines and points

Two-dimensional graphs of three-dimensional data:

clabel	Contour plot elevation labels
comet3	Animated comet plot
contour	Contour plot
contour3	Three-dimensional contour plot
contourc	Contour plot computation (used by contour)
image	Display image
imagesc	Scale data and display as image
pcolor	Pseudocolor (checkerboard) plot
quiver	Quiver plot
slice	Volumetric slice plot

Surface and mesh plots:

mesh	Three-dimensional mesh surface
meshc	Combination mesh/contour plot
meshgrid	Generate x and y arrays
meshz	Three-dimensional mesh with zero plane
slice	Volumetric visualization plot
surf	Three-dimensional shaded surface
surfc	Combined surf/contour plot
surfl	Shaded surface with lighting
trimesh	Triangular mess plot
trisurf	Triangular surface plot
waterfall	Waterfall plot

TABLE G.4.2   MATLAB functions for three-dimensional graphics (Continuing →)

The graphics display is shown in Figure G.4.1(*b*).

• Paint a polygon in black color:

```
>> x =[0.0 1.0 1.0]; y=[0.0 0.0 1.0]; c='k';
>> fill (x,y,c)
```

The graphics display is shown in Figure G.4.1(*c*).

Three-dimensional objects:

cylinder	Generate a cylinder
sphere	Generate a sphere

Graph appearance:

axis	Axis scaling and appearance
caxis	Pseudocolor axis scaling
colormap	Color lookup table
hidden	Mesh hidden line removal mode
shading	Color shading mode
view	Graph viewpoint specification
viewmtx	View transformation matrices

Graph annotation:

grid	Grid lines
legend	Add legend to plot
text	Text annotation
title	Graph title
xlabel	x-axis label
ylabel	y-axis label
zlabel	z-axis label for three-dimensional plots

TABLE G.4.2 ($\rightarrow$ Continued) MATLAB functions for three-dimensional graphics.

- mesh plot:

```
>> [x, y] = meshgrid(-1.0:0.10:1.0, -2.0:0.10:2.0);
>> z = sin(pi*x+pi*y);
>> mesh(z)
```

The graphics display is shown in Figure G.4.1(d).

- Kelvin foam:

FSELIB script *tetrakai*, not listed in the text, generates a periodic, space-filling lattice of Kelvin's tetrakaidecahedron (14-faced polyhedron) encountered in liquid and metallic foam. The graphics display is shown in Figure G.4.1(e).

Graphics control:

capture	Screen capture of current figure in UNIX
clf	Clear current figure
close	Abandon figure
figure	Create a figure in a new graph window
gcf	Get handle to current figure
graymon	Set default figure properties for gray-scale monitors
newplot	Determine correct axes and figure for new graph
refresh	Redraw current figure window
whitebg	Toggle figure background color

Axis control:

axes	Create axes at arbitrary position
axis	Control axis scaling and appearance
caxis	Control pseudo-color axis scaling
cla	Clear current axes
gca	Get handle to current axes
hold	Hold current graph
ishold	True if hold is on
subplot	Create axes in tiled positions

Graphics objects:

figure	Create a figure window
image	Create an image
line	Generate a line
patch	Generate a surface patch
surface	Generate a surface
text	Create text
uicontrol	Create user interface control
uimenu	Create user interface menu

TABLE G.4.3   Miscellaneous MATLAB graphics functions (Continuing →)

Graphics operations:

delete	Delete object
drawnow	Flush pending graphics events
findobj	Find object with specified properties
gco	Get handle of current object
get	Get object properties
reset	Reset object properties
rotate	Rotate an object
set	Set object properties

Hard copy and storage:

orient	Set paper orientation
print	Print graph or save graph to file
printopt	Configure local printer defaults

Movies and animation:

getframe	Get movie frame
movie	Play recorded movie frames
moviein	Initialize movie frame memory

Miscellaneous:

ginput	Graphical input from mouse
ishold	Return hold state
rbbox	Rubber-band box for region selection
waitforbuttonpress	Wait for key/button press over figure

TABLE G.4.3 ($\rightarrow$ Continuing $\rightarrow$) Miscellaneous MATLAB graphics functions.

Color controls:

caxis	Pseudocolor axis scaling
colormap	Color lookup table
shading	Color shading mode

Color maps:

bone	Gray-scale with a tinge of blue color map
contrast	Contrast enhancing gray-scale color map
cool	Shades of cyan and magenta color map
copper	Linear copper-tone color map
flag	Alternating RGB and black color map
gray	Linear gray-scale color map
hsv	Hue-saturation-value color map
hot	Black-red-yellow-white color map
jet	Variation of HSV color map (no wrap)
pink	Pastel shades of pink color map
prism	Prism-color color map
white	All-white monochrome color map

Color map functions:

brighten	Brighten or darken color map
colorbar	Display color map as color scale
hsv2rgb	Hue-saturation-value to RGB equivalent
rgb2hsv	RGB to hue-saturation-value conversion
rgbplot	Plot color map
spinmap	Spin color map

Lighting models:

diffuse	Diffuse reflectance
specular	Specular reflectance
surfl	Three-dimensional shaded surface with lighting
surfnorm	Surface normals

TABLE G.4.3   ($\rightarrow$ Continued) Miscellaneous MATLAB graphics functions.

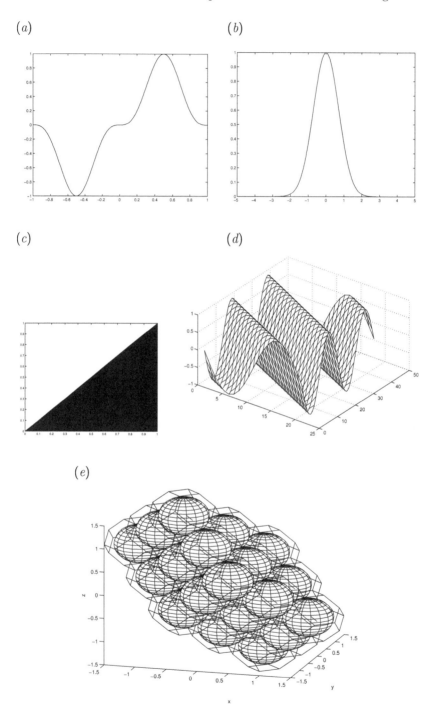

FIGURE G.4.1   An assortment of two- and three-dimensional graphic displays generated by MATLAB.

References

[1] ABRAMOWITZ, M. & STEGUN, I. A. (1970) *Handbook of Mathematical Functions.* Dover.

[2] APPELL, P. (1881) Sur des polynômes de deux variables analogues aux polynômes de Jacobi. *Arch. Math. Phys.* **66**, 238–245.

[3] AREF, H., NEWTON, P. K., STREMLER, M. A., TOKIEDA, T. & VAINCHTEIN, D. L. (2003) Vortex crystals. *Adv. Appl. Mech.* **39**, 1–79.

[4] BLYTH, M. G. & POZRIKIDIS, C. (2005) A Lobatto interpolation grid over the triangle. *IMA J. Appl. Math.* **71**, 153–169.

[5] BOS, L. (1983) Bounding the Lebesgue function for Lagrange interpolation in a simplex. *J. Approx. Theory* **38**, 43–59.

[6] BOS, L., TAYLOR, M. A. & WINGATE, B. A. (2000) Tensor product Gauss–Lobatto points are Fekete points for the cube. *Math. & Comput.* **70**, 1543–1547.

[7] BRIANI, M., SOMMARIVA, A. & VIANELLO, M. (2012) Computing Fekete and Lebesgue points: Simplex, square, disk. *J. Comp. Appl. Math.* **236**, 2477–2486.

[8] CHEN, Q. & BABUŠKA, I. (1995) Approximate optimal points for polynomial interpolation of real functions in an interval and in a triangle. *Comput. Meth. Appl. Mech. Engrg.* **128**, 405–417.

[9] CHEN, Q. & BABUŠKA, I. (1996) The optimal symmetrical points for polynomial interpolation of real functions in the tetrahedron. *Comput. Meth. Appl. Mech. Engrg.* **137**, 89–94.

[10] CHIHARA, T. S. (1978) *An Introduction to Orthogonal Polynomials.* Gordon & Breach.

[11] CIARLET, P. G. (1978) *The Finite Element Method for Elliptic Problems.* North-Holland.

[12] CLOUGH, R. W. & FELIPPA, C. A. (1968) A refined quadrilateral element for analysis of plate bending, *Matrix Methods in Structural Mechanics, Proc. Second Conf.*, pp. 399–440.

[13] COX, S. J. & MCCARTHY, C. M. (1998) The shape of the tallest column. *SIAM J. Math. Anal.* **29**, 547–554.

[14] DAHLQUIST, G. & BJÖRCK, Å. (1974) *Numerical Methods.* Prentice Hall.

[15] DAVIS, P. J. (1975) *Interpolation and Approximation.* Dover.

[16] DEMMEL, J. W. (1997) *Applied Numerical Linear Algebra.* SIAM.

[17] DONEA, J. & HUERTA, A. (2003) *Finite Element Methods for Flow Problems.* Wiley.

[18] DOSSOU, K. & PIERRE, R. (2003) A Newton-GMRES approach for the analysis of the postbuckling behavior of the solutions of the von Kármán equations. *SIAM J. Sci. Comput.* **24**, 1994–2012.

[19] DUBINER, M. (1991) Spectral methods on triangles and other domains. *J. Sci. Comput.* **6**, 345–390.

[20] FEJÉR, L. (1932) Lagrangesche interpolation und die zugehörigen konjugierten punkte. *Mathematische Annalen* **106**, 1–55.

[21] FEJÉR, L. (1932) Bestimmung derjenigen abszissen eines intervalles für welche die quadratsumme der grundfunktionen der Lagrangeschen interpolation im invervalle $[-1,1]$ ein möglichst kleines maximum besitzt. *Ann Scuola Norm. Sup. Pisa Sci. Fis. Mt. Ser. II* **1**, 263–273.

[22] FREUD, G. (1966) *Orthogonal Polynomials.* Pergamon.

[23] GALLAGHER, R. H. (1975) *Finite Element Analysis Fundamentals.* Prentice Hall.

[24] GLOWINSKI, R. (1984) *Numerical Methods for Nonlinear Variational Problems.* Springer.

[25] GRESHO, P. M. & SANI, R. L. (1998) *Incompressible Flow and the Finite Element Method, Volume I, Advection–Diffusion.* Wiley.

[26] HARBANI, Y. & ENGELMAN, M. (1979) Out-of-core solution of linear equations with non-symmetric coefficient matrix. *Computers & Fluids* **7**, 13–31.

[27] HESTENES, M. & STIEFEL, E. (1952) Methods of conjugate gradients for solving linear systems. *J. Res. Natl. Bur, Stand.* **49**, 409–436.

[28] HESTHAVEN, J. S. (1998) From electrostatics to almost optimal nodal sets for polynomial interpolation in a simplex. *SIAM J. Numer. Anal.* **35**, 655–676.

[29] HESTHAVEN, J. S. & TENG, C. H. (2000) Stable spectral methods on tetrahedral elements, *SIAM J. Sci. Comput.* **21**, 2352–2380.

[30] HRABOK, M. M. & HRUDEY, T. M. (1984) A review and catalog of plate bending finite elements. *Computers & Structures* **19**, 479–495.

[31] HOOD, P. (1976) Frontal solution program for unsymmetric matrices. *Int. J. Num. Meth. Eng.* **10**, 379–399.

[32] KARNIADAKIS, G. E. & SHERWIN, S. J. (1999) *Spectral/hp Element Methods for CFD*. Oxford University Press.

[33] KEAST, P. (1986) Moderate degree tetrahedral quadrature formulas. *Comp. Meth. Appl. Mech. Eng.* **43**, 349–353.

[34] KOORNWINDER, T. (1975) Two-variable analogues of the classical orthogonal polynomials. In: Askey, R. A. (Eds) *Theory and Application of Special Functions*. Academic Press.

[35] KRYLOV, V. I. (1962) *Approximate Calculation of Integrals*. Macmillan.

[36] KWAK, S. & POZRIKIDIS, C. (1998) Adaptive triangulation of evolving, closed or open surfaces by the advancing-front method. *J. Comp. Phys.* **145**, 61–88.

[37] KWON, Y. W. & BANG, H. (2000) *The Finite Element Method Using MATLAB*. Chapman & Hall/CRC.

[38] LUO, H. & POZRIKIDIS, C. (2006) A Lobatto interpolation grid in the tetrahedron. *IMA J. Appl. Math.* **71**, 298–313.

[39] MCFARLAND, D., SMITH, B. L. & BERNHART, W. D. (1972) *Analysis of Plates*. Spartan Books.

[40] MORLEY, L. S. D. (1968) The triangular equilibrium problem in the solution for plate bending problems. *Aero. Quart.* **19**, 149–169.

[41] NOVOSHILOV, V. V. (1961) *Theory of Elasticity*. Israel Program for Scientific Translations.

[42] PATERA, A. T. (1984) A spectral element method for fluid dynamics. Laminar flow in a channel expansion. *J. Comp. Phys.* **54**, 468–488.

[43] PERSSON, P.-O. & STRANG, G. (2004) A simple mesh generator in MATLAB. *SIAM Rev.* **46**, 329–345.

[44] PEYRET, R. (2002) *Spectral Methods for Incompressible Viscous Flow*. Springer.

[45] POPOV, E. P. (1991) *Engineering Mechanics of Solids*. Second Edition, Prentice Hall.

[46] POZRIKIDIS, C. (2011) *Introduction to Theoretical and Computational Fluid Dynamics*. Second Edition, Oxford University Press.

[47] POZRIKIDIS, C. (2008) *Numerical Computation in Science and Engineering.* Second Edition, Oxford University Press.

[48] POZRIKIDIS, C. (2002) *A Practical Guide to Boundary-Element Methods with the Software Library BEMLIB.* Chapman & Hall/CRC.

[49] POZRIKIDIS, C. (2004) A finite-element method for interfacial surfactant transport with application to the flow-induced deformation of a viscous drop. *J. Eng. Math.* **49**, 163–180.

[50] PRORIOL, J. (1957) Sur une famille de polynomes á deux variables orthogonaux dans un triangle. *Comptes Rendus de'l Académie des Sciences Paris* **245**, 2459–2461.

[51] REBAY, S. (1993) Efficient unstructured mesh generation by means of Delaunay triangulation and Bower–Watson algorithm. *J. Comp. Phys.* **106**, 125–138.

[52] RIVLIN, T. J. (1969) *An Introduction to the Approximation of Functions.* Dover.

[53] RIVLIN, T. J. (1974) *The Chebyshev Polynomials.* Wiley.

[54] SAAD, Y. & SCHULTZ, M. (1986) GMRES: A generalized minimal residual algorithm for solving nonsymmetric linear systems, *SIAM J. Sci. Statist. Comput.* **7**, 856–869.

[55] DE SAMPIO, P. A. B. (1990) A Petrov–Galerkin/modified operator formulation for convection–diffusion problems. *Int. J. Num. Meth. Eng.* **30**, 331–347.

[56] SANSONE, G. (1959) *Orthogonal Functions.* Reprinted by Dover (1991).

[57] SCHWAB, CH. (1998) *p and hp-Finite Element Methods.* Oxford University Press.

[58] SCHWARZ, H. R. (1988) *Finite Element Methods.* Academic Press.

[59] SHERWIN, S. J. & KARNIADAKIS, G. E. (1995) A triangular spectral element method; applications to the incompressible Navier–Stokes equations. *Comp. Meth. Appl. Mech. Eng.* **123**, 189–229.

[60] SHERWIN, S. J. & KARNIADAKIS, G. E. (1995) A new triangular and tetrahedral basis for high-order (*hp*) finite element methods. *Int. J. Num. Meth. Eng.* **38**, 3775–3802.

[61] SHERWIN, S. J. & KARNIADAKIS, G. E. (1996) Tetrahedral *hp* finite elements: Algorithms and flow simulations. *J. Comp. Phys.* **124**, 14–45.

[62] SHEWCHUK, J. R. (2002) Refinement algorithms for triangular mesh generation. *Computational Geometry: Theory and Applications.* **22**, 21–74.

[63] STROUD, A. H. & SECREST, D. (1966) *Gaussian Quadrature Formulas.* Prentice Hall.

[64] SUETIN, P. K. (1999) *Orthogonal Polynomials in Two Variables.* Gordon & Breach. (Translation of original Russian edition published by Nauka, Moscow, 1988.)

[65] SZABÓ, B. & BABUŠKA, I. (1991) *Finite Element Analysis.* Wiley.

[66] SZEGÖ, G. (1975) *Orthogonal Polynomials.* Fourth Edition, American Mathematical Society, Providence.

[67] TAYLOR, M. A., WINGATE, B. A. & VINCENT, R. E. (2000) An algorithm for computing Fekete points in the triangle. *SIAM J. Numer. Anal.* **38**, 1707–1720.

[68] TIMOSHENKO, S. P. & GERE, J. M. (1961) *Theory of Elastic Stability.* First Edition, McGraw-Hill.

[69] TIMOSHENKO, S. & WOINOWSKY-KRIEGER, S. (1959) *Theory of Plates and Shells.* McGraw-Hill.

[70] YU, C.-C. & HEINRICH, J. C. (1986) Petrov–Galerkin methods for the time-dependent convective transport equation. *Int. J. Num. Meth. Eng.* **23**, 883–901.

[71] WILSON, E. L. (1974) The static condensation algorithm. *Int. J. Num. Meth. Eng.* **8**, 198–203.

[72] ZHANG, S. (2004) A new family of stable mixed finite elements for the 3D Stokes equations. *Math. Comp.* **74**, 543–554.

Index

DistMesh, 764
HCT_ebsm, 508
HCT_psi, 500
HCT_sys, 500
bajanakis, 777
beam_sys, 131
beam, 136
bend_HCT, 512
bos, 399
buckle_tree_sys, 148
buckle_tree, 148
cen3, 282
ces3, 282
cg_dr, 729
cg, 729
cvt6_D, 560
cvt6_u_cn, 567
cvt6, 553
delaunay3, 605
delaunay, 276
det, 193, 307
discr_any, 228
discr_lob_dr, 174
discr_lob, 174
dispersion, 110
eam3, 267
edam6, 335
edges6_3d, 413
edges6, 325
edm3, 263, 286
edm6_abc, 331
edm6, 328
edm_lob_tbl, 193
edm_lob, 193
edm_modal_lob, 219
edmm6, 335

edmm_t10, 632
edmq, 74
eigen, 145
elm6_3d_abc, 413
elm6_abc, 322
elm6_interp, 328
elm_line1_dr, 6
elm_line1, 4
elm_line2_dr, 7
elm_line2, 6
elm_line3_dr, 7
elm_line3, 6
elmt10_interp, 631
emm3, 293, 304
emm_lob, 183
emm_lump, 191
emm_modal_lob, 221
emm_t4, 595
emm, 185
esmm6, 445
fdc, 73
fekete, 405
gauss_tetra24, 629
gauss_trgl, 328
gel_dr, 715
gel, 715
herm10_inv, 485
hexa8, 665
hlm3_n, 303
hlm_lob_sys, 206
hlm_lob, 206
hlml_sys, 46
hlml, 46
int_lob, 183
inv, 193, 485, 494
jacobi, 393

lagrange1, 174

lagrange_psi, 156, 166

lagrange, 172, 176

lapl3_dn_sqr, 313

lapl3_dn, 308

lapl3_d, 286

lapl3_eig, 293

lapl6_d_L, 363

lapl6_d_sc, 363

lapl6_d_ss, 363

lapl6_d, 359

lapl6_eig, 365

laplt10_d, 651

laplt4_d, 608

lobatto_graphs, 166

lobatto, 168, 183

morley_inv, 494

morley_psi, 496

node_fekete, 406

nodes_tetra, 622

nodes, 401

patch, 289, 359

plot_3, 289

plot_6, 359

plot_t10, 651

plot_t4, 608

proj_eig, 96

proriol_graph, 393

proriol_ortho1, 395

proriol_ortho, 395

proriol, 393

psa6M, 460

psa6_stress, 455

psa6, 445

psi_10, 486

psi_lob, 404

psi_q12, 420

psi_q2x2, 423

psi_q4, 318

psi_q8, 418

psi_t3, 258

psi_t6, 324

psi_uni, 401

pstrna6, 467

quiver, 457, 460, 560

rand, 279

scd3_d, 298

scd6_d_rc, 374

scd6_vel, 374

scdl_sys, 115

scdl, 115

sdl_robin, 35

sdl_sys_robin, 35

sdl_sys, 28

sdl, 28

sdq_beta_sys, 73

sdq_beta, 73

sdq_cnd_sys, 68

sdq_cnd, 68

sdq_modal_sys, 82

sdq_modal, 82

sdq_sys, 62

sds_any_sys, 231

sds_any, 231

sds_lob_cnd, 208

sds_lob_sys_cnd, 208

sds_lob_sys, 198

sds_lob, 198

sds_modal, 224

see_elm3, 282

tetra10_cube8, 632

tetra10_sphere8, 632

tetra4_cube12, 598

tetra4_cube8, 598

tetra4_cube_dr, 598

tetra4_delaunay_cube, 605

tetra4_sphere12, 598

tetra4_sphere8_dr, 598

tetra4_sphere8, 598

tetra4_sub12_dr, 583

tetra4_sub12, 583

tetra4_sub8_dr, 583

tetra4_sub8, 579

tetra4_vis, 583

tetra4, 575

tetrakai, 784

tetramesh, 605

thomas, 27

trgl3_delaunay_sqr, 279

trgl3_delaunay, 276

trgl3_disk, 270
trgl3_icos, 316
trgl3_octa, 316
trgl3_sqr, 274
trgl6_L, 347
trgl6_disk, 340
trgl6_icos, 415
trgl6_octa, 415
trgl6_rc, 350, 374
trgl6_sc, 347, 445
trgl6_sqr, 347, 445, 554
trgl6_ss, 347
triangle, 764
trimesh, 279, 289, 359, 457, 460
udl_sys_cn, 101
udl_sys_lump, 103
udl_sys, 96
udl, 96
uds__lob_fe, 240
uds_lob_cn, 235
uds_lob_sys, 233
voronoi, 276

advection
    matrix, 106
    one-dimensional, 105, 114
    three-dimensional, 573
    two-dimensional, 250, 261
AFM, 763
Airy
    equation, 146
    stress function, 438, 439, 526
ALE, 765
alternating tensor, 317, 676
angular frequency, 109
Appell polynomials, 387
    generalized, 391
assembly of a matrix, 25

Babuška–Brezzi condition, 765
back-substitution, 710
bandwidth, 282
barycentric
    coordinates
        for a tetrahedron, 577

for a triangle, 258
    weights, 741
beam
    bending, 120
    buckling, 140
    prismatic, 120
BEM, xxiv
bending
    modulus for a plate, 478
    moment, 471
        tensor, 471
    of a beam, 120
    of a plate, 434, 470
Bessel function, 146, 293
beta function, 267, 597
biconjugate gradients, 733
biharmonic equation, 439, 478, 500, 506
bilinear interpolation, 317
binomial, 51
biquadratic interpolation, 418
body force, 431, 541
    conservative, 438
boundary
    -fitted mesh, 765
    condition
        convection, 34, 35, 44
        Dirichlet, 2, 246, 570
        mixed, 34, 35, 44
        Neumann, 2, 246, 570
        Robin, 34, 35, 44
    flag, 248
brick element, 664
Broydon's method, 120
bubble modes
    for a hexahedron, 672
    for a quadrilateral, 427
    for a tetrahedron, 663
    for a triangle, 401, 410
    in one dimension, 82, 212
buckling
    of a beam, 140
    of a plate, 525
built-in
    beam, 124, 135
    plate, 476

Burgers's equation, 113
    viscous, 119

C^0 continuity, 4, 160
C^1 continuity, 4, 123
C^∞ continuity, 4
cantilever, 124, 132
cardinal interpolation function, 3
Cauchy
    equation of motion, 541
    relations, 740, 749
    stress tensor, 429, 541
cavity flow, 553
cell Péclet number, 114
CFD, 566
Chebyshev
    abscissas, 756
    integration quadrature, 699
    interpolation, 755
    nodal base, 178
    polynomials, 684, 703
        completed, 178
        of the second kind, 179, 700
Christoffel–Darboux formula, 692
clamped
    beam, 124, 135
    plate, 476
combinatorial, 380, 700
compatibility condition
    for Poisson's equation, 307
    in elasticity, 432
condensation of nodes, 67, 206, 387
conductivity
    matrix, 13, 53
    thermal, 1
conformal mesh, 765
connectivity matrix, 25
    generalized, 279
    in one dimension, 52
    in two dimensions, 248
conservative body force, 438
consistency analysis, 103
consistent formulation, 91
constitutive equation
    for a Newtonian fluid, 542

for a solid, 431
convection
    –diffusion, 114, 297, 374
    number, 112
    one-dimensional, 105
cork, 435
Cramer's rule, 269, 328, 592, 745
Crank–Nicolson method, 101, 564
creeping flow, 547
curl of a vector, 678
curvature, 122

damping matrix, 107
decomposition, LU, 711
del operator, 677
Delaunay
    tessellation, 598
    triangulation, 274, 763
delta function, 17, 132
density, 87, 541
derivative
    material, 542
    tangential, 314
determinant of a matrix, 714
differentiation matrix, 159, 742
diffusion
    matrix, 13, 53
        in one dimension, 88
        in two dimensions, 263
    number, 92
diffusivity
    hyper, 103
    thermal, 87, 114, 245, 570
dilatation, 432
Dirac delta function, 17, 132
Dirichlet boundary condition, 2, 135, 246,
        308, 570
discontinuous Galerkin method, 765
divergence
    of a vector function, 678
    theorem, 247, 572, 679
Duffy's transformation, 265

eigenvalue, 93, 144
    of the Laplacian operator, 293, 365

elasticity
    linear, 123, 432
    modulus, 122, 433
    theory, 429
element
    compatible, 479, 497
    conforming, 479, 497
    face, 574
    HCT, 498
    Hermitian, 126, 479
    hexahedral, 664
    high-order triangular, 396
    incompatible, 492
    infinite, 766
    labeling, 282
    macro, 498
    Morley, 493
    nodal set, 154
    non-conforming, 492
    one-dimensional, 3
    quadrilateral
      12-node, 419
      eight-node, 417
      four-node, 317
      high-order, 416
    serendipity, 418
    singular, 766
    subdivision, 270, 340
    tetrahedral, 574, 620
    three-dimensional, 571
    triangular
      four-node, ten-dof, 483
      Hermitian, 482
      six-node, 321
      six-node, 21-dof, 497
      three-node, 257
    two-dimensional, 243
Erdös theorem, 754
error function, 18
Euler
    –Bernoulli beam, 120
    –Lagrange equation, 37
    complex exponential formula, 481
    explicit method, 91
    forward method, 90, 112

Eulerian formulation, 765

face, 574
factorial, 264, 380, 594
FDM, xxiv
Fekete points
    for a tetrahedron, 625, 626
    in one dimension, 738, 746, 760
    on a quadrilateral, 423
    on a triangle, 386, 405
FEM, xxiii
Fick's law, 1
finite element
    analysis, 429
    method, 1
fixed
    beam, 124
    plate, 476
flag, boundary, 248
FLOPS, 709
flow
    free-surface, 547
    Navier–Stokes, 543
    quasi-steady, 547
    Stokes, 547
    viscous, 541
fluid
    flow, 541
    viscosity, 542
flux
    one-dimensional, 2
    three-dimensional, 570
    two-dimensional, 246
free
    -surface flow, 544, 547
    -surface problem, 765
    edge, 476
    end, 124
frequency, angular, 109
FVM, xxiv

Galerkin
    finite element method, 7
      discontinuous, 765
    projection, 7

Gamma function, 392
Gauss
    –Jordan reduction, 721
    –Seidel method, 723
    divergence theorem, 247, 572, 679
    elimination, 27, 709
    quadrature, 694
        for a hexahedron, 666
        for a triangle, 328
geometrical
    node, 153
    stiffness matrix, 142
GFEM, xxiii, 7
GFLOPS, 709
global
    interpolation function, 3
    node, 25
glossary, 765
GMRES, 733
gradient
    of a scalar, 677
    of a vector, 678
    operator, 245, 570
    surface, 314
Gram
    –Schmidt method, 684, 690
    polynomials, 687
Green's second identity, 40, 256
grid
    generation, 763
        for six-node triangles, 340
        for three-node triangles, 270
    structured, 763
    unstructured, 763
GUI, 198

h-refinement, 51
HCT triangle, 498
heat
    capacity, 87
    transfer coefficient, 34
Helmholtz equation, 302
    in one dimension, 45, 74, 206
    variational formulation, 46
Hermite

equation, 744
interpolation, 749
polynomials, 744
Hermitian
    element, 126
    triangle, 482
hexadiagonal matrix, 131
hexahedral elements
    eight-node, 664
    high-order, 667
    spectral, 668
hierarchical expansion
    in one dimension, 213
    in two dimensions, 409
Hooke's law, 122, 125, 432
hp
    -refinement, 51, 153
    -spectral element method, 766
Hsieh–Clough–Tocher element, 498
hydrodynamic pressure, 543
hyperdiffusivity, 103, 104
hypotenuse, 322

incompressible
    fluid, 542
    solid, 433
index notation, 675
inexact integration, 185, 265
infinite element, 766
inner vector product, 675, 676
integral
    over a six-node triangle, 326
    over a three-node triangle, 260
integration quadrature
    Chebyshev, 699
    for a hexahedron, 666
    for a triangle, 328
    Gauss–Chebyshev, 696
    Gauss–Legendre, 696
    Gaussian, 694
    Lobatto, 697
    standard, 696
interpolation
    bilinear, 317
    biquadratic, 418

functions
  for a four-node tetrahedron, 576
  for a six-node triangle, 323
  for a three-node triangle, 258
  for an eight-node hexahedron, 666
  global, 3
  isoparametric, 106, 248, 249, 326
  of functions, 735
  super-parametric, 628
inverse of a matrix, 485, 494, 721
isoparametric interpolation, 106, 248, 249,
    326

Jackson's theorems, 753
Jacobi
  method, 723
  polynomials, 410, 663, 684, 706
Jacobian matrix, 260, 326
  in three dimensions, 578

Kirchhoff theory of plates, 477
Kronecker delta, 3, 675

labeling, 282
Lagrange
  generating polynomial, 741
  interpolation, 155, 740
    Hermitian, 749
  plate bending equation, 478
  polynomials, 422, 740
    generalized, 749
Lagrangian formulation, 766
Lamé constants, 432
Laplace
  equation, 308
  operator, 570
Laplacian
  matrix, 13
  of a scalar, 678
  operator, 245
    eigenvalues, 293, 365, 612, 614,
      651
Lebesgue
  constant, 752, 753
    in one dimension, 161
    in two dimensions, 387

function, 753
  in one dimension, 161
  in three dimensions, 626
  in two dimensions, 387, 406
Legendre polynomials, 174, 684, 686, 688,
    700
Leibniz high-order product differentiation
    rule, 381
linear system, 709
  FEM solvers, 734
  iterative solution, 722, 725
LINUX, 767
Lipschitz condition, 754
Lobatto
  grid on a triangle, 402
  integration quadrature, 697
  interpolation, 756
  modal expansion
    in one dimension, 216
  polynomials, 702, 738
    completed, 172, 177, 698
    orthogonality of, 168
  quadrature, 182
lookup table, 168
LU decomposition, 711

macro-element, 498
mapping
  of a curved triangle, 322
  of a flat triangle, 257
  of a hexahedron, 666
  of a tetrahedron, 576
mass
  lumping, 91, 185, 241, 265
  matrix, 14, 23, 105
    Hermitian, 131
    in one dimension, 88
    in two dimensions, 261
material derivative, 542
mathematical supplement, 675
MATLAB primer, 767
matrix
  inverse, 721
  pentadiagonal, 55
  tridiagonal, 14

MDE, 103
mesh, 763
    conformal, 765
    structured, 763
    unstructured, 763
metric coefficient
    for a six-node triangle, 326
    for a three-node triangle, 260
    volume, 578, 666
MFLOPS, 709
minimax error, 752, 753, 756
mixed
    boundary condition, 34, 35, 44
    finite element expansion, 560
modal expansion
    for a Hermite triangle, 486
    for a hexahedron, 668
    in one dimension, 78, 212
        Lobatto, 216
    on a quadrilateral, 424
    on the triangle, 407
modified differential equation, 103
modulus
    of bending, 478
    of continuity, 754
    of elasticity, 122, 433
moment
    of inertia, 122
    twisting, 473
Morley triangle, 493
mortar finite element method, 766
moving-boundary problems, 766

nabla operator, 677
Navier
    –Stokes equation, 543, 562
    equation, 433
Neumann boundary condition, 2, 246, 302,
        308, 570
neutral surface, 121
Newton
    interpolation, 739
    method, 120
    second law, 541
    third law, 430

Newtonian fluid, 542
node
    condensation, 67, 206
    differentiation matrix, 159, 742
    geometrical, 153
    interpolation, 153
    labeling, 282

ODE, 2
orthogonal polynomials, 683, 738
    over a tetrahedron, 616
    over a triangle, 387, 392
orthonormal polynomials, 683, 692
OS, 767
outer vector product, 676

p-refinement, 51, 153
Péclet number, 114, 298, 374
    cell, 114
Pascal
    pyramid, 619, 621
    triangle, 381
patch test, 766
pentadiagonal matrix, 55
Petrov–Galerkin method, 117, 766
pivoting, 711
plane
    strain analysis, 439
    stress analysis, 434, 443
plate
    bending, 434, 470, 500
    buckling, 525
    circular, 480
point vortex, 744
Poisson
    equation, 307
        compatibility condition, 307
        for the pressure, 566
    ratio, 433
polynomial
    interpolation
        in one dimension, 736
        in three dimensions, 760
        in two dimensions, 760
    orthogonal, 683

over a tetrahedron, 616
over a triangle, 387, 392
orthonormal, 683, 692
roots, 168
post-processing, 455
potential function, 438
pressure, 542
matrix, 553
Poisson equation, 566
prismatic
beam, 120
element, 571, 620
projection
Galerkin, 7
matrix, 93, 723
Proriol polynomials, 392

quadrature
Chebyshev, 699
for a hexahedron, 666
for a triangle, 328
Gaussian, 694
Lobatto, 697
quadrilateral element
12-node, 419
eight-node, 417
four-node, 317
serendipity, 418
spectral, 423
quasi
-linearity, 105
-steady flow, 547

Radau polynomials, 684
random numbers, 279
rate-of-deformation tensor, 542
Rayleigh
–Ritz method, 38, 253, 256
variational formulation, 35
Reissner–Mindlin plate, 479
repeated index summation convention, 675
Reynolds number, 547
Rodrigues formula, 388
Runge
–Kutta method, 90

effect, 156, 754
function, 156, 172

Saint Venant equation, 525
SCILAB, 768
serendipity element, 418
shear
flow, 437
tension, 471
simply supported
beam, 124
edge, 476
singular element, 766
skew-symmetric matrix, 108
skyline storage, 734
smearing, 20
solenoidal vector field, 678
SOR, 724
spectral
nodes
for a hexahedron, 668
for a quadrilateral, 423
for a tetrahedron, 625
for a triangle, 386, 402, 405
in one dimension, 165
radius, 723
spectral radius, 94
square triangulation, 274, 347, 445
stiffness matrix, 129
geometrical, 142
Stokes
flow, 547, 765
unsteady, 500
theorem, 681
strain
plane, 439
tensor, 431
stress
in a fluid, 541
in a solid, 429
normal, 470
recovery, 455
shear, 470
transverse, 470
strong formulation, 38

structural mechanics, 120
Sturm–Liouville operator, 37
subdivision, 270, 340
successive
    over-relaxation, 724
    subdivisions, 270, 340, 763
    substitutions, 722
super-parametric interpolation, 628
SUPG, 766
surface
    elements, 314, 413
    gradient, 314
    metric, 260

tension, 471
    tensor, 471
tensor, 429
tessellation, 763
tetrahedral elements
    four-node, 574
    high-order, 620
    spectral, 625
TFLOPS, 709
thermal
    conductivity, 1
    diffusivity, 87
Thomas algorithm
    for a pentadiagonal system, 57
    for a tridiagonal system, 25
time discretization
    explicit, 91
    implicit, 101
Timoshenko beam theory, 125
trace, 675
traction, 430
transverse shear
    force, 121
    tension, 471
triangle
    grid generation, 270, 340
    Hermitian, 482
    high-order, 380, 396
    in three dimensions, 314, 413
    Morley, 493
    Pascal, 381

six-node, 321
spectral, 402, 405
standard
    equilateral, 400
    right, 257
    three-node, 257
triangulation
    of a disk, 270, 340
    of a rectangle with a hole, 350
    of a square, 274, 347, 445
        with a hole, 347
    of a T-shaped domain, 359
    of an L-shaped domain, 347
tridiagonal matrix, 14
triple scalar vector product, 677
twisting moment, 473

UNIX, 767
upwind differencing, 117
Uzawa algorithm, 553

Vandermonde matrix, 736, 745
    generalized, 162, 214, 384, 738, 746
    in two dimensions, 405
variation, 37, 255
variational formulation, 35, 253
vector
    identities, 678
    product
        dot, 676
        inner, 675, 676
        outer, 676
        triple scalar, 677
velocity, 105, 542
viscosity, 437, 542
    kinematic, 565
volume metric coefficient, 578, 666
von Kármán
    equation, 525
    plate, 479

wave number, 109
weak formulation, 4, 38
weighted residuals, 35, 38

Young modulus, 433